CAMBRIDGE STUDIES IN
ADVANCED MATHEMATICS 71

ANALYSIS IN INTEGER AND FRACTIONAL DIMENSIONS

Already published

1 W.M.L. Holcombe *Algebraic automata theory*
2 K. Peterson *Ergodic theory*
3 P.T. Johnstone *Stone spaces*
4 W.H. Schikhof *Ultrametric calculus*
5 J.-P. Kahane *Some random series of functions, 2nd edition*
6 H. Cohn *Introduction to the construction of class fields*
7 J. Lambek & P.J. Scott *Introduction to higher-order categorical logic*
8 H. Matsumura *Commutative ring theory*
9 C.B. Thomas *Characteristic classes and the cohomology of finite groups*
10 M. Aschbacher *Finite group theory*
11 J.L. Alperin *Local representation theory*
12 P. Koosis *The logarithmic integral I*
13 A. Pietsch *Eigenvalues and s-numbers*
14 S.J. Patterson *An introduction to the theory of the Riemann zeta-function*
15 H.J. Baues *Algebraic homotopy*
16 V.S. Varadarajan *Introduction to harmonic analysis on semisimple Lie groups*
17 W. Dicks & M. Dunwoody *Groups acting on graphs*
18 L.J. Corwin & F.P. Greenleaf *Representations of nilpotent Lie Groups and their applications*
19 R. Fritsch & R. Piccinini *Cellular structures in topology*
20 H. Klingen *Introductory lectures on Siegal modular forms*
21 P. Koosis *The logarithmic integral II*
22 M.J. Collins *Representations and characters of finite groups*
24 H. Kunita *Stochastic flows and stochastic differential equations*
25 P. Wojtaszczyk *Banach spaces for analysts*
26 J.E. Gilbert & M.A.M. Murray *Clifford algebras and Dirac operators in harmonic analysis*
27 A. Frohlich & M.J. Taylor *Algebraic number theory*
28 K. Goebal & W.A. Kirk *Topics in metric fixed point theory*
29 J.F. Humphreys *Reflection groups and Coxeter groups*
30 D.J. Benson *Representations and cohomology I*
31 D.J. Benson *Representations and cohomology II*
32 C. Allday & V. Puppe *Cohomological methods in transformation groups*
33 C. Soulé et al *Lectures on Arakelov geometry*
34 A. Ambrosetti & G. Prodi *A primer of nonlinear analysis*
35 J. Palis & F. Takens *Hyperbolicity, stability and chaos at homoclinic bifurcations*
36 M. Auslander, I. Reiten & S.O. Smalø *Representation theory of Artin algebras*
37 Y. Meyer *Wavelets and operators I*
38 C. Weibel *An introduction to homological algebra*
39 W. Bruns & J. Herzog *Cohen-Macaulay rings*
40 V. Snaith *Explicit Brauer induction*
41 G. Laumon *Cohomology of Drinfield modular varieties I*
42 E.B. Davies *Spectral theory and differential operators*
43 J. Diestel, H. Jarchow & A. Tonge *Absolutely summing operators*
44 P. Mattila *Geometry of sets and measures in Euclidean spaces*
45 R. Pinsky *Positive harmonic functions and diffusion*
46 G. Tenenbaum *Introduction to analytic and probabilistic number theory*
47 C. Peskine *An algebraic introduction to complex projective geometry I*
48 Y. Meyer & R. Coifman *Wavelets and operators II*
49 R. Stanley *Enumerative combinatorics I*
50 I. Porteous *Clifford algebras and the classical groups*
51 M. Audin *Spinning tops*
52 V. Jurdjevic *Geometric control theory*
53 H. Voelklein *Groups as Galois groups*
54 J. Le Potier *Lectures on vector bundles*
55 D. Bump *Automorphic forms*
56 G. Laumon *Cohomology of Drinfield modular varieties II*
57 D.M. Clark & B.A. Davey *Natural dualities for the working algebraist*
59 P. Taylor *Practical foundations of mathematics*
60 M. Brodmann & R. Sharp *Local cohomology*
61 J.D. Dixon, M.P.F. Du Sautoy, A. Mann & D. Segal *Analytic pro-p groups, 2nd edition*
62 R. Stanley *Enumerative combinatorics II*
64 J. Jost & X. Li-Jost *Calculus of variations*
68 Ken-iti Sato *Lévy processes and infinitely divisible distributions*

PUBLISHED BY THE PRESS SYNDICATE OF THE UNIVERSITY OF CAMBRIDGE
The Pitt Building, Trumpington Street, Cambridge, United Kingdom

CAMBRIDGE UNIVERSITY PRESS

The Edinburgh Building, Cambridge CB2 2RU, UK
40 West 20th Street, New York, NY 10011-4211, USA
10 Stamford Road, Oakleigh, VIC 3166, Australia
Ruiz de Alarcón 13, 28014 Madrid, Spain
Dock House, The Waterfront, Cape Town 8001, South Africa

http://www.cambridge.org

First published 2001

Printed in the United Kingdom at the University Press, Cambridge

Typeface CMR 10/13 *System* LaTeX [HBA]

A catalogue record for this book is available from the British Library

Library of Congress Cataloguing in Publication data

Blei, R. C. (Ron C.)
Analysis in integer and fractional dimensions/Ron Blei.
p. cm. – (Cambridge studies in advanced mathematics; 71)
Includes bibliographical references and index.
ISBN 0 521 65084 4
1. Harmonic analysis. 2. Functional analysis. 3. Probabilities.
4. Inequalities (Mathematics) I. Title. II. Series.

QA403 .B54 2001
515′.2433–dc21 00-064229

ISBN 0 521 65084 4 hardback

Analysis in Integer and Fractional Dimensions

Ron Blei
University of Connecticut

CAMBRIDGE
UNIVERSITY PRESS

Dedication:

To the memory of my father, Nicholas Blei (1916–1968),
my mother, Isabel Guth Blei (1921–1975),
and my sister, Maya Blei (1952–1982).

Contents

Preface

What the book is about

In 1976 I gave a new proof to the Grothendieck (two-dimensional) inequality. The proof, pushed a little further, yielded extensions of the inequality to higher dimensions. These extensions, in turn, revealed 'Cartesian products in fractional dimensions', and led in a setting of harmonic analysis to the solution of the (so-called) p-Sidon set problem. The solution subsequently gave rise to an index of *combinatorial dimension*, a general measurement of interdependence with connections to harmonic, functional, and stochastic analysis. In 1993 I was ready to tell the story, and began teaching topics courses about this work. The notes for these courses eventually became this book.

Broadly put, the book is about 'dimensionality'. There are several interrelated themes, sub-themes, variations on themes. But at its very core, there is the notion that when we do mathematics – whatever mathematics we do – we start with independent building blocks, and build our constructs. Or, from an observer's viewpoint – not that of a builder – we *assume* existence of building blocks, and study structures we see. In either case, these are the questions: How are building blocks used, or put together? How complex are the constructs we build, or the structures we observe? How do we gauge, or detect, complexity? The answers involve notions of dimension.

The book is a mix of harmonic analysis, functional analysis, and probability theory. Part text and part research monograph, it is intended for students (no age restriction), whose backgrounds include at least one year of graduate analysis: measure theory, some probability theory, and some functional and Fourier analysis. Otherwise, I start discussions at the very beginning, and try to maintain a self-contained format.

Although the book is about specific brands of analysis, it should be accessible, and – I hope – interesting to mathematicians of other persuasions. I try to convey a sense of a 'big picture', with emphasis on historical links and contextual perspectives. And I try very hard to stay focused, not to be encyclopedic, to stick to the story.

The fourteen chapters are described below. Each except the first starts with 'mise en scène' (the setting of a stage), and ends with exercises. Some exercises are routine, filling in missing details, and some are not. There are some exercises (starred) that I do not know how to do. In fact, there are questions throughout the book, not only in the exercise sections, which I did not answer; some are open problems of long standing, and some arise naturally as the tale unfolds. We start at the beginning ('... a very good place to start ...'), and proceed along marked paths, with pauses at the appropriate stops. We go first through integer dimensions, and, en route, collect problems concerning the gaps between integer dimensions. These problems are solved in the last part of the book. Although there is a story here, and readers are encouraged to start at the beginning, the chapters are by and large modular. A savvy reader could select a starting point, and read confidently; all interconnections are clearly posted.

I A Prologue: Mostly Historical

A historical backdrop and flowchart: how it came about, and how it developed. There are very few proofs, and these few are very easy.

II Three Classical Inequalities

Three inequalities: Khintchin's, Littlewood's, and Orlicz's. These, which are equivalent in a precise sense, mark first steps.

III A Fourth Inequality

Grothendieck's fundamental inequality. Three proofs are given; all three are elementary, and all three involve an 'upgraded' Khintchin inequality.

IV Elementary Properties of the Fréchet Variation – an Introduction to Tensor Products

The Fréchet variation is a multi-dimensional extension of the l^1-norm and is at the heart of the matter. Basic properties are observed. The framework of tensor products is a convenient and natural setting for the 'multi-dimensional' mathematics done here.

V The Grothendieck Factorization Theorem

A two-dimensional statement, an equivalent of the Grothendieck inequality, with key applications in harmonic and stochastic analysis (later in the book). A multi-dimensional version is derived, but open questions persist about 'factorizability' in higher dimensions.

VI An Introduction to Multidimensional Measure Theory

A set-function on a Cartesian product of algebras is a Fréchet measure if it is countably additive separately in each coordinate. The theory of Fréchet measures generalizes notions in Chapter IV. Some multi-dimensional properties extend one-dimensional analogs, and some reveal surprises. The emphasis in this chapter is on the predictable properties.

VII An Introduction to Harmonic Analysis

A distinct introduction to a venerable area. Harmonic analysis in the setting $\{-1,1\}^{\mathbb{N}}$, viewed from the ground up, as it starts from independent Rademacher characters and evolves to the full Walsh system. The focus is on measurements of this evolution. In this chapter, measurements calibrate discrete scales of integer dimensions, and involve the Bonami inequalities and the Littlewood inequalities; measurements gauge interdependence and complexity. Questions concerning feasibility of 'continuous' scales are answered in later chapters.

VIII Multilinear Extensions of the Grothendieck Inequality (via $\Lambda(2)$-uniformizability)

Characterizations of Grothendieck-type inequalities in dimensions greater than two. Proofs are cast in a framework of harmonic analysis, and are based, as in Chapter III, on 'upgraded' Khintchin inequalities. Characterizations involve spectral sets that in a later chapter are viewed as Cartesian products in fractional dimensions.

IX Product Fréchet measures

Product *Fréchet* measures are multidimensional versions of product measures. They are as basic and important in the general multidimensional theory as are their analogs in classical one-dimensional frameworks. Feasibility of these products is inextricably tied to Grothendieck-type inequalities.

X Brownian Motion and the Wiener Process

In science at large, Brownian motion broadly refers to phenomena whose measurements appear to fluctuate randomly. The Wiener process, in effect a limit of simple random walks, provides a mathematical model 'in a first approximation' (Wiener) for such phenomena. Framed in a classical probabilistic setting, the Wiener process and subsequent chaos processes are viewed and analyzed from this book's perspective. Among the main themes are: (1) the identification of chaos processes with Fréchet measures; (2) measurements of evolving stochastic interdependence and complexity; (3) measurements of increasing levels of randomness in random walks.

XI Integrators

A continuation of themes in the previous chapter. A generic identification of Fréchet measures with stochastic processes; stochastic integration in a framework of multidimensional measure theory. The Grothendieck factorization theorem and inequality play prominently in the general stochastic setting.

XII A '3/2-dimensional' Cartesian Product

Analysis of the simplest example of a fractionally-dimensional Cartesian product. Dimension is a gauge of interdependence between coordinates.

XIII Fractional Cartesian Products and Combinatorial Dimension

Precise connections between *combinatorial dimension* and exponents of interdependence in frameworks of harmonic analysis and probability theory. Existence of sets with arbitrarily prescribed combinatorial dimensions (fractional Cartesian products, random sets).

XIV The Last Chapter: Leads and Loose Ends

Some applications and assessments of 'fractional-dimensional' analysis in multidimensional measure theory, harmonic analysis, and stochastic analysis. Open questions and future lines.

Conventions and Notations

Whenever possible, I use language of standard graduate courses in analysis and probability theory. Choice of scalars alternates between real and complex scalars, and is appropriately announced. Conventions and notations are introduced as we go along; every now and then, I review them for the reader.

Here are two examples of conventions that may not be standard, and appear frequently. If n is a positive integer, then $[n]$ denotes the set $\{1, \ldots, n\}$. *Independence* – a recurring theme in the book – appears under several guises, and I explicitly distinguish between these. For example, I refer to *statistical* independence (the mainstay notion in classical probability theory), and to *functional* independence (defined in the sequel). And there are other notions of independence.

Acknowledgements

The mathematics in the book benefited from numerous communications over the years, some in writing, some in conversation, and some by collaboration. I thank, in particular, Michael Benedicks, Bela Bollobás, Lennart Carleson, John Fournier, Evarist Giné, Dick Gosselin, Jean-Pierre Kahane, Tom Körner, Sten Kaijser, Jerry Neuwirth, Yuval Peres, Gilles Pisier, Jim Schmerl, Stu Sidney, Per Sjölin, Nick Varopoulos, and Moshe Zakai. (Some citations appear at various points in the text.)

Teaching topics courses was an integral part of the writing project – many thanks to my lively and loyal audiences for the active interest and the useful feedback. Special thanks to my Ph.D. students, who kept the enterprise going: Jay Caggiano, Fuchang Gao, Slaven Stricevic, and Nasser Towghi.

There were a few places in the book where, telling tales and waxing philosophical, I needed help. Warm thanks to my daughter Micaela, son David, and wife Judy for providing a true sounding board, for the good advice on style and tone, and for their love.

The completion of the book has been long overdue. My appreciation to Roger Astley, Miranda Fyfe, and the other good people at Cambridge University Press, for their unbounded patience, and for their excellent editorial work.

Acknowledgements

I

A Prologue: Mostly Historical

1 From the Linear to the Bilinear

At the start and at the very foundation, there is the *Riesz representation theorem.* In original form it is

Theorem 1 (F. Riesz, 1909). *Every bounded, real-valued linear functional α on $\mathrm{C}([a,b])$ can be represented by a real-valued function g of bounded variation on $[a,b]$, such that*

$$\alpha(f) = \int_a^b f\,\mathrm{d}g, \qquad f \in \mathrm{C}([a,b]), \tag{1.1}$$

where the integral in (1.1) *is a Riemann–Stieltjes integral.*

The measure-theoretic version, headlined also the *Riesz representation theorem*, effectively marks the beginning of functional analysis. In general form, it is

Theorem 2 *Let X be a locally compact Hausdorff space. Every bounded, real-valued linear functional on $\mathrm{C}_0(X)$ can be represented by a regular Borel measure ν on X, such that*

$$\alpha(f) = \int_X f\,\mathrm{d}\nu, \qquad f \in \mathrm{C}_0(X). \tag{1.2}$$

And in its most primal form, measure-theoretic (and non-trivial!) details aside, the theorem is simply

Theorem 3 *If α is a real-valued, bounded linear functional on* $c_0(\mathbb{N}) = c_0$, *then*

$$\|\hat{\alpha}\|_1 := \sum_n |\hat{\alpha}(n)| < \infty, \tag{1.3}$$

and

$$\alpha(f) = \sum_n \hat{\alpha}(n) f(n), \quad f \in c_0,$$

where $\hat{\alpha}(n) = \alpha(\mathbf{e}_n)$ $(\mathbf{e}_n(n) = 1,$ *and* $\mathbf{e}_n(j) = 0$ *for* $j \neq n)$.

The proof of Theorem 3 is merely an observation, which we state in terms of the Rademacher functions.

Definition 4 A Rademacher system indexed by a set E is the collection $\{r_x : x \in E\}$ of functions defined on $\{-1,1\}^E$, such that for $x \in E$

$$r_x(\omega) = \omega(x), \quad \omega \in \{-1,1\}^E. \tag{1.4}$$

To obtain the first line in (1.3), note that

$$\sup\left\{\left\|\sum_{n=1}^{N} \hat{\alpha}(n)\, r_n\right\|_\infty : N \in \mathbb{N}\right\} = \|\hat{\alpha}\|_1, \tag{1.5}$$

and to obtain the second, use the fact that finitely supported functions on $\mathbb{N}$ are norm-dense in $c_0(\mathbb{N})$.

Soon after F. Riesz had established his characterization of bounded *linear* functionals, M. Fréchet succeeded in obtaining an analogous characterization in the *bilinear* case. (Fréchet announced the result in 1910, and published the details in 1915 [Fr]; Riesz's theorem had appeared in 1909 [Ri$_F$1].) The novel feature in Fréchet's characterization was a two-dimensional extension of the *total variation* in the sense of Vitali. To wit, if f is a real-valued function on $[a,b] \times [a,b]$, then the *total variation* of f can be expressed as

$$\sup\left\{\left\|\sum_{n,m} \Delta^2 f(x_n, y_m)\, r_{nm}\right\|_\infty : a < \cdots < x_n < \cdots < b,\right.$$
$$\left. a < \cdots < y_m < \cdots < b\right\}, \tag{1.6}$$

where Δ^2 is the 'second difference',

$$\begin{aligned}\Delta^2 f(x_n, y_m) \\ = f(x_n, y_m) - f(x_{n-1}, y_m) + f(x_{n-1}, y_{m-1}) - f(x_n, y_{m-1}),\end{aligned} \tag{1.7}$$

and $\{r_{nm} : (n,m) \in \mathbb{N}^2\}$ is the Rademacher system indexed by $\mathbb{N}^2$. The two-dimensional extension of this one-dimensional measurement is given by:

Definition 5 The Fréchet variation of a real-valued function f on $[a,b] \times [a,b]$ is

$$\|f\|_{F_2} = \sup\left\{ \left\| \sum_{n,m} \Delta^2 f(x_n, y_m)\, r_n \otimes r_m \right\|_\infty : a < \cdots < x_n < \cdots < b, \right.$$
$$\left. < \cdots < y_m < \cdots < b \right\}. \tag{1.8}$$

($r_n \otimes r_m$ is defined on $\{-1,1\}^{\mathbb{N}} \times \{-1,1\}^{\mathbb{N}}$ by

$$r_n \otimes r_m(\omega_1, \omega_2) = \omega_1(n)\omega_2(m),$$

and $\|\cdot\|_\infty$ is the supremum over $\{-1,1\}^{\mathbb{N}} \times \{-1,1\}^{\mathbb{N}}$.)

Based on (1.8), the bilinear analog of Riesz's theorem is

Theorem 6 (Fréchet, 1915). *A real-valued bilinear functional β on* $\mathrm{C}([a,b])$ *is bounded if and only if there is a real-valued function h on $[a,b] \times [a,b]$ with $\|h\|_{F_2} < \infty$, and*

$$\beta(f,g) = \int_a^b \int_a^b f \otimes g \; \mathrm{d}h, \quad f \in \mathrm{C}([a,b]),\ g \in \mathrm{C}([a,b]), \tag{1.9}$$

where the right side of (1.9) *is an iterated Riemann–Stieltjes integral.*

The crux of Fréchet's proof was a construction of the integral in (1.9), a non-trivial task at the start of the twentieth century when integration theories had just begun developing.

Like Riesz's theorem, Fréchet's theorem can also be naturally recast in the setting of locally compact Hausdorff spaces; we shall come to this in good time. At this juncture we will prove only its primal version.

Theorem 7 *If β is a bounded bilinear functional on c_0, and $\beta(\mathbf{e}_m, \mathbf{e}_n) := \hat{\beta}(m,n)$, then*

$$\sup\left\{ \left\| \sum_{m\in S, n\in T} \hat{\beta}(m,n)\; r_m \otimes r_n \right\|_\infty : \textit{finite sets } S\subset \mathbb{N}, T\subset\mathbb{N} \right\}$$
$$:= \|\hat{\beta}\|_{F_2} < \infty, \tag{1.10}$$

and

$$\begin{aligned} \beta(f,g) &= \sum_{m=1}^{\infty}\left(\sum_{n=1}^{\infty} \hat{\beta}(m,n)\; g(n)\right) f(m) \\ &= \sum_{n=1}^{\infty}\left(\sum_{m=1}^{\infty} \hat{\beta}(m,n)\; f(m)\right) g(n), \\ & \qquad f\in \mathrm{c}_0, \quad g\in \mathrm{c}_0. \end{aligned} \tag{1.11}$$

Conversely, if $\hat{\beta}$ is a real-valued function on $\mathbb{N}\times\mathbb{N}$ such that $\|\hat{\beta}\|_{F_2} < \infty$, then (1.11) defines a bounded bilinear functional on c_0.

The key to Theorem 7 is

Lemma 8 *If $\hat{\beta} = (\hat{\beta}(m,n) : (m,n)\in\mathbb{N}^2)$ is a scalar array, then*

$$\|\hat{\beta}\|_{F_2} = \sup\left\{ \left| \sum_{m\in S, n\in T} \hat{\beta}(m,n)\, x_m\, y_n \right| : x_m \in [-1,1], \right.$$
$$\left. y_n\in[-1,1], \textit{ finite sets } S\subset\mathbb{N}, T\subset\mathbb{N} \right\}. \tag{1.12}$$

Proof: The right side obviously bounds $\|\hat{\beta}\|_{F_2}$. To establish the reverse inequality, suppose S and T are finite subsets of $\mathbb{N}$, and $\omega \in \{-1,1\}^{\mathbb{N}}$. Then

$$\begin{aligned} \|\hat{\beta}\|_{F_2} &\geq \left\| \sum_{n\in T, m\in S} \hat{\beta}(m,n)\; r_m\otimes r_n \right\|_\infty, \\ &\geq \sum_{n\in T}\left| \sum_{m\in S} \hat{\beta}(m,n)\; r_m(\omega)\right|. \end{aligned} \tag{1.13}$$

If $y_n \in [-1,1]$ for $n \in T$, then the right side of (1.13) bounds

$$\left|\sum_{n\in T}\left(\sum_{m\in S}\hat{\beta}(m,n)\; r_m(\omega)\right) y_n\right| = \left|\sum_{m\in S}\left(\sum_{n\in T}\hat{\beta}(m,n)\; y_n\right) r_m(\omega)\right|. \tag{1.14}$$

By maximizing the right side of (1.14) over $\omega \in \{-1,1\}^{\mathbb{N}}$, we conclude that $\|\hat{\beta}\|_{F_2}$ bounds

$$\sum_{m\in S}\left|\sum_{n\in T}\hat{\beta}(m,n)\; y_n\right|. \tag{1.15}$$

If $x_m \in [-1,1]$ for $m \in S$, then (1.15) bounds

$$\begin{aligned}&\left|\sum_{m\in S}\left(\sum_{n\in T}\hat{\beta}(m,n)\; y_n\right) x_m\right| \\ &\quad = \left|\sum_{m\in S, n\in T}\hat{\beta}(m,n)\, x_m\, y_n\right|,\end{aligned} \tag{1.16}$$

which implies that $\|\hat{\beta}\|_{F_2}$ bounds the right side of (1.12). □

Proof of Theorem 7: If β is a bilinear functional on c_0, with norm $\|\beta\| := \sup\{|\beta(f,g)| : f \in B_{c_0},\ g \in B_{c_0}\}$, then (because finitely supported functions are norm-dense in c_0)

$$\begin{aligned}\|\beta\| = \sup\Bigg\{&\left|\sum_{m\in S, n\in T}\hat{\beta}(m,n)\; x_m\; y_n\right| : x_m \in [-1,1], \\ & y_n \in [-1,1],\ \text{finite sets } S \subset \mathbb{N}, T \subset \mathbb{N}\Bigg\},\end{aligned}$$

and Lemma 8 implies (1.10).

Let $f \in c_0$ and $g \in c_0$. If $N \in \mathbb{N}$, then let $f_N = f\mathbf{1}_{[N]}$ and $g_N = g\mathbf{1}_{[N]}$. (Here and throughout, $[N] = \{1,\dots,N\}$.) Because $f_N \to f$ and $g_N \to g$ as $N \to \infty$ (convergence in c_0), and β is continuous in each coordinate, we obtain $\beta(f_N, g) \to \beta(f,g)$ and $\beta(f, g_N) \to \beta(f,g)$ as $N \to \infty$, and then obtain (1.11) by noting that $\beta(f_N, g_N) = \Sigma_{m=1}^N \Sigma_{n=1}^N \hat{\beta}(m,n) g(n) f(m)$.

Conversely, if $\hat{\beta}$ is a scalar array on $\mathbb{N} \times \mathbb{N}$, and f and g are finitely supported real-valued functions on $\mathbb{N}$, then define

$$\beta(f,g) = \sum_m \sum_n \hat{\beta}(m,n) g(n) f(m). \tag{1.17}$$

By Lemma 8 and the assumption $\|\hat{\beta}\|_{F_2} < \infty$, β is a bounded bilinear functional on a dense subspace of c_0, and therefore determines a bounded bilinear functional on c_0. The first part of the theorem implies (1.10) and (1.11).

Theorem 7 was elementary, basic, and straightforward – view it as a warm-up. In passing, observe that whereas every bounded linear functional on c_0 obviously extends to a bounded linear functional on l^∞, the analogous fact in two dimensions, that every bounded bilinear functional on c_0 extends to a bounded bilinear functional l^∞ is also elementary, but not quite as easy to verify. This 'two-dimensional' fact, specifically that (1.11) extends to f and g in l^∞, will be verified in a later chapter.

2 A Bilinear Theory

Notably, Fréchet did not consider in his 1915 paper the question whether there exist functions with bounded variation in his sense, but with infinite total variation in the sense of Vitali. Whether *bilinear* functionals on $C([a,b])$ can be distinguished from *linear* functionals on $C([a,b]^2)$ is indeed a basic and important issue (Exercises 1, 2, 4, 8). So far as I can determine, Fréchet never considered or raised it (at least, not in print). Be that as it may, this question led directly to the next advance.

Littlewood began his classic 1930 paper [Lit4] thus: 'Professor P.J. Daniell recently asked me if I could find an example of a function of two variables, of bounded variation according to a certain definition of Fréchet, but not according to the usual definition.' Noting that the problem was equivalent to finding real-valued arrays

$$\hat{\beta} = (\hat{\beta}(m,n) : (m,n) \in \mathbb{N}^2)$$

with $\|\hat{\beta}\|_{F_2} < \infty$ and $\|\hat{\beta}\|_1 = \Sigma_{m,n}|\hat{\beta}(m,n)| = \infty$, Littlewood settled the problem by a quick use of the Hilbert inequality (Exercise 1). He then considered this question: whereas there are $\hat{\beta}$ with $\|\hat{\beta}\|_{F_2} < \infty$ and $\|\hat{\beta}\|_1 = \infty$, and (at the other end) $\|\hat{\beta}\|_{F_2} < \infty$ implies $\|\hat{\beta}\|_2 < \infty$ (Exercise 3), are there $p \in (1,2)$ such that

$$\|\hat{\beta}\|_{F_2} < \infty \Rightarrow \|\hat{\beta}\|_p < \infty?$$

Littlewood gave this precise answer.

Theorem 9 (the 4/3 inequality, 1930).

$$\|\hat{\beta}\|_p < \infty \ \textit{ for all } \ \hat{\beta} \textit{ with } \|\hat{\beta}\|_{F_2} < \infty \ \textit{ if and only if } \ p \geq \frac{4}{3}.$$

To establish 'sufficiency', that $\|\hat{\beta}\|_{F_2} < \infty$ implies $\|\hat{\beta}\|_{4/3} < \infty$, Littlewood proved and used the following:

Theorem 10 (the mixed (l^1, l^2)-norm inequality, 1930). *For all real-valued arrays* $\hat{\beta} = (\hat{\beta}(m,n) : (m,n) \in \mathbb{N}^2)$,

$$\sum_m \left(\sum_n |\hat{\beta}(m,n)|^2 \right)^{\frac{1}{2}} \leq \kappa \|\hat{\beta}\|_{F_2}, \tag{2.1}$$

where $\kappa > 0$ *is a universal constant.*

This mixed-norm inequality, which was at the heart of Littlewood's argument, turned out to be a precursor (if not a catalyst) to a subsequent, more general inequality of Grothendieck. We shall come to Grothendieck's inequality in a little while.

To prove 'necessity', that there exists $\hat{\beta}$ with $\|\hat{\beta}\|_{F_2} < \infty$ and

$$\|\hat{\beta}\|_p = \infty \text{ for all } p < 4/3,$$

Littlewood used the *finite Fourier transform.* (You are asked to work this out in Exercise 4, which, like Exercise 1, illustrates first steps in harmonic analysis.)

Besides motivating the inequalities we have just seen, Fréchet's 1915 paper led also to studies of 'bilinear integration', first by Clarkson and Adams in the mid-1930s (e.g., [ClA]), and then by Morse and Transue in the late 1940s through the mid-1950s (e.g., [Mor]). For their part, firmly believing that the two-dimensional framework was interesting, challenging, *and* important, Morse and Transue launched extensive investigations of what they dubbed *bimeasures*: bounded bilinear functionals on $C_0(X) \times C_0(Y)$, where X and Y are locally compact Hausdorff spaces. In this book, we take a somewhat more general point of view:

Definition 11 Let X and Y be sets, and let $\mathcal{C} \subset 2^X$ and $\mathcal{D} \subset 2^Y$ be algebras of subsets of X and Y, respectively. A scalar-valued set-function μ on $\mathcal{C} \times \mathcal{D}$ is an F_2-measure if for each $A \in \mathcal{C}$, $\mu(A, \cdot)$ is a scalar measure on $(Y, \mathcal{D})$, and for each $B \in \mathcal{D}$, $\mu(\cdot, B)$ is a scalar measure on $(X, \mathcal{C})$.

That bimeasures are F_2-measures is the two-dimensional extension of Theorem 2. (The utility of the more general definition is illustrated in Exercise 8.)

When highlighting the existence of 'true' bounded bilinear functionals, Morse and Transue all but ignored Littlewood's prior work. In their first paper on the subject, underscoring 'the difficult problem which Clarkson and Adams solve ... ', they stated [MorTr1, p. 155]: 'That [the Fréchet variation] can be finite while the classical total variation ... of Vitali is infinite has been shown by example by Clarkson and Adams [in [ClA]].' (In their 1933 paper [ClA], the authors did, in passing, attribute to Littlewood the first such example [ClA, p. 827], and then proceeded to give their own [ClA, pp. 837–41]. I prefer Littlewood's simpler example, which turned out to be more illuminating.) The more significant miss by Morse and Transue was a fundamental inequality that would play prominently in the bilinear theory – the same inequality that had been foreshadowed by Littlewood's earlier results.

3 More of the Bilinear

The inequality missed by Morse and Transue first appeared in Grothendieck's 1956 work [Gro2], a major milestone that was missed by most. The paper, pioneering new tensor-theoretic technology, was difficult to read and was hampered by limited circulation. (It was published in a journal carried by only a few university libraries.) The inequality itself, the highlight of Grothendieck's 1956 paper, was eventually unearthed a decade or so later. Recast and reformulated in a Banach space setting, this inequality became the focal point in a seminal 1968 paper by Lindenstrauss and Pelczynski [LiPe]. The impact of this 1968 work was decisive. Since then, the inequality, which Grothendieck himself billed as the 'théorème fondamental de la théorie metrique des produits tensoriels' has been reinterpreted and broadly applied in various contexts of analysis. It has indeed become recognized as a fundamental cornerstone.

Theorem 12 (the Grothendieck inequality). *If $\hat{\beta} = (\hat{\beta}(m,n) : (m,n) \in \mathbb{N}^2)$ is a real-valued array, and $\{\mathbf{x}_n\}$ and $\{\mathbf{y}_n\}$ are finite subsets in B_{l^2}, then*

$$\left|\sum_{n,m} \hat{\beta}(m,\ n)\langle \mathbf{x}_m, \mathbf{y}_n\rangle\right| \leq \kappa_G\ \|\hat{\beta}\|_{F_2}, \tag{3.1}$$

where B_{l^2} is the closed unit ball in l^2, $\langle\cdot,\cdot\rangle$ denotes the usual inner product in l^2, and $\kappa_G > 1$ is a universal constant.

Restated (via Lemma 8), the inequality in (3.1) has a certain aesthetic appeal:

$$\sup\left\{\left|\sum_{m\in S,n\in T}\hat{\beta}(m,n)\langle\mathbf{x}_m,\mathbf{y}_n\rangle\right| : \mathbf{x}_m \in l^2, \mathbf{y}_n \in l^2,\right.$$

$$\left.\|\mathbf{x}_m\|_2 \le 1,\ \|\mathbf{y}_n\|_2 \le 1,\ \text{finite } S \subset \mathbb{N},\ T \subset \mathbb{N}\right\}$$

$$\le \kappa_G \sup\left\{\left|\sum_{m\in S,n\in T}\hat{\beta}(m,n)x_m y_n\right| : x_m \in \mathbb{R},\right.$$

$$\left. y_n \in \mathbb{R},\ |x_m| \le 1,\ |y_n| \le 1,\ \text{finite } S \subset \mathbb{N},\ T \subset \mathbb{N}\right\}. \tag{3.2}$$

So stated, the inequality says that products of scalars on the right side of (3.2) can be replaced, up to a universal constant, by the dot product in a Hilbert space. In this light, a question arises whether one can replace the dot product on the left side of (3.1) with, say, the dual action between vectors in the unit balls of l^p and $l^q, 1/p + 1/q = 1$ and $p \in [1, 2)$. The answer is *no* (Exercise 6).

Grothendieck did not explicitly write what had led him to his 'théorème fondamental', but did remark [Gro2, p. 66] that Littlewood's mixed-norm inequality (Theorem 10) was an instance of it (Exercise 5). The actual motivation not withstanding, the historical connections between Grothendieck's inequality, Morse's and Transue's bimeasures, Littlewood's inequality(ies), and Fréchet's 1915 work are apparent in this important consequence of Theorem 12.

Theorem 13 (the Grothendieck factorization theorem). *Let X be a locally compact Hausdorff space. If β is a bounded bilinear functional on $\mathrm{C}_0(X)$ (a bimeasure on $X \times X$), then there exist probability measures ν_1 and ν_2 on the Borel field of X such that for all $f \in \mathrm{C}_0(X), g \in \mathrm{C}_0(X)$,*

$$|\beta(f,g)| \le \kappa_{\mathrm{G}}\|\beta\|\|f\|_{\mathrm{L}^2(\nu_1)}\|g\|_{\mathrm{L}^2(\nu_2)}, \tag{3.3}$$

where $\kappa_G > 0$ is a universal constant, and

$$\|\beta\| = \sup\{|\beta(f,g)| : (f,g) \in B_{\mathrm{C}_0(X)} \times B_{\mathrm{C}_0(X)}\}.$$

This 'factorization theorem', which can be viewed as a two-dimensional surrogate for the 'one-dimensional' Radon–Nikodym theorem, has a far-reaching impact. A case for it will be duly made in this book.

4 From Bilinear to Multilinear and Fraction-linear

Up to this point we have focused on the bilinear theory. As our story unfolds in chapters to come, we will consider questions about extending 'one-dimensional' and 'two-dimensional' notions to other dimensions: higher as well as fractional. Some answers will be predictable and obvious, but some will reveal surprises. In this final section of the prologue, we briefly sketch the backdrop and preview some of what lies ahead.

The multilinear Fréchet theorem in its simplest guise is a straightforward extension of Theorem 7:

Theorem 14 *An n-linear functional β on c_0 is bounded if and only if $\|\hat{\beta}\|_{F_n} < \infty$, where $\hat{\beta}(k_1, \ldots, k_n) = \beta(\mathbf{e}_{k_1}, \ldots, \mathbf{e}_{k_n})$ and*

$$\|\hat{\beta}\|_{F_n} = \sup\left\{ \left\| \sum_{k_1 \in T_1, \ldots, k_n \in T_n} \hat{\beta}(k_1, \ldots, k_n) r_{k_1} \otimes \cdots \otimes r_{k_n} \right\|_\infty : \text{finite sets } T_1 \subset \mathbb{N}, \ldots, T_n \subset \mathbb{N} \right\}. \tag{4.1}$$

Moreover, the n-linear action of β on c_0 is given by

$$\beta(f_1, \ldots, f_n) = \sum_{k_1} \cdots \left(\sum_{k_n} \hat{\beta}(k_1, \ldots, k_n) f_n(k_n) \right) \cdots f_1(k_1), \quad (f_1, \ldots, f_n) \in c_0 \times \cdots \times c_0. \tag{4.2}$$

Though predictable, the analogous general measure-theoretic version requires a small effort. (The proof is by induction.)

The extension of Littlewood's 4/3-inequality to higher (integer) dimensions is not altogether obvious. (So far that I know, Littlewood himself never addressed the issue.) This extension, needed in a harmonic-analytic context, was stated and first proved by G. Johnson and G. Woodward in [JWo]:

Theorem 15

$$\|\hat{\beta}\|_p < \infty \text{ for all } n\text{-arrays } \hat{\beta} \text{ with } \|\hat{\beta}\|_{F_n} < \infty \quad \text{if and only if } p \geq \frac{2n}{n+1}.$$

'One half' of this theorem could be found also in [Da, p. 33]. For his purpose in [Da], Davie called on Littlewood's mixed-norm inequality

(Theorem 10), but did not need the 4/3-inequality. Nevertheless, he stated the latter, and remarked in passing without supplying proof that 'it [was] not hard to extend Littlewood's result' to obtain

$$\|\hat{\beta}\|_{2n/(n+1)} \leq 3^{\frac{n-1}{2}} n^{\frac{n+1}{2n}} \|\hat{\beta}\|_{F_n}. \tag{4.3}$$

(Davie did not state that (4.3) was optimal.)

Davie's paper is interesting in our context not only for its connection with Littlewood's inequalities, but also for a discussion therein of a seemingly unrelated, then-open question concerning multidimensional extensions of the von-Neumann inequality. This particular question was subsequently answered in the negative by N. Varopoulos, who, en route, demonstrated that there was no general *trilinear* Grothendieck-type inequality. The latter result concerning feasibility of Grothendieck-type inequalities in higher dimensions is a crucial part of our story here, indeed leading back to questions about extensions of Littlewood's 4/3-inequality. I will not dwell here or anywhere else in the book on the original problem concerning the von-Neumann inequality. But I shall state here the question, not only for its role as a catalyst, but also because an interesting related problem remains open. It is worth a small detour.

The von-Neumann inequality asserts that if T is a contraction on a Hilbert space and p is a complex polynomial in one variable, then

$$\|p(T)\| \leq \|p\|_\infty := \sup\{|p(z)| : |z| \leq 1\}, \tag{4.4}$$

where $\|\cdot\|$ above denotes the operator norm. The two-dimensional extension of (4.4) asserts that if T_1 and T_2 are commuting contractions on a Hilbert space, and p is a complex polynomial in two variables, then

$$\|p(T_1, T_2)\| \leq \|p\|_\infty := \sup\{|p(z_1, z_2)| : |z_1| \leq 1, |z_2| \leq 1\}. \tag{4.5}$$

(These inequalities can be found in [NF, Chapter 1].) The question whether

$$\|p(T_1, \ldots, T_n)\| \leq \|p\|_\infty,$$

where $n \geq 3$, $T_1, \ldots, T_n$ are commuting contractions on a Hilbert space, and p is a complex polynomial in n variables, was resolved in the negative in [V4]. But a question remains open: for integers $n \geq 3$, are there $K_n > 0$ such that if $T_1, \ldots, T_n$ are commuting contractions on a Hilbert space, and p is a complex polynomial in n variables, then

$$\|p(T_1, \ldots, T_n)\| \leq K_n \|p\|_\infty? \tag{4.6}$$

Let us return to the general $2n/n+1$-inequality in Theorem 15. The arguments used to prove Littlewood's inequality(ies) start from the observation that Rademacher functions are independent in the basic sense manifested by (1.5). The analogous observation in a Fourier-analysis setting is that the lacunary exponentials $\{e^{i3^m x}: m \in \mathbb{N}\}$ on $[0, 2\pi) := \mathbf{T}$ are independent in a like sense. Specifically, if $\Sigma_m \hat{\alpha}(m)\, e^{i3^m x}$ is the Fourier series of a continuous function on $\mathbf{T}$, then $\Sigma_m |\hat{\alpha}(m)| < \infty$ (cf. (1.5)). This phenomenon had been noted first by S. Sidon in 1926 [Si1], and later gave rise to a general concept whose systematic study was begun by Walter Rudin in his classic 1960 paper [RU1]:

Definition 16 $F \subset \mathbb{Z}$ is a Sidon set if

$$f \in \mathrm{C}_F(\mathbf{T}) \Rightarrow \hat{f} \in l^1(F), \tag{4.7}$$

where $\mathrm{C}_F(\mathbf{T}) := \{f \in \mathrm{C}(\mathbf{T}) : \hat{f}(m) = 0 \text{ for } m \notin F\}$.

Note that the counterpoint to Sidon's theorem (asserting that $\{3^k : k \in \mathbb{N}\}$ is a Sidon set) is that Placherel's theorem is otherwise optimal; that is,

$$\hat{f} \in l^p(\mathbb{Z}) \text{ for } \textit{all } f \in \mathrm{C}(\mathbf{T}) \Leftrightarrow p \geq 2. \tag{4.8}$$

These two 'extremal' properties – Sidon's theorem at one end, and (4.8) at the other – lead naturally to a question: for arbitrary $p \in (1, 2)$, are there $F \subset \mathbb{Z}$ such that

$$\hat{f} \in l^q(F) \text{ for } \textit{all } f \in \mathrm{C}_F(\mathbf{T}) \Leftrightarrow q \geq p? \tag{4.9}$$

To make matters concise, we define the *Sidon exponent* of $F \subset \mathbb{Z}$ by

$$\sigma_F = \inf\{p : \|\hat{f}\|_p < \infty \text{ for all } f \in \mathrm{C}_F(\mathbf{T})\}. \tag{4.10}$$

(Two situations could arise: either $\|\hat{f}\|_{\sigma_F} < \infty$ for all $f \in \mathrm{C}_F(\mathbf{T})$, or there exists $f \in \mathrm{C}_F(\mathbf{T})$ with $\|\hat{f}\|_{\sigma_F} = \infty$. Later in the book we will distinguish between these two scenarios.) Let $E = \{3^k : k \in \mathbb{N}\}$, and define for integers, $n \geq 1$

$$E_n = \{\pm 3^{k_1} \pm \cdots \pm 3^{k_n} : (k_1, \ldots, k_n) \in \mathbb{N}^n\}. \tag{4.11}$$

Transported to a context of Fourier analysis, Theorem 15 implies

$$\hat{f} \in l^q(E_n) \text{ for } \textit{all } f \in \mathrm{C}_{E_n}(\mathbf{T}) \Leftrightarrow q \geq \frac{2n}{n+1}. \tag{4.12}$$

In particular,

$$\sigma_{E_n} = 2\Big/\left(1 + \frac{1}{n}\right), \quad n \in \mathbb{N}, \tag{4.13}$$

which leads to the p-Sidon set problem (see (4.9)): for *arbitrary* $p \in (1,2)$, are there $F \subset \mathbb{Z}$ such that $\sigma_F = p$? The resolution of this problem – it so turned out – followed a resolution of a seemingly unrelated problem, that of extending the Grothendieck inequality to higher dimensions.

The Grothendieck inequality (Theorem 12) is a general assertion about bounded bilinear forms on a Hilbert space: in Theorem 12, replace l^2 by a Hilbert space H, and the inner product $\langle \cdot, \cdot \rangle$ in l^2 by a bounded bilinear form on H. A question arises: is there $K > 0$ such that for all bounded trilinear functionals β on c_0, all bounded trilinear forms A on a Hilbert space H, and all finite subsets $\{\mathbf{x}_n\} \subset B_H$, $\{\mathbf{y}_n\} \subset B_H$, and $\{\mathbf{z}_n\} \subset B_H$,

$$\left| \sum_{k,n,m} \hat{\beta}(m,n,k)\ A(\mathbf{x}_k, \mathbf{y}_m, \mathbf{z}_n) \right| \leq K \|\hat{\beta}\|_{F_3}? \tag{4.14}$$

(Here and throughout, B_X denotes the closed unit ball of a normed linear space X.) The question was answered in the negative by Varopoulos [V4], who demonstrated the following. For $H = l^2(\mathbb{N}^2)$, and $\varphi \in l^\infty(\mathbb{N}^3)$, define

$$A_\varphi(\mathbf{x}, \mathbf{y}, \mathbf{z}) = \sum_{k,m,n} \varphi(k,m,n)\ \mathbf{x}(k,m)\ \mathbf{y}(m,n)\ \mathbf{z}(k,n),$$

$$(\mathbf{x}, \mathbf{y}, \mathbf{z}) \in l^2(\mathbb{N}^2) \times l^2(\mathbb{N}^2) \times l^2(\mathbb{N}^2), \tag{4.15}$$

which, by Cauchy–Schwarz, is a bounded trilinear form on H with norm $\|\varphi\|_\infty$. By use of probabilistic estimates, Varopoulos proved the existence of φ for which there was no $K > 0$ such that (4.14) would hold with $A = A_\varphi$ and all bounded trilinear functionals β on c_0. But a question remained: were there *any* $\varphi \in l^\infty(\mathbb{N}^3)$ for which A_φ would satisfy (4.14) for all bounded trilinear functionals β on c_0?

In 1976 I gave a new proof of the Grothendieck inequality [Bl3]. The proof, cast in a harmonic-analysis framework, was extendible to multidimensional settings, and led eventually to characterizations of *projectively bounded* forms [Bl4]. (Projectively bounded forms are those that satisfy Grothendieck-type inequalities, as in (4.14).) We illustrate this characterization in the case of the trilinear forms in (4.15). Choose and fix an arbitrary two-dimensional enumeration of $E = \{3^k : k \in \mathbb{N}\}$, say $E = \{m_{ij} : (i,j) \in \mathbb{N}^2\}$ (any enumeration will do), and consider

$$E^{\frac{3}{2}} := \{(m_{ij}, m_{jk}, m_{ik}) : (i,j,k) \in \mathbb{N}^3\}. \tag{4.16}$$

We then have

Theorem 17 *For $\varphi \in l^\infty(\mathbb{N}^3)$, the trilinear form A_φ is projectively bounded if and only if there exists a regular Borel measure μ on $\mathbf{T}^3$ such that*

$$\hat{\mu}(m_{ij}, m_{jk}, m_{ik}) = \varphi(i,j,k), \quad (i,j,k) \in \mathbb{N}^3. \tag{4.17}$$

Therefore, the question whether there exist φ such that A_φ is not projectively bounded becomes the question: is $E^{3/2}$ a Sidon set in $\mathbb{Z}^3$? The answer is *no*.

In the course of verifying that $E^{\frac{3}{2}}$ is not a Sidon set, certain combinatorial features of it come to light, suggesting that $E^{3/2}$ is a '3/2-fold' Cartesian product of E. Indeed, following this cue, we arrive at a 6/5-inequality [Bl5], which, in effect, is a '3/2-linear' extension of the Littlewood (bilinear) 4/3-inequality. For a scalar 3-array $\hat{\beta} = (\hat{\beta}(i,j,k) : (i,j,k) \in \mathbb{N}^3)$, define (the '3/2-linear' version of the Fréchet variation)

$$\|\hat{\beta}\|_{F_{3/2}} = \sup\left\{ \left\| \sum_{i\in S, j\in T, k\in U} \hat{\beta}(i,j,k)\; r_{ij} \otimes r_{jk} \otimes r_{ik} \right\|_\infty : \right.$$
$$\left. \text{finite sets } S \subset \mathbb{N}, T \subset \mathbb{N}, U \subset \mathbb{N} \right\}. \tag{4.18}$$

(Rademacher systems in (4.18) are indexed by $\mathbb{N}^2$.) The 6/5-inequality is

Theorem 18

$$\|\hat{\beta}\|_p < \infty \textit{ for all 3-arrays } \hat{\beta} \textit{ with } \|\hat{\beta}\|_{F_{3/2}} < \infty$$
$$\textit{if and only if } p \geq 6/5.$$

Transporting this inequality to a setting of Fourier analysis, we let

$$E_{3/2} = \{\pm\, m_{ij} \pm m_{jk} \pm m_{ik} : (i,j,k) \in \mathbb{N}^3\}, \tag{4.19}$$

where $\{m_{ij} \colon (i,j) \in \mathbb{N}^2\}$ is an enumeration of $\{e^{2\pi i 3^k t} \colon k \in \mathbb{N}\}$, and obtain that

$$\sigma_{E_{3/2}} = \frac{6}{5} = 2\Big/\left(1 + 1\Big/\left(\frac{3}{2}\right)\right) \quad \text{(cf. (4.12))}. \tag{4.20}$$

The assertion in (4.20) is a precise link between the harmonic-analytic index $\sigma_{E_{3/2}}$ and the 'dimension' 3/2, a purely combinatorial index. This link naturally suggests a formula relating the harmonic-analytic index of a general 'fractional Cartesian product' to its underlying dimension, and thus the solution of the p-Sidon set problem. This (and much more) will be detailed in good time. The prologue is over. Let us begin.

Exercises

1. i. (*The Hilbert inequality*). Prove that if $(a_n) \in B_{l^2}$ and $(b_n) \in B_{l^2}$ are finitely supported sequences, then

$$\left| \sum_{m \neq n} a_n b_m/(m-n) \right| \leq K,$$

where K is a universal constant.

ii. Applying the Hilbert inequality, reproduce Littlewood's proof of the assertion (on p. 164 of [Li]) that there exist $\hat{\beta} = (\hat{\beta}(m,n) : (m,n) \in \mathbb{Z}^2)$ such that $\|\hat{\beta}\|_{F_2} < \infty$ but $\|\hat{\beta}\|_1 = \infty$.

iii. Compute the infimum of the ps such that $\|\hat{\beta}\|_p < \infty$, where $\hat{\beta}$ is the array obtained in ii.

2. Here are two other proofs, using probability theory, that there exist arrays $\hat{\beta} = (\hat{\beta}(m,n) : (m,n) \in \mathbb{N}^2)$ with $\|\hat{\beta}\|_{F_2} < \infty$ and $\|\hat{\beta}\|_1 = \infty$.

i. (a) Let $\{X_n : n \in \mathbb{N}\}$ be a system of statistically independent standard normal variables on a probability space $(X, \mathfrak{A}, \mathbb{P})$. Show that for every positive integer N, there exists a finite partition $\{A_m : m = 1, \dots, 2^N\}$ of $(X, \mathfrak{A})$ such that if $\hat{\beta}_N(m,n) = \frac{1}{n}\mathbf{E}\mathbf{1}_{A_m} X_n$ for $n = 1, \dots, N$ and $m = 1, \dots, 2^N$, and $\hat{\beta}_N(m,n) = 0$ for all other $(n,m) \in \mathbb{N}^2$ ($\mathbf{E}$ denotes expectation, and $\mathbf{1}$ denotes an indicator function), then

$$\|\hat{\beta}_N\|_{F_2} \leq D \text{ and } \|\hat{\beta}_N\|_1 \geq D \log N,$$

where $D > 0$ is an absolute constant.

(b) Use (a) to produce $\hat{\beta} = (\hat{\beta}(m,n) : (m,n) \in \mathbb{N}^2)$ such that $\|\hat{\beta}\|_{F_2} < \infty$ but $\|\hat{\beta}\|_1 = \infty$ (cf. Exercise 4 iv below). What can be said about $\|\hat{\beta}\|_p$ for $p > 1$?

ii. (a) For each $N > 0$, define

$$\hat{\beta}_N(\omega, n) = r_n(\omega)/N^{\frac{1}{2}} 2^N, \quad \omega \in \{-1,1\}^N, \ n \in [N].$$

Prove that $\|\hat{\beta}_N\|_{F_2} \leq 1$. Compute $\|\hat{\beta}_N\|_p$ for $p \geq 1$.

(b) Use (a) to produce $\hat{\beta} = (\hat{\beta}(m,n)\colon (m,n) \in \mathbb{N}^2)$ such that $\|\hat{\beta}\|_{F_2} < \infty$, $\|\hat{\beta}\|_1 = \infty$, and $\|\hat{\beta}\|_p < \infty$ for all $p > 1$.

(Do you see similarities between the constructions in Parts i and ii? Do you see a similarity between the construction in Part ii and Exercise 4 below?)

3. Verify that if $\hat{\beta} = (\hat{\beta}(m,n) : (m,n) \in \mathbb{N}^2)$ is a scalar array then $\|\hat{\beta}\|_2 \leq \|\hat{\beta}\|_{F_2}$.
4. For $N \in \mathbb{N}$, let $\mathbb{Z}_N = [N]$ (a compact Abelian group with addition modulo N). Consider the characters

$$\chi_n(k) = \mathrm{e}^{2\pi \mathrm{i} kn/N}, \; n \in \mathbb{Z}_N, \; k \in \mathbb{Z}_N,$$

and the Haar measure

$$\nu\{k\} = \frac{1}{N}, \quad k \in \mathbb{Z}_N.$$

For $f \in l^\infty(\mathbb{Z}_N)$, define the transform of f by

$$\hat{f}(n) = \sum_{k \in \mathbb{Z}_N} f(k)\overline{\chi_n(k)}\, \nu(k).$$

i. (Orthogonality of characters) For $m \in \mathbb{Z}_N$ and $n \in \mathbb{Z}_N$, prove

$$\sum_{k \in \mathbb{Z}_N} \chi_m(k)\, \overline{\chi_n(k)}\, \nu(k) = \begin{cases} 1 & \text{if } m = n \\ 0 & \text{otherwise.} \end{cases}$$

ii. (Inversion formula, Parseval's formula, Plancherel's theorem) Prove that for $f \in l^\infty(\mathbb{Z}_N)$,

$$f(n) = \sum_{k \in \mathbb{Z}_N} \hat{f}(k)\, \chi_k(n), \quad n \in \mathbb{Z}_N.$$

Conclude that if $f \in l^\infty(\mathbb{Z}_N)$ and $g \in l^\infty(\mathbb{Z}_N)$, then

$$\sum_{k \in \mathbb{Z}_N} f(k)\, \overline{g(k)}\, \nu(k) = \sum_{k \in \mathbb{Z}_N} \hat{f}(k)\, \overline{\hat{g}(k)},$$

and that if $f \in \mathrm{L}^2(\mathbb{Z}_N, \nu)$, then

$$\|f\|_{\mathrm{L}^2(\mathbb{Z}_N,\nu)} = \|\hat{f}\|_{l^2(\mathbb{Z}_N)}.$$

iii. Prove that the 2-array $(\frac{\mathrm{e}^{2\pi\mathrm{i}(mn/N)}}{\sqrt{N}} : (m,n) \in \mathbb{Z}_N \times \mathbb{Z}_N)$ represents an isometry of $l^2(\mathbb{Z}_N)$. Define

$$\hat{\beta}(m,n) = \begin{cases} \frac{\mathrm{e}^{2\pi\mathrm{i}(mn/N)}}{N^{3/2}} & \text{if } (m,n) \in \mathbb{Z}_N \times \mathbb{Z}_N \\ 0 & \text{otherwise,} \end{cases}$$

and verify that $\|\hat{\beta}\|_{F_2} \leq 1$.

iv. Prove there exists a scalar array $\hat{\beta}$ with $\|\hat{\beta}\|_{F_2} < \infty$ and

$$\|\hat{\beta}\|_p = \infty \text{ for all } p < 4/3.$$

5. Prove that Littlewood's mixed norm inequality (Theorem 10) is an instance of the Grothendieck inequality (Theorem 12).
6. Let $\hat{\beta}$ be the scalar array defined in Exercise 4 iii. Let $q \in (2, \infty)$ and evaluate

$$\sup\left\{\left|\sum_{m\in S, n\in T} \hat{\beta}(m,n)\langle \mathbf{e}_m, \mathbf{y}_n\rangle\right| : \right.$$
$$\left. \mathbf{y}_n \in B_{l^q},\ \text{finite sets } S \subset \mathbb{N}, T \subset \mathbb{N}\right\}.$$

What does your computation say about an extension of Littlewood's mixed norm inequality (Theorem 10)? In particular, prove that the inner product in Grothendieck's inequality cannot be replaced by the dual action between vectors in the unit balls of l^p and l^q, $1/p + 1/q = 1$ and $p \in [1, 2)$.
7. Prove that β is a bounded n-linear functional on c_0 if and only if

$$\sum_{k_1,\ldots,k_n} \hat{\beta}(k_1,\ldots,k_n)\mathrm{e}^{\mathrm{i}3^{k_1}x_1}\cdots\mathrm{e}^{\mathrm{i}3^{k_n}x_n}$$

represents a continuous function on $\mathbf{T}^n$.
8. This exercise, providing yet another example of a function with bounded Fréchet variation and infinite total variation, is a prelude to the 'probabilistic' portion of the book.

A stochastic process $\mathrm{W} = \{\mathrm{W}(t) : t \in [0, \infty)\}$ defined on a probability space $(\Omega, \mathfrak{A}, \mathbb{P})$ is a *Wiener process* if it satisfies these properties:

(a) for $0 \le s < t < \infty, \mathrm{W}(t) - \mathrm{W}(s)$ is a normal r.v. with mean zero and variance $t - s$;

(b) for $0 \le t_0 < t_1 \cdots < t_m < \infty$, $\mathrm{W}(t_k) - \mathrm{W}(t_{k-1})$, $k = 1, \ldots, m$, are independent.

Let $\mathfrak{J}$ denote the algebra generated by the intervals

$$\{(s,t] : 0 \le s < t \le 1\},$$

and let μ_{W} be the set-function on $\mathfrak{A} \times \{(s,t] : 0 \le s < t \le 1\}$ defined by

$$\mu_{\mathrm{W}}(A, (s,t]) = \mathbf{E}\mathbf{1}_A(\mathrm{W}(t) - \mathrm{W}(s)), \quad A \in \mathfrak{A}, \quad 0 \le s < t \le 1.$$

i. Extend μ_W by additivity to $\mathfrak{A} \times \mathfrak{J}$.

ii. Prove that μ_W is an F_2-measure on $\mathfrak{A} \times \mathfrak{J}$ which is uniquely extendible to an F_2-measure on $\mathfrak{A} \times \mathfrak{B}$, where $\mathfrak{B}$ is the Borel field in [0,1].

iii Prove that μ_W cannot be extended to a measure on the σ-algebra generated by $\mathfrak{A} \times \mathfrak{B}$.

Hints for Exercises in Chapter I

1. i. Here is an outline of a proof using elementary Fourier analysis. First, compute the Fourier coefficients of $h(x) = x$ on $\mathbf{T}$. Let $f(x) = \Sigma_n \hat{f}(n)\,\mathrm{e}^{inx}$ and $g(x) = \Sigma_n \hat{g}(n)\,\mathrm{e}^{inx}$ be trigonometric polynomials, and observe that

$$\left|\int_{\mathbf{T}} x f(x) g(x) \mathrm{d}x\right| \leq \pi \|f\|_{\mathrm{L}^2} \|g\|_{\mathrm{L}^2}$$
$$= \pi \left(\sum_n |\hat{f}(n)|^2\right)^{\frac{1}{2}} \left(\sum_n |\hat{g}(n)|^2\right)^{\frac{1}{2}}.$$

To prove the Hilbert inequality, use spectral analysis of fg, and apply Parseval's formula to the integral on the left side.

ii. Littlewood let $a_n = b_n = 1/\sqrt{|n|}(\log|n|)^{\alpha}$ for $n \in \mathbb{N}$, where $1/2 < \alpha < 1$, and then defined $\hat{\beta}(m,n) = a_n b_m/(m-n)$ for $n \neq m$.

2. For $N > 0$, consider $E_i = \{X_i > 0\}$, $i \in [N]$, and then for $s = (s_1, \ldots, s_N) \in \{-1, 1\}^N$, let

$$A_s = E_1^{s_1} \cap E_2^{s_2} \ldots \cap E_k^{s_k},$$

where $E_i^{s_i} = E_i$ if $s_i = 1$, and $E_i^{s_i} = (E_i)^c$ if $s_i = -1$.

3. Cf. Plancherel's theorem.

7. This exercise involves basic notions that are covered at length in Chapter VII.

8. See Remark iv in Chapter VI § 2.

II

Three Classical Inequalities

1 Mise en Scène: Rademacher Functions

Rademacher functions r_n, $n \in \mathbb{N}$, are used here and throughout the book as basic building blocks – there are none more basic! Their original definition, by H. Rademacher in [R, p. 130], was this: if $x \in [0,1]$, and $\Sigma_{n=1}^{\infty} b_n(x)/2^n$ is its binary expansion, then

$$r_n(x) = 1\text{–}2b_n(x), \quad n \in \mathbb{N}. \tag{1.1}$$

(To remove ambiguity for dyadic rationals x, take $b_n(x) = 1$ for infinitely many ns.) This definition, still fairly pervasive throughout the literature, is sometimes restated as

$$\begin{aligned} r_n(x) &= \text{sign } (\sin 2^n \pi x), \quad n \in \mathbb{N}, \\ & \qquad x \in [0,1], \quad x \neq \text{dyadic rational.} \end{aligned} \tag{1.1$'$}$$

For example, see [Zy2, p. 6], [LiTz, p. 24], [Kah3, p. 1], [Hel, p. 170].

In our setting, a *Rademacher system indexed by a set E* will mean a collection of functions $\{r_e : e \in E\}$, defined on $\{-1,1\}^E$ by

$$r_e(\omega) = \omega(e), \quad e \in E, \ \omega \in \{-1,1\}^E. \tag{1.2}$$

While the definitions in (1.1) and (1.2) are equivalent (Exercise 1), I prefer the definition in (1.2) because it makes transparent underlying structures that are germane to these functions. In this book, except for occasional exercises and historical notes, elements of Rademacher systems will always be functions whose domains are Cartesian products of $\{-1,1\}$. Eventually we will distinguish between various underlying indexing sets E, but in the beginning (and for a long while until further notice) we shall use the generic indexing $E = \mathbb{N}$. We will denote $\{-1,1\}^{\mathbb{N}}$ by Ω.

Analysis involving $\{r_n : n \in \mathbb{N}\}$ ultimately rests on two elementary observations, each separately expressing a basic property of independence. The first, which we shall formalize later as *functional independence*, is (in this case) merely a restatement of the product structure of Ω:

$$\text{for every choice of signs } \epsilon_n = \pm 1 \ (n \in \mathbb{N}), \text{ there exists } \omega \in \Omega \text{ such that } r_n(\omega) = \epsilon_n \text{ for every } n \in \mathbb{N}. \tag{1.3}$$

This implies that for all finitely supported $\{a_n\} \subset \mathbb{R}$

$$\sup \left\{ \left| \sum_n a_n \, r_n(\omega) \right| \ : \omega \in \Omega \right\} := \left\| \sum_n a_n \, r_n \right\|_\infty = \sum_n |a_n|, \tag{1.4}$$

and therefore (Exercise 9),

$$\sup \left\{ \sum_n |a_n| : \text{finitely supported } \{a_n\} \subset \mathbb{C}, \left\| \sum_n a_n \, r_n \right\|_\infty = 1 \right\} := c_1 \le 2. \tag{1.5}$$

The second observation is based on a statistical structure that we introduce in Ω. Let us assume that a Rademacher function does not 'favor' either one of the two points in its range $\{-1,1\}$, and then consider Ω as the probability space $(\Omega, \mathfrak{A}, \mathbb{P})$, where $\mathfrak{A}$ is the σ-algebra generated by $\{r_n : n \in \mathbb{N}\}$, and the probability measure $\mathbb{P}$ is determined by

$$\mathbb{P}\{r_n = \epsilon_n \text{ for } n \in F\} = \left(\frac{1}{2}\right)^{|F|}, \quad F \subset \mathbb{N}, \ \epsilon_n = \pm 1 \text{ for } n \in F. \tag{1.6}$$

(Here and throughout, $|\cdot|$ denotes cardinality.) The Rademacher functions thus become *statistically independent symmetric random variables* on $(\Omega, \mathfrak{A}, \mathbb{P})$. In particular, for all $m \in \mathbb{N}$, $n_1 \in \mathbb{N}, \ldots, n_m \in \mathbb{N}$,

$$\mathbf{E}\, r_{n_1} \cdots r_{n_m} = \begin{cases} 1 & |\{l : j = n_l\}| \text{ is even for all } j \in \mathbb{N}, \\ 0 & \text{otherwise,} \end{cases} \tag{1.7}$$

where $\mathbf{E}$ denotes *expectation* (Exercise 1). We shall make strong use of the relations in (1.7). Notice that the instance $m = 2$ is the statement that $\{r_n : n \in \mathbb{N}\}$ is an orthonormal set in $L^2(\Omega, \mathbb{P})$; i.e.,

$$\left(\sum_n |a_n|^2\right)^{\frac{1}{2}} = \left(\mathbf{E}\left|\sum_n a_n r_n\right|^2\right)^{\frac{1}{2}}. \tag{1.8}$$

2 The Khintchin L^1–L^2 Inequality

If $f \in L^2(\Omega, \mathbb{P})$, then

$$\|f\|_{L^2} \geq \|f\|_{L^1}. \tag{2.1}$$

The reverse inequality is generally false; there exist $f \in L^2(\Omega, \mathbb{P})$ such that $\|f\|_{L^1} = 1$ and $\|f\|_{L^2} = \infty$. However, the L^1 and L^2 norms *are* equivalent on the span of the Rademacher system. The latter assertion – widely known as the 'Khintchin inequality' – is among the important tools in modern analysis.

Theorem 1 (the Khintchin L^1–L^2 inequality). *There exists $\kappa_K > 0$ such that for every scalar sequence (a_n) with finite support,*

$$\kappa_K \, \mathbf{E}\left|\sum_n a_n \, r_n\right| \geq \left(\sum_n |a_n|^2\right)^{\frac{1}{2}}. \tag{2.2}$$

Proof: The instance $m = 4$ in (1.7) implies $\mathbf{E}\, r_{n_1} r_{n_2} r_{n_3} r_{n_4} = 1$ whenever two pairs of indices assume the same value, and otherwise, $\mathbf{E}\, r_{n_1} r_{n_2} r_{n_3} r_{n_4} = 0$. Because there are three ways that two pairs of indices can assume the same value ($\{n_1 = n_2, \ n_3 = n_4\}$, $\{n_1 = n_3, \ n_2 = n_4\}$, and $\{n_1 = n_4, \ n_2 = n_3\}$), we obtain

$$\begin{aligned} \mathbf{E}\left|\sum_n a_n \, r_n\right|^4 &= \sum_{n_1,n_2,n_3,n_4} a_{n_1} \, a_{n_2} \, \overline{a}_{n_3} \, \overline{a}_{n_4} \, \mathbf{E}\, r_{n_1} \, r_{n_2} \, r_{n_3} \, r_{n_4} \\ &\leq 3 \sum_{n,m} |a_n|^2 |a_m|^2 = 3\left(\sum_n |a_n|^2\right)^2. \end{aligned} \tag{2.3}$$

Write

$$\mathbf{E}\left|\sum_n a_n\, r_n\right|^2 = \mathbf{E}\left|\sum_n a_n\, r_n\right|^{\frac{2}{3}} \left|\sum_n a_n\, r_n\right|^{\frac{4}{3}}. \tag{2.4}$$

Apply Hölder's inequality with exponents 3/2 and 3 to the right side of (2.4), and obtain

$$\left(\mathbf{E}\left|\sum_n a_n\, r_n\right|^2\right)^{\frac{1}{2}} \le \left(\mathbf{E}\left|\sum_n a_n\, r_n\right|\right)^{\frac{1}{3}} \left(\mathbf{E}\left|\sum_n a_n\, r_n\right|^4\right)^{\frac{1}{6}}. \tag{2.5}$$

Substitute (2.3) in (2.5), and deduce (2.2) (via (1.8)) with $\kappa_K = \sqrt{3}$. □

Remark (notes, mainly historical). In the argument above, the Khintchin L^1–L^2 inequality is derived from the L^2–L^4 inequality in (2.3), which, in turn, is derived from the instance $m = 4$ in (1.7). The full strength of (1.7) (derived from the statistical independence of the Rademacher system) implies L^2–L^{2m} inequalities for all $m \in \mathbb{N}$ (Exercise 3). These L^2–L^{2m} inequalities had been first stated in 1923 by A. Khintchin [Kh1, p. 112], en route to his law of the iterated logarithm, and rediscovered in 1930 by Paley and Zygmund in their study of random series [PaZy1]. Inequalities similar to Khintchin's, involving functions that would later be dubbed *Steinhaus*, were published in 1925 and 1926 by J.E. Littlewood [Lit1], [Lit2], [Lit3], who evidently was unaware of Khintchin's inequalities. We shall return to these matters in Chapter VII.

Khintchin himself did not state the L^1–L^2 inequality that today bears his name. It had been stated first by Littlewood in [Lit3], in a form somewhat different from Theorem 1, and later rephrased by him in [Lit4], in the form of Theorem 1, in order to derive the mixed-norm inequality that we will see in the next section. The proof above of Theorem 1 is Littlewood's; indeed, his 'Rademacher functions' are more akin to (1.2) than to (1.1) (see [Lit4, pp. 169–70]). Be that as it may, the L^1–L^2 inequality involving the Rademacher system today bears Khintchin's name.

An equivalent formulation of the Khintchin L^1–L^2 inequality – that all functions in the L^1-closure of the linear span of the Rademacher system are square-integrable – was proved in 1930 by S. Kacmarz and H. Steinhaus [KaSte, Théorème 8]. Kacmarz and Steinhaus knew, by way of functional analysis that had just then begun developing, that this assertion was equivalent to the L^1–L^2 inequality [KaSte, p. 246] (Exercise 2). But their proof of Théorème 8, and thus the inequality, was rather circuitous. Three years later, W. Orlicz [Or] stated the L^1–L^2 inequality without proof, citing Khintchin's 1923 paper [Kh1] and Paley's and Zygmund's 1930 work [PaZy1]. The paper by Paley and Zygmund contained the full system of L^2–L^q inequalities for all $q > 2$, but not an explicit statement of the L^1–L^2 inequality. The proof of the L^1–L^2 inequality was apparently relegated to folklore sometime between 1930 and 1933. More will be said of Orlicz's paper [O] in the next section.

3 The Littlewood and Orlicz Mixed-norm Inequalities

We define the Fréchet variation of a scalar array

$$\beta = (\beta(m,n) : (m,n) \in \mathbb{N}^2)$$

to be

$$\|\beta\|_{F_2} = \sup \left\{ \left\| \sum_{m \in S, n \in T} \beta(m,n)\, r_m \otimes r_n \right\|_{L^\infty} : \text{finite sets } S \subset \mathbb{N},\ T \subset \mathbb{N} \right\}, \tag{3.1}$$

where $r_m \otimes r_n$ is the function on $\Omega \times \Omega$ defined by

$$r_m \otimes r_n(\omega_1, \omega_2) = r_m(\omega_1) r_n(\omega_2)$$

for $(\omega_1, \omega_2) \in \Omega \times \Omega$. We denote by $F_2(\mathbb{N}, \mathbb{N})$ the class of all scalar arrays β indexed by $\mathbb{N}^2$ such that $\|\beta\|_{F_2} < \infty$.

Theorem 2 (Littlewood's (l^1, l^2)-mixed norm inequality). *There exists $\kappa_L > 0$ such that for every $\beta \in F_2(\mathbb{N}, \mathbb{N})$,*

$$\kappa_L \|\beta\|_{F_2} \geq \sum_m \left(\sum_n |\beta(m,n)|^2 \right)^{\frac{1}{2}}, \tag{3.2}$$

where $\kappa_L \leq c_1 \kappa_K$. ($c_1$ is defined in (1.5), *and κ_K is the constant in* (2.2).)

Proof: We can assume that β has finite support. By an application of (1.5), for all $\omega \in \Omega$,

$$c_1 \left\| \sum_{m,n} \beta(m,n)\, r_m \otimes r_n \right\|_{L^\infty} \geq \sum_m \left| \sum_n \beta(m,n)\, r_n(\omega) \right| . \tag{3.3}$$

Therefore, by taking *expectation* on both sides of (3.3), we obtain

$$c_1 \left\| \sum_{m,n} \beta(m,n)\, r_m \otimes r_n \right\|_{L^\infty} \geq \sum_m \mathbf{E} \left| \sum_n \beta(m,n)\, r_n \right| . \tag{3.4}$$

An application of (2.2) implies

$$c_1 \kappa_K \left\| \sum_{m,n} \beta(m,n)\, r_m \otimes r_n \right\|_{L^\infty} \geq \sum_m \left(\sum_n |\beta(m,n)|^2 \right)^{\frac{1}{2}} . \tag{3.5}$$

□

Theorem 3 (the Orlicz (l^2, l^1)-mixed norm inequality). *There exists $\kappa_0 > 0$ such that for every $\beta \in F_2(\mathbb{N}, \mathbb{N})$,*

$$\kappa_0 \|\beta\|_{F_2} \geq \left(\sum_n \left(\sum_m |\beta(m,n)| \right)^2 \right)^{\frac{1}{2}} , \tag{3.6}$$

and $\kappa_0 \leq \kappa_L$.

Proof: By the triangle inequality in l^2,

$$\sum_m \left(\sum_n |\beta(m,n)|^2 \right)^{\frac{1}{2}} \geq \left(\sum_n \left(\sum_m |\beta(m,n)| \right)^2 \right)^{\frac{1}{2}} , \tag{3.7}$$

which, by Theorem 2, implies (3.6) with $\kappa_0 \leq \kappa_L$. □

Remark (what Orlicz did). In his 1933 paper [Or] (cited in the Remark in the previous section), Orlicz established for $p \geq 1$, that if

$$\sum_n f_n \text{ is unconditionally convergent in } L^p([0,1], \mathfrak{m}) \tag{3.8}$$

($\mathfrak{m}$ = Lebesgue measure), then for all $N > 0$,

$$\int_0^1 \left(\sum_{n=1}^N |f_n|^2 \right)^{\frac{p}{2}} dx \leq K, \tag{3.9}$$

and therefore,

$$\sum_{n=1}^{\infty}\left(\int_0^1 |f_n|^p \mathrm{d}x\right)^{\frac{2}{p}} < \infty. \tag{3.10}$$

(See Exercise 5.) The implication (3.9) ⇒ (3.10) is nowadays routine, via the generalized Minkowski inequality (Exercise 4), but was not routine during the 1930s. Orlicz established the implication (3.8) ⇒ (3.9) by an application of the Khintchin L^1–L^2 inequality, and then gave a detailed proof of the implication (3.9) ⇒ (3.10). Littlewood's mixed-norm inequality (3.2) is in effect an instance of the implication (3.8) ⇒ (3.9), and (3.6) is an instance of the implication (3.8) ⇒ (3.10). Working in a context different from Littlewood's, Orlicz was apparently unaware of the work in [Lit4].

4 The Three Inequalities are Equivalent

In §2 we proved the Khintchin L^1–L^2 inequality. In §3 we used it to deduce Littlewood's mixed-norm inequality, and then applied the latter to obtain the Orlicz mixed-norm inequality. The three inequalities are in fact equivalent; any one is derivable from the other. To show this, it suffices to prove

Theorem 4 *If* (3.6) *holds for all* $\beta \in F_2(\mathbb{N},\mathbb{N})$, *then for every finitely supported sequence of scalars* (a_n),

$$\kappa_0 \,\mathbf{E}\left|\sum_n a_n r_n\right| \geq \left(\sum_n |a_n|^2\right)^{\frac{1}{2}}. \tag{4.1}$$

Proof: Let $(a_n : n \in [N])$ be a scalar sequence, and assume

$$\mathbf{E}\left|\sum_{n=1}^{N} a_n r_n\right| = 1. \tag{4.2}$$

This implies that for every choice of signs $\epsilon_n = \pm 1,\ n = 1, \ldots, N$,

$$\mathbf{E}\left|\sum_{n=1}^{N} \epsilon_n a_n r_n\right| = 1 \tag{4.3}$$

(Exercise 6). Rewrite (4.3) as

$$\left(\frac{1}{2}\right)^N \sum_{\omega\in\{-1,1\}^N} \left|\sum_{n=1}^{N} r_n(u)a_n r_n(\omega)\right| = 1, \quad u \in \Omega, \tag{4.4}$$

which implies (Exercise 7)

$$\left|\left(\frac{1}{2}\right)^N \sum_{\omega\in\{-1,1\}^N} \sum_{n=1}^{N} a_n\, r_n(\omega) r_n(u) r_\omega(v)\right| \le 1, \quad (u,v) \in \Omega^2. \tag{4.5}$$

Define

$$\beta(\omega, n) = a_n\, r_n(\omega)/2^N, \quad \omega \in \{-1,1\}^N,\ n \in \{1,\ldots,N\}. \tag{4.6}$$

After re-indexing, we substitute (4.6) in (3.6), and obtain, by (4.5),

$$\left(\sum_{n=1}^{N}\left(\sum_{\omega\in\{-1,1\}^N} |a_n\, r_n(\omega)|/2^N\right)^2\right)^{\frac{1}{2}} = \left(\sum_{n=1}^{N}|a_n|^2\right)^{\frac{1}{2}} \le \kappa_0. \tag{4.7}$$

□

5 An Application: Littlewood's 4/3-inequality

Littlewood used his mixed-norm inequality (Theorem 2) to prove that the F_2-variation of a scalar array bounds, up to a constant, its $l^{4/3}$-norm. He also demonstrated that this assertion was sharp; that the exponent 4/3 could not, in general, be replaced by exponents $p < 4/3$. (We have already commented in Chapter I on Littlewood's original motivation behind this inequality; that it was a question raised by Daniell (of the *Daniell integral* fame) concerning existence of functions on $[0,1]^2$ 'of bounded variation according to a certain definition of Fréchet, but not according to the usual definition' [Lit4, p. 164].)

Littlewood's 4/3-inequality, which we state and prove below, extends to two dimensions the elementary, one-dimensional inequality in (1.5). Later in the book, this inequality will be further extended to higher dimensional frameworks, including framework of 'fractional dimension'.

Theorem 5

i. *There exists* $1 < \lambda \leq (\kappa_0 \kappa_L)^{\frac{1}{2}}$ *such that for all* $\beta \in F_2(\mathbb{N}, \mathbb{N})$,

$$\|\beta\|_{4/3} \leq \lambda \, \|\beta\|_{F_2}. \tag{5.1}$$

(κ_L *and* κ_0 *are the respective constants in Theorem 2 and Theorem 3.)*

ii. *There exist* $\beta \in F_2(\mathbb{N}, \mathbb{N})$ *such that* $\|\beta\|_p = \infty$ *for all* $p < 4/3$.

Proof:

i. Let $\beta \in F_2(\mathbb{N}, \mathbb{N})$, and write

$$\sum_{n,m} |\beta(n,m)|^{\frac{4}{3}} = \sum_n \left(\sum_m |\beta(n,m)|^{\frac{2}{3}} |\beta(n,m)|^{\frac{2}{3}} \right). \tag{5.2}$$

In the sum over m, apply Hölder's inequality with exponents 3 and 3/2, and then in the sum over n, apply Hölder's inequality with exponents 3/2 and 3. The result is

$$\sum_{n,m} |\beta(n,m)|^{\frac{4}{3}} \leq \left(\sum_n \left(\sum_m |\beta(n,m)|^2 \right)^{\frac{1}{2}} \right)^{\frac{2}{3}}$$

$$\left(\sum_n \left(\sum_m |\beta(n,m)| \right)^2 \right)^{\frac{1}{3}}. \tag{5.3}$$

To the left factor on the right side of (5.3), apply Littlewood's (l^1, l^2)-mixed norm inequality (Theorem 2), and to the right factor apply Orlicz's (l^2, l^1)-mixed norm inequality (Theorem 3). The result is

$$\left(\sum_{n,m} |\beta(n,m)|^{\frac{4}{3}} \right)^{\frac{3}{4}} \leq (\kappa_0 \kappa_L)^{\frac{1}{2}} \|\beta\|_{F_2}. \tag{5.4}$$

ii. Let n be a positive integer. Define

$$\beta_n(j,k) = (1/n)^{\frac{3}{2}} \, e^{2\pi i jk/n}, \quad (j,k) \in [n] \times [n], \tag{5.5}$$

and $\beta_n(j,k) = 0$ for $(j,k) \notin [n] \times [n]$. Then,

$$\|\beta_n\|_{F_2} \leq 1 \quad \text{and} \quad \|\beta_n\|_p = n^{(\frac{2}{p} - \frac{3}{2})}, \tag{5.6}$$

which implies the assertion (Exercise 8).

□

Remark (about Littlewood's original arguments). The proof above of (5.1), which is somewhat different from the argument in [Lit4], was shown to me by Sten Kaijser during my visit to Uppsala University in the fall of 1977. Littlewood's original argument implied the constant $2\kappa_{\mathrm{L}}$ in place of $(\kappa_0\kappa_{\mathrm{L}})^{\frac{1}{2}}$ on the right side of (5.1). The value of the best constant is unknown. (Best constants will be discussed briefly in the next section.)

The proof above that (5.1) is sharp (Theorem 5 ii) is Littlewood's original argument. It is based on the fact that the Gauss matrix

$$(1/\sqrt{n})\ \mathrm{e}^{2\pi \mathrm{i} jk/n}, \quad (j,k) \in [n] \times [n], \tag{5.7}$$

determines an isometry of $l^2([n])$ (Exercise 8). Indeed, the essence of Theorem 5 ii is the existence of matrices $(\beta(j,k) : (j,k) \in [n] \times [n])$ such that $|\beta(j,k)| = 1$ for $(j,k) \in [n] \times [n]$, *and* $(1/\sqrt{n})\beta$ represents an operator on $l^2([n])$ with norm independent of n.

6 General Systems and Best Constants

Analysis of $\{-1,1\}^{\mathbb{N}}$ can be generalized in a framework consisting of a set E, a collection of sets $\{D_e : e \in E\}$, and a system of projections $\{\chi_e : e \in E\}$ defined by

$$\chi_e(x) = x(e), \quad x \in \prod_{e \in E} D_e. \tag{6.1}$$

Specifically, we consider these generalizations of the Rademacher system. Let $m \geq 2$ be an integer, and let $\mathbf{T}_m$ denote the set of mth-roots of unity,

$$\mathbf{T}_m = \{\ \mathrm{e}^{2\pi \mathrm{i} j/m} : j = 0, \dots, m-1\}, \quad (\mathrm{i}\ = \sqrt{-1}). \tag{6.2}$$

Denote the infinite product $(\mathbf{T}_m)^{\mathbb{N}}$ by Ω_m, and let

$$S_m := \{\chi_n^m : n \in \mathbb{N}\}$$

be the system of projections from Ω_m onto $\mathbf{T}_m$. The *functional independence* property (cf. (1.3)) is obvious:

$$\text{if } u_n \in \mathbf{T}_m \text{ for } n \in \mathbb{N}, \text{ then there exists } x \in \Omega_m \text{ such that } \chi_n^m(x) = u_n \text{ for } n \in \mathbb{N}. \tag{6.3}$$

Lemma 6 (cf. (1.5)). *For every integer $m \geq 2$,*

$$c_m := \sup\left\{\sum_n |a_n| : \text{finitely supported } \{a_n\} \subset \mathbb{C}, \left\|\sum_n a_n \chi_n^m\right\|_{L^\infty} = 1\right\} < \infty. \tag{6.4}$$

Proof: It can be assumed that $m \geq 3$. Let $\{a_1, \ldots, a_N\} \subset \mathbb{C}$. By applying (6.3), find $x \in \Omega_m$ such that

$$a_n \chi_n^m(x) = |a_n|\, e^{i\varphi_n}, \quad \varphi_n \in [0, 2\pi/m],\ n = 1, \ldots, N. \tag{6.5}$$

Then, by projecting $a_n \chi_n^m(x)$ on the line bisecting the angle $2\pi/m$, we obtain

$$\left|\sum_{n=1}^{N} a_n\ \chi_n^m(x)\right| \geq (\cos \pi/m) \sum_{n=1}^{N} |a_n|. \tag{6.6}$$

□

It is obvious that $\lim_{m\to\infty} c_m = 1$ (by (6.6)), but the values of the c_m appear to be unknown (Exercise 9).

We will consider also the 'limiting case' $m = \infty$: in (6.1), let $E = \mathbb{N}$ and

$$D_n = \mathbf{T}_\infty := \{e^{2\pi i t} : t \in [0,1]\}, \quad n \in \mathbb{N}, \tag{6.7}$$

and denote the corresponding system of projections by

$$S_\infty := \{\chi_n^\infty : n \in \mathbb{N}\}.$$

(The notation $\mathbf{T}_\infty$ for the *circle group* is temporary, and is used only in this section.) Clearly, for all finitely supported $\mathbb{C}$-valued sequences (a_n),

$$\left\|\sum_n a_n\ \chi_n^\infty\right\|_{L^\infty} = \sum_n |a_n|\ ; \tag{6.8}$$

i.e., $c_\infty = \lim_{m\to\infty} c_m = 1$.

Like the case $\Omega = \{-1,1\}^{\mathbb{N}}$ ($m = 2$), we view Ω_m for each $m > 2$ (including $m = \infty$) as a product of uniform probability spaces. That is, let $\mathfrak{A}_m$ be the σ-algebra generated by S_m; endow $\mathbf{T}_m$ with the uniform probability measure (Lebesgue measure in the case $m = \infty$), and let $\mathbb{P}_m$ be the resulting infinite product measure on $\mathfrak{A}_m$ (cf. (1.6)). If $m < \infty$, then S_m is a system of statistically independent $\mathbf{T}_m$-valued random variables on $(\Omega_m, \mathfrak{A}_m, \mathbb{P}_m)$ such that for every $\chi \in S_m$,

$$\mathbf{E}(\chi)^j = \begin{cases} 0 & \text{if } j = 1, \ldots, m-1 \\ 1 & \text{if } j = m. \end{cases} \tag{6.9}$$

Similarly, if $m = \infty$, then S_∞ is a system of statistically independent $\mathbf{T}_\infty$-valued random variables on $(\Omega_\infty, \mathfrak{A}_\infty, \mathbb{P}_\infty)$ such that for every $\chi \in S_\infty$ and $j \in \mathbb{N}$,

$$\mathbf{E}(\chi)^j = 0. \tag{6.10}$$

The system S_∞ played key roles in Littlewood's 1925 and 1926 papers [Lit1], [Lit2], [Lit3], and in Steinhaus's 1930 paper [Ste]. Elements of S_∞ were dubbed *Steinhaus functions* by Salem and Zygmund [SaZy2], and this term has held (see [Kah3, p. 2]).

Khintchin-type L^1–L^2 inequalities and resulting mixed-norm inequalities can be obtained by arguments very similar to the proofs of Theorems 1, 2, and 3 (Exercise 10). These inequalities assert that the constants defined below are finite:

$$\kappa_{\mathrm{K}}(m) = \sup\left\{ \|(a_n)\|_2 : \text{ finitely supported scalar sequence } (a_n), \quad \mathbf{E}\left|\sum_n a_m\, \chi_n^m\right| = 1\right\}; \tag{6.11}$$

$$\kappa_{\mathrm{L}}(m) = \sup\left\{ \sum_i \left(\sum_j |\beta(i,j)|^2\right)^{\frac{1}{2}} : \text{finitely supported scalar arrays } \beta, \quad \left\|\sum_{i,j} \beta(i,j)\chi_i^m \otimes \chi_j^m\right\|_{\mathrm{L}^\infty} = 1\right\}; \tag{6.12}$$

$$\kappa_{\mathrm{O}}(m) = \sup\left\{ \left(\sum_i \left(\sum_j |\beta(i,j)|\right)^2\right)^{\frac{1}{2}} : \text{finitely supported scalar arrays } \beta, \quad \left\|\sum_{i,j} \beta(i,j)\, \chi_i^m \otimes \chi_j^m\right\|_{\mathrm{L}^\infty} = 1 \right. \tag{6.13}$$

As in the case of the Rademacher system, the L^1–L^2 and the respective mixed-norm inequalities stated above are equivalent (see Theorem 4). Precisely, we have

$$\kappa_{\mathrm{K}}(m) \le \kappa_0(m) \le \kappa_{\mathrm{L}}(m) \le c_m\ \kappa_{\mathrm{K}}(m) < \infty. \tag{6.14}$$

Remark (what is, and what is not known about the constants). Let $\kappa_{\mathrm{K}}^{\mathbb{R}}(m), \kappa_{\mathrm{L}}^{\mathbb{R}}(m)$, and $\kappa_0^{\mathbb{R}}(m)$ be the respective constants defined by the right sides of (6.11), (6.12), and (6.13) where the scalar field is $\mathbb{R}$. In [Lit2], Littlewood obtained $\kappa_{\mathrm{K}}(2) \le \sqrt{3}$ (proof of Theorem 1), and left open the problem of determining $\kappa_{\mathrm{K}}(2)$ (see [Hal]). S. Szarek was the first to show, in his Master's thesis [Sz], that $\kappa_{\mathrm{K}}^{\mathbb{R}}(2) = \kappa_{\mathrm{K}}(2) = \sqrt{2}$. Subsequent proofs establishing $\kappa_{\mathrm{K}}(2) = \sqrt{2}$ (increasing in simplicity, but none trivial) can be found in [H2], [To], and [LatO]. It is thus evident, from (1.4) and the proofs of Theorems 1, 2, 3, and 4, that

$$\sqrt{2} = \kappa_{\mathrm{L}}^{\mathbb{R}}(2) = \kappa_0^{\mathbb{R}}(2). \tag{6.15}$$

At the other end, in the case $m = \infty$, J. Sawa computed $\kappa_{\mathrm{K}}(\infty) = 2/\sqrt{\pi}$ [Saw] (also a Master's thesis). Therefore, by (6.8) and (6.14),

$$2/\sqrt{\pi} = \kappa_0(\infty) = \kappa_{\mathrm{L}}(\infty). \tag{6.16}$$

The values of $\kappa_{\mathrm{K}}(m)$ for $3 \le m < \infty$, and the values of $\kappa_{\mathrm{L}}(m)$ and $\kappa_0(m)$ for $2 \le m < \infty$ are unknown.

Exercises

1. (*measure theory warm-up*).
   i. Verify that $\mathbb{P}$ defined by (1.6) determines a probability measure on $(\Omega, \mathfrak{A})$, where $\mathfrak{A}$ is the σ-algebra generated by $\{r_n : n \in \mathbb{N}\}$.
   ii. Verify that $\{r_n : n \in \mathbb{N}\}$ is a system of statistically independent symmetric random variables on $(\Omega, \mathfrak{A}, \mathbb{P})$, and deduce (1.7).
   iii. Verify that if we use Rademacher's original definition of his functions [R], stated in (1.1), then $\{r_n : n \in \mathbb{N}\}$ is a system of statistically independent symmetric random variables on $([0,1], \mathfrak{B}, \mathfrak{m})$, where $\mathfrak{B}$ is the Borel field of [0,1], and $\mathfrak{m}$ is the Lebesgue measure.
2. (*functional analysis warm-up*). Let (X, ν) be a finite measure space, and suppose $\{f_n : n \in \mathbb{N}\}$ is an orthonormal set in $L^2(X, \nu)$. Prove that the following two assertions are equivalent:

(a) there exists $\kappa > 0$ such that for every finite sequence of scalars (a_n),

$$\kappa \mathbf{E} \left| \sum_n a_n f_n \right| \geq \left(\sum_n |a_n|^2 \right)^{\frac{1}{2}};$$

(b) the L^1-closure and the L^2-closure of the linear span of

$$\{f_n : n \in \mathbb{N}\}$$

are equal.

3. i. Prove that for every integer $n > 1$ and finitely supported scalar sequence (a_n)

$$\left(\mathbf{E} \left| \sum_k a_k r_k \right|^{2n} \right)^{\frac{1}{2n}} \leq \left(\frac{(2n)!}{2^n n!} \right)^{\frac{1}{2n}} \left(\sum_k |a_k|^2 \right)^{\frac{1}{2}}$$

$$\leq \sqrt{n} \left(\sum_k |a_k|^2 \right)^{\frac{1}{2}}.$$

(After you establish this, compare your proof with the one on p. 112 of Khintchin's 1923 paper [K1], and the one on pp. 340–2 of Paley's and Zygmund's 1930 paper [PaZyl].)

ii. Prove that for all $p > 2$ and all finitely supported scalar sequences (a_n)

$$\left(\mathbf{E} \left| \sum_n a_n r_n \right|^p \right)^{\frac{1}{p}} \leq \sqrt{p} \left(\sum_n |a_n|^2 \right)^{\frac{1}{2}},$$

and conclude that for all $p > 0$, the L^p-closure and the L^2-closure of the linear span of $\{r_n : n \in \mathbb{N}\}$ are equal.

4. (*the generalized Minkowski inequality*). Let (X, μ) and (Y, ν) be measure spaces. Suppose f is a measurable function on $X \times Y$ and $p \in [1, \infty)$. Prove that

$$\left(\int_X \left(\int_Y |f(x,y)| \, \nu(dx) \right)^p \mu(dy) \right)^{\frac{1}{p}} \leq \int_Y \left(\int_X |f(x,y)|^p \mu(dy) \right)^{\frac{1}{p}} d\nu(dx).$$

(Compare your proof with Orlicz's argument in [Or, p. 36], which established a special case of this inequality.)

5. In this exercise you will reproduce, in a slightly more general form, the main results in [O]. (Compare your proofs with Orlicz's original arguments.) Let V be a normed linear space and let $(x_n : n \in \mathbb{N})$ be a sequence of elements in V. The series $\Sigma_n x_n$ is said to be *unconditionally convergent* if $\Sigma_n x_{n'}$ is convergent in V for every rearrangement $\{x_{n'}\}$ of $\{x_n\}$.

   i. Prove that if $\Sigma_n x_n$ is unconditionally convergent, then there exist $K > 0$ such that for all integers $N > 0$ and all choices of signs $\epsilon_n = \pm 1$ for $n \in [N]$,

   $$\left\| \sum_{n=1}^{N} \epsilon_n \, x_n \right\|_V \leq K.$$

   ii. Let (X, μ) be a measure space. Prove that if $\Sigma_n f_n$ is unconditionally convergent in $\mathrm{L}^p(X, \mu)$ for $p \geq 1$, then for all $N \in \mathbb{N}$

   $$\int_X \left(\sum_{n=1}^{N} |f_n(x)|^2 \right)^{\frac{p}{2}} \mu(\mathrm{d}x) \leq K,$$

   where K is an absolute constant. From this, deduce Littlewood's (l^1, l^2)-mixed-norm inequality (Theorem 2).

   iii. Let (X, μ) be a measure space. Prove that if $\Sigma_n f_n$ is unconditionally convergent in $\mathrm{L}^p(X, \mu)$ for $p \in [1, 2]$, then

   $$\sum_{n=1}^{\infty} \left(\int_X |f_n(x)|^p \mu(\mathrm{d}x) \right)^{\frac{2}{p}} < \infty.$$

   From this, deduce the Orlicz (l^2, l^1)-mixed-norm inequality (Theorem 3).

6. (*probability theory warm-up*). Prove that for all $\{a_n\} \subset \mathbb{C}$, $N \in \mathbb{N}$, and choices of signs $\epsilon_n = \pm 1$, $n \in [N]$,

$$\sum_{n=1}^{N} a_n r_n \sim_d \sum_{n=1}^{N} \epsilon_n a_n r_n$$

($\sim_d$ means 'has the same distribution as').

(This exercise verifies the implication (4.2) $\Rightarrow$ (4.3), which can be deduced also via harmonic analysis (Chapter VII), from the translation invariance of the Haar measure and functional independence of the Rademacher system.)

7. Verify that (4.4) implies (4.5).
8. (*harmonic analysis warm-up*). For positive integers n, consider the Abelian group $\mathbb{Z}_n = \{0, \dots, n-1\}$ with integer addition mod n. The *characters* of $\mathbb{Z}_n$ are the functions

$$\chi_j : \mathbb{Z}_n \to \{z \in \mathbb{C} : |z| = 1\}, \quad j \in \mathbb{Z}_n,$$

defined by

$$\chi_j(k) = \mathrm{e}^{2\pi \mathrm{i} jk/n}, \quad k \in \mathbb{Z}_n.$$

To underscore that $\{\chi_j : j \in \mathbb{Z}_n\}$ forms the dual group of $\mathbb{Z}_n$, we shall denote its underlying indexing set by $\hat{\mathbb{Z}}_n$. The transform $\hat{f}$ of $f \in l^\infty(\mathbb{Z}_n)$ is

$$\hat{f}(j) = \left(\frac{1}{n}\right) \sum_{k \in \mathbb{Z}_n} f(k)\, \overline{\chi_j(k)}, \quad j \in \hat{\mathbb{Z}}_n.$$

i. (*orthogonality of characters*). For $j \in \hat{\mathbb{Z}}_n$ and $k \in \hat{\mathbb{Z}}_n$, prove

$$\left(\frac{1}{n}\right) \sum_{l \in \mathbb{Z}_n} \chi_j(l)\, \overline{\chi_k(l)} = \begin{cases} 1 & \text{if } j = k \\ 0 & \text{otherwise.} \end{cases}$$

ii. For $f \in l^\infty(\mathbb{Z}_n)$, prove *(the inversion formula)*

$$f(l) = \sum_{k \in \hat{\mathbb{Z}}_n} \hat{f}(k)\, \chi_k(l), \quad l \in \mathbb{Z}_n.$$

Conclude that if $g \in l^\infty(\mathbb{Z}_n)$, then *(Parseval's formula)*

$$\left(\frac{1}{n}\right) \sum_{j \in \mathbb{Z}_n} f(j)\, \overline{g(j)} = \sum_{k \in \hat{\mathbb{Z}}_n} \hat{f}(k)\, \overline{\hat{g}(k)},$$

and therefore *(Plancherel's formula)*,

$$\|f\|_{L^2(\mathbb{Z}_n,\nu)} = \|\hat{f}\|_{l^2(\hat{\mathbb{Z}}_n)},$$

where ν is the Haar measure on $\mathbb{Z}_n$, i.e., the uniform probability measure on $\mathbb{Z}_n$.

iii. Prove that $((1/\sqrt{n})\ e^{2\pi i jk/n} : (j,k) \in \hat{\mathbb{Z}}_n \times \mathbb{Z}_n)$ represents an isometry of $l^2([n])$. That is, if $\mathbf{x} = (x_j) \in B_{l^2(\mathbb{Z}_n)}$ and $\mathbf{y} = (y_j) \in B_{l^2(\hat{\mathbb{Z}}_n)}$, then

$$\langle \mathbf{x}, \mathbf{y} \rangle = \sum_{j,k} x_j\ \overline{y_k} = (1/\sqrt{n}) \sum_{j,k} e^{2\pi i jk/n} x_j\ \overline{y_k}.$$

Conclude that $\|\beta_n\|_{F_2} \leq 1$, where β_n is defined by (5.5).

iv. Prove that there exist $\beta \in F_2(\mathbb{N},\mathbb{N})$ such that $\|\beta\|_{F_2} < \infty$ and $\|\hat{\beta}\|_p = \infty$ for all $p < 4/3$.

9*. What are the values of c_n (defined in (1.5) and (6.4)) for $n \geq 1$?

10. Verify the Khintchin L^1–L^2 inequalities for S_m, $m \geq 3$ and $m = \infty$.

11. Verify $\kappa_K(m) \leq \kappa_O(m)$.

12. i. Prove that for every $m \geq 2$ and $n \geq 2$ (including $m = \infty$ and $n = \infty$) there exist $d_{m,n} > 0$ such that for every finitely supported scalar array β,

$$d_{m,n} \left\| \sum_{j,k} \beta(j,k) \chi_j^m \otimes \chi_k^m \right\|_{L^\infty} \geq \left\| \sum_{j,k} \beta(j,k) \chi_j^n \otimes \chi_k^n \right\|_{L^\infty}.$$

ii.* Let the $d_{m,n}$ denote the best constants in the inequalities in i. Note the relations between $d_{m,n}$, $\kappa_L(m)$, $\kappa_O(m)$, $\kappa_L(n)$ and $\kappa_O(n)$. For example,

$$d_{m,2}\kappa_L(2) \geq \kappa_L(m).$$

Can you prove $d_{m,2}\ \kappa_L(2) = \kappa_L(m)$?

Hints for Exercises in Chapter II

1. i. Mimic the construction of the Lebesgue measure on [0,1] by following these steps (see [Roy, Chapter 3]).

Step 1 For $F \subset \mathbb{N}$ finite, and $\epsilon_j = \pm 1$ for $j \in F$, define the cylinder set

$$O = O(\{\epsilon_j\}_{j \in F}) = \{\omega \in \Omega : \omega(j) = \epsilon_j,\ j \in F\}.$$

Observe that cylinder sets are both open and closed in Ω. Define

$$\mathbb{P}(O) = \left(\frac{1}{2}\right)^{|F|}.$$

Denote the class of cylinder sets by $\mathfrak{C}$. For $A \subset \Omega$, define (the *outer probability measure* $\mathbb{P}^*$ of A)

$$\mathbb{P}^*(A) = \inf\left\{\sum_n \mathbb{P}(O_n) : \{O_n : n \in \mathbb{N}\} \subset \mathfrak{C},\ \cup_n O_n \supset A\right\}.$$

For $O \in \mathfrak{C}$, prove that $\mathbb{P}(O) = \mathbb{P}^*(O)$ (see [Roy, Proposition 1, Chapter 3]).

Step 2 Prove that $\mathbb{P}^*$ is countably subadditive (see [Roy, Proposition 2, Chapter 3]).

Step 3 Following Carathéodory, say that $E \subset \Omega$ is *measurable* if for each $A \subset \Omega$,

$$\mathbb{P}^*(A) = \mathbb{P}^*(A \cap E) + \mathbb{P}^*(A \cap E^c).$$

Denote the class of measurable subsets of Ω by $\mathfrak{M}$. Prove that $\mathfrak{M}$ is a σ-algebra (see [Roy, Theorem 10, Chapter 3]).

Step 4 Prove that $\mathfrak{C} \subset \mathfrak{M}$ (see [Roy, Theorem 12, Chapter 3]).

Step 5 Let $\mathfrak{A}$ denote the σ-algebra generated by $\mathfrak{C}$, and then let $\mathbb{P} = \mathbb{P}^*|_{\mathfrak{A}}$. Prove that $(\Omega, \mathfrak{A}, \mathbb{P})$ is a probability space.

3. i. Review the derivation of (2.3).
   ii. See Exercise 2.
4. Use L^p–L^q duality.
5. See [LiTz, pp. 15–16].
6. Use the 'characteristic function' method, the statistical independence of the Rademacher system, and the symmetry of its elements.

8. iv. Extend the definition of β_n in (5.5) to $\mathbb{Z} \times \mathbb{Z}$ by writing $\beta_m(j,k) = 0$ for all negative integers j and k. For $(j,k) \in \mathbb{N} \times \mathbb{N}$, define

$$\beta(j,k) = \sum_{m=1}^{\infty} (1/m^2)\beta_{2^m}(j - 2^m + 1,\ k - 2^m + 1),$$

and then verify that $\beta \in F_2(\mathbb{N},\mathbb{N})$ and $\|\beta\|_p = \infty$ for every $p < 4/3$.)

11. See the proof of Theorem 4.

III

A Fourth Inequality

1 Mise en Scène: Does the Khintchin L^1–L^2 Inequality Imply the Grothendieck Inequality?

Grothendieck's *théorème fondamental de la théorie metrique des produits tensoriels* appeared first in 1956, in a setting of topological tensor products [Gro2, p. 59], and has resurfaced since that time in various contexts under different guises. Its first reformulation was an elementary assertion that has become known as the *Grothendieck inequality* [LiPe, p. 275]: if (a_{mn}) is a finitely supported scalar array such that

$$\left|\sum_{m,n} a_{mn}\, z_m\, w_n\right| \leq 1 \tag{1.1}$$

for all scalar sequences (w_m) and (z_n) in $B_{\mathbb{C}}$ (the closed unit disk of $\mathbb{C}$), then for all sequences $(\mathbf{x}_m)$ and $(\mathbf{y}_n)$ of vectors in B_{l^2} (the closed unit ball of l^2),

$$\left|\sum_{m,n} a_{mn} \langle \mathbf{x}_m, \mathbf{y}_n \rangle\right| \leq K, \tag{1.2}$$

where K is a universal constant and $\langle \cdot, \cdot \rangle$ denotes the usual inner product in l^2.

This inequality can be restated concisely in terms of the Fréchet variation defined in (II.3.1), and the norm

$$\|\beta\|_{g_2} := \sup\left\{ \left| \sum_{m\in S, n\in T} \beta(m,n)\langle \mathbf{x}_m, \mathbf{y}_n\rangle \right| : \right.$$

$$\left. \{\mathbf{x}_m\} \subset B_{l^2},\ \{\mathbf{y}_m\} \subset B_{l^2},\ \text{finite sets } S \subset \mathbb{N}, T \subset \mathbb{N} \right\} \tag{1.3}$$

(Exercise 1). Indeed, for all scalar arrays β,

$$\|\beta\|_{F_2} \le \|\beta\|_{g_2}, \tag{1.4}$$

and the opposite inequality (with a constant) is

Theorem 1 (the Grothendieck inequality). *There exists $\kappa_G > 0$ such that for all scalar arrays β,*

$$\|\beta\|_{g_2} \le \kappa_G\ \|\beta\|_{F_2}. \tag{1.5}$$

Later in the book we will use the following transcription of Theorem 1: if H_1 and H_2 are Hilbert spaces, and η is a bounded bilinear functional on $H_1 \times H_2$, then for all $\beta \in F_2(\mathbb{N},\mathbb{N})$, and finite subsets

$$\{\mathbf{x}_m : m \in S\} \subset H_1,\ \{\mathbf{y}_m : m \in S\} \subset H_2,$$

$$\left| \sum_{m\in S, n\in T} \beta(m,n)\ \eta(\mathbf{x}_m, \mathbf{y}_n) \right| \le \kappa_G\ \|\beta\|_{F_2}\ \|\eta\|\ \max\{\|\mathbf{x}_m\|_{H_1}, \|\mathbf{y}_n\|_{H_2}\}, \tag{1.6}$$

where $\|\eta\| = \sup\{|\eta(\mathbf{x},\mathbf{y})| : \mathbf{x} \in B_{H_1}, \mathbf{y} \in B_{H_2}\}$ (Exercise 2).

If you skipped Exercise I.6, then now is the time to observe that Littlewood's mixed-norm inequality (Theorem II.2) is an instance of Grothendieck's: let $\{\mathbf{e}_n\}$ denote the standard basis in l^2, and note that

$$\begin{aligned}
\sum_m \left(\sum_n |\beta(m,n)|^2 \right)^{\frac{1}{2}} &= \sum_m \left\| \sum_n \beta(m,n)\ \mathbf{e}_n \right\|_2 \\
&= \sup\left\{ \sum_m \left| \left\langle \sum_n \beta(m,n)\ \mathbf{e}_n, \mathbf{y}_m \right\rangle \right| : \{\mathbf{y}_m\} \in B_{l^2} \right\} \\
&\le \|\beta\|_{g_2} \le \kappa_G\ \|\beta\|_{F_2},
\end{aligned} \tag{1.7}$$

which, in particular, implies $\kappa_G \geq \kappa_L$. Therefore, by Theorem II.4, each of the three inequalities in the previous chapter – (I.2.2), (I.3.2), and (I.3.6) – is implied by Theorem 1, and this naturally brings up the question: *can the Grothendieck inequality be derived from, say, the Khintchin* L^1–L^2 *inequality?*

In the next two sections we give two elementary proofs of Theorem 1, each based, in essence, on an 'upgraded' Khintchin L^1–L^2 inequality. In §4 we formalize the property expressed by such an inequality, dub it $\Lambda(2)$-*uniformizability*, and in §5 use it explicitly to give a third proof of Theorem 1. The $\Lambda(2)$-*uniformizability* property will be used again, in similar fashion, in the multidimensional framework of Chapter VIII.

2 An Elementary Proof

Because the Grothendieck inequality in the complex case follows, modulo the 'best' κ_G, from the inequality in the real case, we will assume in this and the next section that all elements in l^2 have real-valued coordinates. (No attempt is made here to compute 'best' constants.)

Our first step is to state an alternative representation of the standard dot product $\langle \mathbf{x}, \mathbf{y} \rangle = \Sigma_n \mathbf{x}(n)\, \mathbf{y}(n)$ in l^2. Let

$$D_m = \{(n_1, \ldots, n_m) : n_i \in \mathbb{N},\ n_1 < \cdots < n_m\}, \tag{2.1}$$

and fix a one–one correspondence between $\mathbb{N}$ and $\bigcup_{k=1}^{\infty} D_{2k+1}$. (Any correspondence will do.) Denote the correspondence by

$$n \leftrightarrow (n_1, \ldots, n_{2j+1}), \tag{2.2}$$

where $n \in \mathbb{N}$, and $(n_1, \ldots, n_{2j+1}) \in \bigcup_{k=1}^{\infty} D_{2k+1}$. For a scalar sequence $\mathbf{x} = (\mathbf{x}(n) : n \in \mathbb{N})$, let $\phi\mathbf{x}$ be the sequence whose nth entry is

$$(\phi\mathbf{x})_n = \mathbf{x}(n_1) \cdots \mathbf{x}(n_{2j+1}), \tag{2.3}$$

where $n \leftrightarrow (n_1, \ldots, n_{2j+1})$ (as per (2.2)). Then, for all $\mathbf{x} \in l^2$ and $\mathbf{y} \in l^2$,

$$\begin{aligned} &\langle \phi\mathbf{x}, \phi\mathbf{y} \rangle \\ &= \sum_{j=1}^{m} \sum_{(n_1, \ldots, n_{2j+1}) \in D_{2j+1}} \mathbf{x}(n_1) \cdots \mathbf{x}(n_{2j+1}) \mathbf{y}(n_1) \cdots \mathbf{y}(n_{2j+1}). \end{aligned} \tag{2.4}$$

For each $j \geq 1$,

$$\sum_{(n_1,\ldots,n_{2j+1})\in D_{2j+1}} |\mathbf{x}(n_1)\cdots\mathbf{x}(n_{2j+1})\ \mathbf{y}(n_1)\cdots\mathbf{y}(n_{2j+1})|$$

$$\leq \frac{1}{(2j+1)!} \sum_{(n_1,\ldots,n_{2j+1})\in\mathbb{N}^{2j+1}} |\mathbf{x}(n_1)\cdots\mathbf{x}(n_{2j+1})\ \mathbf{y}(n_1)\cdots\mathbf{y}(n_{2j+1})|$$

$$\leq \frac{1}{(2j+1)!}(\|\mathbf{x}\|_2\ \|\mathbf{y}\|_2)^{2j+1}, \tag{2.5}$$

and therefore,

$$|\langle\phi\mathbf{x},\phi\mathbf{y}\rangle| \leq \sum_{j=1}^{\infty}\frac{1}{(2j+1)!}(\|\mathbf{x}\|_2\ \|\mathbf{y}\|_2)^{2j+1}$$

$$= \sinh(\|\mathbf{x}\|_2\ \|\mathbf{y}\|_2) - \|\mathbf{x}\|_2\ \|\mathbf{y}\|_2. \tag{2.6}$$

Let $\mathbf{A}$ be the function defined on $l^2 \times l^2$ by

$$\mathbf{A}(\mathbf{x},\mathbf{y}) = \langle\mathbf{x},\mathbf{y}\rangle + \langle\phi\mathbf{x},\phi\mathbf{y}\rangle, \quad (\mathbf{x},\mathbf{y}) \in l^2 \times l^2. \tag{2.7}$$

Define $\theta\mathbf{x} = \frac{\phi\mathbf{x}}{\sqrt{\sinh(1)-1}}$, and rewrite (2.7) as

$$\langle\mathbf{x},\mathbf{y}\rangle = \mathbf{A}(\mathbf{x},\mathbf{y}) - (\sinh(1)-1)\ \langle\theta\mathbf{x},\theta\mathbf{y}\rangle. \tag{2.8}$$

Lemma 2 *For all* $\mathbf{x}$ *and* $\mathbf{y}$ *in* B_{l^2},

$$\langle\mathbf{x},\mathbf{y}\rangle = \sum_{k=0}^{\infty}(-1)^k(\sinh(1)-1)^k\ \mathbf{A}(\theta^k\mathbf{x},\ \theta^k\mathbf{y}), \tag{2.9}$$

where the convergence of the series is uniform in B_{l^2}. *(*θ^k *is the* k*th iterate of* θ*.)*

Proof: By iterating (2.8), we obtain for $j = 0,\ldots$

$$\langle\mathbf{x},\mathbf{y}\rangle = \sum_{k=0}^{j}(-1)^k(\sinh(1)-1)^k\ \mathbf{A}(\theta^k\mathbf{x},\theta^k\mathbf{y})$$

$$+ (-1)^{j+1}(\sinh(1)-1)^{j+1}\langle\theta^{j+1}\mathbf{x},\theta^{j+1}\mathbf{y}\rangle. \tag{2.10}$$

By (2.6), θ is a map from B_{l^2} into B_{l^2}, and, therefore, the second term on the right side of (2.10) converges to 0 uniformly in B_{l^2}. □

Lemma 3 *If β is a scalar array, and $\{\mathbf{x}_m : m \in S\}$ and $\{\mathbf{y}_n : n \in \mathbb{N}\}$ are finite subsets of B_{l^2}, then*

$$\left| \sum_{m\in S, n\in T} \beta(m,n)\ \mathbf{A}(\mathbf{x}_m, \mathbf{y}_n) \right| \leq \mathrm{e}\|\beta\|_{F_2}. \tag{2.11}$$

Proof: In the ensuing argument, we assume all vectors have finite support (Exercise 3 ii). By the (statistical) independence of the Rademacher system, for all vectors $\mathbf{x}$ and $\mathbf{y}$,

$$\mathbf{A}(\mathbf{x}, \mathbf{y}) = \mathbf{E}\ \mathscr{Im} \prod_k (1 + \mathrm{i}\mathbf{x}(k) r_k)\ \mathscr{Im} \prod_k (1 + \mathrm{i}\mathbf{y}(k) r_k). \tag{2.12}$$

($\mathrm{i} = \sqrt{-1}$, and $\mathscr{Im}$ denotes the imaginary part; see Exercise 3.) Note that

$$\left\| \prod_k (1 + \mathrm{i}\mathbf{x}(k) r_k) \right\|_{\mathrm{L}^\infty} \leq \exp\left(\frac{1}{2}\|\mathbf{x}\|_2^2\right), \tag{2.13}$$

which follows from $|1+\mathrm{i}\mathbf{x}(k)r_k| = \sqrt{1 + (\mathbf{x}(k))^2}$ and $\log(1+(\mathbf{x}(k))^2) \leq (\mathbf{x}(k))^2$. By (2.12), (2.13), and Lemma I.8,

$$\left| \sum_{m\in S, n\in T} \beta(m,n)\ \mathbf{A}(\mathbf{x}_m, \mathbf{y}_n) \right|$$

$$= \left| \sum_{m\in S, n\in T} \beta(m,n)\ \mathbf{E}\ \mathscr{Im} \prod_k (1 + \mathrm{i}\mathbf{x}_m(k) r_k) \mathscr{Im} \prod_k (1 + \mathrm{i}\mathbf{y}_n(k) r_k) \right|$$

$$\leq \mathbf{E} \left| \sum_{m\in S, n\in T} \beta(m,n)\ \mathscr{Im} \prod_k (1 + \mathrm{i}\mathbf{x}_m(k) r_k) \mathscr{Im} \prod_k (1 + \mathrm{i}\mathbf{y}_n(k) r_k) \right| \leq \mathrm{e}\|\beta\|_{F_2}. \tag{2.14}$$

□

Proof of Theorem 1: Let $\{\mathbf{x}_m : m \in S\}$ and $\{\mathbf{y}_n : n \in T\}$ be finite subsets of B_{l^2}. Then,

$$\left| \sum_{m\in S, n\in T} \beta(m,n) \langle \mathbf{x}_m, \mathbf{y}_n \rangle \right|$$

$$= \left| \sum_{m\in S, n\in T} \beta(m,n) \sum_{k=0}^{\infty} (-1)^k (\sinh(1) - 1)^k \mathbf{A}(\theta^k \mathbf{x}_m, \theta^k \mathbf{y}_n) \right|$$

(by Lemma 2)

$$\leq \sum_{k=0}^{\infty}(\sinh(1)-1)^k \left| \sum_{m\in S, n\in T} \beta(m,n)\, \mathbf{A}(\theta^k \mathbf{x}_m, \theta^k \mathbf{y}_n) \right|$$

$$\leq \frac{\mathrm{e}}{2-\sinh(1)} \|\beta\|_{F_2} \quad \text{(by Lemma 3).} \tag{2.15}$$

3 A Second Elementary Proof

As in the previous proof, we first rewrite the dot product in l^2 in terms of a scalar-valued function defined on $l^2 \times l^2$ (cf. Lemma 2), and then verify that this function satisfies a Grothendieck-type inequality (cf. Lemma 3). To underscore similarities between the arguments in this and the previous section, we use the same notation whenever possible.

We fix a one–one correspondence between $\mathbb{N}$ and $\bigcup_{k=2}^{\infty} \mathbb{N}^k$, and denote it by

$$n \leftrightarrow (n_1, \ldots, n_j), \tag{3.1}$$

where $n \in \mathbb{N}$ and $(n_1, \ldots, n_j) \in \bigcup_{k=2}^{\infty} \mathbb{N}^k$ (cf. (2.2)). If

$$\mathbf{x} = (\mathbf{x}(n) : n \in \mathbb{N})$$

is a scalar sequence, then $\phi\mathbf{x}$ will denote here the sequence whose nth coordinate is

$$(\phi\mathbf{x})_n = \mathbf{x}(n_1)\cdots\mathbf{x}(n_j)/\sqrt{j!}, \tag{3.2}$$

where $n \leftrightarrow (n_1, \ldots, n_j)$ (cf. (2.3)). For $\mathbf{x} \in l^2$ and $\mathbf{y} \in l^2$,

$$\begin{aligned} \langle \phi\mathbf{x}, \phi\mathbf{y} \rangle &= \sum_{j=2}^{\infty} \frac{1}{j!} \sum_{(n_1,\ldots,n_j)\in\mathbb{N}^j} \mathbf{x}(n_1)\cdots\mathbf{x}(n_j)\, \mathbf{y}(n_1)\ldots\mathbf{y}(n_j) \\ &= \sum_{j=2}^{\infty} (\langle \mathbf{x}, \mathbf{y} \rangle)^j / j! = \mathrm{e}^{\langle \mathbf{x}, \mathbf{y} \rangle} - \langle \mathbf{x}, \mathbf{y} \rangle - 1. \end{aligned} \tag{3.3}$$

In particular,

$$\|\phi \mathbf{x}\|_2^2 = \mathrm{e}^{\|\mathbf{x}\|_2^2} - \|\mathbf{x}\|_2^2 - 1 \tag{3.4}$$

(cf. (2.6)). Let $\theta = \frac{\phi}{\sqrt{\mathrm{e}-2}}$, which, by (3.4), maps the unit sphere in l^2 into itself. Define a function $\mathbf{A}$ on $l^2 \times l^2$ by

$$\mathbf{A}(\mathbf{x}, \mathbf{y}) = \mathrm{e}^{\langle \mathbf{x}, \mathbf{y} \rangle} - 1 \tag{3.5}$$

(cf. (2.7)), and rewrite (3.3)

$$\langle \mathbf{x}, \mathbf{y} \rangle = \mathbf{A}(\mathbf{x}, \mathbf{y}) - (\mathrm{e} - 2)\, \langle \theta\mathbf{x}, \theta\mathbf{y} \rangle. \tag{3.6}$$

(cf. (2.8)).

Lemma 4 (Exercise 4). *If* $\|\mathbf{x}\|_2 = \|\mathbf{y}\|_2 = 1$, *then*

$$\langle \mathbf{x}, \mathbf{y} \rangle = \sum_{k=0}^{\infty} (-1)^k (\mathrm{e} - 2)^k\ \mathbf{A}(\theta^k \mathbf{x}, \theta^k \mathbf{y}), \tag{3.7}$$

where the convergence of the series is uniform in the unit sphere of l^2.

Lemma 5 *Let* $\{\mathbf{x}_m : m \in S\}$ *and* $\{\mathbf{y}_n : n \in T\}$ *be finite subsets of the unit sphere in* l^2. *Then,*

$$\left| \sum_{m \in S, n \in T} \beta(m, n)\ \mathbf{A}(\mathbf{x}_m, \mathbf{y}_n) \right| \leq (4\mathrm{e} + 1)\|\beta\|_{F_2}. \tag{3.8}$$

Proof: Let $\{Z_k\colon k \in \mathbb{N}\}$ be a system of (statistically) independent standard normal random variables. Then, for $\mathbf{x}$ and $\mathbf{y}$ in the unit sphere of l^2,

$$\begin{aligned} \mathbf{E}\ \mathrm{e}^{\mathrm{i}\Sigma \mathbf{x}(k) Z_k}\ \mathrm{e}^{-\mathrm{i}\Sigma \mathbf{y}(k) Z_k} &= \prod_k \mathbf{E}\ \mathrm{e}^{\mathrm{i}\{\mathbf{x}(k) - \mathbf{y}(k)\} Z_k} \\ &= \prod_k \mathrm{e}^{-\{\mathbf{x}(k) - \mathbf{y}(k)\}^2/2} = \mathrm{e}^{-1}\mathrm{e}^{\langle \mathbf{x}, \mathbf{y} \rangle}. \end{aligned} \tag{3.9}$$

In this computation we used independence, and that for standard normal random variables Z,

$$\mathbf{E}(\mathrm{i}tZ) = \exp(-t^2/2), \quad t \in \mathbb{R}. \tag{3.10}$$

By (3.5), (3.9), and Lemma I.8,

$$\begin{aligned}
&\left|\sum_{m\in S,n\in T} \beta(m,n)\ \mathbf{A}(\mathbf{x}_m,\mathbf{y}_n)\right| \\
&\quad = \left|\sum_{m\in S,n\in T} \beta(m,n)\ (\mathrm{e}^{\langle \mathbf{x}_m,\mathbf{y}_n\rangle} - 1)\right| \\
&\quad = \left|\sum_{m\in S,n\in T} \beta(m,n)\ (\mathrm{e}\,\mathbf{E}\,\mathrm{e}^{\mathrm{i}\Sigma \mathbf{x}_m(k)Z_k}\ \mathrm{e}^{-\mathrm{i}\Sigma \mathbf{y}_n(k)Z_k} - 1)\right| \\
&\quad \leq \mathrm{e}\,\mathbf{E}\left|\sum_{m\in S,n\in T} \beta(m,n)\ \mathrm{e}^{\mathrm{i}\Sigma \mathbf{x}_m(k)Z_k}\ \mathrm{e}^{-\mathrm{i}\Sigma \mathbf{y}_n(k)Zk}\right| + \|\beta\|_{F_2} \\
&\quad \leq (4\mathrm{e}+1)\|\beta\|_{F_2}.
\end{aligned} \tag{3.11}$$

□

Proof of Theorem 1: Apply Lemmas 4 and 5 (and Exercise 1) exactly as we applied Lemmas 2 and 3 in the previous section.

Remark (a summary). A crucial step in all 'self-contained' proofs of the Grothendieck inequality, including Grothendieck's own argument (Exercise 10), is a representation of the dot product in l^2 by absolutely convergent series of integrals of products of bounded functions. We shall see, as the story unfolds, that feasibility of such representations is the essence of Grothendieck's *théorème fondamental.* The two proofs given here contain, respectively, these two representations: if $\|\mathbf{x}\|_2 = 1$ and $\|\mathbf{y}\|_2 = 1$, then

$$\begin{aligned}
&\langle \mathbf{x},\mathbf{y}\rangle \\
&= \sum_{j=0}^{\infty}(-1)^j(\sinh(1)-1)^j\ \mathbf{E}\ \mathscr{Im}\prod_k(1+\mathrm{i}(\theta^j\mathbf{x})_k r_k)\mathscr{Im}\prod_k(1+\mathrm{i}(\theta^j\mathbf{y})_k\ r_k),
\end{aligned} \tag{3.12}$$

where $\theta = \frac{\phi}{\sqrt{\sinh(1)-1}}$ is defined by (2.3), and

$$\langle \mathbf{x},\mathbf{y}\rangle = \sum_{j=0}^{\infty}(-1)^j(\mathrm{e}-2)^j\mathbf{E}(\mathrm{e}^{\mathrm{i}\Sigma_k(\theta^j\mathbf{x})_k Z_k}\ \mathrm{e}^{-\mathrm{i}\Sigma_k(\theta^j\mathbf{y})_k\ Z_k} - 1), \tag{3.13}$$

where $\theta = \phi/\sqrt{\mathrm{e}-2}$ is defined by (3.2).

Each of these identities is based on an 'upgraded' Khintchin L^1–L^2 inequality (Exercises 8, 9), which we formalize in the next section.

4 $\Lambda(2)$-uniformizability

Definition 6 Let (X,ν) be a finite measure space, and let H be a closed subspace of $\mathrm{L}^2(X,\nu)$.

i. H is a $\Lambda(2)$-*space* if there exists $\kappa > 0$ (a $\Lambda(2)$-constant of H) such that for every $g \in H$ there are $f \in H^\perp$ (the orthogonal complement of H) with the property that $g + f \in \mathrm{L}^\infty(X,\nu)$, and

$$\|g+f\|_{\mathrm{L}^\infty} \leq \kappa \|g\|_{\mathrm{L}^2}. \tag{4.1}$$

ii. H is a *uniformizable* $\Lambda(2)$-*space* if for every $\epsilon > 0$ there exists $\delta = \delta_H(\epsilon) > 0$ (a $\Lambda(2)$-*uniformizing constant* associated with H and ϵ) such that for every $g \in H$ there are $f \in H^\perp$ with the property that $g + f \in \mathrm{L}^\infty(X,\nu)$,

$$\|g+f\|_{\mathrm{L}^\infty} \leq \delta \|g\|_{\mathrm{L}^2}, \tag{4.2}$$

and

$$\|f\|_{\mathrm{L}^2} \leq \epsilon \|g\|_{\mathrm{L}^2}. \tag{4.3}$$

The lemma below equates the $\Lambda(2)$-property with an L^1–L^2 inequality. The argument establishing this equivalence is nowadays routine, but during the 1920s and 1930s, in the early stages of functional analysis, it was news. For example, see Banach's paper [Ban], and also Kacmarz's and Steinhaus's paper [KaSte] and the references therein.

Without loss of generality, we will assume, here and throughout, that the underlying measure space (X,ν) in Definition 6 is a probability space.

Lemma 7 (cf. Exercise II.2; [KaSte, Théorème 10]). *A closed subspace H of $\mathrm{L}^2(X,\nu)$ is a $\Lambda(2)$-space if and only if there exists $\kappa > 0$ such that for all $h \in H$,*

$$\kappa \|h\|_{\mathrm{L}^1} \geq \|h\|_{\mathrm{L}^2}; \tag{4.4}$$

i.e., H is a $\Lambda(2)$-space if and only if H is a closed subspace of $\mathrm{L}^1(X,\nu)$.

Proof: Assume (4.4). Let $g \in H$. By the Cauchy–Schwarz inequality and (4.4),

$$\left|\int g\bar{h}\, \mathrm{d}\nu\right| \leq \kappa \|g\|_{\mathrm{L}^2}\, \|h\|_{\mathrm{L}^1}, \quad h \in H. \tag{4.5}$$

This implies that

$$h \mapsto \int g\bar{h}\, \mathrm{d}\nu, \quad h \in H,$$

determines a bounded linear functional on the L^1-closure of H. Therefore, by the duality $(\mathrm{L}^1)^* = \mathrm{L}^\infty$, and by the Hahn–Banach theorem, there exist $F \in \mathrm{L}^\infty(X, \nu)$ such that $\|F\|_{\mathrm{L}^\infty} \leq \kappa$ and for all $h \in H$,

$$\int (F - g)\bar{h}\, \mathrm{d}\nu = 0. \tag{4.6}$$

This means $F - g \in H^\perp$, as required.

Conversely, suppose H is a $\Lambda(2)$-space. Denote by τ_H the canonical projection from $\mathrm{L}^2(X, \nu)$ onto H. Let $h \in H$ be arbitrary. If $g \in \mathrm{L}^2(X, \nu)$ and $\|g\|_{\mathrm{L}^2} = 1$, then there exist $f \in H^\perp$ such that $\tau_H \bar{g} + f \in \mathrm{L}^\infty(X, \nu)$ and $\|\tau_H \bar{g} + f\|_{\mathrm{L}^\infty} \leq \kappa$. Therefore,

$$\left|\int h\bar{g}\, \mathrm{d}\nu\right| = \left|\int h(\tau_H \bar{g} + f)\, \mathrm{d}\nu\right| \leq \kappa \|h\|_{\mathrm{L}^1}. \tag{4.7}$$

Then,

$$\|h\|_{\mathrm{L}^2} = \sup\left\{\left|\int h\bar{g}\, \mathrm{d}\nu\right| : g \in \mathrm{L}^2(X, \nu), \|g\|_{\mathrm{L}^2} = 1\right\} \leq \kappa\, \|h\|_{\mathrm{L}^1}. \tag{4.8}$$

□

Let $\mathrm{L}^p_R(\Omega, \mathbb{P}) = \mathrm{L}^p_R$ denote the L^p-closure of the span of the Rademacher system R. That L^2_R is a $\Lambda(2)$-space follows, via Lemma 7, from the Khintchin L^1–L^2 inequality, which – we recall – was derived in the last chapter from an L^2–L^p inequality. (See Proof of Theorem II.1.) An L^2–L^p inequality for $p > 2$ also yields, with a bit more work, the $\Lambda(2)$-uniformizability property:

Theorem 8 *If H is a closed subspace of $\mathrm{L}^2(X, \nu)$ and $H \subset \mathrm{L}^p(X, \nu)$ for some $p > 2$, then H is a uniformizable $\Lambda(2)$-space.*

Lemma 9 (Exercise 5). *If H is a closed subspace of $\mathrm{L}^2(X, \nu)$ and $H \subset \mathrm{L}^p(X, \nu)$ for some $p > 2$, then H is a $\Lambda(2)$-space.*

Proof of Theorem 8: The inclusion $H \subset \mathrm{L}^p$ is equivalent to existence of $C_p = C > 0$ such that

$$\|g\|_{\mathrm{L}^p} \leq C\|g\|_{\mathrm{L}^2}, \quad g \in H \tag{4.9}$$

(Exercise 5). Let $\epsilon > 0$ and $g \in H$. Define

$$h = \begin{cases} g & \text{if } |g| \leq \epsilon^{\frac{2}{2-p}} \|g\|_{\mathrm{L}^p} \\ 0 & \text{otherwise.} \end{cases} \tag{4.10}$$

Let $\varphi = g - h$. Then,

$$\int |\varphi|^2 \, \mathrm{d}\nu = \int |\varphi|^{2-p} |\varphi|^p \, \mathrm{d}\nu \leq \epsilon^2 \, \|g\|^2_{\mathrm{L}^p}. \tag{4.11}$$

Let τ_H and $\tau_H^\perp$ denote the canonical projections from L^2 onto H and $H^\perp$, respectively. Write

$$\varphi = \tau_H(\varphi) + \tau_H^\perp(\varphi). \tag{4.12}$$

From (4.9) and (4.11), we obtain

$$\|\tau_H(\varphi)\|_{\mathrm{L}^2} \leq \epsilon C \, \|g\|_{\mathrm{L}^2} \quad \text{and} \quad \|\tau_H^\perp(\varphi)\|_{\mathrm{L}^2} \leq \epsilon C \, \|g\|_{\mathrm{L}^2}. \tag{4.13}$$

By Lemma 9, H is a $\Lambda(2)$-space. Therefore, by (4.13), there exists $f \in H^\perp$ such that

$$\|\tau_H(\varphi) + f\|_{\mathrm{L}^\infty} \leq \kappa \|\tau_H(\varphi)\|_{\mathrm{L}^2} \leq \kappa \, \epsilon \, C \, \|g\|_{\mathrm{L}^2}, \tag{4.14}$$

where κ is a $\Lambda(2)$-constant of H. Observe that

$$h + \tau_H(\varphi) + f = g + f - \tau_H^\perp(\varphi), \tag{4.15}$$

and that $f - \tau_H^\perp(\varphi) \in H^\perp$. From (4.15), (4.10), (4.9), and (4.14), we obtain

$$\begin{aligned} \|g + f - \tau_H^\perp(\varphi)\|_{\mathrm{L}^\infty} &= \|h + \tau_H(\varphi) + f\|_{\mathrm{L}^\infty} \\ &\leq \|h\|_{\mathrm{L}^\infty} + \|\tau_H(\varphi) + f\|_{\mathrm{L}^\infty} \leq (\epsilon^{\frac{2}{2-p}} + \epsilon\kappa) \, C \, \|g\|_{\mathrm{L}^2}. \end{aligned} \tag{4.16}$$

Combining (4.14), (4.13), (4.11) and (4.9), we obtain

$$\|f - \tau_H^\perp(\varphi)\|_{\mathrm{L}^2} = \|f + \tau_H(\varphi) - \varphi\|_{\mathrm{L}^2} \leq \epsilon C(\kappa + 1)\|g\|_{\mathrm{L}^2}. \tag{4.17}$$

□

Corollary 10 L^2_R *is a uniformizable* $\Lambda(2)$*-space. Moreover, for an absolute constant* $0 < K < \infty$*, and all* $p > 2$ *and* $\epsilon > 0$*,*

$$\delta_{\mathrm{L}^2_R}(\epsilon) \leq K\sqrt{p} \, \epsilon^{\frac{2}{2-p}}. \tag{4.18}$$

Proof: To obtain the first assertion, it suffices to apply (II.2.3). To obtain the second, apply the full system of the Khintchin inequalities (Exercise II.3), together with (4.16) and (4.17). □

Remarks:

i **(credits).** A version of Theorem 8 had been communicated to me in a letter by Gilles Pisier, in September 1977, and appeared in [Bl4, Lemma 2.2]. Pisier's original proof was based on complex interpolation. The simpler truncation argument given here (and in [Bl4, Lemma 2.2]) was the result of a conversation I had (at Uppsala) with Per Sjölin, also in September 1977.

ii **(a constructive proof).** The proof of Theorem 8 was nonconstructive. That is, in applying the hypothesis $H \subset \mathrm{L}^p$, we concluded only *existence* of $f \in H^\perp$ such that $\tau_H(\varphi) + f$ satisfied (4.14). Although this suffices to prove the Grothendieck inequality (Exercise 6), we shall describe below an explicit algorithm $\alpha : H \to H^\perp$ (a $\Lambda(2)$-*uniformizing map*) such that for every $g \in H$,

$$\|g + \alpha(g)\|_{\mathrm{L}^\infty} \leq \delta \|g\|_{\mathrm{L}^2} \quad \text{and} \quad \|\alpha(g)\|_{\mathrm{L}^2} \leq \epsilon \ \|g\|_{\mathrm{L}^2}. \tag{4.19}$$

The $\Lambda(2)$-uniformizing map α, which we use in §5 and later again in Chapter VIII in a multidimensional context, is closely related to the maps ϕ defined by (2.3) and (3.2) in the previous two sections.

Fix $\epsilon > 0$ such that that $\epsilon C < 1$, where C is the constant in (4.9). Let $g_1 = g$. Apply the truncation in (4.10) to produce $h_1 = g_1 + \varphi_1$ such that

$$\|h_1\|_{\mathrm{L}^\infty} \leq \epsilon^{\frac{2}{2-p}} C \ \|g_1\|_{\mathrm{L}^2} \quad \text{and} \quad \|\varphi_1\|_2 \leq \epsilon C \ \|g_1\|_{\mathrm{L}^2}. \tag{4.20}$$

We continue recursively. Let $n \geq 1$. Assume that we have produced $h_n \in \mathrm{L}^\infty$, $g_n \in \mathrm{L}^2$, and $\varphi_n \in \mathrm{L}^2$ such that $h_n = g_n + \varphi_n$,

$$\|h_n\|_{\mathrm{L}^\infty} \leq \epsilon^{\frac{2}{2-p}} C(\epsilon C)^{n-1} \|g_1\|_{\mathrm{L}^2}, \tag{4.21}$$

and

$$\|\varphi_n\|_{\mathrm{L}^2} \leq (\epsilon C)^n \|g_1\|_{\mathrm{L}^2}.$$

Define $g_{n+1} = \tau_H(\varphi_n)$, and apply (4.10) to g_{n+1}, thus obtaining $h_{n+1} = g_{n+1} + \varphi_{n+1}$ such that

$$\|h_{n+1}\|_{\mathrm{L}^\infty} \leq \epsilon^{\frac{2}{2-p}} C \|\tau_H(\varphi_n)\|_{\mathrm{L}^2} \leq \epsilon^{\frac{2}{2-p}} C(\epsilon C)^n \|g_1\|_{\mathrm{L}^2} \tag{4.22}$$

and

$$\|\varphi_{n+1}\|_{L^2} \leq \epsilon C \|\tau_H(\varphi_n)\|_{L^2} \leq (\epsilon C)^{n+1} \|g_1\|_{L^2}. \tag{4.23}$$

Observe that

$$\sum_{j=1}^{n} (-1)^{j-1} h_j = g + (-1)^{n-1} \tau_H(\varphi_n) + \sum_{j=1}^{n} (-1)^{j-1} \tau_H^{\perp}(\varphi j). \tag{4.24}$$

Letting $n \to \infty$ in (4.24), we obtain

$$\sum_{j=1}^{\infty} (-1)^{j-1} h_j = g + \sum_{j=1}^{\infty} (-1)^{j-1} \tau_H^{\perp}(\varphi_j). \tag{4.25}$$

The series on the left side converges in $L^\infty(X,\nu)$ (by (4.22)), and the series on the right converges in $L^2(X,\nu)$ (by (4.23)). Moreover,

$$\left\| \sum_{j=1}^{\infty} (-1)^{j-1} h_j \right\|_{L^\infty} \leq (\epsilon^{\frac{2}{2-p}} C/(1-\epsilon C)) \|g\|_{L^2}, \tag{4.26}$$

and

$$\left\| \sum_{j=1}^{\infty} (-1)^{j-1} \tau_H^{\perp}(\varphi_j) \right\|_{L^2} \leq \left(\frac{\epsilon C}{1-\epsilon C} \right) \|g\|_{L^2}. \tag{4.27}$$

The $\Lambda(2)$-uniformizing map $\alpha : H \to H^{\perp}$ is defined by

$$\alpha(g) = \sum_{j=1}^{\infty} (-1)^{j-1} \tau_H^{\perp}(\varphi_j), \quad g \in H. \tag{4.28}$$

The second assertion in Corollary 10, regarding dependence of uniformizing constants on ϵ, remains valid; see (4.26) and (4.27) above.

5 A Representation of an Inner Product in a Hilbert Space

In this section we demonstrate that existence of an infinite-dimensional uniformizable $\Lambda(2)$-space yields a representation of an inner product by an absolutely convergent series of integrals of products of bounded functions. This representation, which implies Theorem 1, is the prototype of representations produced in §2 and §3 (cf. (3.12) and (3.13)).

Assume that $\mathrm{L}^2(X,\nu)$ is separable, and that $H \subset \mathrm{L}^2(X,\nu)$ is an infinite-dimensional uniformizable $\Lambda(2)$-space (e.g., $H = \mathrm{L}^2_R(\Omega, \mathbb{P})$). Fix $0 < \epsilon < 1$, and let $\delta = \delta_H(\epsilon)$ be a uniformizing constant associated with H and ϵ. Following the Remark in the previous section, we have a well-defined map $\alpha : H \to H^{\perp}$ such that if $g \in H$ then

$$\|g + \alpha(g)\|_{\mathrm{L}^\infty} \leq \delta \|g\|_{\mathrm{L}^2} \quad \text{and} \quad \|\alpha(g)\|_{\mathrm{L}^2} \leq \epsilon \|g\|_{\mathrm{L}^2}. \tag{5.1}$$

Define (a mixed-norm space)

$$l^1(\mathrm{L}^\infty) = \left\{ (f_j) : f_j \in \mathrm{L}^\infty(X,\nu), \sum_j \|f_j\|_{\mathrm{L}^\infty} < \infty \right\}. \tag{5.2}$$

Theorem 11 *There exists a map $\Psi : l^2 \to l^1(\mathrm{L}^\infty)$ (i.e., for $\mathbf{x} \in l^2$, $\Psi\mathbf{x} = \{(\Psi\mathbf{x})_j : j \in \mathbb{N}\}$, $(\Psi\mathbf{x})_j \in \mathrm{L}^\infty(X,\nu)$, and $\Sigma_j \|(\Psi\mathbf{x})_j\|_{\mathrm{L}^\infty} < \infty$) such that*

$$\|(\Psi\mathbf{x})_j\|_{\mathrm{L}^\infty} \leq \delta\, \epsilon^{j-1} \|\mathbf{x}\|_2, \quad j = 1, \ldots, \tag{5.3}$$

and

$$\langle \mathbf{x}, \mathbf{y} \rangle = \sum_{j=1}^{\infty} (-1)^{j-1} \int (\Psi\mathbf{x})_j\, \overline{(\Psi\mathbf{y})_j}\, \mathrm{d}\nu. \tag{5.4}$$

Proof: Fix unitary maps $U : l^2 \to H$ and $V : H^{\perp} \to H$ (Exercise 7). For $\mathbf{x} \in l^2$, denote $g_1^{\mathbf{x}} = U\mathbf{x}$, and for $j \in \mathbb{N}$ define

$$g_{j+1}^{\mathbf{x}} = V\alpha(g_j^{\mathbf{x}}) \quad \text{and} \quad (\Psi\mathbf{x})_j = g_j^{\mathbf{x}} + \alpha(g_j^{\mathbf{x}}). \tag{5.5}$$

A recursive application of (5.1) implies (5.3), and

$$\|\alpha(g_j^{\mathbf{x}})\|_{\mathrm{L}^2} = \|g_{j+1}^{\mathbf{x}}\|_{\mathrm{L}^2} \leq \epsilon^j \|\mathbf{x}\|_2, \quad j \in \mathbb{N}. \tag{5.6}$$

For $\mathbf{x} \in l^2$ and $\mathbf{y} \in l^2$, we obtain from (5.5)

$$\langle \mathbf{x}, \mathbf{y} \rangle = \int (\Psi \mathbf{x})_1 \overline{(\Psi \mathbf{y})_1} \, \mathrm{d}\nu - \int g_2^{\mathbf{x}} \overline{g_2^{\mathbf{y}}} \, \mathrm{d}\nu, \tag{5.7}$$

and

$$\int g_j^{\mathbf{x}} \overline{g_j^{\mathbf{y}}} \, \mathrm{d}\nu = \int (\Psi \mathbf{x})_j \overline{(\Psi \mathbf{y})_j} \, \mathrm{d}\nu - \int g_{j+1}^{\mathbf{x}} \overline{g_{j+1}^{\mathbf{y}}} \, \mathrm{d}\nu, \quad j = 2, \ldots .$$

(Because U and V are unitary.) By iterating (5.7), we obtain for $n \in \mathbb{N}$

$$\langle \mathbf{x}, \mathbf{y} \rangle = \sum_{j=1}^{n} (-1)^{j-1} \int (\Psi \mathbf{x})_j \overline{(\Psi \mathbf{y})_j} \, \mathrm{d}\nu + (-1)^n \int g_{n+1}^{\mathbf{x}} \overline{g_{n+1}^{\mathbf{y}}} \, \mathrm{d}\nu. \tag{5.8}$$

By applying (5.6), we obtain

$$\lim_{n \to \infty} \int g_{n+1}^{\mathbf{x}} \overline{g_{n+1}^{\mathbf{y}}} \, \mathrm{d}\nu = 0, \tag{5.9}$$

which implies (5.4). □

Proof of Theorem 1: Let $\{\mathbf{x}_m : m \in S\}$ and $\{\mathbf{y}_n : n \in T\}$ be finite subsets of B_{l^2}. Then,

$$\begin{aligned}
&\left| \sum_{m \in S, n \in T} \beta(m,n) \langle \mathbf{x}_m, \mathbf{y}_n \rangle \right| \\
&= \left| \sum_{m \in S, n \in T} \beta(m,n) \sum_{j=1}^{\infty} (-1)^{j-1} \int (\Psi \mathbf{x}_m)_j \overline{(\Psi \mathbf{y}_n)_j} \, \mathrm{d}\nu \right| \quad \text{(by (5.4))} \\
&\leq \sum_{j=1}^{\infty} \left| \sum_{m \in S, n \in T} \beta(m,n) \int (\Psi \mathbf{x}_m)_j \overline{(\Psi \mathbf{y}_n)_j} \, \mathrm{d}\nu \right| \\
&\leq \sum_{j=1}^{\infty} \int \left| \sum_{m,n} \beta(m,n) (\Psi \mathbf{x}_m)_j \overline{(\Psi \mathbf{y}_n)_j} \right| \mathrm{d}\nu \\
&\leq \sum_{j=1}^{\infty} 4(\delta \, \epsilon^{j-1})^2
\end{aligned} \tag{5.10}$$

(by (5.5)), which implies $\kappa_{\mathrm{G}} \leq \frac{4\delta^2}{1-\epsilon^2}$. □

Remark (fine-tuning). Let $(\epsilon_j)_{j=0}^{\infty}$ be a non-increasing sequence such that $\epsilon_0 = 1$, and $1 > \epsilon_j > 0$ for $j \geq 1$, and let $\delta_j = \delta_H(\epsilon_j)$ be uniformizing constants associated with H and ϵ_j. Then, by applying the Remark in §4, for every $j \in \mathbb{N}$ there exists a $\Lambda(2)$-uniformizing map $\alpha_j : H \to H^{\perp}$ such that for $g \in H$,

$$\|g + \alpha_j(g)\|_{\mathrm{L}^\infty} \leq \delta_j \|g\|_{\mathrm{L}^2} \quad \text{and} \quad \|\alpha_j(g)\|_{\mathrm{L}^2} \leq \epsilon_j \|g\|_{\mathrm{L}^2}. \tag{5.11}$$

By making the appropriate (minor) adjustments in the proof of Theorem 11, we obtain a map $\Psi : l^2 \to l^1(\mathrm{L}^\infty)$ such that

$$\|(\Psi\mathbf{x})_j\|_{\mathrm{L}^\infty} \leq \delta_j (\Pi_{i=0}^{j-1} \epsilon_i) \|\mathbf{x}\|_2, \quad j = 1, \ldots, \tag{5.3$'$}$$

and

$$\langle \mathbf{x}, \mathbf{y} \rangle = \sum_{j=1}^{\infty} (-1)^{j-1} \int (\Psi\mathbf{x})_j \overline{(\Psi\mathbf{y})_j} \, \mathrm{d}\nu. \tag{5.4$'$}$$

Applying this in the proof of Theorem 1, we deduce

$$\kappa_{\mathrm{G}} \leq \sum_{j=1}^{\infty} 4(\delta_j \Pi_{i=1}^{j-1} \epsilon_i)^2. \tag{5.12}$$

6 Comments (Mainly Historical) and Loose Ends

$\Lambda(2)$-uniformizability of L^2_R and of $\mathrm{L}^2_{\{Z_k\}}$

In §2 and §3, we made use of

$$\prod_k (1 + \mathrm{i}\mathbf{x}(k) r_k), \tag{6.1}$$

and

$$\mathrm{e}^{\mathrm{i}\Sigma_k \mathbf{x}(k) Z_k}, \tag{6.2}$$

for real-valued $\mathbf{x} \in B_{l^2}$. The device in (6.1), an L^∞-valued function on l^2, is a variant of a *Riesz product.* The measure-valued version of such products (the case $\mathbf{x} \in B_{l^\infty}$) had appeared first in F. Riesz's 1918 paper [Ri$_\mathrm{F}$2] in a context of lacunary Fourier series, and, independently in Rademacher's 1922 paper [R, p. 137] in a context of Rademacher series. The L^∞-valued version first appeared in Salem's and Zygmund's constructive proof [SaZy1] of a theorem originally proved by Banach [Ban] (an analog of Corollary 9 in the case of lacunary Fourier series). Riesz products and their variants will be used extensively in Chapter VII.

The device in (6.2), also an L^∞-valued function on l^2, is similar to the device in (6.1). Its utility here was through the characteristic function of a normal random variable. (See (3.11).)

In Exercises 8 and 9, you will use (6.1) and (6.2) (separately) to prove that L^2_R and $L^2_{\{Z_k\}}$ (L^2-closure of the span of a system of independent standard normal variables $\{Z_k\}$) are uniformizable $\Lambda(2)$-spaces.

Is the Grothendieck Inequality Equivalent to $\Lambda(2)$-uniformizability?

We have shown that existence of infinite-dimensional uniformizable $\Lambda(2)$-spaces implies the Grothendieck inequality. A question arises: *does the Grothendieck inequality imply the existence of an infinite-dimensional uniformizable $\Lambda(2)$-space?* Indeed, because the corresponding weaker statements – the Littlewood mixed-norm inequality, and that L^2_R is a $\Lambda(2)$-space – are derivable from each other (through Theorem II.4), a tempting guess is that the Grothendieck inequality and $\Lambda(2)$-uniformizability are equivalent in the same sense.

More about the Inequality

Grothendieck's original formulation of his inequality involved tensor norms in a then-new setting of topological tensor products. Recognizing the importance of his discovery, Grothendieck dubbed it *le théorème fondamental de la théorie metrique des produits tensoriels* [Gro2]. Alas, the significance of this 1956 theorem was not immediately apparent. Its importance was underscored twelve years later, in Lindenstrauss's and Pelczynski's seminal 1968 paper [LiPe] cast in a framework of absolutely summing operators. (It is here that I first learned about the *théorème fondamental.*) Avoiding the explicit use of tensors, Lindenstrauss and Pelczynski rewrote Grothendieck's theorem and proof, the key to which was this elementary identity: for $\mathbf{x}$ and $\mathbf{y}$ in the unit sphere S_n in $\mathbb{R}^n$,

$$\operatorname{arc}\cos\langle\mathbf{x},\mathbf{y}\rangle = \frac{\pi}{2}\left(1 - \int_{S_n} \operatorname{sign}\langle\mathbf{x},\mathbf{u}\rangle \operatorname{sign}\langle\mathbf{y},\mathbf{u}\rangle\, \sigma(d\mathbf{u})\right), \tag{6.3}$$

where σ is the normalized rotation-invariant measure on S_n. (The role of arc $\cos\langle x, y\rangle$ is analogous to that of $\mathbf{A}(\mathbf{x},\mathbf{y})$ in §2 and §3.) The next step was to apply the *cosine* Taylor series to both sides of (6.3), and thus obtain a representation of the standard dot product in $\mathbb{R}^n$ in terms of an absolutely convergent series of integrals of bounded functions. You are asked to reproduce these arguments in Exercise 10.

The various proofs and interpretations of the inequality that have since surfaced in different settings of analysis attest to the universality of Grothendieck's result. These proofs will not be surveyed here; a partial account can be found in Pisier's book [Pi3]. The three proofs given in this chapter, based on the notion of $\Lambda(2)$-uniformizability, originated in a proof I gave in 1976 in a framework of harmonic analysis [Bl3].

Possibly the best known open problem regarding the Grothendieck inequality is the computation of the smallest κ_G in (1.5); a discussion can be found in [Pi3]. Possibly the most important problem concerns multidimensional extensions of the inequality; this issue will be visited later in the book.

$\Lambda(p)$-sets

The Khintchin L^2–L^p inequalities for $p > 2$, as well as the L^1–L^2 inequality, were viewed in the 1920s and 1930s as new tools, as means to specific ends. (We have seen applications in Chapter II, in this chapter, and will see more in later chapters.) Eventually however, attention turned to general phenomena exemplified by the Rademacher system. In his 1960 classic paper [Ru1], Walter Rudin introduced the following notion cast in a setting of Fourier analysis on the circle group $[0, 2\pi)$.

Definition 12 Let $p \in (0, \infty)$. $E \subset \mathbb{Z}$ is a $\Lambda(p)$-set if for some $q \in (0, p)$ there exists $0 < K < \infty$ such that for every E-polynomial f (a polynomial with spectrum in E),

$$\|f\|_{L^p} \leq K\|f\|_{L^q}. \tag{6.4}$$

(By an argument similar to the one used in Exercise 4, if (6.4) holds for *some* $q \in (0, p)$, then it holds for *all* $q \in (0, p)$ with constants K depending on q; Exercise 11.) This definition can be naturally recast in the setting of this chapter:

Definition 13 Let (X, ν) be a probability space. A closed subspace H of $L^2(X, \nu)$ is a $\Lambda(p)$-space for $p \in (1, \infty)$ if there exist $0 < K < \infty$ such that for all $f \in H$

$$K\|f\|_{L^1} \geq \|f\|_{L^p} \tag{6.5}$$

(i.e., L^p-closure of H = L^1-closure of H).

The connections between $\Lambda(p)$-sets, $\Lambda(p)$-spaces, and the classical Khintchin inequalities are evident; see Exercises II.2, II.3.

Rudin's paper [Ru1] left an indelible mark on harmonic analysis, and indeed on analysis at large, not only by the theorems in it, but also for raising the 'right' questions, some of which are unanswered to this day. I will state two of these unsolved problems, which seem to have bearing on $\Lambda(2)$-uniformizability.

In [Ru1], Rudin produced for all integers $n > 1$, $\Lambda(2n)$-sets that are not $\Lambda(q)$ for all $q > 2n$. He then raised the question, which became known as the *$\Lambda(p)$-set problem*, whether for $p \notin \{4, 6, \ldots\}$ there exist $\Lambda(p)$-sets that are not $\Lambda(q)$ for all $q > p$. In 1974, Bachelis and Ebenstein proved that if $p \in (1, 2)$ and a spectral set E is a $\Lambda(p)$-set, then E is a $\Lambda(q)$-set for *some* $q > p$ [BaE] (Exercise 12). In 1989, Jean Bourgain demonstrated that for every $p > 2$ there exist $\Lambda(p)$-sets that are $\Lambda(q)$-sets for no $q > p$ [Bour]. In his autobiography, Rudin wrote: '[Bourgain] has told me that he regards [the solution to the $\Lambda(p)$-set problem] as the most difficult problem he has ever solved, and he was quite disappointed that $\Lambda(p)$-sets were not mentioned in the lecture (given by Caffarelli) that described the work for which he won the Fields medal, at the Zürich Congress in 1994' [Ru4, p. 178]. The gap between the Bachelis–Ebenstein theorem and the Bourgain theorem remains an open question:

Problem 14 ('the $\Lambda(2)$-set problem'). Does a subspace H of $L^2(X, \nu)$ exist such that H is a $\Lambda(2)$-space, but H is a $\Lambda(q)$-space for no $q > 2$?

A second open problem concerns the stability of the $\Lambda(2)$ property. It is easy to see that if $p > 2$ and H_1 and H_2 are mutually orthogonal $\Lambda(p)$-spaces, then $H_1 \oplus H_2$ (the L^2-direct sum of H_1 and H_2) is also a $\Lambda(p)$-space. Whether this holds also for $p = 2$ is unknown:

Problem 15 ('the $\Lambda(2)$-set union problem'). Let H_1 and H_2 be mutually orthogonal infinite-dimensional $\Lambda(2)$-spaces. Is $H_1 \oplus H_2$ a $\Lambda(2)$-space?

Proposition 16 (Exercise 13). *Let H_1 and H_2 be mutually orthogonal uniformizable $\Lambda(2)$-spaces. Then, $H_1 \oplus H_2$ is a uniformizable $\Lambda(2)$-space.*

Alas, it is unknown whether every $\Lambda(2)$-space is uniformizable. Notice that a negative answer to Problem 14 would imply, by Theorem 8, that every $\Lambda(2)$-space is uniformizable. But in this case, there would be no need to invoke $\Lambda(2)$-uniformizability to solve Problem 15. For, if we knew every $\Lambda(2)$-space to be $\Lambda(q)$ for some $q > 2$, then we would conclude (by an easy argument) that the direct sum of any two $\Lambda(2)$-spaces is also a $\Lambda(2)$-space.

Exercises

1. Verify that in the definition of $\|\beta\|_{g_2}$ in (1.3), the unit ball of l^2 can be replaced by the unit sphere of l^2.
2. Verify that (1.6) is a restatement of the Grothendieck inequality.
3. i. Verify (the identity in the proof of Lemma 3)

$$\mathbf{A}(\mathbf{x},\mathbf{y}) = \mathbf{E}\, \mathscr{Im} \prod_k (1 + \mathrm{i}\mathbf{x}(k)r_k)\, \mathscr{Im}\left(\prod_k (1 + \mathrm{i}\mathbf{y}(k)r_k)\right).$$

   ii. Prove that we can do without the assumption in the beginning of the proof of Lemma 3, that the $\mathbf{x}_m$ and $\mathbf{y}_n$ have finite support. In particular, show that the infinite product $\prod_k (1+\mathrm{i}\mathbf{x}(k)r_k)$ converges almost surely $(\mathbb{P})$, and represents a function in $\mathrm{L}^\infty(\Omega,\mathbb{P})$ satisfying the estimate in (2.13). Then verify that (2.12) holds for every $\mathbf{x} \in l^2$ and $\mathbf{y} \in l^2$.
4. Prove Lemma 4.
5. (*functional analysis warm-ups*)
   i. Let H be a closed subspace of $\mathrm{L}^2(X,\nu)$, and let $p > 2$. Prove that the inclusion $H \subset \mathrm{L}^p$ is equivalent to existence of $C > 0$ such that for all $g \in H$,

$$\|g\|_{\mathrm{L}^p} \leq C\|g\|_{\mathrm{L}^2}.$$

   ii. Prove Lemma 9.
6. Show that the Grothendieck inequality can be verified by applying Theorem 8, as it stands, without using an explicit $\Lambda(2)$-uniformizing map α.
7. Prove that if $\mathrm{L}^2(X,\nu)$ is infinite-dimensional and $H \subset \mathrm{L}^2(X,\nu)$ is a uniformizable $\Lambda(2)$-space, then $H^\perp$ is infinite-dimensional.
8. Use the device $\prod_n (1 + \mathrm{i}\mathbf{x}(n)r_n)$ to establish directly (by construction) that $\mathrm{L}^2_R(\Omega,\mathbb{P})$ is a uniformizable $\Lambda(2)$-space.

9. i. Let $\{Z_n : n \in \mathbb{N}\}$ be a system of independent standard normal variables on a probability space (X, ν). Let H be the L^2-closure of the linear span of $\{Z_n : n \in \mathbb{N}\}$. Use the map

$$f \longmapsto \frac{\mathrm{e}^{\mathrm{i}\epsilon f} - 1}{\epsilon}, \quad f \in H,$$

to prove that H is a uniformizable $\Lambda(2)$-space.

ii. By applying a theorem proved in this chapter, give a second (faster) proof of the fact that H in Part i is a uniformizable $\Lambda(2)$-space.

10. i. Prove the identity in (6.3).

ii. By applying the *cosine* series to both sides of the identity, obtain a representation of the dot product in $\mathbb{R}^n$, and then prove the Grothendieck inequality.

11. Prove that if (6.4) holds for some $q \in (0, p)$ then it holds for all $q \in (0, p)$.

12.* Prove that if $H \subset \mathrm{L}^2(X, \nu)$ is a $\Lambda(p)$-space for $p \in (1, 2)$, then H is a $\Lambda(q)$-space for some $q > p$.

13. Prove that if H_1 is a uniformizable $\Lambda(2)$-space, and H_2 is a $\Lambda(2)$-space orthogonal to H_1, then $H_1 \oplus H_2$ is a $\Lambda(2)$-space. (Cf. Proposition 16.)

Hints for Exercises in Chapter III

3. i. This involves elementary 'harmonic analysis', which will be formalized in Chapter VII: first expand the product $\prod_k (1 + \mathrm{i}\mathbf{x}(k) r_k)$, take the imaginary part, and use the statistical independence of the Rademacher system.

ii. Show that $\prod_k (1 + \mathrm{i}\mathbf{x}(k) r_k)$ converges almost surely if and only if $\mathrm{e}^{\mathrm{i}\Sigma_k \mathbf{x}(k) r_k}$ converges almost surely. Use the Three Series theorem in classical probability theory.

4. This is a transcription of the proof of Lemma 2.

5. Review the argument verifying (II.2.3) $\Rightarrow$ (II.2.1).

6. See Remark ii in §4. This exercise can be done after reading §5.

10. i. It can be assumed that $\mathbf{x}$ and $\mathbf{y}$ are unit vectors in $\mathbb{R}^2$, and that integration can be performed over S_2. This assumption leads to a proof of (6.3) that is simpler than the original argument used by Grothendieck in his *Resumé*, and later by Lindenstrauss

and Pelczynski in their 1968 paper [LiPe]. This simpler, 'two-dimensional' argument appears in [LiTz, p. 68]. (It was shown to me in 1974 by S. Drury.)

13. Use a geometric series argument to show that if $g \in H_1$ then there exist $G \in \mathrm{L}^\infty(X, \nu)$ such that $\tau_{H_1}(G) = g$ and $\tau_{H_2}(G) = 0$.

IV
Elementary Properties of the Fréchet Variation – an Introduction to Tensor Products

1 Mise en Scène: The Space $F_k(\mathbb{N}, \ldots, \mathbb{N})$

In this chapter we focus on the norm that played prominently in the previous two chapters, and will continue to play prominently throughout the book.

Definition 1 (cf. II.3.1). Let $X_1, \ldots, X_k$ be sets. The F_k-variation (Fréchet variation) of a scalar-valued function β defined on $X_1 \times \cdots \times X_k$ is

$$\|\beta\|_{F_k} = \sup\left\{ \left\| \sum_{x_1 \in S_1, \ldots, x_k \in S_k} \beta(x_1, \ldots, x_k)\, r_{x_1} \otimes \cdots \otimes r_{x_k} \right\|_\infty : \text{finite sets } S_1 \subset X_1, \ldots, S_k \subset X_k \right\}, \tag{1.1}$$

where $\{r_x\}_{x \in X_i}$ are Rademacher systems indexed by X_i, $i \in [k]$. The space of scalar-valued functions β on $X_1 \times \cdots \times X_k$ such that $\|\beta\|_{F_k} < \infty$ is denoted by $F_k(X_1, \ldots, X_k)$. (When $X_1, \ldots, X_k$ are arbitrary, or understood from the context, we write F_k for $F_k(X_1, \ldots, X_k)$.)

The F_k-variation appears in the literature sometimes as the norm of a k-linear functional on c_0, sometimes as the k-fold l^1-injective tensor norm, and sometimes (in harmonic analysis) as the sup-norm of a function with spectrum in a k-fold Cartesian product of Sidon sets. Here, starting from first principles, we begin with the Fréchet variation, and in due course will identify it at the appropriate junctures with the aforementioned norms.

The Banach space $(F_k, \| \cdot \|_{F_k})$ (Exercise 1) is an extension to higher dimensions of l^1 $(= F_1)$, the classical space of absolutely summable functions. Some properties of F_k are routine extensions of those of l^1, but some properties manifest surprising, non-trivial 'multidimensional' features that in the 'one-dimensional' case may be unnoticed, uninteresting, or altogether absent. (We have already encountered one such surprise: the 'two-dimensional' Grothendieck inequality, which extends a rather trivial 'one-dimensional' observation, but itself is anything but trivial!). In this chapter, laying down the groundwork, we derive general basic properties that form the mainstay of the multidimensional setting. We leave surprises to later chapters. After we collect some essential tools, we will introduce basic notions of tensor products. These naturally appear in our context through the characterizations of F_k as the space of bounded *k-linear* functionals on c_0, and (equivalently) as the space of bounded *linear* functionals on the k-fold projective tensor product of c_0. These characterizations, extensions of the simple 'one-dimensional' duality $(c_0)^* = l^1$, are at the very heart of the subject, and will be extremely useful in the course of our work.

A study of the Fréchet variation inevitably involves the analysis of Rademacher systems indexed by $X_1, \ldots, X_k$. In later chapters, we will focus on the underlying indexing sets, and, in particular, will distinguish between various indexing schemes, but for work in this chapter, the generic $X_1 = \cdots = X_k = \mathbb{N}$ will do. As work progresses, we will occasionally use slight alterations of this indexing. Specifically, we will apply the (obvious) observation that if $\beta \in F_k(\mathbb{N}, \ldots, \mathbb{N})$ and $N > 0$, then $\beta\, \mathbf{1}_{[N] \times \mathbb{N} \times \cdots \times \mathbb{N}}$ is an element of $F_{k-1}([N] \times \mathbb{N}, \ldots, \mathbb{N})$ (in Definition 1, $X_1 = [N] \times \mathbb{N}$, $X_2 = \cdots = X_{k-1} = \mathbb{N}$). We will use also instances of the observation, which is easy to verify, that for positive integers $j_1, \ldots, j_m$ such that $j_1 + \cdots + j_m = k$,

$$F_m(\mathbb{N}^{j_1}, \ldots, \mathbb{N}^{j_m}) \subset F_k(\mathbb{N}, \ldots, \mathbb{N}). \tag{1.2}$$

2 Examples

At the very outset, confirming that Fréchet variations fundamentally depend on the underlying dimension, we note that inclusions in (1.2) are proper containments:

$$F_k \subsetneqq F_{k+1}, \quad k = 1, \ldots. \tag{2.1}$$

Indeed, that containments are proper – certainly believable, but not all that easy to verify – is the launching point of the subject. The instance $F_1 \subsetneqq F_2$ was observed first by Littlewood, in the introduction to his classic paper [Lit4, p. 164], through a quick application of the Hilbert inequality. Later in his paper, he proved a sharper assertion by use of a Gauss matrix (Theorem II.5 ii):

$$\beta_n(j,k) = (1/n^{\frac{3}{2}})\, \mathrm{e}^{2\pi \mathrm{i} jk/n}, \quad (j,k) \in [n] \times [n], \tag{2.2}$$

whence

$$\|\beta_n\|_{F_2([n],[n])} \leq 1, \tag{2.3}$$

and

$$\|\beta_n\|_{F_1([n]\times[n])} = \|\beta_n\|_1 = \sqrt{n},$$

which imply $F_1(\mathbb{N} \times \mathbb{N}) \subsetneqq F_2(\mathbb{N}, \mathbb{N})$ (Exercise II.8).

To verify the case $k = 2$, we define

$$\beta_n(j,k,l) = (1/n^2)\, \mathrm{e}^{2\pi \mathrm{i}(j+k)l/n}, \quad (j,k,l) \in [n] \times [n] \times [n], \tag{2.4}$$

and observe that

$$\|\beta_n\|_{F_3([n],[n],[n])} \leq 1, \tag{2.5}$$

and

$$\|\beta_n\|_{4/3} = n^{\frac{1}{4}}.$$

An application of Littlewood's 4/3-inequality (Theorem II.5 i) implies

$$\lambda \|\beta_n\|_{F_2([n]^2,[n])} \geq \|\beta_n\|_{4/3} = n^{\frac{1}{4}}, \tag{2.6}$$

and, therefore, $F_2(\mathbb{N}^2, \mathbb{N}) \subsetneqq F_3(\mathbb{N}, \mathbb{N}, \mathbb{N})$ (Exercise 2 i).

The full statement $F_k \subsetneqq F_{k+1}$ for all $k \geq 1$ is a consequence of the multidimensional extension of Littlewood's 4/3-inequality, which we derive in Chapter VII (Exercise 2 ii): that there exist $\lambda_k > 0$ such that for all $k > 1$, and all $\beta \in F_k$,

$$\lambda_k \|\beta\|_{F_k} \geq \|\beta\|_{2k/(k+1)}, \tag{2.7}$$

and there exist $\beta \in F_k$ such that $\|\beta\|_p = \infty$ for all $p < 2k/(k+1)$.

Remark (type F_k). The inequalities in (2.7) imply a statement sharper than (2.1). Let X be an infinite set, and $k \in \mathbb{N}$. We say that a scalar-valued function β defined on X is of *type* F_k if there exist bijections τ from X^k onto X such that $\beta \circ \tau \in F_k(X, \ldots, X)$. If β is of *type* F_k, then $\beta \in l^2(X)$ (Exercise 3). Clearly, if β is of *type* F_k, then β is of *type* F_{k+1}, and an application of (2.7) implies that there exist β of *type* F_{k+1} that are not of *type* F_k. More about *type* will be said later in the book.

3 Finitely Supported Functions are Norm-dense in $F_k(\mathbb{N}, \ldots, \mathbb{N})$

That finitely supported functions are norm-dense in $F_1(\mathbb{N})$ is easy to verify, but its multidimensional analog, that functions on $\mathbb{N}^k$ with finite support are norm-dense in $F_k(\mathbb{N}, \ldots, \mathbb{N})$, requires more work. To begin, we formalize a basic fact (Exercise 4):

Lemma 2 (cf. Theorem I.8). *For $\beta \in F_k(\mathbb{N}, \ldots, \mathbb{N})$,*

$$2^k \|\beta\|_{F_k} \geq \sup\left\{ \left| \sum_{n_1 \in T_1, \ldots, n_k \in T_k} \beta(n_1, \ldots, n_k)\, z^1_{n_1} \cdots z^k_{n_k} \right| : \right.$$

$$\left. |T_i| < \infty,\ z^i_n \in B_{\mathbb{C}},\ i \in [k],\ n \in \mathbb{N} \right\}. \tag{3.1}$$

The next lemma also formalizes a basic fact. Here and throughout, a *rectangle* in $\mathbb{N}^k$ will be a k-fold Cartesian product of subsets of $\mathbb{N}$; *k-disjoint* rectangles will mean rectangles whose respective edges on the coordinate axes are pairwise disjoint.

Lemma 3 *Let $\{C_j : j \in \mathbb{N}\}$ be a collection of k-disjoint rectangles in $\mathbb{N}^k$, and $S = \cup_{j=1}^{\infty} C_j$. Then, for all $\beta \in F_k(\mathbb{N}, \ldots, \mathbb{N})$,*

$$\|\beta \mathbf{1}_S\|_{F_k} \leq 2^k \|\beta\|_{F_k}. \tag{3.2}$$

Proof: Let $\pi_1, \ldots, \pi_k$ denote the canonical projections from $\mathbb{N}^k$ onto $\mathbb{N}$, and let

$$\pi_1[C_j] = A_{j1},\ \ldots,\ \pi_k[C_j] = A_{jk}, \quad j \in \mathbb{N}.$$

By k-disjointness, $\{A_{j1} : j\in\mathbb{N}\}$, ..., $\{A_{jk} : j\in\mathbb{N}\}$ are collections of pairwise disjoint sets. We will use the $\mathbf{T}_k$-valued random variables χ_j^k $(j \in \mathbb{N})$ on the probability space $(\Omega_k, \mathbb{P}_k)$, which were defined in Chapter II §6. Let $B_1 \subset \mathbb{N}, \ldots, B_k \subset \mathbb{N}$ be finite sets, let $t \in \Omega_k$, and define

$$\xi_n^i(t) = \chi_j^k(t)\ \mathbf{1}_{A_{ji}}(n), \quad n \in \mathbb{N},\ j \in \mathbb{N},\ i \in [k]. \tag{3.3}$$

By (II.6.9),

$$\mathbf{E}\ \xi_{n_1}^1 \cdots \xi_{n_k}^k = \mathbf{1}_S(n_1, \ldots, n_k), \quad (n_1, \ldots, n_k) \in \mathbb{N}^k, \tag{3.4}$$

where $\mathbf{E}$ denotes *expectation* with respect to $\mathbb{P}_k$. Fix $\omega_i \in \Omega\ (= \{-1,1\}^{\mathbb{N}})$, $i \in [k]$, and by Lemma 2 obtain

$$2^k \|\beta\|_{F_k} \geq \left| \sum_{n_1 \in B_1, \ldots, n_k \in B_k} \beta(n_1, \ldots, n_k) \xi_{n_1}^1(t) \cdots \xi_{n_k}^k(t) r_{n_1}(\omega_1) \cdots r_{n_k}(\omega_k) \right|. \tag{3.5}$$

By averaging (3.5) over $t \in \Omega_k$, and applying (3.4), we deduce

$$2^k \|\beta\|_{F_k} \geq \left| \sum_{n_1 \in B_1, \ldots, n_k \in B_k} \beta(n_1, \ldots, n_k) \mathbf{1}_S(n_1, \ldots, n_k) r_{n_1}(\omega_1) \cdots r_{n_k}(\omega_k) \right|. \tag{3.6}$$

Now maximize (3.6) over $B_1 \subset \mathbb{N}, \ldots, B_k \subset \mathbb{N}$, and $\omega_1 \in \Omega, \ldots, \omega_k \in \Omega$. □

The third needed fact is

Lemma 4 *Let $\{C_j : j \in \mathbb{N}\}$ be a collection of k-disjoint rectangles in $\mathbb{N}^k$, and $S = \cup_{j=1}^{\infty} C_j$. Then, for all $\beta \in F_k(\mathbb{N}, \ldots, \mathbb{N})$,*

$$\sum_{j=1}^{\infty} \|\beta \mathbf{1}_{C_j}\|_{F_k} \leq 2^k \|\beta \mathbf{1}_S\|_{F_k}. \tag{3.7}$$

Proof: Let $T_{1j} \times \cdots \times T_{kj} \subset C_j$, $j \in \mathbb{N}$, be finite rectangles. Fix $\omega_{1j} \in \Omega, \ldots, \omega_{kj} \in \Omega$, and $\delta_j \in B_{\mathbb{C}}$, such that

$$\delta_j \sum_{n_1 \in T_{1j}, \ldots, n_k \in T_{kj}} \beta(n_1, \ldots, n_k)\, r_{n_1}(\omega_{1j}) \cdots r_{n_k}(\omega_{kj})$$

$$= \left\| \sum_{n_1 \in T_{1j}, \ldots, n_k \in T_{kj}} \beta(n_1, \ldots, n_k)\, r_{n_1} \otimes \cdots \otimes r_{n_k} \right\|_{\mathrm{L}^\infty}. \tag{3.8}$$

Because the C_j are k-disjoint, we can choose $\omega_1 \in B_{l^\infty}$, $\omega_2 \in \Omega, \ldots, \omega_k \in \Omega$, such that for all $j \in \mathbb{N}$,

$$\begin{aligned} \omega_1(n) &= \delta_j\, r_n(\omega_{1j}), && n \in T_{1j}, \\ \omega_2(n) &= r_n(\omega_{2j}), && n \in T_{2j}, \\ &\vdots \\ \omega_k(n) &= r_n(\omega_{kj}), && n \in T_{kj}. \end{aligned}$$

Let $N \geq 1$, and $A_1 = \cup_{j=1}^N T_{1j}, \ldots, A_k = \cup_{j=1}^N T_{kj}$. By (3.8),

$$\sum_{j=1}^N \left\| \sum_{n_1 \in T_{1j}, \ldots, n_k \in T_{kj}} \beta(n_1, \ldots, n_k)\, r_{n_1} \otimes \cdots \otimes r_{n_k} \right\|_{\mathrm{L}^\infty}$$

$$= \left| \sum_{n_1 \in A_1, \ldots, n_k \in A_k} \beta(n_1, \ldots, n_k)\, \mathbf{1}_S(n_1, \ldots, n_k) \omega_1(n_1) \ldots \omega_k(n_k) \right|$$

$$\leq 2^k \|\beta\, \mathbf{1}_S\|_{F_k}. \tag{3.9}$$

To obtain (3.7), maximize the left side of (3.9) over $T_{1j} \times \cdots \times T_{kj}$. □

Lemmas 3 and 4 imply

Lemma 5 *Let*

$$R_N^k = \{(n_1, \ldots, n_k) : n_1 > N, \ldots, n_k > N\}, \quad N \in \mathbb{N}. \tag{3.10}$$

Then, for all $\beta \in F_k(\mathbb{N}, \ldots, \mathbb{N})$,

$$\lim_{N \to \infty} \|\beta\, \mathbf{1}_{R_N^k}\|_{F_k} = 0. \tag{3.11}$$

Proof: If there exist $\delta > 0$ and $N_j \uparrow \infty$ such that

$$\inf\{\|\beta\, \mathbf{1}_{R_{N_j}^k}\|_{F_k} : j \in \mathbb{N}\} > \delta, \tag{3.12}$$

then there exists a collection of k-disjoint rectangles $\{C_j\}$ such that

$$\inf\{\|\beta\, \mathbf{1}_{C_j}\|_{F_k} : j \in \mathbb{N}\} > \delta, \tag{3.13}$$

which, by Lemmas 3 and 4, contradicts $\|\beta\|_{F_k} < \infty$. □

The main result of the section is

Theorem 6 *Finitely supported functions on* $\mathbb{N}^k$ *are norm-dense in* $F_k(\mathbb{N}, \ldots, \mathbb{N})$. *In particular, if* $\beta \in F_k$, *then*

$$\lim_{N\to\infty} \|\beta\, \mathbf{1}_{[N]^k} - \beta\|_{F_k} = 0. \tag{3.14}$$

Proof (by induction). The case $k = 1$ is the assertion that finitely supported functions are norm-dense in $l^1(\mathbb{N})$.

Let $k > 1$, and assume (the induction hypothesis) that (3.14) holds with $k-1$ in place of k for all $\beta \in F_{k-1}$. Let $\beta \in F_k$, and fix $\epsilon > 0$. By applying Lemma 5, let $N_0 > 0$ be such that for all $N \geq N_0$,

$$\|\beta\, \mathbf{1}_{R^k_N}\|_{F_k} < \epsilon. \tag{3.15}$$

By applying the induction hypothesis, we let $K_0 > N_0$ be such that for all $K \geq K_0$,

$$\|\beta\, \mathbf{1}_{\{n_1\in[N_0],(n_2,\ldots,n_k)\notin[K]^{k-1}\}}\|_{F_{k-1}([N_0]\times\mathbb{N},\ldots,\mathbb{N})}$$
$$\leq \|\beta\, \mathbf{1}_{\{n_1\in[N_0],(n_2,\ldots,n_k)\notin[K]^{k-1}\}}\|_{F_k([N_0],\mathbb{N},\ldots,\mathbb{N})} < \epsilon,$$

$\vdots$

$$\|\beta\, \mathbf{1}_{\{(n_1,\ldots,n_{j-1})\notin\mathbb{N}^{j-1},n_j\in[N_0],(n_{j+1},\ldots,n_k)\notin[K]^{k-j}\}}\|_{F_{k-1}(\mathbb{N},\ldots,[N_0]\times\mathbb{N},\ldots,\mathbb{N})}$$
$$\leq \|\beta\, \mathbf{1}_{\{(n_1,\ldots,n_{j-1})\notin\mathbb{N}^{j-1},n_j\in[N_0],(n_{j+1},\ldots,n_k)\notin[K]^{k-j}\}}\|_{F_k(\mathbb{N},\ldots,[N_0],\mathbb{N},\ldots,\mathbb{N})} < \epsilon,$$

$\vdots$

$$\|\beta\, \mathbf{1}_{\{(n_1,\ldots,n_{k-1})\notin\mathbb{N}^{k-1},n_k\in[N_0]\}}\|_{F_{k-1}(\mathbb{N},\ldots,[N_0]\times\mathbb{N})}$$
$$\leq \|\beta\, \mathbf{1}_{\{(n_1,\ldots,n_{k-1})\notin\mathbb{N}^{k-1},n_k\in[N_0]\}}\|_{F_k(\mathbb{N},\ldots,[N_0],\mathbb{N})} < \epsilon. \tag{3.16}$$

Then, by applying (3.15) and (3.16), we deduce

$$\|\beta\, \mathbf{1}_{[N]^k} - \beta\|_{F_k} \leq (2k+1)\epsilon \tag{3.17}$$

for all $N \geq K_0$ (Exercise 5). □

4 Two Consequences

A Fubini-type Property

That every $\beta \in F_1(\mathbb{N})$ determines a bounded linear functional on $l^\infty(\mathbb{N})$ is obvious. The multidimensional analog, which we prove below, is not quite as trivial. En route, we observe a 'Fubini'-type property that will be used extensively throughout the book.

Corollary 7 *Every $\beta \in F_k(\mathbb{N}, \ldots, \mathbb{N})$ determines a bounded k-linear functional $\hat{\beta}$ on l^∞. Specifically, for all $f_1 \in l^\infty, \ldots, f_k \in l^\infty$, and all permutations τ of $[k]$,*

$$\hat{\beta}(f_1, \ldots, f_k) := \sum_{n_1=1}^{\infty} \cdots \left(\sum_{n_k=1}^{\infty} \beta(n_1, \ldots, n_k) f_k(n_k) \right) \cdots f_1(n_1)$$

$$= \sum_{n_{\tau 1}=1}^{\infty} \cdots \left(\sum_{n_{\tau k}=1}^{\infty} \beta(n_1, \ldots, n_k)\; f_{\tau k}(n_{\tau k}) \right) \cdots f_{\tau 1}(n_{\tau 1}), \tag{4.1}$$

and

$$|\hat{\beta}(f_1, \ldots, f_k)| \leq 2^k \|f_1\|_\infty \cdots \|f_k\|_\infty. \tag{4.2}$$

Proof: We prove the case $k = 2$, and relegate the general case to Exercise 6. Let $\beta \in F_2(\mathbb{N}, \mathbb{N})$. For f and g in l^∞, define

$$\beta_{f \otimes g} = \{\beta(m,n) f(m) g(n) : (m,n) \in \mathbb{N}^2\}. \tag{4.3}$$

By Lemma 2, $\beta_{f\otimes g} \in F_2(\mathbb{N}, \mathbb{N})$, and

$$\|\beta_{f\otimes g}\|_{F_2} \leq 4 \|f\|_\infty \; \|g\|_\infty \; \|\beta\|_{F_2}. \tag{4.4}$$

Therefore, it suffices to prove that if $\beta \in F_2(\mathbb{N}, \mathbb{N})$ then $\Sigma_{n=1}^{\infty} \Sigma_{m=1}^{\infty} \beta(m,n)$ exists, and

$$\sum_{m=1}^{\infty} \sum_{n=1}^{\infty} \beta(m,n) = \sum_{n=1}^{\infty} \sum_{m=1}^{\infty} \beta(m,n). \tag{4.5}$$

For $\omega_1 \in \Omega$, and $N_1 \in \mathbb{N}$,

$$\sum_{m=1}^{N} \left| \sum_{n=1}^{N_1} \beta(m,n)\; r_n(\omega_1) \right| \leq \sum_{m=1}^{\infty} \left| \sum_{n=1}^{N_1} \beta(m,n)\; r_n(\omega_1) \right| \leq 2\; \|\beta\|_{F_2}, \tag{4.6}$$

and therefore,

$$\left|\sum_{m=1}^{\infty}\sum_{n=1}^{N_1}\beta(m,n)\ r_n(\omega_1)\right| = \left|\sum_{n=1}^{N_1}\left(\sum_{m=1}^{\infty}\beta(m,n)\right)\ r_n(\omega_1)\right| \leq 2\ \|\beta\|_{F_2}. \tag{4.7}$$

By maximizing (4.6) over $\omega_1 \in \Omega$, we obtain

$$\sum_{n=1}^{N_1}\left|\sum_{m=1}^{\infty}\beta(m,n)\right| \leq 4\ \|\beta\|_{F_2}. \tag{4.8}$$

Letting $N_1 \to \infty$ in (4.8), we conclude that $\Sigma_{n=1}^{\infty}\Sigma_{m=1}^{\infty}\beta(m,n)$ exists, and

$$\left|\sum_{n=1}^{\infty}\sum_{m=1}^{\infty}\beta(m,n)\right| \leq 4\ \|\beta\|_{F_2}. \tag{4.9}$$

We proceed to verify (4.5). By Theorem 6, for every $\epsilon > 0$ there exists $N \in \mathbb{N}$ so that $\|\beta\mathbf{1}_{[N]\times[N]} - \beta\|_{F_2} < \epsilon$. Therefore, by (4.9),

$$\left|\sum_{n=1}^{\infty}\sum_{m=1}^{\infty}(\beta\mathbf{1}_{[N]\times[N]}(m,n) - \beta(m,n))\right| < 4\epsilon, \tag{4.10}$$

and

$$\left|\sum_{m=1}^{\infty}\sum_{n=1}^{\infty}(\beta\mathbf{1}_{[N]\times[N]}(m,n) - \beta(m,n))\right| < 4\epsilon.$$

Obviously,

$$\sum_{m=1}^{\infty}\sum_{n=1}^{\infty}\beta\mathbf{1}_{[N]\times[N]}(m,n) = \sum_{n=1}^{\infty}\sum_{m=1}^{\infty}\beta\mathbf{1}_{[N]\times[N]}(m,n). \tag{4.11}$$

Therefore, by combining (4.10) and (4.11), we obtain

$$\left|\sum_{m=1}^{\infty}\sum_{n=1}^{\infty}\beta(m,n) - \sum_{n=1}^{\infty}\sum_{m=1}^{\infty}\beta(m,n)\right| < 8\epsilon, \tag{4.12}$$

which implies (4.5). □

$F_k(\mathbb{N},\ldots,\mathbb{N})$ *has the Schur Property*

Recall that a sequence (x_n) in a Banach space X *converges weakly* to $x \in X$ if $f(x_n) \to f(x)$ for all $f \in X^*$. Norm convergence and weak convergence are obviously the same in finite-dimensional spaces, but there are (many!) infinite-dimensional Banach spaces where they are not equivalent. A classical theorem by J. Schur [Shu2] states that weak convergence and norm convergence *are* equivalent in $F_1(\mathbb{N})$. (If weak convergence and norm convergence in a Banach space X are equivalent, then X is said to have the *Schur property.*) Below we prove, by induction, the multidimensional version of Schur's theorem.

We begin with Schur's 'one-dimensional' result:

Theorem 8 ([Shu2]). *If a sequence* $(\alpha_n : n \in \mathbb{N})$ *in* $F_1(\mathbb{N})$ *converges weakly to* $\alpha \in F_1(\mathbb{N})$, *i.e., if*

$$\lim_{n\to\infty} \sum_{m=1}^{\infty} \alpha_n(m) f(m) = \sum_{m=1}^{\infty} \alpha(m) f(m) \quad \textit{for all } f \in l^{\infty}, \tag{4.13}$$

then $\lim_{n\to\infty} \|\alpha_n - \alpha\|_{F_1} = 0$.

Proof: It suffices to prove that if $\alpha_n \to 0$ weakly, then $\|\alpha_n\|_{F_1} \to 0$. Suppose it is false, i.e., suppose $\alpha_n \to 0$ weakly and $\|\alpha_n\|_{F_1} = 1$ for all $n \in \mathbb{N}$. Because finitely supported functions on $\mathbb{N}$ are norm-dense in $F_1(\mathbb{N})$, it can be assumed that each α_n has finite support. For each $K \in \mathbb{N}$, weak convergence and norm convergence are equivalent in $F_1([K])$. Therefore, for each $K \in \mathbb{N}$,

$$\lim_{n\to\infty} \|\mathbf{1}_{[K]}\alpha_n\|_{F_1} = 0. \tag{4.14}$$

Suppose *support* $\alpha_1 = [K]$. Let $n_1 = 1$ and $K_1 = K$. By applying (4.14) with $K = K_1$, fix $n_2 > n_1$ such that

$$\|\mathbf{1}_{[K_1]}\alpha_{n_2}\|_{F_1} < \frac{1}{4}. \tag{4.15}$$

Let $K_2 \in \mathbb{N}$ be such that *support* $\alpha_{n_2} \subset [K_2]$. Continuing recursively, we obtain increasing sequences of integers $1 = n_1 < \cdots < n_m < \cdots$, and $0 = K_0 < K_1 < \cdots < K_m < \cdots$ such that for $i \geq 1$, *support* $\alpha_{n_i} \subset [K_i]$, and

$$\|\mathbf{1}_{[K_{i-1}]}\alpha_{n_i}\|_{F_1} < \frac{1}{4}. \tag{4.16}$$

Denote $E_i = \{K_{i-1} + 1, \ldots, K_i\}$. By (4.16) and the assumption $\|\alpha_{n_i}\|_{F_1} = 1$,

$$\|\mathbf{1}_{E_i}\alpha_{n_i}\|_{F_1} > \frac{3}{4}, \quad i \in \mathbb{N}. \tag{4.17}$$

Let $f \in l^\infty(\mathbb{N})$ be so that $f(j)\, \alpha_{n_i}(j) = |\alpha_{n_i}(j)|$ for $j \in E_i$. Then, by (4.16) and (4.17),

$$\left|\sum_{j=1}^{\infty} \alpha_{n_i}(j)\, f(j)\right| > \frac{1}{2} \quad \text{for all } i \in \mathbb{N}, \tag{4.18}$$

which contradicts the assumption that $\alpha_n \to 0$ weakly. □

The argument above verifying Schur's theorem rests on two facts: (1) $F_1([K])$ satisfies the Schur property for every $K \in \mathbb{N}$; (2) finitely supported elements in $F_1(\mathbb{N})$ are norm-dense in $F_1(\mathbb{N})$. The inductive argument below, which verifies the extension of Schur's theorem, is similar: the induction hypothesis corresponds to (1), and Theorem 6, corresponds to (2).

Theorem 9 *Suppose* $(\beta_n : n \in \mathbb{N})$ *is a sequence in* $F_k(\mathbb{N}, \ldots, \mathbb{N})$, $\beta \in F_k(\mathbb{N}, \ldots, \mathbb{N})$, *and*

$$\lim_{n\to\infty} \hat{\beta}_n(f_1, \ldots, f_k) = \hat{\beta}(f_1, \ldots, f_k)$$
$$\textit{for all } f_1 \in l^\infty, \ldots, f_k \in l^\infty. \tag{4.19}$$

Then, $\lim_{n\to\infty} \|\beta_n - \beta\|_{F_k} = 0$. *(See (4.1) for the meaning of 'hat' in (4.19).)*

Proof (by induction). The case $k = 1$ is Schur's theorem. Let $k > 1$, and suppose the assertion is true in the case $k - 1$. In proving the inductive step, we assume (without loss of generality) that $\beta = 0$, and (by Theorem 6) that each β_n has finite support in $\mathbb{N}^k$. Suppose the inductive step fails; that is,

$$\lim_{n\to\infty} \hat{\beta}_n(f_1, \ldots, f_k) = 0 \quad \text{for all } f_1 \in l^\infty, \ldots, f_k \in l^\infty, \tag{4.20}$$

and $\|\beta_n\|_{F_k} = 1$ for all $n \in \mathbb{N}$. For $N \in \mathbb{N}$, denote

$$\mathbf{J}_N = 1 - \mathbf{1}_{R_N^k}. \tag{4.21}$$

(R_N^k is defined in (3.10).) For each $N \in \mathbb{N}$, $\beta_n\, \mathbf{J}_N \in F_{k-1}$ (Exercise 7). By applying the induction hypothesis, we obtain $\lim_{n\to\infty} \|\beta_n\, \mathbf{J}_N\|_{F_{k-1}} = 0$, which implies

$$\lim_{n\to\infty} \|\beta_n\, \mathbf{J}_N\|_{F_k} = 0. \tag{4.22}$$

Suppose that *support* $\beta_1 \subset [N]^k$. Let $n_1 = 1$ and $N_1 = N$. By (4.22), there exists $n_2 > n_1$ such that

$$\|\beta_{n_2}\, \mathbf{J}_{N_1}\|_{F_k} < \frac{1}{4}. \tag{4.23}$$

Continuing recursively, we obtain increasing sequences

$$n_1 < \cdots < n_m < \cdots,$$

and $0 < N_1 < \cdots < N_m < \cdots$ such that *support* $\beta_{n_i} \subset [N_i]^k$ for $i \geq 1$, and

$$\|\beta_{n_i}\, \mathbf{J}_{N_{i-1}}\|_{F_k} < \left(\frac{1}{4}\right)^k \quad \text{for } i > 1. \tag{4.24}$$

Initialize $N_0 = 0$, and define for $i \in \mathbb{N}$,

$$E_i = \{N_{i-1}+1, \ldots, N_i\}^k. \tag{4.25}$$

Because $\beta_{n_i}(\mathbf{J}_{N_i} + \mathbf{1}_{E_i}) = \beta_{n_i}$ and $\|\beta_{n_i}\|_{F_k} = 1$, we obtain from (4.24) that for all $i \geq 1$,

$$\|\beta_{n_i}\, \mathbf{1}_{E_i}\|_{F_k} \geq 1 - \left(\frac{1}{4}\right)^k. \tag{4.26}$$

By applying (4.26), we let $f_{1i} \in B_{l^\infty}, \ldots, f_{ki} \in B_{l^\infty}$, $i \in \mathbb{N}$, be so that

$$\sum_{(j_1,\ldots,j_k)\in E_i} (\beta_{n_i}\, \mathbf{1}_{E_i})(j_1,\ldots,j_k)\, f_{1i}(j_1)\cdots f_{ki}(j_k) > 1 - \left(\frac{1}{4}\right)^k. \tag{4.27}$$

Moreover, for each $j \in [k]$ and $i \in \mathbb{N}$, f_{ji} is chosen to have support in $\{N_{i-1}+1, \ldots, N_i\}$. Let $f_1 = \Sigma_{i=1}^\infty f_{1i}, \ldots, f_k = \Sigma_{i=1}^\infty f_{ki}$. Then, $f_1 \in B_{l^\infty}, \ldots, f_k \in B_{l^\infty}$. By (4.24) and (4.27),

$$\begin{aligned} &|\hat{\beta}_{n_i}(f_1,\ldots,f_k)| \\ &\geq 1 - \left(\frac{1}{4}\right)^k - \left| \sum_{(j_1,\ldots,j_k)\in E_i} (\beta_{n_i}\, \mathbf{J}_{N_{i-1}})(j_1,\ldots,j_k)\, f_{1i}(j_1)\cdots f_{ki}(j_k) \right| \\ &\geq \frac{1}{2} \end{aligned} \tag{4.28}$$

for all $i \in \mathbb{N}$, which contradicts (4.20). □

Remarks:

i **(historical comments).** That finitely supported functions on $\mathbb{N}^k$ are norm-dense in F_k (Theorem 6) appears to be, variously phrased, part of the folklore of functional analysis; I have not been able to track down a first reference. Its two dimensional prototype, that operators of finite rank are norm-dense in the space of operators from c_0 into l^1, was known to the founding masters, and indeed was implicit in Littlewood's paper [Lit4]. The elementary argument establishing it is known among functional analysts as a 'gliding hump' argument.

That summations in (4.1) can be freely interchanged was first verified by Littlewood in the case $\beta \in F_2(\mathbb{N}, \mathbb{N})$ [Lit4, pp. 167–8].

Like Theorem 6, the extension of Schur's theorem to F_k for $k > 1$ has become folklore. I first saw a version of it (without proof) in [Mey2, (6.2.5)] (Exercise 8). Generalizations of this extension in a framework of topological tensor products appeared in [Lu1], [Lu2].

ii **(a preview).** In a framework of harmonic analysis, the separability of F_k is equivalent to the (so-called) Rosenthal property of k-fold Cartesian products of Sidon sets. This will be explained and discussed in Chapter VII.

5 The Space $V_k(\mathbb{N}, \ldots, \mathbb{N})$

That $l^1(\mathbb{N})$ is the dual space of $c_0(\mathbb{N})$ – a simple instance of the Riesz representation theorem – is at the very foundation of (linear) functional analysis. A question arises: can $F_k(\mathbb{N}, \ldots, \mathbb{N})$ be analogously realized as a dual space? It is here, in answer to this question, that tensor products naturally appear.

Let $X_1, \ldots, X_k$ be sets. Let $f_1 \in c_0(X_1), \ldots, f_k \in c_0(X_k)$, and consider their formal product $f_1 \otimes \cdots \otimes f_k$, which we call an *elementary tensor.* The *k-fold algebraic tensor product* of c_0 is the class of all finite combinations of elementary tensors,

$$c_0 \otimes \cdots \otimes c_0$$

$$:= \left\{ \sum_j f_{1j} \otimes \cdots \otimes f_{kj} : \text{finite sum } \sum_j, \text{ and } f_{ij} \in c_0(X_i) \right\}. \quad (5.1)$$

This construction is completely general: given sets $A_1,\dots,A_k$, we view their respective elements as basic building blocks, 'cement' them, and then define $A_1 \otimes \cdots \otimes A_k$ to be the set of all finite combinations of these 'cemented products'. Aiming further, we of course expect that structures in $A_1,\dots,A_k$ will lead to new structures in $A_1 \otimes \cdots \otimes A_k$. In our present setting, we let $A_1 = \cdots = A_k = c_0$. Specifically, we view members in $c_0 \otimes \cdots \otimes c_0$ as functions on $X_1 \times \cdots \times X_k$,

$$\left(\sum_j f_{1j} \otimes \cdots \otimes f_{kj}\right)(x_1,\dots,x_k) = \sum_j f_{1j}(x_1)\cdots f_{kj}(x_k),$$
$$(x_1,\dots,x_k) \in X_1 \times \cdots \times X_k, \tag{5.2}$$

and then consider two elements in $c_0 \otimes \cdots \otimes c_0$ equivalent if each represents the same function on $X_1 \times \cdots \times X_k$. From now on (slightly abusing notation), we let $c_0 \otimes \cdots \otimes c_0$ denote the resulting set of equivalence class representatives.

For $\tau \in c_0 \otimes \cdots \otimes c_0$, define

$$\|\tau\|_{V_k} = \inf\left\{\sum_j \|f_{1j}\|_\infty \cdots \|f_{jk}\|_\infty : \tau = \sum_j f_{1j} \otimes \cdots \otimes f_{kj}\right\}, \tag{5.3}$$

which defines a norm on $c_0 \otimes \cdots \otimes c_0$. The closure of $c_0 \otimes \cdots \otimes c_0$ under this norm is denoted by $V_k(X_1,\dots,X_k)$. This closure comprises all Cauchy sequences in $(c_0 \otimes \cdots \otimes c_0, \|\cdot\|_{V_k})$ with the usual equivalence relation: two Cauchy sequences are considered equivalent if their difference converges to zero in the V_k-norm.

If $\tau \in c_0 \otimes \cdots \otimes c_0$, then $\|\tau\|_{V_k} \geq \|\tau\|_\infty$. Therefore, if (φ_j) is Cauchy in $(c_0 \otimes \cdots \otimes c_0, \|\cdot\|_{V_k})$, then it is Cauchy in $c_0(X_1 \times \cdots \times X_k)$, and $\varphi_j \to f$ uniformly, where $f \in c_0(X_1 \times \cdots \times X_k)$. This naturally defines an injection from $V_k(X_1,\dots,X_k)$ (Cauchy sequences in $(c_0 \otimes \cdots \otimes c_0, \|\cdot\|_{V_k})$) into $c_0(X_1 \times \cdots \times X_k)$. Indeed, we claim that two Cauchy sequences ϕ and ψ in $(c_0 \otimes \cdots \otimes c_0, \|\cdot\|_{V_k})$ that converge uniformly to the same function on $X_1 \times \cdots \times X_k$ represent the same element in $V_k(X_1,\dots,X_k)$. To verify this, by passing to subsequences we can assume

$$\phi = \left(\sum_{j=1}^n \varphi_j : n \in \mathbb{N}\right), \quad \psi = \left(\sum_{j=1}^n \theta_j : n \in \mathbb{N}\right),$$
$$\varphi_j \in c_0 \otimes \cdots \otimes c_0,\ \theta_j \in c_0 \otimes \cdots \otimes c_0,\ j \in \mathbb{N}, \tag{5.4}$$

$$\sum_{j=1}^{\infty} \|\varphi_j\|_{V_k} < \infty, \quad \sum_{j=1}^{\infty} \|\theta_j\|_{V_k} < \infty, \tag{5.5}$$

and

$$\sum_{j=1}^{\infty} \varphi_j(\mathbf{x}) = \sum_{j=1}^{\infty} \theta_j(\mathbf{x}) \quad \text{for all } \mathbf{x} \in X_1 \times \cdots \times X_k. \tag{5.6}$$

By (5.6),

$$\begin{aligned} \left\| \sum_{j=1}^{n-1} \varphi_j - \sum_{j=1}^{n-1} \theta_j \right\|_{V_k} &= \left\| \sum_{j=n}^{\infty} \varphi_j - \sum_{j=n}^{\infty} \theta_j \right\|_{V_k} \\ &\leq \left\| \sum_{j=n}^{\infty} \varphi_j \right\|_{V_k} + \left\| \sum_{j=n}^{\infty} \theta_j \right\|_{V_k} \\ &\leq \sum_{j=n}^{\infty} \|\varphi_j\|_{V_k} + \sum_{j=n}^{\infty} \|\theta_j\|_{V_k}, \end{aligned} \tag{5.7}$$

which proves the claim, and thus $V_k(X_1, \ldots, X_k) \subset c_0(X_1 \times \cdots \times X_k)$. After we verify the duality $(V_k)^* = F_k$ (Proposition 11 below), we will note that this inclusion is proper.

The following proposition characterizes those $\phi \in c_0(X_1 \times \cdots \times X_k)$ that belong in $V_k(X_1, \ldots, X_k)$.

Proposition 10 (Exercise 10). *If $\phi \in c_0(X_1 \times \cdots \times X_k)$, then*

$$\begin{aligned} \|\phi\|_{V_k} = \inf \Bigg\{ \sum_j \|f_{1j}\|_\infty \cdots \|f_{jk}\|_\infty : \phi(x_1, \ldots, x_k) \\ = \sum_j f_{1j}(x_1) \cdots f_{kj}(x_k), \ f_{ij} \in c_0(X_i) \Bigg\}, \end{aligned} \tag{5.8}$$

and

$$V_k(X_1, \ldots, X_k) = \{\phi \in c_0(X_1 \times \cdots \times X_k) : \|\phi\|_{V_k} < \infty\}.$$

Next we verify that $F_k(X_1, \ldots, X_k)$ is the dual space of $V_k(X_1, \ldots, X_k)$. For convenience, we let $X_1 = \cdots = X_k = \mathbb{N}$. Given $\beta \in F_k(\mathbb{N}, \ldots, \mathbb{N})$, we consider $\hat{\beta}$ (defined in (4.1)) as a scalar-valued function of elementary tensors $f_1 \otimes \cdots \otimes f_k \in c_0 \otimes \cdots \otimes c_0$,

$$\begin{aligned} \hat{\beta}(f_1 \otimes \cdots \otimes f_k) &:= \hat{\beta}(f_1, \ldots, f_k) \\ &= \sum_{n_1=1}^{\infty} \cdots \sum_{n_k=1}^{\infty} \beta(n_1, \ldots, n_k)\, f_k(n_k) \cdots f_1(n_1), \end{aligned} \tag{5.9}$$

and extend $\hat{\beta}$ to $c_0 \otimes \cdots \otimes c_0$ by linearity. That is, let $\tau \in c_0 \otimes \cdots \otimes c_0$, represent it by $\Sigma_j f_{1j} \otimes \cdots \otimes f_{kj}$, and then define

$$\hat{\beta}(\tau) = \sum_j \hat{\beta}(f_{1j} \otimes \cdots \otimes f_{kj}). \tag{5.10}$$

To verify that (5.10) is well-defined, we need to check that if

$$\Sigma_j \varphi_{1j} \otimes \cdots \otimes \varphi_{kj}$$

is any other representation of τ, then

$$\sum_j \hat{\beta}(f_{1j} \otimes \cdots \otimes f_{kj}) = \sum_j \hat{\beta}(\varphi_{1j} \otimes \cdots \otimes \varphi_{kj}). \tag{5.11}$$

To confirm this, observe that if $\Sigma_j g_{1j} \otimes \cdots \otimes g_{kj} \in c_0 \otimes \cdots \otimes c_0$, and

$$\sum_j g_{1j}(n_1) \cdots g_{kj}(n_k) = 0 \quad \text{for all } (n_1, \ldots, n_k) \in \mathbb{N}^k, \tag{5.12}$$

then

$$\sum_j \hat{\beta}(g_{1j} \otimes \cdots \otimes g_{kj}) = 0. \tag{5.13}$$

(Because Σ_j is a finite sum, this follows simply from (5.9).)

If $\tau = \Sigma_j f_{1j} \otimes \cdots \otimes f_{kj}$ and $\beta \in F_k(\mathbb{N},\ldots,\mathbb{N})$, then by Lemma 2,

$$\begin{aligned} |\hat{\beta}(\tau)| &\leq \sum_j |\hat{\beta}(f_{1j} \otimes \cdots \otimes f_{kj})| \\ &\leq 2^k \|\beta\|_{F_k} \sum_j \|f_{1j}\|_\infty \cdots \|f_{kj}\|_\infty. \end{aligned} \tag{5.14}$$

Therefore, $\hat{\beta}$ is continuous on $(c_0 \otimes \cdots \otimes c_0, \|\cdot\|_{V_k})$, and thus extendible to a continuous linear functional on $V_k(\mathbb{N},\ldots,\mathbb{N})$ with norm bounded by $2^k\|\beta\|_{F_k}$.

Conversely, if $\mu \in V_k(\mathbb{N},\ldots,\mathbb{N})^*$, then

$$|\mu(f_1 \otimes \cdots \otimes f_k)| \leq \|\mu\|_{V_k^*} \|f_1 \otimes \cdots \otimes f_k\|_{V_k}.$$

But

$$\|f_1 \otimes \cdots \otimes f_k\|_{V_k} = \|f_1\|_\infty \cdots \|f_k\|_\infty, \tag{5.15}$$

which follows from the definition of the V_k-norm and the triangle inequality (see Exercise 15). Therefore, μ determines (*a fortiori*) a bounded k-linear functional on $c_0 \times \cdots \times c_0$, whose value at $(f_1, \ldots, f_k) \in c_0 \times \cdots \times c_0$ is $\mu(f_1 \otimes \cdots \otimes f_k)$, and whose norm is bounded by $\|\mu\|_{V_k^*}$. We summarize:

Proposition 11 *Every $\beta \in F_k(\mathbb{N},\ldots,\mathbb{N})$ determines a bounded linear functional $\hat{\beta}$ on $V_k(\mathbb{N},\ldots,\mathbb{N})$ such that for $\phi \in V_k(\mathbb{N},\ldots,\mathbb{N})$,*

$$\hat{\beta}(\phi) = \sum_{n_1=1}^{\infty} \cdots \sum_{n_k=1}^{\infty} \beta(n_1,\ldots,n_k)\ \phi(n_1,\ldots,n_k). \tag{5.16}$$

Moreover, $\|\hat{\beta}\|_{V_k^*} \leq 2^k \|\beta\|_{F_k}$.

Conversely, if $\mu \in V_k(\mathbb{N},\ldots,\mathbb{N})^*$, and

$$\beta(n_1,\ldots,n_k) = \mu(\mathbf{e}_{n_1} \otimes \cdots \otimes \mathbf{e}_{n_k})$$

for $(n_1,\ldots,n_k) \in \mathbb{N}^k$, then $\beta \in F_k(\mathbb{N},\ldots,\mathbb{N})$ and $\|\beta\|_{F_k} \leq \|\mu\|_{V_k^*}$. ($\{\mathbf{e}_m\}$ is the standard basis in $c_0(\mathbb{N})$: $\mathbf{e}_m(m) = 1$ and $\mathbf{e}_m(j) = 0$ for $m \neq j$.)

Remarks

i **(examples).** We have already noted that

$$V_k(\mathbb{N},\ldots,\mathbb{N}_k) \subset c_0(\mathbb{N} \times \cdots \times \mathbb{N}) := V_1(\mathbb{N} \times \cdots \times \mathbb{N}).$$

More generally, we note that for all $k \geq 1$,

$$V_{k+1}(\mathbb{N},\ldots,\mathbb{N}) \subset V_k(\mathbb{N} \times \mathbb{N},\ldots,\mathbb{N}), \tag{5.17}$$

and that, like (2.1), these containments are proper.

The Case $k = 1$. Define

$$\varphi_n(j,k) = \mathrm{e}^{-2\pi \mathrm{i} jk/n}/\sqrt{n} \quad \text{for} \quad (j,k) \in [n] \times [n], \tag{5.18}$$

and

$$\varphi_n(j,k) = 0 \text{ for } (j,k) \notin [n] \times [n].$$

Then,

$$\sum_{j,k} \varphi_n(j,k)\beta_n(j,k) = \sqrt{n}, \tag{5.19}$$

where β_n is defined in (2.2). Therefore, by (2.3), and by duality (Proposition 11),

$$\|\varphi_n\|_{V_2} \geq \sqrt{n}, \tag{5.20}$$

which implies $V_2(\mathbb{N},\mathbb{N}) \subsetneq c_0(\mathbb{N} \times \mathbb{N})$ (Exercise 9).

The Case $k = 2$. Define

$$\theta_n(j,k,l) = \mathrm{e}^{-2\pi \mathrm{i}(j+k)l/n}/\sqrt{n} \text{ for } (j,k,l) \in [n] \times [n] \times [n], \tag{5.21}$$

and

$$\theta_n(j,k,l) = 0 \quad \text{for } (j,k,l) \notin [n]\times[n]\times[n].$$

Then,

$$\sum_{j,k,l} \theta_n(j,k,l)\beta_n(j,k,l) = \sqrt{n}, \tag{5.22}$$

where β_n is defined in (2.4). Therefore, by (2.5) and duality,

$$\|\theta_n\|_{V_3(\mathbb{N},\mathbb{N},\mathbb{N})} \geq \sqrt{n}. \tag{5.23}$$

To estimate the $V_2(\mathbb{N}^2,\mathbb{N})$-norm of θ_n, we suppose $\beta \in F_2(\mathbb{N}^2,\mathbb{N})$, $\|\beta\|_{F_2} \leq 1$, and note

$$\left|\sum_{j,k,l} \theta_n(j,k,l)\ \beta(j,k,l)\right| \leq \sum_{j,k}\sum_{l} |\theta_n(j,k,l)\ \beta(j,k,l)|$$

$$\leq \sum_{j,k}\left(\sum_{l} |\beta(j,k,l)|^2\right)^{\frac{1}{2}} \leq \kappa_{\mathrm{L}}, \tag{5.24}$$

where the last inequality follows from Littlewood's (l^1,l^2)-mixed norm inequality (Theorem II.2). Therefore (again by duality),

$$\|\theta_n\|_{V_2(\mathbb{N}^2,\mathbb{N})} \leq \kappa_{\mathrm{L}}. \tag{5.25}$$

By combining (5.23) and (5.25), we obtain $V_3(\mathbb{N},\mathbb{N},\mathbb{N}) \subsetneqq V_2(\mathbb{N}^2,\mathbb{N})$ (Exercise 9).

The General Case. The multi-linear extensions of Littlewood's 4/3-inequality (previewed in §2 and proved in Chapter VII) are equivalent, by duality, to the statement:

$$l^{\frac{2k}{k-1}}(\mathbb{N}^k) \subset V_k(\mathbb{N},\ldots,\mathbb{N}) \quad \text{for all } k > 1, \tag{5.26}$$

and there exist $\varphi \in l^q(\mathbb{N}^k)$ for all $q > \frac{2k}{k-1}$ such that $\varphi \notin V_k(\mathbb{N},\ldots,\mathbb{N})$. In particular, these inequalities imply that the inclusion in (5.17) is proper for every $k \geq 1$.

The proof in Chapter VII that for all $q > 2k/k-1$ there exist $\varphi \in l^q(\mathbb{N}^k)$ and $\varphi \notin V_k(\mathbb{N},\ldots,\mathbb{N})$ is non-constructive. Therefore, the preceding argument, which establishes that (5.19) is proper for *all*

$k \geq 1$, unlike the constructive proofs above in the cases $k = 1, 2$, is indirect. The question whether constructive proofs can be given in the cases $k > 2$ is largely open. (More about this will be said in Chapter VII and Chapter X.)

ii **(type V_k).** As in §2, we notice that (5.26) implies a statement stronger than $V_{k+1} \subsetneqq V_k$. Let X be an infinite set. We say that $\varphi \in c_0(X)$ is of *type* V_k if for all bijections τ from X^k onto X, $\varphi \circ \tau \in V_k(X, \ldots, X)$. If $\varphi \in c_0(X)$ is of *type* V_k, and $\beta \in l^2(X)$ is of *type* F_k (defined in §2), then by Proposition 11, the 'sum' $\sum_x \beta(x)\, \varphi(x)$ is well-defined. This implies that $\varphi \in l^2(X)$ is of *type* V_k for all $k \geq 1$ (cf. Exercise 3). By (5.26), every $\varphi \in l^{2k/k-1}(Y)$ is of *type* V_k, and for every $k \geq 1$ there exist φ of *type* V_k that are not of *type* V_{k+1}. Later in the book, we will extend the notion of 'integer-valued' *type* to '*type* α' for arbitrary $\alpha \in [1, \infty)$.

iii **(convolution algebras – a preview).** It is easy to see that with pointwise multiplication on $\mathbb{N}^k$ both V_k and F_k are Banach algebras (Exercise 11).

It is also easy to see that with the additive structure in $\mathbb{N}$, $F_1(\mathbb{N})$ is a convolution algebra:

$$(\beta_1 \star \beta_2)(n) = \sum_{k=0}^{n-1} \beta_1(n-k)\, \beta_2(k), \tag{5.27}$$

and

$$\|\beta_1 \star \beta_2\|_{F_1} \leq \|\beta_1\|_{F_1} \|\beta_2\|_{F_1}, \quad \beta_1 \in F_1(\mathbb{N}), \beta_2 \in F_1(\mathbb{N}).$$

The analogous convolution structure in $F_2(\mathbb{N}, \mathbb{N})$ is not quite as obvious. For β_1 and β_2 in $F_2(\mathbb{N}, \mathbb{N})$, define

$$(\beta_1 \star \beta_2)(m, n) = \sum_{j=0}^{m-1} \sum_{k=0}^{n-1} \beta_1(m-j, n-k)\, \beta_2(j, k). \tag{5.28}$$

Question: *Is $\beta_1 \star \beta_2 \in F_2(\mathbb{N}, \mathbb{N})$?*

We will verify in the next chapter, by use of the Grothendieck inequality (restated below) and the Grothendieck factorization theorem (proved in Chapter V), that the answer is affirmative.

In Chapter IX we will prove that the corresponding question in the three-dimensional case has a negative answer.

iv **(the dual of F_k is a tilde algebra).** Let $\tilde{V}_k(\mathbb{N},\dots,\mathbb{N}) = \tilde{V}_k$ denote the space of $\varphi \in l^\infty(\mathbb{N}^k)$ for which there exist sequences $(\varphi_j : j \in \mathbb{N})$ in V_k such that

$$\lim_{j\to\infty} \varphi_k(\mathbf{n}) = \varphi(\mathbf{n}), \quad \mathbf{n} \in \mathbb{N}^k, \tag{5.29}$$

and

$$\limsup_{j\to\infty} \|\varphi_j\|_{V_k} < \infty.$$

We norm $\tilde{V}_k$ by

$$\|\varphi\|_{\tilde{V}_k} = \inf\left\{\limsup_{j\to\infty} \|\varphi_j\|_{V_k} : \lim_{j\to\infty} \varphi_k(\mathbf{n}) = \varphi(\mathbf{n}),\ \mathbf{n} \in \mathbb{N}^k\right\}. \tag{5.30}$$

Equipped with $\|\cdot\|_{\tilde{V}_k}$ and pointwise multiplication on $\mathbb{N}^k$, $\tilde{V}_k$ is a Banach algebra. Moreover,

$$\tilde{V}_k(\mathbb{N},\dots,\mathbb{N}) = F_k(\mathbb{N},\dots,\mathbb{N})^*. \tag{5.31}$$

The k-fold projective tensor product of l^∞ is

$$l^\infty \underbrace{\hat{\otimes}\cdots\hat{\otimes}}_{k} l^\infty = \left\{\phi \in l^\infty(\mathbb{N}^k) : \phi(\mathbf{n}) = \sum_{j=1}^{\infty} (f_{1j} \otimes \cdots \otimes f_{kj})(\mathbf{n}),\right.$$

$$\left.\sum_{j=1}^{\infty} \|f_{1j}\|_\infty \cdots \|f_{kj}\|_\infty < \infty\right\}. \tag{5.32}$$

The proper inclusion $l^\infty \hat{\otimes} \cdots \hat{\otimes}\, l^\infty \subsetneqq \tilde{V}_k(\mathbb{N},\dots,\mathbb{N})$ was first noticed by N. Varopoulos [V2]. (In this connection, note that if D is the diagonal $\{(n,\dots,n) : n \in \mathbb{N}\} \subset \mathbb{N}^k$, then $\mathbf{1}_D \in \tilde{V}_k(\mathbb{N},\dots,\mathbb{N})$ but $\mathbf{1}_D \notin l^\infty \hat{\otimes} \cdots \hat{\otimes}\, l^\infty$; Exercise 12).)

Tilde algebras, as such, were first defined (somewhat differently) and studied in [V2] and [KatMc]; detailed discussions of these can be found in [GrMc, Chapters 11, 12]. In [V2], Varopoulos showed that $\tilde{V}_k$ (defined above) was the 'multiplier' algebra of $V_k(\mathbb{N},\dots,\mathbb{N})$,

$$\begin{aligned}&\tilde{V}_k(\mathbb{N},\dots,\mathbb{N})\\ &\quad = \{\phi \in l^\infty(\mathbb{N}^k) : \phi\,\varphi \in V_k(\mathbb{N},\dots,\mathbb{N}) \text{ for all } \varphi \in V_k(\mathbb{N},\dots,\mathbb{N})\}\end{aligned} \tag{5.33}$$

(Exercise 13). In [V2], the right side of (5.33) was denoted by M, and then in [V3] by N. In both [V2] and [V3], the symbol $\tilde{V}_k$ was used to denote the algebra of *uniform* limits of sequences in balls of V_k; that is, in Varopoulos's terminology, $\tilde{V}_k$ meant $N \cap c_0$. In this book, $\tilde{V}_k$ denotes the algebra of *pointwise* limits of sequences in balls in V_k; that is, $\tilde{V}_k$ here denotes the dual space of F_k.

The algebras $\tilde{V}_k$ will be revisited in a harmonic-analytic setting in Chapter VII, and then in a framework of multidimensional Grothendieck-type inequalities in Chapter VIII.

v **(a dual formulation of the Grothendieck inequality).** Let H_1 and H_2 be Hilbert spaces, and suppose η is a bounded bilinear functional on $H_1 \times H_2$. Let $S \subset B_{H_1}$ and $T \subset B_{H_1}$ be finite sets, and let ϕ denote the scalar function on $S \times T$ defined by

$$\phi(\mathbf{x}, \mathbf{y}) = \eta(\mathbf{x}, \mathbf{y}), \quad \mathbf{x} \in S, \ \mathbf{y} \in T. \tag{5.34}$$

By Proposition 11, the Grothendieck inequality – as stated in (III.1.6) – is equivalent to

$$\|\phi\|_{V_2(S,T)} \leq \kappa_G \|\eta\|. \tag{5.35}$$

This, in essence, was Grothendieck's original formulation of his inequality in [Gro2].

6 A Brief Introduction to General Topological Tensor Products

We have noted that the first step in producing a k-fold algebraic tensor product is completely formal: we start with sets $A_1, \ldots, A_k$, consider elementary tensors $a_1 \otimes \cdots \otimes a_k$ where $a_1 \in A_1, \ldots, a_k \in A_k$, and define the algebraic tensor product $A_1 \otimes \cdots \otimes A_k$ to be the set consisting of all finite (formal) combinations of elementary tensors. Let us now take $A_1, \ldots, A_k$ to be normed vector spaces, and think of the formal combinations $\tau = \Sigma_j a_{1j} \otimes \cdots \otimes a_{kj} \in A_1 \otimes \cdots \otimes A_k$ as functions on the k-fold Cartesian product of the respective dual spaces of $A_1, \ldots, A_k$,

$$\begin{aligned} \tau(x_1, \ldots, x_k) &= \sum_j x_1(a_{1j}) \cdots x_k(a_{kj}), \\ (x_1, \ldots, x_k) &\in A_1^* \times \cdots \times A_k^*. \end{aligned} \tag{6.1}$$

As in the case $c_0 \otimes \cdots \otimes c_0$, we identify two elements in $A_1 \otimes \cdots \otimes A_k$ if they determine the same function on $A_1^* \times \cdots \times A_k^*$. We denote the resulting set of equivalence class representatives also by $A_1 \otimes \cdots \otimes A_k$, and refer to it as the *algebraic tensor product* of $A_1, \ldots, A_k$. The norms in $A_1, \ldots, A_k$ give rise to Schatten's *greatest crossnorm* [Sc4],

$$\|\tau\|_{\hat{\otimes}} = \inf\left\{ \sum_j \|a_{1j}\|_{A_1} \cdots \|a_{kj}\|_{A_k} : \tau = \sum_j a_{1j} \otimes \cdots \otimes a_{kj} \right\},$$

$$\tau \in A_1 \otimes \cdots \otimes A_k, \tag{6.2}$$

and the completion of $A_1 \otimes \cdots \otimes A_k$ in this crossnorm is denoted by $A_1 \hat{\otimes} \cdots \hat{\otimes} A_k$. Nowadays, $\|\cdot\|_{\hat{\otimes}}$ is usually called the *projective tensor norm*, and the corresponding completion $A_1 \hat{\otimes} \cdots \hat{\otimes} A_k$ is called the *projective tensor product* of $A_1, \ldots, A_k$.

The identification of bounded *k-linear* functionals on $A_1 \times \cdots \times A_k$ as bounded *linear* functionals on $A_1 \hat{\otimes} \cdots \hat{\otimes} A_k$ generalizes the duality $V_k(\mathbb{N}, \ldots, \mathbb{N})^* = F_k(\mathbb{N}, \ldots, \mathbb{N})$ (Proposition 11):

Proposition 12 (Exercise 14). *Let ξ be a bounded k-linear functional on $A_1 \times \cdots \times A_k$, i.e., ξ is linear in each coordinate, and*

$$\sup\{|\xi(a_1, \ldots, a_k)| : a_1 \in B_{A_k}, \ldots, a_k \in B_{A_k}\} := \|\xi\| < \infty. \tag{6.3}$$

Then,

$$\xi(\tau) = \sum_j \xi(a_{1j}, \ldots, a_{kj}), \quad \tau \in A_1 \otimes \cdots \otimes A_k,$$

$$\tau = \sum_j a_{1j} \otimes \cdots \otimes a_{kj}, \tag{6.4}$$

is a well-defined function on $A_1 \otimes \cdots \otimes A_k$, and determines a bounded linear functional on $A_1 \hat{\otimes} \cdots \hat{\otimes} A_k$.

Conversely, if ξ is a bounded linear functional on $A_1 \hat{\otimes} \cdots \hat{\otimes} A_k$, then ξ determines a bounded k-linear functional on $A_1 \times \cdots \times A_k$, defined by $\xi(a_1, \ldots, a_k) = \xi(a_1 \otimes \cdots \otimes a_k)$.

Remark (greatest and least crossnorms). A norm $\|\cdot\|$ on $A_1 \otimes \cdots \otimes A_k$, such that

$$\|a_1 \otimes \cdots \otimes a_k\| = \|a_1\|_{A_1} \cdots \|a_1\|_{A_k}, \quad a_1 \in A_1, \ldots, a_k \in A_k, \tag{6.5}$$

was dubbed a *crossnorm* by R. Schatten [Sc1, Definition 3.3]. The projective tensor norm defined in (6.2) (cf. (5.15)), and the 'k-linear functional' norm (cf. (6.1))

$$\|\tau\|_{\check{\otimes}} = \sup\{|\tau(x_1,\ldots,x_k)| : (x_1,\ldots,x_k) \in B_{A_1^*} \times \cdots \times B_{A_k^*}\},$$
$$\tau \in A_1 \otimes \cdots \otimes A_k, \tag{6.6}$$

are crossnorms on $A_1 \otimes \cdots \otimes A_k$. The latter norm, defined in (6.6), is known as the *k-fold injective tensor norm*, and the completion of $A_1 \otimes \cdots \otimes A_k$ in this norm is the *injective tensor product* $A_1 \check{\otimes} \cdots \check{\otimes} A_k$.

The projective tensor norm on $A_1 \otimes \cdots \otimes A_k$ and the injective tensor norm on $A_1^* \otimes \cdots \otimes A_k^*$ are dual to each other. This means: for all $\tau \in A_1 \otimes \cdots \otimes A_k$,

$$\|\tau\|_{\hat{\otimes}} = \sup\{|\sigma(\tau)| : \sigma \in A_1^* \otimes \cdots \otimes A_k^*,\ \|\sigma\|_{\check{\otimes}} \leq 1\}, \tag{6.7}$$

and for all $\sigma \in A_1^* \otimes \cdots \otimes A_k^*$,

$$\|\sigma\|_{\check{\otimes}} = \sup\{|\sigma(\tau)| : \tau \in A_1 \otimes \cdots \otimes A_k,\ \|\tau\|_{\hat{\otimes}} \leq 1\}. \tag{6.8}$$

The projective and injective tensor norms are, respectively, the *greatest* and the *least* among crossnorms that are dual to each other, and were so dubbed by R. Schatten [Sc1, §3].

The V_k-norm is the *greatest crossnorm* on $c_0 \otimes \cdots \otimes c_0$, and the F_k-variation is the *least crossnorm* on $l^1 \otimes \cdots \otimes l^1$ (Exercise 15). That $V_k(\mathbb{N},\ldots,\mathbb{N})$ is the same as $c_0 \hat{\otimes} \cdots \hat{\otimes} c_0$ is obvious, and the (dual) assertion

$$F_k(\mathbb{N},\ldots,\mathbb{N}) = l^1 \check{\otimes} \cdots \check{\otimes} l^1 \tag{6.9}$$

is a consequence of the norm-density of the algebraic tensor product $l^1 \otimes \cdots \otimes l^1$ in $F_k(\mathbb{N},\ldots,\mathbb{N})$ (Theorem 6).

7 A Brief Introduction to Projective Tensor Algebras

Let $X_1,\ldots,X_k$ be locally compact Hausdorff spaces, and let $C_0(X_1),\ldots,$ $C_0(X_k)$ be the respective Banach algebras of scalar-valued continuous functions vanishing at infinity, with the usual sup-norm $\|\cdot\|_\infty$ and pointwise multiplication. By the Riesz representation theorem,

$$\Sigma_j f_{1j} \otimes \cdots \otimes f_{kj}$$

and $\Sigma_j \varphi_{1j} \otimes \cdots \otimes \varphi_{kj}$ in $C_0(X_1) \otimes \cdots \otimes C_0(X_k)$ are equivalent in the sense of (6.1) precisely when they determine the same function on $X_1 \times \cdots \times X_k$; that is, when

$$\sum_j f_{1j}(x_1) \cdots f_{kj}(x_k) = \sum_j \varphi_{1j}(x_1) \cdots \varphi_{kj}(x_k),$$

$$(x_1, \ldots, x_k) \in X_1 \times \cdots \times X_k. \tag{7.1}$$

By the Stone–Weierstrass theorem, the algebraic tensor product $C_0(X_1) \otimes \cdots \otimes C_0(X_k)$ is norm-dense in $C_0(X_1 \times \cdots \times X_k)$.

The projective tensor product $C_0(X_1) \hat{\otimes} \cdots \hat{\otimes} C_0(X_k)$ consists of all $\phi \in C_0(X_1 \times \cdots \times X_k)$ such that

$$\|\phi\|_{\hat{\otimes}} := \inf \left\{ \sum_{j=1}^{\infty} \|f_{1j}\|_\infty \cdots \|f_{kj}\|_\infty : \right.$$

$$\left. \phi(x_1, \ldots, x_k) = \sum_{j=1}^{\infty} f_{1j}(x_1) \cdots f_{kj}(x_k), x_1 \in X_1, \ldots, x_k \in X_k \right\} < \infty. \tag{7.2}$$

(Cf. Proposition 10, and the discussion preceding it; Exercise 16.) The projective tensor product $C_0(X_1) \hat{\otimes} \cdots \hat{\otimes} C_0(X_k)$ is a Banach algebra with pointwise multiplication, and is usually denoted by $V_k(X_1, \ldots, X_k)$, and its norm by $\|\cdot\|_{V_k}$. (V is for Varopoulos.) Bounded linear functionals on $V_k(X_1, \ldots, X_k)$, which (by Proposition 12) are bounded k-linear functionals on $C_0(X_1) \times \cdots \times C_0(X_k)$, have sometimes been referred to as *k-measures* and sometimes as *multi-measures* – *bimeasures* for $k = 2$. This terminology will not be used here. In Chapter VI we prove that a bounded k-linear functional on $C_0(X_1) \times \cdots \times C_0(X_k)$ can be represented by a set-function μ defined on the k-fold Cartesian product of the Borel fields in $X_1, \ldots, X_k$, which is a measure separately in each coordinate. Such μ will be called here F_k-measures.

The following is a restatement of the Grothendieck inequality in a framework of projective tensor algebras.

Theorem 13 *If X and Y are locally compact Hausdorff spaces, then (the Grothendieck constant)*

$$\kappa_G := \sup\{\|\tau\|_{V_2} : \tau = \Sigma_j f_j \otimes g_j \in C_0(X) \otimes C_0(Y),$$

$$\|(\Sigma_j |f_j|^2)^{\frac{1}{2}}\|_\infty \|(\Sigma_j |g_j|^2)^{\frac{1}{2}}\|_\infty \leq 1\} < \infty. \tag{7.3}$$

Proof: Fix an integer $n > 0$, and suppose $f_j \in C_0(X)$ and $g_j \in C_0(Y)$ ($j \in [n]$) satisfy

$$\sum_{j=1}^{n} |f_j(x)|^2 \leq 1, \quad \sum_{j=1}^{n} |g_j(y)|^2 \leq 1 \quad \text{for } (x, y) \in X \times Y. \tag{7.4}$$

We need to verify

$$\left\| \sum_{j=1}^{n} f_j \otimes g_j \right\|_{V_2} \leq k, \tag{7.5}$$

where $k > 0$ is an absolute constant. For $x \in X$ and $y \in Y$, define $\mathbf{x} \in l^2$ and $\mathbf{y} \in l^2$ by

$$\begin{aligned} \mathbf{x}(j) &= f_j(x) \quad \text{and} \quad \mathbf{y}(j) = g_j(y), \quad j \in [n], \\ \mathbf{x}(j) &= \mathbf{y}(j) = 0, \quad j > n. \end{aligned} \tag{7.6}$$

Then,

$$\sum_{j=1}^{N} f_j(x) g_j(y) = \langle \mathbf{x}, \mathbf{y} \rangle. \tag{7.7}$$

In (III.3.12), the *expectation* $\mathbf{E}$ in each summand can be realized as a finite sum over a finite uniform probability space. Therefore (in the notation of Chapter III), for $j \geq 0$,

$$\mathbf{E}\, \mathscr{I}m \prod_k (1 + \mathrm{i}(\theta^j \mathbf{x})_k\, r_k)\, \mathscr{I}m \prod_k (1 + \mathrm{i}(\theta^j \mathbf{y})_k\, r_k), \qquad (x, y) \in X \times Y, \tag{7.8}$$

is an element in $C_0(X) \otimes C_0(Y)$, and its V_2-norm is bounded by e (by Lemma III.3). By applying (7.7), (7.8), and Lemma III.2, we obtain

$$\left\| \sum_{j=1}^{N} f_j \otimes g_j \right\|_{V_2} \leq \frac{2\mathrm{e}}{4 - \mathrm{e} + \mathrm{e}^{-1}}. \tag{7.9}$$

□

Remarks

i **(the meaning of (7.4) ⇒ (7.5)).** While every finite sum of elementary tensors $\tau = \Sigma_j f_j \otimes g_j$ is (by definition) in $V_2(X, Y)$, the computation of its V_2-norm is often a non-trivial task. Indeed, the obvious estimate

$$\left\| \sum_j f_j \otimes g_j \right\|_{V_2} \leq \sum_j \|f_j\|_\infty \|g_j\|_\infty \tag{7.10}$$

is generally useless (do you see why?), and an effective estimation of its V_2-norm requires other, more 'efficient' representations of τ. And that is the gist of the Grothendieck inequality: under the assumption in (7.4), while $\Sigma_{j=1}^N \|f_j\|_\infty \|g_j\|_\infty$ could be arbitrarily large, there exist representatives $\Sigma_j \varphi_j \otimes \theta_j$ of τ such that

$$\sum_{j=1}^N \|\varphi_j\|_\infty \ \|\theta_j\|_\infty \leq k, \tag{7.11}$$

where $k > 0$ is independent of N. The computation of κ_G in (7.3) – an open problem to this day – is in effect the problem of finding the 'best' representatives of $\Sigma_j f_j \otimes g_j$ in $C_0(X) \otimes C_0(Y)$.

ii **(more crossnorms).** The implication (7.4) $\Rightarrow$ (7.5) is the statement that two norms on $C_0(X) \otimes C_0(Y)$ are equivalent: the first is the V_2-norm, and the second is also a crossnorm, an instance in a family of crossnorms that will play a key role in the next chapter.

Let $\mathbf{p} = (p_1, \ldots, p_n) \in [1, \infty]^n$. For $\varphi \in C_0(X_1) \otimes \cdots \otimes C_0(X_0)(X_n)$, define

$$\|\varphi\|_{g_{n,\mathbf{p}}} = \inf\left\{ \left\| \left(\sum_k |f_{k1}|^{p_1} \right)^{1/p_1} \right\|_\infty \cdots \left\| \left(\sum_k |f_{kn}|^{p_n} \right)^{1/p_n} \right\|_\infty : \right.$$
$$\left. \varphi = \sum_k f_{k1} \otimes \cdots \otimes f_{kn} \right\}. \tag{7.12}$$

(If (a_k) is a scalar sequence, then $(\Sigma_k |a_k|^\infty)^{1/\infty}$ stands for $\|(a_k)\|_\infty$.) If $\mathbf{p} = (p_1, \ldots, p_n) \in [1, \infty]^n$ is a conjugate vector, which means

$$\frac{1}{p_1} + \cdots + \frac{1}{p_n} \leq 1, \tag{7.13}$$

then $\|\cdot\|_{g_{n,\mathbf{p}}}$ is a crossnorm, and $\|\cdot\|_{g_{n,\mathbf{p}}} \leq \|\cdot\|_{V_n}$. In this framework, the implication (7.4) $\Rightarrow$ (7.5) (the Grothendieck inequality) is the assertion that for $n = 2$ and $\mathbf{p} = (2,2)$, there exists $k > 0$ such that

$$\|\varphi\|_{g_{2,(2,2)}} \leq \|\varphi\|_{V_2} \leq k\|\varphi\|_{g_{2,(2,2)}},$$

$$\varphi \in \mathrm{C}_0(X) \otimes \mathrm{C}_0(Y). \tag{7.14}$$

Moreover, the case $n = 2$ and $\mathbf{p} = (2,2)$ is the only such case (in non-trivial settings) where the g-norm and the V-norm are equivalent (Exercise 17).

8 A Historical Backdrop

Tensor products are historically linked to two major developments in twentieth-century physics: general relativity and quantum mechanics. In the first, the tensor calculus that had been invented by Ricci ([RiLe], [Eis]) provided a natural setting for Einstein's theory of general relativity [Weyl], and in the second, *tensors* and *direct products* were key notions in H. Weyl's mathematical formulation of quantum mechanics, the so-called 'new-physics' [Wey2]. In these two contexts, *tensors* were scalar quantities indexed by $\mathbb{N}^k$, and *direct products* – the precursors of tensor products – were Cartesian products of finite-dimensional Euclidean spaces. In this primal phase, both tensors and direct products appeared as purely algebraic entities, without any 'functional analytic' attributes.

In the next phase, bringing then-new functional analysis to bear on Weyl's constructs, F.J. Murray and J. von Neumann investigated operators defined on direct products of Hilbert spaces (e.g., [MuvN1], [vN2], [Mu2], [MuvN2], [Kad]). Their seminal papers, which 'rank among the masterpieces of analysis in the twentieth century' [Die2, p. 90], were written at Princeton during the 1930s and early 1940s, and were largely motivated by von Neumann's prior mathematical work in Europe. The young von Neumann, a Privatdozent in Berlin and Hamburg during the late 1920s, had then embarked on 'axiomatizing' quantum mechanics [vN1]. Later at Princeton, von Neumann's joint papers with Murray were again motivated by the 'new physics'. Alas, heeding the call of the times, von Neumann shifted his interests during the war years away from direct products of Hilbert spaces to more pressing projects ([Ma], [He]).

The next stage, immediately following Murray's and von Neumann's work, was R. Schatten's study of *crossnorms* on direct products of Banach spaces [Sc1], [Sc2], [Sc3]. While Murray had already previously noted instances of such norms (in [Mu1, Chapter 3]), Schatten was first to view them in their full generality. Specifically, he was first to identify the projective tensor norm as the *greatest crossnorm* and its dual as the *least crossnorm* [Sc1, Lemma 4.2, Theorem 4.1]. He also was first to observe the fundamental duality stated in Proposition 12 [Sc3, Theorem 1.2]. The latter paper [Sc3], arriving at the *Transactions* four days after the war in Europe had officially ended, was quickly followed by two sequels, coauthored with von Neumann [ScvN1], [ScvN2]. This work was summarized in Schatten's 1950 book *A Theory of Cross-Spaces* [Sch4].

Moving to a broader context, motivated by L. Schwartz's distributions [Schw] as well as Schatten's direct products, mathematicians in post-war France, notably A. Grothendieck, focused on products of locally convex spaces [Die1]. It was here that the term *produit tensoriel* was introduced, replacing the term *produit direct* [Bou, Chapter III]. This also was the setting for Grothendieck's groundbreaking Saõ Paulo paper [Gro2], arguably the most significant advance in this subject at that time. An account of Grothendieck's early work appears in [Die1], and a more complete description of his researches in functional analysis can be found in [DiU, pp. 253–60].

Grothendieck's mathematics – its language, notation – had encountered some resistance at first, but attention eventually was drawn to the profound results expressed in it. Twelve years after the appearance of Grothendieck's Saõ Paulo article, Lindenstrauss and Pelczynski wrote in the introduction to their 1968 *Studia* paper: 'Though the theory of tensor products constructed in Grothendieck's paper has its intrinsic beauty we feel that the results of Grothendieck and their corollaries can be more clearly presented without the use of tensor products' [LiPe, p. 275]. Be that as it may, in a paper that has since become a classic, Lindenstrauss and Pelczynski changed Grothendieck's bilinear functionals to linear maps, and then applied his results in their own study of *absolutely summing operators* (Exercises 18, 19).

Around the same time that Grothendieck's results were applied to Banach spaces, N. Varopoulos uncovered, independently, basic connections between tensor algebras and harmonic analysis. Using these in works [Herz], [V1] that won him the 1968 Salem prize, Varopoulos applied tensor-theoretic machinery to solve several outstanding problems in harmonic analysis. The proximity of tensor analysis to harmonic analysis, about which more will be said in Chapter VII, was the crux of the matter, and it was here that tensor products entered a mainstream.

A Brief Critique and a Preview

Historically, the first constructions of tensor products involved products of finite order (in Murray's and von Neumann's 1936 paper [MuvN1]), as well as products of unbounded order (in von Neumann's 1938 paper [vN2]). But then in Schatten's and Grothendieck's works, the focus was on two-fold products; analysis in two dimensions was evidently perceived typical, routinely extendible to higher dimensions [Gro1, pp. 50–1]. (Grothendieck did not consider the problem of extending his 'théorème fondamental' to higher dimensions.) Later yet, a focus on two-dimensional settings indeed became natural in Lindenstrauss's and Pelczynski's subsequent view of bilinear functionals as operators (Exercises 17, 18). But dimensionality is very much part of the story – the *raison d'etre* of this book – and a question arises: *how is 'dimension' of a k-fold product noticed precisely?* This question has already been briefly addressed in §2, by use of Littlewood's 4/3-inequality, and will be addressed again, at length, in later chapters.

Exercises

1. Prove that the Fréchet variation is a norm, and that $(F_k, \|\cdot\|_{F_k})$ is a Banach space.
2. i. Prove (2.5), and then conclude that $F_2(\mathbb{N}\times\mathbb{N},\mathbb{N}) \subsetneqq F_3(\mathbb{N},\mathbb{N},\mathbb{N})$.
   ii.* Prove by explicit constructions that for $k > 2$,
   $$F_k(\mathbb{N},\ldots,\mathbb{N}) \subsetneqq F_{k+1}(\mathbb{N}^2,\ldots,\mathbb{N}).$$
   (Presently, I know how to prove this only by non-constructive arguments; see Chapter VII §11 and Chapter X §5, Remark ii.)
3. Prove that if a scalar-valued function β on a set Y is of *type* F_k (defined at the end of §2) for some k, then $\beta \in l^2(Y)$.
4. Prove Lemma 2. Can you improve the constants on the left side of (3.1)? (Cf. Exercise III.9.)
5. Verify the estimate in (3.17).
6. Prove Corollary 7 in the general case.

7. Referring to the start of the proof of Theorem 9, verify that $\beta_n \mathbf{J}_N \in F_{k-1}$.
8. The following concepts are due to Y. Meyer.

Definition 1 [Mey2, p. 243]. A bounded sequence of vectors $(\mathbf{x}_j : j \in \mathbb{N})$ in a normed linear space V is a *Sidon sequence* if there exists $k > 0$ such that for all $n \in \mathbb{N}$,

$$k \left\| \sum_{j=1}^{n} a_j \, \mathbf{x}_j \right\|_V \geq \sum_{j=1}^{n} |a_j|.$$

Definition 2 [Mey2, pp. 250–1]. A normed linear space V is a *Sidon space* if every bounded sequence of vectors in V either has a Cauchy subsequence or has a Sidon subsequence.

i. Prove that $F_k(\mathbb{N}, \dots, \mathbb{N})$ is a Sidon space (cf. [Mey2, p. 251]).
ii. Verify that every Sidon space has the Schur property (cf. [Lu1, p. 285]).

9. Complete the proofs outlined in Remark i §5: construct (1) $\varphi \in c_0(\mathbb{N}^2)$ such that $\varphi \notin V_2(\mathbb{N}, \mathbb{N})$, and (2) $\varphi \in V_2(\mathbb{N}^2, \mathbb{N})$ such that $\varphi \notin V_3(\mathbb{N}, \mathbb{N}, \mathbb{N})$.
10. Prove Proposition 10.
11. Prove that V_k and F_k with pointwise multiplication on $\mathbb{N}^k$ are Banach algebras.
12. i. Prove that equipped with the $\tilde{V}_k$-norm (defined in (5.32)) and pointwise multiplication, $\tilde{V}_k$ is a Banach algebra and

$$\tilde{V}_k(\mathbb{N}, \dots, \mathbb{N}) = F_k(\mathbb{N}, \dots, \mathbb{N})^*.$$

ii. Verify that the k-fold projective tensor product of l^∞ is

$$\underbrace{l^\infty \hat{\otimes} \cdots \hat{\otimes} l^\infty}_{k} = \Bigg\{ \phi \in l^\infty(\mathbb{N}^k) :$$

$$\phi(\mathbf{n}) = \sum_{j=1}^{\infty} (f_{1j} \otimes \cdots \otimes f_{kj})(\mathbf{n}), \sum_{j=1}^{\infty} \|f_{1j}\|_\infty \cdots \|f_{kj}\|_\infty < \infty \Bigg\}.$$

iii. Let $D = \{(n, n) : n \in \mathbb{N}\}$. Prove that $\mathbf{1}_D \in \tilde{V}_2(\mathbb{N}, \mathbb{N})$ and $\mathbf{1}_D \notin l^\infty \hat{\otimes} l^\infty$.

Below you will verify an observation by Varopoulos [V2], that

$$\tilde{V}_2(\mathbb{N}, \mathbb{N}) \cap c_0(\mathbb{N}^2) \supsetneqq V_2(\mathbb{N}, \mathbb{N}). \tag{E.1}$$

iv. Show there exists a sequence of bidisjoint finite rectangles in $\mathbb{N}^2$, $\{A_j \times B_j : j \in \mathbb{N}\}$, and $\{\phi_j \in c_0(A_j \times B_j) : j \in \mathbb{N}\}$ such that $\|\phi_j\|_{\hat{\otimes}} = 1$ for all $j \in \mathbb{N}$ and $\|\phi_j\|_\infty \downarrow 0$.

v. Let $\phi = \sum_j \mathbf{1}_{A_j \times B_j} \phi_j$, where $\|\phi_j\|_{\hat{\otimes}} = 1$ for all $j \in \mathbb{N}$, and $\|\phi_j\|_\infty \downarrow 0$. Prove that ϕ determines a bounded linear functional on $l^1 \check{\otimes} l^1$, and hence is in $\tilde{V}_2(\mathbb{N},\mathbb{N})$, but $\phi \notin V_2(\mathbb{N},\mathbb{N})$.

vi.* Suppose $\phi \in l^\infty \hat{\otimes} l^\infty \cap c_0(\mathbb{N}^2)$. Is $\phi \in V_2(\mathbb{N},\mathbb{N})$? (Compare with (E.1) above.)

13. Prove that $\tilde{V}_k$ consists of all $\phi \in l^\infty(\mathbb{N}^k)$ such that $\phi\,\varphi \in V_k(\mathbb{N},\ldots,\mathbb{N})$ for all $\varphi \in V_k(\mathbb{N},\ldots,\mathbb{N})$.

14. Prove Proposition 12.

15. Let $c_{00}(\mathbb{N}^k)$ denote the linear space of scalar-valued finitely supported functions on $\mathbb{N}^k$. Obviously,

$$c_{00}(\mathbb{N}^k) = c_{00}(\mathbb{N}) \otimes \cdots \otimes c_{00}(\mathbb{N}).$$

Prove that $\|\cdot\|_{F_k}$ and $\|\cdot\|_{V_k}$ are crossnorms on $c_{00}(\mathbb{N}) \otimes \cdots \otimes c_{00}(\mathbb{N})$, which, respectively, are the smallest and largest among crossnorms on $c_{00}(\mathbb{N}) \otimes \cdots \otimes c_{00}(\mathbb{N})$, each of whose dual norms is also a crossnorm.

(Ideas involving general crossnorms appeared first in [Sc1]. Look at this paper, preferably after you do this exercise. The requirement that dual norms be crossnorms is essential; see [ScvN1, Appendix].)

16. Verify the assertions in the beginning of §7 concerning (7.1) and (7.2).

17. i. Prove that if $\mathbf{p} \in [0,\infty]^n$ is a conjugate vector, then $\|\cdot\|_{g_{n,\mathbf{p}}}$ is a crossnorm, and $\|\cdot\|_{g_{n,\mathbf{p}}} \le \|\cdot\|_{V_n}$ on $C_0(X_1) \otimes \cdots \otimes C_0(X_n)$.

ii. Verify that if (p,q) is a conjugate vector and $2 < q \le \infty$, then the norms $\|\cdot\|_{g_{2,(p,q)}}$ and $\|\cdot\|_{V_2}$ are not equivalent.

In Exercises 18 and 19 below, you will note the equivalence between a notion stated in the language of *p-summing operators* [LiPe], and a notion stated in terms of tensor products.

Let A and B be normed linear spaces. If ξ is a bounded bilinear functional on $A \times B$, then associate with ξ the bounded linear map T_ξ from A into B^* such that for $a \in A$, $T_\xi a$ in B^* is determined by

$$T_\xi a(b) = \xi(a,b) \quad \text{for all } b \in B. \tag{E.2}$$

If T is a bounded linear map from A into B, then associate with T the bounded bilinear functional ξ_T on $A \times B^*$ such that

$$\xi_T(a, b^*) = b^*(T, a), \quad a \in A,\ b^* \in B^*. \tag{E.3}$$

Then, $\|\xi\| = \|T_\xi\|$ and $\|T\| = \|\xi_T\|$ ($\|T\| = \sup\{\|T_a\|_B : \|a\|_A = 1\}$, and $\|\xi\|$ is defined in (6.3)).

Definition 1 [LiPe, Definition 3.2]. Let A and B be normed linear spaces, and let T be a bounded linear map from A into B. For $p \geq 1$, define

$$\alpha_p(T) = \sup\Bigg\{\left(\sum_{i=1}^{n} \|T_{a_i}\|^p\right)^{\frac{1}{p}} : \sup\Bigg\{\left(\sum_{i=1}^{n} |a^*(a_i)|^p\right)^{\frac{1}{p}} : a^* \in A^*, \|a^*\| \leq 1\Bigg\} \leq 1, a_i \in A,\ i \in [n],\ n \in \mathbb{N}\Bigg\}. \tag{E.4}$$

If $\alpha_p(T) < \infty$, then T is said to be *p-absolutely* summing.

As indicated in [LiPe, p. 284], this definition had been foreshadowed by [Gro1, Définition 8, p. 160], [Sap], and framed by Pietsch [Pie] in the general form stated above. It is easy to see that for $p_1 \leq p_2$,

$$\alpha_{p_1}(T) \geq \alpha_{p_2}(T). \tag{E.5}$$

Definition 2 Let A and B be normed linear spaces, and let ξ be a bounded bilinear functional on $A \times B$. For $p \geq 1$ and $1/p\ +\ 1/q = 1$, define

$$\gamma_p(\xi) = \sup\Bigg\{ |\xi(\Sigma_j a_j \otimes b_j)| : \Sigma_j a_j \otimes b_j \in A \otimes B, \quad \sup_{x \in A^*, \|x\|=1} \Sigma_j |a_j(x)|^p \leq 1, \quad \sup_{y \in B^*, \|y\|=1} \Sigma_j |b_j(y)|^q \leq 1\Bigg\}. \tag{E.6}$$

18. i. Let A and B be normed linear spaces. Let T be a bounded linear map from A into B, and let ξ_T be the corresponding bilinear functional on $A \times B^*$ defined in (E.3). Prove that for all $p \geq 1$, $\alpha_p(T) = \gamma_p(\xi_T)$. Conclude that if ξ is a bounded bilinear functional on $A \times B$ and T_ξ is the corresponding operator from A into B^* defined by (E.2), then $\gamma_p(\xi) = \alpha_p(T_\xi)$.

    ii. Let K_1 and K_2 be locally compact Hausdorff spaces. Prove that every bounded linear map from $C_0(K_1)$ into $M(K_2)$ (Borel measures on K_2) is 2-absolutely summing.

19. i. Verify that Littlewood's mixed-norm inequality (Theorem II.2) is equivalent to the statement: *the injection from l^1 into l^2 is a 1-absolutely summing map.*

    ii. Verify that the Grothendieck inequality (Theorem III.1) is equivalent to the statement: *every bounded linear map from l^1 into l^2 is 1-absolutely summing.*

    iii. Prove that if $q \in [1, 2]$, then every bounded linear map from l^∞ into l^q is 2-absolutely summing.

    iv. What is an equivalent formulation of Orlicz's inequality (Theorem II.3) in terms of bounded linear maps from l^∞ into l^q?

    v. Can you deduce the Grothendieck inequality from the statement that every bounded linear map from l^∞ into l^2 is 2-absolutely summing?

(Following Exercises 17 and 18 (or before, if hints are needed), read in [LiPe]; specifically, see Theorems 4.1 and 4.3 therein.)

Hints for Exercises in Chapter IV

1. A normed linear space is complete if and only if every absolutely summable series therein is summable.

2. i. Using harmonic analysis on $\mathbb{Z}_n$ (cf. Exercise II.8), note that for all scalar-valued functions f, g and h on $\mathbb{Z}_n$,

$$\sum_{(j,k,l)\in\mathbb{Z}_n\times\mathbb{Z}_n\times\mathbb{Z}_n} (1/n^2)\, e^{2\pi i(j+k)l/n} f(j)g(k)h(l) = \sum_{l\in\mathbb{Z}_n} \hat{f}(l)\hat{g}(l)h(l).$$

8. i. **An outline of a proof in the case $F_1 = l^1$.**

Suppose that the sequence (α_i) in l^1 has no limit points. Without loss of generality, assume the α_i are finitely supported and $\|\alpha_i\|_1 = 1$ for all $i \in \mathbb{N}$.

There exist a subsequence (α_{i_n}) and $\delta > 0$ such that if $F_n =$ support (α_{i_n}) and $T_n = F_n \backslash \bigcup_{j=1}^{n-1} F_j$, where $\backslash$ denotes complementation, then

$$\|\alpha_{i_n} \mathbf{1}_{T_n}\|_1 \geq \delta. \tag{E.7}$$

The proof of this claim, which is a 'gliding hump', uses a diagonalization argument.

Show the following. If (x_i) is a Sidon sequence in the unit ball of a Banach space X, and $(y_i) \subset X$ such that $\|x_i - y_i\| < 1/2$ for all $i \in \mathbb{N}$, then (y_i) is a Sidon sequence. Therefore, if we can find $\delta \in (1/2, 1]$ such that (E.7) holds, then we are done. Otherwise, suppose the 'best' δ in (E.7) is in the interval $(1/2^m, 1/2^{m-1}]$ for $m > 1$. We can assume $\|\alpha_{i_n} \mathbf{1}_{F_n \backslash F_1}\|_1 < 1/2^{m-1}$ for all $n \geq 2$. By 'rescaling' the argument in the case $\delta \in (1/2, 1]$, $(\alpha_{i_n} \mathbf{1}_{F_n \backslash F_1})$ is a Sidon sequence. Because we can find a convergent subsequence of $(\alpha_{i_n} \mathbf{1}_{F_1})$, we are done.

Now prove that $F_k(\mathbb{N}, \ldots, \mathbb{N})$ is a Sidon space for all $k \geq 1$ by induction on k.

12. This can be verified by proving that $\mathbf{1}_D$ is not continuous on $\beta\mathbb{N} \times \beta\mathbb{N}$, where $\beta\mathbb{N}$ is the Stone Čech compactification of $\mathbb{N}$. (This proof was shown to me by F. Gao.)
14. You need to verify that ξ in (6.4) is well-defined. Here is the proof in the case $k = 2$. Suppose A and B are normed linear spaces and $\{a_j\} \subset A$ and $\{b_j\} \subset B$ are finite subsets such that

$$\sum_j a_j(x)\, b_j(y) = 0, \quad (x, y) \in A^* \times B^*.$$

Fix Hamel bases $\{\mathbf{e}_u : u \in U\}$ in A and $\{\mathbf{f}_v : v \in V\}$ in B. Let $S \subset U$ and $T \subset V$ be finite sets such that

$$a_j = \sum_{u \in S} a_{ju}\, \mathbf{e}_u, \quad b_j = \sum_{v \in T} b_{jv}\, \mathbf{f}_v.$$

Observe that

$$\sum_j a_j(x)\, b_j(y) = \sum_{u\in S, v\in T} \left(\sum_j a_{ju}\, b_{jv} \right) \mathbf{e}_u(x)\mathbf{f}_v(y) = 0,$$
$$(x,y) \in A^* \times B^*,$$

which implies that for each $u \in S$ and $v \in T$, $\Sigma_j a_{ju}\, b_{jv} = 0$. Then,

$$\xi(a_j, b_j) = \sum_{u\in S, v\in T} a_{ju}\, b_{jv}\, \xi(\mathbf{e}_u, \mathbf{f}_v),$$

and therefore,

$$\sum_j \xi(a_j, b_j) = \sum_{u\in S, v\in T} \left(\sum_j a_{ju}\, b_{jv} \right) \xi(\mathbf{e}_u, \mathbf{f}_v) = 0.$$

17. ii. For arbitrary $N > 0$, you need to produce β in the unit ball of $F_2(\mathbb{N},\mathbb{N})$ and $\varphi \in V_2(\mathbb{N},\mathbb{N})$ such that $\|\varphi\|_{g_{2,(p,q)}} \leq 1$ and $|\hat{\beta}(\varphi)| > N$. Let β_n be defined by (2.2). If $(\mathbf{x}_j)$ and $(\mathbf{y}_k)$ are finite sequences in the respective unit balls of l^p and l^q, then define

$$\varphi(j,k) = \langle \mathbf{x}_j, \mathbf{y}_k \rangle := \sum_m \mathbf{x}_j(m)\, \mathbf{y}_k(m), \quad (j,k) \in \mathbb{N}^2,$$

and note that $\|\varphi\|_{g_{2,(p,q)}} \leq 1$. Put $\mathbf{x}_j = \mathbf{e}_j$, where $\mathbf{e}_j(j) = 1$, and $\mathbf{e}_j(m) = 0$ for $m \neq j$. Then,

$$\beta_n(\varphi) = \sum_{j,k} \beta_n(j,k) \sum_m \mathbf{e}_j(m)\, y_k(m) = \sum_k \sum_j \beta_n(j,k)\, y_k(j),$$

and therefore,

$$\sup\{|\beta_n(\varphi)| : \{\mathbf{y}_k\} \subset B_{l^q}\} = \sum_k \left(\sum_j |\beta_n(j,k)|^p \right)^{\frac{1}{p}} = n^{1/p-1/2}.$$

V

The Grothendieck Factorization Theorem

1 Mise en Scène: Factorization in One Dimension

In this chapter we prove that *if X and Y are locally compact Hausdorff spaces, then every bounded bilinear functional on* $C_0(X) \times C_0(Y)$ *determines a bounded bilinear functional on a product of two Hilbert spaces* [Gro2, pp. 59–62]. Known as the *Grothendieck factorization theorem*, it is a two-dimensional extension of

Theorem 1 *If X is a locally compact Hausdorff space, and ξ is a bounded linear functional on* $C_0(X)$*, then there exists a probability measure ν on the Borel field in X so that*

$$|\xi(f)| \leq \|\xi\| \|f\|_{L^2(\nu)}, \quad f \in C_0(X), \tag{1.1}$$

where $\|\xi\| = \sup\{|\xi(f)| : \|f\|_\infty \leq 1\}$.

Proof: By the Riesz representation theorem, there exist a Borel measure μ_ξ on X and a Borel-measurable function φ, such that $\|\mu_\xi\|_M = \|\xi\|$, $|\varphi| = 1$ on X, and

$$\xi(f) = \int_X f(x)\, \varphi(x)\, |\mu_\xi|(dx), \quad f \in C_0(X), \tag{1.2}$$

where $|\mu_\xi|$ is the total variation measure. By Cauchy–Schwarz,

$$|\xi(f)| \;\leq\; \|f\|_{L^2(|\mu_\xi|)}\, (\|\mu_\xi\|_M)^{\frac{1}{2}}, \quad f \in C_0(X). \tag{1.3}$$

To obtain (1.1), replace μ_ξ in (1.3) by $\nu = |\mu_\xi|/\|\xi\|$. □

2 An Extension to Two Dimensions

Theorem 2 *If X and Y are locally compact Hausdorff spaces, and ξ is a bounded bilinear functional on $\mathrm{C}_0(X) \times \mathrm{C}_0(Y)$, then there exist probability measures ν_1 and ν_2 on the respective Borel fields of X and Y such that*

$$|\xi(f,g)| \leq \kappa_{\mathrm{G}} \|\xi\| \, \|f\|_{\mathrm{L}^2(\nu_1)} \|g\|_{\mathrm{L}^2(\nu_2)},$$

$$f \in \mathrm{C}_0(X), \; g \in \mathrm{C}_0(Y), \tag{2.1}$$

where $\|\xi\| = \sup\{|\xi(f,g)| : \|f\|_\infty \leq 1, \|g\|_\infty \leq 1\}$, and κ_{G} is defined in (IV.7.3).

Proof: To start, we consider the direct sum $\mathrm{C}_0(X) \oplus \mathrm{C}_0(Y)$, wherein linear structure and norm are given by

$$a(f_1, g_1) + b(f_2, g_2) = (af_1 + bf_2, ag_1 + bg_2), \quad a \in \mathbb{C}, \; b \in \mathbb{C}, \tag{2.2}$$

and

$$\|(f,g)\| = \max\{\|f\|_\infty, \; \|g\|_\infty\}. \tag{2.3}$$

By the Riesz representation theorem,

$$(\mathrm{C}_0(X) \oplus \mathrm{C}_0(Y))^* \; = \; \mathrm{M}(X) \oplus \mathrm{M}(Y), \tag{2.4}$$

and

$$\|(\mu_1, \mu_2)\| \; = \; \|\mu_1\|_{\mathrm{M}} + \|\mu_2\|_{\mathrm{M}}, \quad (\mu_1, \mu_2) \in (\mathrm{C}_0(X) \oplus \mathrm{C}_0(Y))^*. \tag{2.5}$$

We assume (without loss of generality) $\|\xi\| = 1$, and consider two sets. The first is

$$W_\xi = \{(\Sigma_k |f_k|^2, \; \Sigma_k |g_k|^2) : \; \Sigma_k f_k \otimes g_k \in \; \mathrm{C}_0(X) \otimes \mathrm{C}_0(Y), \\ |\xi(\Sigma_k f_k \otimes g_k)| \; \geq \kappa_{\mathrm{G}}\}, \tag{2.6}$$

which is convex in $\mathrm{C}_0(X) \oplus \mathrm{C}_0(Y)$ (Exercise 1), and the second is

$$O = \{(f,g) \; \in \; \mathrm{C}_0(X) \oplus \mathrm{C}_0(Y) : \\ \sup\{f(x), \; g(y) : \; (x,y) \in X \times Y\} \; < \; 1\}, \tag{2.7}$$

which is both open and convex in $\mathrm{C}_0(X) \oplus \mathrm{C}_0(Y)$.

By Theorem IV.13 and Proposition IV.12, W_ξ and O are disjoint. Therefore, by the Hahn–Banach theorem (e.g., [Tr, Proposition 18.1]), there exists $(\mu_1, \mu_2) \in \mathrm{M}(X) \oplus \mathrm{M}(Y)$ such that

$$\int_X f(x)\, \mu_1(\mathrm{d}x) + \int_Y g(y)\, \mu_2(\mathrm{d}y) < 1 \quad \text{for all } (f,g) \in O, \tag{2.8}$$

and

$$\int_X f(x)\, \mu_1(\mathrm{d}x) + \int_Y g(y)\, \mu_2(\mathrm{d}y) > 1 \quad \text{for all } (f,g) \in W_\xi. \tag{2.9}$$

Because O contains all $(f,g) \in \mathrm{C}_0(X) \oplus \mathrm{C}_0(Y)$ such that $f \leq 0$ and $g \leq 0$, (2.8) implies that μ_1 and μ_2 are non-negative measures. Because O contains the real-valued functions in the open unit ball in $\mathrm{C}_0(X) \oplus \mathrm{C}_0(Y)$, (2.8) implies also

$$\|\mu_1\|_{\mathrm{M}} + \|\mu_2\|_{\mathrm{M}} \leq 1. \tag{2.10}$$

Now take arbitrary $f \in \mathrm{C}_0(X)$ and $g \in \mathrm{C}_0(Y)$, and suppose $\xi(f,g)$ is non-zero. Then,

$$\frac{\kappa_{\mathrm{G}}}{|\xi(f,g)|}(|f|^2, |g|^2) \in W_\xi. \tag{2.11}$$

Therefore, by (2.10) and the definition of W_ξ,

$$|\xi(f,g)| \leq \kappa_{\mathrm{G}} \left(\int_X |f(x)|^2\, \mu_1(\mathrm{d}x) + \int_Y |g(y)|^2\, \mu_2(\mathrm{d}y) \right). \tag{2.12}$$

For all $c > 0$ and $d > 0$, (2.12) can be rewritten as

$$|\xi(f,g)| \leq \kappa_{\mathrm{G}} \left(\frac{c}{d} \|f\|^2_{\mathrm{L}^2(\mu_1)} + \frac{d}{c} \|g\|^2_{\mathrm{L}^2(\mu_2)} \right). \tag{2.13}$$

In particular, this implies $\mu_1 \neq 0$ and $\mu_2 \neq 0$. In (2.13), put $c = \|g\|_{\mathrm{L}^2(\mu_2)}$ and $d = \|f\|_{\mathrm{L}^2(\mu_1)}$, and obtain

$$|\xi(f,g)| \leq 2\kappa_{\mathrm{G}}\, \|f\|_{\mathrm{L}^2(\mu_1)} \|g\|_{\mathrm{L}^2(\mu_2)}. \tag{2.14}$$

Define probability measures $\nu_1 = \mu_1 / \|\mu_1\|_{\mathrm{M}}$ and $\nu_2 = \mu_2 / \|\mu_2\|_{\mathrm{M}}$. From (2.10) we obtain

$$(\|\mu_1\|_{\mathrm{M}}\, \|\mu_2\|_{\mathrm{M}})^{\frac{1}{2}} \leq \frac{1}{2} \tag{2.15}$$

(by the 'arithmetic–geometric mean' inequality), and therefore from (2.14),

$$|\xi(f,g)| \leq \kappa_{\mathrm{G}} \|f\|_{\mathrm{L}^2(\nu_1)}\, \|g\|_{\mathrm{L}^2(\nu_2)}. \tag{2.16}$$

□

Remark (a historical note). Dubbing Theorem 2 a *factorization theorem* reflects the view of bounded bilinear functionals ξ on $C_0(X) \times C_0(Y)$ as bounded linear maps T_ξ from $C_0(X)$ into $M(Y)$. (See (IV.E.2) and (IV.E.3).) Restated in the language of linear maps, Theorem 2 asserts that T_ξ and the restriction of its adjoint to $C_0(Y)$ can be 'factored', respectively, through $L^2(X, \nu_1)$ and $L^2(Y, \nu_2)$. Specifically, this means there exists a bounded linear map

$$\tilde{T}_\xi \; : \; L^2(X, \nu_1) \to M(Y), \tag{2.17}$$

such that $T_\xi = \tilde{T}_\xi I$, where I is the canonical inclusion map from $C_0(X)$ into $L^2(X, \nu_1)$. A similar statement holds regarding the restriction of the adjoint $(T_\xi)^*$ to $C_0(Y)$.

Theorem 2 suggests a general question: if A, B and H are Banach spaces, and ξ is a bounded bilinear functional on $A \times B$, then is there a bounded linear map V: $A \to H$ such that

$$\xi(Va, b), \quad (a, b) \in A \times B, \tag{2.18}$$

determines a bounded bilinear functional on $V[A] \times B$? Equivalently, if T_ξ is the bounded linear map from A into $(B)^*$ determined by ξ, then can T_ξ be 'factored' throught H? That is, are there bounded linear maps $V : A \to H$ and $U : H \to (B)^*$ such that $T_\xi = UV$?

This question, with emphasis on Hilbert spaces H, was first studied by Grothendieck in [Gro2]. Eleven years later, working in a framework of Banach spaces, A. Pietsch observed a general connection between *factorization* and *p-summing operators* [Pie, Theorem 2], known today as the Pietsch factorization theorem. (See Exercises IV.18, IV.19, and Exercise 2 in this chapter.) Indeed, the proof above of Theorem 2 uses a key idea from the proof of Pietsch's theorem.

Factorization, as such, was pioneered by Lindenstrauss and Pelczynski in their 1968 classic *Studia* paper [LiPe]. (See Chapter IV §8.) A concise survey of progress in this area up to the early 1980s can be found in [Pi3]; a detailed and more recent treatment can be found in [DiJTon].

3 An Application

In this section we answer the question stated in the previous chapter (Remark ii §5) regarding convolution in $F_2(\mathbb{N}, \mathbb{N})$. The answer to this question will be further amplified in the more general setting of Chapter IX.

Theorem 3 *If $\beta_1 \in F_2(\mathbb{N},\mathbb{N})$, $\beta_2 \in F_2(\mathbb{N},\mathbb{N})$, and for $(m,n) \in \mathbb{N}^2$*

$$(\beta_1 \star \beta_2)(m,n) = \sum_{j=0}^{m-1}\sum_{k=0}^{n-1} \beta_1(m-j,\ n-k)\beta_2(j,k), \tag{3.1}$$

then $\beta_1 \star \beta_2 \in F_2(\mathbb{N},\mathbb{N})$, and

$$\|\beta_1 \star \beta_2\|_{F_2} \leq 4\ (\kappa_{\mathrm{G}})^2 \|\beta_1\|_{F_2}\ \|\beta_2\|_{F_2}. \tag{3.2}$$

The key to the theorem is the following

Lemma 4 *Suppose $\beta \in F_2(\mathbb{N},\mathbb{N})$, and let φ_1 and φ_2 be scalar-valued functions with finite supports in $\mathbb{N}^2$. Define*

$$\phi(j,k) = \sum_{m,n} \beta(m,n)\ \varphi_1(m,j)\ \varphi_2(n,k), \quad (j,k) \in \mathbb{N}^2. \tag{3.3}$$

Then,

$$\|\phi\|_{V_2} \leq 4\ (\kappa_{\mathrm{G}})^2 \|\beta\|_{F_2}\ \|\varphi_1\|_\infty\ \|\varphi_2\|_\infty. \tag{3.4}$$

Proof: Because β is a bounded bilinear functional on $\mathrm{c}_0(\mathbb{N}) \times \mathrm{c}_0(\mathbb{N})$, by Theorem 2 there exist probability measures ν_1 and ν_2 on $\mathbb{N}$ such that for all $f \in \mathrm{c}_0(\mathbb{N})$ and $g \in \mathrm{c}_0(\mathbb{N})$,

$$\begin{aligned} |\hat{\beta}(f,g)| &:= \left| \sum_{m,n} \beta(m,n)\ f(m)\ g(n) \right| \\ &\leq 4\ \kappa_{\mathrm{G}}\ \|\beta\|_{F_2}\ \|f\|_{\mathrm{L}^2(\nu_1)}\ \|g\|_{\mathrm{L}^2(\nu_2)}. \end{aligned} \tag{3.5}$$

That is, $\hat{\beta}$ determines a bilinear functional on $\mathrm{L}^2(\mathbb{N},\nu_1) \times \mathrm{L}^2(\mathbb{N},\nu_2)$, with norm bounded by $4\kappa_{\mathrm{G}}\|\beta\|_{F_2}$.

Assume $\|\varphi_1\|_\infty = \|\varphi_2\|_\infty = 1$, and view $\{\varphi_1(\cdot,j) : j \in \mathbb{N}\}$ and $\{\varphi_2(\cdot,k) : k \in \mathbb{N}\}$ as finite sets in the respective unit balls of $\mathrm{L}^2(\mathbb{N},\nu_1)$ and $\mathrm{L}^2(\mathbb{N},\nu_2)$. Then, by applying the Grothendieck inequality (as stated in (IV.5.37)) to (3.5), we obtain

$$\|\phi\|_{V_2} \leq 4(\kappa_{\mathrm{G}})^2\ \|\beta\|_{F_2}. \tag{3.6}$$

□

Proof of Theorem 3: For arbitrary finite sets $S \subset \mathbb{N}$ and $T \subset \mathbb{N}$, and for arbitrary $\omega_1 \in \{-1,1\}^N$ and $\omega_2 \in \{-1,1\}^N$, we estimate

$$\left| \sum_{m \in S, n \in T} \left(\sum_{j=0}^{m-1}\sum_{k=0}^{n-1} \beta_1(m-j,n-k)\ \beta_2(j,k) \right) r_m(\omega_1)\ r_n(\omega_2) \right|. \tag{3.7}$$

To this end, rewrite (3.7) as

$$\left| \sum_{j,k} \beta_2(j,k) \left(\sum_{m,n} \beta_1(m-j, n-k)\, \mathbf{1}_{[m]}(j+1) \right.\right.$$
$$\left.\left. \cdot \mathbf{1}_S(m)\, r_m(\omega_1)\, \mathbf{1}_{[n]}(k+1) \mathbf{1}_T(n)\, r_n(\omega_2) \right) \right|. \tag{3.8}$$

We change indices in the second sum to $u = m - j$ and $v = n - k$, and rewrite (3.8) as

$$\left| \sum_{j,k} \beta_2(j,k) \left(\sum_{u,v} \beta_1(u,v)\, \mathbf{1}_{[u+j]}(j+1) \right.\right.$$
$$\left.\left. \cdot \mathbf{1}_S(u+j)\, r_{u+j}(\omega_1)\, \mathbf{1}_{[v+k]}(k+1) \mathbf{1}_T(v+k)\, r_{v+k}(\omega_2) \right) \right|. \tag{3.9}$$

The second sum (over u and v) is a function (in j and k) with finite support, which we denote by ϕ. By applying Lemma 4 with $\beta_1 = \beta$, and φ_1, φ_2 defined by

$$\begin{aligned} \varphi_1(u,j) &= \mathbf{1}_{[u+j]}(j+1)\, \mathbf{1}_S(u+j)\, r_{u+j}(\omega_1), \\ \varphi_2(v,k) &= \mathbf{1}_{[v+k]}(k+1)\, \mathbf{1}_T(v+k)\, r_{v+k}(\omega_2), \\ & (u,j) \in \mathbb{N}^2, \quad (v,k) \in \mathbb{N}^2, \end{aligned} \tag{3.10}$$

we obtain $\|\phi\|_{V_2} \leq 4(\kappa_G)^2 \|\beta_1\|_{F_2}$. By duality (Proposition IV.11), we conclude that (3.9) is bounded by $4(\kappa_G)^2 \|\beta_1\|_{F_2}\, \|\beta_2\|_{F_2}$, and obtain the theorem. □

4 The g-norm

The key step in the proof of Theorem 2, that $W_\xi \cap O = \emptyset$, is the application of the Grothendieck inequality (Theorem IV.13). A natural question is whether the Grothendieck inequality is implied by Theorem 2. The answer is *yes*.

We assume Theorem 2 with $\kappa > 0$ in place of κ_G, and proceed to deduce the Grothendieck inequality from it. Suppose $f_j \in C_0(X)$ and

$g_j \in \mathrm{C}_0(Y)$ for $j = 1, \ldots, n$, and

$$\left\| \sum_{j=1}^{n} |f_j|^2 \right\|_\infty = \left\| \sum_{j=1}^{n} |g_j|^2 \right\|_\infty = 1. \tag{4.1}$$

That is, $\|\Sigma_{j=1}^{n} f_j \otimes g_j\|_{g_2,(2,2)} \leq 1$. (For definition of the g-norm, see (IV.7.12).) If ξ is a bounded bilinear functional on $\mathrm{C}_0(X) \times \mathrm{C}_0(Y)$, then

$$\begin{aligned} \left| \xi\left(\sum_{j=1}^{n} f_j \otimes g_j \right) \right| &\leq \sum_{j=1}^{n} |\xi(f_j \otimes g_j)| \\ &\leq \kappa\|\xi\| \left(\sum_{j=1}^{n} \|f_j\|_{\mathrm{L}^2(\nu_1)} \, \|g_j\|_{\mathrm{L}^2(\nu_2)} \right) \\ &\leq \kappa\|\xi\| \left(\sum_{j=1}^{n} \|f_j\|^2_{\mathrm{L}^2(\nu_1)} \right)^{\frac{1}{2}} \left(\sum_{j=1}^{n} \|g_j\|^2_{\mathrm{L}^2(\nu_2)} \right)^{\frac{1}{2}} \leq \kappa\|\xi\|. \end{aligned} \tag{4.2}$$

(The second inequality follows from Theorem 2; the third from Cauchy–Schwarz, and the fourth from the generalized Minkowski inequality and (4.1).) Therefore, by duality (Proposition IV.13),

$$\left\| \sum_{j=1}^{n} f_j \otimes g_j \right\|_{V_2} \leq \kappa, \tag{4.3}$$

which implies Theorem IV.13.

The equivalence of Theorem 2 and the Grothendieck inequality is an instance of a general relation between 'factorizability' (in the sense of Theorem 2) and the g-norm. The following is essentially a restatement of the Pietsch factorization theorem [Pie] (Exercise 2).

Proposition 5 *Let X and Y be locally compact Hausdorff spaces, and let ξ be a bounded bilinear functional on $\mathrm{C}_0(X) \times \mathrm{C}_0(Y)$. If $p \in [1, \infty)$, and $q \geq p/p - 1$, then the following two assertions are equivalent.*

i. *There exist $\gamma > 0$ and probability measures ν_1 and ν_2 on the respective Borel fields of X and Y such that*

$$\begin{gathered} |\xi(f, g)| \leq \gamma\|f\|_{\mathrm{L}^p(\nu_1)} \, \|g\|_{\mathrm{L}^q(\nu_2)}, \\ f \in \mathrm{C}_0(X), g \in \mathrm{C}_0(Y). \end{gathered} \tag{4.4}$$

ii. *There exist* $\kappa > 0$ *such that*

$$|\xi(\varphi)| \leq \kappa\|\varphi\|_{g_2,(p,q)}, \quad \varphi \in \mathrm{C}_0(X) \otimes \mathrm{C}_0(Y). \tag{4.5}$$

Proof: We deal first with the case $p \in (1, \infty)$.

i $\Rightarrow$ ii. Let $\varphi = \Sigma_k f_k \otimes g_k$ be an arbitrary member of $\mathrm{C}_0(X) \otimes \mathrm{C}_0(Y)$. Then,

$$\begin{aligned}
\left|\xi\left(\sum_k f_k \otimes g_k\right)\right| &\leq \sum_k |\xi(f_k \otimes g_k)| \\
&\leq \gamma \sum_k \|f_k\|_{\mathrm{L}^p(\nu_1)} \|g_k\|_{\mathrm{L}^q(\nu_2)} \\
&\leq \gamma \left(\sum_k \|f_k\|^p_{\mathrm{L}^p(\nu_1)}\right)^{\frac{1}{p}} \left(\sum_k \|g_k\|^q_{\mathrm{L}^q(\nu_2)}\right)^{\frac{1}{q}} \\
&\leq \gamma \int_X \left(\sum_k |f_k(x)|^p\right)^{\frac{1}{p}} \nu_1(\mathrm{d}x) \int_Y \left(\sum_k |g_k(y)|^q\right)^{\frac{1}{q}} \nu_2(\mathrm{d}y).
\end{aligned} \tag{4.6}$$

(The second inequality is by (4.4); the third is by Hölder, and the fourth is by the generalized Minkowski inequality.) Because ν_1 and ν_2 are probability measures,

$$|\xi(\varphi)| \leq \gamma \left\|\left(\sum_k |f_k|^p\right)^{\frac{1}{p}}\right\|_\infty \left\|\left(\sum_k |g_k|^q\right)^{\frac{1}{q}}\right\|_\infty, \tag{4.7}$$

which implies (4.5) with $\kappa \leq \gamma$.

ii $\Rightarrow$ i. The argument follows the outline of the proof of Theorem 2. Let $W_\xi(p,q)$ be the set in $\mathrm{C}_0(X) \oplus \mathrm{C}_0(Y)$ defined by

$$\begin{aligned}
W_\xi(p,q) = \{(\Sigma_k |f_k|^p, \Sigma_k |g_k|^q) :& \\
\Sigma_k f_k \otimes g_k \in \mathrm{C}_0(X) \otimes \mathrm{C}_0(Y), &|\xi(\Sigma_k f_k \otimes g_k)| \geq \kappa\}.
\end{aligned} \tag{4.8}$$

Because $1/p + 1/q \leq 1$, $W_\xi(p,q)$ is convex in $\mathrm{C}_0(X) \oplus \mathrm{C}_0(Y)$ (Exercise 1). The definition of O is the same as in (2.7). By (4.5), the sets $W_\xi(p,q)$ and O are disjoint, and therefore there exist non-negative measures $\mu_1 \in \mathrm{M}(X)$ and $\mu_2 \in \mathrm{M}(Y)$ that satisfy (2.8), (2.9), and (2.10). If $f \in \mathrm{C}_0(X)$ and $g \in \mathrm{C}_0(Y)$ are arbitrary and $\xi(f,g) \neq 0$, then

$$\frac{\kappa}{|\xi(f,g)|}(|f|^p, |g|^q) \in W_\xi(p,q). \tag{4.9}$$

Therefore,

$$|\xi(f,g)| \leq \kappa(\|f\|^p_{L^p(\mu_1)} + \|g\|^q_{L^q(\mu_2)}). \tag{4.10}$$

For every $c > 0$ and $d > 0$, (4.10) can be rewritten as

$$|\xi(f,g)| \leq \kappa((c/d^{\frac{p}{q}})\ \|f\|^p_{L^p(\mu_1)} + (d/c^{\frac{q}{p}})\ \|g\|^q_{L^q(\mu_2)}). \tag{4.11}$$

By putting $c = \|g\|_{L^q(\mu_2)}$ and $d = \|f\|_{L^p(\mu_1)}$, and by defining

$$\nu_1 = \mu_1/\|\mu_1\|_M,\ \nu_2 = \mu_2/\|\mu_2\|_M,$$

we obtain (4.4) from (4.11) with $\gamma \leq 2\kappa$ (Exercise 3).

The proof in the case $p = 1$ and $q = \infty$ is similar (Exercise 4). □

5 The *g*-norm in the Multilinear Case

Following linear and bilinear factorizations through Hilbert spaces (Theorems 1, 2), the question arises whether bounded *trilinear* functionals on $C_0(X) \times C_0(Y) \times C_0(Z)$ can be similarly factored. The answer is *no*. Let $X = Y = [0, 2\pi)$, and $Z = \mathbb{Z}$. Define

$$\xi(f,g,h) = \sum_{n\in\mathbb{Z}} \hat{f}(n)\ \hat{g}(n)\ h(n),$$
$$(f,g,h) \in C(X) \times C(Y) \times C_0(Z). \tag{5.1}$$

(The 'hat' above denotes the usual Fourier transform.) Then, ξ is a bounded trilinear functional on $C(X) \times C(Y) \times C_0(Z)$, and for all probability measures ν_1, ν_2, and ν_3 on the respective Borel fields of X, Y, and Z, and all $1 \leq p < \infty$, $1 \leq q < \infty$, and $1 \leq r < \infty$ (Exercise 5),

$$\sup\{|\xi(f,g,h)| : \|f\|_{L^p(\nu_1)} = \|g\|_{L^q(\nu_2)} = \|h\|_{L^r(\nu_3)} = 1\} = \infty. \tag{5.2}$$

While Theorem 2 does not extend (in the obvious way) to higher dimensions, the connection between the *g*-norm and factorizability in two dimensions (Proposition 5) extends (essentially verbatim) to the multidimensional setting.

Proposition 6 (Exercise 6). *Let $X_1, \ldots, X_n$ be locally compact Hausdorff spaces, and let ξ be a bounded n-linear functional on $C_0(X_1) \times \cdots \times C_0(X_n)$. Let $\mathbf{p} = (p_1, \ldots, p_n) \in [1,\infty]^n$ be a conjugate vector; i.e., $1/p_1 + \cdots + 1/p_n \leq 1$. Then, the following are equivalent.*

i. *There exist* $\gamma > 0$ *and probability measures* $\nu_1, \ldots, \nu_n$ *on the respective Borel fields of* $X_1, \ldots, X_n$ *such that*

$$|\xi(f_1, \ldots, f_n)| \leq \gamma \|f_1\|_{L^{p_1}(\nu_1)} \cdots \|f_n\|_{L^{p_n}(\nu_n)},$$
$$(f_1, \ldots, f_n) \in C_0(X_1) \times \cdots \times C_0(X_n). \tag{5.3}$$

ii. *There exists* $\kappa > 0$ *such that for all* $\varphi \in C_0(X_1) \otimes \cdots \otimes C_0(X_n)$,

$$|\xi(\varphi)| \leq \kappa \|\varphi\|_{g_n,\mathbf{p}}. \tag{5.4}$$

Let $\mathbf{p} \in [1, \infty]^n$ be a conjugate vector, and let

$$G_{n,\mathbf{p}} = G_{n,\mathbf{p}}(X_1, \ldots, X_n)$$

denote the closure of $C_0(X_1) \otimes \cdots \otimes C_0(X_n)$ in the $g_{n,\mathbf{p}}$-(cross)norm. (See Remark ii in Chapter IV §7.) For example, $G_{2,(2,2)} = V_2$ (by the Grothendieck inequality), and $G_{2,(p,q)} \subsetneq V_2$ for $p < 2$, $q \geq p/p - 1$, and infinite X_1 and X_2 (Exercise IV.17 ii). Because $G_{n,\mathbf{p}}$ is norm-dense in $V_n(X_1, \ldots, X_n)$ and $\|\cdot\|_{g_n,\mathbf{p}} \leq \|\cdot\|_{V_n}$ (Exercise IV.17 i), we have $(G_{n,\mathbf{p}})^* \subset V_n(X_1, \ldots, X_n)^*$ (Proposition IV.12). Proposition 6 implies the following characterization of $(G_{n,\mathbf{p}})^*$.

Corollary 7 *Suppose* $\xi \in V_n(X_1, \ldots, X_n)^*$. *Then,* $\xi \in [G_{n,\mathbf{p}}]^*$ *if and only if there exist probability measures* $\nu_1, \ldots, \nu_n$ *on the respective Borel fields of* $X_1, \ldots, X_n$ *so that*

$$\|\xi\|_{(\nu_j),\mathbf{p}} := \sup\{|\xi(f_1 \otimes \cdots \otimes f_n)| : \|f_j\|_{L^{p_j}(\nu_j)} \leq 1, \ j \in [n]\} < \infty. \tag{5.5}$$

Remark (an open question). In Chapter VIII, we prove a multilinear extension of the Grothendieck inequality, asserting that there exist $\kappa_n > 0$ such that for all n-linear functionals ξ on $C_0(X_1) \times \cdots \times C_0(X_n)$ and for all $\varphi \in C_0(X) \otimes \cdots \otimes C_0(K_n)$,

$$|\xi(\varphi)| \leq \kappa_n \|\xi\| \, \|\varphi\|_{g_n,(2,\ldots,2)}. \tag{5.6}$$

However, this inequality cannot be applied in Proposition 6, because $\mathbf{p} = (2, \ldots, 2)$ is not a conjugate vector. The following is open.

Question: *Let* ξ *be a bounded* n*-linear functional on* $C_0(X_1) \times \cdots \times C_0(X_n)$. *Is there a conjugate vector* $\mathbf{p}$ *and* $\kappa = \kappa(\xi, \mathbf{p}) > 0$ *such that for all* $\varphi \in C_0(X_1) \otimes \cdots \otimes C_0(X_n)$,

$$|\xi(\varphi)| \leq \kappa \|\varphi\|_{g_{n,\mathbf{p}}}? \tag{5.7}$$

(For example, in the case of the trilinear functional ξ in (5.1), $\mathbf{p} = (2, 2, \infty)$ and $\kappa(\xi, \mathbf{p}) = 1$.)

Exercises

1. Let ξ be a bounded bilinear functional on $C_0(X) \times C_0(Y)$. Verify that if $\mathbf{p} \in [1, \infty] \times [1, \infty]$ is a conjugate vector, then $W_\xi(\mathbf{p})$ (defined in the proof of Proposition 5) is a convex subset of the direct sum $C_0(X) \oplus C_0(Y)$.
2. The following is Pietsch's factorization theorem [Pie], as stated in [LiPe, Proposition 3.1]. (For definitions, see Exercises IV.18 and IV.19).)

   Let A and B be Banach spaces, and let T be a bounded linear map from A into B. Let E^* denote the weak* closure of the set of extreme points of the unit ball in B^*. Prove that T is *p-absolutely summing* if and only if there exists $\kappa > 0$ and a probability measure μ on the Borel field of E^* such that for all $x \in A$

   $$\|Tx\|_B \leq \kappa \|x\|_{L^p(E^*, \mu)}, \tag{E.1}$$

   where x on the right is viewed as a continuous function on E^*.
3.* In verifying the implication i $\Rightarrow$ ii in Proposition 5, we conclude that $\kappa \leq \gamma$, and in verifying ii $\Rightarrow$ i we obtain $\gamma \leq 2\kappa$. Improve the latter inequality; e.g., can you prove $\gamma = \kappa$?
4. Prove Proposition 5 in the case $p = 1$ and $q = \infty$.
5. Let $X = Y = [0, 2\pi)$, $Z = \mathbb{Z}$. Define

   $$\xi(f, g, h) = \sum_{n \in \mathbb{Z}} \hat{f}(n)\, \hat{g}(n)\, h(n), \qquad (f, g, h) \in C(X) \times C(Y) \times C_0(\mathbb{Z}).$$

   Verify ξ is a bounded trilinear functional on $C(X) \times C(Y) \times C_0(\mathbb{Z})$ (cf. Exercise IV.2 i), and show that for all probability measures ν_1, ν_2 and ν_3 on X, Y, and $\mathbb{Z}$, and all $1 \leq p < \infty, 1 \leq q < \infty$, and $1 \leq r < \infty$,

   $$\sup\{|\xi(f, g, h)| : \|f\|_{L^p(\nu_1)} = \|g\|_{L^q(\nu_2)} = \|h\|_{L^r(\nu_3)} = 1\} = \infty.$$
6. Prove Proposition 6.

Hints for Exercises in Chapter V

2. Consider

$$W = \left\{ g \in \mathrm{C}(E^*) : \; g \; = \; a_p(T) \sum_{i=1}^{n} |x_i|^p, \; \sum_{i=1}^{n} \|Tx_i\|_B^p = 1 \right\},$$

where the x_i are viewed as continuous functions on E^*, and then observe that W is convex and disjoint from

$$O \; = \; \{g \in \mathrm{C}(E^*) : \; \sup\{g(t) : t \in E^*\} < 1\}.$$

Apply the argument as in the proof of Proposition 5.

4. To obtain i $\Rightarrow$ ii, instead of applying in (4.6) the generalized Minkowski inequality, interchange the order of summation and integration. To obtain ii $\Rightarrow$ i, replace $\mathrm{C}_0(X) \oplus \mathrm{C}_0(Y)$ by $\mathrm{C}_0(X)$.

VI

An Introduction to Multidimensional Measure Theory

1 Mise en Scène: Fréchet Measures

In this chapter we outline rudiments of multidimensional measure theory, a general framework that we use and further develop in later chapters.

Usual 'one-dimensional' measure theory starts with a set X, an algebra of subsets $C \subset 2^X$, and a scalar measure on C. The multi-dimensional theory starts, similarly, with a scalar-valued set-function on a Cartesian product of algebras:

Definition 1 Let $X_1, \ldots, X_n$ be sets, and let $C_1 \subset 2^{X_1}, \ldots, C_n \subset 2^{X_n}$ be algebras. A scalar-valued set-function μ defined on the Cartesian product $C_1 \times \cdots \times C_n$ is an F_n*-measure* if for each $k \in [n]$ and every $A_j \in C_j$, $j \neq k$,

$$\mu(\ldots, A_j, \ldots, \underset{\substack{\uparrow \\ k\text{th coordinate}}}{A}, \ldots), \quad A \in C_k, \tag{1.1}$$

is a scalar measure on C_k. Such μ will be generically called *Fréchet measures*, or F-measures.

The space of F_n-measures on $C_1 \times \cdots \times C_n$ is denoted by $F_n(C_1, \ldots, C_n)$. If $C_i = 2^{X_i}$, then $F_n(\cdot\cdot, 2^{X_i}, \cdot\cdot)$ is denoted by $F_n(\cdot\cdot, X_i, \cdot\cdot)$.

The space $F_n(C_1, \ldots, C_n)$ is a generalization of $F_n(\mathbb{N}, \ldots, \mathbb{N})$, which was defined in Chapter IV (Definition IV.1). That Definition 1 above is consistent with the definition in Chapter IV §1 follows from a basic theorem that we will soon establish.

By and large, properties of Fréchet measures depend on the ambient dimension. This will become apparent as the theory unravels. We will encounter two kinds of properties as we learn the subject. The first kind comprises properties that extend those of scalar measures, more or less the way we expect, and the second consists of surprises, distinct 'multi-dimensional' characteristics that are not easily guessed. In this chapter, we focus on general properties of the first kind. The more exotic features of Fréchet measures will be revealed in later chapters.

2 Examples

i **('trivial' examples).** Let $\sigma(C_1 \times \cdots \times C_n)$ denote the σ-algebra generated by $C_1 \times \cdots \times C_n$, and observe that every F_1-measure (a scalar measure) on $\sigma(C_1 \times \cdots \times C_n)$ is *a fortiori* an F_n-measure on $C_1 \times \cdots \times C_n$.

ii **('true' examples).** We note that the definition of $F_n(\mathbb{N}_1, \ldots, \mathbb{N}_n)$ in Chapter IV is subsumed by Definition 1 above. For $n = 1$, if $\beta \in F_1(\mathbb{N})$ according to Definition IV.1, then define

$$\mu_\beta(A) = \sum_j \beta(j)\, \mathbf{1}_A(j), \quad A \subset \mathbb{N}, \tag{2.1}$$

and observe that μ_β is countably additive on $2^{\mathbb{N}}$. Conversely, if $\mu \in F_1(\mathbb{N})$ according to Definition 1, and $\beta_\mu(j) = \mu\{j\}$ for $j \in \mathbb{N}$, then $\beta_\mu \in l^1(\mathbb{N})$; i.e., $\beta_\mu \in F_1(\mathbb{N})$ according to Definition IV.1. If $n > 1$, and $\beta \in F_n(\mathbb{N}, \ldots, \mathbb{N})$ according to Definition IV.1, then let

$$\begin{aligned} &\mu_\beta(A_1, \ldots, A_n) \\ &\quad = \sum_{j_1} \left(\cdots \left(\sum_{j_n} \beta(j_1, \ldots, j_n) \mathbf{1}_{A_1}(j_1) \cdots \mathbf{1}_{A_n}(j_n) \right) \cdots \right), \\ &\qquad A_1 \subset \mathbb{N}, \ldots, A_n \subset \mathbb{N}, \end{aligned} \tag{2.2}$$

which is well-defined by Corollary IV.7. Moreover, summations on the right side of (2.2) can be interchanged without affecting the left side (also by Corollary IV.7). Therefore, to verify that $\mu_\beta \in F_n(\mathbb{N}, \ldots, \mathbb{N})$ according to Definition 1, it suffices to check that μ_β is countably additive in the nth coordinate. Suppose $\{B_j\}$ is a countable collection of pairwise disjoint subsets of $\mathbb{N}$. Then

$$\{\mu(A_1, \ldots, A_{n-1}, B_j)\}_j$$

is absolutely summable, and therefore (by the case $n = 1$),

$$\sum_{j=1}^{\infty} \mu_\beta(A_1, \ldots, A_{n-1}, B_j) = \mu_\beta\left(A_1, \ldots, A_{n-1}, \cup_{j=1}^{\infty} B_j\right). \qquad (2.3)$$

The converse ($\mu \in F_n(\mathbb{N}, \ldots, \mathbb{N})$ according to Definition 1 implies

$$\{\mu(\{j_1\}, \ldots, \{j_n\}) : (j_1, \ldots, j_n) \in \mathbb{N}^n\} \in F_n(\mathbb{N}, \ldots, \mathbb{N}) \qquad (2.4)$$

according to Definition IV.1) will be an instance of a general theorem that we prove in the next section (Exercise 1).

Suppose $(X_1, C_1), \ldots, (X_n, C_n)$ are measurable spaces, and each C_j is infinite. Fix countably infinite subsets $E_1 \subset X_1, \ldots, E_n \subset X_n$, and assume that for each $j \in [n]$, there exists a collection $\{B_x : x \in E_j\}$ of pairwise disjoint elements in C_j such that $x \in B_x$. Let $\beta \in F_n(E_1, \ldots, E_n)$, and define (via Corollary IV.7)

$$\mu_\beta(A_1, \ldots, A_n) = \sum_{x_1 \in E_1, \ldots, x_n \in E_n} \beta(x_1, \ldots, x_n)\delta_{x_1}(A_1) \cdots \delta_{x_n}(A_n), \quad A_1 \in C_1, \ldots, A_n \in C_n. \qquad (2.5)$$

($\delta_x(A) = 1$ for $x \in A$, and $\delta_x(A) = 0$ for $x \notin A$.) Then, $\mu_\beta \in F_n(C_1, \ldots, C_n)$. In particular, if $\beta \notin l^1(E_1 \times \cdots \times E_n)$, then μ_β does not extend to a scalar measure on $\sigma(C_1 \times \cdots \times C_n)$. We similarly verify that for all $n \geq 3$ (Exercise 2),

$$F_{n-1}(\sigma(C_1 \times C_2), \ldots, C_n) \subsetneqq F_n(C_1, \ldots, C_n). \qquad (2.6)$$

iii **(examples from harmonic analysis).** In the ensuing discussion we assume the reader has (at least) some knowledge of harmonic analysis. Let $\mathfrak{B}$ denote the usual Borel field of the circle group $\mathbf{T} = [0, 2\pi)$. Let Λ be an infinite subset of $\mathbb{Z}$, and define

$$\mu_\Lambda(A, B) = \sum_{k \in \Lambda} \hat{\mathbf{1}}_A(k)\hat{\mathbf{1}}_B(k), \quad A \in \mathfrak{B},\ B \in \mathfrak{B}. \qquad (2.7)$$

Then, μ_Λ determines a bounded bilinear functional on $\mathrm{L}^2(\mathbf{T}, \mathfrak{m}) \times \mathrm{L}^2(\mathbf{T}, \mathfrak{m})$, which, slightly abusing notation, we write as

$$\mu_\Lambda(f, g) = \sum_{k \in \Lambda} \hat{f}(k)\, \hat{g}(k), \quad (f, g) \in \mathrm{L}^2(\mathbf{T}, m) \times \mathrm{L}^2(\mathbf{T}, m), \qquad (2.8)$$

where $\mathfrak{m}$ is a normalized Lebesgue measure on $(\mathbf{T}, \mathfrak{B})$. Observe that $\mu_\Lambda \in F_2(\mathfrak{B}, \mathfrak{B})$ (Exercise 3 i). Also observe that the transform $\hat{\mu}_\Lambda$ is

$$\hat{\mu}_\Lambda(n, m) \; := \; \mu_\Lambda(\mathrm{e}^{in\cdot}, \mathrm{e}^{im\cdot}) \; = \; \begin{cases} 1 & n = m, \quad n \in \Lambda \\ 0 & \text{otherwise.} \end{cases} \tag{2.9}$$

By the characterization of idempotent measures (in [Ru3, Chapter 3]), μ_Λ defined in (2.7) is extendible to a scalar measure on $\sigma(\mathfrak{B} \times \mathfrak{B})$ if and only if Λ is an element in the coset ring of $\mathbb{Z}$. For instance, if $\Lambda = \{3^k : \; k \in \mathbb{N}\}$ (not in the coset ring), then μ_Λ does not determine an F_1-measure on $\sigma(\mathfrak{B} \times \mathfrak{B})$ (Exercise 3 ii).

Here is a variation on this theme. Define

$$\mu(A, B, C) \; = \; \sum_{k \in \mathbb{Z}} \hat{\mathbf{1}}_A(k) \; \hat{\mathbf{1}}_B(k) \; \mathbf{1}_C(k),$$
$$A \in \mathfrak{B}, \; B \in \mathfrak{B}, \; C \subset \mathbb{Z}. \tag{2.10}$$

Then, $\mu \in F_3(\mathfrak{B}, \mathfrak{B}, \mathbb{Z})$, but μ is neither in $F_2(\sigma(\mathfrak{B} \times \mathfrak{B}), \mathbb{Z})$ nor in $F_2(\mathfrak{B}, \sigma(\mathfrak{B} \times 2^{\mathbb{Z}}))$. Note (in Exercise 4) that the transform $\hat{\mu}$ defined on $\mathbb{Z} \times \mathbb{Z} \times \mathbf{T}$ is

$$\hat{\mu}(n, m, t) := \mu(\mathrm{e}^{in\cdot}, \mathrm{e}^{im\cdot}, \mathrm{e}^{it\cdot}) = \begin{cases} \mathrm{e}^{itn} & n = m, \, t \in \mathbf{T} \\ 0 & \text{otherwise.} \end{cases} \tag{2.11}$$

iv **(an example from stochastic analysis).** Consider a Wiener process $\mathrm{W} = \{\mathrm{W}(t) : t \in [0,1]\}$ on a probability space $(\Omega, \mathfrak{A}, \mathbb{P})$. (See Chapter X §2.) For $A \in \mathfrak{A}$, consider

$$\mathbf{E} \; \mathbf{1}_A \, \mathrm{W}(t), \quad t \in [0,1], \tag{2.12}$$

which defines a continuous function with total variation bounded by 1 (Exercise 5 i). Let $\mathfrak{B}$ denote the Borel field in [0,1]. Then, for each $A \in \mathfrak{A}$, there exists a scalar measure $\mu(A, \cdot)$ on $\mathfrak{B}$, such that for all $J = [s,t] \subset [0,1]$,

$$\mu(A, J) = \mathbf{E} \; \mathbf{1}_A \; \Delta\mathrm{W}(J) := \mathbf{E} \; \mathbf{1}_A \; (\mathrm{W}(t) - \mathrm{W}(s)). \tag{2.13}$$

Moreover, for each $B \in \mathfrak{B}$, $\mu(\cdot, B)$ is a scalar measure on $\mathfrak{A}$ and $\mu(\cdot, B) \ll \mathbb{P}$, which can be verified via the extension theorem in §4 (Exercise 5 ii). Therefore, $\mu \in F_2(\mathfrak{A}, \, \mathfrak{B})$.

We claim μ is not extendible to an F_1-measure on $\sigma(\mathfrak{A} \times \mathfrak{B})$. Let n be a positive integer, and let $J_k = [k/n, \; k+1/n), k = 0, \ldots, n-1$.

Then,

$$\sup\left\{\sum_{i,k} |\mu(A_i, J_k)| : \{A_i\}_i \subset \mathfrak{A}, \Sigma_i \mathbf{1}_{A_i} \leq 1\right\}$$

$$= \sup\left\{\sum_{i,k} \epsilon_{ik} \mathbf{E}\, \mathbf{1}_{A_i} \Delta \mathrm{W}(J_k) : \{A_i\}_i \subset \mathfrak{A},\ \Sigma_i \mathbf{1}_{A_i} \leq 1,\ \epsilon_{ik} = \pm 1\right\}$$

$$= \sup\left\{\sum_{k=0}^{n-1} \mathbf{E}\left(\sum_i \epsilon_{ik}\, \mathbf{1}_{A_i}\right) \Delta \mathrm{W}(J_k) : \{A_i\}_i \subset \mathfrak{A},\ \Sigma_i \mathbf{1}_{A_i} \leq 1,\ \epsilon_{ik} = \pm 1\right\}$$

$$= \sum_{k=0}^{n-1} \mathbf{E}|\Delta \mathrm{W}(J_k)| = \sum_{k=0}^{n-1} \sqrt{2/\pi n} = \sqrt{2n/\pi}. \tag{2.14}$$

(For $k = 0, \ldots, n-1$, $\Delta \mathrm{W}(J_k)$ is Gaussian with mean zero and variance $1/n$.) Because n is arbitrary, the 'total variation' of μ is infinite, and hence our claim.

The F_2-measure in (2.13) – the so-called *white noise* – exemplifies a general correspondence between Fréchet measures and stochastic integrators, which we analyze in Chapter X and Chapter XI. In the 'one-dimensional' case, this correspondence reduces to the usual identification (through the Radon–Nikodym theorem) of random variables in $\mathrm{L}^1(\Omega, \mathfrak{A}, \mathbb{P})$ with scalar measures that are absolutely continuous with respect to $\mathbb{P}$.

3 The Fréchet Variation

If C is an algebra of sets, then a *C-partition* will mean a collection of pairwise disjoint elements in C. If $C_1, \ldots, C_n$ are algebras of sets, then a *$(C_1 \times \cdots \times C_n)$-grid* will mean an n-fold Cartesian product of finite $C_1, \ldots, C_n$-partitions. If $C_1, \ldots, C_n$ are arbitrary or understood from the context, then we will refer simply to *partitions* and *grids*.

Recall that a Rademacher system indexed by a set τ is the collection of functions r_α, $\alpha \in \tau$, defined on $\{-1, +1\}^\tau$ by

$$r_\alpha(\omega) = \omega(\alpha), \quad \alpha \in \tau,\ \omega \in \{-1, +1\}^\tau.$$

If $\tau_1, \ldots, \tau_n$ are indexing sets, and $(\alpha_1, \ldots, \alpha_n) \in \tau_1 \times \cdots \times \tau_n$, then $r_{\alpha_1} \otimes \cdots \otimes r_{\alpha_n}$ denotes the function on $\{-1, +1\}^{\tau_1} \times \cdots \times \{-1, +1\}^{\tau_n}$, whose value at $(\omega_1, \ldots, \omega_n)$ is $r_{\alpha_1}(\omega_1) \cdots r_{\alpha_n}(\omega_n)$.

Throughout, $X_1, \ldots, X_n$ are sets, and $C_1 \subset 2^{X_1}, \ldots, C_n \subset 2^{X_n}$ are algebras.

Definition 2 The F_n-variation of $\mu \in F_n(C_1, \ldots, C_n)$ is

$$\|\mu\|_{F_n(C_1,\ldots,C_n)} = \|\mu\|_{F_n}$$
$$= \sup\left\{ \left\| \sum_{(E_1,\ldots,E_n)\in\gamma} \mu(E_1,\ldots,E_n) r_{E_1} \otimes \cdots \otimes r_{E_n} \right\|_{L^\infty} : \text{grid } \gamma \right\}. \tag{3.1}$$

($r_{E_1}, \ldots, r_{E_n}$ in (3.1) are elements of Rademacher systems indexed by the partitions whose Cartesian product is the grid γ.) F_n-variations will be generically referred to as Fréchet variations.

For $n = 1$, $F_1(C)$ is the space of scalar measures, and the F_1-variation is (equivalent to) the *total variation* norm: for $\mu \in M(X, C)$ $(= F_1(C))$,

$$\|\mu\|_{F_1} \leq \|\mu\|_{\text{total variation}} \leq 2\|\mu\|_{F_1}. \tag{3.2}$$

A classical theorem – the cornerstone of the one-dimensional theory – asserts that if μ is an F_1-measure on a measurable space (X, C), then the F_1-variation of μ is finite; e.g., [DunSchw, Corollary III.4.6]. (*Measurable space* (X, C) means that C is a σ-algebra.) Specifically, $(F_1(C), \|\cdot\|_{F_1})$ is a Banach space. In this section we establish the multi-dimensional version of this theorem: if $C_1, \ldots, C_n$ are σ-algebras, then $\|\cdot\|_{F_n}$ is a norm and $(F_n(C_1, \ldots, C_n), \|\cdot\|_{F_n})$ is a Banach space.

The proof is based on a measure-theoretic 'uniform boundedness' principle [Ni], a basic device that has over the years enjoyed several proofs (e.g., [DunSchw, Theorem IV.9.8], [DiU, p. 33]). The argument verifying it below was shown to me by S.J. Sidney.

Theorem 3 (The Nikodym boundedness principle [Ni]). *Let* $(X, \mathfrak{A})$ *be a measurable space, and let* $\mathfrak{F}$ *be a family of scalar measures on* $\mathfrak{A}$. *If* $\sup\{|\mu(A)| : \mu \in \mathfrak{F}\} < \infty$ *for every* $A \in \mathfrak{A}$, *then*

$$\sup\{|\mu(A)| : A \in \mathfrak{A}, \mu \in \mathfrak{F}\} < \infty;$$
$$\text{i.e., } \sup\{\|\mu\|_{F_1} : \mu \in \mathfrak{F}\} < \infty. \tag{3.3}$$

Proof: For $A \in \mathfrak{A}$, define

$$m(A) = \sup\{|\mu(A)| : \mu \in \mathfrak{F}\}, \tag{3.4}$$

and

$$M_A = \sup\{|m(B)| : B \in \mathfrak{A}, B \subset A\}.$$

Claim: *Suppose $B \in \mathfrak{A}$ and $M_B = \infty$. Then, for every $c > 0$, there exists $A \in \mathfrak{A}$ such that $A \subset B$, $m(A) > c$, and $M_{B\setminus A} = \infty$.*

Proof of Claim: Let $D \in \mathfrak{A}$ and $\mu \in \mathfrak{F}$ be such that $D \subset B$ and $|\mu(D)| > c + m(B)$. Then, either $M_D = \infty$ or $M_{B\setminus D} = \infty$. Note that

$$\begin{aligned} m(B \setminus D) &\geq |\mu(B) - \mu(D)| \geq |\mu(D)| - |\mu(B)| \\ &> c + m(B) - |\mu(B)| > c. \end{aligned}$$

If $M_D = \infty$, then define $A = B\setminus D$, and if $M_{B\setminus D} = \infty$, then define $A = D$.

Suppose the theorem is false; that is, assume

$$m(A) < \infty \quad \text{for all } A \in \mathfrak{A}, \tag{3.5}$$

and $M_X = \infty$. By the claim, we choose inductively $A_1, A_2, \ldots$, pairwise disjoint members of $\mathfrak{A}$, and $\{\mu_k\} \subset \mathfrak{F}$ such that

$$\begin{aligned} &\text{(i)} \lim_{k\to\infty} m(A_k) = \infty; \ \text{(ii)} \lim_{k\to\infty} |\mu_k(A_k)| = \infty; \\ &\text{(iii)} \ M_X\setminus(A_1 \cup \ldots \cup A_k) = \infty \text{ for } k \in \mathbb{N}. \end{aligned} \tag{3.6}$$

We initialize $k_1 = 1$. Because $\Sigma_{k=1}^{\infty}|\mu(A_k)| < \infty$ for each $\mu \in \mathfrak{F}$, we can select inductively (by applying (3.6) (ii)) an increasing sequence of integers (k_j) such that

$$|\mu_{k_{j+1}}(A_{k_{j+1}})| > j + 1 + \Sigma_{i=1}^{j} m(A_{k_i}), \tag{3.7}$$

and

$$\sum_{i=k_{j+1}}^{\infty} |\mu_{k_j}(A_i)| < 1. \tag{3.8}$$

Let $A = \cup_j A_{k_j}$. Then, for every $j \geq 2$,

$$\begin{aligned} m(A) &\geq |\mu_{k_j}(A)| \\ &\geq |\mu_{k_j}(A_{k_j})| - \sum_{i=1}^{j-1} |\mu_{k_j}(A_{k_i})| - \sum_{i=j+1}^{\infty} |\mu_{k_j}(A_{k_i})| \\ &> j + \sum_{i=1}^{j-1} m(A_{k_i}) - \sum_{i=1}^{j-1} |\mu_{k_j}(A_{k_i})| - \sum_{i=k_{j+1}}^{\infty} |\mu_{k_i}(A_i)| > j - 1, \end{aligned}$$

which contradicts (3.5). □

An F_n-measure μ is said to be *bounded* if

$$\sup\{|\mu(E_1,\ldots,E_n)| : (E_1,\ldots,E_n) \in C_1 \times \cdots \times C_n\} < \infty. \tag{3.9}$$

Corollary 4 *If $C_1,\ldots,C_n$ are σ-algebras, then every F_n-measure on $C_1 \times \cdots \times C_n$ is bounded.*

Proof (by induction). The case $n = 1$ is classical. Let $n > 1$, and assume the assertion in the case $n-1$. Let $\mu \in F_n(C_1,\ldots,C_n)$. Then, for every $A \in C_n$,

$$\mu(\cdot,\ldots,\cdot,A) \in F_{n-1}(C_1,\ldots,C_{n-1}). \tag{3.10}$$

By the induction hypothesis for $A \in C_n$,

$$\sup\{|\mu(A_1,\ldots,A_{n-1},A)| : (A_1,\ldots,A_{n-1}) \in C_1 \times \cdots \times C_{n-1}\} < \infty. \tag{3.11}$$

By applying Theorem 3 with $\mathfrak{F} = C_1 \times \cdots \times C_{n-1}$, we deduce

$$\sup\{|\mu(A_1,\ldots,A_n)| : \ (A_1,\ldots,A_n) \in C_1 \times \cdots \times C_n\} < \infty. \tag{3.12}$$

□

The main theorem of this section is

Theorem 5 *If $(X_1, C_1),\ldots,(X_n, C_n)$ are measurable spaces and*

$$\mu \in F_n(C_1,\ldots,C_n),$$

then $\|\mu\|_{F_n} < \infty$.

A key to this theorem is the following elementary

Lemma 6 *Let $(a_{k_1\ldots k_n})$ be a scalar array. For every $N \in \mathbb{N}$, there exist $T_1 \subset [N],\ldots,T_n \subset [N]$ such that*

$$\left|\sum_{(k_1,\ldots,k_n)\in T_1\times\cdots\times T_n} a_{k_1\ldots k_n}\right| \geq \left(\frac{1}{4}\right)^n \left\|\sum_{(k_1,\ldots,k_n)\in[N]^n} a_{k_1\ldots k_n} r_{k_1} \otimes \cdots \otimes r_{k_n}\right\|_{L^\infty}. \tag{3.13}$$

Proof (by induction on n). The case $n = 1$ is the statement that for every set of scalars $\{a_k : k \in [N]\}$ there exists $T \subset [N]$ such that

$$\left|\sum_{k\in T} a_k\right| \geq \frac{1}{4}\sum_{k=1}^{N} |a_k|. \tag{3.14}$$

If $n > 1$ and $\{a_{k_1\dots k_n}\} \subset \mathbb{C}$, then let $\omega_1 \in \{-1,1\}^N, \dots, \omega_n \in \{-1,1\}^N$ be such that

$$\begin{aligned}&\left\|\sum_{(k_1,\dots,k_n)\in[N]^n} a_{k_1\dots k_n} r_{k_1} \otimes \cdots \otimes r_{k_n}\right\|_{L^\infty} \\ &= \left|\sum_{(k_1,\dots,k_n)\in[N]^n} a_{k_1\dots k_n} r_{k_1}(\omega_1)\cdots r_{k_n}(\omega_n)\right|. \end{aligned}\tag{3.15}$$

By the case $n = 1$, there exists $T_1 \subset [N]$ such that (3.15) is bounded by

$$4\left|\sum_{k_1\in T_1}\left(\sum_{(k_2,\dots,k_n)\in[N]^{n-1}} a_{k_1\dots k_n} r_{k_2}(\omega_2)\cdots r_{k_n}(\omega_n)\right)\right|. \tag{3.16}$$

Interchange summations in (3.16), and then, by applying the induction hypothesis, obtain $T_2 \subset [N], \dots, T_n \subset [N]$ such that (3.16) is bounded by

$$4^{n-1}\left|\sum_{(k_2,\dots,k_n)\in T_2\times\cdots\times T_n}\left(4\sum_{k_1\in T_1} a_{k_1\dots k_n}\right)\right|.$$

□

Proof of Theorem 5: Assume the assertion is false. Then, for every $c > 0$ there exist $\mu \in F_n(C_1, \dots, C_n)$ and a grid $\tau_1 \times \cdots \times \tau_n$ such that

$$\left\|\sum_{(A_1,\dots,A_n)\in\tau_1\times\cdots\times\tau_n} \mu(A_1,\dots,A_n)\, r_{A_1}\otimes\cdots\otimes r_{A_n}\right\|_{L^\infty} > c. \tag{3.17}$$

By Lemma 6, there exist $T_1 \subset \tau_1, \dots, T_n \subset \tau_n$ such that

$$\left|\sum_{A_1\in T_1,\dots,A_n\in T_n} \mu(A_1\times\cdots\times A_n)\right| > c/2^n. \tag{3.18}$$

If $D = \cup\{A : A \in T_1\} \times \cdots \times \cup\{A : A \in T_n\}$, then $|\mu(D)| > c/2^n$, and this contradicts Corollary 4.

Corollary 7 (Exercise 6). *If* $C_1, \ldots, C_n$ *are* σ*-algebras, then* $(F_n(C_1, \ldots, C_n), \|\cdot\|_{F_n})$ *is a Banach space.*

4 An Extension Theorem

We have shown in the previous section that the Fréchet variation of an F-measure on a Cartesian product of σ-algebras is finite. In this section we prove the converse: if the Fréchet variation of an F-measure μ on a Cartesian product of algebras is finite, then μ determines an F-measure on the Cartesian product of the corresponding σ-algebras. This generalizes a classical, 'one-dimensional' theorem, known in the literature as the Carathéodory–Hahn–Jordan theorem (e.g., [Du1, Theorem 5.6.3]).

Theorem 8 *Let* $C_1, \ldots, C_n$ *be algebras of sets in* $X_1, \ldots, X_n$*, respectively, and let* $\mu \in F_n(C_1, \ldots, C_n)$*. Then,* μ *is uniquely extendible to an* F_n*-measure* $\tilde{\mu}$ *on* $\sigma C_1 \times \cdots \times \sigma C_n$ *if and only if* $\|\mu\|_{F_n} < \infty$*. Moreover,*

$$\|\mu\|_{F_n(C_1,\ldots,C_n)} = \|\tilde{\mu}\|_{F_n(\sigma C_1,\ldots,\sigma C_n)} \tag{4.1}$$

($\sigma C :=$ σ*-algebra generated by* C*).*

Proof: Necessity follows from Theorem 5.

Sufficiency is proved by induction on n. The case $n = 1$ is the Carathéodory–Hahn–Jordan theorem. Let $n > 1$, and assume the assertion in the case $n - 1$. Let μ be an F_n-measure on $C_1 \times \cdots \times C_n$ such that $\|\mu\|_{F_n} < \infty$. Then, for each $(A_2, \ldots, A_n) \in C_2 \times \cdots \times C_n$, $\mu(\cdot, A_2, \ldots, A_n)$ is extendible to a scalar measure on σC_1. Denote this extension also by μ. Observe that

$$\begin{aligned} &\sup\{|\mu(A_1, A_2, \ldots, A_n)| : (A_1, A_2, \ldots, A_n) \in \sigma C_1 \times C_2 \times \cdots \times C_n\} \\ &\quad < \|\mu\|_{F_n(C_1,\ldots,C_n)}. \end{aligned} \tag{4.2}$$

Claim: *For each* $A \in \sigma C_1, \mu(A, \cdot, \ldots, \cdot) \in F_{n-1}(C_2, \cdots, C_n)$.

Proof of Claim: Let

$$S = \{A \in \sigma C_1 : \mu(A, \cdot, \ldots, \cdot) \in F_{n-1}(C_2, \ldots, C_n)\}. \tag{4.3}$$

Then, S is an algebra containing C_1. To show that S is a σ-algebra, we need to verify that if $\{E_i\}$ is a countable collection of pairwise disjoint elements in S, then $\mu(\cup_i E_i, \cdot, \ldots, \cdot)$ is countably additive in each of the $n-1$ coordinates. We prove that $\mu(\cup_i E_i, \cdot, \ldots, \cdot)$ is countably additive in the nth coordinate. (The argument establishing countable additivity in the other $n-2$ coordinates is identical.) Denote $E = \cup_i E_i$. Let $\{A_j\}$ be a collection of pairwise disjoint elements in C_n such that $\cup_j A_j \in C_n$, and proceed to verify

$$\mu(E, \ldots, \cup_j A_j) = \sum_{j=1}^{\infty} \mu(E, \ldots, A_j). \tag{4.4}$$

Because μ is a scalar measure in the first coordinate,

$$\mu(E, \ldots, \cup_j A_j) = \sum_{i=1}^{\infty} \mu(E_i, \ldots, \cup_j A_j). \tag{4.5}$$

Because $E_i \in S$ for all $i \in \mathbb{N}$,

$$\sum_{i=1}^{\infty} \mu(E_i, \ldots, \cup_j A_j) = \sum_{i=1}^{\infty} \left(\sum_{j=1}^{\infty} \mu(E_i, \ldots, A_j) \right). \tag{4.6}$$

By (4.2) and Lemma 6,

$$\sup \left\{ \left\| \sum_{(E_1, \ldots, E_n) \in \gamma} \mu(E_1, \ldots, E_n)\, r_{E_1} \otimes \cdots \otimes r_{E_n} \right\|_{L^\infty} : (\sigma C_1 \times C_2 \times \cdots \times C_n) - \text{grid } \gamma \right\} \leq \|\mu\|_{F_n}. \tag{4.7}$$

(The last line requires a small argument that you are asked to provide in Exercise 7.) By (4.7) and Corollary IV.7, we can reverse summations on the right side of (4.6). Therefore, because μ is a measure in the first coordinate, we obtain

$$\begin{aligned} \sum_{i=1}^{\infty} \left(\sum_{j=1}^{\infty} \mu(E_i, \ldots, A_j) \right) &= \sum_{j=1}^{\infty} \left(\sum_{i=1}^{\infty} \mu(E_i, \ldots, A_j) \right) \\ &= \sum_{j=1}^{\infty} \mu(E, \ldots, A_j), \end{aligned} \tag{4.8}$$

which proves (4.4), and completes the proof of the claim.

The induction hypothesis together with the claim imply that for each $A \in \sigma C_1$, $\mu(A, \cdot, \ldots, \cdot)$ is extendible to an F_{n-1}-measure $\tilde{\mu}$ on $\sigma C_2 \times \cdots \times \sigma C_n$. The assertion that $\mu(\cdot, A_2, \ldots, A_n)$ is a scalar measure on σC_1 for all $(A_2, \ldots, A_n) \in \sigma C_2 \times \cdots \times \sigma C_n$ is verified by mimicking the proof of the claim above (Exercise 8).

Uniqueness and (4.1) are verified by induction (Exercise 9). □

5 Integrals with Respect to F_n-measures

For a σ-algebra $C \subset 2^X$, let $L^\infty(C)$ denote the space of bounded scalar-valued C-measurable functions on X. Equipped with the supremum norm $\|\cdot\|_\infty$ and pointwise multiplication, $L^\infty(C)$ is a commutative Banach algebra. Observe also that C-simple functions on X are norm-dense in $L^\infty(C)$. Throughout this section we assume $n \geq 2$, and that for $i = 1, \ldots, n$, $C_i \subset 2^{X_i}$ is a σ-algebra.

Lemma 9 *For $\mu \in F_n(C_1, \ldots, C_n)$ and $f \in L^\infty(C_n)$, define*

$$\mu_f(A_2, \ldots, A_n) = \int_{X_1} f(x)\, \mu(\mathrm{d}x, A_2, \ldots, A_n),$$

$$(A_2, \ldots, A_n) \in C_2 \times \cdots \times C_n. \tag{5.1}$$

Then, $\mu_f \in F_{n-1}(C_2, \ldots, C_n)$ and

$$\|\mu_f\|_{F_{n-1}} \leq 2\|f\|_\infty \, \|\mu\|_{F_n}. \tag{5.2}$$

Proof: It suffices to prove the case $n = 2$. The finite additivity of $\mu_f(\cdot)$ is obvious. We need to show that for a countable family $\{A_k\}$ of pairwise disjoint sets in C_2,

$$\int_{X_1} f(x)\, \mu\,(\mathrm{d}x,\, \cup_{k=1}^\infty A_k) = \sum_{k=1}^\infty \int_{X_1} f(x)\, \mu(\mathrm{d}x, A_k). \tag{5.3}$$

Let (φ_j) be a sequence of simple functions on (X_1, C_1) converging uniformly to f. Then,

$$\int_{X_1} \varphi_j(x)\, \mu\,(\mathrm{d}x,\, \cup_{k=1}^m A_k)$$

$$\xrightarrow[j\to\infty]{} \int_{X_1} f(x)\, \mu(\mathrm{d}x, \cup_{k=1}^m A_k), \quad m \in \mathbb{N}, \tag{5.4}$$

and

$$\int_{X_1} \varphi_j(x)\, \mu\,(\mathrm{d}x, \cup_{k=1}^{\infty} A_k) \underset{j\to\infty}{\longrightarrow} \int_{X_1} f(x)\, \mu\,(\mathrm{d}x, \cup_{k=1}^{\infty} A_k). \tag{5.5}$$

The convergence in (5.4) is uniform in m (cf. Exercise 6). Therefore (e.g., [Ru2, Theorem 7.11]),

$$\begin{aligned} &\lim_{j\to\infty} \lim_{m\to\infty} \int_{X_1} \varphi_j(x)\, \mu(\mathrm{d}x, \cup_{k=1}^{m} A_k) \\ &\quad = \lim_{m\to\infty} \lim_{j\to\infty} \int_{X_1} \varphi_j(x)\, \mu(\mathrm{d}x, \cup_{k=1}^{m} A_k). \end{aligned} \tag{5.6}$$

The lemma holds in the case of simple functions, and therefore, by (5.5), the left side of (5.6) equals $\int_{X_1} f(x)\, \mu(\mathrm{d}x, \cup_{k=1}^{\infty} A_k)$. By (5.4) and finite additivity, the right side of (5.6) equals $\sum_{k=1}^{\infty} \int_{X_1} f(x)\, \mu(\mathrm{d}x, A_k)$. This proves (5.3).

The estimate for simple f

$$\|\mu_f\|_{F_1} \leq 2\|f\|_\infty\, \|\mu\|_{F_2} \quad \text{(cf. (3.2))} \tag{5.7}$$

implies the same in the general case. (For real-valued f,

$$\|\mu_f\|_{F_1} \leq \|f\|_\infty \|\mu\|_{F_2}.$$

For complex-valued f, the constant 2 in (5.7) can be made smaller; see Exercise II.9.) □

Let $\mu \in F_n(C_1, \ldots, C_n)$, and $f_1 \in \mathrm{L}^\infty(C_1), \ldots, f_n \in \mathrm{L}^\infty(C_n)$. For $j \in [n-1]$, we obtain $\mu_{f_1\otimes\cdots\otimes f_j} \in F_{n-j}(C_{j+1}, \ldots, C_n)$ by recursion: if $j = 1$, then $\mu_{f_1} \in F_{n-1}(C_2, \ldots, C_n)$ is provided by Lemma 9; if $1 < j < n$, then we apply Lemma 9 to f_j and $\mu_{f_1\otimes\cdots\otimes f_{j-1}}$, and thus obtain $\mu_{f_1\otimes\cdots\otimes f_j} \in F_{n-j}(C_{j+1}, \ldots, C_n)$,

$$\begin{aligned} &\mu_{f_1\otimes\cdots\otimes f_j}(A_{j+1}, \ldots, A_n) \\ &\quad = \int_{X_j} f_j(x)\, \mu_{f_1\otimes\cdots\otimes f_{j-1}}(\mathrm{d}x, A_{j+1}, \ldots, A_n), \\ &\quad (A_{j+1}, \ldots, A_n) \in C_{j+1} \times \cdots \times C_n. \end{aligned} \tag{5.8}$$

For $j = n$, $\mu_{f_1\otimes\cdots\otimes f_n}$ is the integral

$$\begin{aligned} \mu_{f_1\otimes\cdots\otimes f_n} &:= \int f_1 \otimes \cdots \otimes f_n\, \mathrm{d}\mu \\ &= \int_{X_n} f_n(x)\, \mu_{f_1\otimes\cdots\otimes f_{n-1}}(\mathrm{d}x). \end{aligned} \tag{5.9}$$

The recursive construction that ends with (5.9) yields the n-fold iterated integral

$$\int f_1 \otimes \cdots \otimes f_n \, \mathrm{d}\mu := \int_{X_1} f_1(x_1) \Bigg(\ldots \Bigg(\int_{X_{n-1}} f_{n-1}(x_{n-1}) \\ \Bigg(\int_{X_n} f_n(x_n) \mu(\cdot, \cdots, \mathrm{d}x_n) \Bigg) (\mathrm{d}x_{n-1}) \Bigg) \cdots \Bigg) (\mathrm{d}x_1), \tag{5.10}$$

and the estimate

$$\left| \int f_1 \otimes \cdots \otimes f_n \, \mathrm{d}\mu \right| \leq 2^n \|f_1\|_\infty \cdots \|f_n\|_\infty \|\mu\|_{F_n}. \tag{5.11}$$

The recursive definition of $\int f_1 \otimes \cdots \otimes f_n \, \mathrm{d}\mu$ does not depend on the order of the steps leading to (5.9). That is, it does not depend on the order of the iterated integrals in (5.10). To verify this, it suffices to check a 'Fubini'-type property (cf. Corollary IV.7) in the case $n = 2$:

Theorem 10 *Let $\mu \in F_2(C_1, C_2)$, $f_1 \in \mathrm{L}^\infty(C_1)$, and $f_2 \in \mathrm{L}^\infty(C_2)$. Then,*

$$\begin{aligned} \int f_1 \, \mathrm{d}\mu_{f_2} &= \int_{X_1} f_1(x_1) \left(\int_{X_2} f_2(x_2) \, \mu(\cdot, \mathrm{d}x_2) \right) (\mathrm{d}x_1) \\ &= \int_{X_2} f_2(x_2) \left(\int_{X_1} f_1(x_1) \, \mu(\mathrm{d}x_1, \cdot) \right) (\mathrm{d}x_2) \\ &= \int f_2 \, \mathrm{d}\mu_{f_1}. \end{aligned} \tag{5.12}$$

Proof: The assertion clearly holds in the case of simple functions. Therefore, if (θ_k) is a sequence of simple functions converging uniformly to f_2, then, by Lemma 9,

$$\begin{aligned} &\lim_{k\to\infty} \int_{X_1} f_1(x_1) \left(\int_{X_2} \theta_k(x_2) \, \mu(\cdot, \mathrm{d}x_2) \right) (\mathrm{d}x_1) \\ &= \lim_{k\to\infty} \int_{X_2} \theta_k(x_2) \left(\int_{X_1} f_1(x_1) \, \mu(\mathrm{d}x_1, \cdot) \right) (\mathrm{d}x_2) \\ &= \int_{X_2} f_2(x_2) \left(\int_{X_1} f_1(x_1) \, \mu(\mathrm{d}x_1, \cdot) \right) (\mathrm{d}x_2). \end{aligned} \tag{5.13}$$

Also by Lemma 9, $\mu_{\theta_k} \xrightarrow[k\to\infty]{} \mu_{f_2}$ in the $F_1(C_1)$-norm. Therefore,

$$\lim_{k\to\infty} \int_{X_1} f_1(x_1) \left(\int_{X_2} \theta_k(x_2)\, \mu(\cdot, \mathrm{d}x_2) \right) (\mathrm{d}x_1)$$
$$= \int_{X_1} f_1(x_1) \left(\int_{X_2} f_2(x_2)\, \mu(\mathrm{d}x_2, \cdot) \right) (\mathrm{d}x_1), \tag{5.14}$$

which, combined with (5.13), proves the assertion. □

Remark (approximations by simple functions). If

$$\varphi_1 \in \mathrm{L}^\infty(C_1), \ldots, \varphi_n \in \mathrm{L}^\infty(C_n)$$

are simple functions, then the integral $\int \varphi_1 \otimes \cdots \otimes \varphi_n \, \mathrm{d}\mu$ is a finite sum. This implies

Proposition 11 (Exercise 10). *If $\mu \in F_n(C_1, \ldots, C_n)$, and $f_1 \in \mathrm{L}^\infty(C_1), \ldots, f_n \in \mathrm{L}^\infty(C_n)$, then for all $\epsilon > 0$ there exist simple functions $\varphi_1 \in \mathrm{L}^\infty(C_1), \ldots, \varphi_n \in \mathrm{L}^\infty(C_n)$ such that*

$$\left| \int f_1 \otimes \cdots \otimes f_n \, \mathrm{d}\mu - \int \varphi_1 \otimes \cdots \otimes \varphi_n \, \mathrm{d}\mu \right| < \epsilon. \tag{5.15}$$

6 The Projective Tensor Algebra $V_n(C_1, \ldots, C_n)$

Consider the algebraic tensor product $\mathrm{L}^\infty(C_1) \otimes \cdots \otimes \mathrm{L}^\infty(C_n)$ under the equivalence determined by pointwise evaluation on $X_1 \times \cdots \times X_n$. (See Chapter IV §6, §7.) Let $V_n(C_1, \ldots, C_n)$ be the closure of $\mathrm{L}^\infty(C_1) \otimes \cdots \otimes \mathrm{L}^\infty(C_n)$ in the projective tensor norm, which is defined in (IV.7.2) (where C_0-functions are replaced by bounded measurable functions). Let $S(C)$ denote the space of scalar-valued C-measurable simple functions on (X, C) equipped with the supremum norm. Then (Exercise 11 i),

$$V_n(C_1, \ldots, C_n) = S(C_1) \hat{\otimes} \cdots \hat{\otimes} S(C_n). \tag{6.1}$$

If $\sum_j \varphi_{1j} \otimes \cdots \otimes \varphi_{nj} \in S(C_1) \otimes \cdots \otimes S(C_n)$ and

$$\sum_j \varphi_{1j}(x_1) \cdots \varphi_{nj}(x_n) = 0$$
$$\text{for all } (x_1, \ldots, x_n) \in X_1 \times \cdots \times X_n, \tag{6.2}$$

then for every $\mu \in F_n(C_1, \ldots, C_n)$,

$$\sum_j \int \varphi_{1j} \otimes \cdots \otimes \varphi_{nj} \, \mathrm{d}\mu = 0 \tag{6.3}$$

(Exercise 11 ii). In particular, this implies that if $\phi \in S(C_1) \otimes \cdots \otimes S(C_n)$, and $\phi = \sum_j \varphi_{1j} \otimes \cdots \otimes \varphi_{nj}$, then

$$\int \phi \, \mathrm{d}\mu := \sum_j \int \varphi_{1j} \otimes \cdots \otimes \varphi_{nj} \, \mathrm{d}\mu \tag{6.4}$$

does not depend on the pointwise representation of ϕ by elements in $S(C_1) \otimes \cdots \otimes S(C_n)$. Moreover,

$$\left| \int \phi \, \mathrm{d}\mu \right| \leq 2^n \|\phi\|_{\hat{\otimes}} \|\mu\|_{F_n}. \tag{6.5}$$

Therefore, by passing to limits, we conclude that the integral $\int \phi \, \mathrm{d}\mu$ exists for all $\phi \in V_n(C_1, \ldots, C_n)$ and $\mu \in F_n(C_1, \ldots, C_n)$, and that it satisfies (6.5).

The verification that $\int f_1 \otimes \cdots \otimes f_n \, \mathrm{d}\mu$ defined in the previous section is the same as the integral defined above is relegated to Exercise 12.

Remark (a problem). We noted that all functions in $V_n(C_1, \ldots, C_n)$ are canonically integrable with respect to all $\mu \in F_n(C_1, \ldots, C_n)$. How to characterize functions on $X_1 \times \cdots \times X_n$ that are 'canonically integrable' with respect to a given $\mu \in F_n(C_1, \ldots, C_n)$ (that is, how to describe '$\mathrm{L}^1(\mu)$') is an open (-ended) question. This problem, which in the one-dimensional case is resolved by the classical Radon–Nikodym theorem, is in essence the question: how do we differentiate, in a Radon–Nikodym sense, in dimensions greater than one (Exercise 13)?

7 A Multilinear Riesz Representation Theorem

In the previous section we observed that the integral with respect to $\mu \in F_n(C_1, \ldots, C_n)$ determines a bounded n-linear functional on $\mathrm{L}^\infty(C_1) \times \cdots \times \mathrm{L}^\infty(C_n)$. In this section we prove a converse.

Let $X_1, \ldots, X_n$ be locally compact Hausdorff spaces, and let $\mathfrak{A}_1, \ldots, \mathfrak{A}_n$ be their respective Borel fields. Recall that a scalar-valued function ξ on $\mathrm{C}_0(X_1) \times \cdots \times \mathrm{C}_0(X_n)$ is a bounded n-linear functional if it is linear in each coordinate and

$$\|\xi\| := \sup\{\{|\xi(f_1, \ldots, f_n)|\} : \|f_1\|_\infty \leq 1, \ldots, \|f_n\|_\infty \leq 1\} < \infty. \tag{7.1}$$

(Generality is not sacrificed here, for we can view elements in the commutative Banach algebra $L^\infty(C)$ as continuous functions on the maximal ideal space of $L^\infty(C)$.)

Theorem 12 *If ξ is a bounded n-linear functional on $C_0(X_1) \times \cdots \times C_0(X_n)$, then there exists a unique $\mu_\xi \in F_n(\mathfrak{A}_1, \ldots, \mathfrak{A}_n)$ such that*

$$\xi(f_1, \ldots, f_n) = \int f_1 \otimes \cdots \otimes f_n \, d\mu_\xi, \qquad (f_1, \ldots, f_n) \in C_0(X_1) \times \cdots \times C_0(X_n), \tag{7.2}$$

and

$$\|\mu_\xi\|_{F_n} \leq \|\xi\| \leq 2^n \|\mu_\xi\|_{F_n}. \tag{7.3}$$

Proof: The proof is by induction on n. The case $n = 1$ is the classical Riesz representation theorem. We will prove here the case $n = 2$, which is typical (Exercise 14).

Denote $X_1 = X$, $X_2 = Y$, $\mathfrak{A}_1 = \mathfrak{A}$ and $\mathfrak{A}_2 = \mathfrak{B}$.

Step 1 If $f \in C_0(X)$, then $\xi(f, \cdot) := \xi_f(\cdot)$ is a bounded linear functional on $C_0(Y)$, and hence, by the Riesz representation theorem, there exists a unique regular measure μ_{ξ_f} on $\mathfrak{B}$ such that

$$\xi_f(g) = \int g(y) \, \mu_{\xi_f}(dy), \quad g \in C_0(Y), \tag{7.4}$$

and

$$\|\mu_{\xi_f}\|_{F_1(\mathfrak{B})} \leq \|\xi\|. \tag{7.5}$$

Step 2 If $B \in \mathfrak{B}$, then

$$f \mapsto \mu_{\xi_f}(B), \; f \in C_0(X), \tag{7.6}$$

defines a bounded linear functional on $C_0(X)$ with norm bounded by $\|\xi\|$. (Boundedness is implied by (7.5), and linearity follows from the linearity of ξ.) Therefore, for each $B \in \mathfrak{B}$, there exists a regular measure $\mu_\xi(\cdot, B)$ on $\mathfrak{A}$ such that

$$\mu_{\xi_f}(B) = \int f(x) \, \mu_\xi(dx, B), \tag{7.7}$$

and $\|\mu_\xi(\cdot, B)\|_{F_1(\mathfrak{A})} \leq \|\xi\|$. More generally, if $\theta = \Sigma_k a_k \ \mathbf{1}_{B_k}$ is a simple function in the unit ball of $\mathrm{L}^\infty(\mathfrak{B})$, then $\Sigma_k a_k \ \mu_\xi(\cdot, B_k)$ is the representing measure of the functional

$$f \mapsto \int \theta(y) \ \mu_{\xi,f}(\mathrm{d}y) = \int f(x) \Sigma_k a_k \ \mu_\xi(\mathrm{d}x, B_k), \quad f \in \mathrm{C}_0(X), \tag{7.8}$$

and $\|\Sigma_k a_k \ \mu_\xi(\cdot, B_k)\|_{F_1(\mathfrak{A})} \leq \|\xi\|$.

Step 3 Let $\{A_j\} \subset \mathfrak{A}$ and $\{B_k\} \subset \mathfrak{B}$ be countable collections of pairwise disjoint sets, and let $\beta = (\mu_\xi(A_j \times B_k) : (j, k) \in \mathbb{N}^2)$. Then,

$$\|\beta\|_{F_2(\mathbb{N},\mathbb{N})} \leq \|\xi\|. \tag{7.9}$$

Proof of Step 3: Fix arbitrary finite sets $S \subset \mathbb{N}$ and $T \subset \mathbb{N}$, and fix arbitrary $\omega \in \{-1, 1\}^{\mathbb{N}}$ and $\eta \in \{-1, 1\}^{\mathbb{N}}$. In (7.8), put

$$f = \Sigma_{j \in S} r_j(\omega) \mathbf{1}_{A_j} \quad \text{and} \quad \theta = \Sigma_{k \in T} r_k(\eta) \ \mathbf{1}_{B_k}.$$

Then,

$$\begin{aligned} &\left| \int f(x) \Sigma_k a_k \ \mu_\xi(\mathrm{d}x, B_k) \right| \\ &= \left| \sum_{j \in S, k \in T} \mu_\xi(A_j, B_k) \ r_j(\omega) \ r_k(\eta) \right| \leq \|\xi\|. \end{aligned} \tag{7.10}$$

Step 4 If $\{B_k\}$ is a countable collection of pairwise disjoint sets in $\mathfrak{B}$, then $\Sigma_{k=1}^\infty \mu_\xi(\cdot, B_k) \in F_1(\mathfrak{A})$.

Proof of Step 4: Let $\{A_j\}$ be a countable collection of pairwise disjoint sets in $\mathfrak{A}$. Because $\mu_\xi(\cdot, B_k)$ is a measure for each $k \in \mathbb{N}$, we have

$$\sum_{k=1}^\infty \mu_\xi(\cup_j A_j, B_k) = \sum_{k=1}^\infty \left(\sum_{j=1}^\infty \mu_\xi(A_j, B_k) \right). \tag{7.11}$$

By Step 3 and Corollary IV.7,

$$\sum_{k=1}^\infty \left(\sum_{j=1}^\infty \mu_\xi(A_j \times B_k) \right) = \sum_{j=1}^\infty \left(\sum_{k=1}^\infty \mu_\xi(A_j \times B_k) \right). \tag{7.12}$$

Step 5 Let $\{B_k\}$ be a countable collection of pairwise disjoint sets in $\mathfrak{B}$. Then,

$$\sum_{k=1}^{\infty} \mu_\xi(\cdot, B_k) = \mu_\xi(\cdot, \cup_k B_k).$$

Proof of Step 5: Let $f \in C_0(X)$. By Step 4,

$$\sum_{k=1}^{n} \int f(x)\ \mu_\xi(\mathrm{d}x, B_k) \xrightarrow[n\to\infty]{} \int f(x) \Sigma_{k=1}^{\infty} \mu_\xi(\mathrm{d}x, B_k), \tag{7.13}$$

and by Step 1,

$$\begin{aligned}\sum_{k=1}^{n} \int f(x)\ \mu_\xi(\mathrm{d}x, B_k) &= \sum_{k=1}^{n} \mu_{\xi_f}(B_k) \xrightarrow[n\to\infty]{} \mu_{\xi_f}\left(\cup_{k=1}^{\infty} B_k\right) \\ &= \int f(x)\ \mu_\xi\left(\mathrm{d}x, \cup_{k=1}^{\infty} B_k\right).\end{aligned} \tag{7.14}$$

Because f is arbitrary, this implies Step 5.

We now put the steps together. By Step 2, $\mu_\xi(\cdot \times B) \in F_1(\mathfrak{A})$ for each $B \in \mathfrak{B}$. By Step 5, $\mu_\xi(A \times \cdot) \in F_1(\mathfrak{B})$ for each $A \in \mathfrak{A}$; that is, $\mu_\xi \in F_2(\mathfrak{A} \times \mathfrak{B})$. By the definition of integration with respect to $\mu_\xi \in F_2(\mathfrak{A} \times \mathfrak{B})$, Steps 1 and 2 imply

$$\xi(f,g) = \int f \otimes g\ \mathrm{d}\mu_\xi, \quad (f,g) \in C_0(X) \times C_0(Y), \tag{7.15}$$

and $\|\xi\| \le 4\|\mu_\xi\|_{F_2}$. The reverse inequality $\|\xi\| \ge \|\mu_\xi\|_{F_2}$ follows from (7.9). Uniqueness ($\mu_\xi = 0$ implies $\xi = 0$) follows from uniqueness in the case $n = 1$. □

Theorem 12 implies a characterization of $V_n(X_1, \ldots, X_n)^*$, which generalizes the characterization in Proposition IV.11. To see this, we first observe (as in Chapter IV, in the case $X_1 = \cdots = X_n = \mathbb{N}$) that every $\mu \in F_n(\mathfrak{A}_1, \ldots, \mathfrak{A}_n)$ determines a bounded linear functional ξ_μ on $V_n(X_1, \ldots, X_n)$: if $\phi \in V_n(X_1, \ldots, X_n)$ and $\phi = \sum_j f_{1j} \otimes \cdots \otimes f_{nj}$ such that $\sum_j \|f_{1j}\|_\infty \cdots \|f_{nj}\|_\infty < \infty$, then

$$\xi_\mu(\phi) = \sum_j \int f_{1j} \otimes \cdots \otimes f_{nj}\ \mathrm{d}\mu, \tag{7.16}$$

and $\|\xi_\mu\| \leq 2^n \|\mu\|_{F_n}$. (See §6.) In the opposite direction, if

$$\xi \in V_n(X_1, \ldots, X_n)^*,$$

then by Theorem 12, there exists $\mu_\xi \in F_n(\mathfrak{A}_1, \ldots, \mathfrak{A}_n)$ such that the linear action of ξ on $V_n(X_1, \ldots, X_n)$ is given by (7.16) with $\mu_\xi = \mu$, and $\|\mu_\xi\|_{F_n} \leq \|\xi\|$. We summarize:

Theorem 13 (cf. Proposition IV.11). *If $X_1, \ldots, X_n$ are locally compact Hausdorff spaces with respective Borel fields $\mathfrak{A}_1, \ldots, \mathfrak{A}_n$, then,*

$$V_n(X_1, \ldots, X_n)^* = F_n(\mathfrak{A}_1, \ldots, \mathfrak{A}_n). \tag{7.17}$$

8 A Historical Backdrop

Bounded bilinear functionals on C([0,1]) had been characterized first by Fréchet in [Fr], and later were dubbed *bimeasures* by Morse and Transue [Mor]. A bilinear Riesz representation-type theorem, identifying these bimeasures as bona fide set-functions, was stated and proved first by Ylinen in [Y1, Theorem 6.6], where they were also called *bimeasures.* In general multidimensional settings, which began attracting attention in the mid-1980s, the terms used were *multimeasures* or *polymeasures* (e.g., [GrY], [Do]), but I prefer F_n*-measures,* mainly because these register the ambient dimension.

Multidimensional measure theory, as such, began with this definition by Fréchet [Fr]: for a scalar-valued function u on $[0,1]^2$, let

$$\|u\| = \sup\left\{\left|\sum_{x_i \in \rho, y_j \in \tau} \Delta^2 u(x_i, y_j)\epsilon_i \delta_j\right| : \right.$$

$$\left. \epsilon_i = \pm 1,\ \delta_j = \pm 1,\ \text{partitions } \rho, \tau\right\}, \tag{8.1}$$

where

$$\rho = \{0 \leq x_1 < \cdots < x_m \leq 1\}, \quad \tau = \{0 \leq y_1 < \cdots < y_n \leq 1\},$$

and the 'second difference' Δ^2 is given by

$$\Delta^2 u(x_i, y_j)$$

$$= u(x_i, y_j) - u(x_{i-1}, y_j) + u(x_{i-1}, y_{j-1}) - u(x_i, y_{j-1}). \tag{8.2}$$

(The connection between (8.1) and Definition 2 in §3 should be evident.) Fréchet constructed in [Fr] a *Riemann–Stieltjes* double integral with respect to such u, and used it to represent bounded bilinear functionals on C([0,1]). This generalized F. Riesz's prior characterization of bounded linear functionals on C([0,1]) [Ri$_f$1]. (See Exercises 15, 16, 17.)

While measure theory (in one dimension) got off to a quick start in the beginning of the twentieth century, the bilinear theory was far slower to develop. Sustained interest in any area requires non-trivial examples, and indeed hardly anything at all had transpired in two dimensions until first Littlewood [Lit4], and then Clarkson and Adams [ClA] produced functions $u = u(x, y)$ with finite variation in the sense of Fréchet, i.e., $\|u\| < \infty$, and infinite variation in the sense of Vitali, i.e.,

$$\sup\left\{\left|\sum_{i,j}\Delta^2 u(x_i, y_j)\epsilon_{ij}\right| : \epsilon_{ij} = \pm 1, \text{ partitions } \rho, \tau\right\} = \infty. \tag{8.3}$$

(See Exercise I.2.) Littlewood's examples and inequalities [Lit4], foreshadowing the probabilistic aspects of the subject, were in hindsight more illuminating than the constructions in [ClA]. Littlewood himself, for reasons unknown, did not pursue this further. On the other hand, Adams and Clarkson (only briefly acknowledging Littlewood's prior examples, and paying no attention to his inequalities [ClA, p. 827, p. 837]) continued to investigate the Fréchet variation largely in a context of then-current integration theories (e.g., [ACl1], [ACl2], [Cl]).

Following the work of Adams and Clarkson, Morse and Transue continued essentially in the same spirit. Recognizing that bilinear functionals were fundamentally different from linear functionals, they concentrated on extending the classical 'one-dimensional' theory to two dimensions. Their work consisted of two series of papers. In the first, staying within Euclidean settings, they made precise an analogy between distribution functions on the line and functions with bounded Fréchet variation on the plane [MorTr1], [MorTr4], and then investigated Stieltjes integral representations of bilinear actions on function spaces [MorTr2], [MorTr3]. In a second series of papers [Mor], [MorTr5], [MorTr6], [MorTr7], they replaced Fréchet's setting [0,1] × [0,1] with a general Cartesian product $K_1 \times K_2$, where K_1 and K_2 were locally compact Hausdorff spaces, and considered *bimeasures* on $K_1 \times K_2$. These, in their context, were scalar-valued functions Λ on $C_0(K_1) \times C_0(K_2)$ such that for each $g \in C_0(K_2)$, $\Lambda(\cdot, g)$ was a bounded linear functional on $C_0(K_1)$, and for each $f \in C_0(K_1)$, $\Lambda(f, \cdot)$ was a bounded linear functional on $C_0(K_2)$. (That every bimeasure on $K_1 \times K_2$ determined an F_2-measure on the two-fold Cartesian product of the respective Borel fields (Theorem 12) was nowhere stated in their work. As far as I can determine, a general bilinear Riesz representation-type theorem first appeared in [Y1].) Like their predecessors Adams and Clarkson, Morse and Transue were guided largely by 'one-dimensional' measure theory, which was then the state of the art. But notably, they were also motivated by a firm belief that 'multi-dimensional' issues were more challenging and interesting than their 'one-dimensional' antecedents. In that respect – I daresay – they had it right.

Morse and Transue recognized at the very outset an important difference between the two-dimensional and the one-dimensional theories; that there was no Hahn-type decomposition $\Lambda = \Lambda^+ - \Lambda^-$ for bimeasures Λ [MorTr6, §10]. To negotiate around this obstacle, in search for a concept of 'absolute integrability' in two dimensions, they considered the following notion. Let Λ be a bimeasure on $K_1 \times K_2$. For p and q positive *l.sc.* (lower semicontinuous) functions on K_1 and K_2, respectively, define

$$\begin{aligned}&\Lambda^*(p, q)\\ &\quad = \sup\{|\Lambda(u, v)| : (u, v) \in C_0(K_1) \times C_0(K_2), |u| \leq p, |v| \leq q\}, \quad (8.4)\end{aligned}$$

and then, for positive functions h and k on K_1 and K_2, respectively, define

$$\begin{aligned}&\Lambda^*(h, k)\\ &\quad = \inf\,\{\Lambda^*(p, q) : h \leq p, k \leq q, p > 0 \text{ and } q > 0 \text{ are } l.sc.\}. \quad (8.5)\end{aligned}$$

Their extensive studies of Λ^*, which they dubbed 'superior integral', were based on a view of it as a two-dimensional extension of the usual Lebesgue integral. (See the survey article [Mor], which previewed [MorTr5], [MorTr6], [MorTr7].) Alas, Morse and Transue were missing basic tools. Like Adams and Clarkson, they paid scant attention to Littlewood's prior work [Lit4], which throughout their papers was mentioned only once [MorTr4, p. 106] (cf. Exercise 18). In this respect, Littlewood's 4/3-inequality (Theorem II.5) could have pointed them to p-variations, which extend the usual total variation norm in the 'one-dimensional' setting. But Morse's and Transue's most significant miss was the natural role of tensor products in the study of bimeasures;

specifically, that bilinear functionals on $C_0(K_1) \times C_0(K_2)$ were *linear* functionals on the projective tensor product $C_0(K_1) \hat{\otimes} C_0(K_2)$. (Ideas involving tensor products had been introduced by von Neumann and Schatten [Sc4] at the very same geographic location, just prior to the researches by Morse and Transue. See Chapter IV §8.) But for this miss, their attention might have been directed to Grothendieck's work in the early and mid-1950s, work that plays prominently in the bilinear setting (see Chapter IX). The Grothendieck factorization theorem, in particular, would have been useful in their investigations of the *superior integral* Λ^* (Exercise 19).

Exercises

1. Prove directly, without invoking the Nikodym boundedness principle (Theorem 3), that if $\mu \in F_n(\mathbb{N}, \ldots, \mathbb{N})$ according to Definition 1, then

$$\{\mu(\{j_1\}, \ldots, \{j_n\}) : (j_1, \ldots, j_n) \in \mathbb{N}^n\} \in F_n(\mathbb{N}, \ldots, \mathbb{N})$$

according to Definition IV.1.

2. Referring to (2.5), prove that if $\beta \notin l^1(E_1 \times \cdots \times E_n)$, then μ_β is not extendible to a scalar measure on $\sigma(C_1 \times \cdots \times C_n)$.

   More generally, assuming results cited in the remark in Chapter IV §2, verify that $F_{n-1}(\sigma(C_1 \times C_2), \ldots, C_n) \subsetneqq F_n(C_1, \ldots, C_n)$ for all $n \geq 3$.

3. This exercise refers to the definition of μ_Λ in (2.8).

   i. Prove that $\mu_\Lambda \in F_2(\mathfrak{B}, \mathfrak{B})$.

   ii. Prove directly, without appealing to the characterization of idempotent measures, that if $\Lambda = \{3^k : k \in \mathbb{N}\}$, then μ_Λ does not determine an F_1-measure on $\sigma(\mathfrak{B} \times \mathfrak{B})$.

4. This exercise refers to the definition of μ in (2.10).

   i. Prove that $\mu \in F_3(\mathfrak{B}, \mathfrak{B}, \mathbb{Z})$, but $\mu \notin F_2(\sigma(\mathfrak{B} \times \mathfrak{B}), \mathbb{Z})$ and $\mu \notin F_2(\mathfrak{B}, \sigma(\mathfrak{B} \times 2^{\mathbb{Z}}))$.

   ii. Prove that $\hat{\mu}$ (given in (2.11)) is not in $\tilde{V}_2(\mathbb{Z}, (\mathbb{Z} \times \mathbf{T}))$; that is, $\hat{\mu}$ is not the pointwise limit on $\mathbb{Z} \times \mathbb{Z} \times \mathbf{T}$ of a uniformly bounded sequence in $V_2(\mathbb{Z}, (\mathbb{Z} \times \mathbf{T}))$. This also implies the assertion in i above, that $\mu \notin F_2$. Do you see why?

5. Let $W = \{W(t) : t \in [0,1]\}$ be a Wiener process on a probability space $(\Omega, \mathfrak{A}, \mathbb{P})$.

   i. For each $A \in \mathfrak{A}$, define

   $$f_A(t) = \mathbf{E}\, \mathbf{1}_A W(t), \quad t \in [0,1].$$

   Prove that f_A is a continuous function of bounded variation on [0,1], with total variation bounded by 1.

   ii. Prove that μ defined in (2.13) determines an F_2-measure on $\mathfrak{A} \times \mathfrak{B}$.

6. Prove Corollary 6.
7. Referring to (4.7), prove that

   $$\|\mu\|_{F_n(\sigma C_1, C_2, \ldots, C_n)} \leq \|\mu\|_{F_n(C_1, C_2, \ldots, C_n)}.$$

8. Referring to the end of the proof of Theorem 8, let

   $$S = \{(A_2, \ldots, A_n) \in \sigma C_2 \times \cdots \times \sigma C_n : \mu(\cdot, A_2, \ldots, A_n) \in F_1(C_1)\}.$$

   Prove that $S = \sigma C_2 \times \cdots \times \sigma C_n$.

9. This exercise refers to the end of the proof of Theorem 8.

   i. Prove the uniqueness of the extension. That is, show that if $\mu_1 \in F_n(\sigma C_1, \ldots, \sigma C_n)$, $\mu_2 \in F_n(\sigma C_1, \ldots, \sigma C_n)$, and $\mu_1 = \mu_2$ on $C_1 \times \cdots \times C_n$, then $\mu_1 = \mu_2$ on $\sigma C_1 \times \cdots \times \sigma C_n$.

   ii. Prove (4.1).

10. Verify Proposition 11.
11. Let $V_n(C_1, \ldots, C_n)$ be the projective tensor algebra $L^\infty(C_1) \hat{\otimes} \cdots \hat{\otimes}\, L^\infty(C_n)$, where $(X_1, C_1), \ldots, (X_n, C_n)$ are measurable spaces, and let $S(C)$ denote the space of scalar-valued C-measurable simple functions on (X, C) equipped with the supremum norm.

    i. Prove that

    $$V_n(C_1, \ldots, C_n) = S(C_1) \hat{\otimes} \cdots \hat{\otimes} S(C_n).$$

    ii. Prove that if $\Sigma_j \varphi_{1j} \otimes \cdots \otimes \varphi_{nj} \in S(C_1) \otimes \cdots \otimes S(C_n)$ and $\Sigma_j \varphi_{1j} \otimes \cdots \otimes \varphi_{nj} = 0$, then for all $\mu \in F_n(C_1, \ldots, C_n)$,

    $$\sum_j \int \varphi_{1j} \otimes \cdots \otimes \varphi_{nj}\, d\mu = 0.$$

12. Verify that the integrals $\int f_1 \otimes \cdots \otimes f_n d\mu$ defined in §5 and §6 are equal.

13. Let $(X_1, C_1), \ldots, (X_n, C_n)$ denote measurable spaces. Let $\mu \in F_n(C_1, \ldots, C_n)$ and $\phi \in \mathrm{L}^\infty(C_1) \otimes \cdots \otimes \mathrm{L}^\infty(C_n)$. Define a set function $\int \phi \; \mathrm{d}\mu$ by

$$\left(\int \phi \; \mathrm{d}\mu\right)(E_1, \ldots, E_n) = \int \mathbf{1}_{E_1 \times \cdots \times E_n} \phi \; \mathrm{d}\mu,$$
$$(E_1, \ldots, E_n) \in C_1 \times \cdots \times C_n.$$

   i. Verify that $\int \phi \; \mathrm{d}\mu$ determines an F_n-measure on $C_1 \times \cdots \times C_n$, and

$$\left\| \int \phi \; \mathrm{d}\mu \right\|_{F_n} \leq \|\phi\|_{V_n} \|\mu\|_{F_n}.$$

   ii.* Denote by $LF_n(\mu)$ the closure of

$$\left\{ \int \phi \; \mathrm{d}\mu : \phi \in \mathrm{L}^\infty(C_1) \otimes \cdots \otimes \mathrm{L}^\infty(C_n) \right\}$$

   in $F_n(C_1, \ldots, C_n)$. Can you, somehow, associate with every $\lambda \in LF_n(\mu)$ a function f defined on $X_1 \times \cdots \times X_n$, such that for $(E_1, \ldots, E_n) \in C_1 \times \cdots \times C_n$,

$$\lambda(E_1, \ldots, E_n) = \int \mathbf{1}_{E_1 \times \cdots \times E_n} f \; \mathrm{d}\mu$$

   makes sense?

14. Supply the details in the general inductive step in the proof of Theorem 12.

15. In this exercise, by applying results of this chapter, you will construct the double Stieltjes integrals obtained by Fréchet in his 1915 paper [Fr].

   Let k be a real-valued function on $[0,1]^2$ such that $\|k\| < \infty$ (see (8.1)). Let $u \in \mathrm{C}([0,1])$ and $v \in \mathrm{C}([0,1])$. First prove that

$$\int_0^1 u(t) \; \mathrm{d}_t k(s,t), \quad s \in [0,1],$$

   is a function of bounded variation on $[0,1]$, where d_t denotes Stieltjes integration in t. Then show

$$\int_0^1 u(t) \; \mathrm{d}_t \int_0^1 v(s) \; \mathrm{d}_s k(s,t) = \int_0^1 v(s) \; \mathrm{d}_s \int_0^1 v(t) \; \mathrm{d}_t k(s,t).$$

16. In this exercise you will obtain the extension to two dimensions of the usual formula relating a Riemann–Stieltjes integral to a Lebesgue–Stieltjes integral.

    Suppose k is a real-valued function on $[0,1]^2$ which is left-continuous in each variable separately, and satisfies $\|k\| < \infty$. Prove that there exists $\mu \in F_2(\mathfrak{B}, \mathfrak{B})$, where $\mathfrak{B}$ denotes the usual Borel field in $[0,1]^2$, such that

$$k(s,t) = \mu((0,s],(0,t]),$$

    and

$$\int u \otimes v \, \mathrm{d}\mu = \int_0^1 u(s) \, \mathrm{d}_s \int_0^1 v(t) \, \mathrm{d}_t k(s,t)$$
$$= \int_0^1 v(t) \, \mathrm{d}_t \int_0^1 v(s) \, \mathrm{d}_s k(s,t)$$

    for all $(u,v) \in \mathrm{C}([0,1]) \times \mathrm{C}([0,1])$, where the integral on the left side is defined in §5, and the integrals on the right side are the iterated Stieltjes integrals obtained in Exercise 15.

17. In the first part of this exercise you will deduce Fréchet's representation of bounded bilinear functionals on C([0,1]), and in the second part you will deduce a result obtained in [MorTr1]; see [Mor, p. 346].

    i. Let Λ be a real-valued bounded bilinear functional on C([0,1]). Prove there exists a real-valued function k on $[0,1]^2$ such that $\|k\| < \infty$, and

$$\Lambda(u,v) = \int_0^1 u(s) \, \mathrm{d}_s \int_0^1 v(t) \, \mathrm{d}_t k(s,t)$$

$$= \int_0^1 v(t) \, \mathrm{d}_t \int_0^1 v(s) \, \mathrm{d}_s k(s,t)$$

    for $(u,v) \in \mathrm{C}([0,1]) \times \mathrm{C}([0,1])$.

    ii. Prove that k (in i) can be chosen so that it is left-continuous in each variable separately. Then verify

$$\|k\| = \|\Lambda\| := \sup\{|\Lambda(u,v)| : u \text{ and } v \text{ in the unit ball of } \mathrm{C}([0,1])\}.$$

18. By applying results in Chapter IV, you will derive here a result of Morse and Transue concerning the Fréchet variation; see [Mor, Theorem 2.1]. To facilitate comparison with Morse's and Transue's work, I will adopt here their notation.

    Let k be a real-valued function defined on $[0,1] \times [0,1]$, and denote (temporarily) its Fréchet variation $\|k\|$ by $P(k)$. For $s \in [0,1]$, let $X_s = (0, s] \times [0,1]$. Denote the Fréchet variation of the restriction of k to X_s by $P(k, X_s)$. Prove that if $P(k) < \infty$, then $\lim_{s\to 0^+} P(k, X_s) = 0$.

Morse and Transue billed the theorem above as one of the deep results in the bilinear theory [Mor, p. 351]. Although this specific theorem was not explicitly stated in Littlewood's 1930 paper [Lit4], its proof was implicit there; see [Lit4, pp. 167–8].

19. Let K_1 and K_2 be locally compact Hausdorff spaces with respective Borel fields $\mathfrak{B}_1$ and $\mathfrak{B}_2$. Let Λ^* be defined by (8.4) and (8.5).
    i. Does Λ^* determine an F_2-measure on $\mathfrak{B}_1 \times \mathfrak{B}_2$?
    ii. Show that there exist probability measures ν on $(K_1, \mathfrak{B}_1)$ and λ on $(K_2, \mathfrak{B}_2)$ such that $\Lambda^* \ll \nu \times \lambda$, where $\nu \times \lambda$ is the product measure, and $\Lambda^* \ll \nu \times \lambda$ means that if $\nu \times \lambda(A \times B) = 0$ for $A \in \mathfrak{B}_1$ and $B \in \mathfrak{B}_2$, then $\Lambda^*(\mathbf{1}_A, \mathbf{1}_B) = 0$.

Hints for Exercises in Chapter VI

1. Assume that the assertion is false, and apply Lemma 6.
3. ii. Use the Khintchin L^1–L^2 inequality; specifically, that the transform of every measure on $(\mathbf{T} \times \mathbf{T}, \sigma(\mathfrak{B} \times \mathfrak{B}))$ with spectrum in $\{(3^k, 3^k) : k \in \mathbb{N}\}$ is square-summable.
4. i. If $\mu \in F_2$, then μ can be 'factored' in the sense of Chapter V. Use an argument similar to the one used in Exercise V.5.
   ii. Cf. Remark iv in Chapter IV §5.
5. i. For hints, browse through Chapters X and XI.
   ii. This can be shown by applying Theorem 8, but it also can be proved directly.

6. The case $n = 1$ is classical. It is clear that $\|\cdot\|_{F_n}$ is a norm for all integers $n > 0$. To verify that $(F_n, \|\cdot\|_{F_n})$ is a Banach space for $n > 1$, first note that if $\mu \in F_n$, then for all $(A_1, \ldots, A_{n-1}) \in C_1 \times \cdots \times C_{n-1}$,

$$\|\mu(A_1, \ldots, A_{n-1}, \cdot)\|_{F_1(C_n)} \leq \|\mu\|_{F_n},$$

which follows from the definition of the Fréchet variation. Observe the same regarding $F_1(C_j)$ for $j = 1, \ldots, n-1$. Now suppose (μ_k) is Cauchy in $(F_n, \|\cdot\|_{F_n})$. Then, there exists a scalar-valued set function μ on $C_1 \times \cdots \times C_n$ such that for $A_1 \in C_1, \ldots, A_n \in C_n$

$$\lim_{k\to\infty} \mu_k(A_1, \ldots, A_n) = \mu(A_1, \ldots, A_n)$$

By applying this observation and the case $n = 1$, conclude that $\mu \in F_n$.

VII

An Introduction to Harmonic Analysis

1 Mise en Scène: Mainly a Historical Perspective

A recurring construct in previous chapters was based on this simple blueprint:

given sets $E_1, \ldots, E_n$ and $x_1 \in E_1, \ldots, x_n \in E_n$, form products $x_1 \otimes \cdots \otimes x_n$, and consider the class $E_1 \otimes \cdots \otimes E_n$ comprising all linear combinations of such products.

At the very outset, if nothing is known or assumed about the 'building blocks' $x_1, \ldots, x_n$, then their product $x_1 \otimes \cdots \otimes x_n$ is merely a formal object, and not much more can be said. If something *is* known about $E_1, \ldots, E_n$, then meaning could be ascribed to $x_1 \otimes \cdots \otimes x_n$, and analysis of $E_1 \otimes \cdots \otimes E_n$ would proceed accordingly. In our specific context, we considered Rademacher functions and their products. We considered the set of independent functions $R = \{r_k\}$ on $\Omega = \{-1, 1\}^{\mathbb{N}}$, and viewed the elements in the n-fold $R \otimes \cdots \otimes R$ as functions on Ω^n. An underlying theme has been that Rademacher functions are basic objects from which all else is constructed, a notion that can be formulated effectively in a framework of harmonic analysis. And that is our purpose in this chapter: to learn and analyze this framework, as it is built from the ground up.

Loosely put, *harmonic analysis* is about representing general phenomena in terms of familiar phenomena. The subject's beginnings – in the mid-eighteenth century, about ninety years after the invention of the calculus – were rooted in the notion that arbitrary functions could be represented by series of sines and cosines. This idea, which had appeared first in D. Bernoulli's solution to the vibrating string problem [Be], encountered some initial resistance. The contested points were largely conceptual: what is an *arbitrary function*, and what is its *representation*? Historical accounts of speculations about these and related issues can be found in [Dug]. (See also [GraR, pp. 243–53], and the introduction in [Cars].)

The official debut of *Fourier analysis* is usually marked by Joseph Fourier's extensive use of trigonometric series in his researches of heat, which had appeared first in an 1807 Memoir, then an 1811 Prize Essay, and finally in his 1822 classic *Théorie Analytique de la Chaleur*. While Fourier's priority in the use of trigonometric series was indeed challenged by some of his contemporaries ([Her, pp. 318–21], [GraR, pp. 243–53]), it was Fourier's name that eventually became associated with the subject. For so it was, largely through his insight and tenacity, that an elusive eighteenth-century idea became the foundation of a thriving enterprise. Detailed accounts of Fourier's work and unusual life story, cast in the tumultuous French political scene in and out of academe, can be found in [GraR] and [Her].

The vast subject emanating from the study of trigonometric series falls under the headings *Fourier analysis* and *harmonic analysis*: the first, a homage to Fourier, refers to the classical theory as well as areas close to it; the second, honoring a plucked string, refers to a much wider field including Fourier analysis and related mathematics. A broad survey that traces ideas of harmonic analysis in various contexts, some fairly far flung, can be found in [Mac].

Framed in a neo-classical setting, Fourier analysis begins with the *Fourier–Stieltjes transform* $\hat{\mu}$ of a Borel measure μ on the *circle group* $\mathbf{T} := [0, 2\pi)$,

$$\hat{\mu}(n) = \int_{\mathbf{T}} \mathrm{e}^{-\mathrm{i}tn} \mu(\mathrm{d}t), \quad n \in \mathbb{Z}. \tag{1.1}$$

A central problem is to characterize these measures in terms of their transforms. Specifically, how is $\mu \in \mathrm{M}(\mathbf{T})$ reconstructed from its Fourier–Stieltjes series

$$S[\mu] = \sum_{n \in \mathbb{Z}} \hat{\mu}(n) \mathrm{e}^{\mathrm{i}nt}? \tag{1.2}$$

More generally, how are properties of μ reflected by properties of $\hat{\mu}$? The normalized Lebesgue measure on $\mathbf{T}$ plays here a fundamental role, primarily because the exponentials $\mathrm{e}^{\mathrm{i}nt}$, $n \in \mathbb{Z}$, form a complete orthonormal set with respect to it. This (in a nutshell) is the foundation of the classical theory (Exercise 1). Detailed accounts of Fourier analysis from varying viewpoints in (neo-)classical settings can be found in [Kat],

[Ko], and [Zy2]; a concise account of select highlights can be found in [Hel]. All four were written by grandmasters.

Ideas underlying Fourier analysis extend well beyond trigonometric series. During the 1930s, a general locally compact group was proposed as a generic framework for harmonic analysis at large [PaWi2], [We]. In the commutative section of this framework, a compact Hausdorff space G equipped with a continuous Abelian group operation stands for the circle group $\mathbf{T}$; continuous characters of G, i.e., continuous functions

$$\gamma : G \mapsto \{z \in \mathbb{C} : |z| = 1\}, \tag{1.3}$$

such that

$$\gamma(x \cdot y) = \gamma(x)\gamma(y), \quad (x, y) \in G \times G,$$

correspond to exponentials, and *Haar measures* on G (positive translation-invariant Borel measures) stand for Lebesgue measure. The main problems here, like those in the classical setting, focus on representing objects (e.g., measures) defined on G in terms of the characters of G. An accessible account of harmonic analysis on general Abelian groups, also told by a grandmaster, can be found in [Ru3]. Further studies, detailing some of the researches on the subject's frontiers, can be found in [GrMc].

In this chapter we outline rudiments of harmonic analysis in $G = \Omega := \{-1, +1\}^{\mathbb{N}}$. This primal setting, equipped with minimal and transparent structures, is considerably simpler than the circle group $\mathbf{T}$. Yet, it illustrates effectively the workings of general principles. Starting with the Rademacher system $R = \{r_j : j \in \mathbb{N}\}$, a set of basic characters on Ω, we learn about $\hat{\Omega}$ from the ground up. We view the full character group $\hat{\Omega}$ as an increasing union of k-fold products of R,

$$\hat{\Omega} = \bigcup_{k=1}^{\infty} \{r_{j_1} \cdots r_{j_k} : (j_1, \ldots, j_k) \in \mathbb{N}^k\} \cup \{r_0\}, \tag{1.4}$$

and analyze the evolving complexity of its constituent systems

$$\{r_{j_1} \ldots r_{j_k} : (j_1, \ldots, j_k) \in \mathbb{N}^k\},$$

as it depends on k. The increasing complexity of these systems is the main theme of this chapter.

Focusing on the group Ω, we begin from first principles. To underscore the generality of what is done here, further along we shall allude also to general compact Abelian groups, and recast results in that setting. We expect that readers eventually become familiar with the material found in (at least) the first two chapters in [Ru3].

2 The Setup

A Compact Abelian Group and its Dual

To start, we equip $\Omega := \{-1,+1\}^{\mathbb{N}}$ with the usual Tychonoff product topology, and obtain a compact Hausdorff space, wherein the resulting Borel field, denoted here by $\mathfrak{A}$, is the σ-algebra generated by the Rademacher functions. For $\omega = (\omega_n) \in \Omega$ and $\eta = (\eta_n) \in \Omega$, we define the product $\omega \cdot \eta \in \Omega$ by

$$(\omega \cdot \eta)_n = \omega_n \, \eta_n, \quad n \in \mathbb{N}, \tag{2.1}$$

and obtain, from the definition of the product topology, that $(\omega, \eta) \mapsto \omega \cdot \eta$ is a continuous function from $\Omega \times \Omega$ onto Ω. Endowed with these structures, Ω becomes a compact Abelian group.

We denote by $\hat{\Omega}$ the set of continuous characters of Ω. That is, $\hat{\Omega}$ comprises all continuous non-zero scalar-valued functions χ on Ω such that

$$\chi(\omega \cdot \eta) = \chi(\omega) \, \chi(\eta), \quad (\omega, \eta) \in \Omega \times \Omega. \tag{2.2}$$

Let $R = \{r_n : n \in \mathbb{N}\}$ be the usual system of Rademacher functions on Ω, and let r_0 denote the function on Ω that is identically 1. Define

$$W = \{r_{j_1} \cdots r_{j_k} : j_1 > \cdots > j_k \geq 0, \ k \in \mathbb{N}\}. \tag{2.3}$$

Then, W is a countable Abelian group under pointwise multiplication of functions, and r_0 is its group-identity.

Proposition 1 $\hat{\Omega} = W$.

Proof: It is clear that $W \subset \hat{\Omega}$.
We prove the reverse inclusion. Let e_0 be the group-identity in Ω, i.e., $e_0(n) = 1$ for all $n \in \mathbb{N}$. If χ is a continuous character of Ω, then,

$$\chi(\omega)^2 = \chi(\omega^2) = \chi(e_0) = 1, \quad \omega \in \Omega. \tag{2.4}$$

Therefore, χ takes values in $\{-1,+1\}$. For $n \geq 1$, define $e_n \in \Omega$ by

$$e_n(m) = \begin{cases} -1 & \text{if } n = m \\ 1 & \text{if } n \neq m. \end{cases} \tag{2.5}$$

Then, $\lim_{n\to\infty} e_n = e_0$ (in the product topology). Because χ is continuous, there exists $N > 0$ such that $\chi(e_n) = 1$ for all $n \geq N$. Let $F = \{n : \chi(e_n) = -1\}$. Observe that $gp\{e_n\}$ (= group generated by the e_n) is dense in Ω, and that

$$\chi(\omega) = \prod_{n\in F} r_n(\omega), \quad \omega \in gp\{e_n\}. \tag{2.6}$$

Therefore,

$$\chi = \prod_{n\in F} r_n. \tag{2.7}$$

□

Remark (a leitmotif). The Rademacher system R is a basic independent set from which $\hat{\Omega}$ is synthesized. When referring to *independence*, we need to make precise what we mean by it. In Chapter II §1 we noted that R is *functionally* independent, a notion that will be revisited later in this chapter. We noted also that R is *statistically* independent with respect to a probability measure that will soon be recalled. And now, with group multiplication in W, we notice that the Rademacher system is also *algebraically* independent. This means: if

$$r_{j_1}\cdots r_{j_k} = r_0, \quad 0 \leq j_1 < \cdots < j_k, \tag{2.8}$$

then $k = 1$ and $j_1 = 0$. That is, every $w \in W$ can be represented uniquely as a product of distinct elements in R.

Convolution

Next we observe that the multiplicative structure in Ω naturally gives rise to a multiplicative structure in the space of Borel measures $\mathrm{M}(\Omega)$.

If f is a measurable function on Ω and $\omega \in \Omega$, then let f_ω denote the function defined by $f_\omega(\eta) = f(\eta\cdot\omega)$, $\eta \in \Omega$. Fix Borel measures μ and ν on Ω. If $f \in \mathrm{C}(\Omega)$, then $\omega \mapsto \int_\Omega f_\omega(\eta)\ \mu(\mathrm{d}\eta)$ defines a continuous function on Ω, and

$$f \mapsto \int_\Omega \left(\int_\Omega f_\omega(\eta)\ \mu(\mathrm{d}\eta)\right)\ \nu(\mathrm{d}\omega), \quad f \in \mathrm{C}(\Omega), \tag{2.9}$$

defines a continuous linear functional on $\mathrm{C}(\Omega)$, whose norm is bounded by $\|\nu\|_M\|\mu\|_M$. By the Riesz representation theorem, there exists a unique Borel measure $\nu \star \mu$ on Ω – dubbed *convolution* of ν and μ – such that

$$\int_\Omega f(\omega)\nu \star \mu(\mathrm{d}\omega) = \int_\Omega \left(\int_\Omega f_\omega(\eta)\ \mu(\mathrm{d}\eta) \right) \nu(\mathrm{d}\omega). \tag{2.10}$$

Then, $\nu \star \mu = \mu \star \nu$ and $\|\nu \star \mu\|_M \leq \|\nu\|_M\ \|\mu\|_M$. Thus, $\mathrm{M}(\Omega)$ equipped with convolution and the total variation norm becomes a commutative Banach algebra (Exercise 2).

Proposition 2 (Exercise 3). *If* $\mu \in \mathrm{M}(\Omega)$, $\nu \in \mathrm{M}(\Omega)$, *and* $A \in \mathfrak{A}$, *then*

$$\nu \star \mu(A) = \int_\Omega \mu(A \cdot \omega)\ \nu(\mathrm{d}\omega) = \int_\Omega \nu(A \cdot \omega)\ \mu(\mathrm{d}\omega). \tag{2.11}$$

Transforms

For $\mu \in \mathrm{M}(\Omega)$, define its W-transform $\hat{\mu}$ by

$$\hat{\mu}(w) = \int_\Omega w(\eta)\ \mu(\mathrm{d}\eta), \quad w \in \hat{\Omega} \ \ (\text{cf. } (1.1)). \tag{2.12}$$

Proposition 3 (Exercise 4). *For all* μ *and* ν *in* $\mathrm{M}(\Omega)$,

$$(\mu \star \nu)\hat{}\,(w) = \hat{\mu}(w)\ \hat{\nu}(w), \quad w \in \hat{\Omega}.$$

The W-series of $\mu \in \mathrm{M}(\Omega)$ is

$$S[\mu] = \sum_{w \in \hat{\Omega}} \hat{\mu}(w)\ w \quad (\text{cf. } (1.2)), \tag{2.13}$$

which, at this juncture, is merely a formal object. A question naturally arises: in what sense does $S[\mu]$ represent μ?

Haar Measure

The question concerning representations of $\mu \in \mathrm{M}(\Omega)$ is of special interest when μ is absolutely continuous with respect to a *Haar measure*, a positive translation-invariant measure on $(\Omega, \mathfrak{A})$. Existence of such measures in general, by no means obvious, is guaranteed by a basic theorem due to A. Haar [Ha2]. In our specific setting, the normalized Haar measure on Ω is the probability measure $\mathbb{P}$ defined in Chapter II §1. That is,

$$\mathbb{P}(A \cdot \omega) = \mathbb{P}(A), \quad A \in \mathfrak{A},\ \omega \in \Omega, \tag{2.14}$$

and $\mathbb{P}$ is the only probability measure with this property (Exercise 5). To underscore connections between probability theory and harmonic analysis on Ω, we will denote integration with respect to $\mathbb{P}$ – here and throughout the chapter – by $\mathbf{E}$ (*expectation*).

By a standard application of the Radon–Nikodym theorem, absolutely continuous measures with respect to $\mathbb{P}$ can be naturally identified with elements of $\mathrm{L}^1(\Omega, \mathbb{P})$. The W-transform of $f \in \mathrm{L}^1(\Omega, \mathbb{P})$ is the transform of the measure $f\mathrm{d}\mathbb{P}$,

$$\hat{f}(w) = \mathbf{E}\ wf, \tag{2.15}$$

and the convolution of f and g in $\mathrm{L}^1(\Omega, \mathbb{P})$ is the Radon–Nikodym derivative with respect to $\mathbb{P}$ of the convolution $(f\mathrm{d}\mathbb{P}) \star (g\mathrm{d}\mathbb{P})$. We summarize.

Proposition 4 (Exercise 6). *If $f \in \mathrm{L}^1(\Omega, \mathbb{P})$ and $\nu \in \mathrm{M}(\Omega)$, then $(f\mathrm{d}\mathbb{P}) \star \nu \ll \mathbb{P}$, and*

$$f \star \nu := \frac{\mathrm{d}(f\mathrm{d}\mathbb{P} \star \nu)}{\mathrm{d}\mathbb{P}} = \int_\Omega f_\omega\ \nu(\mathrm{d}\omega), \tag{2.16}$$

where the integral on the right side is the element in $\mathrm{L}^1(\Omega, \mathbb{P})$ represented almost everywhere ($\mathbb{P}$) by the function $\int_\Omega f_\omega(\eta)\ \nu(\mathrm{d}\omega)$, $\eta \in \Omega$.

In particular, if $f \in \mathrm{L}^1(\Omega, \mathbb{P})$ and $g \in \mathrm{L}^1(\Omega, \mathbb{P})$, then for almost all $\eta \in (\Omega, \mathbb{P})$,

$$f \star g(\eta) := \frac{\mathrm{d}(f\mathrm{d}\mathbb{P} \star g\mathrm{d}\mathbb{P})}{\mathrm{d}\mathbb{P}}(\eta) = \mathbf{E}_\omega f_\eta g, \tag{2.17}$$

and

$$(f \star g)\hat{}\ (w) = \mathbf{E}_\eta\{\mathbf{E}_\omega f_\eta(\omega) g(\omega)\ w(\eta)\} = \hat{f}(w)\ \hat{g}(w), \quad w \in \hat{\Omega}. \tag{2.18}$$

($\mathbf{E}_\eta$ and $\mathbf{E}_\omega$ denote integrations with respect to $\mathbb{P}(\mathrm{d}\eta)$ and $\mathbb{P}(\mathrm{d}\omega)$, respectively.)

The importance of the Haar measure stems from the orthogonality relations: if w and w' are characters on Ω, then

$$\mathbf{E}\ w\ w' = \begin{cases} 1 & w = w' \\ 0 & w \neq w'. \end{cases} \tag{2.19}$$

(To verify (2.19), write w and w' as products of Rademacher functions, and then apply (II.1.7).) This implies that if f is a W-polynomial, i.e., $f = \Sigma_{w \in F}\ a_w w$ where F is a finite subset of $\hat{\Omega}$, then

$$\hat{f}(w) = \begin{cases} a_w & w \in F \\ 0 & w \notin F. \end{cases} \tag{2.20}$$

In particular, if f and g are W-polynomials, then

$$\mathbf{E}fg = \sum_{w} \hat{f}(w)\, \hat{g}(w), \tag{2.21}$$

and therefore,

$$(\mathbf{E}|f|^2)^{\frac{1}{2}} := \|f\|_{L^2} = \|\hat{f}\|_2 = \left(\sum_{w\in F} |a_w|^2\right)^{\frac{1}{2}}. \tag{2.22}$$

The formula in (2.21) is known as *Parseval's relation.* In the next section we will verify that (2.22) holds for all $f \in L^2(\Omega, \mathbb{P})$; specifically, that $f \mapsto \hat{f}$ determines a unitary equivalence between $L^2(\Omega, \mathbb{P})$ and $l^2(\hat{\Omega})$.

3 Elementary Representation Theory

In this section we make precise how μ is represented by $S[\mu]$. The results and methods used to derive them are classical, and indeed typical of results and methods in any setting, not only Ω.

Definition 5 *A summability kernel* $(k_n : n \in \mathbb{N})$ on Ω is a sequence of scalar-valued continuous functions on Ω with these properties:

$$\mathbf{E}\, k_n = \hat{k}_n(r_0) = 1, \quad n \in \mathbb{N}; \tag{3.1}$$

$$\sup_{n\in\mathbb{N}} \|k_n\|_{L^1} < \infty; \tag{3.2}$$

$$\lim_{n\to\infty} \mathbf{E}\, |k_n \mathbf{1}_{V^c}| = 0 \quad \text{for every neighbourhood } V \text{ of } e_0. \tag{3.3}$$

(V^c denotes the complement of V.)

Proposition 6 (Exercise 7). *If* $(k_n : n \in \mathbb{N})$ *is a summability kernel on* Ω*, and* B *is any one of the spaces* $C(\Omega)$, $L^p(\Omega, \mathbb{P})$ *for* $1 \le p < \infty$, *then for all* $f \in B$,

$$\lim_{n\to\infty} (k_n \star f) = f, \quad \textit{convergence in the } B\textit{-norm.} \tag{3.4}$$

The key to the proposition is

Lemma 7 (Exercise 7). *If* $B = C(\Omega)$, *or* $B = L^p(\Omega, \mathbb{P})$ *for* $1 \le p < \infty$, *then for all* $f \in B$,

$$\|f\|_B = \|f_\omega\|_B, \quad \omega \in \Omega, \tag{3.5}$$

and

$$\lim_{\omega \to e_0} f_\omega = f, \quad \textit{convergence in the } B\textit{-norm.} \tag{3.6}$$

In order to obtain $\mu \in \mathrm{M}(\Omega)$ as a limit of partial sums of $S[\mu]$, we verify that *Riesz products* form a summability kernel:

Proposition 8 *Let*

$$R_n = \prod_{j=1}^{n} (1 + r_j), \quad n \in \mathbb{N} \text{ (cf. (III.6.1))}. \tag{3.7}$$

Then, $(R_n : n \in \mathbb{N})$ *is a summability kernel on* Ω.

Proof: By the statistical independence of $\{r_n\}$,

$$\hat{R}_n(r_0) = \mathbf{E} \prod_{j=1}^{n} (1 + r_j) = \prod_{j=1}^{n} \mathbf{E}\ (1 + r_j) = 1, \tag{3.8}$$

which verifies (3.1). To verify (3.2), note that $R_n \geq 0$, and therefore $\|R_n\|_{\mathrm{L}^1} = \hat{R}_n(r_0) = 1$ for all $n \in \mathbb{N}$.

To verify (3.3), let V be a neighbourhood of e_0, and, without loss of generality, assume $V = \{(1, \ldots, 1, \omega_{k+1}, \ldots) : (\omega_{k+1}, \ldots) \in \Omega\}$. Then, for all $\omega \notin V$ there exist $j \in [k]$ such that $1 + r_j(\omega) = 0$. Therefore, $\mathbf{E}\ R_n \mathbf{1}_{V^c} = 0$ for all $n \geq k$. ☐

For each $n \in \mathbb{N}$, define

$$W(n) = \{r_{j_1} \cdots r_{j_n} : 0 \leq j_1 \leq \cdots \leq j_n \leq n\}, \tag{3.9}$$

which is the support of $\hat{R}_n$ (cf. (3.7)). Note that $W(n) \subset W(n+1)$, and $\cup_n W(n) = W$. For $\mu \in \mathrm{M}(\Omega)$, we consider the partial sums (cf. (1.2))

$$\sum_{w \in W(n)} \hat{\mu}(w)\ w, \quad n \in \mathbb{N}. \tag{3.10}$$

Corollary 9

i. *If* $B = \mathrm{C}(\Omega)$ *or* $B = \mathrm{L}^p(\Omega, \mathbb{P})$ *for* $1 \leq p < \infty$, *then for all* $f \in B$,

$$\sum_{w \in W(n)} \hat{f}(w)\ w \xrightarrow[n \to \infty]{} f, \quad \textit{convergence in the } B\textit{-norm.} \tag{3.11}$$

ii. *If* $f \in \mathrm{L}^\infty(\Omega, \mathbb{P})$, *then*

$$\sum_{w \in W(n)} \hat{f}(w)\, w \xrightarrow[n\to\infty]{} f, \quad \textit{weak}^* \textit{ convergence in } \mathrm{L}^\infty(\Omega, \mathbb{P}). \tag{3.12}$$

iii. *If* $\mu \in \mathrm{M}(\Omega)$, *then*

$$\sum_{w \in W(n)} \hat{\mu}(w)\, w \xrightarrow[n\to\infty]{} \mu, \quad \textit{weak}^* \textit{ convergence in } \mathrm{M}(\Omega). \tag{3.13}$$

Proof: We expand the Riesz product R_n,

$$R_n = 1 + \sum_{1\le j\le n} r_j + \cdots + \sum_{1\le j_1<\cdots<j_k\le n} r_{j_1}\cdots r_{j_k} + \cdots + r_1\cdots r_n, \tag{3.14}$$

and note that $\hat{R}_n = \mathbf{1}_{W(n)}$. By Proposition 3, for $\mu \in \mathrm{M}(\Omega)$,

$$R_n \star \mu = \sum_{w \in W(n)} \hat{\mu}(w)\, w. \tag{3.15}$$

Therefore, Part i follows from Propositions 6 and 8.

To prove Part ii, observe that if $f \in \mathrm{L}^\infty(\Omega, \mathbb{P})$ and $g \in \mathrm{L}^1(\Omega, \mathbb{P})$, then

$$\mathbf{E}\; g(R_n \star f) = \mathbf{E}\;(g \star R_n) f. \tag{3.16}$$

Therefore, by Part i,

$$\mathbf{E}\; g(R_n \star f) \xrightarrow[n\to\infty]{} \mathbf{E}\; gf. \tag{3.17}$$

The proof of Part iii is similar: if $\mu \in \mathrm{M}(\Omega)$ and $f \in \mathrm{C}(\Omega)$, then

$$\int_\Omega f(\omega)\,(R_n \star \mu)(\mathrm{d}\omega) = \int_\Omega f \star R_n(\omega)\, \mu(\mathrm{d}\omega), \tag{3.18}$$

and therefore (again by Part i),

$$\int_\Omega f(\omega)(R_n \star \mu)(\mathrm{d}\omega) \xrightarrow[n\to\infty]{} \int_\Omega f(\omega)\, \mu(\mathrm{d}\omega). \tag{3.19}$$

□

Corollary 10 *If* $\mu \in \mathrm{M}(\Omega)$, *and* $\hat{\mu}(w) = 0$ *for all* $w \in W$, *then* $\mu = 0$. *In particular,* W *is a complete orthonormal system in* $\mathrm{L}^2(\Omega, \mathbb{P})$, *and*

$$\|f\|_{\mathrm{L}^2} = \left(\sum_{w \in W} |\hat{f}(w)|^2 \right)^{\frac{1}{2}}, \quad f \in \mathrm{L}^2(\Omega, \mathbb{P}). \tag{3.20}$$

The formula in (3.20), widely known as *Plancherel's Theorem*, is at the very foundation of classical harmonic analysis (see Exercise 1).

For $\mu \in \mathrm{M}(\Omega)$, we define the *spectrum of* μ to be the support of its W-transform:

$$\text{spect}\ \mu := \{w \in W : \hat{\mu}(w) \neq 0\}. \tag{3.21}$$

Measures with finite spectrum are naturally identified as continuous functions on Ω, and will be called W-polynomials. (Polynomials with spectrum in $E \subset W$ will be called E-polynomials.) Corollary 9 implies that W-polynomials are norm-dense in $\mathrm{C}(\Omega)$ and $\mathrm{L}^p(\Omega, \mathbb{P})$ for $p \in [1, \infty)$, and weak*-dense in $\mathrm{L}^\infty(\Omega, \mathbb{P})$ and $\mathbf{M}(\Omega)$.

Corollary 11 (Exercises 8, 26). *If* $f \in \mathrm{L}^1(\Omega, \mathbb{P})$ *then* $\hat{f} \in \mathrm{c}_0(W)$.

Remarks:

i **(a word of caution).** The assertion in Proposition 6 is false in the instances $B = \mathrm{L}^\infty(\Omega, \mathbb{P})$ and $B = \mathrm{M}(\Omega)$. For, if $f \in \mathrm{L}^\infty(\Omega, \mathbb{P})$ and $\sum_{w \in W(n)} \hat{f}(w) \to f$ in the L^∞-norm, then f represents a continuous function on Ω. Similarly, if $\mu \in \mathrm{M}(\Omega)$ and $\sum_{w \in W(n)} \hat{\mu}(w)w \to \mu$ in the $\mathrm{M}(\Omega)$-norm, then μ is absolutely continuous with respect to $\mathbb{P}$. Indeed, the gist of the proposition below is that Lemma 7, the key to Proposition 6, fails in these two cases:

Proposition 12 (Exercise 9).

(1) $f \in \mathrm{L}^\infty(\Omega, \mathbb{P})$ *is in* $\mathrm{C}(\Omega)$ *if and only if* $\lim_{\omega \to e_0} \|f_\omega - f\|_{\mathrm{L}^\infty} = 0$.
(2) $\mu \in \mathrm{M}(\Omega)$ *is in* $\mathrm{L}^1(\Omega, \mathbb{P})$ *if and only if* $\lim_{\omega \to e_0} \|\mu_\omega - \mu\|_{\mathrm{M}} = 0$ (μ_ω *is defined by* $\mu_\omega(A) = \mu(\omega{\cdot}A)$).

ii **(a preview).** Corollary 11 (usually referred to as the Riemann–Lebesgue lemma) states that $f \mapsto \hat{f}$ is a continuous injection from $\mathrm{L}^1(\Omega, \mathbb{P})$ into $\mathrm{c}_0(W)$. This injection is not surjective. If it were, then $\mathrm{L}^1(\Omega, \mathbb{P})$ would be isomorphic to $\mathrm{c}_0(W)$, and $\mathrm{L}^\infty(\Omega, \mathbb{P})$ would be isomorphic to $l^1(W)$, which is impossible. (l^1 is separable, whereas L^∞ is not!) Of particular interest are spectral sets $E \subset W$ with the property that for all $\varphi \in \mathrm{c}_0(E)$ there exist $f \in \mathrm{L}^1(\Omega, \mathbb{P})$ such that $\hat{f}|_E = \varphi$. The latter property, known as *Sidonicity*, is a statement of *functional independence*, and will be discussed at length later in the chapter.

4 Some History

Elements of R were originally defined as functions on the interval [0,1] [R]; see (II.1.2). This view of R, still fairly common, implies that elements of W also could be viewed as functions on [0,1]. (See (2.3).) These same functions on [0,1] were introduced first by J. Walsh [W], essentially via the Haar system [Ha1], without mention of the Rademacher system. In his work [W], Walsh noted that 'the chief interest of the set φ [comprising his newly discovered functions] lies in its similarity to the usual (e.g., sine, cosine, Sturm–Liouville, Legendre) sets of orthogonal functions, ...' [W, p. 5]. He proved that φ was complete, and observed some local properties of φ-series that strongly resembled properties of classical Fourier series.

R.E.A.C. Paley was the first to notice that Walsh's functions were the products of Rademacher's functions, and that they could be naturally ordered according to this scheme: let $n = 2^{m_1} + \cdots + 2^{m_k}$ be the binary expansion of a positive integer n, and define the nth Walsh function to be

$$w_n = r_{m_1+1} \cdots r_{m_k+1}. \tag{4.1}$$

Using this ordering, Paley investigated basic similarities between the classical Fourier series and the series $\Sigma_n a_n w_n$, and, en route, discovered also new properties, which effectively foreshadowed the concept of martingales [Pa] (Exercise 11). A student of Littlewood, Paley was 25 years old when his paper [Pa] was published. He died a year later in a skiing accident in the Canadian Rockies [Har].

That Walsh functions on [0,1] can be viewed as *characters* of Ω was observed first by N. Fine [Fi1], [Fi2]. (From here on, we will refer to elements in W as *Walsh characters* and to their correspondents on [0,1] as *Walsh functions*, and to elements in R as *Rademacher characters* and to their correspondents on [0,1] as *Rademacher functions.*) This equivalent view of Walsh's functions is based on the measure-preserving map σ from $(\Omega, \mathfrak{A}, \mathbb{P})$ onto $([0,1], \mathfrak{B}, \mathfrak{m})$ defined by

$$\sigma(\omega) = \sum_{n=1}^{\infty} (1 - \omega(n))/2^{n+1}, \quad \omega \in \Omega \tag{4.2}$$

($\mathfrak{B}$ = Borel field in [0,1], $\mathfrak{m}$ = Lebesgue measure), and the fact that if w is a Walsh function on [0,1], then $w \circ \sigma$ is a Walsh character on

Ω (Exercise 10). Therefore, results about W that involve only measure-theoretic properties can be easily shuttled between $([0,1], \mathfrak{B}, \mathfrak{m})$ and $(\Omega, \mathfrak{A}, \mathbb{P})$. For example, the completeness of Walsh functions in $L^2([0,1], \mathfrak{m})$, first observed in [W], can be quickly obtained from Corollary 10, which asserts completeness of Walsh characters in $L^2(\Omega, \mathbb{P})$.

The bare-bone representation theory outlined in the previous section is but a very small portion of a very large body of results that deal with convergence of Fourier series and other Fourier-type series. (See Zygmund's treatise [Zy2].) Indeed, Corollary 9 is but one of several resolutions of the 'representation' problem. The two theorems below, established about forty years apart, are prominent examples of fundamental results in this area.

Theorem 13 ([Ri$_\text{m}$, p. 230], [Zy2, Vol. I, p. 253], Exercise 12). *For all* $f \in L^p(\mathbf{T}, m)$, $p \in (1, \infty)$,

$$\sum_{j=-n}^{n} \hat{f}(j)\, e^{ijt} \xrightarrow[n\to\infty]{} f, \quad \textit{convergence in the } L^p\textit{-norm}. \tag{4.3}$$

Theorem 14 [Car], [Hu]. *For all* $f \in L^p(\mathbf{T}, m)$, $p \in (1, \infty]$,

$$\sum_{j=-n}^{n} \hat{f}(j)\, e^{ijt} \xrightarrow[n\to\infty]{} f(t) \quad \textit{for almost all } t\ (\mathfrak{m}). \tag{4.4}$$

Theorem 13 is due to M. Riesz (F. Riesz's brother), and is standard fare in books on classical harmonic analysis (e.g. [Kat, Chapter III], [Hel, Chapter 5]). Theorem 14, proved first by Lennart Carleson in the case $p = 2$, settled a long-standing problem concerning pointwise representation of a function f by its Fourier series $S[f]$. This problem, in essence going back to the time of Fourier, had been unresolved prior to Carleson's theorem even in the case $f \in C(T)$; see [Zy2, Vol. I, *preface*]. The analog of Theorem 13 for Walsh series was obtained by Paley [Pa], and analogs of Theorem 14 were obtained by P. Billard [Bi] and P. Sjölin [Sj].

5 Analysis of Walsh Systems: A First Step

For $k \in \mathbb{N}$, we consider the k-fold products of elements in $\{r_n : n \in \mathbb{N}\}$,

$$W_k = \{r_{n_1} \cdots r_{n_k} : 0 \le n_1 \le \cdots \le n_k\}. \tag{5.1}$$

We refer to W_k as the Walsh system of order k, and will continue to use R to denote the Rademacher system. Clearly, $W_k \subset W_{k+1}$ and

$$\hat{\Omega} = W = \bigcup_{k=0}^{\infty} W_k. \tag{5.2}$$

It is evident – certainly in a heuristic sense – that W_1 $(= R \cup \{r_0\})$ is the 'least' complex system, and, at the other end, W is the 'most' complex. It is also apparent that the complexity of W_k increases as k increases. A fundamental question arises: how can we gauge precisely, starting with W_1, the evolving complexity of W_k?

We begin by observing a property enjoyed by every W_k, but (obviously) not by W. Specifically, we will prove that for all $k \in \mathbb{N}$, if a bounded measurable function has spectrum in W_k, then the function is necessarily continuous.

Throughout, we use the following notation. If Space(Ω) denotes a subspace of $\mathrm{M}(\Omega)$, and $E \subset W$, then

$$\mathrm{Space}_E(\Omega) = \mathrm{Space}_E = \{\nu \in \ \mathrm{Space}(\Omega) : \ \mathrm{spect}\ \nu \subset E\}. \tag{5.3}$$

(This notation is used also in the general setting, where a compact Abelian group G and its dual $\hat{G}$ stand for Ω and W, respectively.) We will verify by induction on k that

$$\mathrm{L}^{\infty}_{W_k} = \mathrm{C}_{W_k}, \quad k \in \mathbb{N}. \tag{5.4}$$

The case $k = 1$ is

Proposition 15 *If $f \in \mathrm{L}^{\infty}_{R}$, then*

$$\sum_{w \in R} |\hat{f}(w)| = \|\hat{f}\|_1 \leq 2\ \|f\|_{\mathrm{L}^{\infty}}. \tag{5.5}$$

In particular, $\mathrm{L}^{\infty}_{W_1} = \mathrm{C}_{W_1}$.

Proof: Let $f \in \mathrm{L}^{\infty}_{R}$. Fix $\omega \in \Omega$, $n \in \mathbb{N}$, and consider the Riesz product

$$F_\omega = \prod_{j=1}^{n} (1 + r_j(\omega)\ r_j). \tag{5.6}$$

As in the proof of Proposition 8,

$$\|F_\omega\|_{\mathrm{L}^1} = \hat{F}_\omega(r_0) = 1. \tag{5.7}$$

Also observe

$$\mathbf{1}_R\, \hat{F}_\omega(w) = \begin{cases} w(\omega) & \text{if } w \in \{r_1, \dots, r_n\} \\ 0 & \text{otherwise.} \end{cases} \tag{5.8}$$

Therefore,

$$\left| \sum_{j=1}^{n} \hat{f}(r_j)\, r_j(\omega) \right| = |\mathbf{E}\, f F_\omega| \leq \|f\|_{\mathrm{L}^\infty} \|F_\omega\|_{\mathrm{L}^1} \leq \|f\|_{\mathrm{L}^\infty}. \tag{5.9}$$

Then by maximizing (5.9) over $\omega \in \Omega$ and $n \in \mathbb{N}$, we obtain (5.5) (cf. (II.1.4), (II.1.5)). By (5.5), $\sum_{j=1}^{n} \hat{f}(r_j)\, r_j \to f$ uniformly on Ω, implying that $f \in \mathrm{C}(\Omega)$. Therfore, $\mathrm{L}^\infty_{W_1} \subset \mathrm{C}(\Omega)$. □

The counterpoint to Proposition 15 is

Proposition 16 (Exercise 13). *For all* $f \in \mathrm{L}^\infty(\Omega)$,

$$\|f\|_{\mathrm{L}^\infty} \geq \|f\|_{\mathrm{L}^2} = \|\hat{f}\|_2 := \left(\sum_{w \in W} |\hat{f}(w)|^2 \right)^{\frac{1}{2}}. \tag{5.10}$$

Moreover, (5.10) *is best possible: there exist* $f \in \mathrm{C}(\Omega)$ *such that*

$$\|\hat{f}\|_p := \left(\sum_{w \in W} |\hat{f}(w)|^p \right)^{\frac{1}{p}} = \infty \quad \text{for all } p < 2. \tag{5.11}$$

The first assertion in Proposition 16 follows easily from Plancherel's theorem (cf. (3.20)). The proof of the second part, that (5.10) is best possible, is not quite as easy (Exercise 13). Notice the gap between the l^1-norm in (5.5) and the l^2-norm in (5.10). Later in this chapter, we will fill and calibrate this gap with Walsh systems of increasing order.

Remark (how it began). The properties exemplified by Proposition 15 first attracted attention in the classical setting $\mathbf{T} = [0, 2\pi)$. Consider the space of continuous functions on $\mathbf{T}$ with absolutely convergent Fourier series, commonly denoted as

$$A(\mathbf{T}) = \{f \in \mathrm{C}(\mathbf{T}) : \hat{f} \in l^1(\mathbb{Z})\}, \tag{5.12}$$

and normed by

$$\|f\|_A = \|\hat{f}\|_1, \quad f \in A(\mathbf{T}). \tag{5.13}$$

Equipped with the A-norm and pointwise multiplication on $\mathbf{T}$ (convolution on $\mathbb{Z}$), $A(\mathbf{T})$ is a Banach algebra. Notice that

$$A(\mathbf{T}) \subsetneq \mathrm{C}(\mathbf{T}) \subsetneq \mathbf{L}^\infty(\mathbf{T}, \mathfrak{m}). \tag{5.14}$$

(The quickest way to obtain the proper inclusions in (5.14) is to argue that if $A(\mathbf{T}) = \mathrm{C}(\mathbf{T})$, then $\mathrm{C}(\mathbf{T}) = \mathrm{L}^\infty(\mathbf{T}, \mathfrak{m})$; see Exercise 14.) Addressing an issue implicit in (5.14), S. Sidon was the first to observe infinite sets $E \subset \mathbb{Z}$ such that

$$A_E(\mathbf{T}) = \mathrm{C}_E(\mathbf{T}) = \mathrm{L}_E^\infty(\mathbf{T}, \mathfrak{m}) \quad \text{[Si1], [Si2]}. \tag{5.15}$$

Specifically, by using the products introduced by F. Riesz [Ri$_\mathrm{f}$2], Sidon proved that if $E = \{\lambda_j\} \subset \mathbb{Z}^+,\ \lambda_1 < \cdots < \lambda_j < \lambda_{j+1} < \cdots$, and

$$q_E := \inf\{\lambda_j / \lambda_{j-1} : j \in \mathbb{N}\} > 1, \tag{5.16}$$

then E satisfies (5.15). Sets $E \subset \mathbb{Z}^+$ with $q_E > 1$ are sometimes called *Hadamard* sets [Zy2, Vol. 1, p. 208], and sometimes *lacunary* sets [Kat, Chapter V]; we shall use the latter term. In general, spectral subsets E of $\hat{G}$ (the dual of a compact Abelian group G) such that

$$A_E(G) = \mathrm{C}_E(G) \tag{5.17}$$

are called *Sidon* sets [Ru1, p. 204]. (See Remark ii in §3, and also Exercise 15 i.)

In our setting, $A(\Omega) = \{f \in \mathrm{C}(\Omega) : \hat{f} \in l^1(W)\}$, and (5.5) becomes the statement

$$A_R(\Omega) = \mathrm{C}_R(\Omega) = \mathrm{L}_R^\infty(\Omega, \mathbb{P}), \tag{5.18}$$

i.e., $R \subset \hat{\Omega}$ is a Sidon set. Indeed, the proof of Proposition 15 is nearly identical to the argument in [Si2] verifying that a lacunary set satisfies (5.15). (The use of algebraic independence of R in the proof of Proposition 15 is analogous to the use of lacunarity in Sidon's proof.)

Here is an interesting aside. If $E \subset \mathbb{Z}^+$ is lacunary (and hence Sidon in $\mathbb{Z}$), and $\{w_j : j \in \mathbb{N}\}$ is the Paley ordering of W (defined in (4.1)), then $\{w_j : j \in E\}$ is Sidon in W. This was established by G. Morgenthaler [Mo, §7], also by use of Riesz products. I do not know the answer to this question: if $E \subset \mathbb{Z}^+$ is Sidon, then is $\{w_j : j \in E\}$ a Sidon set in W? (See Exercise 19.)

Note the implication

$$A_E = \mathrm{C}_E \Rightarrow \mathrm{C}_E = \mathrm{L}^\infty_E, \quad E \subset \mathbb{Z} \text{ or } E \subset W \text{ (Exercise 15 ii).} \tag{5.19}$$

(This implication works easily and equally well in any Abelian group setting.) Demonstrating that the reverse implication is not true, Haskell Rosenthal was first to observe non-Sidon sets $E \subset \mathbb{Z}$ such that $\mathrm{C}_E(\mathbf{T}) = \mathrm{L}^\infty_E(\mathbf{T}, \mathfrak{m})$ [Ro]. Spectral sets E (in any group) such that $\mathrm{C}_E = \mathrm{L}^\infty_E$ are thus called *Rosenthal sets.* In the next section we verify that W_k is Rosenthal for every $k \in \mathbb{N}$, and in Exercise 16 you will verify that W_2 (and therefore W_k for every $k \geq 2$) is not Sidon. In §12 we will observe the same phenomenon in every discrete Abelian group.

6 W_k is a Rosenthal Set

We first do the groundwork. Let $R_0 = \{r_0\}$, and for $k \in \mathbb{N}$, define

$$R_k = \{r_{n_1} \cdots r_{n_k} : 0 < n_1 < \cdots < n_k\} \tag{6.1}$$

(k-fold products of *distinct* Rademacher functions). Then, $R_j \cap R_k = \emptyset$ for $j \neq k$, and

$$W_k = \bigcup_{j=0}^{k} R_j \tag{6.2}$$

(k-fold products of Rademacher functions). For $i = 0, \ldots,$ and $m = i+k, i+k+1, \ldots,$ define

$$T^k_{m,i} = \{r_{n_1} \cdots r_{n_k} : i < n_1 < \cdots < n_k \leq m\}. \tag{6.3}$$

For convenience, we denote $T^k_{m,0}$ by T^k_m.

Lemma 17 *For all* $f \in \mathrm{L}^\infty_{R_k}$,

$$\|f\|_{\mathrm{L}^\infty} = \sup \left\{ \left\| \sum_{w \in T^k_m} \hat{f}(w)\, w \right\|_{\mathrm{L}^\infty} : m \in \mathbb{N} \right\}. \tag{6.4}$$

Proof: Fix $\omega \in \Omega$ and $m \in \mathbb{N}$. Consider the Riesz product (cf. (5.6))

$$F_\omega = \prod_{j=1}^{m} (1 + r_j(\omega)\, r_j). \tag{6.5}$$

The spectral analysis of F_ω implies that if $r_{n_1} \cdots r_{n_k} \in T_m^k$, then (cf. (5.8))

$$\hat{F}_\omega(r_{n_1} \cdots r_{n_k}) = r_{n_1}(\omega) \cdots r_{n_k}(\omega). \tag{6.6}$$

Thorofore (because $\|F_\omega\|_{\mathrm{L}^1} = \hat{F}_\omega(r_0) = 1$),

$$\left| \sum_{w \in T_m^k} \hat{f}(w)\, w(\omega) \right| = |\mathbf{E} f F_\omega| \leq \|f\|_{\mathrm{L}^\infty}, \tag{6.7}$$

which implies that the right side of (6.4) is bounded by $\|f\|_{\mathrm{L}^\infty}$.

To verify the reverse inequality, let g be a W-polynomial; that is, for some $N \in \mathbb{N}$,

$$\operatorname{spect} g \subset \{r_{n_1} \cdots r_{n_N} : 0 \leq n_1 \leq \cdots \leq n_N \leq N\}. \tag{6.8}$$

Then,

$$\begin{aligned} |\mathbf{E} f g| &= \left| \sum_{w \in T_N^k} \hat{f}(w)\, \hat{g}(w) \right| = \left| \mathbf{E} g \sum_{w \in T_N^k} \hat{f}(w)\, w \right| \\ &\leq \|g\|_{\mathrm{L}^1} \sup \left\{ \left\| \sum_{w \in T_m^k} \hat{f}(w)\, w \right\|_{\mathrm{L}^\infty} : m \in \mathbb{N} \right\}, \end{aligned} \tag{6.9}$$

which implies, by the density of W-polynomials in $\mathrm{L}^1(\Omega, \mathbb{P})$, that $\|f\|_{\mathrm{L}^\infty}$ is bounded by the right side of (6.4). □

For $k \in \{2, 3, \ldots\}$ and $i \in \mathbb{N}$, define

$$L_{i,k} = \{w \in R_k : w = r_i r_{j2} \cdots r_{jk} : i < j_2 < \cdots < j_k\}. \tag{6.10}$$

The $L_{i,k}$ are pairwise disjoint, and $R_k = \bigcup_{i=1}^\infty L_{i,k}$.

Lemma 18 *For $k \in \{2, 3, \ldots\}$ and $i \in \mathbb{N}$, there exist $\mu_i \in \mathrm{M}(\Omega)$ such that $\|\mu_i\|_{\mathrm{M}} = 1$ and*

$$\hat{\mu}_i(w) = \begin{cases} 1 & \textit{if } w \in L_{i,k} \\ 0 & \textit{if } w \in R_k \backslash L_{i,k}. \end{cases} \tag{6.11}$$

Proof: Define $e_i \in \Omega$ by $e_i(i) = -1$, and $e_i(j) = 1$ for $j \neq i$. Consider the Riesz product (cf. Exercises 17 and 18)

$$\rho = \prod_{j=i}^{\infty}(1 + e_i(j)\ r_j). \tag{6.12}$$

Then, $\|\rho\|_{\mathrm{M}} = 1$, and for $w \in R_k$,

$$\hat{\rho}(w) = \begin{cases} -1 & w \in L_{i,k} \\ 1 & w \in R_k \backslash L_{i,k}. \end{cases} \tag{6.13}$$

Then, $\mu_i = \rho \star \rho - \rho/2$ satisfies (6.11). □

A subset $E \subset R$ is said to be the *generating set* for $D \subset R_k$ if E is the smallest set such that

$$D \subset \{w_1 \cdots w_k : w_1 \in E, \ldots, w_k \in E\}. \tag{6.14}$$

Subsets $D_j \subset R_k$, $j = 1, \ldots,$ are said to be *strongly disjoint* if their respective generating sets are pairwise disjoint.

For the proof of the lemma below, we recall the framework in Chapter II §5. Let $\mathbf{T}_k$ denote the set of kth roots of unity (a subset of $\{z \in \mathbb{C} : |z| = 1\}$), and let $\Omega_k = (\mathbf{T}_k)^{\mathbb{N}}$. Let $\mathbb{P}_k$ denote the probability measure on Ω_k, which is the infinite product of the uniform probability measure on $\mathbf{T}_k$. For $n \in \mathbb{N}$, let χ_n be the projection from Ω_k onto the nth coordinate. Then, $\{\chi_n : n \in \mathbb{N}\}$ is a system of statistically independent $\mathbf{T}_k$-valued random variables on $(\Omega_k, \mathbb{P}_k)$ such that $(\chi_n)^k = 1$ and $\mathbf{E}(\chi_n)^j = 0$, $j \in [k-1]$ and $n \in \mathbb{N}$. (See (II.6.10).)

Lemma 19 *Suppose $\{D_j : j \in \mathbb{N}\}$ is a collection of finite and strongly disjoint subsets of R_k. Then, there exists $c_k > 0$ (depending only on k) such that for all $f \in \mathrm{L}^{\infty}_{R_k}$ and $N \in \mathbb{N}$,*

$$c_k \left\| \sum_{j=1}^{N} \sum_{w \in D_j} \hat{f}(w)\ w \right\|_{\mathrm{L}^\infty} \leq \|f\|_{\mathrm{L}^\infty}. \tag{6.15}$$

Proof: Assume $k \geq 2$. Let $\gamma_n = \chi_n + \overline{\chi}_n/2$, and observe that for every $n \in \mathbb{N}$,

$$\mathbf{E}\ (\gamma_n)^i = \begin{cases} c_k & \text{if } i = k \\ 0 & \text{if } i = 1, \ldots, k-1, \end{cases} \tag{6.16}$$

where

$$c_k = \begin{cases} \frac{1}{2^{k-1}} & k \text{ odd} \\ \binom{k}{k/2} \frac{1}{2^k} + \frac{1}{2^{k-1}} & k \text{ even.} \end{cases} \tag{6.17}$$

Let E_j be the generating set of D_j, $j \in \mathbb{N}$. Fix $s \in \Omega_k$, and define

$$H_s = \prod_{i \in E_1} (1+\gamma_1(s)r_i) \cdots \prod_{i \in E_j} (1+\gamma_j(s)r_i) \cdots \prod_{i \in E_N} (1+\gamma_N(s)r_i). \tag{6.18}$$

Then, H_s is a non-negative W-polynomial such that

$$\|H_s\|_{\mathrm{L}^1(\Omega,\mathbb{P})} = \hat{H}_s(r_0) = 1. \tag{6.19}$$

The spectral analysis of H_s implies the following. If $j \in [\mathrm{N}]$ and $w \in D_j$, then

$$\hat{H}_s(w) = \gamma_j(s)^k. \tag{6.20}$$

Otherwise, if $j \in [\mathrm{N}]$, $w \in R_k$, and $w \notin D_j$, then either $\hat{H}_s(w) = 0$ or

$$\hat{H}_s(w) = \gamma_1(s)^{i_1} \cdots \gamma_N(s)^{i_N}, \tag{6.21}$$

where $0 \leq i_n < k$ and $\Sigma_{n=1}^{N} i_n = k$.

We now 'average' H_s over $s \in \Omega_k$ (with respect to $\mathbb{P}_k$),

$$F = \int_{\Omega_k} H_s \, \mathbb{P}_k(\mathrm{d}s). \tag{6.22}$$

The function F is a non-negative W-polynomial such that

$$\hat{F}(w) = \mathbf{E}\, w \int_{\Omega_k} H_s \, \mathbb{P}_k(\mathrm{d}s) = \int_{\Omega_k} \hat{H}_s(w) \, \mathbb{P}_k(\mathrm{d}s), \quad w \in W. \tag{6.23}$$

Therefore, by (6.19), $\|F\|_{\mathrm{L}^1} = \hat{F}(r_0) = 1$. By (6.16), (6.20), (6.21), and the (statistical) independence of the γ_n, if $w \in R_k$ then

$$\hat{F}(w) = \begin{cases} c_k & w \in D_j, \quad j \in [\mathrm{N}] \\ 0 & \text{otherwise.} \end{cases} \tag{6.24}$$

Therefore, for $f \in \mathrm{L}^\infty_{R_k}$,

$$c_k \left\| \sum_{j=1}^{N} \sum_{w \in D_j} \hat{f}(w)\, w \right\|_{\mathrm{L}^\infty} = \|f \star F\|_{\mathrm{L}^\infty} \leq \|f\|_{\mathrm{L}^\infty}. \tag{6.25}$$

□

Lemma 20 *Suppose $D_j \subset R_k$, $j \in \mathbb{N}$, are finite and strongly disjoint. Then, for $\beta \in l^\infty(R_k)$ and $N \in \mathbb{N}$,*

$$\sum_{j=1}^{N} \left\| \sum_{w \in D_j} \beta(w)\ w \right\|_{\mathrm{L}^\infty} \leq 4 \left\| \sum_{j=1}^{N} \sum_{w \in D_j} \beta(w)\ w \right\|_{\mathrm{L}^\infty} . \tag{6.26}$$

Proof: Because the D_j are strongly disjoint, for $\omega_j \in \Omega$ and $j \in [\mathrm{N}]$ there exists $\omega \in \Omega$ such that

$$\sum_{w \in D_j} \beta(w)\ w(\omega_j) = \sum_{w \in D_j} \beta(w)\ w(\omega), \quad j \in [\mathrm{N}] \tag{6.27}$$

(Exercise 20 i). For $j \in [\mathrm{N}]$, let ω_j be such that

$$\left\| \sum_{w \in D_j} \beta(w)\ w \right\|_{\mathrm{L}^\infty} = \left| \sum_{w \in D_j} \beta(w)\ w(\omega_j) \right| . \tag{6.28}$$

By applying (6.27) and Lemma VI.6 (in the case $n = 1$), we choose $S \subset [\mathrm{N}]$ so that

$$\begin{aligned} \sum_{j=1}^{N} \left| \sum_{w \in D_j} \beta(w)\ w(\omega_j) \right| &= \sum_{j=1}^{N} \left| \sum_{w \in D_j} \beta(w)\ w(\omega) \right| \\ &\leq 4 \left| \sum_{j \in S} \sum_{w \in D_j} \beta(w)\ w(\omega) \right| \leq 4 \left\| \sum_{j \in S} \sum_{w \in D_j} \beta(w)\ w \right\|_{\mathrm{L}^\infty} . \end{aligned} \tag{6.29}$$

Let $\mu \in \mathrm{M}(\Omega)$ (a Riesz product) be such that $\|\mu\|_{\mathrm{M}} = 1$, and

$$\hat{\mu}(w) = \begin{cases} 1 & w \in D_j \text{ and } j \in S \\ 0 & w \in D_j \text{ and } j \notin S \end{cases} \tag{6.30}$$

(Exercise 20 ii), and obtain

$$\begin{aligned} \left\| \sum_{j \in S} \sum_{w \in D_j} \beta(w)\ w \right\|_{\mathrm{L}^\infty} &= \left\| \mu \star \left(\sum_{j=1}^{N} \sum_{w \in D_j} \beta(w)\ w \right) \right\|_{\mathrm{L}^\infty} \\ &\leq \left\| \sum_{j=1}^{N} \sum_{w \in D_j} \beta(w)\ w \right\|_{\mathrm{L}^\infty} . \end{aligned} \tag{6.31}$$

We obtain (6.26) from (6.28), (6.29), and (6.31). □

Theorem 21 *For all $k \in \mathbb{N}$, $\mathrm{L}^\infty_{R_k} = \mathrm{C}_{R_k}$.*

Proof (by induction). The case $k = 1$ is Proposition 15. Let $k > 1$ and assume $\mathrm{L}^\infty_{R_{k-1}} = \mathrm{C}_{R_{k-1}}$. Let $f \in \mathrm{L}^\infty_{R_k}$, and define

$$f_m = \sum_{m<n_1\ldots<n_k} \hat{f}(r_{n_1}\cdots r_{n_k})\, r_{n_1}\cdots r_{n_k}, \quad m \in \mathbb{N}. \tag{6.32}$$

By Lemma 18,

$$f - f_m \in \mathrm{L}^\infty_{L_{1,k}\cup\cdots\cup L_{m,k}}, \tag{6.33}$$

and by the induction hypothesis, $f - f_m \in \mathrm{C}(\Omega)$ (Exercise 21). Therefore, to conclude that $f \in \mathrm{C}(\Omega)$, it suffices to verify the following.

Claim: $\lim_{m\to\infty} \|f_m\|_{\mathrm{L}^\infty} = 0$.

Proof: Suppose the claim is false. Then, by Lemma 17, there exist $\delta > 0$ and increasing sequences of integers (l_j) and (m_j) such that for all $j \in \mathbb{N}$, $l_j < m_j < l_{j+1}$ and

$$\left\| \sum_{w\epsilon T^k_{m_j,l_j}} \hat{f}(w)\, w \right\|_{\mathrm{L}^\infty} > \delta. \tag{6.34}$$

The sets $T^k_{m_j,l_j}$ $(j \in \mathbb{N})$ are strongly disjoint, and therefore by Lemma 19 and Lemma 20,

$$\sum_{j=1}^{N} \left\| \sum_{w\epsilon T^k_{m_j,l_j}} \hat{f}(w)\, w \right\|_{\mathrm{L}^\infty} \leq 4 \left\| \sum_{j=1}^{N} \sum_{w\epsilon T^k_{m_j,l_j}} \hat{f}(w)\, w \right\|_{\mathrm{L}^\infty}$$
$$\leq (4/c_k)\|f\|_{\mathrm{L}^\infty}, \tag{6.35}$$

which contradicts (6.34). □

Finally, we need the following lemma in order to 'piece together' Theorem 21 for R_j, $j = 1, \ldots, k$, and conclude that W_k is a Rosenthal set.

Lemma 22 *Let $j \in [k]$. There exist $\mu_j \in \mathrm{M}(\Omega)$ such that*

$$\hat{\mu}_j(w) = \begin{cases} 1 & \text{if } w \in R_j \\ 0 & \text{if } w \in W_k \backslash R_j. \end{cases} \tag{6.36}$$

Proof: Consider the Riesz product

$$\rho = \prod_{n=1}^{\infty} \left(1 + \frac{1}{2} r_n\right). \tag{6.37}$$

Then, $\|\rho\|_{\mathrm{M}} = 1$, and for $i \in [k]$,

$$\hat{\rho}(w) = 1/2^i \quad w \in R_i. \tag{6.38}$$

Let P be a real-valued polynomial of degree k defined on [0,1], such that $P(0) = 0$ and

$$P(1/2^i) = \begin{cases} 1 & \text{if } i = j \\ 0 & \text{if } i \neq j \text{ and } i \in [k]. \end{cases} \tag{6.39}$$

Write $P(x) = \sum_{n=1}^{k} a_n \, x^n$ ($x \in [0,1]$), and then define the measure

$$\mu_j = P(\rho) = \sum_{n=1}^{k} a_n \, \rho^n, \tag{6.40}$$

where ρ^n is the n-fold convolution of ρ. By (6.38) and (6.39), $\hat{\mu}_j$ satisfies (6.36). □

Theorem 23 $\mathrm{L}^{\infty}_{W_k} = \mathrm{C}_{W_k}$.

Proof: Let $f \in \mathrm{L}^{\infty}_{W_k}$. By Lemma 22, we can write $f = f_1 + \cdots + f_k$, where $f_j \in \mathrm{L}^{\infty}_{R_j}$ for $j = 0, \ldots, k$. By Theorem 21, $f \in \mathrm{C}_{W_k}$. □

7 Restriction Algebras

In this section we collect preliminaries concerning algebras of restrictions of transforms. These algebras will play prominently in the rest of the chapter (see Exercise 15).

For $F \subset W$, consider

$$\begin{aligned} B(F) \;&:=\; \mathrm{M}(\Omega)\hat{}/\{\hat{\mu} : \mu \in \mathrm{M},\ \hat{\mu}|_F = 0\} \\ &= \{\varphi \in l^{\infty}(F) : \exists\ \mu \in \mathrm{M}(\Omega) \text{ such that } \hat{\mu}|_F = \varphi\}. \end{aligned} \tag{7.1}$$

($\hat{\mu}|_F$ is the restriction of $\hat{\mu}$ to F.) The $B(F)$-norm of $\varphi \in l^{\infty}(F)$ is the quotient norm,

$$\|\varphi\|_{B(F)} = \inf\{\|\mu\|_{\mathrm{M}} : \mu \in \mathrm{M}(\Omega),\ \hat{\mu}|_F = \varphi\}. \tag{7.2}$$

Similarly, define

$$A(F) := \mathrm{L}^1(\Omega,\mathbb{P})\hat{}/\{\hat{f} : f \in \mathrm{L}^1(\Omega,\mathbb{P}), \hat{f}|_F = 0\}$$
$$= \{\varphi \in \mathrm{c}_0(F) : \exists\, f \in \mathrm{L}^1(\Omega,\mathbb{P}) \text{ such that } \hat{f}|_F = \varphi\}. \qquad (7.3)$$

The $A(F)$-norm of $\varphi \in \mathrm{c}_0(F)$ is

$$\|\varphi\|_{A(F)} = \inf\{\|f\|_{\mathrm{L}^1} : f \in \mathrm{L}^1(\Omega,\mathbb{P}), \hat{f}|_F = \varphi\}. \qquad (7.4)$$

Equipped with pointwise multiplication and these quotient norms, $B(F)$ and $A(F)$ are Banach algebras.

The following proposition is a summary of basic properties. Its proof, a mix of functional and harmonic analysis, is left to the reader.

Proposition 24 (Exercise 22). *Let $F \subset W$.*

i. $\mathrm{C}_F(\Omega)^* = B(F)$. *Specifically, $\varphi \in B(F)$ defines the functional*

$$f \mapsto \int_\Omega f\, \mathrm{d}\mu = \lim_{k\to\infty} \sum_{w\in F\cap W_k} \hat{f}(w)\, \varphi(w), \quad f \in \mathrm{C}_F(\Omega), \qquad (7.5)$$

where $\mu \in \mathrm{M}(\Omega)$ and $\hat{\mu}|_F = \varphi$. Conversely, if $\varphi \in \mathrm{C}_F(\Omega)^*$, then there exist $\mu \in \mathrm{M}(\Omega)$ such that

$$\hat{\mu}|_F(w) = \varphi(w), \quad w \in F, \qquad (7.6)$$

and the action of φ on $\mathrm{C}_F(\Omega)$ is given by (7.5).

ii. $A(F)^* = \mathrm{L}^\infty_F(\Omega,\mathbb{P})$. *Specifically, $f \in \mathrm{L}^\infty_F(\Omega,\mathbb{P})$ defines the functional*

$$\varphi \mapsto \mathbf{E} fg = \lim_{k\to\infty} \sum_{w\in F\cap W_k} \hat{f}(w)\, \hat{g}(w), \quad \varphi \in A(F), \qquad (7.7)$$

where $g \in \mathrm{L}^1(\Omega,\mathbb{P})$ and $\hat{g}|_F = \varphi$. Conversely, if $\nu \in A(F)^*$ then $\sum_{w\in F} \nu(w) w$ is the W-series of $f \in \mathrm{L}^\infty_F(\Omega,\mathbb{P})$, whose action on $A(F)$ is given by (7.7).

iii. *$A(F)$ is an isometrically closed subalgebra of $B(F)$. Moreover, finitely supported functions on F are norm-dense in $A(F)$.*

iv. *Let $\varphi \in l^\infty(F)$. Then, $\varphi \in B(F)$ if and only if there exists a sequence of finitely supported functions $(\varphi_j : j \in \mathbb{N})$ on F such that*

$$\lim_{j\to\infty} \varphi_j(w) = \varphi(w), \quad w \in F, \qquad (7.8)$$

and

$$\sup\{\|\varphi_j\|_{B(F)} : j \in \mathbb{N}\} < \infty.$$

In particular, for $\varphi \in l^\infty(F)$,

$$\|\varphi\|_{B(F)} = \inf\left\{\limsup_{j\to\infty} \|\varphi_j\|_{B(F)} : \{\varphi_j\} \subset A(F),\right.$$

$$\left.\lim_{j\to\infty} \varphi_j(w) = \varphi(w) \textit{ for } w \in F\right\}. \tag{7.9}$$

8 Harmonic Analysis and Tensor Analysis

In this section we identify restriction algebras involving products of Rademacher systems with the tensor algebras defined in Chapter IV. We first consider the n-fold *Cartesian* product R^n (a subset of W^n), and then will transport results to the n-fold product R_n (a subset of W).

Proving grounds will be the compact Abelian group $\Omega^n = \Omega \times \cdots \times \Omega$, whose normalized Haar measure is the n-fold product measure $\mathbb{P}^n = \mathbb{P} \times \cdots \times \mathbb{P}$, and whose dual group is $W^n = W \times \cdots \times W$. Characters on Ω^n are elementary tensors

$$w_1 \otimes \cdots \otimes w_n, \quad w_j \in W,\ j \in [n]. \tag{8.1}$$

Because Ω can be naturally identified with the n-fold Cartesian product $\Omega \times \cdots \times \Omega$, the analysis of Ω^n can be carried out within Ω proper. To be precise, let $P_1, \dots, P_n$ be pairwise disjoint infinite subsets of $\mathbb{N}$, and denote $\Omega^{(i)} = \{-1,1\}^{P_i}$, $i \in [n]$. Then, following a bijection between P_i and $\mathbb{N}$, we identify $\Omega^{(i)}$ with Ω, and write (slightly abusing notation)

$$\Omega = \Omega^{(1)} \times \cdots \times \Omega^{(n)} = \Omega^n.$$

Again (only slightly) abusing notation, we let P_i denote also the corresponding system of Rademacher characters ($j \leftrightarrow r_j$ for $j \in P_i$), and then let $W^{(i)}$ be the subgroup of $\hat{\Omega}$ generated by $P_i \cup \{r_0\}$. Then,

$$\{w^{(1)} \cdots w^{(n)} : w^{(1)} \in W^{(1)}, \dots, w^{(n)} \in W^{(n)}\} = W. \tag{8.2}$$

The aforementioned bijections between P_i and $\mathbb{N}$ (identifying Ω with $\Omega^{(i)}$) also give rise to bijections between W and $W^{(i)}$, $i \in [n]$. We denote these bijections by $\tau_i : W^{(i)} \mapsto W$, $i \in [n]$, and obtain a one–one map τ from W onto $W \times \cdots \times W$,

$$w \mapsto \tau(w) = (\tau_1(w^{(1)}), \dots, \tau_n(w^{(n)})), \quad w \in W, \tag{8.3}$$

where $w = w^{(1)} \cdots w^{(n)}$, and $w^{(1)} \in W^{(1)}, \ldots, w^{(n)} \in W^{(n)}$. The spaces $\mathrm{L}^p(\Omega, \mathbb{P})$ and $\mathrm{C}(\Omega)$ are identified, respectively, with $\mathrm{L}^p(\Omega^n, \mathbb{P}^n)$ and $\mathrm{C}(\Omega^n)$: if $f \in \mathrm{L}^p(\Omega, \mathbb{P})$, then

$$\sum_{w \in W} \hat{f}(w)\, \tau(w) \tag{8.4}$$

is the W^n-series of an element in $\mathrm{L}^p(\Omega^n, \mathbb{P}^n)$ with the same norm; similarly, if $f \in \mathrm{C}(\Omega)$, then (8.4) represents a continuous function on Ω^n with the same norm (Exercise 23).

The identification of restriction algebras as tensor algebras extends to higher dimensions the observation

$$A(R) = \mathrm{c}_0(R) \quad (R \text{ is a Sidon set; cf. Exercise 15}). \tag{8.5}$$

Specifically, for every $\varphi \in \mathrm{c}_0(\mathbb{N})$ there exist $f \in \mathrm{L}^1(\Omega, \mathbb{P})$ with the property that $\hat{f}(r_j) = \varphi(j)$ for all $j \in \mathbb{N}$. The extension of (8.5) is

$$A(R^n) = \mathrm{c}_0(R) \,\hat{\otimes} \cdots \hat{\otimes}\, \mathrm{c}_0(R) = V_n(R, \ldots, R). \tag{8.6}$$

To establish it, we first identify finitely supported $\beta \in F_n(R, \ldots, R)$ as R^n-polynomials,

$$f_\beta = \sum_{(w_1, \ldots, w_n) \in R^n} \beta(w_1, \ldots, w_n)\, w_1 \otimes \cdots \otimes w_n, \tag{8.7}$$

and obtain (from definitions)

$$\|\beta\|_{F_n} = \|f_\beta\|_{\mathrm{L}^\infty}. \tag{8.8}$$

Next we will require the duality $V_n(R, \ldots, R)^* = F_n(R, \ldots, R)$ (Proposition IV.11); that is, $\beta \in F_n(R, \ldots, R)$ is the bounded linear functional $\hat{\beta}$ on V_n given by

$$\hat{\beta}(\varphi) = \sum_{(w_1, \ldots, w_n) \in R^n} \beta(w_1, \ldots, w_n)\, \varphi(w_1, \ldots, w_n),$$

$$\varphi \in V_n(R, \ldots, R), \tag{8.9}$$

and

$$\|\beta\|_{F_n} \leq \|\hat{\beta}\|_{(V_n)^*} \leq 2^n \|\beta\|_{F_n}. \tag{8.10}$$

Finally, we will use the observation that R^n is a Rosenthal set in W^n, which can be verified as follows. Let $P = P_1 \cdots P_n$, where $P_1, \ldots, P_n$ are pairwise disjoint infinite subsets of the Rademacher system R.

Through (8.3), identify P with R^n, and then through (8.4), identify $\mathrm{L}^\infty_P(\Omega, \mathbb{P})$ with $\mathrm{L}^\infty_{R^n}(\Omega^n, \mathbb{P}^n)$, and $\mathrm{C}_P(\Omega)$ with $\mathrm{C}_{R^n}(\Omega^n)$. Because $P \subset R_n$ is a Rosenthal set (Theorem 21), we conclude

$$\mathrm{L}^\infty_{R^n}(\Omega^n, \mathbb{P}^n) = \mathrm{C}_{R^n}(\Omega^n). \tag{8.11}$$

Proposition 25

i. $V_n(R, \dots, R) = A(R^n)$. *In particular, for all* $\varphi \in \mathrm{c}_0(R^n)$,

$$\|\varphi\|_{V_n(R,\dots,R)} \le \|\varphi\|_{A(R^n)} \le 2^n \|\varphi\|_{V_n(R,\dots,R)}. \tag{8.12}$$

ii. $\beta \mapsto f_\beta$ *determines an isometry from* $F_n(R, \dots, R)$ *onto* $\mathrm{C}_{R^n}(\Omega^n)$.

iii. $F_n(R, \dots, R)^* = B(R^n)$. *In particular,* (a) *for* $\varphi \in B(R^n)$, *there exists* $(\varphi_k) \subset V_n(R, \dots, R)$ *such that*

$$\lim_{k\to\infty} \varphi_k(w_1, \dots, w_n) = \varphi(w_1, \dots, w_n), \quad (w_1, \dots, w_n) \in R^n, \tag{8.13}$$

and

$$\sup_{k\in\mathbb{N}} \|\varphi_k\|_{V_n} \le \|\varphi\|_{B(R^n)}; \tag{8.14}$$

(b) *conversely, if* $\varphi \in l^\infty(R^n)$, *and if there exists* (φ_k) *in the unit ball of* $V_n(R, \dots, R)$ *such that* $\lim_{k\to\infty} \varphi_k = \varphi$ *(pointwise on* R^n*), then* $\varphi \in B(R^n)$ *and* $\|\varphi\|_{B(R^n)} \le 2^n$.

Proof:

i. Let φ be a finitely supported function on $R \times \cdots \times R$. By applying duality, (8.8), (8.10), and Proposition 24, we obtain

$$\begin{aligned} &\|\varphi\|_{V_n(R,\dots,R)} \\ &\le \sup\left\{ \left| \sum_{(w_1,\dots,w_n)\in R^n} \varphi(w_1,\dots,w_n)\beta(w_1,\dots,w_n) \right| : \|\beta\|_{F_n} \le 1 \right\} \\ &= \sup\left\{ \left| \sum_{(w_1,\dots,w_n)\in R^n} \varphi(w_1,\dots,w_n)\, \hat{f}(w_1,\dots,w_n) \right| : \right. \\ &\qquad\qquad \left. R^n\text{-polynomials } f, \|f\|_{\mathrm{L}^\infty} \le 1 \right\} \\ &= \|\varphi\|_{A(R^n)}. \end{aligned} \tag{8.15}$$

Similarly, by (8.8) and (8.10),

$$\|\varphi\|_{A(R^n)} \le 2^n \|\varphi\|_{V_n(R,\dots,R)}. \tag{8.16}$$

Part i follows from norm-density of finitely supported functions in $V_n(R,\ldots,R)$ and $A(R^n)$.

ii. By duality and Part i, $A(R^n)^* = \mathrm{L}^\infty_{R^n} = V_n^* = F_n$, and, because R^n is Rosenthal, $F_n(R,\ldots,R)$ is canonically isometric to $\mathrm{C}_{R^n}(\Omega^n)$ ((8.7) and (8.8)).

iii. The first assertion follows from duality (Proposition 24) and Part ii. To verify the assertion in (a), use weak* density of finitely supported functions on R^n in $B(R^n)$ ($\mathrm{C}_{R^n}(\Omega^n)^*$), i.e., for $\varphi \in B(R^n)$, there exists a sequence (φ_k) of finitely supported functions on $R \times \cdots \times R$ converging pointwise to φ, and $\|\varphi_k\|_{B(R^n)} \leq \|\varphi\|_{B(R^n)}$ (Proposition 24 iv). Each φ_k is in $V_n(R,\ldots,R)$, and by (8.8),

$$\begin{aligned}\|\varphi_k\|_{V_n} &\leq \sup\left\{\left|\sum_{(w_1,\ldots,w_n)\in R^n} \varphi_k(w_1,\ldots,w_n)\beta(w_1,\ldots,w_n)\right| : \right.\\ &\qquad\qquad \left. \beta \in F_n(R,\ldots,R), \|\beta\|_{F_n} \leq 1\right\}\\ &= \sup\left\{\left|\sum_{(w_1,\ldots,w_n)\in R^n} \varphi_k(w_1,\ldots,w_n)\, \hat{f}(w_1,\ldots,w_n)\right| : \right.\\ &\qquad\qquad \left. f \in \mathrm{C}_{R^n}(\Omega^n),\ \|f\|_{\mathrm{L}^\infty} \leq 1\right\}\\ &= \|\varphi_k\|_{B(R^n)}. \end{aligned} \tag{8.17}$$

To verify (b), note that the unit ball of $V_n(R,\ldots,R)$ is weak* dense in the unit ball of its second dual, which is $F_n(R,\ldots,R)^*$, and then apply Part ii together with the equality $\mathrm{C}_{R^n}(\Omega^n)^* = B(R^n)$. □

Having the tensor-theoretic representations of $\mathrm{C}_{R^n}(\Omega)$ and $A(R^n)$, we proceed to the analogous representations of $\mathrm{C}_{R_n}(\Omega)$ and $A(R_n)$.

An n-array $\beta \in l^\infty(\mathbb{N}^n)$ is *symmetric* if for every $\tau \in \mathrm{per}[n]$ (permutations of $[n]$),

$$\beta(i_1,\ldots,i_n) = \beta(i_{\tau 1},\ldots,i_{\tau n}), \quad (i_1,\ldots,i_n) \in \mathbb{N}^n; \tag{8.18}$$

it is said to *vanish on diagonals* if $\beta(j_1,\ldots,j_n) = 0$ for all $(j_1,\ldots,j_n) \in \mathbb{N}^n$ such that $|\{j_1,\ldots,j_n\}| < n$ (i.e., at least two of the n coordinates have the same value). Define $F_{n\sigma}(\mathbb{N},\ldots,\mathbb{N})$ to be

$$\{\beta \in F_n(\mathbb{N},\ldots,\mathbb{N}) : \beta \text{ symmetric and vanishes on diagonals}\}, \tag{8.19}$$

and note that it is a closed subalgebra of $F_n(\mathbb{N}, \ldots, \mathbb{N})$ (pointwise multiplication on $\mathbb{N}^n$ and F_n-norm). For $F \subset \mathbb{N}^n$, we denote the algebra of restrictions to F of elements in $V_n(\mathbb{N}, \ldots, \mathbb{N})$ by

$$V_n|_F = V_n(\mathbb{N}, \ldots, \mathbb{N})/\{\varphi \in V_n : \varphi|_F = 0\}, \tag{8.20}$$

and endow it with the quotient norm. Consider

$$D_n = \{(j_1, \ldots, j_n) \in \mathbb{N}^n : 0 < j_1 < \cdots < j_n\}, \tag{8.21}$$

and identify it (canonically) with $R_n = \{r_{j_1} \cdots r_{j_n} : (j_1, \ldots, j_n) \in D_n\}$. If f is an R_n-polynomial, then define $\beta_f \in F_{n\sigma}(\mathbb{N}, \ldots, \mathbb{N})$ by

$$\beta_f(j_1, \ldots, j_n) = \begin{cases} \hat{f}(r_{j_1} \cdots r_{j_n}) & \text{if } |\{j_1, \ldots, j_n\}| = n \\ 0 & \text{otherwise.} \end{cases} \tag{8.22}$$

Theorem 26

i. *The linear map* $f \mapsto \beta_f$ *in* (8.22) *determines an isomorphism from* $\mathrm{C}_{R_n}(\Omega)$ *onto* $F_{n\sigma}(\mathbb{N}, \ldots, \mathbb{N})$. *In particular, for* $f \in \mathrm{C}_{R_n}(\Omega)$,

$$\|f\|_{\mathrm{L}^\infty} \leq (1/n!)\ \|\beta_f\|_{F_n} \leq (2\mathrm{e})^n\ \|f\|_{\mathrm{L}^\infty}. \tag{8.23}$$

ii. $A(R_n) = V_n|_{D_n}$. *In particular, for* $\varphi \in \mathrm{c}_0(D_n)$,

$$\|\varphi\|_{A(R_n)} \leq \|\varphi\|_{V_n|_{D_n}} \leq (2\mathrm{e})^n\ \|\varphi\|_{A(R_n)}. \tag{8.24}$$

We need three lemmas. The first is a polarization device.

Lemma 27 (The Mazur–Orlicz identity [MazOr1, p. 63]). *Suppose* $\beta \in F_n(\mathbb{N}, \ldots, \mathbb{N})$ *is symmetric, and* $\varphi_1 \in l^\infty(\mathbb{N}), \ldots, \varphi_n \in l^\infty(\mathbb{N})$. *Then,*

$$\hat{\beta}(\varphi_1, \ldots, \varphi_n) = \frac{1}{n!}\ \mathbf{E}\ r_1 \cdots r_n \hat{\beta}(\Sigma_{j=1}^n r_j\ \varphi_j, \ldots, \Sigma_{j=1}^n r_j\ \varphi_j). \tag{8.25}$$

(If $\beta \in F_n(\mathbb{N}, \ldots, \mathbb{N})$, *then* $\hat{\beta}$ *denotes the corresponding* n*-linear functional on* $l^\infty(\mathbb{N})$.*)*

Proof: By the symmetry of β,

$$\hat{\beta}(\varphi_1, \ldots, \varphi_n) = \hat{\beta}(\varphi_{\tau 1}, \ldots, \varphi_{\tau n}), \quad \tau \in \text{per}[n]. \tag{8.26}$$

By the linearity of $\hat{\beta}$,

$$\begin{aligned} &\mathbf{E}\ r_1 \cdots r_n\ \hat{\beta}(\Sigma_{j=1}^n r_j\ \varphi_j, \ldots, \Sigma_{j=1}^n r_j\ \varphi_j) \\ &\quad = \sum_{(j_1, \ldots, j_n) \in [n]^n} \hat{\beta}(\varphi_{j_1}, \ldots, \varphi_{j_n})\ \mathbf{E}\ r_1 \cdots r_n\ r_{j_1} \cdots r_{j_n}. \end{aligned} \tag{8.27}$$

Note that

$$\mathbf{E}\, r_1 \cdots r_n\, r_{j_1} \cdots r_{j_n} = \begin{cases} 1 & \{j_1, \ldots, j_n\} = [n] \\ 0 & \text{otherwise.} \end{cases} \tag{8.28}$$

Therefore, by (8.26),

$$\begin{aligned} &\mathbf{E}\, r_1 \cdots r_n\, \hat{\beta}\left(\Sigma_{j=1}^n r_j\, \varphi_j, \ldots, \Sigma_{j=1}^n r_j\, \varphi_j\right) \\ &= \sum_{\tau \in \operatorname{per}[n]} \hat{\beta}(\varphi_{\tau 1}, \ldots, \varphi_{\tau n}) = n!\, \hat{\beta}(\varphi_1, \ldots, \varphi_n). \end{aligned} \tag{8.29}$$

□

The second lemma formalizes a fact used extensively in previous sections.

Lemma 28 *If $\theta \in l^\infty(\mathbb{N})$ is $\mathbb{R}$-valued, then there exists $\mu \in \mathrm{M}(\Omega)$ such that*

$$\hat{\mu}(r_{j_1} \cdots r_{j_n}) = \theta(i_1) \cdots \theta(i_n), \quad r_{j_1} \cdots r_{j_n} \in R_n, \tag{8.30}$$

and

$$\|\mu\|_{\mathrm{M}} \leq \|\theta\|_\infty^n.$$

Proof: The required measure is the Riesz product

$$\mu = \|\theta\|_\infty^n \prod_{j=1}^{\infty} \left(1 + \frac{\theta(j)}{\|\theta\|_\infty}\, r_j\right). \tag{8.31}$$

□

The third lemma is a consequence of the preceding two.

Lemma 29 *Let $\varphi_1 \in l^\infty(\mathbb{N}), \ldots, \varphi_n \in l^\infty(\mathbb{N})$. There exist $\mu \in \mathrm{M}(\Omega)$ such that*

$$\hat{\mu}(r_{j_1} \cdots r_{j_n}) = \varphi_1(j_1) \cdots \varphi_n(j_n), \quad r_{j_1} \cdots r_{j_n} \in R_n, \tag{8.32}$$

and

$$\|\mu\|_{\mathrm{M}} \leq (2\mathrm{e})^n\, \|\varphi_1\|_\infty \cdots \|\varphi_n\|_\infty. \tag{8.33}$$

Proof: Assume (without loss of generality) that $\|\varphi_j\|_\infty = 1$ for each $j \in [n]$. Assume also that the φ_j are real-valued. We argue by duality

(Proposition 24). Let f be an R_n-polynomial and let β_f be defined by (8.22). By symmetry,

$$\hat{\beta}_f(\varphi_1, \ldots, \varphi_n) = n! \sum_{j_1 < \cdots < j_n} \hat{f}(r_{j_1} \cdots r_{j_n})\, \varphi_1(j_1) \cdots \varphi_n(j_n). \tag{8.34}$$

By Lemma 28, if $\theta \in l^\infty(\mathbb{N})$ is real-valued, then

$$\begin{aligned} |\hat{\beta}_f(\theta, \ldots, \theta)| &= \left| n! \sum_{j_1 < \cdots < j_n} \hat{f}(r_{j_1} \cdots r_{j_n})\, \theta(j_1) \cdots \theta(j_n) \right| \\ &\leq n!\, \|\theta\|_\infty^n\, \|f\|_\infty. \end{aligned} \tag{8.35}$$

By (8.34) and Lemma 27,

$$\begin{aligned} &(n!)^2 \sum_{j_1 < \cdots < j_n} \hat{f}(r_{j_1} \cdots r_{j_n})\, \varphi_1(j_1) \cdots \varphi_n(j_n) \\ &\quad = \mathbf{E}\, r_1 \cdots r_n\, \hat{\beta}_f(\Sigma_{j=1}^n r_j\, \varphi_j, \ldots, \Sigma_{j=1}^n\, r_j\, \varphi_j). \end{aligned} \tag{8.36}$$

By (8.35), for every $\omega \in \Omega$,

$$|\hat{\beta}_f(\Sigma_{j=1}^n r_j(\omega)\, \varphi_j, \ldots, \Sigma_{j=1}^n\, r_j(\omega)\, \varphi_j)| \leq n^n\, n!\, \|f\|_{\mathrm{L}^\infty}. \tag{8.37}$$

Therefore, by (8.36),

$$\begin{aligned} &\left| \sum_{j_1 < \cdots < j_n} \hat{f}(r_{j_1} \cdots r_{j_n})\, \varphi_1(j_1) \cdots \varphi_n(j_n) \right| \\ &\quad \leq \frac{n^n}{n!}\, \|f\|_{\mathrm{L}^\infty} \leq \mathrm{e}^n\, \|f\|_{\mathrm{L}^\infty}. \end{aligned} \tag{8.38}$$

Because f is arbitrary, this implies the lemma for real-valued φ_j. The complex case follows by treating separately real and imaginary parts. □

Proof of Theorem 26:

i. If f is an R_n-polynomial, then,

$$\|f\|_{\mathrm{L}^\infty} \leq (1/n!)\, \|\beta_f\|_{F_n} \leq \mathrm{e}^n \|f\|_{\mathrm{L}^\infty}, \tag{8.39}$$

where the inequality on the right follows from Lemma 29 (see (8.38)), and the inequality on the left is obvious. Therefore, because finitely supported elements in $F_{n\sigma}$ are norm-dense in $F_{n\sigma}$ (Theorem IV.6), the linear map $f \mapsto \beta_f$ determines an isomorphism from $C_{R_n}(\Omega)$ onto $F_{n\sigma}(\mathbb{N},\ldots,\mathbb{N})$.

ii. The assertion follows by duality from (8.39). □

Remarks:

i **(R_n is Rosenthal $\Leftrightarrow$ F_n is separable).** Proposition 25 i implies (by duality)

$$L^{\infty}_{R^n}(\Omega^n, \mathbb{P}^n) = F_n(R,\ldots,R), \tag{8.40}$$

which (because R^n is Rosenthal) implies

$$C_{R^n}(\Omega^n) = F_n(R,\ldots,R) \quad \text{(Proposition 25 ii)}. \tag{8.41}$$

This proves that finitely supported functions on $R \times \cdots \times R$ are norm-dense in $F_n(R,\ldots,R)$. That F_n is separable – we recall – was verified from first principles in Chapter IV (Theorem IV.6).

The argument is reversible: in (8.40), apply the fact that finitely supported functions are norm-dense in $F_n(R,\ldots,R)$ (Theorem IV.6), and conclude that R^n is a Rosenthal set in W^n. With additional effort we can, similarly, obtain that R_n (and hence W_n) is Rosenthal in W: following the proof of Theorem 26, deduce

$$L^{\infty}_{R_n}(\Omega, \mathbb{P}) = F_{n\sigma}(\mathbb{N},\ldots,\mathbb{N}), \tag{8.42}$$

and then, by using norm-density of finitely supported symmetric functions in $F_{n\sigma}(\mathbb{N},\ldots,\mathbb{N})$, conclude that $L^{\infty}_{R_n}(\Omega,\mathbb{P}) = C_{R_n}(\Omega)$. (See Exercises 24 and 25.)

ii **(tilde algebras).** Recall (Chapter IV §5, Remark iii) that the *tilde algebra* $\tilde{V}_n(\mathbb{N},\ldots,\mathbb{N}) = \tilde{V}_n$ comprises all $\varphi \in l^{\infty}(\mathbb{N}^n)$ for which there exist $\{\varphi_k : k \in \mathbb{N}\} \subset V_n$ such that

$$\lim_{k\to\infty} \varphi_k(\mathbf{j}) = \varphi(\mathbf{j}), \quad \mathbf{j} \in \mathbb{N}^n, \tag{8.43}$$

and

$$\limsup_{k\to\infty} \|\varphi_k\|_{V_n} < \infty.$$

The norm in $\tilde{V}_n$ is

$$\|\varphi\|_{\tilde{V}_n} = \inf\left\{\limsup_{k\to\infty} \|\varphi_k\|_{V_n} : \lim_{k\to\infty} \varphi_k(\mathbf{j}) = \varphi(\mathbf{j}), \quad \mathbf{j} \in \mathbb{N}^n\right\}. \tag{8.44}$$

By Proposition 25 iii, the tilde algebra $\tilde{V}_n$ is the Banach algebra of restrictions of transforms to R^n:

$$\tilde{V}_n(\mathbb{N}, \ldots, \mathbb{N}) = B(R^n). \tag{8.45}$$

We will use this equality in the next chapter, in characterizations of multilinear Grothendieck-type inequalities.

iii **(credits).** Connections between harmonic analysis and tensor analysis were discovered by Varopoulos [Herz]. (See Chapter IV §8.) The equivalences between restriction algebras and tensor algebras in Proposition 25 were brought to light in [V3].

An isomorphism similar to $\mathrm{C}_{R_n}(\Omega) = F_{n\sigma}(\mathbb{N}, \ldots, \mathbb{N})$ (Theorem 26 i) was first shown by A.M. Davie in [Da, Lemma 2.1] by a combinatorial device analogous to the Mazur–Orlicz identity in Lemma 27. (See Exercise 27.) We will revisit the isomorphism $\mathrm{C}_{R_n}(\Omega) = F_{n\sigma}(\mathbb{N}, \ldots, \mathbb{N})$ in Chapter XI §5.

9 Bonami's Inequalities: A Measurement of Complexity

Thus far we have shown that $\mathrm{L}^\infty_{W_n}(\Omega, \mathbb{P}) = \mathrm{C}_{W_n}(\Omega)$ for all $n \in \mathbb{N}$, a property that distinguishes Walsh systems of finite order from the full Walsh system. In this and the next two sections we will distinguish between the W_n themselves.

We begin with the Khintchin inequalities, which follow from the statistical independence of R (Exercise II.3): for all R-polynomials f,

$$\|f\|_{\mathrm{L}^p} \leq \sqrt{p}\, \|f\|_{\mathrm{L}^2}, \quad p > 2. \tag{9.1}$$

In Chapter X we will interpret the constants' growth $\mathcal{O}(\sqrt{p})$ in (9.1) as yet another manifestation of independence. Here we verify that this constants' growth is best possible:

Lemma 30 (cf. Exercise 28). *For every positive integer k,*

$$2\left\|\sum_{j=1}^{k} r_j\right\|_{\mathrm{L}^k} \geq k = \sqrt{k}\left\|\sum_{j=1}^{k} r_j\right\|_{\mathrm{L}^2}. \tag{9.2}$$

Proof: Consider the Riesz product (cf. (3.7))

$$F_k = \prod_{j=1}^{k}(1 + r_j), \tag{9.3}$$

and observe that

$$\mathbf{E}\, F_k \sum_{j=1}^{k} r_j = k. \tag{9.4}$$

Also note that

$$\|F_k\|_{\mathrm{L}^1} = 1 \quad \text{and} \quad \|F_k\|_{\mathrm{L}^2} = 2^{k/2}. \tag{9.5}$$

Therefore, for every $q \in (1, 2)$,

$$\|F_k\|_{\mathrm{L}^q} \leq 2^{k/p}, \quad \frac{1}{p} + \frac{1}{q} = 1 \text{ (Exercise 29)}. \tag{9.6}$$

In (9.4) apply Hölder's inequality with exponents $q = k/(k-1)$ and $p = k$, use (9.6), and conclude

$$k = \mathbf{E}\, F_k \sum_{j=1}^{k} r_j \leq 2 \left\| \sum_{j=1}^{k} r_j \right\|_{\mathrm{L}^k}. \tag{9.7}$$

$\square$

Proposition 31 (n-dimensional Khintchin inequalities; Exercise 31). *For all $n \in \mathbb{N}$, and all $f \in \mathrm{L}^2_{R^n}(\Omega^n, \mathbb{P}^n)$,*

$$\|f\|_{\mathrm{L}^p} \leq p^{n/2}\, \|f\|_{\mathrm{L}^2}, \quad p > 2. \tag{9.8}$$

Moreover, for all $k \subset \mathbb{N}$,

$$2^n \left\| \sum_{i_1,\ldots,i_n=1}^{k} r_{i_1} \otimes \cdots \otimes r_{i_n} \right\|_{\mathrm{L}^k} \geq k^n$$

$$= k^{n/2} \left\| \sum_{i_1,\ldots,i_n=1}^{k} r_{i_1} \otimes \cdots \otimes r_{i_n} \right\|_{\mathrm{L}^2}; \tag{9.9}$$

that is, the constants' growth $\mathscr{O}(p^{n/2})$ in (9.8) is best possible.

Proof: (by induction on n). The case $n = 1$ is (9.1) and Lemma 30. Let $n > 1$, and let f be an R^n-polynomial. Then,

$$\begin{aligned}
\|f\|_{\mathrm{L}^p}^p &= \mathbf{E}\left|\sum_{i_1,\dots,i_n} \hat{f}(r_{i_1},\dots,r_{i_n})\, r_{i_1}\otimes\cdots\otimes r_{i_n}\right|^p \\
&= \mathbf{E}_{\omega_2\dots\omega_n}\left(\mathbf{E}_{\omega_1}\left|\sum_{i_1}\left(\sum_{i_2,\dots,i_n} \hat{f}(r_{i_1},\dots,r_{i_n}) r_{i_2}(\omega_2)\cdots r_{i_n}(\omega_n)\right) r_{i_1}(\omega_1)\right|^p\right) \\
&\le p^{p/2}\,\mathbf{E}_{\omega_2\cdots\omega_n}\left(\sum_{i_1}\left|\sum_{i_2,\dots,i_n} \hat{f}(r_{i_1},\dots,r_{i_n}) r_{i_2}(\omega_2)\cdots r_{i_n}(\omega_n)\right|^2\right)^{p/2} \\
&\le p^{p/2}\left(\sum_{i_1}\left(\mathbf{E}_{\omega_2\dots\omega_n}\left|\sum_{i_2,\dots,i_n} \hat{f}(r_{i_1},\dots,r_{i_n}) r_{i_2}(\omega_2)\cdots r_{i_n}(\omega_n)\right|^p\right)^{2/p}\right)^{p/2} \\
&\le p^{p/2}\, p^{p(n-1)/2}\left(\sum_{i_1,\dots,i_n} |\hat{f}(r_{i_1},\dots,r_{i_n})|^2\right)^{p/2} \\
&= p^{pn/2}\, \|f\|_{\mathrm{L}^2}^p.
\end{aligned} \tag{9.10}$$

The first inequality in (9.10) is a consequence of (9.1); the second is a consequence of the generalized Minkowski inequality via the interchange of $\mathbf{E}_{\omega_2\dots\omega_n}$ and $\sum_{i_1}$; and the third inequality follows from the induction hypothesis.

To obtain (9.9), note that

$$\left\|\sum_{i_1,\dots,i_n=1}^{k} r_{i_1}\otimes\cdots\otimes r_{i_n}\right\|_{\mathrm{L}^k(\Omega^n)} = \left\|\sum_{j=1}^{k} r_j\right\|_{\mathrm{L}^{kn}(\Omega)}^n \ge \left\|\sum_{j=1}^{k} r_j\right\|_{\mathrm{L}^k}^n, \tag{9.11}$$

and then apply Lemma 30. □

Next, by using $A(R_n) = V_n|_{D_n}$ (Theorem 26 ii), we transport the n-dimensional Khintchin inequalities involving R^n to the setting W_n. These inequalities were established first by Aline Bonami [Bon1], [Bon2], by combinatorial methods.

Theorem 32 (Bonami's inequalities). *For all integers $n > 1$, and all $f \in L^2_{W_n}(\Omega, \mathbb{P})$,*

$$\|f\|_{L^p} \leq 2\, e^n p^{n/2}\, \|f\|_{L^2}, \quad p > 2. \tag{9.12}$$

Moreover, the constants' growth in (9.12) is optimal: for all $\alpha < n/2$,

$$\sup\{\|f\|_{L^p}/p^\alpha : p > 2, f \in B_{L^2_{W_n}}\} = \infty. \tag{9.13}$$

Proof: Let f be an R_n-polynomial,

$$f = \sum_{i_1 < \cdots < i_n} \hat{f}(r_{i_1} \cdots r_{i_n})\, r_{i_1} \cdots r_{i_n}. \tag{9.14}$$

Fix $\mathbf{s} = (s_1, \ldots, s_n) \in \Omega^n$, and define

$$f_{\mathbf{s}} = \sum_{i_1 < \cdots < i_n} \hat{f}(r_{i_1} \cdots r_{i_n})\, r_{i_1}(s_1) \cdots r_{i_n}(s_n)\, r_{i_1} \cdots r_{i_n}. \tag{9.15}$$

By Theorem 26, there exist $F_{\mathbf{s}} \in L^1(\Omega, \mathbb{P})$ such that

$$\hat{F}_{\mathbf{s}}(r_{i_1} \cdots r_{i_n}) = r_{i_1}(s_1) \cdots r_{i_n}(s_n), \quad r_{i_1} \cdots r_{i_n} \in \text{spect } f, \tag{9.16}$$

and

$$\|F_{\mathbf{s}}\|_{L^1} \leq e^n. \tag{9.17}$$

(In the application of Theorem 26, the factor 2^n in (8.24) can be dropped because the right side of (9.16) is real-valued.) Observe that

$$f = F_s \star f_{\mathbf{s}}, \tag{9.18}$$

and therefore,

$$\|f\|^p_{L^p} = \|F_s \star f_{\mathbf{s}}\|^p_{L^p} \leq \|F_s\|^p_{L^1} \|f_{\mathbf{s}}\|^p_{L^p} \leq e^{np} \|f_{\mathbf{s}}\|^p_{L^p}. \tag{9.19}$$

By integrating (9.19) over $\mathbf{s} \in (\Omega^n, \mathbb{P}^n)$, and then interchanging integrations, we obtain

$$\|f\|^p_{L^p} \leq e^{np}\, \mathbf{E}_\omega \left(\mathbf{E}_{\mathbf{s}} \left| \sum_{i_1 < \cdots < i_n} \hat{f}(r_{i_1} \cdots r_{i_n}) r_{i_1}(\omega) \cdots r_{i_n}(\omega)\, r_{i_1}(s_1) \cdots r_{i_n}(s_n) \right|^p \right). \tag{9.20}$$

An application in (9.20) of the n-dimensional Khintchin inequalities implies

$$\|f\|^p_{L^p} \le e^{np} p^{np/2} \left(\sum_{i_1<\cdots<i_n} |\hat{f}(r_{i_1}\cdots r_{i_n})|^2 \right)^{p/2}$$

$$= e^{np}\, p^{np/2}\, \|f\|_{L^2}, \tag{9.21}$$

which verifies (9.12) in the case $f \in L^2_{R_n}(\Omega, \mathbb{P})$.

Let $f \in L^2_{W_n}(\Omega, \mathbb{P})$, and write $f = \Sigma^n_{j=0} f_j$, where $f_j \in L^2_{R_j}(\Omega, \mathbb{P})$ for $j = 0, \ldots, n$. Then, by (9.21) (applied to each f_j) and the Cauchy–Schwarz inequality,

$$\|f\|_{L^p} \le \sum_{j=0}^{n} \|f_j\|_{L^p} \le \sum_{j=0}^{n} e^j\, p^{j/2} \|f_j\|_{L^2}$$

$$\le \left(\sum_{j=0}^{n} e^{2j}\, p^j \right)^{\frac{1}{2}} \|f\|_{L^2} \le 2e^n p^{n/2}\, \|f\|_{L^2}. \tag{9.22}$$

To verify (9.13), notice that R_n contains a copy of the n-fold Cartesian product R^n: let $P_1, \ldots, P_n$ be infinite pairwise disjoint subsets of R, consider $P_1 \cdots P_n$, and apply (9.9). □

Remarks:

i **(history).** The case $p = 2m$, $m \in \mathbb{N}$, in (9.1) was proved first by A. Khintchin in his classic 1923 paper [Kh1, pp. 111–12]. Khintchin needed this to deduce exponential tail-probability estimates for the distribution of the deviation from $n/2$ of the number of 1s among the first n digits in the binary expansion of a random point in ([0,1], *Lebesgue measure*). He concluded from these estimates that if $\mu(n)$ denotes the aforementioned deviation, then

$$\mu(n) = \mathscr{O}((n \log\log n)^{\frac{1}{2}}) \quad \text{almost surely.} \tag{9.23}$$

This was the first half of Khintchin's celebrated *law of the iterated logarithm*. A year later he published the full statement:

$$\overline{\lim_{n\to\infty}} \left(\frac{\mu(n)}{(2n \log\log n)^{1/2}} \right) = 1 \quad \text{almost surely [Kh2].} \tag{9.24}$$

The first half of the law follows from Khintchin's inequalities, while the second half corresponds to Lemma 30. K. Chung called this *law of the iterated logarithm* 'a crowning achievement in classical probability theory' [Ch, p. 231].

In a 1926 paper [Lit3], interfacing complex and harmonic analysis, Littlewood established inequalities nearly identical to (9.1), wherein, in place of Rademacher functions, he used independent $\mathbf{T}_\infty$-valued random variables uniformly distributed in [0,1] [Lit3, Lemma 3]. ($\mathbf{T}_\infty := \{z \in \mathbb{C} : |z| = 1\}$; see (II.6.7).) These random variables were later dubbed 'Steinhaus functions' by Salem and Zygmund [SaZY1, p. 285]; see Chapter II §6. The two sets of inequalities, involving separately the Rademacher system and the Steinhaus system, turn out to be equivalent: each is derivable from the other. (This will be shown in §12). Unaware of Khintchin's prior work, Littlewood obtained his inequalities by a combinatorial argument very similar to Khintchin's, but going a little further, Littlewood deduced the L^2–L^p inequalities for *all* $p > 2$, and, by use of a convexity argument [Lit3, Lemma 2], derived also the L^1–L^p inequalities for all $p \in (1, 2]$ (cf. Chapter II).

The inequalities in (9.1) for the Rademacher functions were (re)proved in a joint 1930 paper by Paley and Zygmund [PaZy1, Lemma 2], who, like Littlewood, were unaware of Khintchin's 1923 inequalities. Paley and Zygmund did not state the connection between Littlewood's inequalities involving the Steinhaus system [Lit3, Lemma 2], which they knew, and those involving the Rademacher functions in their own paper. (Three decades later, Zygmund, without citing [Kh1], called (9.1) 'a classical result of the Calculus of Probability' [Zy2, p. 380].) These inequalities, including the L^1–L^p inequalities for $p \in (1, 2]$ are commonly known today as the *Khintchin inequalities.*

Theorem 32 implies

$$\mathrm{L}^1_{W_n}(\Omega, \mathbb{P}) = \mathrm{L}^p_{W_n}(\Omega, \mathbb{P}), \quad n \in \mathbb{N},\ p \in (1, \infty). \tag{9.25}$$

This property of a spectral set, that every L^1-function with spectrum therein is in L^p for $p > 1$, was first noted by Sidon for lacunary sets in $\mathbb{Z}^+$ and $p = 2$ [Si3], and independently by Zygmund [Zy1], also in the case of lacunary sets but for *all* $p > 1$. In this regard, Littlewood's 1926 paper [Lit3] was crucial for the Khintchin-type inequalities needed by Zygmund (cf. [Zy1, p. 140]); see Exercise 30.

(This very same 1926 paper [Lit3] led also to Littlewood's 1930 paper [Lit4], much about which has been said in previous chapters.) The various studies during the 1940s and 1950s of lacunarity *vis à vis* the property in (9.25) were eventually recounted and summarized in Zygmund's 1959 treatise [Zy2]. Among these studies, of particular interest (to us) is Salem's and Zygmund's constructive proof [SaZy1] of Banach's 1930 theorem, that if $F \subset \mathbb{Z}^+$ is lacunary and $\varphi \in l^2(F)$, then there exist $f \in \mathrm{C}(\mathbf{T})$ such that $\hat{f}|_F = \varphi$ [Ban, p. 212]. Indeed, the same device (an L^∞-type Riesz product) used to construct such an $f \in \mathrm{C}(\mathbf{T})$ in [SaZy1] was used in Chapter III, in the proof that $\mathrm{L}^2_R(\Omega, \mathbb{P})$ is a *uniformizable* $\Lambda(2)$-space; see (III.6.1).

The first systematic study of the property exemplified by (9.25) appeared in Walter Rudin's 'Trigonometric series with gaps' [Ru1], arguably among the most influential works in harmonic analysis in the latter half of the twentieth century. (Some of its highlights have already been discussed in Chapter III §6.) In that paper Rudin introduced the notion of $\Lambda(p)$-sets, casting it in the setting of $\mathbb{Z}$ (Definition III.12). The same notion, of course, can be viewed equally well in any Abelian group setting [Ru3]. (This notion was further generalized by A. Figà-Talamanca and D. Rider in a framework of non-Abelian compact groups [FigRid].)

Definition 33 Let G be a compact Abelian group. A spectral set $F \subset \hat{G}$ is a $\Lambda(p)$-set for $p \in (1, \infty)$ if

$$\mathrm{L}^1_F(G) = \mathrm{L}^p_F(G). \tag{9.26}$$

Equivalently, $F \subset \hat{G}$ is a $\Lambda(p)$-set if there exists $k_p > 0$ ($\Lambda(p)$-constant) such that for all F-polynomials g,

$$\|g\|_{\mathrm{L}^p} \leq k_p \, \|g\|_{\mathrm{L}^1}. \tag{9.27}$$

(In Rudin's original formulation, the range of p was $(0, \infty)$; here, for our purposes, it suffices to consider $p \in (1, \infty)$.)

Underscoring the significance of the constants in (9.27), Rudin demonstrated $k_p = \mathcal{O}(\sqrt{p})$ for all Sidon sets. He proved this growth to be optimal, and then raised a fundamental question: does $k_p = \mathcal{O}(\sqrt{p})$ characterize Sidonicity? The question was answered in the affirmative seventeen years later by Gilles Pisier [P1].

Following Rudin's 1960 paper [Ru1], the next major advance was Aline Bonami's landmark work [Bon1], [Bon2] (her Ph.D. dissertation) dealing with non-Sidon $\Lambda(p)$-sets and associated growths of $\Lambda(p)$-constants. Her work was motivated by the following question: if G is a compact Abelian group and $q \in (1, 2]$, then which $\varphi \in l^\infty(\hat{G})$ have the property that for all $f \in \mathrm{L}^q(G)$,

$$\sum_{\gamma \in \hat{G}} \varphi(\gamma)\ |\hat{f}(\gamma)|^2 < \infty? \tag{9.28}$$

Notice that $F \subset \hat{G}$ is $\Lambda(p)$ if and only if (9.28) is satisfied with $\varphi = \mathbf{1}_F$ for all $f \in \mathrm{L}^q(G)$, $1/q + 1/p = 1$ (Exercise 31). In her 1968 paper [Bon1], working mainly in the group $\{-1, 1\}^{\mathbb{N}}$, Bonami proved Theorem 32 by an intricate combinatorial argument, producing k_ps somewhat sharper than those obtained in the proof above (but with same growth). In her 1970 article [Bon2], which has become a classic, she generalized and extended this result.

ii **(a measurement of complexity).** Bonami's inequalities suggest the following measurement. For $F \subset W$, let

$$\eta_F(a) = \sup\{\|f\|_{\mathrm{L}^p}/p^a : p > 2, f \in B_{\mathrm{L}^2_F(\Omega,\mathbb{P})}\}, \quad a > 0, \tag{9.29}$$

and then define

$$\delta_F = \inf\{a : \eta_F(a) < \infty\}. \tag{9.30}$$

If F is finite, then $\delta_F = 0$, in which case η_F is the relevant measurement. If $\delta_F = \alpha$ and $\eta_F(\alpha) < \infty$, then we say that $\delta_F = \alpha$ *exactly*; otherwise, if $\eta_F(\alpha) = \infty$, then we say that $\delta_F = \alpha$ *asymptotically*. If F is infinite, then $\delta_F \in [1/2, \infty]$. Observe that $\delta_W = \infty$, and if $F \subset W$ is infinite and statistically independent, then $\delta_F = 1/2$ *exactly* (cf. (9.1) and (9.2)). Bonami's inequalities state

$$\delta_{W_n} = \frac{n}{2} \quad \textit{exactly}, \quad n \in \mathbb{N}. \tag{9.31}$$

I view δ_F as a measurement of *complexity*, a notion I have thus far used in a heuristic sense. Making this precise in Chapter X, we will interpret the δ-scale as a gauge of statistical *interdependence* in a probabilistic context.

iii **(a preview).** Bonami's inequalities, as stated in (9.31), naturally lead to the question whether the δ-scale is 'continuous': *for arbitrary* $x \in (1/2, \infty)$, *are there* $F \subset W$ *such that* $\delta_F = x$? In Chapter XIII, we will resolve this and related questions by constructing Walsh systems of 'non-integer order'.

10 The Littlewood $2n/(n+1)$-Inequalities: Another Measurement of Complexity

In the previous section we detected the complexity of W_n by a measurement that conveys a degree of *statistical interdependence*; we shall revisit this measurement in Chapter X. Next, we will detect the complexity of W_n by yet another measurement, which, in effect, marks a degree of *functional interdependence.*

We begin with the basic property of the Rademacher system that we have already used and highlighted several times (e.g., (II.1.5), Remark in §2, Proposition 15, Remark in §5):

$$\|\hat{f}\|_1 := \sum_{w \in W_1} |\hat{f}(w)| \leq c_1 \|f\|_{L^\infty}, \quad \text{for all } f \in C_{W_1}(\Omega), \tag{10.1}$$

where c_1 denotes the best constant in the inequality (Exercise II.9). Obviously, $\|\hat{f}\|_1 \geq \|f\|_{L^\infty}$ for all $f \in C(\Omega)$, and therefore $\|\hat{f}\|_1$ cannot be replaced in (10.1) by $\|\hat{f}\|_p$ where $p < 1$. The theorem below extends these two observations to Walsh systems of finite order.

Theorem 34 *For all* $n \in \mathbb{N}$, *and all* $f \in C_{W_n}(\Omega)$,

$$\|\hat{f}\|_{2n/(n+1)} := \left(\sum_{w \in W_n} |\hat{f}(w)|^{\frac{2n}{n+1}} \right)^{\frac{n+1}{2n}} \leq c_n \|f\|_\infty, \tag{10.2}$$

where $c_n > 0$ *depend only on* n.

Moreover, (10.2) *is sharp: there exist* $f \in C_{W_n}(\Omega)$ *such that* $\|\hat{f}\|_p = \infty$ *for all* $p < 2n/n+1$.

The theorem has two parts: the inequality in (10.2), which we prove in this section, and the optimality of (10.2), which we will prove in the next section. To prove (10.2), we need n-linear extensions of Littlewood's (bilinear) 4/3-inequality [Lit4] (Chapter II §5), and to this end, we obtain below n-linear extensions of Littlewood's and Orlicz's bilinear mixed-norm inequalities.

Lemma 35 (cf. Theorems II.2, II.3). *For all integers $n \geq 1$, and $f \in \mathrm{C}_{R^n}(\Omega^n)$,*

$$c_1(\sqrt{2})^{n-1}\ \|f\|_{\mathrm{L}^\infty} \geq \left\{ \sum_{(w_2,\dots,w_n)\in R^{n-1}} \left(\sum_{w_1\in R} |\hat{f}(w_1,\dots,w_n)| \right)^2 \right\}^{\frac{1}{2}}. \tag{10.3}$$

Proof: The case $n = 1$ is (10.1). Let $n \geq 2$, and note (an extension of Littlewood's (l^1, l^2)-mixed norm inequality):

$$\begin{aligned} &c_1\ \|f\|_{\mathrm{L}^\infty} \\ &\geq \left\| \sum_{w_1\in R} \left| \sum_{(w_2,\dots,w_n)\in R^{n-1}} \hat{f}(w_1,\dots,w_n)\ w_2 \otimes \cdots \otimes w_n \right| \right\|_{\mathrm{L}^\infty(\Omega^{n-1})} \\ &\geq \sum_{w_1\in R} \mathbf{E}_{\omega_2\dots\omega_n} \left| \sum_{(w_2,\dots,w_n)\in R^{n-1}} \hat{f}(w_1,\dots,w_n)\ w_2(\omega_1)\cdots w(\omega_n) \right| \\ &\geq (1/\sqrt{2})^{n-1} \sum_{(w_2,\dots,w_n)\in R^{n-1}} \left(\sum_{w_1\in R} |\hat{f}(w_1,\dots,w_n)|^2 \right)^{\frac{1}{2}}, \end{aligned} \tag{10.4}$$

where the first inequality follows from (10.1), and the third follows from the $(n-1)$-dimensional Khintchin L^1–L^2 inequality (Exercise 32). An application of the generalized Minkowski inequality to (10.4) implies (10.3) (an extension of Orlicz's (l^2, l^1)-mixed norm inequality). □

Theorem 36 (the Littlewood $2n/(n+1)$-inequality). *For all $n \in \mathbb{N}$, and all $f \in \mathrm{C}_{R^n}(\Omega^n)$,*

$$\|\hat{f}\|_{2n/n+1} \leq 2^{\frac{n+1}{2}}\ \|f\|_{\mathrm{L}^\infty}. \tag{10.5}$$

Proof: (by induction on n). The case $n = 1$ is (10.1). Let $n > 1$, and define

$$c_n = \sup\{\|\hat{f}\|_{2n/n+1} : f \in B_{\mathrm{C}_{R^n}(\Omega^n)}\}. \tag{10.6}$$

Assume that $c_{n-1} < \infty$. Let $f \in \mathrm{C}_{R^n}(\Omega^n)$ be a polynomial. Then,

$$
\begin{aligned}
&(c_{n-1}\|f\|_{\mathrm{L}^\infty})^{\frac{2n-2}{n}} \\
&\quad \geq \sup_{\omega\in\Omega} \sum_{(w_2,\dots,w_n)\in R^{n-1}} \left|\sum_{w_1\in R} \hat{f}(w_1,\dots,w_n)\, w_1(\omega)\right|^{\frac{2n-2}{n}} \\
&\quad \geq \sum_{(w_2,\dots,w_n)\in R^{n-1}} \mathbf{E}_\omega \left|\sum_{w_1\in R} \hat{f}(w_1,\dots,w_n)\, w_1(\omega)\right|^{2-\frac{2}{n}} \\
&\quad \geq (1/2)^{\frac{n-1}{n}} \sum_{(w_2,\dots,w_n)\in R^{n-1}} \left(\sum_{w_1\in R} |\hat{f}(w_1,\dots,w_n)|^2\right)^{\frac{n-1}{n}} .
\end{aligned}
\tag{10.7}
$$

The first inequality in (10.7) is a consequence of the induction hypothesis; the second is obvious, and the third follows from the Khintchin $\mathrm{L}^{2-2/n}$–L^2 inequality. (In the latter inequality, we use the L^1–L^2 Khintchin constant, which is $\sqrt{2}$; see Remark i below.) Next, we write

$$
\begin{aligned}
&\sum_{(w_1,\dots,w_n)\in R^n} |\hat{f}(w_1,\dots,w_n)|^{\frac{2n}{n+1}} \\
&\quad = \sum_{(w_2,\dots,w_n)\in R^{n-1}} \sum_{w_1\in R} |\hat{f}(w_1,\dots,w_n)|^{\frac{2n-2}{n+1}} |\hat{f}(w_1,\dots,w_n)|^{\frac{2}{n+1}} .
\end{aligned}
\tag{10.8}
$$

On the right side of (10.8), first apply Hölder's inequality to the sum over w_1 with exponents $(n+1)/(n-1)$ and $(n+1)/2$, and then apply Hölder's inequality with exponents $(n+1)/n$ and $n+1$ to the sum over $w_2, \dots, w_n$. These two applications of Hölder's inequality result in

$$
\begin{aligned}
&\sum_{(w_1,\dots,w_n)\in R^n} |\hat{f}(w_1,\dots,w_n)|^{\frac{2n}{n+1}} \\
&\quad \leq \left\{ \sum_{w_2,\dots,w_n} \left(\sum_{w_1} |\hat{f}(w_1,\dots,w_n)|^2 \right)^{\frac{n-1}{n}} \right\}^{\frac{n}{n+1}} \\
&\qquad \cdot \left\{ \sum_{w_2,\dots,w_n} \left(\sum_{w_1} |\hat{f}(w_1,\dots,w_n)| \right)^2 \right\}^{\frac{1}{n+1}} .
\end{aligned}
\tag{10.9}
$$

Apply (10.3) and (10.7) to (10.9), thus obtaining

$$\|\hat{f}\|_{2n/(n+1)} \leq (c_1)^{\frac{1}{n}} (2c_{n-1})^{1-\frac{1}{n}} \|f\|_{L_\infty}. \tag{10.10}$$

Using the estimate $c_1 \leq 2$ in (10.10), we obtain

$$(c_n)^n/(c_{n-1})^{n-1} \leq 2^n, \tag{10.11}$$

which (by 'telescoping') implies $c_n \leq 2^{(n+1)/2}$. □

Remarks:

i **(better constants).** To obtain the estimate $c_n \leq 2^{(n+1)/2}$, we used the L^p–L^2 inequality

$$\kappa_p \left(\mathbf{E} \left| \sum_{j=1}^{N} a_j \, r_j \right|^p \right)^{\frac{1}{p}} \geq \left(\sum_{j=1}^{N} |a_j|^2 \right)^{\frac{1}{2}}, \quad N \in \mathbb{N}, \tag{10.12}$$

which we applied in (10.7) with $p = 2-2/n$ and the estimate $\kappa_p \leq \sqrt{2}$ (the Khintchin L^1–L^2 constant). It was shown by U. Haagerup [H1] that the best constants κ_p in (10.12) satisfy

$$\kappa_p \leq 2^{\frac{1}{p}-\frac{1}{2}}, \; p \in (0, 2]. \tag{10.13}$$

More precisely, Haagerup showed that there exists p_0, whose approximate value is 1.847..., such that (i) if $p \leq p_0$ then $\kappa_p = 2^{1/p-1/2}$, and (ii) if $p > p_0$ then κ_p (computed in [H1]) is strictly less than $2^{1/p-1/2}$. Using (10.13) in the third line of (10.7), we deduce

$$c_n \leq \sqrt{2} \; (c_1)^{\frac{1}{n}} (c_{n-1})^{\frac{n-1}{n}}, \tag{10.14}$$

thus obtaining an estimate sharper than (10.5),

$$c_n \leq c_1 (\sqrt{2})^{(\frac{n+1}{2}-\frac{1}{n})}, n \geq 1. \tag{10.15}$$

The value of c_n, $n \in \mathbb{N}$, is unknown.

ii **(history in brief).** The 4/3-inequality first appeared in Littlewood's 1930 paper [Lit4]. Its first application, since its discovery by Littlewood, was in R. Edward's and K. Ross's 1974 work [ERos] on *p-Sidon sets.* (More will be said about p-Sidon sets in the next section.) The n-linear extension of Littlewood's inequality was first noted by A.M. Davie, who had no use for it [Da, (2.2)]. It was also noted, independently, by G. Johnson and G. Woodward in their paper on p-Sidon sets [JWo], which followed Edwards's and Ross's work [ERos]. In Davie's paper, the inequality was stated without proof, and in Johnson's and Woodward's paper it was derived by an n-dimensional version of Littlewood's original argument. The inductive argument above, starting from the case $n = 1$, is different from theirs, and yields sharper constants.

iii **(Theorem 36 is best possible).** In his paper [Lit4] Littlewood observed, by use of the Gauss matrix, that the 4/3-inequality (the case $n = 2$) was sharp; see Chapter II §5. In order to highlight the harmonic analysis implicit in Littlewood's original argument, we recast it below in the framework of this chapter.

Lemma 37 *For every $k \in \mathbb{N}$, there exists a $\{-1,1\}$-valued 2-array β indexed by $[2^k] \times [2^k]$, such that*

$$\|\beta\|_{F_2} \leq 2^{3k/2}. \tag{10.16}$$

Proof: Consider the compact Abelian group $\{-1,1\}^k$, and denote it by $\Omega(k)$. Its dual group is

$$\{r_{j_1} \cdots r_{j_k} : 0 \leq j_1 \leq \cdots \leq j_k \leq k\}, \tag{10.17}$$

which we denote by $W(k)$. (We have already encountered $W(k)$ in the beginning of the chapter; see (3.9).) Define

$$\beta(\omega, w) = w(\omega), \quad w \in W(k), \quad \omega \in \Omega(k). \tag{10.18}$$

Let f be an arbitrary scalar-valued function on $\Omega(k)$, $\|f\|_\infty \leq 1$, and consider its transform

$$\hat{f}(w) = (1/2^k) \sum_{\omega \in \Omega(k)} f(\omega)\, \beta(\omega, w), \quad w \in W(k). \tag{10.19}$$

By Plancherel's theorem, if g is a scalar-valued function on $W(k)$ such that

$$\sum_{w \in W(k)} |g(w)|^2 \leq 1,$$

then

$$\left| \sum_{w \in W(k)} \hat{f}(w) g(w) \right| = (1/2^k) \left| \sum_{w \in W(k)} \sum_{\omega \in \Omega(k)} \beta(\omega, w) f(\omega) g(w) \right| \leq 1. \tag{10.20}$$

If h is a scalar-valued function on $W(k)$, $\|h\|_\infty \leq 1$, then put $g = (1/\sqrt{2^k})h$ in (10.20). Enumerate $\Omega(k)$ and $W(k)$ by $[2^k]$ (any enumeration will do), and deduce (10.16). $\square$

Corollary 38 *There exist $f \in C_{R^2}(\Omega^2)$ such that $\|\hat{f}\|_p = \infty$ for all $p < 4/3$.*

Proof: Let $A_k \subset R$ ($k \in \mathbb{N}$) be pairwise disjoint sets such that $|A_k| = 2^k$. Fix k and take (any) two enumerations of A_k: one by $\Omega(k)$, and the other by $W(k)$. Then, by applying Lemma 37, produce $f_k \in C_{A_k \times A_k}(\Omega^2)$ such that for $(w_1, w_2) \in A_1 \times A_2$

$$\|f_k\|_{L^\infty} \leq 1, \quad \text{and} \quad |\hat{f}_k(w_1, w_2)| = 1/2^{3k/2}. \tag{10.21}$$

Note

$$\|\hat{f}_k\|_p^p = 2^{kp(\frac{2}{p} - \frac{3}{2})}, \quad p > 0. \tag{10.22}$$

Define

$$f = \sum_{k=1}^{\infty} f_k/k^2, \tag{10.23}$$

and conclude that $f \in C_{R^2}(\Omega^2)$, and $\|\hat{f}\|_p = \infty$ for all $p < 4/3$. $\square$

The 2-array β in (10.18) is analogous to the Gauss matrix in Littlewood's proof (cf. (II.5.5)). A similar construction, producing the same effect, can be given in the case $n = 3$. Fix an integer $k > 1$, and define a $\{-1, 1\}$-valued 3-array β_2 by

$$\beta_2(\omega_1, \omega_2, w_3) = w_3(\omega_1)\ w_3(\omega_2), \qquad w_3 \in W(k), \ \ \omega_1 \in \Omega(k), \ \omega_2 \in \Omega(k). \tag{10.24}$$

Let f and g be scalar-valued functions on $\Omega(k)$, $\|f\|_\infty \leq 1$, $\|g\|_\infty \leq 1$, and let h be a scalar-valued function on $W(k)$, $\|h\|_\infty \leq 1$. Then,

$$\begin{aligned}(1/2^{2k})&\left|\sum_{\omega_1 \in \Omega(k), \omega_2 \in \Omega(k), w_3 \in W(k)} \beta_2(\omega_1, \omega_2, w_3) f(\omega_1)\ g(\omega_2)\ h(w_3)\right| \\ &= \left|\sum_{w \in W(k)} \hat{f}(w)\ \hat{g}(w)\ h(w)\right| \leq \|\hat{f}\|_2\ \|\hat{g}\|_2\ \|h\|_\infty \\ &\leq \|f\|_\infty\ \|g\|_\infty\ \|h\|_\infty. \end{aligned} \tag{10.25}$$

The equality above follows by inversion of the transform; the first inequality follows by Cauchy–Schwarz, and the second inequality follows by Plancherel. We conclude

$$\|(1/2^{2k})\ \beta_2\|_{F_3} \leq 1, \quad \|(1/2^{2k})\ \beta_2\|_p = 2^{k(\frac{3}{p}-2)} \quad \text{for } p > 0, \tag{10.26}$$

and proceed, as in the proof of Corollary 38, to construct $f \in \mathrm{C}_{R^3}(\Omega^3)$ such that $\|\hat{f}\|_p = \infty$ for all $p < 3/2$.

These are the only explicit constructions I know. In the next section we establish by an indirect argument that the $2n/(n+1)$-inequalities are optimal for *all* $n \in \mathbb{N}$. In Chapter X we will give a direct proof, based on random constructions. Otherwise, for $n \geq 4$, I do not know any 'deterministic' constructions verifying that the Littlewood $2n/(n+1)$-inequalities are best possible.

We conclude the section with the proof of the first half of Theorem 34:

Proof of (10.2). Let $f \in \mathrm{C}_{R_n}(\Omega)$ be a polynomial, and define $\tilde{f} \in \mathrm{C}_{R^n}(\Omega^n)$ by

$$(\tilde{f})\hat{}\,(w_1, \ldots, w_n) = \hat{f}(w_1 \cdots w_n), \quad (w_1, \ldots, w_n) \in R^n. \tag{10.27}$$

($\tilde{f}$ corresponds to $\beta_f \in F_n(R, \ldots, R)$ in (8.22).) By Theorem 26,

$$\|\hat{f}\|_{\mathrm{L}^\infty} \leq n!\ (2\mathrm{e})^n\ \|f\|_{\mathrm{L}^\infty}. \tag{10.28}$$

Then, by Theorem 36,

$$(n!)^{\frac{n+1}{2n}}\ \|\hat{f}\|_{2n/n+1} = \|(\tilde{f})\hat{}\,\|_{2n/n+1} \leq (\sqrt{2})^{n+1}\ n!\ (2\mathrm{e})^n\ \|f\|_{\mathrm{L}^\infty}. \tag{10.29}$$

By Lemma 22, if $f \in \mathrm{C}_{W_n}(\Omega)$, then $f = \Sigma_{j=0}^n f_j$ where $f_j \in \mathrm{C}_{R_j}(\Omega)$, and there exists a constant $K_n > 0$ such that $K_n \|\Sigma_{j=0}^n\ f_j\|_{\mathrm{L}^\infty} \geq \|f_k\|_{\mathrm{L}^\infty}$ for $k = 0, 1, \ldots, n$. By applying (10.29) to each f_j, we obtain

$$\|\hat{f}\|_{2n/n+1} \leq K_n (n+1)^{\frac{n+1}{2n}} (n!)^{\frac{n-1}{2n}} (\sqrt{2})^{n+1} (2\mathrm{e})^n\ \|f\|_{\mathrm{L}^\infty}. \tag{10.30}$$

11 p-Sidon Sets

In this section, the inequalities in (10.2) are shown to be sharp. The arguments will be carried out in a framework of *p-Sidon sets*, described below, and work will continue in the setting of Ω. (In the next section, all that has and will have been done – the Rosenthal property, Bonami's inequalities, and Littlewood's inequalities – will be officially transcribed to the general Abelian group setting.)

Proposition 39 (Exercise 33). *The following are equivalent for* $F \subset W$, $p \in [1,2]$, *and* $1/p+1/q=1$: (i) $\mathrm{L}^\infty_F(\Omega)\hat{} \subset l^p(F)$; (ii) $\mathrm{C}_F(\Omega)\hat{} \subset l^p(F)$; (iii) *there exist* $\zeta > 0$ *such that for all* F*-polynomials* g,

$$\|\hat{g}\|_p \;\le\; \zeta \|g\|_{\mathrm{L}^\infty}; \tag{11.1}$$

(iv) $l^q(F) \subset A(F)$; (v) *there exist* $\zeta > 0$ *such that for all finitely supported* $\varphi \in A(F)$,

$$\|\varphi\|_{A(F)} \le \zeta \|\varphi\|_q. \tag{11.2}$$

Definition 40 For $F \subset W$ and $t \in (0,\infty)$, let

$$\zeta_F(t) = \sup\{\|\hat{g}\|_t : g \in B_{\mathrm{C}_F(\Omega)}\}, \tag{11.3}$$

and define the *Sidon exponent* of F to be

$$\sigma_F = \inf\{t : \zeta_F(t) < \infty\}. \tag{11.4}$$

If $\sigma_F = p$, then F is said to be *p-Sidon.* If $\zeta_F(\sigma_F) < \infty$, then

$$\sigma_F = p \quad \textit{exactly} \quad (F \text{ is } \textit{exactly } p\text{-Sidon}), \tag{11.5}$$

and if $\zeta_F(\sigma_F) = \infty$, then

$$\sigma_F = p \quad \textit{asymptotically} \quad (F \text{ is } \textit{asymptotically } p\text{-Sidon}). \tag{11.6}$$

Remarks:

i **(about the terminology).** The statement that F is Sidon, in the sense of §5, is equivalent to the statement that F is exactly 1-Sidon in the sense of Definition 40. We shall use both terms 'Sidon' and 'exactly 1-Sidon' interchangeably.

According to Definition 40, the assertion 'F is p-Sidon' means:

$$(\mathrm{C}_F)\hat{} \subset l^t \text{ for all } t > p, \text{ and there exist } f \in \mathrm{C}_F \\ \text{such that } \hat{f} \notin l^t \text{ for all } t < p.$$

In Edwards's and Ross's paper [ERos], the assertion 'F is p-Sidon' meant that F satisfied (any one of) the properties in Proposition 39. For our purposes, I prefer Definition 40.

We have shown, so far, that W_1 is exactly 1-Sidon (easily!), that W_2 is exactly 4/3-Sidon, and that W_3 is exactly 3/2-Sidon ($n = 2$ and $n = 3$ in Theorem 34, Remark iii in §10). We have also shown that for all $n \geq 1$,

$$\zeta_{W_n}\left(\frac{2n}{n+1}\right) < \infty. \tag{11.7}$$

In this section we prove

$$\zeta_{W_n}(t) = \infty \text{ for all } t < 2n/(n+1),$$

and thus verify that W_n is *exactly* $(2n/(n+1))$-Sidon. To this end, we use

Theorem 41 *Let $F \subset W$ and $t \geq 1$. For all F-polynomials g and all $s \geq 1$,*

$$\|g\|_{\mathrm{L}^s} \leq \zeta_F(t)\ \sqrt{s}\ \|\hat{g}\|_{2t/(3t-2)}. \tag{11.8}$$

Proof: We write $g = \sum_{\gamma \in F} a_\gamma \gamma$ (an F-polynomial), and assume

$$\sum_{\gamma \in F} |a_\gamma|^{2t/(3t-2)} = 1.$$

For $u \in \{-1, 1\}^F$, define

$$g_u = \sum_{\gamma \in F} (a_\gamma)^\alpha\ r_\gamma(u)\ \gamma, \tag{11.9}$$

where $\alpha = t/(3t-2)$, and $\{r_\gamma : \gamma \in F\}$ is the Rademacher system indexed by F. By assumption, $\sum_{\gamma \in F} |(a_\gamma)^{1-\alpha}|^{t/(t-1)} = 1$, and therefore, by Proposition 39, there exist $\mu_u \in \mathrm{M}(\Omega)$ such that

$$\|\mu_u\|_{\mathrm{M}}\ \leq\ \zeta_F(t) \tag{11.10}$$

and

$$\hat{\mu}_u(\gamma) = (a_\gamma)^{1-\alpha}\ r_\gamma(u), \quad \gamma \in F.$$

For such measures μ_u,

$$g_u \star \mu_u = g, \tag{11.11}$$

and therefore

$$\|g\|_{\mathrm{L}^s}^s = \|g_u \star \mu_u\|_{\mathrm{L}^s}^s\ \leq\ \zeta_f(t)^s\ \|g_u\|_{\mathrm{L}^s}^s. \tag{11.12}$$

By integrating (11.12) with respect to the Haar measure on $\{-1,1\}^F$, applying Fubini's theorem, and the Khintchin inequalities, we obtain

$$
\begin{aligned}
\|g\|_{\mathrm{L}^s}^s &\le \zeta_F(t)^s\, \mathbf{E}_u\, \mathbf{E}_\omega \left|\sum_{\gamma\in F} (a_\gamma)^\alpha\, r_\gamma(u)\, \gamma(\omega)\right|^s \\
&= \zeta_F(t)^s\, \mathbf{E}_\omega\, \mathbf{E}_u \left|\sum_{\gamma\in F} (a_\gamma)^\alpha\, r_\gamma(u)\, \gamma(\omega)\right|^s \\
&\le \zeta_F(t)^s\, (\sqrt{s})^s \left(\sum_{\gamma\in F} |a_\gamma|^{2\alpha}\right)^{\frac{s}{2}} = \zeta_F(t)^s\, (\sqrt{s})^s. \qquad (11.13)
\end{aligned}
$$

□

Corollary 42 *For all $n \in \mathbb{N}$,*

$$\zeta_{W_n}(t) = \infty, \quad t < \frac{2n}{n+1}, \tag{11.14}$$

and (therefore) W_n is exactly $2n/(n+1)$-Sidon.

Proof: Fix $t < 2n/(n+1)$. Let $k > 1$ be an arbitrary integer, and let

$$g = \sum_{i_1,\ldots,i_n=1}^{k} r_{i_1} \otimes \cdots \otimes r_{i_n}. \tag{11.15}$$

By (9.9),

$$2^n\, \|g\|_{\mathrm{L}^k(\Omega^n)} \ge k^n = \sqrt{k}\, \|\hat{g}\|_{2t/(3t-2)} k^{\frac{n}{t}-\frac{n}{2}-\frac{1}{2}}. \tag{11.16}$$

By combining (11.16) and (11.8), we obtain

$$\zeta_{W_n}(t) \ge 2^{-n} k^{\frac{2n-tn-t}{2t}}, \tag{11.17}$$

By assumption, $n/t - n/2 - 1/2 > 0$, and therefore the right side of (11.17) is an unbounded function of k, which implies (11.14).

By combining (11.7) with (11.14), we obtain $\sigma_{W_n} = 2n/(n+1)$ exactly.

□

Remarks:

ii **(alternative proofs).** In Chapter X we produce by random constructions $f \in \mathrm{C}_{W_n}$ such that $\|\hat{f}\|_p = \infty$ for all $p < 2n/(n+1)$

(Remark i in Chapter X §5). I do not know any deterministic constructions of such $f \in C_{W_n}$ for $n > 3$. (See Remark iii in §10.)

iii **(who did what and how).** The key idea in the proof of Theorem 41 is due to Rudin, who showed in [Ru1] that $\Lambda(s)$-constants' growth of every Sidon set is $\mathscr{O}(\sqrt{s})$ (the case $t = 1$ in (11.8)).

The first allusion to p-Sidonicity – without dubbing it so – was in a 1972 work by M. Bożejko and T. Pytlik [BozPy], where Theorem 41 was obtained by extending Rudin's aforementioned argument in the case $t = 1$; see [BozPy, Theorem 2]. Two years after Bożejko's and Pytlik's paper had appeared, Edwards and Ross published their own study of 'p-Sidon sets' [ERos]. In it they showed that if E and F are disjoint spectral sets such that E and F are infinite, and $E \cup F$ is dissociate (e.g., $E \cup F$ lacunary in $\mathbb{Z}$, or $E \cup F = R$ in W), then $E \cdot F$ is exactly 4/3-Sidon. (*Dissociate sets* will be defined in the next section.) To verify $(C_{E \cdot F})\hat{} \subset l^{4/3}$, Edwards and Ross applied Littlewood's 4/3-inequality, and to verify that there exist $f \in C_{E \cdot F}$ such that $\hat{f} \notin l^t$ for all $t < 4/3$, they used the (indirect) proof of Corollary 42 in the case $n = 2$. (Although in this case, to show the latter, a direct construction would have done just as well; see Remark iii in §10.) The n-fold version of Edwards' and Ross's result – that $E_1 \cdots E_n$ is exactly $(2n/(n+1))$-Sidon whenever $E_1, \ldots, E_n$ are pairwise disjoint infinite spectral sets whose union is dissociate – was established by G. Johnson and G. Woodward [JWo]. They deduced $\zeta_{E_1 \cdots E_n}(2n/(n+1)) < \infty$ by extending Littlewood's 4/3-inequality, and obtained $\zeta_{E_1 \cdots E_n}(t) = \infty$ for $t < 2n/(n+1)$ via the proof of Corollary 42 above. The requirement that the E_i be disjoint was removed in [Bl2].

Lemma 30, which provides a crucial step in the proof of Corollary 42, is due to A. Bonami [Bon1, Théorème 1]. As far as I can determine, its first use in an argument like the proof of Corollary 42 appeared in [Fig, Lemma].

iv **(applications and a preview).** We can now expeditiously verify that for all $n \in \mathbb{N}$,

$$F_n(\mathbb{N}^2, \ldots, \mathbb{N}) \subsetneqq F_{n+1}(\mathbb{N}, \ldots, \mathbb{N}),$$

and (equivalently via duality)

$$V_{n+1}(\mathbb{N}, \ldots, \mathbb{N}) \subsetneqq V_n(\mathbb{N}^2, \ldots, \mathbb{N}). \tag{11.18}$$

(See Remarks in Chapter IV §2 and §5.) The argument is this. By Corollary 42, there exist scalar-valued functions β on $\mathbb{N}^{n+1}$ such that $\beta \in F_{n+1}(\mathbb{N}, \ldots, \mathbb{N})$ and $\|\beta\|_p = \infty$ for all $p < 2(n+1)/(n+2)$. But, again by Corollary 42, such β cannot be in $F_n(\mathbb{N}^2, \ldots, \mathbb{N})$.

More generally, Corollary 42 implies that if Y is any countably infinite set, then there exist $\beta \in l^2(Y)$ of type F_{n+1} but not of type F_n; equivalently, there exist $\varphi \in c_0(Y)$ of type V_n but not of type V_{n+1}. (For definitions of *type*, see Chapter IV §2 and §5.)

The focus of the projective tensor algebra V_n is on the question: can a function in n independent variables be represented as an absolutely convergent series of n-fold elementary tensors, each of whose n factors is a function, respectively, of each of the independent variables. This question is an instance of a more general problem. Let

$$U = \{S_j : j = 1, \ldots, n\} \tag{11.19}$$

be a cover of $[m]$; that is, $S_j \subset [m]$, and $\cup\{S_j : j \in [n]\} = [m]$. For $S \subset [m]$, let π_S denote the canonical projection from $\mathbb{N}^m (= \mathbb{N}^{[m]})$ onto $\mathbb{N}^S$; that is,

$$\pi_S(i_1, \ldots, i_m) = (i_k : k \in S), \quad (i_1, \ldots, i_m) \in \mathbb{N}^m. \tag{11.20}$$

(For convenience, subsets $S \subset [m]$ are enumerated in increasing order: $S = (k_1, k_2, \ldots)$, $0 < k_1 < k_2 < \cdots \leq m$.) We consider those $\varphi \in c_0(\mathbb{N}^m)$ that can be represented (pointwise on $\mathbb{N}^m$) as

$$\varphi = \sum_\alpha \theta_\alpha^{(1)} \circ \pi_{S_1} \cdots \theta_\alpha^{(n)} \circ \pi_{S_n},$$

$$\theta_\alpha^{(1)} \in c_0(\mathbb{N}^{S_1}), \ldots, \theta_\alpha^{(n)} \in c_0(\mathbb{N}^{S_n}), \quad \alpha \in \mathbb{N}, \tag{11.21}$$

where

$$\sum_\alpha \|\theta_\alpha^{(1)}\|_\infty \cdots \|\theta_\alpha^{(n)}\|_\infty < \infty. \tag{11.22}$$

We define $\|\varphi\|_{V_U}$ to be the infimum of (11.22) over representations of φ by (11.21), and let $V_U(\mathbb{N}^n)$ denote the algebra consisting of all $\varphi \in c_0(\mathbb{N}^n)$ with $\|\varphi\|_{V_U} < \infty$. Notice that in (11.21), a function φ in m independent variables is represented by an absolutely convergent series of n-fold elementary tensors, each of whose n factors is, respectively, a function of one of n *interdependent* variables. In Chapter XII and Chapter XIII we will gauge the degree of this interdependence by the 'combinatorial dimension' of a 'fractional Cartesian product' based on U. Indeed, by using this gauge, we will distinguish between the *algebras* V_U that are based on different covers U of $[m]$.

To illustrate the issues that arise here, let us consider the case $m = 3$, and covers

$$U_1 = \{(1), (2), (3)\},\ U_2 = \{(1,2), (3)\},\ U_3 = \{(1,3), (2)\},$$
$$U_4 = \{(2,3), (1)\},\ U_5 = \{(1,2), (2,3), (1,3)\},$$
$$U_6 = \{(1,2,3)\}. \tag{11.23}$$

Then,

$$V_{U_1} \subset V_{U_j} \subset V_{U_5} \subset V_{U_6}\ (= c_0(\mathbb{N}^3)), \quad j = 2,3,4. \tag{11.24}$$

Whereas each of V_{U_2}, V_{U_3}, and V_{U_4} is isomorphic to $V_2(\mathbb{N}, \mathbb{N})$, they *are* distinct in the sense that there exists $\varphi \in V_{U_2}$ such that $\varphi \notin V_{U_j}, j = 3, 4$. (Do you see why?) Note also (an instance of (11.18)) that $V_{U_1} \subsetneqq V_{U_j}, j = 2,3,4$. Questions concerning the remaining inclusions $V_{U_j} \subset V_{U_5} \subset V_{U_6}, j = 2,3,4$, lead to new issues that will be resolved in Chapter XII by the use of a Littlewood-type inequality in 'dimension' 3/2.

v **(a measurement of complexity).** Corollary 42 provides, in effect, a calibration of Plancherel's theorem. If $f \in L^\infty(\Omega, \mathbb{P})$, then $\|f\|_{L^\infty} \geq \|\hat{f}\|_2$, which generally cannot be improved (Proposition 16): there exist $f \in C(\Omega)$ such that $\|\hat{f}\|_p = \infty$ for all $p < 2$. But if $f \in L^\infty_{W_n} (= C_{W_n})$, then $c_n\|f\|_{L^\infty} \geq \|\hat{f}\|_{2n/(n+1)}$, which also cannot be improved (Corollary 42): there are $f \in C_{W_n}(\Omega)$ such that $\|\hat{f}\|_p = \infty$ for all $p < 2n/(n+1)$. In this connection, $\sigma_{W_n} = 2n/(n+1)$ registers the complexity of W_n.

By 'complexity' I mean – roughly speaking – the 'engineering effort' needed to construct a set by using basic 'building blocks'. (See the first comment in Section 1 of this chapter.) We also could think of complexity – again heuristically speaking – as 'interdependence' between elements in a given set. The distinction between these two 'characterizations' is that in the first we are handed 'building blocks' and we build, whereas in the second we are given the set and observe the interdependence of its elements. In either case, we want measurements of 'effort', or 'interdependence'. In our specific context, I view ζ_F and the Sidon exponent σ_F as these measurements.

To make this precise, let us consider a completely general notion of independence. Suppose X, Y, and Z are sets (with no *a priori* structures). Suppose f is a function from X onto Y, and g a function from X onto Z. Speaking heuristically, when we say that f and g are *independent*, or that there is no interaction between them, we mean essentially this: for any $x \in X$, knowledge of $f(x)$ implies no information about $g(x)$, and knowledge of $g(x)$ implies no information about $f(x)$. Translating this into mathematics we say that f and g are *functionally independent* if for every $y \in Y$ and $z \in Z$ there exists $x \in X$ such that $f(x) = y$ and $g(x) = z$. (Convince yourself that the latter indeed conveys independence.)

Definition 43 Let X be a set, $\{Y_\alpha : \alpha \in A\}$ a family of sets, and let f_α be a function from X onto Y_α, $\alpha \in A$. The system $\{f_\alpha : \alpha \in A\}$ is *functionally independent on* X if

$$\text{for all } (y_\alpha : \alpha \in A) \in \prod_{\alpha \in A}^{\infty} Y_\alpha, \text{ there exists } x \in X \quad \text{such that } f_\alpha(x) = y_\alpha \text{ for all } \alpha \in A. \tag{11.25}$$

Functional independence appears under different guises in various contexts (Exercise 34). In our context it appears as 1-Sidonicity. To see this, consider characters γ of a compact Abelian group G as functions on $\mathrm{M}(G)$,

$$\mu \mapsto \hat{\mu}(\gamma), \quad \mu \in \mathrm{M}(G), \tag{11.26}$$

and recall that $F \subset \hat{G}$ is exactly 1-Sidon if and only if there exists $\zeta > 0$ such that for every $(y_\gamma : \gamma \in F)$ in $B_{l^\infty(F)}$, there exist measures μ in the ζ-ball of $\mathrm{M}(G)$ such that for all $\gamma \in F$,

$$\hat{\mu}(\gamma) = y_\gamma. \tag{11.27}$$

In this light,

$F \subset \hat{G}$ is exactly 1-Sidon if and only if there exists $\zeta > 0$ such that F is functionally independent on the ζ-ball of $\mathrm{M}(G)$.

Now in place of $B_{l^\infty(F)}$ in the discussion above, consider the more 'restrictive' $B_{l^q(F)}$ for $q \in (2, \infty)$. Then, the subsequent measurement $\zeta_F(q)$ and the index σ_F will register an increasing level of complexity in F, above that of functional independence.

vi **(more about functional independence vis-à-vis Sidonicity).** Functional independence is tied to the underlying domain of the functions in question. (We expect this from any notion of independence, however it is defined.) Suppose $\{f_\alpha : \alpha \in A\}$ is a system of functions from X onto Y, that $X' \subset X$, and that each member of $\{f_\alpha : \alpha \in A\}$ also maps X' onto Y. If $\{f_\alpha : \alpha \in A\}$ is functionally independent on X', then it is *a fortiori* functionally independent on X, but the reverse implication need not hold, and thus the question: what is the smallest domain X' on which $\{f_\alpha : \alpha \in A\}$ is functionally independent? (Smaller X' convey 'stronger' independence.) In a context of harmonic analysis, every finite spectral set F is Sidon, and therefore functionally independent on sufficiently large balls of measures. The size of these balls can be viewed as an estimate on the 'amount' of functional independence in F. Indeed, the smallest ζ such that F is functionally independent on the ζ-ball of $\mathrm{M}(G)$ is the Sidon constant $\zeta_F(1)$ of F.

Sidon sets in general Abelian groups Γ are (by definition) functionally independent on balls of measures. In W, the Rademacher system is functionally independent on Ω (extreme points in $B_{\mathrm{M}(\Omega)}$), which conveys a 'stronger' notion of independence. In $\mathbb{Z}$, while obviously no two exponentials can be functionally independent on $\mathbf{T}$ (extreme points in $B_{M(\mathbf{T})}$), if $E \subset \mathbb{Z}$ is sufficiently lacunary (e.g., $q_E \geq 3$), then E is 'almost' functionally independent on $\mathbf{T}$. (See Exercise 15 iii.) Specifically, this implies: if $E \subset \mathbb{Z}$ is lacunary ($q_E > 1$), then for some $\zeta > 1, E$ is functionally independent on the ζ-ball of the space of *discrete* measures on $\mathbf{T}$ [Mé]. This, again, conveys a notion of independence 'stronger' than 1-Sidonicity. Indeed, there exist Sidon sets in $\mathbb{Z}$ that are not even finite unions of lacunary sets [Ru1]. In general, the 'strongest' independence property that can be ascribed to Sidon sets is that *exact* 1-Sidon sets are functionally independent on balls of *continuous* measures. (This follows from the proof in [Dru] that a finite union of Sidon sets is Sidon.)

vii **(the p-Sidon set problem).** Corollary 42 leads to the question: are there spectral sets F with $\sigma_F = p$ for *arbitrary* $p \in (1,2)$? This question, like the corresponding problem concerning the δ-scale (Remark ii §9), will be resolved in Chapter XIII by constructions of Walsh systems of 'fractional order'.

viii **(two open questions).** Rudin's question [Ru1, Section 3] whether $\Lambda(s)$-constants' growth $\mathscr{O}(\sqrt{s})$ is equivalent to Sidonicity was affirmatively answered by Pisier [Pi1]. (See also [MaPi].) Notably, Pisier's theorem in a general group setting had been deduced first in Ω [Bon2,

p. 350], from a purely combinatorial characterization of Sidonicity in W [MM]:

$$F \subset W \text{ Sidon} \quad \Leftrightarrow \quad F = \text{finite union of}$$
$$\text{algebraically independent sets (Remark in §2).} \qquad (11.28)$$

(Whether there are 'purely combinatorial' characterizations of Sidon sets in general groups is an open (-ended) problem of long standing; e.g., see [Pi2].) Two questions arise:

1. Is there a characterization of p-Sidon sets in W that is analogous to (11.28)?
2. Does $\mathscr{O}(\sqrt{s})$ in (11.8) characterize p-Sidonicity? That is, does the implication

$$\sup\{\|g\|_{L^s}/\sqrt{s} : s > 0, \text{ spect } g \subset F, \|\hat{g}\|_{2t/(3t-2)} = 1\} < \infty$$
$$\Rightarrow \zeta_F(t) < \infty \qquad (11.29)$$

hold for all $t \in (1, 2)$?

ix **(the p-Sidon set union problem).** It is unknown whether for arbitrary subsets F_1 and F_2 of a discrete Abelian group Γ,

$$\sigma_{F_1 \cup F_2} = \max\{\sigma_{F_1}, \sigma_{F_2}\}. \qquad (11.30)$$

The question whether (11.30) holds in the instance $\sigma_{F_1} = \sigma_{F_2} = 1$ *exactly* had been first raised by Rudin [Ru1, Remark 2.5 (2)], well before the notion of p-Sidonicity was framed, and was answered in the affirmative by S. Drury [Dru]. The answer to Rudin's union question subsequently provided, via D. Rider's reformulation of Drury's solution [Rid], one of the two key ingredients in Pisier's solution (seven years later) to yet another of Rudin's problems, whether $\mathscr{O}(\sqrt{p})$ growth of $\Lambda(p)$-constants characterized Sidonicity. (The other ingredient was *metric entropy* [MaPi].) Notably, the Sidon set union problem was first solved in the simplest setting $W = \hat{\Omega}$ [MM], some three years prior to its resolution in an arbitrary group setting, but the p-Sidon set union problem is still unresolved even in W.

12 Transcriptions

In the last section of this chapter we indicate how Walsh systems and their properties arise in a general setting.

We let G denote a compact Abelian group with normalized Haar measure $\mathfrak{m}$, and let $\Gamma(=\hat{G})$ denote its discrete dual group. The reader familiar with material in the first two chapters of [Ru3] should easily be able to transcribe rudiments (definitions and the like) from Ω to G. Specifically, definitions and preliminaries involving restriction algebras in §7, definitions of the Rosenthal property in §5, and indices involving $\Lambda(q)$-sets and p-Sidon sets in §9 and §11 can be recast essentially verbatim in the general setting.

Dissociate Sets – Definition and Examples

Following Hewitt and Ross [HewRos, 37.12], we say $E \subset \Gamma$ is *dissociate* if E satisfies the following property: for every $n \in \mathbb{N}$, if $\gamma_1, \ldots, \gamma_n$ are distinct elements in E, and

$$(\gamma_1)^{k_1} \cdots (\gamma_n)^{k_n} = \text{identity element of } \Gamma, \tag{12.1}$$

where $k_j \in \{-2,-1,0,1,2\}$ for $j \in [n]$, then

$$(\gamma_1)^{k_1} = \cdots = (\gamma_n)^{k_n} = \text{identity element of } \Gamma. \tag{12.2}$$

This purely algebraic property manifests independence. In a harmonic-analytic context, we view elements of dissociate sets as basic 'building blocks' in $\hat{G}$ much the same way we view Rademacher characters in $\hat{\Omega}$. (Clearly, the Rademacher system is dissociate.) Every infinite discrete Abelian group contains an infinite dissociate set (e.g., see [HewRos, Theorem 37.18]).

Besides the Rademacher system, other canonical examples of dissociate sets are: lacunary sets $E \subset \mathbb{Z}^+$ such that $q_E \geq 3$ (§5), and the systems $S_m, m \geq 3$, defined in Chapter II §6. The systems S_m are generalizations of R: for $m \geq 3$, $\Omega_m = (\mathbf{T}_m)^{\mathbb{N}}$ is a compact Abelian group ($\mathbf{T}_m$ = mth roots of unity with the uniform probability measure), whose discrete dual is generated by S_m, just as $\hat{\Omega}$ is generated by R. We have already noted in Chapter II §6 that S_m is a system of *statistically independent* $\mathbf{T}_m$-valued random variables on Ω_m. It is also *functionally independent* on Ω_m (see (II.6.3) and Remark iv §11), and *algebraically independent* in $\hat{\Omega}_m$; the latter means that for $E = S_m$ and $\Gamma = \hat{\Omega}_m$, if (12.1) is assumed with

$$k_j \in \{-m, -m+1, \ldots, 0, \ldots, m-1, m\}, \tag{12.3}$$

then (12.2) holds. The limiting case $m = \infty$ is of particular interest to us. In this case, Ω_∞ is the infinite product $\mathbf{T}^{\mathbb{N}}$, endowed with the usual product topology and coordinate-wise addition; the Haar measure is the infinite product of the normalized Lebesgue measure in each coordinate, and $(\mathbf{T}^{\mathbb{N}})\hat{} = \oplus\, \mathbb{Z}$ (= integer-valued sequences with finitely many non-zero terms) equipped with coordinate-wise integer addition. The action of $\mathbf{n} = (n_k) \in \oplus\, \mathbb{Z}$ on $\mathbf{t} = (t_k) \in \mathbf{T}^{\mathbb{N}}$ is given by

$$\mathbf{t} \mapsto \exp(2\pi\mathrm{i}\langle \mathbf{n}, \mathbf{t}\rangle) = \exp(2\pi\mathrm{i}\Sigma_k n_k\, t_k). \tag{12.4}$$

The system $S_\infty := S$, known as the Steinhaus system (Chapter II §6, Remark i §9), can be identified with the canonical basis $\{\mathbf{e}_j : j \in \mathbb{N}\}$ in $\oplus\, \mathbb{Z}$ $(\mathbf{e}_j(j) = 1$ and $\mathbf{e}_j(n) = 0$ for $n \neq j)$: the jth element of $S = \{\mathbf{e}_j : j \in \mathbb{N}\}$ is the character χ_j on $\mathbf{T}^{\mathbb{N}}$ given by

$$\chi_j(\mathbf{t}) = \exp(2\pi\mathrm{i}\Sigma_k \mathbf{e}_j(k)\, t_k) = \exp(2\pi\mathrm{i} t_j), \quad \mathbf{t} = (t_k) \in \mathbf{T}^{\mathbb{N}}. \tag{12.5}$$

The decisive advantage of the Steinhaus system, which we will utilize in Chapter XII and Chapter XIII, is that it is *fully* algebraically independent ($m = \infty$ in (12.3)), and therefore easier to handle than S_m, $2 \leq m < \infty$.

Riesz Products

Throughout the section, E will denote a dissociate set in Γ. We assume E is countably infinite, and enumerate $E = \{\gamma_j : j \in \mathbb{N}\}$. We denote the identity element in Γ by γ_0; i.e., $\gamma_0 = \mathbf{1}_G$.

For real-valued $\theta \in B_{l^\infty(\mathbb{N})}$, and $N \in \mathbb{N}$, consider the finite Riesz product

$$R_N = \prod_{j=1}^{N} \left(\gamma_0 + \theta(j)\frac{\gamma j + \bar{\gamma} j}{2}\right), \tag{12.6}$$

whose spectral analysis implies

$$\hat{R}_N(\gamma_0) = 1, \tag{12.7}$$

and, because $R_N \geq 0$,

$$\|R_N\|_{\mathrm{L}^1(G)} = \int_G R_N(x)\, \mathfrak{m}(\mathrm{d}x) = \hat{R}_N(\gamma_0) = 1. \tag{12.8}$$

Also, denote

$$D_N = \{(\gamma_{i_1})^{\epsilon_1} \cdots (\gamma_{i_k})^{\epsilon_k} : 0 < i_1 < \cdots < i_k \leq N, k \in [N], \epsilon_j = \pm 1, j \in [k]\}, \tag{12.9}$$

and obtain

$$\hat{R}_N(\chi) = \begin{cases} (1/2^k)\ \theta(i_1) \cdots \theta(i_k) & \chi \in D_N, \quad \chi = (\gamma_{i_1})^{\epsilon_1} \ldots (\gamma_{i_k})^{\epsilon_k} \\ 0 & \chi \notin D_N, \ \chi \neq \gamma_0. \end{cases} \tag{12.10}$$

By (12.8) and (12.10), there exists a positive Borel measure $\mu \in \mathrm{M}(G)$ such that $\mu = \text{weak}^* \lim_{N\to\infty} R_N$, $\|\mu\|_{\mathrm{M}} = \hat{\mu}(\gamma_0) = 1$, and

$$\begin{aligned} &\hat{\mu}(\gamma_{i_1} \cdots \gamma_{i_k}) = (1/2^k)\ \theta(i_1) \cdots \theta(i_k), \\ &0 < i_1 < \cdots < i_k, \ k \in \mathbb{N}. \end{aligned} \tag{12.11}$$

We denote

$$\mu = \prod_{j=1}^{\infty} \left(\gamma_0 + \theta(j) \frac{\gamma j + \bar{\gamma} j}{2} \right), \tag{12.12}$$

and call it a *Riesz product.* This measure implies the analog of Lemma 28, where G replaces Ω and E replaces R.

Restriction Algebras and Tensor Algebras

For $n \in \mathbb{N}$, consider (the analog of R_n)

$$E_n = \{\gamma_{i_1} \cdots \gamma_{i_n} : 0 < i_1 < \cdots < i_n\}. \tag{12.13}$$

By use of Riesz products and polarization (Lemma 27), we obtain that if f is an E_n-polynomial,

$$f = \sum_{i_1,\ldots,i_n} \hat{f}(\gamma_{i_1} \cdots \gamma_{i_n})\ \gamma_{i_1} \cdots \gamma_{i_n}, \tag{12.14}$$

and

$$\beta_f(i_1,\ldots,i_n) = \begin{cases} \hat{f}(\gamma_{i_1}\cdots\gamma_{i_n}) & \text{if } |\{i_1,\ldots,i_n\}| = n \\ 0 & \text{otherwise,} \end{cases} \tag{12.15}$$

then $f \mapsto \beta_f$ determines an isomorphism from C_{E_n} onto $F_{n\sigma}(\mathbb{N},\ldots,\mathbb{N})$. This implies

$$A(E_n) = V_n|_{D_n} \quad \text{(cf. Theorem 26)}, \tag{12.16}$$

where D_n is the tetrahedral set defined in (8.21).

The Rosenthal Property

The dual of $A(E_n)$ is $\mathrm{L}^\infty_{E_n}(G,\mathfrak{m})$ (Proposition 24), and that of $V_n|_{D_n}$ is $F_{n\sigma}(\mathbb{N},\ldots,\mathbb{N})$ (Chapter IV §5). Therefore, by (12.16), $\mathrm{L}^\infty_{E_n}(G,\mathfrak{m}) = F_{n\sigma}(\mathbb{N},\ldots,\mathbb{N})$. Scalar-valued functions with finite support on $\mathbb{N}^n$ are norm-dense in $F_{n\sigma}(\mathbb{N},\ldots,\mathbb{N})$ (Theorem IV.6), and therefore

$$\mathrm{L}^\infty_{E_n}(G,\mathfrak{m}) = \mathrm{C}_{E_n}(G) \text{ (cf. Exercise 25, and Remark i§8).}$$

Bonami's Inequalities

Transcribed to the general setting, Theorem 32 asserts

$$\delta_{E_n} = \frac{n}{2} \text{ exactly}, \quad n \in \mathbb{N}. \tag{12.17}$$

(For definitions see Remark ii §9.) A key to the transcription is

Proposition 44 *Let G and G' be compact Abelian groups with respective Haar measures $\mathfrak{m}$ and $\mathfrak{m}'$, and respective dual groups Γ and Γ'. If $E = \{\gamma_j : j \in \mathbb{N}\}$ is dissociate in Γ, and $F = \{\chi_j : j \in \mathbb{N}\}$ is dissociate in Γ', then*

$$\eta_{E_n}(a) \leq 12^n \; \eta_{F_n}(a), \quad a > 0. \tag{12.18}$$

Proof: (cf. Proof of Theorem 32): Let

$$f = \sum_{i_1,\ldots,i_n} \hat{f}(\gamma_{i_1}\cdots\gamma_{i_n})\gamma_{i_1}\cdots\gamma_{i_n}$$

be an E_n-polynomial. For each $x \in G'$, consider

$$f_x = \sum_{i_1,\ldots,i_n} \hat{f}(\gamma_{i_1} \cdots \gamma_{i_n})\, \bar{\chi}_{i_1}(x) \cdots \bar{\chi}_{i_n}(x)\, \gamma_{i_1} \cdots \gamma_{i_n}. \tag{12.19}$$

By (12.16), for each $x \in G'$ there exist $\mu_x \in \mathrm{M}(G)$ such that

$$\hat{\mu}_x(\gamma_{i_1} \cdots \gamma_{i_n}) = \bar{\chi}_{i_1}(x) \cdots \bar{\chi}_{i_n}(x), \quad \gamma_{i_1} \cdots \gamma_{i_n} \in \text{ spect } f \tag{12.20}$$

and

$$\|\mu_x\|_{\mathrm{M}(G)} \leq 12^n, \tag{12.21}$$

where 12^n in (12.21) is a composite of polarization constants and Riesz product-norms. Then,

$$f = f_x \star \mu_x, \tag{12.22}$$

and

$$\|f\|^p_{\mathrm{L}^p(G)} \leq \|\mu_x\|^p_{\mathrm{M}(G)} \|f_x\|^p_{\mathrm{L}^p(G)} \leq 12^{np}\ \|f_x\|^p_{\mathrm{L}^p(G)}, \quad p > 2. \tag{12.23}$$

By integrating both sides of (12.23) over G' with respect to $\mathfrak{m}'$, applying Fubini's theorem, and the definition of $\eta_{F_n}(a)$, we deduce

$$\|f\|_{\mathrm{L}^p(G)} \leq 12^n\ \eta_{F_n}(a)\ p^a \|\hat{f}\|_2, \tag{12.24}$$

which implies (12.18). □

We prove (12.17) via Proposition 44 and Bonami's inequalities (Theorem 32). Notice that Proposition 44 implies that the Khintchin inequalities (obtained in [Kh1]), and the Khintchin-type inequalities involving the Steinhaus system (obtained in [Lit3]) are equivalent in the sense that any one set of inequalities can be derived from the other. (See Remark i §9.)

p-Sidonicity

The transcription of Corollary 42,

$$\sigma_{E_n} = \frac{2n}{n+1} \text{ exactly}, \quad n \in \mathbb{N}, \tag{12.25}$$

follows from (12.16) and Proposition 39 (rephrased in the general setting).

Exercises

1. Define the Fourier transform of $f \in \mathrm{L}^1(\mathbf{T}, \mathfrak{m})$ ($\mathfrak{m}$ = normalized Lebesgue measure) by

$$\hat{f}(n) = (1/2\pi) \int_{\mathbf{T}} \mathrm{e}^{-int} f(t) \mathfrak{m}(\mathrm{d}t), \quad n \in \mathbb{Z}.$$

   Prove that if $f \in \mathrm{L}^2(\mathbf{T}, \mathfrak{m})$, then

$$\left(\sum_{n \in \mathbb{Z}} |\hat{f}(n)|^2 \right)^{\frac{1}{2}} = \|f\|_{\mathrm{L}^2},$$

   thus verifying that the partial sums of the Fourier series $\Sigma_n \hat{f}(n)\, \mathrm{e}^{int}$ converge to f in $\mathrm{L}^2(\mathbf{T}, \mathfrak{m})$.
2. Verify that convolution in $\mathrm{M}(\Omega)$ is commutative, and

$$\|\mu \star \nu\|_{\mathrm{M}} \leq \|\mu\|_{\mathrm{M}}\, \|\nu\|_{\mathrm{M}}, \quad \mu \in \mathrm{M}(\Omega),\ \nu \in \mathrm{M}(\Omega).$$

3. Prove Proposition 2.
4. Prove Proposition 3.
5. Prove that the probability measure $\mathbb{P}$, which is defined in Chapter II §1, is the unique normalized Haar measure on Ω.
6. Prove Proposition 4.
7. i. Prove Lemma 7.
   ii. Prove Proposition 6.
8. Prove Corollary 11.
9. Prove Proposition 12.
10. Show that the binary digit expansion of $x \in [0,1]$ gives rise to the measure-preserving map $\sigma : (\Omega, \mathfrak{A}, \mathbb{P}) \mapsto ([0,1], \mathfrak{B}, \mathfrak{m})$ defined in (4.2). Verify that if w is a Walsh function on $[0,1]$, then $w \circ \sigma$ is the corresponding Walsh character on Ω.
11. Paley's work in [Pa] foreshadowed the theory of martingales. If you are familiar with martingales, then this exercise is merely a historical note. Otherwise, I urge you to learn about them (at the very least for the purpose of this exercise); e.g., [Tu, Chapter 7], [Bu], [Ga].

Let $\{w_j : j \in \mathbb{N}\}$ be Paley's enumeration of the Walsh functions. For $f \in \mathrm{L}^1([0,1], \mathfrak{B}, \mathfrak{m})$, let

$$f_n = \sum_{j=2^n}^{2^{n+1}-1} \hat{f}(w_j)\, w_j, \quad n \in \mathbb{N}.$$

Observe that if σ is the measure-preserving map in Exercise 10, then

$$f_n \circ \sigma = \sum_{w \in W(n+1) \setminus W(n)} (f \circ \sigma)\hat{}(w)\, w,$$

where $W(n)$ is defined in (3.9).

i. Prove that $(\Sigma_{j=1}^n f_j : n \in \mathbb{N})$ is a martingale sequence.
ii. Deduce from the Burkholder–Gundy martingale inequalities that for all $p \in (1, \infty)$,

$$B_p \|\Sigma_n |f_n|^2\|_{\mathrm{L}^p} \le \|f\|_{\mathrm{L}^p}^2 \le C_p \|\Sigma_n |f_n|^2\|_{\mathrm{L}^p},$$

where $B_p > 0$ and $C_p > 0$ are numerical constants depending only on p; compare this with [Pa, Theorem V].
iii. Prove that if $f \in \mathrm{L}^1([0,1], \mathfrak{B}, \mathfrak{m})$, then $\Sigma_{j=1}^n f_j \to f$ a.e. $(\mathfrak{m})$ on $[0,1]$.

12. i. Prove that Theorem 13 is equivalent to the following assertion: for $f \in \mathrm{L}^p(\mathbf{T}, \mathfrak{m})$,

$$\left\| \sum_{j=-n}^{n} \hat{f}(j)\, \mathrm{e}^{\mathrm{i}jt} \right\|_{\mathrm{L}^p} \le C_p \|f\|_{\mathrm{L}^p}, \quad n \in \mathbb{N}, \tag{E.1}$$

where $C_p > 0$ depends only on p.
ii. Show that (E.1) is false for $p = 1$, and false also if we replace $\mathrm{L}^p(\mathbf{T}, \mathfrak{m})$ by $\mathrm{C}(\mathbf{T})$.

13. i. Show that for all $1 \le p < 2$ and all $K > 0$, there exist W-polynomials g such that $\|g_j\|_{\mathrm{L}^\infty} \le 1$, and $\|\hat{g}_j\|_p > K$.
ii. Show that there exist $f \in \mathrm{C}(\Omega)$ such that $\|\hat{f}\|_p = \infty$ for all $p \in [1, 2)$.

14. Prove that $A(\mathbf{T}) \subsetneq \mathrm{C}(\mathbf{T}) \subsetneq \mathrm{L}^\infty(\mathbf{T}, \mathfrak{m})$.

In the exercises that follow, G denotes a general metrizable compact Abelian group with a normalized Haar measure $\mathfrak{m}$, and Γ denotes its countable discrete dual. The first two chapters in Rudin's book [Ru3] suffice, but if you have not yet read through them, then take $G = \Omega$; all that is needed here has been done in this chapter.

15. The following restriction algebra is officially introduced in §7. Define

$$A(E) = \mathrm{L}^1(G, \mathfrak{m})\hat{}/\{\hat{f} \in \mathrm{L}^1(G, \mathfrak{m})\hat{} : \hat{f} = 0 \text{ on } E\}.$$

$A(E)$ is the quotient $\mathrm{L}^1(G, \mathfrak{m})\hat{}$ *modulo* the ideal
$$\{\hat{f} \in \mathrm{L}^1(G, \mathfrak{m})\hat{} : \hat{f} = 0 \text{ on } E\}.$$

Equipped with the quotient norm

$$\|\hat{f}|_E\|_{A(E)} = \inf\{\|g\|_{\mathrm{L}^1} : g \in \mathrm{L}^1,\ \hat{g}|_E = \hat{f}|_E\},$$

it is the Banach algebra of restrictions of $\hat{f} \in \mathrm{L}^1(G, \mathfrak{m})\hat{}$ to E.

i. Show that E is Sidon if and only if $A(E) = \mathrm{c}_0(E)$.
ii. Show that if E is Sidon, then $\mathrm{C}_E = \mathrm{L}^\infty_E$.
iii. Part ii leads to a simple proof of (5.18): the functional independence property of the Rademacher system in (II.1.3) easily implies $A_R(\Omega) = \mathrm{C}_R(\Omega)$, and this, by ii above, implies (5.18). A similar proof can be given in $\mathbf{T}$, once you verify – as you do below – an 'approximate' functional independence property for a lacunary $E = \{\lambda_j\} \subset \mathbb{Z}^+$ with $q_E \geq 3$. (See (5.16).)

   a. Prove that there exists $0 < \delta < 1$ such that for all $\{\theta_j\} \subset \mathbf{T}$, there exists $x \in \mathbf{T}$ such that

   $$|e^{i\theta_j} - e^{i\lambda_j x}| < \delta, \quad j \in \mathbb{N}.$$

   b. Without use of Riesz products, prove that E is Sidon.

16. You have already noticed, for example in Exercise 15 i, that $E \subset \Gamma$ is a Sidon set if and only if there exists $\zeta > 0$ such that for all E-polynomials f

$$\zeta\|f\|_{\mathrm{L}^\infty} \geq \|\hat{f}\|_1.$$

Use this, in conjunction with any of the constructions in Chapter IV §2, to prove that W_2 (and hence W_k for every $k \geq 2$) is not a Sidon set.

17. Let $\{\mu_n\} \subset \mathrm{M}(G)$, and assume that $\lim_{n\to\infty} \hat{\mu}_n(\gamma) = \phi(\gamma)$ for all $\gamma \in \Gamma$. Prove that there exists $\mu \in \mathrm{M}(G)$ such that $\hat{\mu} = \phi$ if and only if $\sup\{\|\mu_n\| : n \in \mathbb{N}\} < \infty$.

18. For real-valued φ in the unit ball of $l^\infty(\mathbb{N})$, define

$$\mu_n = \prod_{j=1}^{n}(1 + \varphi(j)\; r_j).$$

   i. Show that if $w = r_{j_1} \cdots r_{j_k}$ then $\hat{\mu}_n(w) \to \varphi(j_1) \cdots \varphi(j_k)$ as $n \to \infty$.

   ii. Show there exists $\rho_\varphi \in \mathrm{M}(\Omega)$, which is called a *Riesz product* and is written as an 'infinite product'

$$\rho_\varphi = \prod_{j=1}^{\infty}(1 + \varphi(j)\; r_j), \tag{E.2}$$

   such that if $w \in W$ and $w = r_{j_1} \cdots r_{j_k}$, then

$$\hat{\rho}_\varphi(w) = \varphi(j_1) \cdots \varphi(j_k).$$

   iii. Let $\varphi \in l^\infty(\mathbb{N})$ be $\mathbb{R}$-valued such that $\|\varphi\|_\infty \leq 1$, and let ρ_φ be the corresponding Riesz product. Prove that $\rho_\varphi \ll \mathbb{P}$ if and only if $\varphi \in l^2(\mathbb{N})$.

   iv. Prove that the converse to Corollary 11 is false: there exists $\mu \in \mathrm{M}(\Omega)$ such that $\hat{\mu} \in \mathrm{c}_0(W)$, but μ is not absolutely continuous with respect to $\mathbb{P}$.

19. i. Let $W = \{w_j : j \in \mathbb{N}\}$ be the Walsh system enumerated by Paley's ordering. Prove that if $E \subset \mathbb{Z}^+$ is a lacunary set, then $\{w_j : j \in E\}$ is Sidon in $\hat{\Omega}$.

   ii.* Let $E \subset \mathbb{Z}^+$ be a Sidon set. Is $\{w_j : j \in E\}$ a Sidon set in W?

20. i. Referring to the proof of Lemma 20, verify (6.27).

   ii. Construct the Riesz product $\mu \in \mathrm{M}(\Omega)$ that satisfies (6.30).

21. Verify the first step in the proof of Theorem 21. That is, show that if $f \in \mathrm{L}^\infty_{R_k}(\Omega, \mathbb{P})$ and f_m is defined by (6.32), then $f - f_m \in \mathrm{C}(\Omega)$.

22. Prove Proposition 24.

23. Verify that (8.4) is a W^n-series of $f \in \mathrm{L}^p(\Omega^n, \mathbb{P}^n)$ (resp., $f \in \mathrm{C}(\Omega^n)$) if and only if $f \in \mathrm{L}^p(\Omega, \mathbb{P})$ (resp., $f \in \mathrm{C}(\Omega)$).

24. Suppose $A \subset G$ and $\mathfrak{m}(A) > 0$. Prove that $A \cdot A$ contains a non-empty neighborhood of e_G (identity element of the group G).

25. Prove that $E \subset \hat{G}$ is a *Rosenthal* set if and only if $\mathrm{L}^\infty_E(G, \mathfrak{m})$ is a separable Banach space.

26. i. A spectral set $E \subset \Gamma$ is said to be a *Riesz set* if $\mathrm{M}_E(G) = \mathrm{L}^1_E(G, \mathfrak{m})$. The origin of the definition is a classical result due to the brothers F. and M. Riesz, that $\mathbb{Z}^+$ is a Riesz set in $\mathbb{Z}$ [Hel, p. 106]. The first general study of such sets appeared in [Mey1].

    Prove that E is a Riesz set if and only if $\mathrm{M}_E(G)$ is separable.

    ii.* By using a key result from H^p-theory, R. Dressler and L. Pigno proved in [DrP] that if $E \subset \mathbb{Z}$ is Rosenthal, then E is Riesz. Prove the general result: if $E \subset \Gamma$ is Rosenthal, then E is Riesz, where Γ is a discrete Abelian group. Little is known; even the case $\Gamma = W$ has not yet been resolved.

27. In this exercise you will verify a combinatorial formula (used by Davie in [Da, Lemma 2.1]), and then apply it to prove Lemma 29.

    i. Let $\varphi_1, \ldots, \varphi_n$ be scalar-valued functions defined on $[n]$. For $S \subset [n]$, define $\phi_S = \sum_{l \in S} \varphi_l$. Prove

$$\sum_{\tau \in \mathrm{per}[n]} \varphi_1(\tau 1) \cdots \varphi_n(\tau n) = \sum_{m=1}^{n} (-1)^{n+m} \sum_{\substack{S \subset [n] \\ |S| = m}} \phi_S(1) \cdots \phi_S(n),$$

    where $\mathrm{per}[n]$ denotes the set of permutations of $[n]$.

    ii. Establish Lemma 29 by using Part i.

28. Prove Lemma 30 by a combinatorial argument, showing that for every positive integer k,

$$\left\| \sum_{j=1}^{k} r_j \right\|_{\mathrm{L}^{2k}} = \sum_{\substack{i_j \\ j=1,\ldots,2k}} \mathbf{E}\, r_{i_1} \cdots r_{i_{2k}} \geq k.$$

    (Compare your effort in this exercise with the proof of Lemma 30.)

29. Let (X, ν) be a finite measure space. Let f be a scalar-valued measurable function on X. Show that for every $q \in (1, 2)$

$$\|f\|_{\mathrm{L}^q(\nu)} \leq (\|f\|_{\mathrm{L}^1(\nu)})^{1-\frac{2}{p}} (\|f\|_{\mathrm{L}^2(\nu)})^{\frac{2}{p}}, \qquad \frac{1}{p} + \frac{1}{q} = 1.$$

30. Use the Khintchin inequalities to deduce the results in a and b below (proved first by Littlewood in [Lit3]).

a. Suppose $(a_n) \in B_{l^2(\mathbb{Z})}$. Then for every $p > 2$, there exists $(\alpha_n) \in B_{l^\infty(\mathbb{Z})}$ such that

$$\sum_{n\in\mathbb{Z}} \alpha_n a_n \mathrm{e}^{int} \in \mathrm{L}^p(\mathbf{T}) \quad \text{[Lit3, Theorem 1].}$$

Proof of a.

i. Consider the random series

$$S_\omega = \sum_{n\in\mathbb{Z}} r_n(\omega)\; a_n\; \mathrm{e}^{int}, \quad \omega \in (\Omega, \mathbb{P}),$$

where the Rademacher system above is indexed by $\mathbb{Z}$, and $\Omega = \{-1,1\}^{\mathbb{Z}}$. Let $\{K_j : j \in \mathbb{N}\}$ denote the usual Féjer kernel; see [Kat, Chapter I]. Define

$$A_j = \{\omega \in \Omega : \|S_\omega \star K_j\|_{\mathrm{L}^p(\mathbf{T})} \le 100\; p^{\frac{1}{2}}\}.$$

Prove there exist $c_p > 0$ (depending only on p) such that

$$\mathbb{P}(A_j) > c_p.$$

ii. Prove that there exist $\omega \in \Omega$ and increasing sequences of positive integers (N_j) such that

$$\|S_\omega \star K_{N_j}\|_{\mathrm{L}^p(\mathbf{T})} \le 100\; p^{\frac{1}{2}}, \quad j \in \mathbb{N}.$$

Conclude that $S_\omega \in \mathrm{L}^p(\mathbf{T})$.

b. Suppose $(a_n) \in c_0(\mathbb{Z})$, and $\sum_n \alpha_n a_n \mathrm{e}^{int} \in \mathrm{L}^1(\mathbf{T})$ for every sequence of scalars (α_n) such that $|\alpha_n| = 1,\ n \in \mathbb{N}$. Then, $(a_n) \in l^2(\mathbb{Z})$; [Lit3, Theorem 2].

Proof of b.

i. Prove

$$\sup\left\{\left\|\sum_n r_n(\omega)\; a_n\; \mathrm{e}^{int}\right\|_{\mathrm{L}^1(\mathbf{T})} : \omega \in \Omega\right\} < \infty.$$

ii. Prove b.

31. Let Γ be a discrete Abelian group, $p > 2$, and $1/q + 1/p = 1$. Prove that $F \subset \Gamma$ is a $\Lambda(q)$-set if and only if for all $f \in \mathrm{L}^q(\hat{\Gamma})$,

$$\sum_{\gamma \in \Gamma} \mathbf{1}_F(\gamma)\, |\hat{f}(\gamma)|^2 < \infty.$$

32. i. Prove the n-dimensional Khintchin L^2–L^2 inequality, and show that the best constant in the inequality is $2^{n/2}$.
    ii.* Prove the Khintchin L^1–L^2 inequality in the case W_n. What is the best constant for this inequality?
33. i. Prove Proposition 39.
    ii. Prove that if $p > 1$, then $(\mathrm{C}_F)\hat{} = l^p$ only if F is finite.
    iii.* For $p > 1$, does $(\mathrm{C}_F)\hat{} \subset l^p$ imply that F is a Rosenthal set?
34. Show that the usual notion of *linear independence* of vectors in the framework of linear algebra can be rephrased in terms of *functional independence.*

Hints for Exercises in Chapter VII

1. Use the norm-density of trigonometric polynomials in $\mathrm{C}(\mathbf{T})$ (the Stone–Weierstrass theorem), which implies that $\{e^{int} : n \in \mathbb{Z}\}$ is a complete orthonormal set in $\mathrm{L}^2(\mathbf{T}, \mathfrak{m})$.
3. Establish it first for *cylinders*

$$A(F; (\epsilon_j)) := \{(\omega_j) \in \Omega : \omega_j = \epsilon_j, j \in F\},$$

   where $F \subset \mathbb{N}$ is finite, and $\epsilon_j \in \{-1, 1\}$, $j \in F$. Note that $\mathbf{1}_A$ is continuous on Ω.
4. Apply (2.9).
5. $\mathbb{P}$ is translation-invariant. If μ is a translation-invariant probability measure on Ω and $\mu \neq \mathbb{P}$, then there exists a non-empty cylinder A such that $\mu(A) \neq \mathbb{P}(A)$. This, by translation invariance, leads to a contradiction.
7. i. Verify it first for simple functions.
   ii. First show that

$$\|k_n \star f - f\|_B \;\leq\; \mathbf{E}_\eta\, |k_n(\eta)|\, \|f_\eta - f\|_B,$$

and then split the right side into two integrals: one over a judiciously chosen neighbourhood V of e_0, and the other over V^c. A similar argument in the framework of $\mathbf{T}$ appears in [Kat, p. 10].

8. First note that the assertion is trivial for W-polynomials, and then use the norm-density of W-polynomials in $L^1(\Omega, \mathbb{P})$. This is the analog of the classical Riemann–Lebesgue lemma on $\mathbf{T}$ and $\mathbb{R}$: $f \in L^1(\mathbf{T}, \mathfrak{m}) \Rightarrow \hat{f} \in c_0(\mathbb{Z})$, and $f \in L^1(\mathbb{R}, dx) \Rightarrow \hat{f} \in c_0(\mathbb{R})$; e.g., [Kat, p. 13], [Roy, Exercise 16, p. 94].
9. The assertion in (1) is not as trivial as it appears. The sticking point is that members of $L^\infty(\Omega, \mathbb{P})$ are equivalence classes determined by the $\mathbb{P}$-null sets. Use the observation that if $f \in L^\infty(\Omega, \mathbb{P})$ and $\lim_{\omega \to e_0} \|f_\omega - f\|_{L^\infty} = 0$, then $\{k_n \star f\}$ is equicontinuous. To prove (2), observe that if $\mu \in M(\Omega)$ and $\lim_{\omega \to e_0} \|\mu_\omega - \mu\|_M = 0$, then (by an argument like the one used in Exercise 7) $k_n \star \mu \to \mu$ in $M(\Omega)$.
11. iii. The assertion that $\Sigma_{j=1}^n f_j \to f$ in L^1 follows from Corollary 9. (The statement in iii had been proved first by Kaczmarz [Ka], and extended later by Paley in [Pa, Theorem IX].)
12. ii. To establish the first statement, use

$$\left\| \sum_{j=-n}^{n} e^{ijt} \right\|_{L^1} \sim \log n,$$

together with the *de la Vallée Poussin Kernel*; see [Kat, p. 15, p. 50]. To establish the second statement, use the relation above and duality.
13. i. For real-valued $\varphi \in B_{l^2}$, consider $\prod_{j=1}^n (1 + i\varphi(j) r_j)$; cf. (III.2.13).
    ii. Let (p_j) be a sequence in [1,2) converging to 2, and $(g_j : j \in \mathbb{N})$ a sequence of W-polynomials with pairwise disjoint spectra, such that $\|g_j\|_{L^\infty} \leq 1$ and $\|\hat{g}_j\|_{p_j} > 2^j$ for every $j \in \mathbb{N}$. Let $f = \sum_{j=1}^\infty g_j / j^2$; cf. [Kat, pp. 99–100].
14. The quickest argument I know uses $A(\mathbf{T}) = C(\mathbf{T}) \Rightarrow C(\mathbf{T}) = L^\infty(\mathbf{T}, m)$. A constructive argument is an instance of the preceding exercise.
15. iii. For hints, or to see where results of this type lead, consult [Mé].
17. To show the 'only if' direction, observe that if $\hat{\mu}_n(\gamma) \to \hat{\mu}(\gamma)$ for all $\gamma \in \Gamma$, then $\mu_n \to \mu$ weak* in $M(G)$, and apply a uniform boundedness principle. To prove the other direction, use the Riesz representation theorem. For example, see [Kat, Exercise I.7.9].

18. Browse through Zygmund's treatise [Zy2]; there you will find the proofs (and much more!).
19. i. Assume first $q_E \geq 3$, and verify that the Riesz product

$$\rho_\varphi = \prod_{j \in E} (1 + \varphi(j)\ w_j),$$

where $-1 \leq \varphi(j) \leq 1$ for $j \in E$, is a well-defined measure on Ω, and $\hat{\rho}_\varphi(w_j) = \varphi(j)$ for all $j \in \mathbb{N}$. To this end you need to show that the lacunarity of E implies that $\{w_j : j \in E\}$ is algebraically independent in W. See [Mo, p. 498].
20. i. By assumption, the D_j are generated by mutually disjoint blocks of Rademacher characters, and, therefore, functions spanned by these blocks are 'functionally' independent.
    ii. Apply the hypothesis that the D_j are strongly disjoint; cf. (6.18).
22. Exercise 17 is a special case. See the proof of Proposition 26. The following basic tenet of functional analysis is used (here and throughout): if X is a Banach space and H is a closed subspace of X, then the dual space of the quotient X/H is the annihilator $H^\perp$ of H in X^*, and the dual space of H is $X^*/H^\perp$.
25. One direction is easy. To prove the other direction, let $\{g_j\}$ be a dense subset of L^∞_E. Fix $f \in \mathrm{L}^\infty_E$, and fix $\epsilon > 0$. Define

$$A_j = \{x \in G : \|f_x - g_j\|_{\mathrm{L}^\infty} < \epsilon\},$$

where f_x is the translate of f. Then, $G = \cup A_j$, and therefore there exists j_0 such that $m(A_{j0}) > 0$. Therefore, $A_{j0} \cdot A_{j0}$ contains an open neighborhood of $e_G \in G$. You have just set the stage for an application of Exercise 9. Note that your proof in Exercise 9, properly adapted, works in a setting of general groups.
26. i. Use the strategy of the previous exercise.
30. i. in the proof of a. Apply Fubini's theorem, and the Khintchin L^p–L^2 inequality,

$$\mathbf{E}_\omega\ \|S_\omega \star K_j\|^p_{\mathrm{L}^p(\mathbf{T})} = \int_{\mathbf{T}} \mathbf{E}_\omega |S_\omega \star K_j|^p\ dt \leq p^{p/2}.$$

    ii. in the proof of b. Assume the assertion is false. By use of approximate identities, produce $\omega \in \Omega$ such that

$$\sum_n r_n(\omega)\ a_n e^{int} \notin \mathrm{L}^1(\mathbf{T}).$$

iii. in the proof of b. By i, Fubini's theorem, and the Khintchin L^1–L^2 inequality,

$$\infty > \mathbf{E}_\omega \left\| \sum_n r_n(\omega)\, a_n\, e^{int} \right\|_{L^1(\mathbf{T})}$$

$$= \int_{\mathbf{T}} \left(\mathbf{E}_\omega \left| \sum_n r_n(\omega)\, a_n\, e^{int} \right| \right) dt \geq (1/\sqrt{2}) \left(\sum_n |a_n|^2 \right)^{\frac{1}{2}}.$$

32. i. Induction does it.

33. i. The implication 'iii ⇒ ii' follows by the norm-density of F-polynomials in $C_F(\Omega)$. Conversely, if iii fails, then construct $f \in C_F(\Omega)$ such that $\hat{f} \notin l^p$. The equivalences 'ii ⇔ iv' and 'iii ⇔ iv' follow by duality: $C_F(\Omega)^* = B(F)$, $A(F)^* = L_F^\infty(\Omega)$, and $(l^p)^* = l^q$; see Proposition 24. The implication 'ii ⇒ i' follows from Corollary 9 ii.

VIII

Multilinear Extensions of the Grothendieck Inequality (via $\Lambda(2)$-uniformizability)

1 Mise en Scène: A Basic Issue

The Grothendieck inequality, a fundamental statement about bilinear functionals, can be expressed equivalently in several ways. In previous chapters we have noted:

(a) Grothendieck's original formulation [Gro2, p. 59];
(b) Lindenstrauss's and Pelczynski's restatement of it [LiPe, p. 275] (see Theorem III.1);
(c) Theorem IV.13 (cf. (V.4.3));
(d) The inequality in (IV.5.37), which is akin to Grothendieck's original formulation;
(e) The factorization theorem Theorem V.2.

The equivalences (a) $\Leftrightarrow$ (b) $\Leftrightarrow$ (c) $\Leftrightarrow$ (d) are based on the duality $V_2^* = F_2$, and are easy to verify; each of the four assertions conveys the same phenomenon, generically referred to as the *inequality.* The equivalence of the inequality and (e), a result of convexity (Proposition V.5), is not quite as obvious. We recall:

The Inequality (cf. (IV.5.37))

Let H be a Hilbert space, and let η be a bilinear functional on H. Consider the norms

$$\|\eta\|_{f_2} := \sup\{|\eta(\mathbf{x},\mathbf{y})| : \mathbf{x} \in B_H, \mathbf{y} \in B_H\}, \tag{1.1}$$

and

$$\|\eta\|_{\mathrm{pb}_2} := \sup\{\|\eta|_{E\times F}\|_{\mathrm{V}_2(E,F)} : \text{finite subsets } E \subset B_H, F \subset B_H\}. \tag{1.2}$$

The Grothendieck inequality is the assertion

$$\|\eta\|_{\mathrm{pb}_2} \leq k_{\mathrm{G}}\ \|\eta\|_{f_2}, \tag{1.3}$$

where $k_{\mathrm{G}} > 0$ is a universal constant. The opposite inequality $\|\eta\|_{f_2} \leq \|\eta\|_{\mathrm{pb}_2}$ is obvious. (Here and throughout, $\|\cdot\|_{f_n}$ will denote in the given context the usual norm of an n-linear functional, and $\|\cdot\|_{\mathrm{pb}_n}$ will denote a norm conveying a Grothendieck-type inequality.)

The Factorization Theorem (cf. Theorem IV.2)

Let K_1 and K_2 be locally compact Hausdorff spaces. Let η be a bilinear functional on $\mathrm{C}_0(K_1) \times \mathrm{C}_0(K_2)$, with its (usual) norm

$$\|\eta\|_{f_2} := \sup\{|\eta(f,g)| : f \in B_{\mathrm{C}_0(K_1)}, g \in B_{\mathrm{C}_0(K_2)}\}. \tag{1.4}$$

Given probability measures ν_1 and ν_2 on the respective Borel fields of K_1 and K_2, define

$$\begin{aligned}&\|\eta\|_{(\nu_1,\nu_2)}\\ &\quad = \sup\{|\eta(f,g)| : f \in B_{\mathrm{L}^2(\nu_1)} \cap \mathrm{C}_0(K_1), g \in B_{\mathrm{L}^2(\nu_2)} \cap \mathrm{C}_0(K_2)\},\end{aligned} \tag{1.5}$$

and

$$\|\eta\|_{\phi} = \inf\{\|\eta\|_{(\nu_1,\nu_2)} : \text{ probability measures } \nu_1, \nu_2\}. \tag{1.6}$$

The Grothendieck factorization theorem is the assertion

$$\|\eta\|_{\phi} \leq k_{\mathrm{G}}\ \|\eta\|_{f_2}, \tag{1.7}$$

where $k_{\mathrm{G}} > 0$ is a universal constant. The opposite inequality $\|\eta\|_{f_2} \leq \|\eta\|_{\phi}$ is easy to verify.

The equivalence of thc inequality and the factorization theorem (with the same constants k_{G}) is a 'two-dimensional' phenomenon that does not extend, as such, to dimensions greater than two (see Exercise V.5).

Indeed, questions about multidimensional extensions of the Grothendieck inequality and factorization theorem deal with three ostensibly separate issues: the first concerns extensions of the inequality; the second concerns extensions of the factorization theorem; and the third concerns relationships between multilinear Grothendieck-type inequalities and multilinear factorization. We have already touched on the second and third issues in Chapter V, and will return to them in later chapters. In this chapter, we consider the first issue: extensions of the Grothendieck bilinear inequality to dimensions greater than two.

2 Projective Boundedness

To begin, we formalize a notion that conveys, in effect, a multidimensional Grothendieck inequality:

Definition 1 Let η be an n-linear functional on a Hilbert space H. Define

$$\|\eta\|_{\mathrm{pb}_n} = \sup\{\|\eta|_{F_1\times\cdots\times F_n}\|_{V_n(F_1,\ldots,F_n)} : F_i \subset B_H,\ |F_i| < \infty,\ i \in [n]\}. \tag{2.1}$$

If $\|\eta\|_{\mathrm{pb}_n} < \infty$, then η is said to be *projectively bounded*; otherwise, if $\|\eta\|_{\mathrm{pb}_n} = \infty$, then η is *projectively unbounded.*

Following the identification of tensor algebras as restriction algebras (Chapter VIII §8), we can restate the definition in the language of harmonic analysis. Let η be an n-linear functional on a Hilbert space H. If $E \subset B_H$ and R_E denotes the Rademacher system indexed by E, then let $\phi_{\eta,E}$ be the scalar-valued function on the n-fold product $R_E \times \cdots \times R_E$ defined by

$$\phi_{\eta,E}(r_{x_1},\ldots,r_{x_n}) = \eta(x_1,\ldots,x_n),\ (r_{x_1},\ldots,r_{x_n}) \in R_E^n. \tag{2.2}$$

Proposition 2 (Remark ii in Chapter VII §8; Exercise 1 i). *If η is an n-linear functional on a Hilbert space H, then*

$$\begin{aligned}\left(\frac{1}{2}\right)^n \sup\{\|\phi_{\eta,E}\|_{B(R_E^n)} : E \subset B_H, |E| < \infty\} &\le \|\eta\|_{\mathrm{pb}_n} \\ &\le \sup\{\|\phi_{\eta,E}\|_{B(R_E^n)} :\ E \subset B_H, |E| < \infty\}.\end{aligned} \tag{2.3}$$

($B(R^n_E)$ is the algebra of restrictions to R^n_E of transforms of Borel measures on $\{-1,1\}^E$.)

Proposition 3 (Exercise 1 ii). *An n-linear functional η on a Hilbert space H is projectively bounded if and only if $\phi_{\eta,E} \in B(R^n_E)$ for all countably infinite sets $E \subset B_H$.*

Note that $\|\cdot\|_{\mathrm{pb}_n}$ defines a norm on the space of projectively bounded n-linear functionals on H, and that the resulting normed linear space is a Banach space. For future use, we record also completeness with respect to weak convergence:

Proposition 4 (Exercise 1 iii). *If $(\eta_m : m \in \mathbb{N})$ is a sequence of n-linear functionals on a Hilbert space H such that*

$$\lim_{m\to\infty} \eta_m(\mathbf{x}_1,\ldots,\mathbf{x}_n) = \eta(\mathbf{x}_1,\ldots,\mathbf{x}_n), \ (\mathbf{x}_1,\ldots,\mathbf{x}_n) \in H^n, \tag{2.4}$$

then

$$\|\eta\|_{\mathrm{pb}_n} \le \liminf_{m\to\infty} \|\eta_m\|_{\mathrm{pb}_n}. \tag{2.5}$$

If η is an n-linear functional on H, then

$$\|\eta\|_{f_n} := \sup\{|\eta(\mathbf{x}_1,\ldots,\mathbf{x}_n)| : (\mathbf{x}_1,\ldots,\mathbf{x}_n) \in B^n_H\} \le \|\eta\|_{\mathrm{pb}_n}. \tag{2.6}$$

Our focus is on the opposite inequality. We obviously have that every bounded *linear* functional on H is projectively bounded. That every bounded *bilinear* functional on H is projectively bounded (not quite as obvious) is precisely the Grothendieck inequality. In this chapter we consider the question: *what are the projectively bounded n-linear functionals for $n \ge 3$?* We establish a characterization, and will obtain as a consequence that there exist bounded n-linear functionals that are projectively unbounded.

3 Uniformizable $\Lambda(2)$-sets

The notion of $\Lambda(2)$-uniformizability, the key to the proof of thc (bilinear) Grothendieck inequality in Chapter III, can be rephrased in a framework of harmonic analysis:

Definition 5 (cf. Definition III.6). Let G be a compact abelian group, and let $\hat{G}$ be its dual group. A spectral set $E \subset \hat{G}$ is a uniformizable $\Lambda(2)$-set if for every $0 < \epsilon < 1$ there exists $\delta = \delta_E(\epsilon)$ and a map $\psi : l^2(E) \mapsto \mathrm{L}^\infty(G)$ so that for all $\mathbf{x} \in l^2(E)$,

$$\text{(i)} \quad \psi(\mathbf{x})\hat{}\,(\gamma) = \mathbf{x}(\gamma), \quad \gamma \in E;$$

$$\text{(ii)} \quad \|\psi(\mathbf{x})\|_{\mathrm{L}^\infty} \leq \delta \|\mathbf{x}\|_2;$$

$$\text{(iii)} \quad \|\psi(\mathbf{x})\hat{}\,|_{E^c}\|_2 := \left(\sum_{\gamma \notin E} |\psi(\mathbf{x})\hat{}\,(\gamma)|^2 \right)^{\frac{1}{2}} \leq \epsilon \|x\|_2. \tag{3.1}$$

ϵ and δ are said to be $\Lambda(2)$-*uniformizing constants* of E associated with the $\Lambda(2)$-*uniformizing map* ψ. *(E^c denotes the complement of E in $\hat{G}$.)*

In this chapter we work in the setting $G = \Omega^n$, where $\Omega = \{-1, +1\}^{\mathbb{N}}$, and its dual group $\hat{G} = W^n$, where W is the system of Walsh characters. Here and throughout, we write $\mathrm{L}^p(\Omega^n)$ for $\mathrm{L}^p(\Omega^n, \mathbb{P}^n)$ $(1 \leq p \leq \infty)$, where $\mathbb{P}^n$ is the n-fold product of $\mathbb{P}$ (Haar measure on Ω).

Lemma 6 *For every integer $n > 0$, R^n is a uniformizable $\Lambda(2)$-set.*

Proof: (see Remarks i, iii below). Observe that R^n is a $\Lambda(p)$-set for all $p > 2$ (Definition VII.33, Proposition VII.31), and then apply Theorem III.8 and Remark ii §4. □

Remarks:

i **(a constructive proof of Lemma 6).** To ease notation, we identify $l^2(R^n)$ with $l^2(\mathbb{N}^n)$, and construct a $\Lambda(2)$-uniformizing map ψ on $l^2(\mathbb{N}^n)$.

We first verify Lemma 6 for $n = 1$ (cf. Exercise III.8). Fix $0 < \epsilon < 1$. Let $\mathbf{x} \in l^2(\mathbb{N})$ be a vector with real-valued coordinates, and define

$$\psi_1(\mathbf{x}) = -\mathscr{R}e \, \frac{\mathrm{i}\|\mathbf{x}\|_2}{\epsilon} \prod_{j=1}^{\infty} \left(r_0 + \mathrm{i}\epsilon \frac{\mathbf{x}(j)}{\|\mathbf{x}\|_2} r_j \right) \quad \text{(Exercise 2).} \tag{3.2}$$

Then, $\psi_1(\mathbf{x}) \in \mathrm{L}^\infty(\Omega)$ is a real-valued function such that

$$\|\psi_1(\mathbf{x})\|_{\mathrm{L}^\infty} \leq \frac{\exp(\epsilon^2/2)}{\epsilon} \|\mathbf{x}\|_2. \tag{3.3}$$

The spectral analysis of $\psi_1(\mathbf{x})$ yields

$$\psi_1(\mathbf{x}) = \sum_{j=1}^{\infty} \mathbf{x}(j)\ r_j$$
$$+ \sum_{k=1}^{\infty} (-1)^k (\epsilon/\|\mathbf{x}\|_2)^{2k} \sum_{0<j_1<\cdots<j_{2k+1}} \mathbf{x}(j_1)\cdots\mathbf{x}(j_{2k+1})\ r_{j_1}\cdots r_{j_{2k+1}}, \tag{3.4}$$

and therefore,

$$\psi(\mathbf{x})\hat{}\,(r_j) = \mathbf{x}(j), \quad j \in \mathbb{N}. \tag{3.5}$$

We estimate

$$\sum_{w \notin R} |\psi_1(\mathbf{x})\hat{}\,(w)|^2$$
$$\leq \sum_{k=1}^{\infty} (\epsilon/\|\mathbf{x}\|_2)^{4k} \sum_{0<j_1<\cdots<j_{2k+1}} |\mathbf{x}(j_1)\cdots\mathbf{x}(j_{2k+1})|^2$$
$$\leq (\|\mathbf{x}\|_2/\epsilon)^2 \sum_{k=1}^{\infty} \frac{1}{(2k+1)!}\ \epsilon^{4k+2} \leq \epsilon^4 \|\mathbf{x}\|_2^2. \tag{3.6}$$

By combining (3.3), (3.5), and (3.6), we conclude that R is a uniformizable $\Lambda(2)$-set, and

$$\delta_R(\epsilon) := \delta_1 \leq \frac{2}{\sqrt{\epsilon}} \exp(\epsilon/4). \tag{3.7}$$

We proceed by induction. For $n > 1$, let $\mathbf{x} \in l^2(\mathbb{N}^n)$ have real-valued coordinates, and let

$$\psi_{n-1} : l^2(\mathbb{N}^{n-1}) \mapsto \mathrm{L}^\infty(\Omega^{n-1})$$

be the $\Lambda(2)$-uniformizing map provided by the induction hypothesis. Specifically, assume

$$\|\psi_{n-1}(\mathbf{x}(j,\cdot))\|_{\mathrm{L}^\infty} \leq \delta_{n-1}\ \|\mathbf{x}(j,\cdot)\|_2, \quad j \in \mathbb{N}, \tag{3.8}$$

where $\psi_{n-1}(\mathbf{x}(j,\cdot))$ is a real-valued element in $\mathrm{L}^\infty(\Omega^{n-1})$ such that

$$\psi_{n-1}(\mathbf{x}(j,\mathbf{k}))\hat{}\,(r_{k_1},\ldots,r_{k_{n-1}}) = \mathbf{x}(j,\mathbf{k}),$$
$$\mathbf{k} = (k_1,\ldots,k_{n-1}) \in \mathbb{N}^{n-1}, \tag{3.9}$$

and

$$\sum_{\gamma \notin R^{n-1}} |\psi_{n-1}(\mathbf{x}(j,\cdot))\hat{}\,(\gamma)|^2 \leq \epsilon^2\ \|\mathbf{x}(j,\cdot)\|_2^2. \tag{3.10}$$

To simplify notation, denote

$$\psi_{n-1}(\mathbf{x}(j,\cdot)) = f_j, \quad j \in \mathbb{N}. \tag{3.11}$$

By (3.8),

$$\sum_{j=1}^{\infty} \|f_j\|_{\mathrm{L}^\infty}^2 \preceq (\delta_{n-1})^2 \, \|\mathbf{x}\|_2^2. \tag{3.12}$$

Define

$$\psi_n(\mathbf{x}) = -\mathscr{R}e \, \frac{\mathrm{i}\delta_{n-1}\|\mathbf{x}\|_2}{\epsilon} \prod_{j=1}^{\infty} \left(r_0 + \mathrm{i}\epsilon \frac{f_j}{\delta_{n-1}\|\mathbf{x}\|_2} \otimes r_j \right) \quad \text{(cf. (3.2))}. \tag{3.13}$$

By (3.12), $\psi_n(\mathbf{x}) \in \mathrm{L}^\infty(\Omega^n)$ is real-valued, and

$$\|\psi_n(\mathbf{x})\|_{\mathrm{L}^\infty} \leq \delta_{n-1} \frac{\exp(\epsilon^2/2)}{\epsilon} \|\mathbf{x}\|_2 \quad \text{(cf. (3.3))}. \tag{3.14}$$

The series expansion of $\psi_n(\mathbf{x})$ is

$$\psi_n(\mathbf{x}) = \sum_{j=1}^{\infty} f_j \otimes r_j$$
$$+ \sum_{k=1}^{\infty} (-1)^k \, (\epsilon/\delta_{n-1}\|\mathbf{x}\|_2)^{2k} \sum_{0<j_1<\cdots<j_{2k+1}} f_{j_1} \cdots f_{j_{2k+1}} \otimes r_{j_1} \cdots r_{j_{2k+1}}. \tag{3.15}$$

By an application of (3.9) to the first sum on the right side of (3.15),

$$\psi_n(\mathbf{x})\,\hat{}\,(r_{k_1}, \ldots, r_{k_n}) = \mathbf{x}(\mathbf{k}), \quad \mathbf{k} = (k_1, \ldots, k_n) \in \mathbb{N}^n. \tag{3.16}$$

Next, estimate

$$\|\psi_n(\mathbf{x})\,\hat{}\,|_{(R^n)^c}\|_2^2 \;\leq\; \sum_{j=1}^{\infty} \|\hat{f}_j|_{(R^{n-1})^c}\|_2^2$$
$$+ \sum_{k=1}^{\infty} (\epsilon/\delta_{n-1} \, \|\mathbf{x}\|_2)^{4k} \sum_{0<j_1<\cdots<j_{2k+1}} \|f_{j_1}\|_{\mathrm{L}^2}^2 \cdots \|f_{j_{2k+1}}\|_{\mathrm{L}^2}^2. \tag{3.17}$$

Observe that

$$\sum_{0<j_1<\cdots<j_{2k+1}} \|f_{j_1}\|_{\mathrm{L}^2}^2 \cdots \|f_{j_{2k+1}}\|_{\mathrm{L}^2}^2$$
$$\leq \frac{1}{(2k+1)!} \left(\sum_{j=1}^{\infty} \|f_j\|_{\mathrm{L}^2}^2 \right)^{2k+1}. \tag{3.18}$$

By (3.9), (3.10) and (3.11),

$$\sum_{j=1}^{\infty} \|f_j\|_{\mathrm{L}^2}^2 \leq (1+\epsilon^2)\ \|\mathbf{x}\|_2^2. \tag{3.19}$$

The estimates in (3.17), (3.18), and (3.19) imply

$$\begin{aligned} &\|\psi_n(\mathbf{x})\hat{\ }|_{(R^n)^c}\|_2^2 \\ &\leq \epsilon^2\|\mathbf{x}\|_2^2 + (\delta_{n-1}\ \|\mathbf{x}\|_2/\epsilon)^2 \sum_{k=1}^{\infty} \frac{1}{(2k+1)!}\left(\frac{\epsilon^2+\epsilon^4}{\delta_{n-1}^2}\right)^{2k+1} \\ &\leq (\epsilon^2 + (1+\epsilon^2)(\epsilon^2+\epsilon^4)/\delta_{n-1}^2)\ \|\mathbf{x}\|_2^2. \end{aligned} \tag{3.20}$$

This proves the case where the vectors $\mathbf{x}$ are real-valued. The complex case is obtained by treating separately real and imaginary parts.

Corollary 7 (Exercise 3). *The $\Lambda(2)$-uniformizing map*

$$\psi_n : l^2(\mathbb{N}^n) \mapsto \mathrm{L}^\infty(\Omega^n)$$

*constructed above is weak * continuous. Specifically, if*

$$\lim_{j\to\infty} \mathbf{x}_j(\mathbf{k}) = \mathbf{x}(\mathbf{k}) \ \textit{for}\ \mathbf{x}_j \in B_{l^2(\mathbb{N}^n)},\ \mathbf{x} \in B_{l^2(\mathbb{N}^n)},\ \textit{and}\ \mathbf{k} \in \mathbb{N}^n,$$

then

$$\lim_{j\to\infty} \mathbf{E}\ \psi_n(\mathbf{x}_j)g = \mathbf{E}\ \psi_n(\mathbf{x})g,\ \ g \in \mathrm{L}^1(\Omega^n). \tag{3.21}$$

ii **(about Corollary 7).** Adding to Definition 5 the requirement that $\Lambda(2)$-uniformizing maps $\psi : l^2(E) \to \mathrm{L}^\infty(G)$ be weak* continuous produces an ostensibly stronger notion of $\Lambda(2)$-uniformizability. To wit, the proof of Lemma 6, based on Theorem III.8 and the observation that R^n is a $\Lambda(p)$ set for $p > 2$, does not directly yield the assertion in Corollary 7. I do not know whether $\Lambda(2)$-uniformizability alone, with no further structural conditions, implies the existence of weak* continuous $\Lambda(2)$-uniformizing maps.

iii **(A quick non-constructive proof).** Suppose we do not insist on explicitly defined $\Lambda(2)$-uniformizing maps. That is, suppose Definition 5 is rephrased (cf. Definition III.6): $E \subset \hat{G}$ is a uniformizable $\Lambda(2)$-set if for every $0 < \epsilon < 1$ there exists $\delta = \delta_E(\epsilon)$ such that for every $\mathbf{x} \in l^2(E)$ there exists $f \in \mathrm{L}^\infty(G)$ with these properties:

(i) $\hat{f}(\gamma) = \mathbf{x}(\gamma),\ \gamma \in E$;

(ii) $\|f\|_{\mathrm{L}^\infty} \leq \delta\|\mathbf{x}\|_2$;

(iii) $\|\hat{f}|_{E^c}\|_2 := \left(\sum_{\gamma\notin E} |\hat{f}(\gamma)|^2\right)^{\frac{1}{2}} \leq \epsilon\|x\|_2$.

Then, we obtain that R^n is a uniformizable $\Lambda(2)$-set by taking $f \in L^\infty(\Omega^n)$ that satisfies (i) and (ii) above (the Khintchin L^1–L^2 inequality for R^n), and then convolving it with a Riesz product (Exercise 4).

iv **($\Lambda(2)$- uniformizing constants of R^n)** The constructive proof in Remark i implies that $\delta_{R^n}(\epsilon)$ is $\mathscr{O}(\epsilon^{-n+1/2})$. A modification of the proof implies that for every $k > 0$ there exists $C_k > 0$ such that

$$\delta_{R^n}(\epsilon) \le C_k\, \epsilon^{-(1/k)} \quad \text{(Exercise 5).} \tag{3.22}$$

4 A Projectively Bounded Trilinear Functional

In Chapter III we proved the Grothendieck bilinear inequality by using the $\Lambda(2)$-uniformizability of R, and in this chapter we use the $\Lambda(2)$-uniformizability of R^n to derive analogous inequalities in higher dimensions. To ease our way into the multilinear framework, where notation is inevitably more complicated, we first derive an archetypal trilinear instance. Then, generalizing this instance, we shall identify within certain classes of multilinear functionals those that are projectively bounded.

Let $H = l^2(\mathbb{N}^2)$. For $\mathbf{x} \in H$, $\mathbf{y} \in H$, and $\mathbf{z} \in H$ with finite support, define

$$\eta(\mathbf{x}, \mathbf{y}, \mathbf{z}) = \sum_{i,j,k} \mathbf{x}(i,j)\, \mathbf{y}(j,k)\, \mathbf{z}(i,k), \quad (\mathbf{x}, \mathbf{y}, \mathbf{z}) \in H^3. \tag{4.1}$$

Lemma 8 *η determines a bounded trilinear functional on H. Specifically,*

$$\|\eta\|_{\mathcal{F}_3} := \sup\{|\eta(\mathbf{x}, \mathbf{y}, \mathbf{z})| : (\mathbf{x}, \mathbf{y}, \mathbf{z}) \in B_H \times B_H \times B_H\} = 1. \tag{4.2}$$

Proof: Three successive applications of the Cauchy–Schwarz inequality to the sums over i, j, and k, respectively, imply that for $\mathbf{x} \in H$, $\mathbf{y} \in H$, and $\mathbf{z} \in H$,

$$\sum_{i,j,k} |\mathbf{x}(i,j)\, \mathbf{y}(j,k)\, \mathbf{z}(i,k)| \le \|\mathbf{x}\|_2\, \|\mathbf{y}\|_2\, \|\mathbf{z}\|_2.$$

Denote the canonical basis of H by $\{\mathbf{e}_{ij} : (i,j) \in \mathbb{N}^2\}$ (i.e., $\mathbf{e}_{ij}(k,l) = 1$ if $(i,j) = (k,l)$, and $\mathbf{e}_{ij}(k,l) = 0$ otherwise), and observe that for $(i,j,k) \in \mathbb{N}^3$, $\eta(\mathbf{e}_{ij}, \mathbf{e}_{jk}, \mathbf{e}_{ik}) = 1$. □

By fixing an enumeration $W = \{w_j : j \in \mathbb{N}\}$ of the Walsh characters (any enumeration will do), we view η as a bounded trilinear functional on $l^2(W^2)$, and then, by applying Parseval's formula, we realize it as a bounded trilinear functional on $\mathrm{L}^2(\Omega^2)$:

Lemma 9 *For* $f \in \mathrm{L}^2(\Omega^2),\ g \in \mathrm{L}^2(\Omega^2),\ h \in \mathrm{L}^2(\Omega^2)$,

$$\begin{aligned}\eta(\hat{f}, \hat{g}, \hat{h}) &:= \sum_{i,j,k} \hat{f}(w_i, w_j)\hat{g}(w_j, w_k)\hat{h}(w_i, w_k) \\ &= \int_{\Omega^3} f(\omega_1, \omega_2)\ g(\omega_2, \omega_3)\ h(\omega_1, \omega_3)\mathbb{P}^3(\mathrm{d}\omega_1, \mathrm{d}\omega_2, \mathrm{d}\omega_3).\end{aligned} \tag{4.3}$$

Proof: If f, g, and h are W^2-polynomials, then (4.3) is obtained by three successive applications of Parseval's formula. The general case follows by density of polynomials in L^2. □

Lemma 10 *If E_1, E_2 and E_3 are finite subsets of $B_{\mathrm{L}^\infty(\Omega^2)}$, then*

$$\|\eta|_{E_1 \times E_2 \times E_3}\|_{V_3(E_1, E_2, E_3)} \leq 1. \tag{4.4}$$

Proof: Because E_1, E_2, and E_3 are finite, $\eta|_{E_1 \times E_2 \times E_3}$ is obviously in $V_3(E_1, E_2, E_3)$. We verify (4.4) by duality. Let $\beta \in (V_3)^*$; i.e., β is a bounded trilinear functional on $\mathrm{c}_0(E_1) \times \mathrm{c}_0(E_2) \times \mathrm{c}_0(E_3)$, and $\|\beta\|_{V_3^*} = \|\beta\|_{\mathcal{F}_3}$ (Chapter IV §5). The evaluation of β at $\eta|_{E_1 \times E_2 \times E_3}$ is

$$\beta(\eta|_{E_1 \times E_2 \times E_3}) = \sum_{(f,g,h) \in E_1 \times E_2 \times E_3} \beta(\delta_f \otimes \delta_g \otimes \delta_h)\ \eta(\hat{f}, \hat{g}, \hat{h}), \tag{4.5}$$

where δ_f, δ_g, and δ_h are the indicator functions of $\{f\}, \{g\}$, and $\{h\}$, respectively. Apply (4.3), and estimate

$$\begin{aligned}&\left|\sum_{(f,g,h) \in E_1 \times E_2 \times E_3} \beta(\delta_f \otimes \delta_g \otimes \delta_h)\ \eta(\hat{f}, \hat{g}, \hat{h})\right| \\ &= \left|\sum_{(f,g,h)} \beta(\delta_f \otimes \delta_g \otimes \delta_h) \int_{\Omega^3} f(\omega_1, \omega_2) g(\omega_2, \omega_3)\ h(\omega_1, \omega_3)\ \mathbb{P}^3(\mathrm{d}\omega_1, \mathrm{d}\omega_2, \mathrm{d}\omega_3)\right| \\ &\leq \int_{\Omega^3} \left|\sum_{(f,g,h)} \beta(\delta_f \otimes \delta_g \otimes \delta_h)\ f(\omega_1, \omega_2) g(\omega_2, \omega_3)\ h(\omega_1, \omega_3)\right| \mathbb{P}^3(\mathrm{d}\omega_1, \mathrm{d}\omega_2, \mathrm{d}\omega_3) \\ &\leq \|\beta\|_{V_3^*}.\end{aligned} \tag{4.6}$$

The last estimate is obtained as follows. Fix $(\omega_1, \omega_2, \omega_3) \in \Omega^3$. Write

$$\sum_{(f,g,h)\in E_1\times E_2\times E_3} \beta(\delta_f \otimes \delta_g \otimes \delta_h) f(\omega_1, \omega_2)\ g(\omega_2, \omega_3)\ h(\omega_1, \omega_3)$$
$$= \beta(t \otimes u \otimes v), \tag{4.7}$$

where

$$t(f) = f(\omega_1, \omega_2),\ u(g) = g(\omega_2, \omega_3),$$
$$v(h) = h(\omega_1, \omega_3),\ (f, g, h) \in E_1 \times E_2 \times E_3,$$

and obtain (4.6) by integrating the inequality $|\beta(t \otimes u \otimes v)| \leq \|\beta\|_{V_3^*}$ over Ω^3. □

Next, we verify that η is projectively bounded by representing $\eta(\mathbf{x}, \mathbf{y}, \mathbf{z})$ as a convergent series whose summands have the form $\eta(\hat{f}_{\mathbf{x}}, \hat{g}_{\mathbf{y}}, \hat{h}_{\mathbf{z}})$, where $(f_{\mathbf{x}}, g_{\mathbf{y}}, h_{\mathbf{z}}) \in \mathrm{L}^\infty(\Omega^2) \times \mathrm{L}^\infty(\Omega^2) \times \mathrm{L}^\infty(\Omega^2)$. This series representation will be deduced from the $\Lambda(2)$-uniformizability of R^2. To start, enumerate the complement of the Rademacher system in W,

$$W \backslash R := R^{\mathrm{c}} = \{\chi_k : k \in \mathbb{N}\}. \tag{4.8}$$

(Any enumeration will do.) Choose a one–one correspondence between the Rademacher system $R = \{r_k : k \in \mathbb{N}\}$ and R^{c},

$$r_k \leftrightarrow \chi_k, \quad r_k \in R,\ \chi_k \in R^{\mathrm{c}}. \tag{4.9}$$

(Any correspondence will do.) We denote the two-point set $\{0, 1\}$ by D. For $\mathbf{s} = (u, t) \in D \times D$, denote $|\mathbf{s}| = \max\{u, t\}$, and for $\mathbf{s} \in (D \times D)^n$ define

$$|\mathbf{s}| = \sum_{j=1}^{n} |\mathbf{s}(j)| \tag{4.10}$$

($\mathbf{s} = (\mathbf{s}(1), \ldots, \mathbf{s}(n))$, and $\mathbf{s}(j) \in D \times D$ for $j = 1, \ldots, n$). Fix

$$0 < \epsilon < 1/7.$$

By Lemma 6 (the case $n = 2$), we can choose a $\Lambda(2)$-uniformizing map

$$\psi_2 := \psi : l^2(\mathbb{N}^2) \mapsto \mathrm{L}^\infty(\Omega^2) \tag{4.11}$$

associated with ϵ and $\delta_{R^2}(\epsilon) = \delta$ (as per Definition 5). We identify $l^2(\mathbb{N}^2)$ with $l^2(R^2)$.

Let $\mathbf{x} \in B_H$. For $\mathbf{s} \in D \times D$ and $(j,k) \in \mathbb{N}^2$, define

$$\mathbf{x_s}(j,k) = \psi(\mathbf{x})\hat{}\,(\gamma_j, \gamma_k)_{\mathbf{s}}, \tag{4.12}$$

where

$$(\gamma_j, \gamma_k)_{\mathbf{s}} = \begin{cases} (r_j, r_k) & \text{if } \mathbf{s} = (0,\ 0) \\ (r_j, \chi_k) & \text{if } \mathbf{s} = (0,\ 1) \\ (\chi_j, r_k) & \text{if } \mathbf{s} = (1,\ 0) \\ (\chi_j, \chi_k) & \text{if } \mathbf{s} = (1,\ 1). \end{cases} \tag{4.13}$$

We proceed recursively. Suppose $n > 1$, and that $\mathbf{x_s} \in l^2(\mathbb{N}^2)$ has been obtained for every $\mathbf{s} \in (D \times D)^{n-1}$. For $\mathbf{s} \in (D \times D)^n$, write $\mathbf{s} = (\mathbf{s}_2, \mathbf{s}_1)$, where $\mathbf{s}_2 \in (D \times D)^{n-1}$ and $\mathbf{s}_1 \in D \times D$, and let

$$\mathbf{x_s} = (\mathbf{x}_{\mathbf{s}_2})_{\mathbf{s}_1}, \tag{4.14}$$

where $(\mathbf{x}_{\mathbf{s}_2})_{\mathbf{s}_1}$ is defined by (4.12) with $\mathbf{x}_{\mathbf{s}_2}$ in place of $\mathbf{x}$. That is,

$$\mathbf{x_s}(j,k) = \psi(\mathbf{x}_{\mathbf{s}_2})\hat{}\,(\gamma_j, \gamma_k)_{\mathbf{s}_1}, \quad (j,k) \in \mathbb{N}^2. \tag{4.15}$$

Lemma 11 *For* $\mathbf{x} \in B_H$, $\mathbf{s} \in (D \times D)^n$, *and* $n \geq 1$, *let* $\mathbf{x_s} \in l^2(\mathbb{N}^2)$ *be defined recursively by* (4.12) *and* (4.14). *Then,*

$$\mathbf{x}_{(\mathbf{s},(0,0))} = \mathbf{x_s}; \tag{4.16}$$

$$\psi(\mathbf{x_s})\hat{}\,(r_j, r_k) = \mathbf{x_s}(j,k), \quad (j,k) \in \mathbb{N}^2; \tag{4.17}$$

$$\|\mathbf{x_s}\|_2 \leq \epsilon^{|\mathbf{s}|}; \tag{4.18}$$

$$\|\psi(\mathbf{x_s})\|_{\mathrm{L}^\infty} \leq \delta\ \epsilon^{|\mathbf{s}|}; \tag{4.19}$$

$$\|\mathbf{x_s}\|_2 \leq \epsilon^{|\mathbf{s}|}. \tag{4.20}$$

Proof: By (4.15) and (3.1) (i),

$$\mathbf{x}_{(\mathbf{s},(0,0))}(j,k) = \psi(\mathbf{x_s})\hat{}\,(r_j, r_k) = \mathbf{x_s}(j,k), \quad (j,k) \in \mathbb{N}^2, \tag{4.21}$$

which verifies (4.16) and (4.17).

The statement in (4.18) is proved by induction on n. The case $n = 1$ follows from (4.12), (3.1) (i) and (3.1) (ii). Let $n > 1$ and $\mathbf{s} = (\mathbf{s}_2, \mathbf{s}_1)$, where $\mathbf{s}_2 \in (D \times D)^{n-1}$ and $\mathbf{s}_1 \in D \times D$. If $\mathbf{s} = (\mathbf{s}_2, (0,0))$, then $|\mathbf{s}| = |\mathbf{s}_2|$; by (4.16) and the induction hypothesis,

$$\|\mathbf{x}_{(\mathbf{s}_2,(0,0))}\|_2 = \|\mathbf{x}_{\mathbf{s}_2}\|_2 \leq \epsilon^{|\mathbf{s}_2|} = \epsilon^{|\mathbf{s}|}. \tag{4.22}$$

For any other $\mathbf{s}$, by (4.15), (3.1) (iii), and the induction hypothesis,

$$\|\mathbf{x}_{\mathbf{s}}\|_2 = \left(\sum_{(j,k)\in\mathbb{N}^2} |\psi(\mathbf{x}_{\mathbf{s}_2})\hat{\ }(\gamma_j, \gamma_k)_{\mathbf{s}_1}|^2 \right)^{\frac{1}{2}}$$

$$\leq \epsilon \,\|\mathbf{x}_{\mathbf{s}_2}\|_2 \leq \epsilon\, \epsilon^{|\mathbf{s}_2|} = \epsilon^{|\mathbf{s}|}. \tag{4.23}$$

The statements (4.19) and (4.20) follow, by (4.18), from (3.1) (ii) and (iii). □

Let π_a, π_b and π_c denote the projections from D^3 onto D^2 defined by

$$\pi_a(t,u,v) = (u,t), \quad \pi_b(t,u,v) = (u,v),$$

$$\pi_c(t,u,v) = (t,v), \quad (t,u,v) \in D^3. \tag{4.24}$$

For integers $k \geq 1$, let $\mathbf{t} \in (D^3)^k$, and write $\mathbf{t} = (\mathbf{t}(j) : j \in [k])$, where $\mathbf{t}(j) \in D^3$. Define

$$\boldsymbol{\pi}_a\mathbf{t} := (\pi_a\mathbf{t}(1), \ldots, \pi_a\mathbf{t}(k)); \tag{4.25}$$

$\boldsymbol{\pi}_b\mathbf{t}$ and $\boldsymbol{\pi}_c\mathbf{t}$ are defined similarly. Denote $\tilde{D}^3 = D^3 \sim (0,0,0)$.

Theorem 12 *Let η be the trilinear functional on H^3 defined in* (4.1). *Let $\psi : l^2(\mathbb{N}^2) \mapsto \mathrm{L}^\infty(\Omega^2)$ be a $\Lambda(2)$-uniformizing map for R^2 with uniformizing constants $0 < \epsilon < 1/7$ and $\delta_{R^2}(\epsilon) = \delta$. Then, for $(\mathbf{x},\mathbf{y},\mathbf{z}) \in B_H \times B_H \times B_H$,*

$$\eta(\mathbf{x},\mathbf{y},\mathbf{z}) = \sum_{k=0}^{\infty}(-1)^k \sum_{\mathbf{t}\in(\tilde{D}^3)^k} \eta(\psi(\mathbf{x}_{\boldsymbol{\pi}_a\mathbf{t}})\hat{\ },\ \psi(\mathbf{y}_{\boldsymbol{\pi}_b\mathbf{t}})\hat{\ },\ \psi(\mathbf{z}_{\boldsymbol{\pi}_c\mathbf{t}})\hat{\ }), \tag{4.26}$$

where the series on the right side converges uniformly in B_H^3.

Proof: Let $(\mathbf{x},\mathbf{y},\mathbf{z}) \in B_H^3$. By the first identity in (4.3),

$$\eta(\psi(\mathbf{x})\hat{\ }, \psi(\mathbf{y})\hat{\ }, \psi(\mathbf{z})\hat{\ }) = \sum_{(w_1,w_2,w_3)\in W^3} \psi(\mathbf{x}\hat{\ }(w_1,w_2)\ \psi(\mathbf{y})\hat{\ }(w_2,w_3)\ \psi(\mathbf{z})\hat{\ }(w_1,w_3). \tag{4.27}$$

By splitting the sum in (4.27) into a sum over R^3 and a sum over its complement $(R^3)^{\mathrm{c}}$, we obtain

$$\begin{aligned}\sum_{(i,j,k)\in\mathbb{N}^3} &\psi(\mathbf{x})\hat{}\,(r_i,r_j)\,\psi(\mathbf{y})\hat{}\,(r_j,r_k)\,\psi(\mathbf{z})\hat{}\,(r_i,r_k)\\ &= \eta(\psi(\mathbf{x})\hat{}\,,\psi(\mathbf{y})\hat{}\,,\psi(\mathbf{z})\hat{}\,)\\ &- \sum_{(\gamma_1,\gamma_2,\gamma_3)\in(R^3)^{\mathrm{c}}} \psi(\mathbf{x})\hat{}\,(\gamma_1,\gamma_2)\,\psi(\mathbf{y})\hat{}\,(\gamma_2,\gamma_3)\,\psi(\mathbf{z})\hat{}\,(\gamma_1,\gamma_3). \end{aligned}\tag{4.28}$$

By (3.1) (i), for all $(i,j)\in\mathbb{N}^3$,

$$\begin{aligned}\psi(\mathbf{x})\hat{}\,(r_i,r_j) = \mathbf{x}(i,j), \quad \psi(\mathbf{y})\hat{}\,(r_i,r_j) = \mathbf{y}(i,j),\\ \psi(\mathbf{z})\hat{}\,(r_i,r_j) = \mathbf{z}(i,j).\end{aligned}\tag{4.29}$$

By the definitions in (4.12) and (4.24),

$$\begin{aligned}\sum_{(\gamma_1,\gamma_2,\gamma_3)\in(R^3)^{\mathrm{c}}} &\psi(\mathbf{x})\hat{}\,(\gamma_1,\gamma_2)\,\psi(\mathbf{y})\hat{}\,(\gamma_2,\gamma_3)\,\psi(\mathbf{x})\hat{}\,(\gamma_1,\gamma_3)\\ &= \sum_{\mathbf{s}\in\tilde{D}^3} \eta(\mathbf{x}_{\pi_a\mathbf{s}},\mathbf{y}_{\pi_b\mathbf{s}},\mathbf{z}_{\pi_c\mathbf{s}}).\end{aligned}\tag{4.30}$$

Combining (4.29) and (4.30), we rewrite (4.28) as

$$\eta(\mathbf{x},\mathbf{y},\mathbf{z}) = \eta(\psi(\mathbf{x})\hat{}\,,\psi(\mathbf{y})\hat{}\,,\psi(\mathbf{z})\hat{}\,) - \sum_{\mathbf{s}\in\tilde{D}^3} \eta(\mathbf{x}_{\pi_a\mathbf{s}},\mathbf{y}_{\pi_b\mathbf{s}},\mathbf{z}_{\pi_c\mathbf{s}}).\tag{4.31}$$

Claim: *For $n\geq 1$,*

$$\begin{aligned}\eta(\mathbf{x},\mathbf{y},\mathbf{z}) &= \sum_{k=0}^{n-1}(-1)^k \sum_{\mathbf{t}\in(\tilde{D}^3)^k} \eta(\psi(\mathbf{x}_{\pi_a\mathbf{t}})\hat{}\,,\psi(\mathbf{y}_{\pi_b\mathbf{t}})\hat{}\,,\psi(\mathbf{z}_{\pi_c\mathbf{t}})\hat{}\,)\\ &\quad +(-1)^n \sum_{\mathbf{t}\in(\tilde{D}^3)^n} \eta(\mathbf{x}_{\pi_a\mathbf{t}},\ \mathbf{y}_{\pi_b\mathbf{t}},\ \mathbf{z}_{\pi_c\mathbf{t}}).\end{aligned}\tag{4.32}$$

Proof of Claim (by induction). The case $n=1$ is the statement in (4.31). Let $n\geq 1$ and assume (4.32). Let $\mathbf{t}\in(\tilde{D}^3)^n$. By applying (4.31) with $\mathbf{x}_{\pi_a\mathbf{t}}$ in place of $\mathbf{x}$, $\mathbf{y}_{\pi_b\mathbf{t}}$ in place of $\mathbf{y}$, and $\mathbf{z}_{\pi_c\mathbf{t}}$ in place of $\mathbf{z}$, we obtain

$$\begin{aligned}\eta(\mathbf{x}_{\pi_a\mathbf{t}},\ \mathbf{y}_{\pi_b\mathbf{t}},\ \mathbf{z}_{\pi_c\mathbf{t}}) &= \eta(\psi(\mathbf{x}_{\pi_a\mathbf{t}})\hat{}\,,\psi(\mathbf{y}_{\pi_b\mathbf{t}})\hat{}\,,\psi(\mathbf{z}_{\pi_c\mathbf{t}})\hat{}\,)\\ &\quad - \sum_{\mathbf{s}\in\tilde{D}^3} \eta((\mathbf{x}_{\pi_a\mathbf{t}})_{\pi_a\mathbf{s}},(\mathbf{y}_{\pi_b\mathbf{t}})_{\pi_b\mathbf{s}},(\mathbf{z}_{\pi_c\mathbf{t}})_{\pi_c\mathbf{s}}).\end{aligned}\tag{4.33}$$

The recursion in (4.14) and the definition in (4.25) imply

$$(\mathbf{x}_{\boldsymbol{\pi}_a\mathbf{t}})_{\boldsymbol{\pi}_a\mathbf{s}} = \mathbf{x}_{\boldsymbol{\pi}_a(\mathbf{t},\mathbf{s})},\ (\mathbf{y}_{\boldsymbol{\pi}_b\mathbf{t}})_{\boldsymbol{\pi}_b\mathbf{s}} = \mathbf{y}_{\boldsymbol{\pi}_b(\mathbf{t},\mathbf{s})}, (\mathbf{z}_{\boldsymbol{\pi}_c\mathbf{t}})_{\boldsymbol{\pi}_c\mathbf{s}} = \mathbf{z}_{\boldsymbol{\pi}_c(\mathbf{t},\mathbf{s})},$$

where $\mathbf{t} \in (\tilde{D}^3)^n$ and $\mathbf{s} \in \tilde{D}^3$. Therefore, the second term on the right side of (4.32) can be rewritten as

$$\begin{aligned}
&(-1)^n \sum_{\mathbf{t}\in(\tilde{D}^3)^n} \eta(\mathbf{x}_{\boldsymbol{\pi}_a\mathbf{t}}, \mathbf{y}_{\boldsymbol{\pi}_b\mathbf{t}}, \mathbf{z}_{\boldsymbol{\pi}_c\mathbf{t}}) \\
&= (-1)^n \sum_{\mathbf{t}\in(\tilde{D}^3)^n} \Bigg(\eta(\psi(\mathbf{x}_{\boldsymbol{\pi}_a\mathbf{t}})\hat{\ }, \psi(\mathbf{y}_{\boldsymbol{\pi}_b\mathbf{t}})\hat{\ }, \psi(\mathbf{z}_{\boldsymbol{\pi}_c\mathbf{t}})\hat{\ }) \\
&\qquad - \sum_{\mathbf{s}\in\tilde{D}^3} \eta(\mathbf{x}_{\boldsymbol{\pi}_a(\mathbf{t},\mathbf{s})}, \mathbf{y}_{\boldsymbol{\pi}_b(\mathbf{t},\mathbf{s})}, \mathbf{z}_{\boldsymbol{\pi}_c(\mathbf{t},\mathbf{s})}) \Bigg) \\
&= (-1)^n \sum_{\mathbf{t}\in(\tilde{D}^3)^n} \eta(\psi(\mathbf{x}_{\boldsymbol{\pi}_a\mathbf{t}})\hat{\ }, \psi(\mathbf{y}_{\boldsymbol{\pi}_b\mathbf{t}})\hat{\ }, \psi(\mathbf{z}_{\boldsymbol{\pi}_c\mathbf{t}})\hat{\ }) \\
&\quad + (-1)^{n+1} \sum_{\mathbf{t}\in(\tilde{D}^3)^{n+1}} \eta(\mathbf{x}_{\boldsymbol{\pi}_a\mathbf{t}}, \mathbf{y}_{\boldsymbol{\pi}_b\mathbf{t}}, \mathbf{z}_{\boldsymbol{\pi}_c\mathbf{t}}).
\end{aligned} \tag{4.34}$$

By (4.34) and the induction hypothesis, we obtain (4.32) with $n+1$ in place of n, and the claim follows.

Lemma 8 and (4.18) imply that for all $n > 0$ and $\mathbf{t} \in (\tilde{D}^3)^n$,

$$\begin{aligned}
|\eta(\mathbf{x}_{\boldsymbol{\pi}_a\mathbf{t}}, \mathbf{y}_{\boldsymbol{\pi}_b\mathbf{t}}, \mathbf{z}_{\boldsymbol{\pi}_c\mathbf{t}})| &\le \|\mathbf{x}_{\boldsymbol{\pi}_a\mathbf{t}}\|_2 \ \|\mathbf{y}_{\boldsymbol{\pi}_b\mathbf{t}}\|_2 \ \|\mathbf{z}_{\boldsymbol{\pi}_c\mathbf{t}}\|_2 \\
&\le \epsilon^{|\boldsymbol{\pi}_a\mathbf{t}|} \ \epsilon^{|\boldsymbol{\pi}_b\mathbf{t}|} \ \epsilon^{|\boldsymbol{\pi}_c\mathbf{t}|}.
\end{aligned} \tag{4.35}$$

For all $\mathbf{t} \in (\tilde{D}^3)^n$,

$$|\boldsymbol{\pi}_a\mathbf{t}| + |\boldsymbol{\pi}_b\mathbf{t}| + |\boldsymbol{\pi}_c\mathbf{t}| \ge n, \tag{4.36}$$

and therefore, by (4.35),

$$\sum_{\mathbf{t}\in(\tilde{D}^3)^n} |\eta(\mathbf{x}_{\boldsymbol{\pi}_a\mathbf{t}}, \mathbf{y}_{\boldsymbol{\pi}_b\mathbf{t}}, \mathbf{z}_{\boldsymbol{\pi}_c\mathbf{t}})| \le (7\epsilon)^n. \tag{4.37}$$

The assertion in (4.26) is obtained by applying (4.37) in (4.31), and letting $n \to \infty$. □

Corollary 13 *η is projectively bounded. In particular,*

$$\|\eta\|_{\mathrm{pb}_3} \le \frac{\delta^3}{1 - 7\epsilon}, \tag{4.38}$$

where ϵ and $\delta = \delta_{R^2}(\epsilon)$ are $\Lambda(2)$-uniformizing constants of R^2.

Proof: We estimate $\|\eta|_{E_1\times E_2\times E_3}\|_{V_3(E_1,E_2,E_3)}$, where E_1, E_2 and E_3 are arbitrary finite subsets of B_H. For $k \geq 0$ and $\mathbf{t} \in (\tilde{D}^3)^k$, define

$$\begin{aligned} E_{1\mathbf{t}} &= \{\psi(\mathbf{x}_{\boldsymbol{\pi}_a\mathbf{t}})\hat{}: x \in E_1\},\ E_{2\mathbf{t}} = \{\psi(\mathbf{y}_{\boldsymbol{\pi}_b\mathbf{t}})\hat{}: y \in E_2\},\\ E_{3\mathbf{t}} &= \{\psi(\mathbf{z}_{\boldsymbol{\pi}_c\mathbf{t}})\hat{}: z \in E_3\}. \end{aligned} \tag{4.39}$$

Then, by (4.19) and Lemma 10,

$$\|\eta|_{E_{1\mathbf{t}}\times E_{2\mathbf{t}}\times E_{3\mathbf{t}}}\|_{V_3(E_{1\mathbf{t}},E_{2\mathbf{t}},E_{3\mathbf{t}})} \leq \delta^3 \epsilon^{|\boldsymbol{\pi}_a\mathbf{t}|}\ \epsilon^{|\boldsymbol{\pi}_b\mathbf{t}|}\ \epsilon^{|\boldsymbol{\pi}_c\mathbf{t}|}. \tag{4.40}$$

Therefore, by an application of (4.40) to (4.26),

$$\begin{aligned} &\|\eta|_{E_1\times E_2\times E_3}\|_{V_3(E_1,E_2,E_3)}\\ &\leq \sum_{k=0}^{\infty}\sum_{\mathbf{t}\in(\tilde{D}^3)^k} \|\eta|_{E_{1\mathbf{t}}\times E_{2\mathbf{t}}\times E_{3\mathbf{t}}}\|_{V_3(E_{1\mathbf{t}},E_{2\mathbf{t}},E_{3\mathbf{t}})}\\ &\leq \delta^3 \sum_{k=0}^{\infty}\sum_{\mathbf{t}\in(\tilde{D}^3)^k} \epsilon^{|\boldsymbol{\pi}_a\mathbf{t}|}\ \epsilon^{|\boldsymbol{\pi}_b\mathbf{t}|}\ \epsilon^{|\boldsymbol{\pi}_c\mathbf{t}|} \leq \frac{\delta^3}{1-7\epsilon}, \end{aligned} \tag{4.41}$$

which implies the estimate in (4.38). □

5 A Characterization

We build on the main result of the previous section. Let $H = l^2(\mathbb{N}^2)$ and $\varphi \in l^2(\mathbb{N}^3)$. Define

$$\begin{aligned} \eta_\varphi(\mathbf{x},\mathbf{y},\mathbf{z}) &= \sum_{i,j,k} \varphi(i,j,k)\ \mathbf{x}(i,j)\ \mathbf{y}(j,k)\mathbf{z}(i,k),\\ &\qquad (\mathbf{x},\mathbf{y},\mathbf{z}) \in H^3. \end{aligned} \tag{5.1}$$

Then, η_φ is a well-defined bounded trilinear functional on H, and $\|\eta_\varphi\|_{\mathcal{F}_3} \leq \|\varphi\|_\infty$ (see Lemma 8). In this section we answer the question: *when is η_φ projectively bounded?*

We consider the Rademacher system indexed by $\mathbb{N}^2$, whose underlying domain is $\Omega = \{-1,1\}^{\mathbb{N}^2}$. We denote it by

$$R = \{r_{ij} : (i,j) \in \mathbb{N}^2\}, \tag{5.2}$$

and define (a subset of R^3)

$$R^U = \{(r_{ij}, r_{jk}, r_{ik}) : (i,j,k) \in \mathbb{N}^3\}. \tag{5.3}$$

Theorem 14 *The trilinear functional η_φ on H defined in* (5.1) *is projectively bounded if and only if $\varphi \in B(R^U)$.*

The assertion that a scalar-valued φ defined on $\mathbb{N}^3$ is in $B(R^U)$ means there exist $\mu \in \mathrm{M}(\Omega^3)$ such that

$$\varphi(i,j,k) = \hat{\mu}(r_{ij}, r_{jk}, r_{ik}), \ (i,j,k) \in \mathbb{N}^3, \tag{5.4}$$

and its $B(R^U)$-norm is the infimum of $\|\mu\|_{\mathrm{M}}$ over all such μ; see Chapter VII §7 for definitions and basic facts.

To prove necessity ($\varphi \in B(R^U) \Rightarrow \|\eta_\varphi\|_{\mathrm{pb}_3} < \infty$), we identify $B(R^U)$ as a *tilde algebra.* (See Remark ii in Chapter VII §8.) Let $V_U(\mathbb{N}^3)$ denote the set of $\varphi \in c_0(\mathbb{N}^3)$ that can be written as

$$\begin{aligned} \varphi(i,j,k) \ &= \ \sum_{m=1}^{\infty} \alpha_m \ \theta_{m1}(i,j) \ \theta_{m2}(j,k) \ \theta_{m3}(i,k), \\ & \quad\ (i,j,k) \in \mathbb{N}^3, \end{aligned} \tag{5.5}$$

where $\theta_{ml} \in B_{c_0(\mathbb{N}^2)}$ for $m \in \mathbb{N}$ and $l \in \{1,2,3\}$, and $\sum_{m=1}^{\infty} |\alpha_m| < \infty$. The norm in $V_U(\mathbb{N}^m)$ is

$$\|\varphi\|_{V_U} = \inf \left\{ \sum_{m=1}^{\infty} |\alpha_m| : \ \text{representations of } \varphi \text{ by (5.5)} \right\} \tag{5.6}$$

(cf. (VII.8.21) and (VII.8.22)). Then, $\tilde{V}_U(\mathbb{N}^3)$ is the algebra of pointwise limits of sequences uniformly bounded in $V_U(\mathbb{N}^3)$. Elements in $\tilde{V}_U(\mathbb{N}^3)$ are normed by

$$\|\varphi\|_{\tilde{V}_U} = \inf \left\{ \sup_m \|\varphi_m\|_{V_U} : \ \lim_{m\to\infty} \varphi_m(\mathbf{n}) = \varphi(\mathbf{n}), \ \mathbf{n} \in \mathbb{N}^3 \right\}. \tag{5.7}$$

The algebra $V_U(\mathbb{N}^3)$ can be realized also as the algebra of restrictions of elements in $V_3(\mathbb{N}^2, \ \mathbb{N}^2, \ \mathbb{N}^2)$ to

$$\mathbb{N}^U := \{((i,j),(j,k),(i,k)) : (i,j,k) \in \mathbb{N}^3\} \subset \mathbb{N}^2 \times \mathbb{N}^2 \times \mathbb{N}^2. \tag{5.8}$$

Similarly, we identify $\tilde{V}_U(\mathbb{N}^3)$ as the algebra of restrictions of elements in $\tilde{V}_3(\mathbb{N}^2, \mathbb{N}^2, \mathbb{N}^2)$ to $\mathbb{N}^U$. ($\tilde{V}_U(\mathbb{N}^3)$ is *complemented* in $\tilde{V}_3(\mathbb{N}^2, \mathbb{N}^2, \mathbb{N}^2)$; see Exercise 6.) By Proposition VII.24, $V_3(\mathbb{N}^2, \mathbb{N}^2, \mathbb{N}^2)$ and $\tilde{V}_3(\mathbb{N}^2, \mathbb{N}^2, \mathbb{N}^2))$ are canonically isomorphic to $A(R^3)$ and $B(R^3)$, respectively. This implies

Lemma 15 (Exercise 7).

i. $V_U(\mathbb{N}^3)$ *is canonically isomorphic to* $A(R^U)$. *Specifically, if* $\varphi \in V_U(\mathbb{N}^3)$, *then there exists* $f \in \mathrm{L}^1(\Omega^3)$ *such that*

$$\varphi(i,j,k) = \hat{f}(r_{ij}, r_{jk}, r_{ik}), \quad (i,j,k) \in \mathbb{N}^3, \tag{5.9}$$

and

$$\|f\|_{\mathrm{L}^1} \leq 2^3 \; \|\varphi\|_{V_U}. \tag{5.10}$$

Conversely, if $\varphi \in A(R^U)$, *then there exists a representation of* φ,

$$\varphi(r_{ij}, r_{jk}, r_{ik}) = \sum_{m=1}^{\infty} \alpha_m \; \theta_{m1}(i,j) \; \theta_{m2}(j,k)\theta_{m3}(i,k), \quad (i,j,k) \in \mathbb{N}^3, \tag{5.11}$$

such that $\theta_{ml} \in B_{\mathrm{c}_0(\mathbb{N}^2)}$ *for* $m \in \mathbb{N}$ *and* $l \in \{1,2,3\}$, *and*

$$\sum_{m=1}^{\infty} |\alpha_m| \leq \|\varphi\|_{A(R^U)}. \tag{5.12}$$

ii. $\tilde{V}_U(\mathbb{N}^3)$ *is canonically isomorphic to* $B(R^U)$. *Specifically, if* $\varphi \in \tilde{V}_U(\mathbb{N}^m)$, *then there exists* $\mu \in \mathrm{M}(\Omega^3)$ *such that*

$$\varphi(i,j,k) = \hat{\mu}(r_{ij}, r_{jk}, r_{ik}), \quad (i,j,k) \in \mathbb{N}^3, \tag{5.13}$$

and

$$\|\mu\|_{\mathrm{M}} \leq 2^3 \; \|\varphi\|_{\tilde{V}_U}. \tag{5.14}$$

Conversely, if $\varphi \in B(R^U)$, *then there exists a sequence* $(\varphi_k : k \in \mathbb{N})$ *in* $V_U(\mathbb{N}^3)$ *such that*

$$\varphi(r_{ij}, r_{jk}, r_{ik}) = \lim_{k\to\infty} \varphi_k(i,j,k), \quad (i,j,k) \in \mathbb{N}^3, \tag{5.15}$$

and

$$\sup\{\|\varphi_k\|_{V_U} : k \in \mathbb{N}\} \leq \|\varphi\|_{B(R^U)}. \tag{5.16}$$

Proof of Theorem 14: We first prove that if $\varphi \in B(R^U)$, then η_φ is projectively bounded.

Step 1 Suppose

$$\varphi(i,j,k) = \theta_1(i,j) \; \theta_2(j,k) \; \theta_3(i,k), \quad (i,j,k) \in \mathbb{N}^3, \tag{5.17}$$

where $\theta_l \in B_{c_0(\mathbb{N}^2)}$, $l \in \{1,2,3\}$. Let E_l be a finite subset of B_H, $l \in \{1,2,3\}$. Let

$$F_l = \{\mathbf{x}\theta_l : \mathbf{x} \in E_l\}, \quad l \in \{1,2,3\}, \tag{5.18}$$

where

$$(\mathbf{x}\theta_l)(i,j) = \mathbf{x}(i,j)\, \theta_l(i,j), \quad (i,j) \in \mathbb{N}^2. \tag{5.19}$$

Then, $F_1 \times F_2 \times F_3 \subset B_H$. By Corollary 13,

$$\|\eta_\varphi\|_{V_3(E_1,E_2,E_3)} = \|\eta\|_{V_3(F_1,F_2,F_3)} \leq K, \tag{5.20}$$

where $K > 0$ is an absolute constant (cf. (4.38)). This implies

$$\|\eta_\varphi\|_{\mathrm{pb}_3} \leq K. \tag{5.21}$$

Step 2 Suppose $\varphi \in V_U(\mathbb{N}^3)$. Write

$$\varphi = \sum_{m=1}^{\infty} \alpha_m\, \varphi_m, \tag{5.22}$$

where $\varphi_m \in B_{c_0(\mathbb{N}^3)}$ is an 'elementary tensor' of the type defined in (5.17), and $\sum_{m=1}^{\infty} |\alpha_m| < \infty$. Then,

$$\eta_\varphi = \sum_{m=1}^{\infty} \alpha_m\, \eta_{\varphi_m}, \tag{5.23}$$

and by Step 1,

$$\|\eta_\varphi\|_{\mathrm{pb}_3} \leq \sum_{m=1}^{\infty} \alpha_m \|\eta_{\varphi_m}\|_{\mathrm{pb}_3} \leq K \sum_{m=1}^{\infty} |\alpha_m|. \tag{5.24}$$

Step 3 Suppose $\varphi \in B(R^U)$. Then, by Lemma 15, there exists $\{\varphi_k : k \in \mathbb{N}\} \subset V_U(\mathbb{N}^3)$ such that

$$\varphi(i,j,k) = \lim_{k\to\infty} \varphi_k(i,j,k), \quad (i,j,k) \in \mathbb{N}^3, \tag{5.25}$$

and

$$\sup\{\|\varphi_k\|_{V_U}\colon k \in \mathbb{N}\} \leq \|\varphi\|_{B(R^U)}. \tag{5.26}$$

By (5.26) and Step 2,

$$\|\eta_{\varphi_k}\|_{\mathrm{pb}_3} \leq K\|\varphi\|_{B(R^U)}, \quad k \in \mathbb{N}. \tag{5.27}$$

By (5.25), $\lim_{k\to\infty} \eta_{\varphi_k}(\mathbf{x},\mathbf{y},\mathbf{z}) = \eta_\varphi(\mathbf{x},\mathbf{y},\mathbf{z}), (\mathbf{x},\mathbf{y},\mathbf{z}) \in B_H \times B_H \times B_H$. Then, by (5.27) and Proposition 4 (Exercise 1 iii),

$$\|\eta_\varphi\|_{\mathrm{pb}_3} \leq K\|\varphi\|_{B(R^U)}. \tag{5.28}$$

Next we prove the sufficiency part in Theorem 14 ($\|\eta_\varphi\|_{\mathrm{pb}_3} < \infty \Rightarrow \varphi \in B(R^U)$). Let $E = \{\mathbf{e}_{ij}: (i,j) \in \mathbb{N}\}$ be the basis in $H = l^2(\mathbb{N}^2)$,

$$\mathbf{e}_{ij}(k,l) = \begin{cases} 1 & i = k, \quad j = l \\ 0 & \text{otherwise.} \end{cases} \tag{5.29}$$

Define (cf. (2.2))

$$\begin{aligned} \phi_{\eta_\varphi,E}(r_{i_1 j_1},\ r_{i_2 j_2},\ r_{i_3 j_3}) &= \eta_\varphi(\mathbf{e}_{i_1 j_1}, \mathbf{e}_{i_2 j_2}, \mathbf{e}_{i_3 j_3}), \\ (r_{i_1 j_1},\ r_{i_2 j_2},\ r_{i_3 j_3}) &\in R^3. \end{aligned} \tag{5.30}$$

By Proposition 3,

$$\|\phi_{\eta_\varphi,E}\|_{B(R^3)} \leq \|\eta_\varphi\|_{\mathrm{pb}_3}, \tag{5.31}$$

and by (5.29),

$$\phi_{\eta_\varphi,E}(r_{i_1 j_1},\ r_{i_2 j_2},\ r_{i_3 j_3}) = \begin{cases} \varphi(i,j,k) & \text{if} \quad i_1 = i_3 = i, \\ & \qquad j_1 = i_2 = j, \\ & \qquad j_2 = j_3 = k \\ 0 & \text{otherwise,} \end{cases} \tag{5.32}$$

which imply $\|\varphi\|_{B(R^U)} \leq \|\eta_\varphi\|_{\mathrm{pb}_3}$.

Corollary 16 (cf. Exercise 6). $\mathbf{1}_{R^U} \in B(R^3)$.

Proof: Let $\varphi(i,j,k) = 1$ for $(i,j,k) \in \mathbb{N}^3$. Then, $\phi_{\eta_\varphi,E} = \mathbf{1}_{R^U}$, where $\phi_{\eta_\varphi,E}$ is defined in (5.30), and computed in (5.32). By Corollary 13, η_φ is projectively bounded, and by (5.31), $\mathbf{1}_{R^U} \in B(R^3)$. □

6 Projectively Unbounded Trilinear Functionals

To verify existence of bounded trilinear functionals that are projectively unbounded, we use the implication $\|\eta_\varphi\|_{\mathrm{pb}_3} < \infty \Rightarrow \varphi \in B(R^U)$ in Theorem 14 (the easy direction).

Theorem 17 *R^U is not Sidon; i.e., there exists $\varphi \in l^\infty(\mathbb{N}^3)$ such that $\varphi \notin B(R^U)$.*

Proof: We will prove that the $\Lambda(q)$ constants' growth of R^U is no better than $\mathcal{O}(q^{3/4})$, and thus deduce, from the instance $t = 1$ in Theorem VII.41, that R^U is not a Sidon set.

For $m \in \mathbb{N}$, let

$$f_m = \sum_{(i,j,k)\in[m]^3} r_{ij} \otimes r_{jk} \otimes r_{ik}, \tag{6.1}$$

and consider the Riesz product

$$R_m = \prod_{(i,j)\in[m]^2} (1 + r_{ij}) \otimes \prod_{(i,j)\in[m]^2} (1 + r_{ij}) \otimes \prod_{(i,j)\in[m]^2} (1 + r_{ij}). \tag{6.2}$$

Then, $\|R_m\|_{\mathrm{L}^1} = 1$,

$$\|R_m\|_{\mathrm{L}^2} = 2^{3m^2} \quad \text{(cf. (VII.9.5))}, \tag{6.3}$$

and therefore,

$$\|R_m\|_{\mathrm{L}^p} \leq 2^{3m^2/q} \tag{6.4}$$

for $p \in (1, 2)$ and $1/q + 1/p = 1$ (Exercise VII.29). Also, because $\hat{R}_m = 1$ on the support of $\hat{f}$,

$$|\mathbf{E}\; R_m\; f_m| = m^3 = m^{\frac{3}{2}} \|f_m\|_{\mathrm{L}^2}. \tag{6.5}$$

By applying Hölder's inequality to the left side of (6.5) with

$$p = m^2/(m^2 - 1) \quad \text{and} \quad q = m^2,$$

and applying (6.4), we obtain

$$m^{\frac{3}{2}} \|f_m\|_{\mathrm{L}^2} \leq 2^3\; \|f_m\|_{\mathrm{L}^{m^2}}, \tag{6.6}$$

which implies that the $\Lambda(q)$ constants' growth of R^U is no better than $\mathscr{O}(q^{3/4})$. $\square$

Remarks:

i **(a preview).** In Chapter XII we will prove that the $\Lambda(q)$ constants' growth of R^U is precisely $\mathscr{O}(q^{3/4})$, and thus obtain that R^U is, in essence, a '3/2-dimensional fractional Cartesian product'. (Cf. Bonami's inequalities in Chapter VII §9.) To wit, the spectral set

R^U will foreshadow a general notion of a *fractional Cartesian product* and a measurement of *combinatorial dimension*, which we formally introduce in Chapter XIII.

ii **(credits).** That a Grothendieck-type inequality need not hold in dimensions greater than two was shown first by Varopoulos in [V4, Proposition 4.2]. This discovery underscored that analysis in dimensions greater than two is fundamentally different from analysis in two dimensions. Specifically, Varopoulos proved by an application of the Kahane–Salem–Zygmund probabilistic estimates that there exists $\varphi \in l^\infty(\mathbb{N}^3)$ such that η_φ fails a trilinear Grothendieck-type inequality. A result in the opposite direction, demonstrating (non-trivial) trilinear functionals that *do* satisfy a Grothendieck-type inequality, appeared first in [Bl3]. The characterization in Theorem 14 appeared in [Bl4]. The proof here of Varopoulos's original result, that there exist projectively unbounded η_φ, is different from the proof in [V4].

7 The General Case

For a set Y and a positive integer m, let $\pi_1, \ldots, \pi_m$ denote the usual projections from Y^m onto Y,

$$\pi_i(y_1, \ldots, y_m) = y_i, \quad (y_1, \ldots, y_m) \in Y^m, \ i \in [m]. \tag{7.1}$$

If $S \subset [m]$ and $S \neq \emptyset$, then π_S will denote the projection from Y^m onto Y^S defined by

$$\pi_S(y_1, \ldots, y_m) = (y_i : i \in S), \quad (y_1, \ldots, y_m) \in y^m. \tag{7.2}$$

Here and throughout, in keeping with standard notation, we write Y^m for $Y^{[m]}$. For $A \subset Y^m$, let $\pi_S[A] = \{\pi_S(\mathbf{y}) : \mathbf{y} \in A\}$, and for $\mathbf{y} \in Y^S$, let $\pi_S^{-1}\{\mathbf{y}\} = \{\mathbf{u} \in Y^m : \pi_S(\mathbf{u}) = \mathbf{y}\}$.

Let $U = \{S_1, \ldots, S_n\}$ be a cover of $[m]$ consisting of non-empty ordered subsets of $[m]$, such that every $j \in [m]$ appears in at least two elements of U; that is,

$$\bigcup_{p=1}^{n} S_p = [m], \tag{7.3}$$

and

$$|\{p : j \in S_p\}| \geq 2, \quad j \in [m]. \tag{7.4}$$

(We allow repetitions in U; e.g., $U = \{(1),(1)\}$.) For each $p \in [n]$, let H_p be the Hilbert space $l^2(\mathbb{N}^{S_p})$. That is, we take a separable Hilbert space and index its basis by $\mathbb{N}^{S_p}$. For $\varphi \in l^\infty(\mathbb{N}^m)$, define

$$\eta_{\varphi,U}(\mathbf{x}_1,\dots,\mathbf{x}_n) = \sum_{l\in\mathbb{N}^m} \varphi(l)\, \mathbf{x}_1(\pi_{S_1} l)\cdots \mathbf{x}_n(\pi_{S_n} l), \tag{7.5}$$

for $\mathbf{x}_1 \in H_1, \dots, \mathbf{x}_n \in H_n$ with finite support. Note that the trilinear functional η_φ in (5.1) is the instance $U = \{(1,2),(2,3),(1,3)\}$ in (7.5).

Lemma 18 (cf. Lemma 8). *If $\varphi \in l^\infty(\mathbb{N}^m)$, then $\eta_{\varphi,U}$ determines a bounded n-linear functional on $H_1 \times \cdots \times H_n$, and*

$$\|\eta_{\varphi,U}\|_{f_n} := \sup\{|\eta_{\varphi,U}(\mathbf{x}_1,\dots,\mathbf{x}_n)| : \mathbf{x}_p \in B_{H_p}, p \in [n]\} \le \|\varphi\|_\infty. \tag{7.6}$$

Proof: By Hölder's inequality, it suffices to prove (7.6) in the case $\eta_{\varphi,U} := \eta_U$, where $\varphi \equiv 1$. In this case we denote $\eta_{\varphi,U} := \eta_U$. The proof is by induction on m. For $m = 1$,

$$\left|\sum_{l\in\mathbb{N}} \mathbf{x}_1(l)\cdots\mathbf{x}_n(l)\right| \le \|\mathbf{x}_1\|_n \cdots \|\mathbf{x}_n\|_n \le \|\mathbf{x}_1\|_2 \cdots \|\mathbf{x}_n\|_2. \tag{7.7}$$

Let $m > 1$, and write

$$\begin{aligned}&\left|\sum_{l\in\mathbb{N}^m} \mathbf{x}_1(\pi_{S_1} l)\cdots\mathbf{x}_n(\pi_{S_n} l)\right| \\ &\qquad = \left|\sum_{k\in\mathbb{N}} \sum_{l\in\pi_n^{-1}\{k\}} \mathbf{x}_1(\pi_{S_1} l)\cdots\mathbf{x}_n(\pi_{S_n} l)\right|.\end{aligned} \tag{7.8}$$

For $k \in \mathbb{N}$, the sum $\sum_{l\in\pi_n^{-1}\{k\}}$, after relabeling coordinates, is a sum over $\mathbb{N}^{m-1}$. Therefore, by the induction hypothesis,

$$\left|\sum_{l\in\pi_n^{-1}\{k\}} \mathbf{x}_1(\pi_{S_1} l)\cdots\mathbf{x}_n(\pi_{S_n} l)\right| \le \prod_{p=1}^{n}\left(\sum_{\mathbf{j}\in\pi_{S_p}[\pi_n^{-1}\{k\}]} |\mathbf{x}_p(\mathbf{j})|^2\right)^{\frac{1}{2}}. \tag{7.9}$$

If $S_p = (m)$ for some p, then

$$\left(\sum_{\mathbf{j} \in \pi_{S_p}[\pi_n^{-1}\{k\}]} |\mathbf{x}_p(\mathbf{j})|^2 \right)^{\frac{1}{2}} = |\mathbf{x}_p(k)|. \tag{7.10}$$

Also, if $m \notin S_p$, then

$$\left(\sum_{\mathbf{j} \in \pi_{S_p}[\pi_n^{-1}\{k\}]} |\mathbf{x}_p(\mathbf{j})|^2 \right)^{\frac{1}{2}} = \|\mathbf{x}_p\|_2. \tag{7.11}$$

By applying (7.9) to (7.8), we obtain

$$\left| \sum_{l \in \mathbb{N}^m} \mathbf{x}_1(\pi_{S_1} l) \cdots \mathbf{x}_n(\pi_{S_n} l) \right| \leq \sum_{k \in \mathbb{N}} \prod_{p=1}^{n} \left(\sum_{\mathbf{j} \in \pi_{S_p}[\pi_n^{-1}\{k\}]} |\mathbf{x}_p(\mathbf{j})|^2 \right)^{\frac{1}{2}}$$

$$= \prod_{\{p:\ m \notin S_p\}} \|\mathbf{x}_p\|_2 \sum_{k \in \mathbb{N}} \prod_{\{p:\ m \in S_p\}} \left(\sum_{\mathbf{j} \in \pi_{S_p}[\pi_n^{-1}\{k\}]} |\mathbf{x}_p(\mathbf{j})|^2 \right)^{\frac{1}{2}}. \tag{7.12}$$

By the case $m = 1$,

$$\sum_{k \in \mathbb{N}} \prod_{\{p:\ m \in S_p\}} \left(\sum_{\mathbf{j} \in \pi_{S_p}[\pi_n^{-1}\{k\}]} |\mathbf{x}_p(\mathbf{j})|^2 \right)^{\frac{1}{2}}$$

$$\leq \prod_{\{p:\ m \in S_p\}} \left(\sum_{k \in \mathbb{N}} \sum_{\mathbf{j} \in \pi_{S_p}[\pi_n^{-1}\{k\}]} |\mathbf{x}_p(\mathbf{j})|^2 \right)^{\frac{1}{2}}$$

$$= \prod_{\{p:\ m \in S_p\}} \|\mathbf{x}_p\|_2. \tag{7.13}$$

By combining (7.12) and (7.13), we obtain (7.6). □

For $S \subset [m]$, let R_S denote the Rademacher system indexed by $\mathbb{N}^S$, and let $\Omega_S = \{-1,+1\}^{\mathbb{N}^S}$ denote its underlying domain. Define (cf. (5.3))

$$R^U = \{(r_{\pi_{S_1} l}, \dots, r_{\pi_{S_n} l}) : l \in \mathbb{N}^m\} \subset R_{S_1} \times \cdots \times R_{S_n}. \tag{7.14}$$

Following the correspondence $l \leftrightarrow (r_{\pi_{S_1} l}, \dots, r_{\pi_{S_n} l})$ between $\mathbb{N}^m$ and R^U, we say that a scalar function φ on $\mathbb{N}^m$ is in $B(R^U)$ if there exists $\mu \in \mathrm{M}(\Omega_{S_1} \times \cdots \times \Omega_{S_n})$ such that for all $l \in \mathbb{N}^m$,

$$\varphi(l) = \hat{\mu}(r_{\pi_{S_1} l}, \dots, r_{\pi_{S_n} l}) \quad \text{(cf. (5.4))}. \tag{7.15}$$

The main result of this chapter is

Theorem 19 (cf. Theorem 14). $\|\eta_{U,\varphi}\|_{\mathrm{pb}_n} < \infty \Leftrightarrow \varphi \in B(R^U)$.

In the next section we prove that $\eta_{U,\varphi}$ is projectively bounded for $\varphi \equiv 1$, and use this result in §9 to establish the general implication $\varphi \in B(R^U) \Rightarrow \|\eta_{U,\varphi}\|_{\mathrm{pb}_n} < \infty$. The converse, that $\|\eta_{U,\varphi}\|_{\mathrm{pb}_n} < \infty$ implies $\varphi \in B(R^U)$, will also be verified in §9.

8 $\varphi \equiv 1$

We write η_U for $\eta_{U,\varphi}$. In order to prove that $\|\eta_U\|_{\mathrm{pb}_n} < \infty$, we will obtain a representation of η_U generalizing the representation in Theorem 12, and to this end, we first generalize the representation in Lemma 9 (the case $U = \{(1,2),(2,3),(1,3)\}$). The key observation, which we generalize below, is simply that the representation in Lemma 9 consists of three successive convolutions.

For $p \in [n]$, following an enumeration of the Walsh system W by $\mathbb{N}$, we identify $l^2(\mathbb{N}^{S_p})$ with $l^2(W^{S_p})$. By Plancherel's theorem, we realize η_U as an n-linear functional on $\mathrm{L}^2(\Omega^{S_1}, \mathbb{P}^{S_1}) \times \cdots \times \mathrm{L}^2(\Omega^{S_n}, \mathbb{P}^{S_n})$:

$$\eta_U(\hat{f}_1, \dots, \hat{f}_n) := \sum_{\mathbf{w}=(w_1,\dots,w_m) \in W^m} \hat{f}_1(\pi_{S_1}\mathbf{w}) \cdots \hat{f}_n(\pi_{S_n}\mathbf{w}),$$

$$f_1 \in \mathrm{L}^2(\Omega^{S_1}, \mathbb{P}^{S_1}), \dots, f_n \in \mathrm{L}^2(\Omega^{S_n}, \mathbb{P}^{S_n}). \tag{8.1}$$

(For $S \subset [m]$, $\mathbb{P}^S$ denotes the $|S|$-fold product measure $\mathbb{P} \times \cdots \times \mathbb{P}$ on Ω^S.) If $\xi \in \Omega$, and f is scalar-valued on Ω^S, then we let $f|_{\omega_i=\xi}$ denote the restriction of f to $\pi_i^{-1}\{\xi\}$. If $f \in \mathrm{L}^1(\Omega^S)$ and $j \in S$, then $\hat{f}^j$ will denote the W-transform of f with respect to the coordinate indexed by j. (For $w \in W$, $\hat{f}^j(w)$ is a function on $\Omega^{S\setminus\{j\}}$.) Define

$$l_1 = \max\{p : 1 \in S_p\}, \ldots, l_m = \max\{p : m \in S_p\}, \tag{8.2}$$

and

$$A_1 = \{p : 1 \in S_p\}\setminus\{l_1\}, \ldots, A_m = \{p : m \in S_p\}\setminus\{l_m\}. \tag{8.3}$$

Let $f_1 \in \mathrm{L}^2(\Omega^{S_1}), \ldots, f_n \in \mathrm{L}^2(\Omega^{S_n})$. For $j \in [m]$ and $k \in A_j$, let $\xi_{kj} \in \Omega$. For $p = 1, \ldots, n$, denote

$$f_p^{(m)} = \begin{cases} f_{l_m}|_{\omega_m = \Pi_{k \in A_m} \xi_{km}} & p = l_m \\ f_p & m \notin S_p \\ f_p|_{\omega_m = \xi_{pm}} & p \in A_m. \end{cases} \tag{8.4}$$

We continue recursively: for $1 \le i < m$, if $f_1^{(i+1,\ldots,m)}, \ldots, f_n^{(i+1,\ldots,m)}$ are functions on $\Omega^{S_1\setminus\{i+1,\ldots,m\}}, \ldots, \Omega^{S_n\setminus\{i+1,\ldots,m\}}$, respectively, then for $p = 1, \ldots, n$,

$$f_p^{(i,\ldots,m)} = \begin{cases} f_{l_i}^{(i+1,\ldots,m)}|_{\omega_i = \Pi_{k \in A_i} \xi_{ki}} & p = l_i \\ f_p^{(i+1,\ldots,m)} & i \notin S_p \\ f_p^{(i+1,\ldots,m)}|_{\omega_i = \xi_{pi}} & p \in A_i. \end{cases} \tag{8.5}$$

This recursion ends with $f_p^{(1,\ldots,m)}$, $p = 1, \ldots, n$, which we view as functions on $\Omega^{A_1} \times \cdots \times \Omega^{A_m}$. To record for future use that these functions depend on U, we write

$$f_p^{(1,\ldots,m)} := T_U f_p, \quad p = 1, \ldots, n. \tag{8.6}$$

Lemma 20 (cf. Lemma 9). *If $f_1 \in \mathrm{L}^2(\Omega^{S_1}), \ldots, f_n \in \mathrm{L}^2(\Omega^{S_n})$, then*

$$\begin{aligned} \eta_U(\hat{f}_1, \ldots, \hat{f}_n) &:= \sum_{\mathbf{w}=(w_1,\ldots,w_m) \in W^m} \hat{f}_1(\pi_{S_1}\mathbf{w}) \cdots \hat{f}_n(\pi_{S_n}\mathbf{w}) \\ &= \int_{\xi \in \Omega^{A_1} \times \cdots \times \Omega^{A_m}} \left(\prod_{p=1}^n T_U f_p \right)(\xi)\ (\mathbb{P}^{A_1} \times \cdots \times \mathbb{P}^{A_m})(\mathrm{d}\xi). \end{aligned} \tag{8.7}$$

Proof: We prove (8.7) by induction on m. If $m = 1$, then $l_1 = n$ and $A_1 = [n-1]$. In this case the right side of (8.7) is the n-fold convolution of $f_1, \dots, f_n$ evaluated at the identity element e_0 of Ω:

$$\int_{\Omega^{n-1}} f_1(\xi_1) \cdots f_{n-1}(\xi_{n-1}) f_n(\Pi_{k=1}^{n-1} \xi_k) \, \mathbb{P}^{n-1}(d\xi_1, \dots, d\xi_{n-1})$$
$$= (f_1 \star \cdots \star f_n)(e_0) = \sum_{w \in W} \hat{f}_1(w) \cdots \hat{f}_n(w). \tag{8.8}$$

Let $m > 1$, and assume (8.7) in the case $m - 1$. The right side of (8.7) is an integral over $\Omega^{A_1} \times \cdots \times \Omega^{A_{m-1}}$ with respect to $\mathbb{P}^{A_1} \times \cdots \times \mathbb{P}^{A_{m-1}}$ of

$$\prod_{\{p:\ m \notin S_p\}} T_U f_p \left(\int_{\Omega^{A_m}} T_U f_{l_m} \prod_{\{p:\ p \in A_m\}} T_U f_p \, d\mathbb{P}^{A_m} \right). \tag{8.9}$$

In particular,

$$\int_{\Omega^{A_m}} T_U f_{l_m} \prod_{\{p:\ p \in A_m\}} T_U f_p \, d\mathbb{P}^{A_m}$$

is the multi-fold convolution on Ω of f_{l_m} with f_p ($p \in A_m$), evaluated at $e_0 \in \Omega$. Let U' be the cover of $[m-1]$ obtained by replacing each $S \in U$ with $S' = S \backslash \{m\}$, and let $T_{U'}$ be defined by (8.6), where U is replaced by U'. Then,

$$\int_{\Omega^{A_m}} T_U f_{l_m} \prod_{\{p:\ p \in A_m\}} T_U f_p \, d\mathbb{P}^{A_m}$$
$$= \sum_{w \in W} \prod_{\{p:\ m \in S_p\}} T_{U'} \hat{f}_p^m(w). \tag{8.10}$$

If $m \notin S_p$, then

$$T_U f_p = T_{U'} f_p. \tag{8.11}$$

Following a substitution of (8.10) and (8.11) in (8.9), we rewrite the right side of (8.7) as

$$\sum_{w \in W} \int_{\Omega^{A_1} \times \cdots \times \Omega^{A_{m-1}}} \left(\prod_{\{p:\ m \notin S_p\}} T_{U'} f_p \right)$$
$$\left(\prod_{\{p:\ m \in S_p\}} T_{U'} \hat{f}_p^m(w) \right) d\mathbb{P}^{A_1} \times \cdots \times d\mathbb{P}^{A_{m-1}}. \tag{8.12}$$

For each $w \in W$, apply the induction hypothesis to each of the summands in (8.12), thus obtaining that the right side of (8.7) equals

$$\sum_{w_m \in W} \sum_{\mathbf{w}=(w_1,\ldots,w_{m-1}) \in W^{m-1}} \prod_{\{p:\ m \notin S_p\}} \hat{f}_\alpha(\pi_{S_p}\mathbf{w}) \prod_{\{p:\ m \in S_p\}} (\hat{f}_p^m(w_m))\hat{\ }(\pi_{S'_p}\mathbf{w})$$

$$= \sum_{\mathbf{w}=(w_1,\ldots,w_m) \in W^m} \hat{f}_1(\pi_{S_1}\mathbf{w}) \cdots \hat{f}_n(\pi_{S_n}\mathbf{w}). \qquad (8.13)$$

□

Corollary 21 (cf. Corollary 10). *If $E_1, \ldots, E_n$ are finite subsets of the respective unit balls of* $\mathrm{L}^\infty(\Omega^{S_1}), \ldots, \mathrm{L}^\infty(\Omega^{S_n})$, *then,*

$$\|\eta_U|_{E_1 \times \cdots \times E_n}\|_{V_n(E_1,\ldots,E_n)} \leq 1. \qquad (8.14)$$

Proof: If $(f_1, \ldots, f_n) \in E_1 \times \cdots \times E_n$, then $T_U f_1, \ldots, T_U f_n$ are in the respective unit balls of $\mathrm{L}^\infty(\Omega^{S_p})$, $p = 1, \ldots, n$. (See (8.6).) The estimate in (8.14) follows by duality. (Review the argument used to prove Corollary 10.) □

We enumerate R^c by $\mathbb{N}$ (as in (4.8)), where R is the Rademacher system R indexed by $\mathbb{N}$ and R^c denotes its complement in W, and (as in (4.9)) we fix a one–one correspondence between R and R^c. Let $D = \{0, 1\}$, and $S \subset [m]$. For $(s_i\colon s_i \in D,\ i \in S) = \mathbf{s}$ in D^S, define $|\mathbf{s}| = \max\{s_i : i \in S\}$, and for $(\mathbf{s}(j) : \mathbf{s}(j) \in D^S, j \in [k]) = \mathbf{s}$ in $(D^S)^k$, define

$$|\mathbf{s}| = \sum_{j=1}^{k} |\mathbf{s}(j)| \quad \text{(cf. (4.10))}. \qquad (8.15)$$

Let

$$\psi_S : l^2(\mathbb{N}^S) \mapsto \mathrm{L}^\infty(\Omega^S) \qquad (8.16)$$

be a $\Lambda(2)$-uniformizing map of R^S provided by Lemma 8, associated with uniformizing constants $0 < \epsilon < 1$ (to be specified later) and $\delta_{R^S}(\epsilon) = \delta$. ($l^2(R^S)$ is identified with $l^2(\mathbb{N}^S)$; R is the Rademacher system indexed by $\mathbb{N}$, and Lemma 8 is invoked in the case $n = |S|$.) For $j \in \mathbb{N}$ and $s \in D$, define

$$v_j(s) = \begin{cases} r_j & \text{if } s = 0 \\ \chi_j & \text{if } s = 1. \end{cases} \qquad (8.17)$$

For $\boldsymbol{l} = (l_j : j \in S) \in \mathbb{N}^S$ and $\mathbf{s} = (s_j : j \in S) \in D^S$, define

$$(\gamma_{\boldsymbol{l}})_s = (v_{l_j}(\mathbf{s}_j) : j \in S) \quad \text{(cf. (4.13))}. \tag{8.18}$$

Let $\mathbf{x} \in B_{l^2(\mathbb{N}^S)}$. For $\boldsymbol{l} \in \mathbb{N}^S$ and $\mathbf{s} \in D^S$, define

$$x_s(\boldsymbol{l}) = \psi_S(\mathbf{x})\hat{\ }(\gamma_{\boldsymbol{l}})_s \quad \text{(cf. (4.12))}. \tag{8.19}$$

We proceed recursively. For $n > 1$, assume $x_s \in l^2(\mathbb{N}^S)$ has been obtained for every $\mathbf{s} \in (D^S)^{n-1}$. Let $\mathbf{s} \in (D^S)^n$, write $\mathbf{s} = (\mathbf{s}_2, \mathbf{s}_1)$ where $\mathbf{s}_2 \in (D^S)^{n-1}$ and $\mathbf{s}_1 \in D^S$, and define

$$\mathbf{x_s} = (\mathbf{x}_{\mathbf{s}_2})_{\mathbf{s}_1} \quad \text{(cf. (4.14))}. \tag{8.20}$$

Specifically, (8.20) means that for $\boldsymbol{l} \in \mathbb{N}^S$,

$$\mathbf{x_s}(\boldsymbol{l}) = \psi(\mathbf{x}_{\mathbf{s}_2})\hat{\ }(\gamma_{\boldsymbol{l}})_{\mathbf{s}_1} \quad \text{(cf. (4.15))}. \tag{8.21}$$

The lemma below generalizes Lemma 11.

Lemma 22 (Exercise 8). *For $\mathbf{x} \in B_{l^2(\mathbb{N}^S)}$, let $\mathbf{x_s} \in l^2(\mathbb{N}^S)$ be defined recursively by* (8.18) *and* (8.20). *Then,*

$$\mathbf{x}_{(\mathbf{s},(0,\ldots,0))} = \mathbf{x_s}; \tag{8.22}$$

$$\psi(\mathbf{x_s})\hat{\ }(\pi_S(r_{j_1}, \ldots, r_{j_m})) = \mathbf{x_s}(\pi_S(j_1, \ldots, j_m)), \quad (j_1, \ldots, j_m) \in \mathbb{N}^m; \tag{8.23}$$

$$\|\mathbf{x_s}\|_2 \le \epsilon^{|\mathbf{s}|}; \tag{8.24}$$

$$\|\psi(\mathbf{x_s})\|_{\mathrm{L}^\infty} \le \delta\, \epsilon^{|\mathbf{s}|}; \tag{8.25}$$

$$\|\mathbf{x_s}\|_2 \le \epsilon^{|\mathbf{s}|}. \tag{8.26}$$

The theorem below generalizes Theorem 12.

Theorem 23 *Let $U = \{S_1, \ldots, S_n\}$ be a cover of $[m]$ satisfying* (7.3) *and* (7.4), *and let η_U be the n-linear functional on $H_1 \times \cdots \times H_n$ defined by* (7.5) *with $\varphi \equiv 1$, where $H_p = l^2(\mathbb{N}^{S_p})$, $p = 1, \ldots, n$. Let*

$$\psi_p : l^2(\mathbb{N}^{S_p}) \mapsto \mathrm{L}^\infty(\Omega^{S_p}), \quad p = 1, \ldots, n, \tag{8.27}$$

be $\Lambda(2)$-uniformizing maps associated with $0 < \epsilon < 1/(2^m - 1)$ and $\delta_{R^{S_p}}(\epsilon) = \delta_p$. Then, for $(\mathbf{x}^{(1)}, \ldots, \mathbf{x}^{(n)}) \in B_{H_1} \times \cdots \times B_{H_n}$,

$$\eta_U(\mathbf{x}^{(1)}, \ldots, \mathbf{x}^{(n)}) = \sum_{k=0}^{\infty} (-1)^k \sum_{\mathbf{t} \in (\tilde{D}^m)^k} \eta_U(\psi(\mathbf{x}^{(1)}_{\pi_{S_1}\mathbf{t}})\hat{}, \ldots, \psi(\mathbf{x}^{(n)}_{\pi_{S_n}\mathbf{t}})\hat{}), \tag{8.28}$$

where series on the right side converges uniformly in $B_{H_1} \times \cdots \times B_{H_n}$. (For $S \subset [m]$ and $\mathbf{t} = (\mathbf{t}(1), \ldots, \mathbf{t}(k)) \in (D^m)^k$,

$$\boldsymbol{\pi}_S \mathbf{t} = (\pi_S \mathbf{t}(1), \ldots, \pi_S \mathbf{t}(k)).) \tag{8.29}$$

Proof: (cf. Proof of Theorem 12). By (8.1),

$$\begin{aligned} &\eta_U(\psi(\mathbf{x}^{(1)})\hat{}, \ldots, \psi(\mathbf{x}^{(n)})\hat{}) \\ &= \sum_{\mathbf{w} \in W^m} \psi(\mathbf{x}^{(1)})\hat{}(\pi_{S_1}\mathbf{w}) \cdots \psi(\mathbf{x}^{(n)})\hat{}(\pi_{S_n}\mathbf{w}). \end{aligned} \tag{8.30}$$

Split the sum on the right side of (8.30) into a sum over R^m and a sum over its complement:

$$\begin{aligned} &\sum_{\mathbf{w} \in R^m} \psi(\mathbf{x}^{(1)})\hat{}(\pi_{S_1}\mathbf{w}) \cdots \psi(\mathbf{x}^{(n)})\hat{}(\pi_{S_n}\mathbf{w}) \\ &= \eta_U(\psi(\mathbf{x}^{(1)})\hat{}, \ldots, \psi(\mathbf{x}^{(n)})\hat{}) \\ &\quad - \sum_{\mathbf{w} \in (R^m)^c} (\psi(\mathbf{x}^{(1)})\hat{}(\pi_{S_1}\mathbf{w}) \cdots \psi(\mathbf{x}^{(n)})\hat{}(\pi_{S_n}\mathbf{w})). \end{aligned} \tag{8.31}$$

By the definition in (8.18), and by (8.23) (the case $\mathbf{s} = (0, \ldots, 0)$), we rewrite (8.31) as

$$\eta_U(\mathbf{x}^{(1)}, \ldots, \mathbf{x}^{(n)}) = \eta_U(\psi(\mathbf{x}^{(1)})\hat{}, \ldots, \psi(\mathbf{x}^{(n)})\hat{}) - \sum_{\mathbf{s} \in \tilde{D}^m} \eta(\mathbf{x}^{(1)}_{\pi_{S_1}\mathbf{s}}, \ldots, \mathbf{x}^{(n)}_{\pi_{S_n}\mathbf{s}}). \tag{8.32}$$

By iterating the identity in (8.32), we obtain for $N \geq 1$,

$$\begin{aligned} \eta_U(\mathbf{x}^{(1)}, \ldots, \mathbf{x}^{(n)}) = &\sum_{k=1}^{N-1} \sum_{\mathbf{t} \in (\tilde{D}^m)^k} \eta_U(\psi(\mathbf{x}^{(1)}_{\pi_{S_1}\mathbf{t}})\hat{}, \ldots, \psi(\mathbf{x}^{(n)}_{\pi_{S_n}\mathbf{t}})\hat{}) \\ &+ (-1)^N \sum_{\mathbf{t} \in (\tilde{D}^m)^N} \eta_U(\mathbf{x}^{(1)}_{\pi_{S_1}\mathbf{t}}, \ldots, \mathbf{x}^{(n)}_{\pi_{S_n}\mathbf{t}}) \end{aligned} \tag{8.33}$$

(cf. Claim in the proof of Theorem 12). By Lemma 15 and (8.24), for all $N \geq 1$ and $\mathbf{t} \in (\tilde{D}^m)^N$ (cf. (4.35)),

$$\begin{aligned} |\eta_U(\mathbf{x}^{(1)}_{\pi_{S_1}\mathbf{t}}, \ldots, \mathbf{x}^{(n)}_{\pi_{S_n}\mathbf{t}})| &\leq \|\mathbf{x}^{(1)}_{\pi_{S_1}\mathbf{t}}\|_2 \cdots \|\mathbf{x}^{(n)}_{\pi_{S_n}\mathbf{t}}\|_2 \\ &\leq \epsilon^{|\pi_{S_1}\mathbf{t}|} \cdots \epsilon^{|\pi_{S_n}\mathbf{t}|}. \end{aligned} \tag{8.34}$$

For $\mathbf{t} \in (\tilde{D}^m)^N$,

$$\sum_{p=1}^{n} |\pi_{S_p}\mathbf{t}| \geq N, \tag{8.35}$$

and therefore, by (8.34),

$$\sum_{\mathbf{t}\in(\tilde{D}^m)^N} |\eta_U(\mathbf{x}^{(1)}_{\pi_{S_1}\mathbf{t}}, \ldots, \mathbf{x}^{(n)}_{\pi_{S_n}\mathbf{t}})| \leq ((2^m-1)\epsilon)^N \quad \text{(cf. (4.37))}. \tag{8.36}$$

The theorem follows by applying (8.36) in (8.33), and letting $N \to \infty$. □

Corollary 24 (cf. Corollary 13). *Let ϵ and δ_p, $p = 1, \ldots, n$, be the $\Lambda(2)$-uniformizing constants in Theorem 20. Then,*

$$\|\eta_U\|_{\mathrm{pb}_n} \leq \frac{\delta_1 \cdots \delta_m}{1-(2^m-1)\epsilon}. \tag{8.37}$$

Proof: Apply Corollary 21 and (8.26) to the representation of η_U in (8.28). □

9 Proof of Theorem 19

We first identify $B(R^U)$ as a tilde algebra. Let $V_U(\mathbb{N}^m)$ denote the space of those $\varphi \in c_0(\mathbb{N}^m)$ that can be represented as

$$\varphi(\boldsymbol{l}) = \sum_{i=1}^{\infty} \alpha_i\, \theta_{i1}(\pi_{S_1}\boldsymbol{l}) \cdots \theta_{in}(\pi_{S_n}\boldsymbol{l}),$$

$$\boldsymbol{l} \in \mathbb{N}^m,\ \theta_{ip} \in B_{c_0(\mathbb{N}^{S_p})} \text{ for } i \in \mathbb{N} \text{ and } p \in [n],\ \sum_{i=1}^{\infty}|\alpha_i| < \infty, \tag{9.1}$$

and norm it by

$$\|\varphi\|_{V_U} = \inf\left\{\sum_{i=1}^{\infty} |\alpha_i| : \text{ representations of } \varphi \text{ by } (9.1)\right\}. \tag{9.2}$$

The space $V_U(\mathbb{N}^m)$ is the algebra of restrictions of elements in $V_n(\mathbb{N}^{S_1}, \ldots, \mathbb{N}^{S_n})$ to $\mathbb{N}^U$, where

$$\mathbb{N}^U = \{(\pi_{S_1}\boldsymbol{l}, \ldots, \pi_{S_n}\boldsymbol{l}) : \boldsymbol{l} \in \mathbb{N}^m\} \subset \mathbb{N}^{S_1} \times \cdots \times \mathbb{N}^{S_n}. \tag{9.3}$$

That is, $V_U(\mathbb{N}^m)$ is the quotient $V_n(\mathbb{N}^{S_1}, \ldots, \mathbb{N}^{S_n})$ modulo the ideal

$$\{\varphi \in V_n(\mathbb{N}^{S_1}, \ldots, \mathbb{N}^{S_n}) : \varphi(\mathbf{j}) = 0 \text{ for all } \mathbf{j} \in \mathbb{N}^U\}. \tag{9.4}$$

(In Exercise 9 you are asked to prove that $V_n(\mathbb{N}^{S_1}, \ldots, \mathbb{N}^{S_n})$ is isomorphic to the direct sum of $V_U(\mathbb{N}^m)$ with the ideal in (9.4).)

We define $\tilde{V}_U(\mathbb{N}^m)$ to be the algebra of pointwise limits of bounded sequences in $V_U(\mathbb{N}^m)$, normed by

$$\|\varphi\|_{\tilde{V}_U} = \inf\left\{\sup_j \|\varphi_j\|_{V_U} : \lim_{j\to\infty} \varphi_j(\boldsymbol{l}) = \varphi(\boldsymbol{l}), \boldsymbol{l} \in \mathbb{N}^m\right\} \quad \text{(cf. (5.7))}. \tag{9.5}$$

In the language of harmonic analysis, $V_U(\mathbb{N}^m)$ and $\tilde{V}_U(\mathbb{N}^m)$ are restriction algebras:

Lemma 25 (cf. Lemma 15; Exercise 10).

i. $V_U(\mathbb{N}^m) = A(R^U)$. *Specifically, if* $\varphi \in V_U(\mathbb{N}^m)$, *then there exists* $f \in L^1(\Omega^m)$ *such that*

$$\varphi(\boldsymbol{l}) = \hat{f}(r_{\pi_{S_1}\boldsymbol{l}}, \ldots, r_{\pi_{S_n}\boldsymbol{l}}), \quad \boldsymbol{l} \in \mathbb{N}^m, \tag{9.6}$$

and

$$\|f\|_{L^1} \leq 2^n \, \|\varphi\|_{V_U}. \tag{9.7}$$

Conversely, if $\varphi \in A(R^U)$, then there exists a representation

$$\varphi(r_{\pi_{S_1}\boldsymbol{l}}, \ldots, r_{\pi_{S_n}\boldsymbol{l}}) = \sum_{k=1}^{\infty} \alpha_k \; \theta_{k1}(\pi_{S_1}\boldsymbol{l}) \cdots \theta_{kn}(\pi_{S_n}\boldsymbol{l}), \quad \boldsymbol{l} \in \mathbb{N}^m, \tag{9.8}$$

such that $\theta_{kp} \in B_{c_0(\mathbb{N}^{S_p})}$ $(k \in \mathbb{N},\ p \in [n])$, and

$$\sum_{m=1}^{\infty} |\alpha_m| \leq \|\varphi\|_{A(R^U)}. \tag{9.9}$$

ii. $\tilde{V}_U(\mathbb{N}^m) = B(R^U)$. *Specifically, if $\varphi \in \tilde{V}_U(\mathbb{N}^m)$, then there exists $\mu \in \mathrm{M}(\Omega^m)$ such that*

$$\varphi(\boldsymbol{l}) = \hat{\mu}(r_{\pi_{S_1}\boldsymbol{l}}, \ldots, r_{\pi_{S_n}\boldsymbol{l}}), \quad \boldsymbol{l} \in \mathbb{N}^m, \tag{9.10}$$

and

$$\|\mu\|_{\mathrm{M}} \leq 2^n \; \|\varphi\|_{\tilde{V}_U}. \tag{9.11}$$

Conversely, if $\varphi \in B(R^U)$, then there exists a sequence $(\varphi_j : j \in \mathbb{N})$ in $V_U(\mathbb{N}^m)$ such that

$$\varphi(r_{\pi_{S_1}\boldsymbol{l}}, \ldots, r_{\pi_{S_n}\boldsymbol{l}}) = \lim_{j\to\infty} \varphi_j(\boldsymbol{l}), \quad \boldsymbol{l} \in \mathbb{N}^m, \tag{9.12}$$

and

$$\sup\{\|\varphi_j\|_{V_U} : j \in \mathbb{N}\} \leq \|\varphi\|_{B(R^U)}. \tag{9.13}$$

Proof of Theorem 19: Suppose $\varphi \in B(R^U)$. As in the proof of Theorem 14, we verify $\|\eta_{U,\varphi}\|_{\mathrm{pb}_n} < \infty$ in three steps.

Step 1 If

$$\varphi(\boldsymbol{l}) = \theta_1(\pi_{S_1}\boldsymbol{l}) \cdots \theta_n(\pi_{S_n}\boldsymbol{l}), \quad \boldsymbol{l} \in \mathbb{N}^m, \tag{9.14}$$

where $\theta_p \in B_{c_0(\mathbb{N}^{S_p})}, p \in [n]$, then, by Corollary 24,

$$\|\eta_{U,\varphi}\|_{\mathrm{pb}_n} \leq \frac{\delta_1 \cdots \delta_m}{1 - (2^m - 1)\epsilon}. \tag{9.15}$$

Step 2 If $\varphi \in V_U(\mathbb{N}^m)$, then, by Step 1,

$$\|\eta_{U,\varphi}\|_{\mathrm{pb}_n} \leq \left(\frac{\delta_1 \cdots \delta_m}{1 - (2^m - 1)\epsilon}\right) \|\varphi\|_{V_U}. \tag{9.16}$$

Step 3 If $\varphi \in B(R^U)$, then, by Lemma 25 and Proposition 4,

$$\|\eta_{U,\varphi}\|_{\mathrm{pb}_n} \leq \left(\frac{\delta_1 \cdots \delta_m}{1-(2^m-1)\epsilon}\right) \|\varphi\|_{B(R^U)}. \tag{9.17}$$

Conversely, we verify that if $\|\eta_{U,\varphi}\|_{\mathrm{pb}_n} < \infty$, then $\varphi \in B(R^U)$. Let

$$E_p = \{e_{\mathbf{k}} : \mathbf{k} \in \mathbb{N}^{S_p}\} \tag{9.18}$$

be the standard basis in $l^2(\mathbb{N}^{S_p})$, $p \in [n]$. Define

$$\phi_{\eta_{U,\varphi}}(r_{\mathbf{k}_1}, \ldots, r_{\mathbf{k}_n}) = \eta_{U,\varphi}(e_{\mathbf{k}_1}, \ldots, e_{\mathbf{k}_n}), \quad \mathbf{k}_p \in \mathbb{N}^{S_p},\ p \in [n]. \tag{9.19}$$

By assumption, $\phi_{\eta_{U,\varphi}} \in B(R_{S_1} \times \cdots \times R_{S_n})$; in particular,

$$\|\phi_{\eta_{U,\varphi}}\|_{B(R_{S_1}\times\cdots\times R_{S_n})} \leq 2^n\ \|\eta_{U,\varphi}\|_{\mathrm{pb}_n}. \tag{9.20}$$

By the definition of $\eta_{U,\varphi}$,

$$\phi_{\eta_{U,\varphi}}(r_{\mathbf{k}_1}, \ldots, r_{\mathbf{k}_n}) = \begin{cases} \varphi(\boldsymbol{l}) & \text{if } \pi_{S_p}\boldsymbol{l} = \mathbf{k}_p, p \in [n], \boldsymbol{l} \in \mathbb{N}^m \\ 0 & \text{otherwise,} \end{cases} \tag{9.21}$$

which implies

$$\|\varphi\|_{B(R^U)} \leq \|\phi_{\eta_{U,\varphi}}\|_{B(R_{S_1}\times\cdots\times R_{S_n})} \leq 2^n\ \|\eta_{U,\varphi}\|_{\mathrm{pb}_n}. \tag{9.22}$$

Remarks:

i **(upgrading Theorem 19).** When $\varphi \in B(R^U)$, an integral representation of $\eta_{U,\varphi}$ based on (8.28) implies a property ostensibly stronger than $\|\eta_{U,\varphi}\|_{\mathrm{pb}_n} < \infty$ (Exercise 11 ii). To see this, we first restate the definition of tilde algebras (see Chapter III §8, Remark iii). Let $K_1, \ldots, K_n$ be compact Hausdorff spaces, and let $\tilde{V}_n(K_1, \ldots, K_n)$ consist of those $f \in \mathrm{L}^\infty(K_1 \times \cdots \times K_n)$ for which there exist sequences $(\varphi_k : k \in \mathbb{N})$ in $V_n(K_1, \ldots, K_n)$, such that

$$\lim_{k\to\infty} \varphi_k(t_1, \ldots, t_n) = f(t_1, \ldots, t_n), \qquad (t_1, \ldots, t_n) \in K_1 \times \cdots \times K_n \tag{9.23}$$

and

$$\limsup_{k\to\infty} \|\varphi_k\|_{V_n} < \infty.$$

($L^\infty(K_1 \times \cdots \times K_n)$ is the space of all bounded Borel-measurable functions on $K_1 \times \cdots \times K_n$, and the definition of $V_n(K_1, \ldots, K_n)$ can be found in Chapter IV §7.) We norm $\tilde{V}_n(K_1, \ldots, K_n)$ by

$$\begin{aligned}\|f\|_{V_n} &= \inf\Big\{ \limsup_{k\to\infty} \|\varphi_k\|_{V_n} : \\ &\qquad \lim_{k\to\infty} \varphi_k = f \text{ pointwise on } K_1 \times \cdots \times K_n \Big\}. \end{aligned} \tag{9.24}$$

We consider the case $n = 2$, which is archetypal. We take K to be the unit ball in l^2 equipped with the weak topology, and let η be defined by

$$\eta(\mathbf{x}, \mathbf{y}) = \sum_{j=1}^{\infty} \mathbf{x}(j)\ \mathbf{y}(j), \quad (\mathbf{x}, \mathbf{y}) \in l^2 \times l^2. \tag{9.25}$$

(Note that η is not continuous on $K \times K$.) Let ψ be the $\Lambda(2)$-uniformizing map in (3.2). Then, the function on $K \times K$ defined by

$$\begin{aligned}\eta(\psi(\mathbf{x})\hat{\ }, \psi(\mathbf{y})\hat{\ }) &= \int_\Omega \psi(\mathbf{x})(\omega)\ \psi(\mathbf{y})(\omega)\ \mathbb{P}(d\omega) \\ &:= \mathbf{E}\ \psi(\mathbf{x})\ \psi(\mathbf{y}), \quad (\mathbf{x}, \mathbf{y}) \in K \times K, \end{aligned} \tag{9.26}$$

is in $\tilde{V}_2(K, K)$. Therefore, by applying the representation of η in Theorem 23 in the (simplest) instance $U = \{(1), (1)\}$ (see Chapter III), we obtain

$$\eta|_{K^2} \in \tilde{V}_2(K, K) \quad \text{(Exercise 11 i)}. \tag{9.27}$$

In the multidimensional framework, we have

Theorem 26 $\eta_{U,\varphi}|_{B_{H_1} \times \cdots \times B_{H_n}} \in \tilde{V}_n(B_{H_1}, \ldots, B_{H_n}) \Leftrightarrow \varphi \in B(R^U)$.

ii **(a characterization?).** Let η be a bounded trilinear functional on a Hilbert space H. Let $E = \{\mathbf{e}_i : i \in \mathbb{N}\}$ be a basis in H, and write for $(\mathbf{x}, \mathbf{y}, \mathbf{z}) \in H^3$ (Exercise 12)

$$\eta(\mathbf{x}, \mathbf{y}, \mathbf{z}) = \sum_{i,j,k} a_{ijk}\ \mathbf{x}(i)\ \mathbf{y}(j)\ \mathbf{z}(k), \tag{9.28}$$

where

$$a_{ijk} = \eta(\mathbf{e}_i, \mathbf{e}_j, \mathbf{e}_k), \quad (i,j,k) \in \mathbb{N}^3, \tag{9.29}$$

and

$$\begin{aligned}
&\sum_{i,j,k} a_{ijk}\ \mathbf{x}(i)\ \mathbf{y}(j)\ \mathbf{z}(k) \\
&\quad = \lim_{N\to\infty} \sum_{(i,j,k)\in[N]^3} a_{ijk}\ \mathbf{x}(i)\ \mathbf{y}(j)\ \mathbf{z}(k) \\
&\quad = \sum_{i=1}^{\infty}\sum_{j=1}^{\infty}\sum_{k=1}^{\infty} a_{ijk}\ \mathbf{x}(i)\ \mathbf{y}(j)\ \mathbf{z}(k).
\end{aligned} \tag{9.30}$$

By Proposition 3, if η is projectively bounded, then

$$(a_{ijk} : (i,j,k) \in \mathbb{N}^3) \in B(R^3),$$

where R is the Rademacher system indexed by $\mathbb{N}$; that is, there exists $\mu \in \mathrm{M}(\Omega^3)$ such that

$$\hat{\mu}(r_i, r_j, r_k) = a_{ijk}, \quad (i,j,k) \in \mathbb{N}^3, \tag{9.31}$$

$$\|\mu\|_{\mathrm{M}} \leq 2^3 \|\eta\|_{\mathrm{pb}_3}.$$

I do not know whether the converse holds:

Problem (Exercise 13). Suppose η is a bounded trilinear functional on H and

$$\eta|_{E^3} \in B(R^3). \tag{9.32}$$

Is η projectively bounded?

Here is a plausible approach to an affirmative answer (?). Suppose there exists $\mu \in \mathrm{M}(\Omega^3)$ such that $\hat{\mu}|_E = \eta|_{E^3}$. Let $(\mathbf{x}, \mathbf{y}, \mathbf{z}) \in B_H^3$. By $\Lambda(2)$-uniformizability, we obtain $f_x \in \mathrm{L}^\infty(\Omega)$, $f_y \in \mathrm{L}^\infty(\Omega)$, $f_z \in \mathrm{L}^\infty(\Omega)$ with L^∞-norms bounded by an absolute constant, so that

$$\hat{f}_x(r_i) = \mathbf{x}(i),\ \hat{f}_y(r_i) = \mathbf{y}(i),\ f_z(r_i) = \mathbf{z}(i),\ i \in \mathbb{N}, \tag{9.33}$$

and $\|\hat{f}_x|_{R^c}\|_2$, $\|\hat{f}_y|_{R^c}\|_2$, $\|\hat{f}_z|_{R^c}\|_2$ are 'small'. Observe

$$\begin{aligned}\eta(\mathbf{x},\mathbf{y},\mathbf{z}) &= \int_{\Omega^3} f_x \otimes f_y \otimes f_z d\mu \\ &\quad - \sum_{(w_1,w_2,w_3)\in W^3\sim R^3} \hat{f}_x(w_1)\ \hat{f}_y\ (w_2)\hat{f}_z(w_3)\ \hat{\mu}(w_1,w_2,w_3).\end{aligned} \tag{9.34}$$

(It is not difficult to make sense of the second term on the right side of (9.34).) The main obstacle is the feasibility of a recursion based on (9.34). Specifically, can $\eta(\mathbf{x},\mathbf{y},\mathbf{z})$ be represented by an absolutely convergent series whose summands are multiples of $\int_{\Omega^3} f_x \otimes f_y \otimes f_z d\mu$?

iii **(a preview).** Let $\eta_{U,\varphi}$ be the trilinear functional considered in §5:

$$\begin{aligned}\eta_{U,\varphi}(\mathbf{x},\mathbf{y},\mathbf{z}) &= \sum_{i,j,k} \varphi(i,j,k)\ \mathbf{x}(i,j)\ \mathbf{y}(j,k)\ \mathbf{z}(i,k), \\ \mathbf{x} &\in l^2(\mathbb{N}^2),\ \mathbf{y} \in l^2(\mathbb{N}^2),\ \mathbf{z} \in l^2(\mathbb{N}^2),\end{aligned} \tag{9.35}$$

where $\varphi\in l^\infty(\mathbb{N}^3)$, and $U=\{(1,2),(2,3),(1,3)\}$. The same functional can be viewed also as a bounded 4-linear functional on $l^2(\mathbb{N})\times l^2(\mathbb{N})\times l^2(\mathbb{N}^2)\times l^2(\mathbb{N}^2)$:

$$\begin{aligned}\eta_{U',\varphi}(\mathbf{w},\mathbf{x},\mathbf{y},\mathbf{z}) &= \sum_{i,j,k} \varphi(i,j,k)\ \mathbf{w}(i)\ \mathbf{x}(j)\ \mathbf{y}(j,k)\ \mathbf{z}(i,k), \\ \mathbf{w} \in l^2(\mathbb{N}),\ \mathbf{x} &\in l^2(\mathbb{N}),\ \mathbf{y} \in l^2(\mathbb{N}^2),\ \mathbf{z} \in l^2(\mathbb{N}^2),\end{aligned} \tag{9.36}$$

where $U' = \{(1),(2),(2,3),(1,3)\}$. (See (7.5).) If $\|\eta_{U',\varphi}\|_{\mathrm{pb}_4} < \infty$, then $\|\eta_{U,\varphi}\|_{\mathrm{pb}_3} < \infty$ (Exercise 14). A question naturally arises: does the converse hold? The answer will become evident in Chapter XIII, as a corollary to the solution of the p-Sidon set problem (Chapter VII §11, Remark vii).

Exercises

1. i. Verify Proposition 2.
   ii. Verify Proposition 3.
   iii. Verify Proposition 4.

2. Let $\mathbf{x} \in l^2(\mathbb{N})$ have real-valued coordinates, and consider

$$F_n = \prod_{j=1}^{n} (r_0 + \mathrm{i}\mathbf{x}(j)\ \mathbf{r}_j), \quad n \in \mathbb{N}. \tag{E.1}$$

   i. Show that $\|F_n\|_{\mathrm{L}^\infty} \leq \exp(\frac{1}{2}\|\mathbf{x}\|_2)$.

   ii. Show that if $k \in [n]$ and $0 < j_1 < \cdots < j_k \leq n$, then

$$\hat{F}_n(r_{j_1} \ldots r_{j_k}) = \mathrm{i}^k\ \mathbf{x}(j_1) \ldots \mathbf{x}(j_k).$$

   Otherwise, if

$$w \notin \{r_{j_1} \cdots r_{j_k} : 0 < j_1 < \cdots < j_m \leq n,\ m \in [n]\} \cup \{r_0\}$$

   (cf. (VII.3.9)), then $\hat{F}_n(w) = 0$.

   iii. Prove there exists $\psi(\mathbf{x}) \in \mathrm{L}^\infty(\Omega, \mathbb{P})$, denoted as the infinite product

$$\psi(\mathbf{x}) := \prod_{j=1}^{n} (r_0 + \mathrm{i}\mathbf{x}(j)\ r_j) \tag{E.2}$$

   (L^∞-Riesz product), such that $\psi(\mathbf{x})\hat{}\,(r_0) = 1$ and

$$\psi(\mathbf{x})\hat{}\,(r_{j_1} \cdots r_{j_k}) = \mathrm{i}^k\ \mathbf{x}(j_1) \cdots \mathbf{x}(j_k), \\ 0 < j_1 < \cdots < j_k,\ k \in \mathbb{N}.$$

3. Verify Corollary 7.
4. Recall that a subset F of a discrete Abelian group $\hat{G}$ is a $\Lambda(2)$-set if there exists $k > 0$ such that for all $f \in \mathrm{L}^2_E(G)$,

$$\|f\|_{\mathrm{L}^2} \leq k\|f\|_{\mathrm{L}^1} \tag{E.3}$$

   (Definitions III.13, VII.33). The assertion that $R \subset W$ is a $\Lambda(2)$-set is the classical L^1–L^2 Khintchin inequality (Chapter II); that $R^n \subset W^n$ is a $\Lambda(2)$-set for all $n \in \mathbb{N}$ follows by induction and Minkowski's inequality (Exercise VII.32).

   i. Prove that $F \subset \hat{G}$ is a $\Lambda(2)$-set if and only if for every $\varphi \in l^2(F)$ there exists $f \in \mathrm{L}^\infty(G)$ such that

$$\hat{f}|_F = \varphi, \tag{E.4}$$

   and

$$\|f\|_{\mathrm{L}^\infty} \leq k\|\varphi\|_2.$$

ii. Prove that R^n is $\Lambda(2)$-uniformizable by applying i, and the Riesz product

$$\prod_{j=1}^{\infty}\left(r_0+\left(\frac{1}{2}\right)r_j\right)\otimes\cdots\otimes\prod_{j=1}^{\infty}\left(r_0+\left(\frac{1}{2}\right)r_j\right).$$

iii.* Is every $\Lambda(2)$-set $\Lambda(2)$-uniformizable?

5. i. Show that for every integer $m>1$ there exists $\mu\in \mathrm{M}(\Omega)$ such that

$$\hat{\mu}|_R = 1 \text{ and } \hat{\mu}|_{R_j} = 0 \text{ for } j=2,\ldots,m.$$

ii. Fix an integer $m>1$. In the definition of ψ_1 in (3.2), replace ϵ by $\epsilon^{1/2m}$, and then convolve the resulting ψ_1 by $\mu\in\mathrm{M}(\Omega)$ such that $\hat{\mu}|_R=1$, and $\hat{\mu}|_{R_j}=0$ for $j=2,\ldots,2m$. Deduce that $\delta_R(\epsilon)$ is $\mathscr{O}(\epsilon^{1/2m})$.

iii. Prove that for every $k>0$ there exists $C_k>0$ such that

$$\delta_{R^n}(\epsilon)\le C_k\,\epsilon^{-(1/k)}.$$

6. Let $U=\{(1,2),(2,3),(1,3)\}$. Prove that if $\varphi\in V_3(\mathbb{N}^2,\mathbb{N}^2,\mathbb{N}^2)$, then $\mathbf{1}_{\mathbb{N}^U}\varphi$ is also in $V_3(\mathbb{N}^2,\mathbb{N}^2,\mathbb{N}^2)$. In particular,

$$\|\mathbf{1}_{\mathbb{N}^U}\varphi\|_{V_3(\mathbb{N}^2,\mathbb{N}^2,\mathbb{N}^2)}\le\|\varphi\|_{V_3(\mathbb{N}^2,\mathbb{N}^2,\mathbb{N}^2)}.$$

7. i. Prove Lemma 15.

ii.* Find the 'optimal' constant in (5.14). For example, can you show

$$\|f\|_{\mathrm{L}^1}\le 2^{3/2}\,\|\varphi\|_{V_U}?$$

8. Prove Lemma 22.
9. Prove

$$\begin{aligned}&V_n(\mathbb{N}^{S_1},\ldots,\mathbb{N}^{S_n})\\&\quad= V_U(\mathbb{N}^m)\oplus\{\varphi\in V_n(\mathbb{N}^{S_1},\ldots,\mathbb{N}^{S_n}):\varphi(\mathbf{j})=0,\ \mathbf{j}\in\mathbb{N}^U\},\end{aligned}\tag{E.5}$$

where $\oplus$ denotes a Banach algebra direct sum.

10. i. Verify Lemma 25.

ii.* Can the constants' growths in (9.7) and (9.11), which depend on n, be improved?

11. i. Let η be the usual dot product in l^2 (defined in (9.26)). Prove that $\eta|_{K\times K}\in\tilde{V}_2(K,K)$, where K is the unit ball in l^2 equipped with the weak topology.

ii.* Obtain $\eta|_{K\times K} \in \tilde{V}_2(K,K)$ directly from the Grothendieck inequality. More generally, is it true that if η is a projectively bounded n-linear functional on a Hilbert space H, and K is the unit ball of H equipped with the weak topology, then $\eta_{|K^n} \in \tilde{V}_n(K,\dots,K)$?

12. Verify (9.30).
13. Verify that an affirmative answer to the open problem in Remark ii §9 implies Theorem 19.
14. Prove that if $\|\eta_{U',\varphi}\|_{\mathrm{pb}_4} < \infty$, then $\|\eta_{U,\varphi}\|_{\mathrm{pb}_3} < \infty$, where $\eta_{U',\varphi}$ is defined in (9.36) and $\eta_{U,\varphi}$ is defined in (9.35).

Hints for Exercises in Chapter VIII

1. i. See definitions and Remark ii in Chapter VII §8.

   ii. Necessity follows from Proposition 2 and Exercise VII.17. Conversely, if $\|\eta\|_{\mathrm{pb}_n} = \infty$, then for every $K > 0$ there exists a finite set $T \subset B_H$ such that

$$\|\eta\|_{V_n(T,\dots,T)} \geq K.$$

   By taking arbitrarily large Ks and corresponding Ts, produce $E \subset B_H$ such that $\phi_{\eta,E} \notin B(R^n)$. Apply Proposition 2 and the 'dual' version of Lemma VII.20.

   iii. See Proposition 3 and Exercise VII.17.

2. Review the appropriate sections in Chapter III.
3. It suffices to verify (3.21) for W^n-polynomials. If $\mathbf{x}_j \to \mathbf{x}$ weakly, then $\lim_{j\to\infty} \psi_n(\mathbf{x}_j)\hat{}\,(\gamma) = \psi_n(\mathbf{x})\hat{}\,(\gamma)$ for $\gamma \in W^n$.
4. i. Consider the restriction algebra

$$\mathrm{L}^\infty(G)\hat{}\,/\{\hat{f} : f \in \mathrm{L}^\infty(G), \hat{f}|_F = 0\} := Q^\infty(F),$$

   and verify that $\mathrm{L}^1_F(G)^* = Q^\infty(F)$; cf. Proposition VII.24.

   iii.* See Chapter III §6.

5. i. See Lemma VII.22.
6. It suffices to prove that the indicator function $\mathbf{1}_{\mathbb{N}^U}$ is in $\tilde{V}_3(\mathbb{N}^2,\mathbb{N}^2,\mathbb{N}^2)$. To show this, consider the Riesz product

$$\prod_{i,j}(r_0 + r_i(\xi_1)r_j(\xi_2)r_{ij}) \otimes \prod_{i,j}(r_0 + r_i(\xi_2)r_j(\xi_3)r_{ij}) \otimes \prod_{i,j}(r_0 + r_i(\xi_1)r_j(\xi_3)r_{ij}),$$

$$(\xi_1,\xi_2,\xi_3) \in \Omega^3,$$

   and average it over $(\xi_1,\xi_2,\xi_3) \in \Omega^3$ (cf. Lemma 9, Exercise VI.12 iii).

7. Recall the following. If $\phi \in c_0(\mathbb{N}^2 \times \mathbb{N}^2 \times \mathbb{N}^2)$ and

$$\begin{aligned}\phi((i_1,i_2),(j_1,j_2),(k_1,k_2)) &= \sum_{m=1}^{\infty} \alpha_m\, \theta_{m1}(i_1,i_2)\, \theta_{m2}(j_1,j_2)\, \theta_{m3}(k_1,k_2),\\ &((i_1,i_2),(j_1,j_2),(k_1,k_2)) \in \mathbb{N}^2 \times \mathbb{N}^2 \times \mathbb{N}^2,\end{aligned}$$

where $\sum_{m=1}^{\infty} |\alpha_m| < \infty$, and $\theta_{mp} \in B_{c_0(\mathbb{N}^2)}$ for $m \in \mathbb{N}$ and $p \in \{1,2,3\}$, then there exists $f \in L^1(\Omega^3)$ such that

$$\begin{aligned}\hat{f}(r_{i_1j_1}, r_{i_2j_2}, r_{i_3j_3}) &= \phi((i_1,j_1),(i_2,j_2),(i_3,j_3)),\\ &((i_1,j_1),(i_2,j_2),(i_3,j_3)) \in \mathbb{N}^2 \times \mathbb{N}^2 \times \mathbb{N}^2,\end{aligned}$$

and

$$\|f\|_{L^1} \leq 2^3 \sum_{m=1}^{\infty} |\alpha_m|.$$

Conversely, if $\phi \in A(R^3)$, then there exists a representation

$$\begin{aligned}\phi(r_{i_1j_1}, r_{i_2j_2}, r_{i_3j_3}) &= \sum_{m=1}^{\infty} \alpha_m\, \theta_{m1}(i_1,j_1)\, \theta_{m2}(i_2,j_2)\, \theta_{m3}(i_3,j_3),\\ &((i_1,j_1),(i_2,j_2),(i_3,j_3)) \in \mathbb{N}^2 \times \mathbb{N}^2 \times \mathbb{N}^2,\end{aligned}$$

such that $\theta_{mp} \in B_{c_0(\mathbb{N}^2)}$ for $m \in \mathbb{N}$ and $p \in \{1,2,3\}$, and

$$\sum_{m=1}^{\infty} |\alpha_m| \leq \|\phi\|_{A(R^3)}.$$

Now obtain $A(R^U)$ as a quotient of $A(R^3)$, and $V_U(\mathbb{N}^3)$ as a quotient of $V_3(\mathbb{N}^2,\mathbb{N}^2,\mathbb{N}^2)$.

8. Review the proof of Lemma 11.
9. Let $\varphi \in V_n(\mathbb{N}^{S_1},\ldots,\mathbb{N}^{S_n})$, and write $\varphi = \mathbf{1}_{\mathbb{N}^U}\varphi + (\varphi - \mathbf{1}_{\mathbb{N}^U}\varphi)$. Show that $\mathbf{1}_{\mathbb{N}^U} \in \tilde{V}_n(\mathbb{N}^{S_1},\ldots,\mathbb{N}^{S_n})$. You can establish this by generalizing the argument used in Exercise 6; you can use the systems discussed in Chapter II §6, or the 'fractional' convolution operation defined in (8.7). You can obtain (E.5) also as a byproduct of the proof of the 'easy' direction in Theorem 19; cf. Corollary 16. However, the proof generalizing Exercise 6 is more direct and yields better constants.

10. i. See Exercise 7.

    ii. This problem may become more tractable when we show in Chapter XIII that $\mathbb{N}^U$ is a 'fractional Cartesian product' with 'combinatorial dimension' dim $\mathbb{N}^U$. A reasonable conjecture is that best constants in (9.7) and (9.11) are bounded by $2^{\dim \mathbb{N}^U}$. Exercise 7 ii is an instance of this problem.

11. i. For each $n \geq 1$, consider

$$\eta_n(\mathbf{x}, \mathbf{y}) = \sum_{j=1}^{n} \mathbf{x}(j)\ \mathbf{y}(j), \quad (\mathbf{x}, \mathbf{y}) \in K \times K,$$

    and prove that $\|\eta_n\|_{V_2(K,K)} \leq c$, where c is an absolute constant. To this end, use Theorem 23 in the case $m = 1$, $n = 2$, and $U = \{(1), (1)\}$. See Theorem IV.13.

    ii. In the two-dimensional case, use the Grothendieck factorization theorem. I do not know the answer to the question in the n-dimensional case for $n \geq 3$.

IX
Product Fréchet Measures

1 Mise en Scène: A Basic Question

Product measures pervade analysis from the foundations up. In our context for example, in harmonic analysis they are the key to convolution, and in probability theory they underlie the notion of statistical independence. And of course there are other examples in various other settings. The feasibility of product measures is guaranteed by this classical result:

Theorem 1 *If μ is a scalar measure on a measurable space $(X, \mathfrak{A})$ and ν is a scalar measure on a measurable space $(Y, \mathfrak{B})$, then*

$$\mu \times \nu(A \times B) = \mu(A)\ \nu(B), \quad A \in \mathfrak{A},\ B \in \mathfrak{B}, \tag{1.1}$$

determines a scalar measure on the product space $(X \times Y, \sigma(\mathfrak{A} \times \mathfrak{B}))$.

A basic question arises: are products of Fréchet measures also feasible?

Definition 2 Let $(X_1, \mathfrak{A}_1), \dots, (X_n, \mathfrak{A}_n), (Y_1, \mathfrak{B}_1), \dots, (Y_n, \mathfrak{B}_n)$ be measurable spaces. For $\mu \in F_n(\mathfrak{A}_1, \dots, \mathfrak{A}_n)$ and $\nu \in F_n(\mathfrak{B}_1, \dots, \mathfrak{B}_n)$, define

$$\begin{aligned} &\mu \times \nu((A_1, B_1), \dots, (A_n, B_n)) = \mu(A_1, \dots, A_n)\ \nu(B_1, \dots, B_n), \\ &(A_1, \dots, A_n) \in \mathfrak{A}_1 \times \cdots \times \mathfrak{A}_n, \\ &(B_1, \dots, B_n) \in \mathfrak{B}_1 \times \cdots \times \mathfrak{B}_n. \end{aligned} \tag{1.2}$$

If $\mu \times \nu$ determines an F_n-measure on $\sigma(\mathfrak{A}_1 \times \mathfrak{B}_1) \times \cdots \times \sigma(\mathfrak{A}_n \times \mathfrak{B}_n)$, then we write $\mu \times \nu \in F_n(\sigma(\mathfrak{A}_1 \times \mathfrak{B}_1), \dots, \sigma(\mathfrak{A}_n \times \mathfrak{B}_n))$, and refer to $\mu \times \nu$ as a *product F-measure.*

That product F_1-measures are always feasible is a straightforward matter (Theorem 1), and that product F_2-measures are always feasible (not quite as obvious) follows from the Grothendieck inequality and factorization theorem. In dimensions $n > 2$, the absence of a universal n-linear Grothendieck inequality implies existence of F_n-measures μ and ν such that $\mu \times \nu \notin F_n$. Indeed, the main lesson in this chapter is that product F-measures are inextricably linked to Grothendieck-type inequalities.

2 A Preview

In this section, we illustrate in a simple setting the connection between Grothendieck-type inequalities and product Fréchet measures.

For a scalar array $A = (a_{ij})$ of finite rank, define

$$\|A\|_{f_{2,p}} = \sup\left\{\left|\sum_{i,j} a_{ij}\ s_i\ t_j\right| : (s_i) \in B_{l^p},\ (t_j) \in B_{l^p}\right\},$$
$$p \in [2, \infty]. \tag{2.1}$$

For scalar arrays $A = (a_{ij})$ and $B = (b_{ij})$ with finite rank, we define the tensor product

$$(A \otimes B)_{imjn} = a_{ij}\ b_{mn}, \quad (i,j,m,n) \in \mathbb{N}^4, \tag{2.2}$$

and view it as a bilinear functional acting on scalar-valued functions defined on $\mathbb{N}^2$: for $x = (x_{im})$ and $y = (y_{jn})$,

$$(A \otimes B)(x,y) = \sum_{i,j,m,n} a_{ij}\ b_{mn}\ x_{im}\ y_{jn}. \tag{2.3}$$

(In a context of multilinear algebra, $A \otimes B$ is sometimes called a *Kronecker product*; e.g., see [L, Chapter 12].) For $p \in [2, \infty]$,

$$\|A \otimes B\|_{f_{2,p}} = \sup\left\{\left|\sum_{i,j,m,n} a_{ij}\ b_{mn}\ x_{im}\ y_{jn}\right| : \right.$$
$$\left. (x_{im}) \in B_{l^p(\mathbb{N}^2)},\quad (y_{jn}) \in B_{l^p(\mathbb{N}^2)}\right\}. \tag{2.4}$$

We note two relations involving these norms. The first is elementary, but the second relation requires the intervention of the Grothendieck theorems. (See Exercise 1 for other relations.)

Theorem 3 *If A and B are matrices with finite rank, then*

$$\|A \otimes B\|_{f_{2,2}} \leq \|A\|_{f_{2,2}} \|B\|_{f_{2,2}}, \tag{2.5}$$

and

$$\|A \otimes B\|_{f_{2,\infty}} \leq \kappa_G^2 \, \|A\|_{f_{2,\infty}} \|B\|_{f_{2,\infty}}, \tag{2.6}$$

where κ_G is the Grothendieck constant. (To underscore in this section that the case $p = \infty$ is but instance, we use the notation $\|\cdot\|_{f_{2,\infty}}$. Elsewhere in the book, $\|\cdot\|_{f_2}$ stands for $\|\cdot\|_{f_{2,\infty}}$.)

Proof: To verify (2.5), let $x = (x_{im}) \in B_{l^2(\mathbb{N}^2)}$ be arbitrary, and estimate

$$\sum_{j,n} \left| \sum_{i,m} a_{ij} \, b_{mn} \, x_{im} \right|^2 = \sum_j \left(\sum_n \left| \sum_m b_{mn} \sum_i a_{ij} \, x_{im} \right|^2 \right)$$

$$\leq \|B\|_{f_{2,2}}^2 \sum_j \left| \sum_{i,m} a_{ij} \, x_{im} \right|^2 \leq \|B\|_{f_{2,2}}^2 \|A\|_{f_{2,2}}^2. \tag{2.7}$$

To prove (2.6), we first deduce from the Grothendieck factorization theorem that there exist probability measures ν_1 and ν_2 on $\mathbb{N}$ such that for all $h \in \mathrm{L}^2(\mathbb{N}, \nu_1)$ and $g \in \mathrm{L}^2(\mathbb{N}, \nu_2)$,

$$\left| \sum_{i,j} a_{ij} h(i) \, g(j) \right| \leq \kappa_G \|A\|_{f_{2,\infty}} \|h\|_{\mathrm{L}^2(\nu_1)} \|g\|_{\mathrm{L}^2(\nu_2)}. \tag{2.8}$$

That is, A defines a bilinear functional on $\mathrm{L}^2(\mathbb{N}, \nu_1) \times \mathrm{L}^2(\mathbb{N}, \nu_1)$ with norm bounded by $\kappa_G \|A\|_{f_{2,\infty}}$. Next, we let $(x_{im}) \in B_{c_0(\mathbb{N}^2)}$ and $(y_{jn}) \in B_{c_0(\mathbb{N}^2)}$ be arbitrary, and consider the subsets $\{h_m\}$ and $\{g_n\}$ of the respective unit balls in $\mathrm{L}^2(\mathbb{N}, \nu_1)$ and $\mathrm{L}^2(\mathbb{N}, \nu_2)$, where

$$h_m(i) = x_{im}, \quad g_n(j) = y_{jn}, \quad (i, j, m, n) \subset \mathbb{N}^2. \tag{2.9}$$

Then, by the Grothendieck inequality,

$$\left| \sum_{i,j,m,n} a_{ij} \, b_{mn} \, x_{im} \, y_{jn} \right| = \left| \sum_{m,n} b_{mn} \, A(h_m, g_n) \right|$$

$$\leq \kappa_G^2 \|B\|_{f_{2,\infty}} \|A\|_{f_{2,\infty}}. \tag{2.10}$$

□

Remark (an overview). The assertion in (2.5) is immediate from the definition of $\|\cdot\|_{f_{2,\infty}}$, but going up a notch, if we consider $A \otimes B$ as a bilinear functional acting on the 'mixed norm' space

$$l^\infty(l^2) = \left\{(x_{ij}) : \sup_j \Sigma_i |x_{ij}|^2 < \infty\right\}, \tag{2.11}$$

then an assertion analogous to (2.5) is a restatement of the Grothendieck inequality (Exercise 1 i).

The statement in (2.6) requires the Grothendieck factorization theorem *and* the Grothendieck inequality. The argument verifying it is essentially the same as the argument we use in a later section to prove the analogous assertion in the general measurable setting.

Next we show that Theorem 3 cannot be extended in the obvious way to dimensions greater than two. For a 3-array $A = (a_{ijk})$ of finite rank,

$$\|A\|_{f_{3,p}} = \sup\left\{\left|\sum_{i,j,k} a_{ijk}\ s_i\ t_j\ u_k\right| : (s_i) \in B_{l^p}, (t_j) \in B_{l^p},\ (u_k) \in B_{l^p}\right\},$$
$$p \in [2,\infty]. \tag{2.12}$$

For $A = (a_{ijk})$ and $B = (b_{ijk})$, define

$$(A \otimes B)_{i_1 j_1 k_1 i_2 j_2 k_2} = a_{i_1 j_1 k_1}\ b_{i_2 j_2 k_2}, \quad (i_1, j_1, k_1, i_2, j_2, k_2) \in \mathbb{N}^6, \tag{2.13}$$

and for A and B with finite rank,

$$\|A \otimes B\|_{f_{3,p}} = \sup\left\{\left|\sum_{i_1,j_1,k_1,i_2,j_2,k_2} a_{i_1 j_1 k_1} b_{i_2 j_2 k_2} x_{i_1 i_2} y_{j_1 j_2} z_{k_1 k_2}\right| :\right.$$
$$\left.(x_{ij}) \in B_{l^p(\mathbb{N}^2)},\ (y_{ij}) \in B_{l^p(\mathbb{N}^2)},\ (z_{ij}) \in B_{l^p(\mathbb{N}^2)}\right\}. \tag{2.14}$$

Theorem 4

i. *For every $K > 0$, there exist 3-arrays A and B such that*

$$\|A\|_{f_{3,2}} \leq 1,\ \|B\|_{f_{3,2}} \leq 1,\ \textit{and}\ \|A \otimes B\|_{f_{3,2}} > K.$$

ii. *For every $K > 0$, there exist 3-arrays A and B such that*

$$\|A\|_{f_{3,\infty}} \leq 1,\ \|B\|_{f_{3,\infty}} \leq 1,\ \textit{and}\ \|A \otimes B\|_{f_{3,\infty}} > K.$$

Proof: For Part i see Exercise 2.

To prove Part ii, we first note that Theorem VIII.14 and Theorem VIII.17 imply existence of trilinear functionals η on $l^2(\mathbb{N}) \times l^2(\mathbb{N}) \times l^2(\mathbb{N})$ such that $\|\eta\|_{\mathcal{F}_{3,\infty}} \leq 1$, and

$$\|\eta|_{E^3}\|_{\tilde{V}_3(E,E,E)} = \infty, \tag{2.15}$$

where $E = \{\mathbf{e}_i\}$ is an orthonormal basis of l^2. Write $\phi_N(i,j,k) = \eta(\mathbf{e}_i, \mathbf{e}_j, \mathbf{e}_k)$ for $(i,j,k) \in [2^N] \times [2^N] \times [2^N]$ and $N \in \mathbb{N}$, and conclude from (2.15) that for every $K > 0$ there exists $N \in \mathbb{N}$ such that

$$\|\phi_N\|_{V_3([2^N],[2^N],[2^N])} > K. \tag{2.16}$$

Next we view η as a trilinear functional on $\mathrm{C}(\Omega_N) \times \mathrm{C}(\Omega_N) \times \mathrm{C}(\Omega_N)$, where Ω_N is the compact Abelian group $\{-1,1\}^N$: enumerate $\hat{\Omega}_N = \{w_j : j \in [2^N]\}$, and define

$$A(f,g,h) := \sum_{(i,j,k)\in[2^N]^3} \eta(\mathbf{e}_i, \mathbf{e}_j, \mathbf{e}_k)\ \hat{f}(w_i)\ \hat{g}(w_j)\ \hat{h}(w_k),$$

$$(f,g,h) \in \mathrm{C}(\Omega_N) \times \mathrm{C}(\Omega_N) \times \mathrm{C}(\Omega_N). \tag{2.17}$$

Then,

$$\begin{aligned} |A(f,g,h)| &\leq \|\hat{f}\|_{\mathrm{L}^2} \|\hat{g}\|_{\mathrm{L}^2} \|\hat{h}\|_{\mathrm{L}^2} \\ &\leq \|f\|_{\mathrm{C}(\Omega_N)} \|g\|_{\mathrm{C}(\Omega_N)} \|h\|_{\mathrm{C}(\Omega_N)}, \end{aligned} \tag{2.18}$$

which implies $\|A\|_{\mathcal{F}_{3,\infty}} \leq 1$. Enumerate $\Omega_N = \{\omega_i : i \in [2^N]\}$, and let

$$\begin{aligned} a_{ijk} &= A(\mathbf{1}_{\{\omega_i\}}, \mathbf{1}_{\{\omega_j\}}, \mathbf{1}_{\{\omega_k\}}), \\ &\qquad (i,j,k) \in [2^N] \times [2^N] \times [2^N], \\ a_{ijk} &= 0, \quad (i,j,k) \notin [2^N] \times [2^N] \times [2^N]. \end{aligned} \tag{2.19}$$

Define

$$x_{ij} = w_i(\omega_j), \quad (i,j) \in [2^N] \times [2^N], \tag{2.20}$$

whence the characters $w_i \in \hat{\Omega}_N$ $(i \in [2^N])$ can be written as

$$w_i = \sum_j x_{ij}\ \mathbf{1}_{\{\omega_j\}}, \tag{2.21}$$

and

$$\eta(\mathbf{e}_{i_1}, \mathbf{e}_{i_2}, \mathbf{e}_{i_3}) = A(w_{i_1}, w_{i_2}, w_{i_3})$$
$$= \sum_{j_1,j_2,j_3} a_{j_1j_2j_3}\ x_{i_1j_1}\ x_{i_2j_2}\ x_{i_3j_3}. \tag{2.22}$$

Then, by (2.16) (via duality), there exist $B = (b_{ijk})$ so that $\|B\|_{f_{3,\infty}} = 1$, and

$$\left| \sum_{i_1,i_2,i_3} b_{i_1i_2i_3} \sum_{j_1,j_2,j_3} a_{j_1j_2j_3}\ x_{i_1j_1}\ x_{i_2j_2}\ x_{i_3j_3} \right| > K, \tag{2.23}$$

which implies $\|A \otimes B\|_{f_{3,\infty}} > K$. □

3 Projective Boundedness

Projective boundedness, viewed in Chapter VIII in a framework of Hilbert spaces, can be considered also in a framework of F_n-measures:

Definition 5 (cf. Definition VIII.1). Let $(X_1, \mathfrak{A}_1), \ldots, (X_n, \mathfrak{A}_n)$ be measurable spaces. For $\mu \in F_n(\mathfrak{A}_1, \ldots, \mathfrak{A}_n)$, $F_1 \subset B_{\mathrm{L}^\infty(\mathfrak{A}_1)}, \ldots, F_n \subset B_{\mathrm{L}^\infty(\mathfrak{A}_n)}$, let

$$\phi_\mu(f_1, \ldots, f_n) = \int f_1 \otimes \cdots \otimes f_n \, \mathrm{d}\mu,$$
$$(f_1, \ldots, f_n) \in F_1 \times \cdots \times F_n, \tag{3.1}$$

and define

$$\|\mu\|_{\mathrm{pb}_n} = \sup\{\|\phi_\mu\|_{V_n(F_1,\ldots,F_n)} : F_i \subset B_{\mathrm{L}^\infty(\mathfrak{A}_i)}, |F_i| < \infty,\ i \in [n]\}. \tag{3.2}$$

If $\|\mu\|_{\mathrm{pb}_n} < \infty$, then μ is said to be *projectively bounded.* The class of projectively bounded F_n-measures is denoted by

$$PBF_n = PBF_n(\mathfrak{A}_1, \ldots, \mathfrak{A}_n).$$

A projectively bounded form in a Hilbert space setting (defined in Chapter VIII) conveys, in effect, a general Grothendieck-type inequality. Every projectively bounded functional on a Hilbert space can be naturally realized as a projectively bounded Fréchet measure (Exercise 3, cf. (2.17)). For this reason we sometimes refer (somewhat loosely) to projective boundedness as a Grothendieck-type inequality.

The linear space PBF_n equipped with $\|\cdot\|_{\mathrm{pb}_n}$ is a Banach space, and $(PBF_n, \|\cdot\|_{\mathrm{pb}_n}) \subset (F_n, \|\cdot\|_{F_n})$ is a norm-decreasing inclusion (Exercise 4). In one dimension we obviously have $PBF_1 = F_1$, but in higher dimensions,

$$PBF_2 = F_2, \tag{3.3}$$

and

$$PBF_n \subsetneqq F_n, \quad n \geq 3, \tag{3.4}$$

which have been previewed in the previous section, are not quite as obvious.

The theorem below is the link between product Fréchet measures and Grothendieck-type inequalities.

Theorem 6 *Let $(X_1, \mathfrak{A}_1), \ldots, (X_n, \mathfrak{A}_n), (Y_1, \mathfrak{B}_1), \ldots, (Y_n, \mathfrak{B}_n)$ be measurable spaces. If $\mu \in PBF_n(\mathfrak{A}_1, \ldots, \mathfrak{A}_n)$, then $\mu \times \nu \in F_n$ for all $\nu \in F_n(\mathfrak{B}_1, \ldots, \mathfrak{B}_n)$. Conversely, if $\mathfrak{B}_1, \ldots, \mathfrak{B}_n$ are infinite σ-algebras, and $\mu \times \nu \in F_n$ for every $\nu \in F_n(\mathfrak{B}_1, \ldots, \mathfrak{B}_n)$, then $\mu \in PBF_n(\mathfrak{A}_1, \ldots, \mathfrak{A}_n)$.*

Two lemmas are needed for the proof: the first follows from the extension theorem Theorem VI.8, and the second is essentially a restatement of the projective boundedness property. For $\mu \in F_n(\mathfrak{A}_1, \ldots, \mathfrak{A}_n)$ and $\nu \in F_n(\mathfrak{B}_1, \ldots, \mathfrak{B}_n)$, define $\|\mu \times \nu\|_{F_n}$ to be the supremum of

$$\left\| \sum_{(A_1,\ldots,A_n)\in\boldsymbol{\alpha},(B_1,\ldots,B_n)\in\boldsymbol{\beta}} \mu \times \nu((A_1, B_1), \ldots, (A_n, B_n)) r_{A_1 B_1} \otimes \cdots \otimes r_{A_n B_n} \right\|_\infty \tag{3.5}$$

taken over all grids $\boldsymbol{\alpha}$ of $X_1 \times \cdots \times X_n$ and $\boldsymbol{\beta}$ of $Y_1 \times \cdots \times Y_n$. (The Rademacher systems above are indexed respectively by grids of $X_i \times Y_i$, $i \in [n]$.) Justifying this definition of $\|\mu \times \nu\|_{F_n}$, we note below that (3.5) is precisely the F_n-variation of the product F-measure $\mu \times \nu$, whenever the latter exists:

Lemma 7 *$\mu \times \nu$ determines an F_n-measure on*

$$\sigma(\mathfrak{A}_1 \times \mathfrak{B}_1) \times \cdots \times \sigma(\mathfrak{A}_n \times \mathfrak{B}_n)$$

if and only if $\|\mu\times\nu\|_{F_n} < \infty$. *If* $\mu\times\nu \in F_n(\sigma(\mathfrak{A}_1\times\mathfrak{B}_1),\ldots,\sigma(\mathfrak{A}_n\times\mathfrak{B}_n))$, *then* $\|\mu\times\nu\|_{F_n}$ *in* (3.5) *is the* F_n*-variation of* $\mu\times\nu$.

Proof: Necessity follows from Theorem VI.5.

To prove sufficiency, we first extend by finite additivity the domain of $\mu\times\nu$ to $a(\mathfrak{A}_1\times\mathfrak{B}_1)\times\cdots\times a(\mathfrak{A}_n\times\mathfrak{B}_n)$ ($a(\mathfrak{A}\times\mathfrak{B}) :=$ algebra generated by $(\mathfrak{A}\times\mathfrak{B})$). By Theorem 1,

$$\mu\times\nu \in F_n(a(\mathfrak{A}_1\times\mathfrak{B}_1),\ldots,a(\mathfrak{A}_n\times\mathfrak{B}_n)) \quad \text{(Exercises 5, 6)}. \tag{3.6}$$

Then, because the right side of (3.5) is finite,

$$\|\mu\times\nu\|_{F_n(a(\mathfrak{A}_1\times\mathfrak{B}_1),\ldots,a(\mathfrak{A}_n\times\mathfrak{B}_n))} < \infty. \tag{3.7}$$

By Theorem VI.8, $\mu\times\nu \in F_n(\sigma(\mathfrak{A}_1\times\mathfrak{B}_1),\ldots,\sigma(\mathfrak{A}_n\times\mathfrak{B}_n))$, and its F_n-variation equals the right side of (3.5). □

Let $S(\mathfrak{A})$ denote the space of $\mathfrak{A}$-measurable simple functions on $(X,\mathfrak{A})$. (See Chapter VI §6.)

Lemma 8 (Exercise 7 i). *If* $\mu \in F_n(\mathfrak{A}_1,\ldots,\mathfrak{A}_n)$, *then*

$$\|\mu\|_{\mathrm{pb}_n} = \sup\{\|\phi_\mu\|_{V_n(F_1,\ldots,F_n)} : F_i \subset B_{S(\mathfrak{A}_i)}, |F_i| < \infty, i\in[n]\}. \tag{3.8}$$

(*See* (3.1) *for the definition of* ϕ_μ.)

Proof of Theorem 6: For $\mu \in F_n(\mathfrak{A}_1,\ldots,\mathfrak{A}_n)$, $f_1 \in \mathrm{L}^\infty(\mathfrak{A}_1\times\mathfrak{B}_1),\ldots,$ $f_n \in \mathrm{L}^\infty(\mathfrak{A}_n\times\mathfrak{B}_n)$, let

$$\phi_{f_1\ldots f_n;\mu}(y_1,\ldots,y_n) = \int f_1(x_1,y_1)\cdots f_n(x_n,y_n)\mu(\mathrm{d}x_1,\ldots,\mathrm{d}x_n),$$
$$(y_1,\ldots,y_n)\in Y_1\times\cdots\times Y_n. \tag{3.9}$$

The definition of $\phi_{f_1\ldots f_n;\mu}$ is essentially the same as that of ϕ_μ in (3.1): $Y_1,\ldots,Y_n$ here play the role of $F_1,\ldots,F_n$ in (3.1). By Lemma 8, if $\mathfrak{B}_1,\ldots,\mathfrak{B}_n$ are infinite, then

$$\begin{aligned}\|\mu\|_{\mathrm{pb}_n} &= \sup\{\|\phi_{f_1\ldots f_n;\mu}\|_{V_n(\mathfrak{B}_1,\ldots,\mathfrak{B}_n)} : f_i \in S(\mathfrak{A}_i)\otimes S(\mathfrak{B}_i), \|f_i\|_\infty \le 1,\ i\in[n]\}\\ &\quad \text{(Exercise 7 ii)}.\end{aligned} \tag{3.10}$$

Suppose $\|\mu\|_{\text{pb}_n} < \infty$, and $\nu \in F_n(\mathfrak{B}_1, \ldots, \mathfrak{B}_n)$. Let $\alpha_1, \ldots, \alpha_n, \beta_1, \ldots, \beta_n$ be finite partitions of $X_1, \ldots, X_n$, $Y_1, \ldots, Y_n$, respectively, and let $\boldsymbol{\alpha} = \alpha_1 \times \cdots \times \alpha_n$ and $\boldsymbol{\beta} = \beta_1 \times \cdots \times \beta_n$ be the resulting grids. Fix $\omega_i \in \{-1,1\}^{\alpha_i \times \beta_i}$, and define

$$f_{\omega_i} = \sum_{(A,B)\in\alpha_i\times\beta_i} r_{AB}(\omega_i)\, \mathbf{1}_{A\times B}, \quad i \in [n]. \tag{3.11}$$

By the duality $V_n^* = F_n$,

$$\begin{aligned} &\left| \sum_{(A_1,\ldots,A_n)\in\boldsymbol{\alpha},(B_1,\ldots,B_n)\in\boldsymbol{\beta}} \mu \times \nu((A_1,B_1),\ldots,(A_n,B_n)) r_{A_1B_1}(\omega_1)\cdots r_{A_nB_n}(\omega_n) \right| \\ &\quad = \left| \int \phi_{f_{\omega_1}\cdots f_{\omega_n};\mu}\, d\nu \right| \\ &\quad \le \|\phi_{f_{\omega_1}\cdots f_{\omega_n};\mu}\|_{V_n}\, \|\nu\|_{F_n} \le \|\mu\|_{\text{pb}_n} \|\nu\|_{F_n}, \end{aligned} \tag{3.12}$$

which implies

$$\|\mu \times \nu\|_{F_n} \le \|\mu\|_{\text{pb}_n}\, \|\nu\|_{F_n}. \tag{3.13}$$

By Lemma 7, $\mu \times \nu \in F_n(\sigma(\mathfrak{A}_1 \times \mathfrak{B}_1), \ldots, \sigma(\mathfrak{A}_n \times \mathfrak{B}_n))$.

Conversely, suppose $\mu \times \nu \in F_n(\sigma(\mathfrak{A}_1 \times \mathfrak{B}_1), \ldots, \sigma(\mathfrak{A}_n \times \mathfrak{B}_n))$ for every $\nu \in F_n(\mathfrak{B}_1, \ldots, \mathfrak{B}_n)$. Then, there exists $0 < K < \infty$ such that

$$\|\mu \times \nu\|_{F_n} \le K \|\nu\|_{F_n} \tag{3.14}$$

for all $\nu \in F_n(\mathfrak{B}_1, \ldots, \mathfrak{B}_n)$ (Exercise 8). For $i \in [n]$, let

$$f_i = \sum_{(A,B)\in\alpha_i\times\beta_i} a_{AB}\, \mathbf{1}_{A\times B} \tag{3.15}$$

be simple functions on $X_i \times Y_i$, where α_i is a partition of X_i and β_i is a partition of Y_i. Let $\nu \in F_n(\mathfrak{B}_1, \ldots, \mathfrak{B}_n)$ be arbitrary. By (3.14),

$$\begin{aligned} &\left| \int \phi_{f_1\cdots f_n;\mu}\, d\nu \right| \\ &\quad = \left| \sum_{(A_1,\ldots,A_n)\in\boldsymbol{\alpha},\ (B_1,\ldots,B_n)\in\boldsymbol{\beta}} \mu \times \nu((A_1,B_1),\ldots,(A_n,B_n))\, a_{A_1B_1}\cdots a_{A_nB_n} \right| \\ &\quad \le 2^n \|\mu \times \nu\|_{F_n}\, \|f_1\|_\infty \cdots \|f_n\|_\infty \\ &\quad \le 2^n K\, \|\nu\|_{F_n} \|f_1\|_\infty \cdots \|f_n\|_\infty. \end{aligned} \tag{3.16}$$

By the duality $V_n^* = F_n$, this implies

$$\|\phi_{f_1 \dots f_n;\mu}\|_{V_n} \leq 2^n K \ \|f_1\|_\infty \cdots \|f_n\|_\infty, \tag{3.17}$$

which, by (3.10), proves $\|\mu\|_{\mathrm{pb}_n} < \infty$.

4 Every $\mu \in F_2$ is Projectively Bounded

Theorem 9 (cf. Lemma V.4). *If $(X_1, \mathfrak{A}_1)$ and $(X_2, \mathfrak{A}_2)$ are measurable spaces and $\mu \in F_2(\mathfrak{A}_1, \mathfrak{A}_2)$, then*

$$\|\mu\|_{\mathrm{pb}_2} \leq 4\kappa_{\mathrm{G}}^2 \ \|\mu\|_{F_2} \tag{4.1}$$

(κ_{G} := the Grothendieck constant).

Proof: By Lemma 8, it suffices to verify that if $F_1 \subset B_{S(\mathfrak{A}_1)}$ and $F_2 \subset B_{S(\mathfrak{A}_2)}$ are finite, then

$$\|\phi_\mu\|_{V_2(F_1,F_2)} \leq 4\kappa_{\mathrm{G}}^2 \ \|\mu\|_{F_2}. \tag{4.2}$$

To this end, let D_1 be a finite partition of X_1 and let D_2 be a finite partition of X_2 so that every $f \in F_1$ and $g \in F_2$ can be written as

$$f = \sum_{d \in D_1} f(d) \ \mathbf{1}_d, \quad g = \sum_{d \in D_2} g(d) \ \mathbf{1}_d. \tag{4.3}$$

($f(d)$ and $g(d)$ denote the constant values that f and g assume on d.) Then,

$$\begin{aligned} \phi_\mu(f,g) \quad &= \sum_{d_1 \times d_2 \in D_1 \times D_2} \mu(d_1 \times d_2) \ f(d_1) \ g(d_2), \\ & \qquad f \in F_1, \ g \in F_2, \end{aligned} \tag{4.4}$$

determines a bilinear functional on $\mathrm{C}(D_1) \times \mathrm{C}(D_2)$ such that

$$\|\phi_\mu\|_{f_2} \leq 4 \ \|\mu\|_{F_2}. \tag{4.5}$$

By the Grothendieck factorization theorem, there exist probability measures ν_1 on D_1 and ν_2 on D_2 such that

$$\sup\{|\phi_\mu(f,g)| : f \in B_{\mathrm{L}^2(\nu_1)}, g \in B_{\mathrm{L}^2(\nu_1)}\} \leq 4\kappa_{\mathrm{G}} \ \|\mu\|_{F_2}. \tag{4.6}$$

To obtain (4.2), apply the Grothendieck inequality as stated in (III.1.6), or equivalently in (IV.5.37), with $H_1 = \mathrm{L}^2(D_1, \nu_1)$, $H_2 = \mathrm{L}^2(D_2, \nu_2)$, and $\eta = \phi_\mu$. □

5 There Exist Projectively Unbounded F_3-measures

Theorem 10 *If* $(X, \mathfrak{A})$, $(Y, \mathfrak{B})$, *and* $(Z, \mathfrak{C})$ *are measurable spaces with infinite* $\mathfrak{A}$, $\mathfrak{B}$, *and* $\mathfrak{C}$, *then there exists* $\mu \in F_3(\mathfrak{A}, \mathfrak{B}, \mathfrak{C})$ *with* $\|\mu\|_{\mathcal{PB}_3} = \infty$.

We first prove a quantitative version of this theorem. Let Ω_m denote the finite Abelian group $\{-1,1\}^m$, and let $\hat{\Omega}_m$ denote its dual. Define (a variant of the Gauss matrix)

$$\varphi(\omega, w) = w(\omega), \quad w \in \hat{\Omega}_m,\ \omega \in \Omega_m. \tag{5.1}$$

For $\alpha \in l^2(\hat{\Omega}_m)$ and $\beta \in l^2(\Omega_m)$,

$$\left| \sum_{(w,\omega) \in \hat{\Omega}_m \times \Omega_m} \sqrt{\frac{1}{2^m}} \varphi(w,\omega)\ \alpha(w)\ \beta(\omega) \right| \le \|\alpha\|_2\ \|\beta\|_2. \tag{5.2}$$

$\sqrt{1/2^m}\varphi\ \alpha \otimes \beta$ defines a bilinear functional on $\mathrm{C}(\Omega_m) \times \mathrm{C}(\hat{\Omega}_m)$:

$$\sum_{(w,\omega) \in \hat{\Omega}_m \times \Omega_m} \sqrt{\frac{1}{2^m}} \varphi(w,\omega)\ \alpha(w)\ \beta(\omega)\ f(w)\ g(\omega),$$

$$(f,g) \in \mathrm{C}(\Omega)_m \times \mathrm{C}(\hat{\Omega}_m), \tag{5.3}$$

whence

$$\left\| \sqrt{\frac{1}{2^m}} \varphi\ \alpha \otimes \beta \right\|_{F_2(\hat{\Omega}_m \times \Omega_m)} \le \left\| \sqrt{\frac{1}{2^m}} \varphi\ \alpha \otimes \beta \right\|_{f_2}$$

$$\le \|\alpha\|_2 \|\beta\|_2 \quad \text{(Exercise 9)}. \tag{5.4}$$

Lemma 11 *For all scalar-valued functions* ρ *on* $\hat{\Omega}_m$,

$$\|\rho\|_2 \le \|\rho \cdot \varphi\|_{V_2(\hat{\Omega}_m, \Omega_m)} \le \sqrt{2}\ \|\rho\|_2, \tag{5.5}$$

where

$$\rho \cdot \varphi(w,\omega) = \rho(w)\ \varphi(w,\omega), \quad (w,\omega) \in \hat{\Omega}_m \times \Omega_m.$$

Proof: The inequality on the right side of (5.5) follows from the Littlewood mixed (l^1, l^2)-norm inequality (Exercise 10).

To obtain the inequality on the left side, let $\alpha \in B_{l^2(\hat{\Omega}_m)}$ be arbitrary, and from (5.4) deduce

$$\left|\sum_{w\in\hat{\Omega}_m} \rho(w)\ \alpha(w)\right| = \left|\sum_{(w,\omega)\in\hat{\Omega}_m\times\Omega_m} \rho(w)\ \varphi(\omega,w)\frac{\alpha(w)\ \overline{\varphi(\omega,w)}}{2^m}\right|$$

$$\leq \|\rho\cdot\varphi\|_{V_2(\hat{\Omega}_m,\Omega_m)} \left\|\frac{\alpha\varphi}{2^m}\right\|_{f_2} \leq \|\rho\cdot\varphi\|_{V_2(\hat{\Omega}_m,\Omega_m)}. \tag{5.6}$$

The desired inequality follows by maximizing (5.6) over $\alpha\in B_{l^2(\hat{\Omega}_m)}$. □

For a scalar-valued function ρ on $\hat{\Omega}_m$, define the F_3-measure μ_ρ on $\Omega_m \times \Omega_m \times \hat{\Omega}_m$:

$$\mu_\rho(A,B,C) = \sum_{(w,\omega)\in\hat{\Omega}_m\times\Omega_m} \rho(w)\ \varphi(w,\omega)\ \hat{\mathbf{1}}_A(w)\hat{\mathbf{1}}_B(w)\ \hat{\mathbf{1}}_C(\omega),$$

$$A\subset\Omega_m,\ B\subset\Omega_m,\ C\subset\hat{\Omega}_m. \tag{5.7}$$

The lemma below is a quantitative version of Theorem 10.

Lemma 12 $\|\mu_\rho\|_{F_3} \leq \|\rho\|_\infty$ *and* $\|\mu\|_{\mathrm{pb}_3} \geq \|\rho\|_2$.

Proof: If $f\in \mathrm{C}(\Omega_m)$, $g\in\mathrm{C}(\Omega_m)$, and $h\in\mathrm{C}(\hat{\Omega}_m)$, then

$$\begin{aligned}\left|\int f\otimes g\otimes h\ \mathrm{d}\mu_\rho\right| &= \sum_{(w,\omega)\in\hat{\Omega}_m\times\Omega_m} \rho(w)\ \varphi(w,\omega)\hat{f}(w)\ \hat{g}(w)\ \hat{h}(\omega)\\ &= \left|\sum_{w\in\hat{\Omega}_m}\rho(w)\ \hat{f}(w)\ \hat{g}(w)\ h(w)\right|\\ &\leq \|\rho\|_\infty\ \|\hat{f}\|_2\ \|\hat{g}\|_2\ \|h\|_{\mathrm{C}(\hat{\Omega}_m)}\\ &\leq \|\rho\|_\infty\ \|f\|_{\mathrm{C}(\Omega_m)}\ \|g\|_{\mathrm{C}(\Omega_m)}\ \|h\|_{\mathrm{C}(\hat{\Omega}_m)},\end{aligned} \tag{5.8}$$

which implies $\|\mu_\rho\|_{F_3}\leq\|\rho\|_\infty$. The transform $\hat{\mu}_\rho$ is

$$\hat{\mu}_\rho(w_1,w_2,\omega) = \int w_1\otimes w_2\otimes\omega\ \mathrm{d}\mu_\rho,$$

$$(w_1,w_2,\omega)\in\hat{\Omega}_m\times\hat{\Omega}_m\times\Omega_m. \tag{5.9}$$

By an elementary computation,

$$\hat{\mu}_\rho(w_1,w_2,\omega) = \begin{cases}\rho(w)\ \varphi(w,\omega) & \text{if } w_1=w_2=w\\ 0 & \text{otherwise.}\end{cases} \tag{5.10}$$

In the definition of projective boundedness, let $F_1 = F_2 = \hat{\Omega}_m$ (characters on Ω_m), $F_3 = \Omega_m$ (characters on $\hat{\Omega}_m$), and deduce

$$\|\mu_\rho\|_{\mathrm{pb}_3} \geq \|\hat{\mu}_\rho\|_{V_3(\hat{\Omega}_m,\hat{\Omega}_m,\Omega_m)}. \tag{5.11}$$

Therefore, by (5.10) and Lemma 11,

$$\|\mu_\rho\|_{\mathrm{pb}_3} \geq \|\hat{\mu}_\rho\|_{V_3(\hat{\Omega}_m,\hat{\Omega}_m,\Omega_m)} \geq \|\rho\cdot\varphi\|_{V_2(\hat{\Omega}_m,\Omega_m)} \geq \|\rho\|_2. \tag{5.12}$$

□

Proof of Theorem 10: (Exercise 11). For $m \in \mathbb{N}$, let $A_m \subset X$, $B_m \subset Y, C_m \subset Z$ be finite sets so that $|A_m| = |B_m| = |C_m| = 2^m$, and $\{A_m\}, \{B_m\}, \{C_m\}$ are pairwise disjoint. We identify A_m and B_m with Ω_m, and C_m with $\hat{\Omega}_m$, and then, by applying Lemma 12 with $\rho \equiv 1$, we obtain $\mu_m \in F_3(A_m, B_m, C_m)$ (cf. Chapter VI §2 ii) such that

$$\|\mu_m\|_{F_3} \leq 1 \quad \text{and } \|\mu_m\|_{\mathrm{pb}_3} \geq 2^{\frac{m}{2}}. \tag{5.13}$$

Each such μ_m determines an F_3-measure on $\mathfrak{A} \times \mathfrak{B} \times \mathfrak{C}$, which we denote also by μ_m:

$$\mu_m(A, B, C) = \sum_{a\in A_m, b\in B_m, c\in C_m} \mu_m(a, b, c)\ \mathbf{1}_A(a)\mathbf{1}_B(b)\ \mathbf{1}_C(c),$$

$$(A, B, C) \in \mathfrak{A} \times \mathfrak{B} \times \mathfrak{C}. \tag{5.14}$$

Note that these extensions of the μ_m satisfy (5.13). Let $\mu = \Sigma_m\, \mu_m/m^2$. Then, $\mu \in F_3(\mathfrak{A}, \mathfrak{B}, \mathfrak{C})$, and

$$\|\mu\|_{\mathrm{pb}_3} \geq \|\mu_m\|_{\mathrm{pb}_3}/m^2 \geq 2^{\frac{m}{2}}/m^2, \quad m \in \mathbb{N}. \tag{5.15}$$

6 Projective Boundedness in Topological Settings

Let $X_1, \ldots, X_n$ be locally compact Hausdorff spaces, and let $\mathfrak{B}_1, \ldots, \mathfrak{B}_n$ denote their respective Borel fields. As usual, $V_n(X_1, \ldots, X_n)$ denotes the completion of the (algebraic) tensor product $\mathrm{C}_0(X_1) \otimes \cdots \otimes \mathrm{C}_0(X_n)$ in the projective tensor norm, and, similarly, we let $V_n(\mathfrak{B}_1, \ldots, \mathfrak{B}_n)$ be the projective tensor norm-completion of $S(\mathfrak{B}_1) \otimes \cdots \otimes S(\mathfrak{B}_n)$. (See Chapter VI §6, §7.) If $X_1, \ldots, X_n$ are compact, then

$$\begin{aligned} &\mathrm{C}(X_1 \times \cdots \times X_n) \cap V_n(\mathfrak{B}_1, \ldots, \mathfrak{B}_n) \\ &\quad = V_n(X_1, \ldots, X_n) \quad \text{[S, Theorem 4.3]}. \end{aligned} \tag{6.1}$$

(I do not know whether (6.1) holds with non-compact $X_1, \dots, X_n$ and C_0 in place of C (Exercise 12).)

Let $Y_1, \dots, Y_n$ be locally compact Hausdorff spaces with respective Borel fields $\mathfrak{C}_1, \dots, \mathfrak{C}_n$. For

$$\mu \in F_n(\mathfrak{B}_1, \dots, \mathfrak{B}_n) \text{ and } f \in V_n(X_1 \times Y_1, \dots, X_n \times Y_n),$$

define (cf. (3.1), (3.9))

$$\phi_{f;\mu}(y_1, \dots, y_n) = \int f(x_1, y_1, \dots, x_n, y_n)\ \mu(\mathrm{d}x_1, \dots, \mathrm{d}x_n),$$

$$(y_1, \dots, y_n) \in Y_1 \times \cdots \times Y_n. \tag{6.2}$$

Proposition 13 *If*

$$\mu \in PBF_n(\mathfrak{B}_1, \dots, \mathfrak{B}_n) \text{ and } f \in V_n(X_1 \times Y_1, \dots, X_n \times Y_n),$$

then $\phi_{f;\mu} \in V_n(Y_1, \dots, Y_n)$, *and*

$$\|\phi_{f;\mu}\|_{V_n} \leq \|f\|_{V_n}\ \|\mu\|_{\mathrm{pb}_n}. \tag{6.3}$$

Proof: Let $g = g_1 \otimes \cdots \otimes g_n$ be an elementary tensor in $\mathrm{C_c}(X_1 \times Y_1) \otimes \cdots \otimes \mathrm{C_c}(X_n \times Y_n)$ ($\mathrm{C_c}$:= continuous functions with compact support). Then, $\phi_{g;\mu} \in \mathrm{C_c}(Y_1 \times \cdots \times Y_n)$. If X and Y are locally compact spaces, then $V_2(X, Y)$ is dense in $\mathrm{C}_0(X \times Y)$ (Exercise 13), and, therefore, for each $k \in [n]$ there exist sequences $(\varphi_{jk} : j \in \mathbb{N})$ in $S(\mathfrak{B}_k) \otimes S(\mathfrak{C}_k)$ such that $\lim_{j\to\infty} \varphi_{kj} = g_k$ (uniform norm limit). Denote $\theta_j = \varphi_{j1} \otimes \cdots \otimes \varphi_{jn}$. Then,

$$\lim_{j\to\infty} \phi_{\theta_j;\mu} = \phi_{g;\mu} \quad \text{(uniform norm limit)}. \tag{6.4}$$

Note that $(\phi_{\theta_j;\mu} : j \in \mathbb{N})$ is Cauchy in $V_n(\mathfrak{C}_1, \dots, \mathfrak{C}_n)$, and (Exercise 14)

$$\|\phi_{\theta_j;\mu}\|_{V_n(\mathfrak{C}_1,\dots,\mathfrak{C}_n)} \leq \|\varphi_{j1}\|_\infty \cdots \|\varphi_{jn}\|_\infty\ \|\mu\|_{\mathrm{pb}_n},\ j \in \mathbb{N}. \tag{6.5}$$

Therefore,

$$\phi_{g;\mu} \in \mathrm{C_c}(Y_1 \times \cdots \times Y_n) \cap V_n(\mathfrak{C}_1, \dots, \mathfrak{C}_n).$$

By (6.1), $\phi_{g;\mu} \in V_n(Y_1, \dots, V_n)$, and by (6.5),

$$\|\phi_{g;\mu}\|_{V_n} \leq \|g_1\|_\infty \cdots \|g_n\|_\infty\ \|\mu\|_{\mathrm{pb}_n}. \tag{6.6}$$

Suppose $f \in V_n(X_1 \times Y_1, \ldots, X_n \times Y_n)$, and write

$$f = \sum_k g_k, \tag{6.7}$$

where $g_k = g_{1k} \otimes \cdots \otimes g_{nk}$ are elementary tensors in $\mathrm{C_c}(X_1 \times Y_1) \otimes \cdots \otimes \mathrm{C_c}(X_n \times Y_n)$ $(k \in \mathbb{N})$, and

$$\|f\|_{V_n} \leq (1+\epsilon) \sum_k \|g_{1k}\|_\infty \cdots \|g_{nk}\|_\infty \tag{6.8}$$

(arbitrary $\epsilon > 0$). Then

$$\phi_{f;\mu} = \sum_k \phi_{g_k;\mu}, \tag{6.9}$$

and (6.3) follows from (6.6) and (6.8). □

Corollary 14 *If* $\mu \in PBF_n(\mathfrak{B}_1, \ldots, \mathfrak{B}_n)$, $\nu \in F_n(\mathfrak{C}_1, \ldots, \mathfrak{C}_n)$, *and* $f \in V_n(X_1 \times Y_1, \ldots, X_n \times Y_n)$, *then*

$$\int f \,\mathrm{d}(\mu \times \nu) \tag{6.10}$$

$$= \int \left(\int f(x_1, y_1, \ldots, x_n, y_n) \mu(\mathrm{d}x_1, \ldots, \mathrm{d}x_n) \right) \nu(\mathrm{d}y_1, \ldots, \mathrm{d}y_n).$$

We are naturally led to

Definition 15 Let $X_1, \ldots, X_n$ be locally compact Hausdorff spaces with respective Borel fields $\mathfrak{B}_1, \ldots, \mathfrak{B}_n$. Let $Y_1, \ldots, Y_n$ be locally compact Hausdorff spaces, and denote $\tau = (Y_1, \ldots, Y_n)$. We say that μ in $F_n(\mathfrak{B}_1, \ldots, \mathfrak{B}_n)$ is *τ-projectively bounded* if

$$\|\mu\|_{\tau\mathrm{pb}_n} := \sup \|\phi_{f;\mu}\|_{V_n(Y_1,\ldots,Y_n)} < \infty, \tag{6.11}$$

where the supremum is over elementary tensors f in the unit ball of $\mathrm{C_0}(X_1 \times Y_1) \otimes \cdots \otimes \mathrm{C_0}(X_n \times Y_n)$, and $\phi_{f;\mu}$ is defined by (6.2). The class of τ-projectively bounded F-measures on $\mathfrak{B}_1 \times \cdots \times \mathfrak{B}_n$ is denoted by $\tau PBF_n(\mathfrak{B}_1, \ldots, \mathfrak{B}_n)$.

The proof of Proposition 13 yields

$$PBF_n(\mathfrak{B}_1, \ldots, \mathfrak{B}_n) \subset \tau PBF_n(\mathfrak{B}_1, \ldots, \mathfrak{B}_n). \tag{6.12}$$

I suspect the inclusion is proper, but cannot prove it.

Remarks:

i **(a general iterated integral?).** If $\mu \in PBF_n(\mathfrak{B}_1, \ldots, \mathfrak{B}_n)$, $\nu \in F_n(\mathfrak{C}_1, \ldots, \mathfrak{C}_n)$, and $f \in V_n(\sigma(\mathfrak{B}_1 \times \mathfrak{C}_1), \ldots, (\mathfrak{B}_n \times \mathfrak{C}_n))$, then, by Theorem 6, the integral $\int f \, \mathrm{d}(\mu \times \nu)$ is well-defined. (See Chapter VI §6.) A natural question arises: can f be integrated iteratively, as in the topological setting (Corollary 14), first with respect to μ and then ν? The question reduces to this: for $\mu \in PBF_n(\mathfrak{B}_1, \ldots, \mathfrak{B}_n)$, and $E_1 \in \sigma(\mathfrak{B}_1 \times \mathfrak{C}_1), \ldots, E_n \in \sigma(\mathfrak{B}_1 \times \mathfrak{C}_1)$, is

$$\int \mathbf{1}_{E_1}(x_1, y_1) \ldots \mathbf{1}_{E_n}(x_n, y_n) \ \mu(\mathrm{d}x_1, \ldots, \mathrm{d}x_n),$$
$$(y_1, \ldots, y_n) \in Y_1 \times \cdots \times Y_n, \tag{6.13}$$

an element of $V_n(\mathfrak{C}_1, \ldots, \mathfrak{C}_n)$? (I do not know the answer.)

ii **(is projective boundedness stable under products?).**

Lemma 16 (Exercise 15 i; cf. Lemma 7). *Let $X_1, \ldots, X_n, Y_1, \ldots, Y_n$ be locally compact Hausdorff spaces with respective Borel fields $\mathfrak{B}_1, \ldots, \mathfrak{B}_n, \mathfrak{C}_1, \ldots, \mathfrak{C}_n$, $\mu \in F_n(\mathfrak{B}_1, \ldots, \mathfrak{B}_n)$, and $\nu \in F_n(\mathfrak{C}_1, \ldots, \mathfrak{C}_n)$. Then, $\mu \times \nu \in F_n(\sigma(\mathfrak{B}_1 \times \mathfrak{C}_1), \ldots, \sigma(\mathfrak{B}_n \times \mathfrak{C}_n))$ if and only if*

$$\sup\left\{\left|\int \phi_{f;\mu} \, \mathrm{d}\nu\right| : f = f_1 \otimes \cdots \otimes f_n, f_j \in B_{C_0(X_j \times Y_j)}, \ j \in [n]\right\} < \infty. \tag{6.14}$$

Proposition 17 (Exercise 15 ii; cf. Theorem 6, Corollary 14). *Suppose $X_1, \ldots, X_n, Y_1, \ldots, Y_n$ are locally compact Hausdorff spaces with respective Borel fields $\mathfrak{B}_1, \ldots, \mathfrak{B}_n, \mathfrak{C}_1, \ldots, \mathfrak{C}_n$. Let $\mu \in F_n(\mathfrak{B}_1, \ldots, \mathfrak{B}_n)$.*

i. $\|\mu\|_{\tau\mathrm{pb}_n} < \infty$ *if and only if* $\mu \times \nu \in F_n(\sigma(\mathfrak{B}_1 \times \mathfrak{C}_1), \ldots, \sigma(\mathfrak{B}_n \times \mathfrak{C}_n))$ *for all* $\nu \in F_n(\mathfrak{C}_1, \ldots, \mathfrak{C}_n)$.

ii. *If* $\|\mu\|_{\tau\mathrm{pb}_n} < \infty$, $\nu \in F_n(\mathfrak{C}_1, \ldots, \mathfrak{C}_n)$, $f \in V_n(X_1 \times Y_1, \ldots, X_n \times Y_n)$, *then*

$$\int f \, \mathrm{d}(\mu \times \nu) = \int \left(\int f \, \mathrm{d}\mu\right) \mathrm{d}\nu, \tag{6.15}$$

and

$$\left|\int f \, \mathrm{d}(\mu \times \nu)\right| \leq 2^n \, \|f\|_{V_n} \|\mu\|_{\tau \mathrm{pb}_n} \|\nu\|_{F_n}. \tag{6.16}$$

Proposition 18 (Exercise 15 iii). *Suppose $X_1, \ldots, X_n$ are locally compact Hausdorff spaces with respective Borel fields $\mathfrak{B}_1, \ldots, \mathfrak{B}_n$, and let $\tau = (X_1, \ldots, X_n)$. If μ and ν are τ-projectively bounded F_n-measures on $\mathfrak{B}_1 \times \cdots \times \mathfrak{B}_n$, then $\mu \times \nu$ is τ-projectively bounded.*

I do not know whether the same is true in the measurable setting: Suppose $(X_1, \mathfrak{A}_1), \ldots, (X_n, \mathfrak{A}_n), (Y_1, \mathfrak{B}_1), \ldots, (Y_n, \mathfrak{B}_n)$ are measurable spaces, $\mu \in PBF_n(\mathfrak{A}_1, \ldots, \mathfrak{A}_n)$, and $\nu \in PBF_n(\mathfrak{B}_1, \ldots, \mathfrak{B}_n)$. Is $\mu \times \nu \in PBF_n(\sigma(\mathfrak{A}_1 \times \mathfrak{B}_1), \ldots, \sigma(\mathfrak{A}_n \times \mathfrak{B}_n))$?

7 Projective Boundedness in Topological-group Settings

In this section, $X_1, \ldots, X_n$ are locally compact Abelian groups with respective Borel fields $\mathfrak{B}_1, \ldots, \mathfrak{B}_n$. Let μ be a bounded n-linear functional on $\mathrm{C}_0(X_1) \times \cdots \times \mathrm{C}_0(X_n)$ represented as an F_n-measure μ on $\mathfrak{B}_1 \times \cdots \times \mathfrak{B}_n$ (Theorem VI.12), and define its transform

$$\begin{aligned} \hat{\mu}(\gamma_1, \ldots, \gamma_n) &= \int \gamma_1 \otimes \cdots \otimes \gamma_n \, \mathrm{d}\mu, \\ & \qquad (\gamma_1, \ldots, \gamma_n) \in \hat{X}_1 \times \cdots \times \hat{X}_n. \end{aligned} \tag{7.1}$$

(Notice that the intervention of the multilinear Riesz representation theorem is essential: it provides the extension of the n-linear functional μ to $\mathrm{L}^\infty(\mathfrak{B}_1) \times \cdots \times \mathrm{L}^\infty(\mathfrak{B}_n)$, making possible the evaluation of μ at $(\gamma_1, \ldots, \gamma_n) \in \hat{X}_1 \times \cdots \times \hat{X}_n$.)

Proposition 19 (Exercise 16).

i. *If $\mu \in F_n(\mathfrak{B}_1, \ldots, \mathfrak{B}_n)$, then $\hat{\mu}$ is bounded and uniformly continuous on $\hat{X}_1 \times \cdots \times \hat{X}_n$ separately in each coordinate.*
ii. *If $\mu \in F_2(\mathfrak{B}_1, \mathfrak{B}_2)$, then $\hat{\mu}$ is bounded and uniformly continuous on $X_1 \times X_2$.*

A fundamental issue arises in the harmonic-analytic setting: is convolution in $F_1(\sigma(\mathfrak{B}_1 \times \cdots \times \mathfrak{B}_n))$ extendible to $F_n(\mathfrak{B}_1, \ldots, \mathfrak{B}_n)$? We give two equivalent definitions of convolution in the multidimensional framework (Exercise 17), each mimicking a standard definition in the one-dimensional case.

The first extends the construction in [Ru3, pp. 14–15]. If

$$\mu \in PBF_n(\mathfrak{B}_1, \ldots, \mathfrak{B}_n) \text{ and } \nu \in F_n(\mathfrak{B}_1, \ldots, \mathfrak{B}_n),$$

then, by applying Theorem 6, we define $\mu \star \nu$ to be the F_n-measure

$$\begin{aligned} &\mu \star \nu(E_1, \ldots, E_n) \\ &\quad = \int \mathbf{1}_{E_1}(x_1 + y_1) \cdots \mathbf{1}_{E_n}(x_n + y_n) \mu \times \nu(\mathrm{d}(x_1, y_1), \ldots, \mathrm{d}(x_n, y_n)), \\ &\quad E_1 \in \mathfrak{B}_1, \ldots, E_n \in \mathfrak{B}_n. \end{aligned} \tag{7.2}$$

The second definition extends the construction in [Kat, p. 41]. Suppose $\mu \in \tau PBF_n(\mathfrak{B}_1, \ldots, \mathfrak{B}_n)$, where $\tau = (X_1, \ldots, X_n)$ (Definition 15), and $\nu \in F_n(\mathfrak{B}_1, \ldots, \mathfrak{B}_n)$. (The stipulation $\mu \in \tau PBF_n$ is ostensibly weaker than $\mu \in PBF_n$.) Define a linear functional ϕ on $V_n(X_1, \ldots, X_n)$,

$$\begin{aligned} \phi(f) = \int \left(\int f(x_1 + y_1, \ldots, x_n + y_n)\, \mu(\mathrm{d}x_1, \ldots, \mathrm{d}x_n) \right) \nu(\mathrm{d}y_1, \ldots, \mathrm{d}y_n), \\ f \in V_n(X_1, \ldots, X_n). \end{aligned} \tag{7.3}$$

Then (cf. Proposition 17),

$$|\phi(f)| \leq \|f\|_{V_n} \|\mu\|_{\tau \mathrm{pb}_n} \|\nu\|_{f_n} \tag{7.4}$$

or

$$|\phi(f)| \leq 2^n \|f\|_{V_n} \|\mu\|_{\tau \mathrm{pb}_n} \|\nu\|_{F_n} \tag{7.5}$$

($\|\nu\|_{f_n} :=$ the norm of ν in V_n^*). Define $\mu \star \nu$ to be the F_n-measure on $\mathfrak{B}_1 \times \cdots \times \mathfrak{B}_n$ representing ϕ. This definition of convolution leads to

Definition 20 (Exercise 18). Let $X_1, \ldots, X_n$ be locally compact Abelian groups with respective Borel fields $\mathfrak{B}_1, \ldots, \mathfrak{B}_n$. For

$$\mu \in F_n(\mathfrak{B}_1, \ldots, \mathfrak{B}_n), \quad \text{elementary tensor } f \in \mathrm{C}_0(X_1^2) \otimes \cdots \otimes \mathrm{C}_0(X_n^2),$$

denote

$$\begin{aligned} \Psi_{f;\mu}(y_1, \ldots, y_n) = \int f(x_1 + y_1, \ldots, x_n + y_n) \mu(\mathrm{d}x_1, \ldots, \mathrm{d}x_n), \\ (y_1, \ldots, y_n) \in X_1 \times \cdots \times X_n, \end{aligned} \tag{7.6}$$

and define

$$\|\mu\|_{\mathrm{gpb}_n} = \sup\{\|\Psi_{f;\mu}\|_{V_n} : \text{elementary tensors}$$
$$f \in \mathrm{C}_0(X_1 \times X_1) \otimes \cdots \otimes \mathrm{C}_0(X_n \times X_n), \|f\|_\infty \leq 1\}. \quad (7.7)$$

If $\|\mu\|_{\mathrm{gpb}_n} < \infty$, then $\mu \in F_n(\mathfrak{B}_1, \ldots, \mathfrak{B}_n)$ is said to be g-projectively bounded, and the class of such μ is designated by $gPBF_n(\mathfrak{B}_1, \ldots, \mathfrak{B}_n)$.

If $\mu \in gPBF_n$ and $\nu \in F_n$, then $\mu \star \nu$ is the F_n-measure representing the bounded linear functional in (7.3). Note that

$$(\mu \star \nu)\hat{} = \hat{\mu}\,\hat{\nu} \quad \text{(Exercise 19).} \tag{7.8}$$

Indeed, we can start with (7.8) as the definition if there exists $\lambda \in F_n(\mathfrak{B}_1, \ldots, \mathfrak{B}_n)$ such that $\hat{\lambda} = \hat{\mu}\,\hat{\nu}$ for $\mu \in F_n(\mathfrak{B}_1, \ldots, \mathfrak{B}_n)$ and $\nu \in F_n(\mathfrak{B}_1, \ldots, \mathfrak{B}_n)$, then the convolution $\mu \star \nu$ is defined to be this λ.

Proposition 21 (Exercise 20). *$\mu \in gPBF_n(\mathfrak{B}_1, \ldots, \mathfrak{B}_n)$ if and only if for every $\nu \in F_n(\mathfrak{B}_1, \ldots, \mathfrak{B}_n)$ there exists $\lambda \in F_n(\mathfrak{B}_1, \ldots, \mathfrak{B}_n)$ such that $\hat{\lambda} = \hat{\mu}\,\hat{\nu}$. (Elements of $gPBF_n(\mathfrak{B}_1, \ldots, \mathfrak{B}_n)$ will be called convolvers.)*

By Theorem 9, $gPBF_2(\mathfrak{B}_1, \mathfrak{B}_2) = F_2(\mathfrak{B}_1, \mathfrak{B}_2)$, which implies that convolution in $F_1(\sigma(\mathfrak{B}_1, \mathfrak{B}_2))$ is canonically extendible to $F_2(\mathfrak{B}_1, \mathfrak{B}_2)$. However, the three-dimensional case is fundamentally different (cf. Theorem 10):

Lemma 22 *For all $K > 0$ there exist discrete measures on $X_1 \times X_2 \times X_3$ such that $\|\mu\|_{F_3} \leq 1$ and $\|\mu\|_{\mathrm{gpb}_3} \geq K$.*

Proof: For every $K > 0$, there exist $N > 0$ and $\mu \in F_3([N], [N], [N])$ such that $\|\mu\|_{F_3} \leq 1$ and $\|\mu\|_{\mathrm{pb}_3} \geq K$ (Theorem 4 ii, or Lemma 12). This means: there exist scalar 3-arrays $\mu = \{\mu_{xyz} : (x, y, z) \in [N]^3\}$, and 2-arrays $a \in B_{l^\infty([N]^2)}$, $b \in B_{l^\infty([N]^2)}$, $c \in B_{l^\infty([N]^2)}$, such that

$$\|\mu\|_{F_3} \leq 1, \tag{7.9}$$

and

$$\|\phi_{a,b,c;\mu}\|_{V_3([N],[N],[N])} \geq K, \tag{7.10}$$

where

$$\phi_{a,b,c;\mu}(i,j,k) = \sum_{(x,y,z)\in[N]^3} \mu_{xyz}\ a_{xi}\ b_{yj}\ c_{zk}, \quad (i,j,k) \in [N]^3. \tag{7.11}$$

Suppose F_l and G_l are disjoint and mutually independent N-subsets of X_l, $l = 1,2,3$,

$$F_l = \{s_{jl} : j \in [N]\}, \quad G_l = \{t_{jl} : j \in [N]\}, \quad l = 1,2,3. \tag{7.12}$$

(Mutually independent subsets F and G means that if $x_1 + s_1 = x_2 + s_2$ for (x_1, s_1) and (x_2, s_2) in $F \times G$, then $x_1 = x_2$ and $s_1 = s_2$.) Define a discrete measure on $X_1 \times X_2 \times X_3$ by

$$\mu = \sum_{(x,y,z)\in[N]^3} \mu_{xyz}\ \delta_{(s_{x1},s_{y2},s_{z3})}. \tag{7.13}$$

By (7.9), $\|\mu\|_{F_3(\mathfrak{B}_1,\mathfrak{B}_2,\mathfrak{B}_3)} \leq 1$. Because F_l and G_l are mutually independent ($l = 1,2,3$), there exist $f \in \mathrm{C}_0(X_1)$, $g \in \mathrm{C}_0(X_2)$, and $h \in \mathrm{C}_0(X_3)$ such that for $(i,j) \in [N]^2$,

$$f(s_{i1} + t_{j1}) = a_{ij}, \quad g(s_{i2} + t_{j2}) = b_{ij}, \quad h(s_{i3} + t_{j3}) = c_{ij}. \tag{7.14}$$

Then,

$$\phi_{a,b,c;\mu}(i,j,k) = \Psi_{f\otimes g\otimes h;\mu}(t_{i1}, t_{j2}, t_{k3}), \quad (i,j,k) \in [N]^3, \tag{7.15}$$

where $\phi_{a,b,c;\mu}$ is defined in (7.11) and $\Psi_{f\otimes g\otimes h;\mu}$ is defined in (7.6). By (7.10), $\|\Psi_{f\otimes g\otimes h;\mu}\|_{V_3} \geq K$, and therefore $\|\mu\|_{\mathrm{gpb}_3} \geq K$. □

Corollary 23 (Exercise 21). *If X_1, X_2, and X_3 are infinite locally compact Abelian groups with respective Borel fields $\mathfrak{B}_1, \mathfrak{B}_2$ and $\mathfrak{B}_3$, then convolution in $F_1(\sigma(\mathfrak{B}_1 \times \mathfrak{B}_2 \times \mathfrak{B}_3))$ is not extendible to $F_3(\mathfrak{B}_1, \mathfrak{B}_2, \mathfrak{B}_3)$.*

Remarks:

i **(are containments proper?).** Let $X_1, \ldots, X_n$ be locally compact Abelian groups with respective infinite Borel fields $\mathfrak{B}_1, \ldots, \mathfrak{B}_n$. We have already noted (cf. (6.12))

$$PBF_n(\mathfrak{B}_1,\ldots,\mathfrak{B}_n) \subset \tau PBF_n(\mathfrak{B}_1,\ldots,\mathfrak{B}_n) \subset gPBF_n(\mathfrak{B}_1,\ldots,\mathfrak{B}_n). \tag{7.16}$$

That $PBF_n(\mathfrak{B}_1, \ldots, \mathfrak{B}_n) \subsetneqq gPBF_n(\mathfrak{B}_1, \ldots, \mathfrak{B}_n)$ is a consequence of constructions in the previous chapter. (See the next section.) However, I do not know (and only suspect) that both inclusions in (7.16) are proper.

The Banach spaces $qPBF_n$ and τPBF_n equipped with convolution are Banach algebras (Exercise 22), but I do not know that $PBF_n(\mathfrak{B}_1, \ldots, \mathfrak{B}_n)$ equipped with convolution is also a Banach algebra.

It is easy to verify that

$$F_1(\sigma(\mathfrak{B}_1 \times \cdots \times \mathfrak{B}_n)) \subset PBF_n(\mathfrak{B}_1, \ldots, \mathfrak{B}_n) \tag{7.17}$$

for $n \geq 1$, and, therefore, that every F_1-measure on $\sigma(\mathfrak{B}_1 \times \cdots \times \mathfrak{B}_n)$ is a convolver in $F_n(\mathfrak{B}_1, \ldots, \mathfrak{B}_n)$. In the case $n = 2$, every F_2-measure on $\mathfrak{B}_1 \times \mathfrak{B}_2$ is a convolver in $F_2(\mathfrak{B}_1, \mathfrak{B}_2)(= PBF_2(\mathfrak{B}_1, \mathfrak{B}_2))$, but this does not extend to higher dimensions [BlCag]: there exists $\mu \in F_2(\sigma(\mathfrak{B}_1 \times \mathfrak{B}_2), \mathfrak{B}_3)$ such that $\mu \notin gPBF_3(\mathfrak{B}_1, \mathfrak{B}_2, \mathfrak{B}_3)$ (Exercise 23).

ii **(credits).** The feasibility of convolution of bounded bilinear functionals (bimeasures) on $C_0(X) \times C_0(Y)$, where X and Y are locally compact Abelian groups, was first observed in [GrSch1, §2]. (The key was an answer by Pisier [GrSch1, p. 91] to a question concerning a characterization of Fourier transforms of bimeasures [GrMc, p. 313].) The two definitions of convolution in F_2 based on (7.2) and (7.3) are different from the one in [GrSch1, §2], and resemble the definition in [GiISch].

That convolution could not be extended from $F_1(\sigma(\mathfrak{A} \times \mathfrak{B} \times \mathfrak{C}))$ to the entire space $F_3(\mathfrak{A}, \mathfrak{B}, \mathfrak{C})$, where $\mathfrak{A}, \mathfrak{B}, \mathfrak{C}$ are the Borel fields of locally compact infinite Abelian groups, was shown in [GrSch2, Theorem 6]. The proof in [GrSch2], like the proof here (Exercise 21, 23), was based on a quantitative version of this phenomenon; see [GrSch2, pp. 23–5]. The quantitative versions obtained here are tied to projectively unbounded F_3-measures (Theorem 10), which, in turn, can be obtained either from Lemm 12 (cf. Exercise 23 iv), or from the failure of a trilinear Grothendieck-type inequality (as in the proof of Theorem 4 ii).

Product F-measures had been previewed in the stochastic framework of [Bl6], and appeared in general multidimensional settings in [Bl8]. The convolution of projectively bounded F_n-measures on locally compact Abelian groups was noted also in [Bl8].

8 Examples

If $X_1, \ldots, X_n$ are compact Abelian groups with respective Borel fields $\mathfrak{B}_1, \ldots, \mathfrak{B}_n$, Haar measures $\mathfrak{m}_1, \ldots, \mathfrak{m}_n$, and dual groups $\hat{X}_1, \ldots, \hat{X}_n$, then the evaluation of $\mu \in F_n(\mathfrak{B}_1, \ldots, \mathfrak{B}_n)$ at an $(\hat{X}_1 \times \cdots \times \hat{X}_n)$-trigonometric polynomial f (necessarily an element of $V_n(X_1, \ldots, X_n)$) is

$$\int f \, \mathrm{d}\mu = \sum_{(\hat{x}_1, \ldots, \hat{x}_n) \in \hat{X}_1 \times \cdots \times \hat{X}_n} \hat{f}(\hat{x}_1, \ldots, \hat{x}_n) \, \hat{\mu}(\hat{x}_1, \ldots, \hat{x}_n). \tag{8.1}$$

This representation (Parseval's formula) extends to arbitrary $f \in V_n(X_1, \ldots, X_n)$, *provided* f is convolved with a summability kernel (k_j) in $\mathrm{L}^1(X_1 \times \cdots \times X_n,\ \mathbf{m})$, where $\mathbf{m} = \mathfrak{m}_1 \times \cdots \times \mathfrak{m}_n$ (cf. Definition VII.5):

$$\int f \, \mathrm{d}\mu = \lim_{j \to \infty} \sum_{(\hat{x}_1, \ldots, \hat{x}_n) \in \hat{X}_1 \times \cdots \times \hat{X}_n} \hat{k}_j(\hat{x}_1, \ldots, \hat{x}_n) \hat{f}(\hat{x}_1, \ldots, \hat{x}_n) \, \hat{\mu}(\hat{x}_1, \ldots, \hat{x}_n). \tag{8.2}$$

If μ acts boundedly on $\mathrm{L}^2(X_1 \times \cdots \times X_n, \mathbf{m})$, then (*a fortiori*) it acts boundedly on $V_n(X_1, \ldots, X_n)$, and, in this case, we can dispense with the summability kernel on the right side of (8.2):

$$\int f \, \mathrm{d}\mu = \sum_{(\hat{x}_1, \ldots, \hat{x}_n) \in \hat{X}_1 \times \cdots \times \hat{X}_n} \hat{f}(\hat{x}_1, \ldots, \hat{x}_n) \, \hat{\mu}(\hat{x}_1, \ldots, \hat{x}_n), \tag{8.3}$$

where the sum on the right side is performed iteratively. We use this comment and results of the previous chapter to produce examples of F_n-measures that are convolvers but are not projectively bounded.

Take $n = 3$, and assume that X, Y, and Z are infinite compact Abelian groups with respective Borel fields $\mathfrak{A}$, $\mathfrak{B}$, and $\mathfrak{C}$. Choose countably infinite spectral sets $E \subset \hat{X}$, $F \subset \hat{Y}$, and $G \subset \hat{Z}$, and enumerate them

$$E = \{\hat{x}_{ij} : (i,j) \in \mathbb{N}^2\}, \quad F = \{\hat{y}_{ij} : (i,j) \in \mathbb{N}^2\},$$
$$G = \{\hat{z}_{ij} : (i,j) \in \mathbb{N}^2\}. \tag{8.4}$$

For $\varphi \in l^\infty(\mathbb{N}^3)$, define

$$\mu_\varphi(A,B,C) = \sum_{i,j,k} \varphi(i,j,k)\ \hat{\mathbf{1}}_A(\hat{x}_{ij})\ \hat{\mathbf{1}}_B(\hat{y}_{jk})\ \hat{\mathbf{1}}_C(\hat{z}_{ik}),$$

$$A \in \mathfrak{A},\ B \in \mathfrak{B},\ C \in \mathfrak{C}. \tag{8.5}$$

Then (Exercise 24), $\mu_\varphi \in F_3(\mathfrak{A},\mathfrak{B},\mathfrak{C})$, $\|\mu_\varphi\|_{F_3} \leq \|\varphi\|_\infty$, and

$$\int f\ \mathrm{d}\mu_\varphi = \sum_{i,j,k} \varphi(i,j,k)\ \hat{f}(\hat{x}_{ij},\hat{y}_{jk},\hat{z}_{ik}), \quad f \in V_3(X,Y,Z). \tag{8.6}$$

Consider the spectral subset of $\hat{X} \times \hat{Y} \times \hat{Z}$ (cf. (VIII.5.8))

$$\mathbf{E}^U = \{(\hat{x}_{ij},\hat{y}_{jk},\hat{z}_{ik}) : (i,j,k) \in \mathbb{N}^3\}. \tag{8.7}$$

We redefine (for bookkeeping purposes)

$$\tilde{\varphi}(\hat{x}_{ij},\hat{y}_{jk},\hat{z}_{ik}) = \varphi(i,j,k), \quad (i,j,k) \in \mathbb{N}^3, \tag{8.8}$$

and deduce from (8.6)

$$\hat{\mu}_\varphi = \tilde{\varphi}\ \mathbf{1}_{E^U}. \tag{8.9}$$

Theorem 24

i. *μ_φ is a convolver for every $\varphi \in l^\infty(\mathbb{N}^3)$.*
ii. *There exists $\varphi \in l^\infty(\mathbb{N}^3)$ such that μ_φ is not projectively bounded.*

Proof: Let $\nu \in F_3(\mathfrak{A},\mathfrak{B},\mathfrak{C})$ be arbitrary, and consider

$$\tilde{\phi}(\hat{x}_{ij},\hat{y}_{jk},\hat{z}_{ik}) := \phi(i,j,k) := \hat{\mu}_\varphi(\hat{x}_{ij},\hat{y}_{jk},\hat{z}_{ik})\hat{\nu}(\hat{x}_{ij},\hat{y}_{jk},\hat{z}_{ik}),$$
$$(i,j,k) \in \mathbb{N}^3. \tag{8.10}$$

Now observe that $\hat{\mu}_\varphi\hat{\nu} = \tilde{\phi}\ \mathbf{1}_{E^U} = \hat{\mu}_\phi$. This proves Part i. (See (7.8) and the comment following it.)

To prove Part ii, produce $\varphi \in l^\infty(\mathbb{N}^3)$ such that $\varphi \notin B(R^U)$, as per Theorem VIII.17, and apply Theorem VIII.14. □

Remarks:

i **(a characterization?).** If $X_1, \ldots, X_n$ are compact Abelian groups with respective Borel fields $\mathfrak{B}_1, \ldots, \mathfrak{B}_n$, and $\mu \in PBF_n(\mathfrak{B}_1, \ldots, \mathfrak{B}_n)$, then $\hat{\mu} \in \tilde{V}_n(\hat{X}_1, \ldots, \hat{X}_n)$ (Exercise 25). Is the converse true? This is an open question closely related to the problem in Remark ii, Chapter VIII §9.

ii **(L^2-factorizability and complete boundedness).** A bounded n-linear functional μ on $C_0(X_1) \times \cdots \times C_0(X_n)$, where $X_1, \ldots, X_n$ are locally compact Hausdorff spaces, is said to be L^2-*factorizable* if there exist $0 < K < \infty$ and probability measures $\nu_1, \ldots, \nu_n$ on the respective Borel fields of $X_1, \ldots, X_n$ such that

$$\left| \int f_1 \otimes \cdots \otimes f_n \, d\mu \right| \leq K \, \|f_1\|_{L^2(\nu_1)} \cdots \|f_n\|_{L^2(\nu_n)},$$

$$f_1 \in C_0(X_1), \ldots, f_n \in C_0(X_n). \tag{8.11}$$

Every bounded bilinear functional on $C_0(X_1) \times C_0(X_2)$ is projectively bounded and L^2-factorizable. (See §1.) In higher dimensions, matters are fundamentally different: if $n > 2$, then there exist L^2-factorizable n-linear functionals that are projectively unbounded (Exercise 26). I do not know whether every projectively bounded n-linear functional is L^2-factorizable.

A bounded n-linear functional μ on $C_0(X_1) \times \cdots \times C_0(X_n)$ is said to be *completely bounded* if there exist a Hilbert space H, *-representations $\pi_1 : C_0(X_1) \mapsto \mathscr{B}(H), \ldots, \pi_n : C_0(X_n) \mapsto \mathscr{B}(H)$, $\mathbf{x} \in H$, $\mathbf{y} \in H$, such that

$$\int f_1 \otimes \cdots \otimes f_n \, d\mu = \langle \pi_1(f_1) \ldots \pi_n(f_n)\mathbf{x}, \mathbf{y} \rangle, \tag{8.12}$$

where $\langle \cdot, \cdot \rangle$:= inner product in H, and $\mathscr{B}(H)$:= bounded linear operators on H. (See [ChrEfSin, Corollary 3.2].) If μ is L^2-factorizable, then μ is completely bounded, and for $n > 2$, there exist completely bounded μ that are not L^2-factorizable [Sm]. Following Theorem 24, because μ_φ defined in (8.5) is L^2-factorizable for every $\varphi \in l^\infty(\mathbb{N}^3)$ (cf. Exercise 26), there exist completely bounded n-linear functionals that are projectively unbounded. I do not know whether every projectively bounded n-linear functional is completely bounded.

If $X_1, \ldots, X_n$ are locally compact Abelian groups, and μ, ν are completely bounded n-linear functionals on $C_0(X_1) \times \cdots \times C_0(X_n)$, then the convolution $\mu \star \nu$ exists and is a completely bounded n-linear functional [Y2], [ZSch]. It is unknown whether every completely bounded n-linear functional is a convolver.

Exercises

1. i. For scalar matrices $A = (a_{ij})$ and $B = (b_{ij})$ of finite rank, define

$$\|A \otimes B\|_{f2,(2,\infty)}$$
$$= \sup \left\{ \left| \sum_{i,j,m,n} a_{ij} b_{mn} x_{im} y_{jn} \right| : \sup_m \Sigma_i |x_{im}|^2 \leq 1, \ \sup_n \Sigma_j |y_{jn}|^2 \leq 1 \right\}.$$

Verify that

$$\|A \otimes B\|_{f_2,(2,\infty)} \leq K \|A\|_{f_2,2} \|B\|_{f_2,\infty},$$

where $0 < K < \infty$ is an absolute constant, is equivalent to the Grothendieck inequality.

ii.* Prove or disprove: for all $p \in (2, \infty)$,

$$\|A \otimes B\|_{f_2,p} \leq K_p \ \|A\|_{f_2,p} \ \|B\|_{f_2,p},$$

where $K_p > 0$ depends only on p.

2. Let $A = (a_{ij})$ and $B = (b_{ij})$ be scalar matrices. Define $A \cdot B = (c_{ij})$ by

$$c_{ij} = a_{ij} \ b_{ij}, \quad (i,j) \in \mathbb{N}^2.$$

(The matrix $A \cdot B$ is called the Schur product of A and B [Schu1].)

i. Prove that $\|A \cdot B\|_{f_2,2} \leq \|A\|_{f_2,2} \|B\|_{f_2,2}$.

ii. It is demonstrated in [V4, Proposition 3.1] that the assertion in i cannot be extended to the trilinear case; i.e., for every $k > 0$ there exist scalar 3-arrays $A = (a_{ijk})$ and $B = (b_{ijk})$ such that $\|A\|_{f_3,2} \leq 1, \|B\|_{f_3,2} \leq 1$, and $\|A \cdot B\|_{f_3,2} \geq K$.

Prove Theorem 4 i by this result.

3. i. Let η be a bounded n-linear functional on

$$\mathrm{L}^2(X_1, \nu_1) \times \cdots \times \mathrm{L}^2(X_n, \nu_n),$$

where ν_i is a finite positive measure on $(X_i, \mathfrak{A}_i)$ ($i \in [n]$). Verify that

$$\eta(\mathbf{1}_{A_1}, \ldots, \mathbf{1}_{A_n}), \quad (A_1, \ldots, A_n) \in \mathfrak{A}_1 \times \cdots \times \mathfrak{A}_n, \tag{E.1}$$

defines an F_n-measure on $\mathfrak{A}_1 \times \cdots \times \mathfrak{A}_n$, which we denote also by η. Show that if $f_1 \in \mathrm{L}^\infty(X_1), \ldots, f_n \in \mathrm{L}^\infty(X_n)$, then

$$\eta(f_1, \ldots, f_n) = \int f_1 \otimes \cdots \otimes f_n \, \mathrm{d}\eta.$$

(Left side is the action of η on $\mathrm{L}^2(X_1, \nu_1) \times \cdots \times \mathrm{L}^2(X_n, \nu_n)$, and right side is the integral with respect to the F_n-measure η defined in (E.1).)

Conclude that if η is a projectively bounded n-linear functional on $\mathrm{L}^2(X_1, \nu_1) \times \cdots \times \mathrm{L}^2(X_n, \nu_n)$, according to Definition VIII.1, then η is a projectively bounded F_n-measure, according to Definition 5.

ii.* Can every projectively bounded F_n-measure be realized as a projectively bounded multilinear functional on a Hilbert space?

4. Verify that $\|\cdot\|_{\mathrm{pb}_n}$ defines a norm, that $(PBF_n, \|\cdot\|_{\mathrm{pb}_n})$ is a Banach space, and that $\|\mu\|_{F_n} \le \|\mu\|_{\mathrm{pb}_n}$ for $\mu \in F_n$.
5. Verify (3.6): show that the extension of $\mu \times \nu$ to

$$a(\mathfrak{A}_1 \times \mathfrak{B}_1) \times \cdots \times a(\mathfrak{A}_n \times \mathfrak{B}_n),$$

where $\mu \times \nu$ is defined by (1.2), is an F_n-measure on $a(\mathfrak{A}_1 \times \mathfrak{B}_1) \times \cdots \times a(\mathfrak{A}_n \times \mathfrak{B}_n)$.

6. In this exercise you will verify that Theorem 1 is essential for the proof of Lemma 7. Specifically, you will establish existence of σ-algebras $\mathfrak{A}$ and $\mathfrak{B}$ and a finitely additive positive set-function μ on $a(\mathfrak{A} \times \mathfrak{B})$ such that

$$\mu \in F_2(\mathfrak{A}, \mathfrak{B}), \tag{E.2}$$

(hence μ has finite total variation), but

$$\mu \notin F_1(a(\mathfrak{A} \times \mathfrak{B})). \tag{E.3}$$

(The example was shown to me by J. Schmerl.)

i. Let $\mathfrak{B}$ denote the usual Borel field in [0,1], and let λ denote Lebesgue measure on $\mathfrak{B}$. Let $X \subset [0,1]$ be such that

$$A \cap X \neq \emptyset \text{ and } A \cap X^c \neq \emptyset \text{ for all } A \in \mathfrak{B} \text{ with } \lambda(A) > 0.$$

Establish existence of such sets X, which are necessarily non-measurable.

ii. Let $Y = X^c$, and denote $\mathfrak{B}_X = \{X \cap A : A \in \mathfrak{B}\}$ and $\mathfrak{B}_Y = \{Y \cap A : A \in \mathfrak{B}\}$. If $E \in \mathfrak{B}_X$ and $F \in \mathfrak{B}_Y$, then define $\mu(E, F) = \lambda(A \cap B)$, where $E = X \cap A$ $(A \in \mathfrak{B})$ and $F = X \cap B$ $(B \in \mathfrak{B})$. Verify that μ is well-defined and that $\mu \in F_2(\mathfrak{B}_X, \mathfrak{B}_Y)$.

iii. Let $D = \{(x, x) : x \in [0, 1]\}$. Prove that there exists a collection of pairwise disjoint rectangles $\{I_k \times J_k : k \in \mathbb{N}\}$, such that

$$[0,1] \times [0,1] \backslash D = \bigcup_k I_k \times J_k.$$

Let $E_k = I_k \cap X$ and $F_k = J_k \cap Y$, and observe that $X \times Y = \bigcup_k E_k \times F_k$ and that $\mu(E_k, F_k) = 0$ for all $k \in \mathbb{N}$.

7. i. Prove (Lemma 8): if $\mu \in F_n(\mathfrak{A}_1, \ldots, \mathfrak{A}_n)$, then

$$\|\mu\|_{\mathrm{pb}_n} = \sup\{\|\phi_\mu\|_{V_n(F_1,\ldots,F_n)} : F_i \subset B_{S(\mathfrak{A}_i)}, |F_i| < \infty, i \in [n]\}.$$

ii. Prove (3.10) (in the proof of Theorem 6):

$$\|\mu\|_{\mathrm{pb}_n}$$

$$= \sup\{\|\phi_{f_1\cdots f_n;\mu}\|_{V_n(\mathfrak{B}_1,\ldots,\mathfrak{B}_n)} : f_i \in S(\mathfrak{A}_i) \otimes S(\mathfrak{B}_i), \|f_i\|_\infty \leq 1, i \in [n]\}.$$

8. Verify that if $\mu \times \nu \in F_n(\sigma(\mathfrak{A}_1 \times \mathfrak{B}_1), \ldots, \sigma(\mathfrak{A}_n \times \mathfrak{B}_n))$ for every $\nu \in F_n(\mathfrak{B}_1, \ldots, \mathfrak{B}_n)$, then there exists $K > 0$ such that for every $\nu \in F_n(\mathfrak{B}_1, \ldots, \mathfrak{B}_n)$

$$\|\mu \times \nu\|_{F_n} \leq K \|\nu\|_{F_n}.$$

9. Verify (5.3): if $\alpha \in l^2(\Omega_m)$ and $\beta \in l^2(\hat{\Omega}_m)$, then

$$\left| \sum_{(w,\omega) \in \Omega_m \times \hat{\Omega}_m} \sqrt{\frac{1}{m}} \varphi(w, \omega)\ \alpha(w)\ \beta(\omega) \right| \leq \|\alpha\|_2\ \|\beta\|_2.$$

10. Prove the inequality on the right side of (5.5).

11. i. Verify that the μ_m defined in (5.14) satisfy (5.13).

ii. Verify that if $\mu = \Sigma_m\ \mu_m/m^2$, then $\mu \in F_3(\mathfrak{A} \times \mathfrak{B} \times \mathfrak{C})$, and (5.15) holds.

12.* Let $X_1, \ldots, X_n$ be *locally* compact, non-compact Hausdorff spaces with respective Borel fields $\mathfrak{B}_1, \ldots, \mathfrak{B}_n$. Prove or disprove

$$\mathrm{C}_0(X_1 \times \cdots \times X_n) \cap V_n(\mathfrak{B}_1, \ldots, \mathfrak{B}_n)$$

$$= V_n(X_1, \ldots, X_n) \quad \text{(cf. Exercise IV.12 vi).}$$

13. Verify the following (in the proof of Proposition 13).

i. If g is an elementary tensor in $\mathrm{C_c}(X_1 \times Y_1) \otimes \cdots \otimes \mathrm{C_c}(X_n \times Y_n)$, then $\phi_{g;\mu} \in \mathrm{C_c}(Y_1 \times \cdots \times Y_n)$, where $\phi_{g;\mu}$ is defined in (6.2).

ii. If X and Y are locally compact Hausdorff spaces then $V(X,Y)$ is dense in $\mathrm{C_0}(X \times Y)$.

14. Let $X_1, \ldots, X_n, Y_1, \ldots, Y_n$ be locally compact Hausdorff spaces with respective Borel fields $\mathfrak{B}_1, \ldots, \mathfrak{B}_n, \mathfrak{C}_1, \ldots, \mathfrak{C}_n$, and suppose $\mu \in PBF_n(\mathfrak{B}_1, \ldots, \mathfrak{B}_n)$. For $k \in [n]$, let $(\varphi_{jk}\colon j \in \mathbb{N})$ be Cauchy sequences in $S(\mathfrak{B}_k) \otimes S(\mathfrak{C}_k)$, and denote $\boldsymbol{\varphi}_j = \varphi_{j1} \otimes \cdots \otimes \varphi_{jn}$. Prove that $(\phi_{\boldsymbol{\varphi}j;\mu} : j \in \mathbb{N})$ is Cauchy in $V_n(Y_1, \ldots, Y_n)$.

15. i. Prove that if $X_1, \ldots, X_n, Y_1, \ldots, Y_n$ are locally compact Hausdorff spaces with respective Borel fields $\mathfrak{B}_1, \ldots, \mathfrak{B}_n, \mathfrak{C}_1, \ldots, \mathfrak{C}_n$, $\mu \in F_n(\mathfrak{B}_1, \ldots, \mathfrak{B}_n)$, and $\nu \in F_n(\mathfrak{C}_1, \ldots, \mathfrak{C}_n)$, then $\mu \times \nu \in F_n(\sigma(\mathfrak{B}_1 \times \mathfrak{C}_1), \ldots, \sigma(\mathfrak{B}_n \times \mathfrak{C}_n))$ if and only if

$$\sup\left\{\left|\int \phi_{f;\mu}\, \mathrm{d}\nu\right| : f = f_1 \otimes \cdots \otimes f_n,\ f_j \in B_{\mathrm{C_0}(X_j \times Y_j)} j \in [n]\right\} \tag{E.4}$$

is finite.

ii. Prove Proposition 17.

iii. Prove Proposition 18.

16. i. Prove Proposition 19.

ii.* For $n \geq 3$ and $\mu \in F_n(\mathfrak{B}_1, \ldots, \mathfrak{B}_n)$, is $\hat{\mu}$ jointly continuous on $\hat{X}_1 \times \cdots \times \hat{X}_n$?

17. Verify that convolution defined by (7.2) is the same as the definition of convolution based on (7.3).

18. Prove $\|\cdot\|_{\mathrm{gpb}_n}$ (defined in (7.7)) is a norm, and $(gPBF_n, \|\cdot\|_{\mathrm{gpb}_n})$ is a Banach space.

19. Verify $(\mu \star \nu)\hat{\ } = \hat{\mu}\, \hat{\nu}$ for all $\mu \in gPBF_n$ and $\nu \in F_n$.

20. Prove Proposition 21.

21. Use Lemma 22 and Proposition 21 to show that if X_1, X_2, and X_3 are infinite locally compact Abelian groups, then convolution in $F_1(\sigma(\mathfrak{B}_1 \times \mathfrak{B}_2 \times \mathfrak{B}_3))$ cannot be extended to the entire space $F_3(\mathfrak{B}_1, \mathfrak{B}_2, \mathfrak{B}_3)$.

22. Prove that the Banach spaces $gPBF_n$ and τPBF_n equipped with convolution are Banach algebras.

23. Fix an integer $m > 1$, and define μ in $F_3(\Omega_m, \Omega_m, \hat{\Omega}_m)$ by

$$\mu(A, B, C) = \sum_{\omega \in \Omega_m} \mathbf{1}_A(\omega)\, \mathbf{1}_B(\omega)\, \hat{\mathbf{1}}_C(\omega)/2^{\frac{n}{2}}.$$

$$A \subset \Omega_m, \ B \subset \Omega_m, \ C \subset \hat{\Omega}_m. \tag{E.5}$$

i. Verify that $\mu \in F_2(\Omega_m^2, \hat{\Omega}_m)$ with $\|\mu\|_{F_2} \leq 1$, and therefore $\mu \in F_3(\Omega_m, \Omega_m, \hat{\Omega}_m)$ with $\|\mu\|_{F_3} \leq 1$.

ii. Verify

$$\hat{\mu}(w_1, w_2, \omega) = w_1(\omega)\, w_2(\omega)/2^{\frac{n}{2}}, w_1 \in \hat{\Omega}_m, w_2 \in \hat{\Omega}_m, \ \omega \in \Omega_m. \tag{E.6}$$

iii. Prove that $\|\hat{\mu}\|_{V_3(\hat{\Omega}_m, \hat{\Omega}_m, \Omega_m)} \geq 2^{n/2}$.

iv. Prove that for every $K > 0$ there exists a discrete measure μ with finite support in $X_1 \times X_2 \times X_3$ such that $\|\mu\|_{F_2(\sigma(\mathfrak{B}_1 \times \mathfrak{B}_2), \mathfrak{B}_3)} \leq 1$ and $\|\mu\|_{\mathrm{pb}_g} \geq K$. Conclude that there exist F_2-measures on $\sigma(\mathfrak{B}_1 \times \mathfrak{B}_2) \times \mathfrak{B}_3$ which are not convolvers in $F_3(\mathfrak{B}_1, \mathfrak{B}_2, \mathfrak{B}_3)$.

24. Prove that the set-function μ_φ defined in (8.5) is an F_3-measure on $\mathfrak{A} \times \mathfrak{B} \times \mathfrak{C}$, that $\|\mu_\varphi\|_{F_3} \leq \|\varphi\|_\infty$, and that

$$\int f \, \mathrm{d}\mu_\varphi = \sum_{i,j,k} \varphi(i,j,k)\, \hat{f}(\hat{x}_{ij}, \hat{y}_{jk}, \hat{z}_{ik}), \quad f \in V_3(X, Y, Z).$$

25. Prove that if $X_1, \ldots, X_n$ are compact Abelian groups with respective Borel fields $\mathfrak{B}_1, \ldots, \mathfrak{B}_n$, and $\mu \in PBF_n(\mathfrak{B}_1, \ldots, \mathfrak{B}_n)$, then $\hat{\mu} \in V_n(\hat{X}_1, \ldots, \hat{X}_n)$.

26. Prove there exist L^2-factorizable trilinear functionals that are projectively unbounded.

Hints for Exercises in Chapter IX

1. ii*. Interpolation?
2. i. An instance of Theorem 3.
4. The first assertion is straightforward. So is the second: if (μ_j) is Cauchy in PBF_n, then it is Cauchy in F_n, and hence converges to some $\mu \in F_n$. Then $\mu \in PBF_n$ – by the definition of projective boundedness, and because $V_n = \tilde{V}_n$ when underlying domains are finite.

7. i. Use the norm-density of $S(\mathfrak{A})$ in $\mathrm{L}^\infty(\mathfrak{A})$. The norms in V and $\tilde{V}$ are the same when the underlying sets are finite.
   ii. Index $F_i \subset B_{S^\infty(\mathfrak{A}_i)}$ by elements in $\mathfrak{B}_i$.
8. Uniform boundedness principle.
9. The Plancherel Theorem.
10. Verify the dual formulation of Littlewood's mixed norm inequality: for finite sets E and F, and $\phi \in l^\infty(E \times F)$,

$$\|\phi\|_{V_2(E,F)} \leq \sqrt{2} \max \left\{ \sum_{e \in E} |\phi(e, f)|^2 : f \in F \right\}$$

($\sqrt{2}$ = the Khintchin constant). In this formulation, let $E = W$, $F = \Omega$, and $\phi = \rho \cdot \varphi$.

13. i. Use basic properties of integrals with respect to F-measures.
   ii. Apply the Stone–Weierstrass theorem.
14. Show that if $\boldsymbol{\varphi} = \varphi_1 \otimes \cdots \otimes \varphi_n$ is an elementary tensor such that $\varphi_k \in S(\mathfrak{B}_k) \otimes S(\mathfrak{C}_k)$ $(k \in [n])$, then

$$\|\phi_{\boldsymbol{\varphi};\mu}\|_{V_n} \leq \|\varphi_1\|_\infty \cdots \|\varphi_n\|_\infty \; \|\mu\|_{\mathrm{pb}_n}.$$

15. i. Formally, $\int \phi_{f;\mu} \, \mathrm{d}\nu = \int (\int f \, \mathrm{d}\mu) \, \mathrm{d}\nu$. If (E.4) holds, then

$$\int \left(\int f \mathrm{d}\mu \right) \mathrm{d}\nu$$

   determines a bounded n-linear functional on $\mathrm{C}_0(X_1 \times Y_1) \times \cdots \times \mathrm{C}_0(X_n \times Y_n)$, and hence there is an F-measure on $\sigma(\mathfrak{B}_1 \times \mathfrak{C}_1) \times \cdots \times \sigma(\mathfrak{B}_n \times \mathfrak{C}_n)$, which is $\mu \times \nu$. The converse follows from $\int \phi_{f;\mu} \, \mathrm{d}\nu = \int f \, \mathrm{d}(\mu \times \nu)$.
   ii. See proof of Theorem 6; cf. Corollary 14.
   iii. Apply Lemma 16 and Proposition 17.
16. The first part of the proposition can be obtained from a standard convergence theorem. Use the Grothendieck factorization theorem to prove the second part. (Is the use here of the factorization theorem necessary?)
23. See [BlCag].

   i. Compare with Lemma 11.
   ii. Compare with (VII.10.24).

iii. Use $V_3^* = F_3$, and that if $\beta(w_1, w_2, \omega) = w_1(\omega)\, w_2(\omega)/2^{2n}$, then $\|\beta\|_{F_3} \leq 1$. See (VII.10.25).

iv. Use iii; cf. Lemma 22. See Exercise 21.

24. See Lemma VIII.8 and (VIII.5.1). Observe that

$$\int f \otimes g \otimes h \; \mathrm{d}\mu_\varphi = \sum_{i,j,k} \varphi(i,j,k)\; \hat{f}(\hat{x}_{ij})\; \hat{g}(\hat{y}_{jk})\; \hat{h}(\hat{z}_{ik}),$$

$$f \in \mathrm{C}(X),\; g \in \mathrm{C}(Y),\; h \in \mathrm{C}(Z),$$

and apply Plancherel's theorem.

25. Cf. (5.10) and (5.11).

26. See examples in §8.

X

Brownian Motion and the Wiener Process

1 Mise en Scène: A Historical Backdrop and Heuristics

The Wiener process – a stochastic process with independent Gaussian increments – was originally conceived as a probabilistic model for Brownian movement, and has been, ever since, among the most influential mathematical constructs in the twentieth century. For our purposes, we used it in Chapter VI §2 to produce a canonical example of an F_2-measure that cannot be extended to an F_1-measure. In this chapter and the next, we examine and develop ideas underlying this example.

We begin here with some of the history and heuristics behind Brownian motion and the Wiener process. (In this book, 'Brownian motion' or 'Brownian movement' will refer always to a physical phenomenon, and the 'Wiener process' to Norbert Wiener's mathematical model of it.)

From Brown to Wiener

In the sciences at large, Brownian movement generically refers to haphazard, erratic, difficult-to-predict trajectories of particles. Such movements exhibited by tiny particles suspended in liquid first became known to naturalists in the seventeenth century, soon after the invention of the microscope, and for a long time were thought to be vital – always manifesting life. Refuting that 'vitality' was the cause, the botanist Robert Brown recorded in 1827 that erratic movements, such as those observed by his colleagues and predecessors, were in fact performed by inorganic as well as organic particles. He guessed these particles to be nature's most basic constituents, and referred to them as 'active molecules' [Br]. Brown *almost* got it right. Today it is commonly known that the particles he observed were not bona fide molecules, but that their movements were caused by invisible sub-microscopic molecular activity. In nineteenth-century science, however, the idea that matter was physically constituted from atoms and molecules in perpetual motion, though widely believed, was still an unproven notion, the so-called *atomic–molecular hypothesis*. Indeed, this notion, proposed first by the philosopher Democritus (465–400 B.C.), led to centuries of speculations, with growing numbers of proponents of 'atomism' on the one side, but also with some illustrious opponents on the other. First among the skeptics, in antiquity, was the philosopher Aristotle, and last, in the modern era, was the scientist–philosopher Ernst Mach – the same Mach of the speed-of-sound fame. (Ernst Mach also had doubts about ether, whose existence, like that of atoms, was widely accepted by nineteenth-century physicists...) Aristotle's opposition to 'atomism' stemmed from his belief that we could accept reality only of that which we could experience through our senses. Mach's skepticism was essentially the same: an ardent phenomenonologist, he demanded physical proof that atoms actually existed. And so it was, in this very context, that a plausible explanation for Brownian motion became an important pivotal issue [Ny], [Bru, Chapter 15].

Siding with the atomists, a young Albert Einstein – then a clerk in the Swiss patent office – proposed in a landmark 1905 paper a statistical–mechanical model for Brownian movement based on the assumption that [Ei1, pp. 3–4]

> the suspended particles perform an irregular movement – even if a very slow one – in the liquid, on account of the molecular movement of the liquid.*

After deriving and solving a diffusion equation for the 'suspended particles', Einstein obtained that probability distributions of the 'irregular movement' were Gaussian, and deduced a simple formula relating certain physical constants to the average displacement of a particle. He concluded with the hope [Ei1, p. 18]

> that some enquirer may succeed shortly in solving the problem suggested here [verifying his model, and, specifically, determining atomic and molecular dimensions].

Soon after Einstein's paper had appeared, Jean Perrin provided experimental proof, based on Brownian movement and Einstein's model of it,

* Otherwise, but for his obvious belief in atomic structures, young Einstein was deeply influenced by Ernst Mach's then-maverick ideas about physics, and, in particular, about space–time; see [Ei2, p. 21].

that atoms and molecules were in fact 'real' entities [Pe1], [Pe2], [Ny, Chapter 4]. Einstein was awarded the 1921 Nobel physics prize for the first of his three celebrated 1905 papers (*Annalen der Physik*, Vol. 17), for theoretical work on the photoelectric effect; the citation by the prize committee only obliquely mentioned his third paper (about special relativity), and ignored altogether the second paper (about Brownian movement) [Ber, p. 188]. Perrin was awarded the 1926 Nobel physics prize for his experimental work on Brownian movement and atomic measurements. A detailed account of the key role of Brownian motion in the experimental verification of the atomic–molecular hypothesis is found in [Ny]; a briefer account can be found in [Bru, Chapter 15].

Underscoring the significance of the atomic–molecular hypothesis itself, Richard Feynman offered, somewhat darkly, this tribute [Fey, pp. 1–2].

> If, in some cataclysm, all of scientific knowledge were to be destroyed, and only one sentence passed on to the next generations of creatures, what statement would contain the most information in the fewest words? I believe it is the *atomic hypothesis* (or the *atomic* fact, or whatever you wish to call it) that *all things are made of atoms – little particles that move around in perpetual motion, attracting each other when they are a little distance apart, but repelling upon being squeezed into one another.*

On the mathematical side, the first study of Brownian movement from a purely probabilistic viewpoint appeared in Louis Bachelier's University of Paris doctoral thesis, submitted in 1898 and defended in 1900 [B1]. In his dissertation, Bachelier made no mention of haphazard motions observed in contexts of physical science. Rather, he was inspired by price fluctuations on the Paris stock exchange, which led him to a mathematical model of time-dependent randomness. The model, although flawed, was in hindsight a precursor to later constructs by others; see [B2, Chapter XII–XIV]. Bachelier's work in probability theory, with the Bourse as his laboratory, should have been seminal, but, alas, attracted little notice from his contemporaries. Interesting comments by Paul Lévy about Bachelier's results (*and* Lévy's own) can be found in [Lé4, pp. 97–98]; see also [Man, pp. 392–5].*

* The Bourse *is* of course a physical context. In a biographical sketch of Bachelier [Man, pp. 392–5], Benoit Mandelbrot suggests that focus on the stock market might have tainted Bachelier's mathematics; l'Académie would have been more receptive to Bachelier's ideas had they been cast in then-traditional settings of physical science.

Unlike Bachelier's work, Norbert Wiener's study of Brownian movement was decidedly motivated by physics [Wi2, pp. 131–133]. Citing Einstein and Perrin in his 1923 work on *Differential-Space*, Wiener noted the

> physical explanation of the Brownian movement... that it is due to the haphazard impulses given to the particles by the collisions of the molecules of the fluid in which the particles are suspended [Wi2, p. 133].

He then constructed a model: a Gaussian stochastic process representing the seemingly random movements of these particles. The construction of this process was, at the time, a major mathematical breakthrough. We explain below the motivation behind Wiener's model, and defer its precise definition and construction to the next section.

Heuristics

Say we are observing haphazard movements of particles. Our goal: model these movements. We consider one such generic particle p, which we dub *Brownian*, and suppose it is free. That is, we assume the only forces acting on p are imparted by an ambient environment – all forces are hidden, and all acting in a very complicated manner. Also, to simplify matters, we suppose the particle p is moving continuously along a straight line. (Think of p's position on this line as the x-coordinate of an actual Brownian particle in three-dimensional space.) We let the particle's position at time $t = 0$ be the origin, and ask: what can be said about its position $X = X(t)$ at time $t > 0$?

We concede that we do not know, and are unable to determine the particle's extremely complex dynamics. Our perceptions, based on 'zero knowledge', are that

(i) at any instant, p moves to the right or to the left with equal likelihood, and

(ii) p's trajectories over disjoint time intervals appear unrelated.

These two assumptions about Brownian particles and their trajectories are meant at the very outset in an intuitive sense, and will soon be made precise. Particles and their trajectories about which we assume i and ii will be called *Brownian.* We also presume that

(iii) p's 'statistics' over time intervals of equal length are the same.

The third surmise (time-homogeneity) also stems from 'zero knowledge': knowing (and assuming) nothing about forces acting on p, we imagine that Brownian motion is statistically the same in every time interval of the same length. (In iii, 'statistics' could mean *expected* distance traveled by p, *variance* of its increments, or *probability distributions* of its increments.)

We model these perceptions in a probabilistic framework. To begin, we think of $X(t)$ as a real-valued random variable with finite variance. We assume that $\mathbf{E}X(t) = 0$ (assertion i); that increments over disjoint time intervals are uncorrelated (assertion ii, Exercise 1),

$$\mathbf{E}(X(t_1) - X(s_1))(X(t_2) - X(s_2)) = 0,$$
$$0 \leq s_1 < t_1 \leq s_2 < t_2 < \infty; \tag{1.1}$$

and that the variance of $X(t) - X(s)$ is a function of $t - s$ (assertion iii),

$$\mathrm{Var}(X(t) - X(s)) = v(t - s), \quad 0 \leq s < t \leq \infty, \tag{1.2}$$

where v is a non-negative function on $[0, \infty)$. From (1.1) and (1.2) we conclude that

$$\mathrm{Var}X(t) = ct \text{ for all } t \geq 0, \tag{1.3}$$

where $c \geq 0$ is a numerical constant (Exercise 2).

To derive $\mathrm{Var}X(t) = ct$ for all $t \geq 0$, we used a mild, indeed a minimal interpretation of assertions i, ii, and iii. Let us now apply the more stringent interpretation, that path increments over disjoint time intervals are statistically independent random variables. This, in a probabilistic context, is the most extreme interpretation of assertion ii (and, insofar that modeling 'reality' is our objective, the simplest and most naive...). We fix time $t > 0$, fix an arbitrary integer $n > 0$, and imagine $X(t)$ to be 'approximately' the result of a simple random walk clocked by discrete time $t/n, 2t/n, \ldots, jt/n, \ldots, (n-1)t/n$: at time jt/n, $j = 0, \ldots, n-1$, we imagine the particle moving a distance s_n to the right or to the left with probability $1/2$, the moves are independent, and $s_n = \sqrt{ct/n}$ (cf. (1.3)). Then, $X(t)$ is 'approximately'

$$X_n(t) = \sqrt{ct/n} \sum_{j=1}^{n} r_j, \tag{1.4}$$

where $\{r_j : j \in \mathbb{N}\}$ is the usual Rademacher system. (See Chapter VII.) Taking $n \to \infty$, we obtain by the Central Limit Theorem that X_n converges in distribution to a Gaussian r.v. with mean 0 and variance ct, and think of this limit as $X(t)$.

Remarks:

i **(a Gaussian from the viewpoint of physics).** The perception that a Brownian particle's position is a Gaussian random variable follows, as we have just seen, from the Central Limit Theorem, via statistical analysis based on 'zero knowledge'. This perception is a cornerstone to Wiener's mathematical model of Brownian motion, which Wiener himself viewed as 'a first approximation' [Wi1, p. 295].

A Gaussian model of Brownian movement can be derived also in a context of statistical mechanics from 'idealized' physical principles. This indeed was Einstein's observation in his 1905 paper on Brownian movement. We briefly describe such a derivation (cf. [Re, pp. 483–4]). Let $n(x,t)$ be the linear density of Brownian particles at time $t > 0$ and position x. (For simplicity, as before, consider Brownian movement in one dimension.) Let $J(x,t)$ (*flux*) be the average number of Brownian particles per unit time crossing a point x at time t. Conservation of mass implies

$$\frac{\partial n}{\partial t} = -\frac{\partial J}{\partial x}. \tag{1.5}$$

If we assume also that flux is proportional to the spatial derivative of the density,

$$J = -D\frac{\partial n}{\partial x} \tag{1.6}$$

(constant $D > 0$), then, by combining (1.6) with (1.5), we obtain the diffusion equation

$$\frac{\partial n}{\partial t} = D\frac{\partial^2 n}{\partial x^2} \quad \text{(cf. [Ei1, p. 15, equation (I)])}, \tag{1.7}$$

whose solutions involve Gaussian kernels.

The assertion in (1.6), a simple and 'ideal' assumption on which Einstein's model rests, is analogous to the independence-of-Brownian increments assumption in the statistical context.

ii **(a Gaussian from the viewpoint of information theory).** Like assumptions i and ii, the time-homogeneity in iii stems from 'zero knowledge': with no information about ambient forces acting on the

Brownian particle, we cannot but surmise that Brownian movement is statistically the same in every time interval of the same length. A simple analogy is that a uniform probability measure is a model for random sampling from a finite set. A rigorous justification for this model – usually glossed over in elementary courses – is that the uniform probability measure on a finite set, among all probability measures on the set, has maximum entropy [I3, p. 17]. Here we view the entropy of a distribution, in the sense of Shannon, as a gauge of the 'amount' of information contained in the distribution: greater entropy means less information. Indeed, by applying the maximum entropy method to find the distribution of displacements of a Brownian particle (under a hypothesis that we have no information about the particle's dynamics), we conclude that the distribution is Gaussian [Sh, pp. 56–7].

2 A Mathematical Model for Brownian Motion

By a *stochastic process* (or simply a *process*) we mean a collection of random variables indexed by a prescribed set. The underlying probability space $(\Omega, \mathscr{A}, \mathbb{P})$ will always be complete, the random variables real-valued, and the indexing set until further notice will be the unit interval [0,1]. If $X = \{X(t) : t \in [0,1]\}$ is a stochastic process, and [0,1] denotes a time scale, then we call $X(t), t \in [0,1]$, a *sample-path*; otherwise, we refer to it as a *random function*. We can think of a stochastic process X also as a function of two variables,

$$X = X(t, \omega), \quad t \in [0,1],\ \omega \in \Omega. \tag{2.1}$$

But, unless specifying otherwise, we follow the usual convention, writing $X(t)$ for $X(t, \omega)$. (In a probabilistic context, the sample point $\omega \in \Omega$ is almost always implicit.)

Definition 1 A stochastic process $\mathrm{W} = \{\mathrm{W}(t) : t \in [0,1]\}$ on a probability space $(\Omega, \mathscr{A}, \mathbb{P})$ is a *Wiener process* if

$$\mathrm{W}(0) = 0 \text{ a.s. } (\mathbb{P}); \tag{2.2}$$

$$\text{for } [s,t] \subset [0,1],\ \mathrm{W}(t) - \mathrm{W}(s) \text{ is a Gaussian random variable with mean 0 and variance } t - s; \tag{2.3}$$

for every partition $\{0 \le t_1 < \cdots < t_n \le 1\}$ of [0,1], $W(t_2) - W(t_1), \ldots, W(t_n) - W(t_{n-1})$ are statistically independent random variables. (2.4)

A Wiener process is a model for Brownian movement: the sample space Ω represents the ensemble of all possible paths of a Brownian particle, and $W(t)$ ($t \in [0,1]$) is its ($\mathbb{P}$) random trajectory; properties (2.2), (2.3), and (2.4) reflect the heuristics in the previous section. Notice that continuity of sample paths is not part of the definition; sample-path continuity will be a consequence of the model. (Recall that continuity of Brownian trajectories was built into the heuristics that led to Gaussian distributions.)

First on the agenda is the question: does such a model exist?

A Construction of a Wiener Process

We start with a countably infinite system of independent standard Gaussian variables on a probability space $(\Omega, \mathscr{A}, \mathbb{P})$, and let H be the $L^2(\Omega, \mathbb{P})$-closure of the linear span of this system. Every element in H is Gaussian with mean 0 (Exercise 4 i). We let U be a unitary map from $L^2([0,1], \mathfrak{m})$ onto H, where $\mathfrak{m}$ denotes Lebesgue measure on [0,1], and define

$$W(t) = U\mathbf{1}_{[0,t]}, \quad t \in [0,1]. \tag{2.5}$$

Then, $W=\{W(t)\colon t \in [0,1]\}$ is a Wiener process. Properties (2.2) and (2.3) are evident. Property (2.4) follows from this general fact: if $Y_1, \ldots, Y_n$ are mutually orthogonal Gaussian variables with mean 0 such that every element in the linear span of $\{Y_1, \ldots, Y_n\}$ is Gaussian, then $Y_1, \ldots, Y_n$ are independent (Exercise 4 ii).

This realization of a Wiener process, which is due to S. Kakutani, is a characterization (cf. [Kak1], [Kak3, pp. 241–2]). That is, if W is a Wiener process, then it can be realized as

$$W(t) = U\mathbf{1}_{[0,t]}, \quad t \in [0,1], \tag{2.6}$$

where U is a unitary map from $L^2([0,1], \mathfrak{m})$ onto the L^2-closure of the span of a countably infinite system of independent standard Gaussian variables. We will prove this in the next section, immediately after we define the *Wiener integral.*

Remark (the Wiener space). In §5 we will demonstrate, by using stochastic series representations, that if W is any Wiener process, then sample-paths of W are almost surely continuous. Indeed, we expect sample-path continuity from a model of Brownian movement. And therein lies Wiener's achievement: a construction of a process satisfying (2.2), (2.3), and (2.4), *and* the sample space is $\mathrm{C}_{\mathbb{R}}([0,1])$ (the space of real-valued continuous functions on [0,1]).

Here is an outline of such a construction. For $n \in \mathbb{N}$, $0 \le t_1 < \cdots < t_n \le 1$, and Borel sets $A_1 \subset \mathbb{R}, \ldots, A_{n-1} \subset \mathbb{R}$, define

$$\begin{aligned} &C(t_1, \ldots, t_n; A_1, \ldots, A_{n-1}) \\ &\quad = \{g \in \mathrm{C}_{\mathbb{R}}([0,1]) : g(t_2) - g(t_1) \in A_1, \ldots, g(t_n) - g(t_{n-1}) \in A_{n-1}\}, \end{aligned} \tag{2.7}$$

and then define

$$\begin{aligned} \mathscr{C} = \{&C(t_1, \ldots, t_n; A_1, \ldots, A_{n-1}) : \\ &n \in \mathbb{N},\ 0 \le t_1 < \cdots < t_n \le 1,\ \text{Borel sets } A_1 \subset \mathbb{R}, \ldots, A_{n-1} \subset \mathbb{R}\}. \end{aligned} \tag{2.8}$$

It is easy to verify that $\mathscr{C}$ is closed under finite intersections, and that complements of elements in $\mathscr{C}$ are finite unions of elements in $\mathscr{C}$.

Next, to construct a probability measure on $\sigma\mathscr{C}$ (= σ-algebra generated by $\mathscr{C}$), we consider

$$\mathscr{G}_{\sigma^2}(A) = \frac{1}{\sigma\sqrt{2\pi}} \int_A \exp\left(\frac{-t^2}{2\sigma^2}\right) \mathrm{d}t, \quad \text{Borel set } A \subset \mathbb{R}, \tag{2.9}$$

and define for $C = C(t_1, \ldots, t_n; A_1, \ldots, A_{n-1}) \in \mathscr{C}$,

$$\mathbb{P}_{\mathrm{W}}(C) = \prod_{j=2}^{n} \mathscr{G}_{t_j - t_{j-1}}(A_{j-1}). \tag{2.10}$$

After verifying that $\mathbb{P}_{\mathrm{W}}$ is well-defined and countably additive on $\mathscr{C}$ (which requires some work!) we conclude, by the Carathéodory extension theorem (e.g., [Roy, pp. 295–7]), that $\mathbb{P}_{\mathrm{W}}$ is extendible to a probability measure on $\sigma\mathscr{C}$.

A Wiener process W on the probability space $(\mathrm{C}_{\mathbb{R}}([0,1]), \sigma\mathscr{C}, \mathbb{P}_{\mathrm{W}})$ is defined by

$$\mathrm{W}(t, \omega) = \omega(t) - \omega(0), \quad \omega \in \mathrm{C}_{\mathbb{R}}([0,1]),\ t \in [0,1]. \tag{2.11}$$

Property (2.2) is evident, and properties (2.3) and (2.4) follow from (2.9) and (2.10), respectively.

This construction (detailed, for example, in [Hi, pp. 44–51]) is based on Wiener's original insight: the realization of $C_{\mathbb{R}}([0,1])$ as a product space with a probability measure on it determined by a product of Gaussian measures. It was this view of $C_{\mathbb{R}}([0,1])$ that eluded Paul Lévy in his own quest for 'la fonction du mouvement brownien'; see [Lé4, pp. 97–100]. Wiener dubbed this probability space *differential-space.* He imagined a Brownian path to be synthesized from statistically independent 'differences' sampled from $C_{\mathbb{R}}([0,1])$ (cf. (1.4) and (2.5)); and hence the term *differential-space.* Today $\mathbb{P}_W$ is called the Wiener measure, and $(C_{\mathbb{R}}([0,1]), \sigma\mathscr{C}, \mathbb{P}_W)$ is called the Wiener space.

3 The Wiener Integral

The next question is: *can a function on* [0,1] *be integrated, in some reasonable sense, along sample-paths of a Wiener process?* Notice that the question considered in the previous section – does a Wiener process exist? – can itself be restated as a question about the feasibility of an integral: write (formally)

$$W(t) = \int_0^1 \mathbf{1}_{[0,t]}\, dW,$$

and ask whether the right side exists as a limit, in some appropriate sense, of sums of 'differences' ΔW (cf. (1.4)). (Wiener himself imagined Brownian motion as this limit; see [Wi1, p. 294], and also §12 in this chapter.)

We first observe a natural obstacle to integration in the usual Riemann–Stieltjes sense. For the purpose of the proof below, we take for granted that sample-paths of a Wiener process are almost surely continuous. (This will be verified in §5.)

Proposition 2 *On every subinterval of* [0,1], *almost all* ($\mathbb{P}$) *sample-paths of a Wiener process have unbounded variation.*

Proof: Let W be a Wiener process, and define

$$W_{n,j} = W(j/n) - W((j-1)/n), \; n \in \mathbb{N}, \; j \in [n].$$

We claim that

$$\lim_{n\to\infty}\sum_{j=1}^{n}(\mathrm{W}_{n,j})^2 = 1, \tag{3.1}$$

where the limit is taken in the $\mathrm{L}^2(\Omega, \mathbb{P})$-norm. To verify this, write

$$\mathbf{E}\left(\sum_{j=1}^{n}(\mathrm{W}_{n,j})^2 - 1\right)^2 = \mathbf{E}\left(\sum_{j=1}^{n}\left[(\mathrm{W}_{n,j})^2 - \frac{1}{n}\right]\right)^2. \tag{3.2}$$

The random variables $(\mathrm{W}_{n,j})^2$, $j \in [n]$, are independent and have mean $1/n$. Therefore, the right side of (3.2) equals

$$\sum_{j=1}^{n}\mathbf{E}\left[(\mathrm{W}_{n,j})^2 - \frac{1}{n}\right]^2 = \sum_{j=1}^{n}\left[\mathbf{E}(\mathrm{W}_{n,j})^4 - \left(\frac{1}{n}\right)^2\right]. \tag{3.3}$$

Because the $\mathrm{W}_{n,j}$ are Gaussian with mean 0 and variance $1/n$, we have $\mathbf{E}(\mathrm{W}_{n,j})^4 = 3/n^2$, $j \in [n]$. Therefore,

$$\mathbf{E}\left(\sum_{j=1}^{n}(\mathrm{W}_{n,j})^2 - 1\right)^2 = \frac{2}{n}, \tag{3.4}$$

which implies (3.1).

Because sample-paths of W are almost surely continuous, we conclude from (3.1) that they almost surely have infinite variation over [0,1] (Exercise 5). The same proof applies to any subinterval of [0,1] (Exercise 6). □

Remarks:

i **(what does infinite variation mean?).** According to the proposition above almost all sample-paths in Wiener's model have infinite variation, while a trajectory of a physical particle, no matter how erratic, surely has finite variation. In fact, it can be further shown that, although a physical particle has always finite velocity, almost all sample-paths of W are nowhere differentiable [Wi2, §4], [DvErKak]. This seems paradoxical, but all is in order: in a probabilistic framework, *nowhere differentiability* means *uncertainty.* To wit, the haphazardness perceived in (physical) Brownian movement becomes the assumption underlying the probabilistic model, that a Brownian particle's direction cannot at any instant be predicted; that all directions are equally likely. (Review the heuristics in §1.)

This assumption implies – formally, at least – that there are no tangent lines to the graph $\{(t, W(t)) : t \in [0,1]\}$; and hence *nowhere differentiability.* In this sense, Proposition 2 is a statement about inability to predict Brownian motion, and not about the motion itself.

ii **(quadratic variation).** The left side of (3.1) is the *quadratic variation* of W at $t = 1$. The main step in the proof of Proposition 2, that this variation is finite and non-zero, contains the argument that the F_2-measure associated with W (recalled in Remark iv below) cannot be extended to a bona fide measure (see also Exercise 7.)

In a broader context of stochastic integration, the quadratic variation is key to the *Itô integral*: an integral of a random function (of W) with respect to W [I1]. This integral and this quadratic variation, of which more will be said in §8 and the next chapter, are at the very foundation of *adaptive* stochastic integration (e.g., [Pr]).

According to Proposition 2, we cannot integrate along sample-paths of W in the usual Riemann-Stieltjes sense. Instead, we use a functional-analytic approach. Let f be a step function on [0,1].

$$f = \sum_i a_i \, \mathbf{1}_{J_i}, \tag{3.5}$$

where the J_i are pairwise disjoint subintervals of [0,1]. Define

$$I_{\mathrm{W}}(f) := \sum_j a_j \, \Delta \mathrm{W}(I_j) \tag{3.6}$$

($\Delta \mathrm{W}(J) := \mathrm{W}(t) - \mathrm{W}(s)$, where $J \subset [0,1]$ is an interval with end-points $s \leq t$). The $\Delta \mathrm{W}(J_i)$ are independent Gaussian random variables with mean 0 and variance $\mathfrak{m}(J_i)$, and therefore, $I_{\mathrm{W}}(f)$ is Gaussian with mean 0 and variance

$$\mathbf{E}|I_{\mathrm{W}}(f)|^2 = \sum_i |a_i|^2 \, \mathfrak{m}(J_i) = \int_0^1 |f|^2 \, dt. \tag{3.7}$$

Therefore, (3.6) defines an isometry from the space of step functions in $\mathrm{L}^2([0,1], \mathfrak{m})$ into $\mathrm{L}^2(\Omega, \mathbb{P})$, and because step functions are norm-dense in $\mathrm{L}^2([0,1], \mathfrak{m})$, this isometry is uniquely extendible to $\mathrm{L}^2([0,1], \mathfrak{m})$. Its evaluation at $f \in \mathrm{L}^2([0,1], \mathfrak{m})$ is, by definition, the Wiener integral $I_{\mathrm{W}}(f)$.

Proposition 3 (Exercise 8). *For* $f \in \mathrm{L}^2([0,1], \mathfrak{m})$, $I_{\mathrm{W}}(f)$ *is a Gaussian random variable with mean* 0 *and variance* $\|f\|^2_{\mathrm{L}^2}$. *Moreover, for* *all f and g in* $\mathrm{L}^2([0,1],\mathfrak{m})$,

$$\int_0^1 f(t)g(t)\mathrm{d}t = \mathbf{E}\ I_{\mathrm{W}}(f)I_{\mathrm{W}}(g). \tag{3.8}$$

We now can verify that Kakutani's construction of a Wiener process is a characterization:

Proposition 4 *Let* $\mathrm{W} = \{\mathrm{W}(t) : t \in [0,1]\}$ *be a Wiener process on* $(\Omega, \mathscr{A}, \mathbb{P})$. *Then, there exists a system* $\{X_j : j \in \mathbb{N}\}$ *of independent standard Gaussian r.v. on* $(\Omega, \mathscr{A}, \mathbb{P})$, *and a unitary equivalence* U *from* $\mathrm{L}^2([0,1],\mathfrak{m})$ *onto the* $\mathrm{L}^2(\Omega,\mathbb{P})$*-closure of the linear span of* $\{X_j : j \in \mathbb{N}\}$, *such that*

$$\mathrm{W}(t) = U\mathbf{1}_{[0,t]}, \quad t \in [0,1]. \tag{3.9}$$

Proof: Let $\{\mathbf{e}_j : j \in \mathbb{N}\}$ be an orthonormal basis of $\mathrm{L}^2([0,1],\mathfrak{m})$, and let

$$X_j = I_{\mathrm{W}}(\mathbf{e}_j), \quad j \in \mathbb{N}. \tag{3.10}$$

By Proposition 3 and Exercise 4 ii, the X_j are independent standard Gaussian variables. Define $U\mathbf{e}_j = X_j$. Then,

$$Uf = I_{\mathrm{W}}(f), \quad f \in \mathrm{L}^2([0,1],\mathfrak{m}). \tag{3.11}$$

To verify (3.11), expand $f = \sum_j c_j\, \mathbf{e}_j$, and obtain

$$Uf = \sum_j c_j\, X_j = \sum_j c_j\, I_{\mathrm{W}}(\mathbf{e}_j) = I_{\mathrm{W}}\left(\sum_j c_j\, \mathbf{e}_j\right), \tag{3.12}$$

where each of the series above converges in the respective L^2-spaces. In particular,

$$U\mathbf{1}_{[0,t]} = I_{\mathrm{W}}(\mathbf{1}_{[0,t]}) = \mathrm{W}(t). \tag{3.13}$$

□

Remarks:

iii **(a new integral).** The Wiener integral $I_{\mathrm{W}}(f)$ is a *stochastic integral* of a *deterministic integrand* (an integrand that does not depend on $\omega \in \Omega$) with respect to a *random integrator* (a stochastic process). It appeared first in various guises in Wiener's papers on

Brownian motion, where its construction was not quite as transparent as it is today; see commentaries by K. Itô and J.-P. Kahane [Wi3, pp. 513–19, 558–63].

For every continuous function f there exists a sequence of partitions

$$\pi_k = \{0 \le t_{1,k} < \cdots < t_{n_k,k} = 1\} \tag{3.14}$$

whose mesh goes to 0, such that

$$\sum_{i=1}^{n_k} f(t_{i,k})\,[\mathrm{W}(t_{i,k}) - \mathrm{W}(t_{i-1,k})] \xrightarrow[k\to\infty]{} I_{\mathrm{W}}(f) \text{ almost surely } (\mathbb{P}). \tag{3.15}$$

We cannot conclude the stronger property (integrability in the Riemann–Stieltjes sense), that (3.15) holds for *all* sequences of partitions whose mesh goes to 0.

iv **(an F_2-measure).** We revisit the F_2-measure constructed in Chapter VI §2 iv. Let W be a Wiener process on a probability space $(\Omega, \mathscr{A}, \mathbb{P})$. For $A \in \mathscr{A}$ and $B \in \mathscr{B}$ (Borel field in [0,1]), define

$$\mu_{\mathrm{W}}(A, B) = \mathbf{E}\ \mathbf{1}_A\ I_{\mathrm{W}}(\mathbf{1}_B), \quad A \in \mathscr{A},\ B \in \mathscr{B}. \tag{3.16}$$

(This agrees with the definition of μ in (VI.2.13) for intervals B.) Then, for all $A \in \mathscr{A}$, $\mu_{\mathrm{W}}(A, \cdot)$ is a scalar measure on $([0,1], \mathscr{B})$ that is absolutely continuous with respect to Lebesgue measure (because $\|I_{\mathrm{W}}(\mathbf{1}_B)\|_{\mathrm{L}^2} = (\mathfrak{m}(B))^{1/2}$). Similarly, for all $B \in \mathscr{B}$, $\mu_{\mathrm{W}}(\cdot, B)$ is a scalar measure on $(\Omega, \mathscr{A})$ that is absolutely continuous with respect to $\mathbb{P}$ (by the definition of μ_{W}). In particular, $\mu_{\mathrm{W}} \in F_2(\mathscr{A}, \mathscr{B})$ and $\|\mu_{\mathrm{W}}\|_{F_2} = \sqrt{2/\pi}$ (Exercise 9). Observe that μ_{W} is not extendible to a scalar measure on $\sigma(\mathscr{A} \times \mathscr{B})$ (as per (VI.2.14)).

If f is a bounded measurable function on $([0,1], \mathscr{B})$, then (Exercise 10)

$$I_{\mathrm{W}}(f) = \frac{\mathrm{d}}{\mathrm{d}\mathbb{P}} \int_{[0,1]} f(t)\ \mu_{\mathrm{W}}(\cdot, \mathrm{d}t). \tag{3.17}$$

v **(white noise).** For a Borel set $B \subset [0,1]$, define

$$\Delta W(B) := I_{\mathrm{W}}(\mathbf{1}_B), \tag{3.18}$$

which obviously extends the definition of $\Delta\mathrm{W}(J)$, where J is an interval. If $B_1, \ldots, B_k$ are pairwise disjoint Borel subsets of [0,1], then $\Delta\mathrm{W}(B_1), \ldots, \Delta\mathrm{W}(B_k)$ are independent Gaussian variables with mean zero and variances $\mathfrak{m}(B_1), \ldots, \mathfrak{m}(B_k)$, respectively. The random set-function $\Delta\mathrm{W}(\cdot)$ is sometimes called *white noise* (e.g., [Nu, p. 8]), and I, too, will use this terminology.

vi **(a generalized Wiener process and its associated F_2-measure).** The following set-indexed process was proposed by S. Kakutani [Kak1] as a generalization of the Wiener process. In the definition of W, replace [0,1] by a locally compact Abelian group G, and replace the Lebesgue measure on [0,1] by a Haar measure $\mathfrak{m}$ on G. Then, a *generalized Wiener process* $\Delta\mathrm{W}$ on G is a collection of random variables indexed by the Borel field $\mathscr{B}$ in G,

$$\Delta\mathrm{W} = \{\Delta\mathrm{W}(B) : B \in \mathscr{B}\}, \tag{3.19}$$

with the following properties:

$$\begin{gathered}\text{for every } B \in \mathscr{B},\ \Delta\mathrm{W}(B) \text{ is Gaussian with}\\ \text{mean 0 and variance } \mathfrak{m}(B);\end{gathered} \tag{3.20}$$

$$\begin{gathered}\text{if } \{B_j\} \text{ is a collection of pairwise disjoint sets in } \mathscr{B},\\ \text{then } \{\Delta\mathrm{W}(B_j)\} \text{ is a statistically independent}\\ \text{system, and if } \mathfrak{m}(\cup_j B_j) < \infty, \text{ then}\\ \Delta\mathrm{W}(\cup_j B_j) = \sum_j \Delta\mathrm{W}(B_j) \quad (\text{convergence in } \mathrm{L}^2(\Omega, \mathbb{P})).\end{gathered} \tag{3.21}$$

Kakutani called this process *generalized Brownian motion.* Its construction is nearly identical to the one in §2: let U be a unitary map from $\mathrm{L}^2(G, \mathfrak{m})$ onto the Hilbert space spanned by independent standard Gaussian variables, and define

$$\Delta\mathrm{W}(B) = U\mathbf{1}_B, \quad B \in \mathscr{B}. \tag{3.22}$$

If $f \in \mathrm{L}^2(G, \mathfrak{m})$, then Uf is the 'Wiener integral' of f with respect to $\Delta\mathrm{W}$ (cf. Proposition 4). In the specific case $[0,1]^2$, the process defined by

$$U\mathbf{1}_{[0,s]\times[0,t]}, \quad (s,t) \in [0,1]^2, \tag{3.23}$$

is the (so-called) *Brownian sheet* [CW].

The definition of the F_2-measure $\mu_{\Delta\mathrm{W}}$ associated with a generalized $\Delta\mathrm{W}$ is the same as that of the F_2-measure in (3.16):

$$\mu_{\Delta\mathrm{W}}(A, B) = \mathbf{E}\ \mathbf{1}_A\ U\mathbf{1}_B, \quad A \in \mathscr{A},\ B \in \mathscr{B}. \tag{3.24}$$

If G is infinite, then $\mu_{\Delta W}$ is not extendible to a scalar measure on $\sigma(\mathscr{A} \times \mathscr{B})$ (Exercise 11).

The construction of $\mu_{\Delta W}$ is reversible. That is, we start with the definition: $\mu \in F_2(\mathscr{A}, \mathscr{B})$ is a *Wiener F_2-measure* if

for each $B \in \mathscr{B}$, $\mu(\cdot, B) \ll \mathbb{P}$, and the Radon–Nikodym derivative $\mathrm{d}\mu(\cdot, B)/\mathrm{d}\mathbb{P}$ is a Gaussian r.v. with mean 0 and variance $\mathfrak{m}(B)$; (3.25)

for every collection of pairwise disjoint sets $\{B_j\}$ in $\mathscr{B}$, $\{\mathrm{d}\mu(\cdot, B_j)/\mathrm{d}\mathbb{P}\}$ is an orthogonal system in $\mathrm{L}^2(\Omega, \mathscr{A}, \mathbb{P})$. (3.26)

Then, a generalized Wiener process (in the sense of Kakutani) associated with a Wiener F_2-measure μ is

$$\Delta \mathrm{W} = \{\mathrm{d}\mu(\cdot, B)/\mathrm{d}\mathbb{P} : B \in \mathscr{B}\}. \tag{3.27}$$

(The association between stochastic processes and Fréchet measures, of which (3.27) is an instance, will be studied in the next chapter.)

vii **(stochastic series of** W**).** Given an orthonormal basis $\{\mathbf{e}_j : j \in \mathbb{N}\}$ of $\mathrm{L}^2([0,1], \mathfrak{m})$, we obtain from Proposition 4 a series representation of W,

$$\mathrm{W}(t) = \sum_{j=1}^{\infty} \left(\int_{[0,t]} \mathbf{e}_j(x) \, \mathrm{d}x \right) I_{\mathrm{W}}(\mathbf{e}_j), \tag{3.28}$$

which converges in $\mathrm{L}^2(\Omega, \mathbb{P})$ uniformly for all $t \in [0,1]$ (Exercises 12, 13). Such a series based on the classical trigonometric system was first considered by Wiener [Wi4, p. 570]. (See also [PaZy2, Chapter IX].) In this case, we take the orthonormal basis to be the normalized cosine system, and expand

$$\mathbf{1}_{[0,t]}(s) \sim t + \sqrt{2} \sum_{j=1}^{\infty} \frac{\sin \pi j t}{\pi j} \cos \pi j s. \tag{3.29}$$

From (3.28), we obtain

$$\mathrm{W}(t) \sim t X_0 + \sqrt{2} \sum_{j=1}^{\infty} \frac{\sin \pi j t}{\pi j} X_j, \tag{3.30}$$

where

$$X_j = I_{\mathrm{W}}(\sqrt{2} \cos \pi j s), \quad j \in \mathbb{N}, \tag{3.31}$$

and

$$X_0 = \mathrm{W}(1). \tag{3.32}$$

The series in (3.30), known as the Fourier–Wiener (sine) series [Kah2, p. 150], converges in $\mathrm{L}^2(\Omega, \mathbb{P})$ for all $t \in [0,1]$, and in $\mathrm{L}^2([0,1], \mathfrak{m})$ almost surely ($\mathbb{P}$). After some groundwork in the next two sections, we will establish that a subsequence of partial sums of this series converges uniformly on [0,1] almost surely, and thus conclude that sample-paths of a Wiener process are almost surely continuous.

4 Sub-Gaussian Systems

In this section we formalize a notion of independence based on measurements of decay of tail-probabilities $\mathbb{P}(|X| > x)$ as $x \to \infty$. We start with

Definition 5

i. $X \in \mathrm{L}^2(\Omega, \mathbb{P})$ is a *sub-Gaussian* variable if there exists $0 < A < \infty$ such that

$$\overline{\lim_{x\to\infty}} \exp(Ax^2)\ \mathbb{P}(|X| > x\|X\|_{\mathrm{L}^2}) < \infty. \tag{4.1}$$

ii. $F \subset \mathrm{L}^2(\Omega, \mathbb{P})$ is a *sub-Gaussian* system if every $X \in \mathrm{L}^2_F(\Omega, \mathbb{P})$ is sub-Gaussian. (As in Chapter VII, $\mathrm{L}^2_F(\Omega, \mathbb{P})$ denotes the L^2-closure of the linear span of F.)

Lemma 6 (see [Kah3, p. 82]). *Suppose $X \in \mathrm{L}^2(\Omega, \mathbb{P})$, $\|X\|_{\mathrm{L}^2} \le 1$. The following are equivalent.*

i. *X is sub-Gaussian;*

ii. *there exists $0 < B < \infty$ such that*

$$\sup\{\|X\|_{\mathrm{L}^p}/\sqrt{p} : p > 1\} \le B; \tag{4.2}$$

iii. *there exists $0 < C < \infty$ such that*

$$\overline{\lim_{t\to\infty}} \mathbf{E} \exp(t|X| - Ct^2) < \infty; \tag{4.3}$$

iv. *there exists $0 < D < \infty$ such that*

$$\mathbf{E} \exp(D|X|^2) < \infty. \tag{4.4}$$

Proof: ii ⇒ iii. Estimate

$$\begin{aligned}\mathbf{E}\exp(t|X|) &= \sum_{n=0}^{\infty} t^n\, \mathbf{E}|X|^n/n! \quad \text{(Taylor series expansion of exp)}\\ &\le \sum_{n=0}^{\infty} (B^2t^2)^{\frac{n}{2}} n^{\frac{n}{2}}/n! \quad (p = n \text{ in } (4.2))\\ &\le \exp(8B^2t^2) \quad \left(n^{\frac{n}{2}}/n! < 2^n\Big/\left(\frac{n}{2}\right)!\right). \end{aligned} \tag{4.5}$$

iii ⇒ i. For $t > 0$ and $x > 0$,

$$\mathbb{P}(|X| > x)\exp(tx) \le \mathbf{E}\exp(t|X|), \tag{4.6}$$

and, by (4.3), for sufficiently large $t > 0$,

$$\mathbb{P}(|X| > x) \le \exp(Ct^2 - tx). \tag{4.7}$$

Put $t = x/2C$ in (4.7), and obtain (4.1) with $A = 1/4C$.
i ⇒ iv. Take $D < A$, and, for sufficiently large $k > 0$, estimate

$$\begin{aligned}\mathbf{E}\exp(D|X|^2) &= \int_0^{\infty} \mathbb{P}(\exp(D|X|^2) > x)\, \mathrm{d}x\\ &\le k + \int_k^{\infty} x^{-C/A} \mathrm{d}x < \infty. \end{aligned} \tag{4.8}$$

iv ⇒ i. Estimate

$$\begin{aligned}\mathbb{P}(|X| > x) &\le \exp(-Dx^2)\, \mathbf{E}\exp(D|X|^2)\\ &\le \exp(-Ax^2) \end{aligned} \tag{4.9}$$

for sufficiently large x, where $0 < A < \infty$ is chosen appropriately.
i ⇒ ii. Assume $p > 1$, and estimate

$$\begin{aligned}\mathbf{E}|X|^p &= \int_0^{\infty} \mathbb{P}(|X|^p > x)\, \mathrm{d}x \le K\int_0^{\infty} \exp(-Ax^{2/p})\, \mathrm{d}x\\ &= Kp\sqrt{\pi/A}\int_0^{\infty} (\sqrt{A/\pi})\, x^{p-1} \exp(-Ax^2)\, \mathrm{d}x, \end{aligned} \tag{4.10}$$

for some $K > 0$. The last integral is the $(p-1)$st moment of a Gaussian r.v. with mean 0 and variance $1/2A$. Therefore, $\mathbf{E}|X|^p$ is $\mathscr{O}(p^{(p+1)/2})$, and ii follows. □

Remarks:

i **(an Orlicz norm).** Consider the Orlicz function

$$\phi_1(x) = \exp(x^2) - 1, \quad x > 0, \tag{4.11}$$

and the corresponding Orlicz norm

$$\|Y\|_{\phi_1} = \inf\{\rho > 0 : \mathbf{E}\, \phi_1(|Y|/\rho) \leq 1\}, \quad Y \in \mathrm{L}^0(\Omega, \mathscr{A}). \tag{4.12}$$

We denote the Orlicz space comprising all $Y \in \mathrm{L}^0(\Omega, \mathscr{A})$ such that $\|Y\|_{\phi_1} < \infty$ by $\mathrm{L}_{\phi_1}(\Omega, \mathbb{P})$. (See [LiTz, Vol. II, p. 120].)

By Lemma 6, $F \subset \mathrm{L}^2(\Omega, \mathbb{P})$ is sub-Gaussian if and only if

$$\mathrm{L}^2_F(\Omega, \mathbb{P}) \subset \mathrm{L}_{\phi_1}(\Omega, \mathbb{P}). \tag{4.13}$$

Equivalently, $F \subset \mathrm{L}^2(\Omega, \mathbb{P})$ is sub-Gaussian if and only if there exists $K > 0$ such that

$$\|X\|_{\phi_1} \leq K \|X\|_{\mathrm{L}^2} \tag{4.14}$$

for all $X \in \operatorname{span}(F)$.

ii **(a sub-Gaussian system conveys independence).** Suppose a measurement of a certain quantity, whose true value is μ, is performed N times. Let y_j be the jth measurement, and denote the jth error by

$$x_j := y_j - \mu, \quad j = 1, \ldots, N. \tag{4.15}$$

If the N measurements are independent – in some heuristic sense – then we expect these errors to cancel out, and therefore expect $\Sigma_j x_j$ to be small. Conversely, if the N trials are, to some degree, interdependent – again, in some heuristic sense – then we expect fewer cancellations, and thus expect $\Sigma_j x_j$ to be correspondingly larger (Exercise 14). To make this precise in a 'statistical' setting, suppose this procedure (N repeated measurements) is performed K times, and that K is large. Let Ω be the set of the K outcomes, each an N-tuple $(x_1, \ldots, x_N)$, and let $\mathbb{P}$ be the uniform probability measure on Ω ($\mathbb{P}(x_1, \ldots, x_N) = 1/K$). Consider the projections

$$X_j(x_1, \ldots, x_N) = x_j, \quad j \in [N], \ (x_1, \ldots, x_N) \in \Omega. \tag{4.16}$$

In this scenario, viewing these X_j as random variables, we assess cancellation of errors by the tail-probabilities

$$\mathbb{P}(|\Sigma_j X_j| > x), \quad x > 0, \tag{4.17}$$

and think of (4.17) as a gauge of interdependence of the N measurements: smaller $\mathbb{P}(|\Sigma_j X_j| > x)$ mean less interdependence.

Now consider a general probability space $(\Omega, \mathbb{P})$, and $\{X_j : j \in \mathbb{N}\}$ an orthonormal system in $\mathrm{L}^2(\Omega, \mathbb{P})$. In general, tail-probabilities can be estimated by Chebyshev's inequality,

$$\mathbb{P}\left(\left|\sum_j a_j\, X_j\right| > x\right) \le 1/x^2 \text{ for all } x > 0, \text{and all } (a_j) \in B_{l^2}, \tag{4.18}$$

estimates that could be sharper, but *no sharper* than sub-Gaussian. Precisely, this means that for all $\alpha > 2$,

$$\inf\left\{\varliminf_{x\to\infty} (-1/x^\alpha) \log \mathbb{P}\left(\left|\sum_j a_j\, X_j\right| > x\right) : \Sigma_j |a_j|^2 = 1\right\} = 0. \tag{4.19}$$

(See [Ru1, Theorem 3.4] and Lemma 6.) We view orthonormal sub-Gaussian systems $\{X_j : j \in \mathbb{N}\}$ as independent systems.

Because every finite set of sub-Gaussian variables is obviously a sub-Gaussian system, this notion of independence based on tail-probability estimates needs further fine-tuning. For an orthonormal system $F = \{X_j\}$ (finite or infinite) of sub-Gaussian variables, define (cf. (4.19))

$$c_F := \inf\left\{\varliminf_{x\to\infty} (-1/x^2) \log \mathbb{P}\left(\left|\sum_j a_j\, X_j\right| > x\right) : \Sigma_j |a_j|^2 = 1\right\}. \tag{4.20}$$

F is sub-Gaussian if and only if there exists $K > 0$ such that $K \le c_S$ for all finite subsets $S \subset F$. Equivalently (cf. (VII.9.29)), we define

$$\eta_F := \sup\left\{\varlimsup_{x\to\infty} \|X\|_{\mathrm{L}^p}/\sqrt{p} : X \in \text{span } F, \ \ \|X\|_{\mathrm{L}^2} = 1\right\}, \tag{4.21}$$

and

$$o_F := \sup\{\|X\|_{\phi_1} : X \in \text{span } F, \|X\|_{\mathrm{L}^2} = 1\}. \tag{4.22}$$

By Lemma 6, $c_F > 0 \Leftrightarrow \eta_F < \infty \Leftrightarrow o_F < \infty$. Any one of these constants can be viewed as a gauge of the 'independence' manifested in F.

iii **(examples).**

1. A system of statistically independent standard Gaussian variables $\{X_j : j \in \mathbb{N}\}$ is sub-Gaussian: every element in the L^2-closure of the span of $\{X_j : j \in \mathbb{N}\}$ is Gaussian with mean 0 (Exercise 4 i), and every standard Gaussian variable X satisfies

$$\begin{aligned}\mathbf{E}\exp(s|X|) &\leq \mathbf{E}(\exp(-sX) + \exp(sX))\\ &\leq 2\exp(s^2/2), \quad s > 0. \end{aligned} \tag{4.23}$$

2. The Rademacher system $\{r_j : j \in \mathbb{N}\}$ is sub-Gaussian. This, by Lemma 6, is equivalent to the Khintchin inequalities (Exercise II.4),

$$\left\| \sum_{j=1}^{n} a_j\, r_j \right\|_{\mathrm{L}^p} \leq B\sqrt{p}\left(\sum_{j=1}^{n} |a_j|^2\right)^{\frac{1}{2}}, \quad p > 2. \tag{4.24}$$

   To prove directly that $\{r_j : j \in \mathbb{N}\}$ is sub-Gaussian, observe that if $X = \Sigma_{j=1}^{n} a_j r_j$, and $s \in (-\infty, \infty)$, then (Exercise 15)

$$\begin{aligned}\mathbf{E}\exp(sX) &= \prod_{j=1}^{N} \mathbf{E}\exp(s a_j r_j) = \prod_{j=1}^{N} \cosh(s a_j)\\ &\leq \exp(s^2 \|X\|_{\mathrm{L}^2}^2). \end{aligned} \tag{4.25}$$

   More generally, by the Burkholder–Gundy inequalities, every martingale-difference sequence (X_j) with $\sup \|X_j\|_{\mathrm{L}^\infty} < \infty$ is sub-Gaussian. (See Exercise VII.11.)

3. Every Sidon set is sub-Gaussian. This follows from Lemma 6 and Theorem VII.41. (In Theorem VII.41, let $t = 1$, and replace the Walsh system W by any discrete Abelian group.)

These examples illustrate three ostensibly separate notions of 'independence': *statistical, functional*, and *sub-Gaussian.* The first is at the heart of classical probability theory; the second was discussed in a framework of harmonic analysis in Chapter VII (in connection with Sidonicity), and the third is the essence of this section. Every statistically independent, $L_{\phi_1}(\Omega, \mathbb{P})$-bounded system is sub-Gaussian (Exercise 16), but sub-Gaussian systems need not be statistically independent (e.g., Sidon sets). In a framework of harmonic analysis, every sub-Gaussian system of characters is Sidon [Pi1], and therefore can be viewed as functionally independent; see Chapter VII §11. Whether functionally independent systems can be viewed as sub-Gaussian systems, and how to view sub-Gaussian systems as functionally independent systems are open (-ended) questions.

5 Random Series

The idea of *random series* is, arguably, the single important concept that brought probability theory and classical analysis together. It first appeared at the end of the nineteenth century, in the work of Emil Borel [Bor], but not quite in the form known today ([Kah3, p. 37]). Random series in their present-day guise were pioneered during the 1920s and 1930s by the grandmasters Khintchin, Kolmogorov, Lévy, Littlewood, Rademacher, Steinhaus, Paley, Wiener, and Zygmund (e.g., [PaWiZy]). In the beginning, the focus was mainly on Rademacher series,

$$\sum_j r_j(\omega)\ f_j, \quad \omega \in \{-1, +1\}^{\mathbb{N}}, \tag{5.1}$$

where (f_j) was a prescribed sequence of functions. But it soon became apparent (e.g., [Lit3], [Ste2]) that Rademacher functions could be replaced, to good effect, by the so-called *Steinhaus functions* (see Chapter VII §9, Remark i): uniformly distributed, $\{z \in \mathbb{C} : |z| = 1\}$-valued, statistically independent random variables (Chapter II §6). These functions were dubbed *Steinhaus* in a seminal paper [SaZy2] (dedicated to Steinhaus), where Salem and Zygmund – building on Paley's and Zygmund's previous work – established fundamental properties of series 'randomized' by Rademacher as well as Steinhaus systems.

In a subsequent phase, building on Salem's and Zygmund's work, J.-P. Kahane focused on random series involving statistically independent Gaussian variables [Kah1]. This indeed was a major step – certainly from this chapter's viewpoint – primarily because every Wiener integral can be represented as a random Gaussian series (Exercise 17).

In particular, properties of sample-paths of a Wiener process could then be deduced from general properties of series randomized by statistically independent Gaussian variables.

In this section, building on work of Kahane, Salem and Zygmund, we consider series randomized by sub-Gaussian systems. For applications to a Wiener process, we could just as well take statistically independent Gaussian variables, but the more general approach taken here will become useful at a later point, in the analysis of more general stochastic processes. The lemma and theorem below are the main tools. The estimates in Theorem 8 in the case of statistically independent sub-Gaussian systems are the *Kahane–Salem–Zygmund estimates* [Kah3, pp. 68–9], which are of paramount importance.

In the lemma below, (S, ν) is a finite measure space, and T a linear subspace of $\mathrm{L}^\infty(S, \nu)$. For $0 < u \leq 1$, we consider

$$\rho(T, u) = \inf\{\nu\{|f| \geq u\|f\|_{\mathrm{L}^\infty}\} : f \in T\}. \tag{5.2}$$

Lemma 7 *Let $\{X_j : j \in \mathbb{N}\} \subset \mathrm{L}^2(\Omega, \mathbb{P})$ be an orthonormal sub-Gaussian system. Suppose $\rho(T, u) = \rho > 0$ for some $u \in (0, 1]$. Let $\{f_j\}$ be a finite subset of T such that*

$$\left\| \sum_j |f_j|^2 \right\|_{\mathrm{L}^\infty} \leq 1, \tag{5.3}$$

and consider (the random function) $p = \sum_j f_j \otimes X_j$. Then, the random variable

$$\|p\|_{\mathrm{L}^\infty} = \operatorname*{ess\,sup}_{s \in S} \left| \sum_j f_j(s)\ X_j \right| \tag{5.4}$$

is sub-Gaussian. In particular, for all $0 < A < c_{\{X_j\}}$, there exist $L > 0$ such that

$$\mathbb{P}(\|p\|_{\mathrm{L}^\infty} > x) < \frac{\nu(S)}{\rho} \exp(-Au^2x^2), \quad x > L. \tag{5.5}$$

Proof: By (5.3) and the definition of $c_{\{X_j\}}$ (in (4.20)), there exists $L > 0$ such that for all $s \in S$,

$$\begin{aligned} \mathbb{P}\{\xi : |p(s, \xi)| > x\} &= \mathbf{E}_\omega \mathbf{1}_{\{\xi : |p(s,\xi)| > x\}}(\omega) \\ &< \exp(-Ax^2), \quad x > L. \end{aligned} \tag{5.6}$$

(For clarity's sake, we make explicit the appearance of sample points ξ and ω in Ω.) Integrating (5.6) over S with respect to ν, we obtain

$$\begin{aligned}\mathbf{E}_\omega &\left(\int_S \mathbf{1}_{\{\xi:|p(s,\xi)|>x\}}(\omega)\ \nu(\mathrm{d}s)\right)\\ &= \int_S (\mathbf{E}_\omega \mathbf{1}_{\{\xi:|p(s,\xi)|>x\}}(\omega))\ \nu(\mathrm{d}s)\\ &\le \nu(S)\exp(-Ax^2). \qquad (5.7)\end{aligned}$$

For $\omega \in \Omega$, define

$$I_\omega = \{t \in S : |p(t,\omega)| \ge u\ \|p(\cdot,\omega)\|_{\mathrm{L}^\infty}\}. \qquad (5.8)$$

Then, for all $s \in S$, $\omega \in \Omega$, and $x > L$,

$$\mathbf{1}_{I_\omega}(s)\ \mathbf{1}_{\{\xi:|p(s,\xi)|>x\}}(\omega) \ge \mathbf{1}_{I_\omega}(s)\ \mathbf{1}_{\{\xi:\|p(\cdot,\xi)\|_{\mathrm{L}^\infty}>x/u\}}(\omega). \qquad (5.9)$$

Therefore, by (5.7) and the definition of ρ,

$$\begin{aligned}\nu(S)\exp(-Ax^2) &\ge \mathbf{E}_\omega\left(\int_S \mathbf{1}_{\{\xi:|p(s,\xi)|>x\}}(\omega)\ \nu(\mathrm{d}s)\right)\\ &\ge \mathbf{E}_\omega\left(\int_{I_\omega} \mathbf{1}_{\{\xi:|p(s,\xi)|>x\}}(\omega)\ \nu(\mathrm{d}s)\right)\\ &\ge \rho\ \mathbb{P}\{\xi : \|p(\cdot,\xi)\|_{\mathrm{L}^\infty} > x/u\}, \qquad (5.10)\end{aligned}$$

which (because $\rho > 0$) implies (5.5) (Exercise 18). □

Let $S = [0,1]$, and ν = Lebesgue measure. For $N > 0$, P_N below will stand either for the space of trigonometric polynomials on [0,1] of degree N, or the space of Walsh polynomials of degree N (the span of $\{w_n : n \le N\}$, where $\{w_n : n \in \mathbb{N}\}$ is the Walsh system enumerated in (VII.4.1)).

Theorem 8 *If $\{X_j : j \in \mathbb{N}\} \subset \mathrm{L}^2(\Omega,\mathbb{P})$ is an orthonormal sub-Gaussian system, then there exists $C > 0$ such that for all finite sets $\{f_j\} \subset P_N$,*

$$\mathbb{P}\left\{\left\|\sum_j f_j \otimes X_j\right\|_{\mathrm{L}^\infty} > C\left\|\sum_j |f_j|^2\right\|_{\mathrm{L}^\infty}^{\frac{1}{2}} (\ln N)^{\frac{1}{2}}\right\} \le 1/N. \qquad (5.11)$$

Proof: Let P_N be the space of trigonometric polynomials of degree N. In order to estimate $\rho(P_N, \cdot)$ (defined in (5.2)), we first estimate the modulus of continuity of

$$f(x) = \sum_{j=0}^{N} (a_j \cos 2\pi jx + b_j \sin 2\pi jx). \tag{5.12}$$

For $x \in [0,1]$ and $y \in [0,1]$,

$$\begin{aligned} &|f(x) - f(y)| \\ &\leq \max_{1 \leq j \leq N} \{|\cos 2\pi jx - \cos 2\pi jy|,\ |\sin 2\pi jx - \sin 2\pi jy|\} \sum_{j=0}^{N} (|a_j| + |b_j|) \\ &\leq 2\pi N\ |x-y| \sum_{j=0}^{N} (|a_j| + |b_j|) \\ &\leq 2\pi N\ |x-y|\ \sqrt{2N} \left(\sum_{j=0}^{N} |a_j|^2 + |b_j|^2 \right)^{\frac{1}{2}} \\ &\leq 2\pi\sqrt{2}\ N^{\frac{3}{2}} \|f\|_{L^\infty} |x-y|. \end{aligned} \tag{5.13}$$

Suppose $\|f\|_{L^\infty} = |f(x_0)|$, and let $\delta = 1/4\pi\sqrt{2}\ N^{3/2}$. By (5.13), for all $t \in (x_0 - \delta, x + \delta)$,

$$|f(x_0)| \leq 2\ |f(t)|, \tag{5.14}$$

which implies

$$\rho\left(P_N, \frac{1}{2}\right) \geq 1/2\pi\sqrt{2}\ N^{\frac{3}{2}}. \tag{5.15}$$

Apply Lemma 7 with $u = 1/2$ and $x = C(\ln N)^{\frac{1}{2}}$, for an appropriately chosen $C > 0$.

We now consider the space P_N of Walsh polynomials of degree N, and let $f \in P_N$,

$$f = \sum_{j=0}^{N} a_j w_j. \tag{5.16}$$

Consider the measure preserving equivalence between ([0,1], *Lebesgue measure*) and ($\{-1,1\}^{\mathbb{N}}$, *normalized Haar measure*). (See Chapter VII

§4 and Exercise VII.10.) We view f as a W_K-polynomial defined on $\{-1,1\}^{\mathbb{N}}$, where

$$\log_2(N) - 1 \le K \le \log_2(N), \tag{5.17}$$

and spect $f \subset \{r_{j_1} \cdots r_{j_K} : 0 \le j_1 \le \cdots \le j_K \le K\}$. For $u \in \{-1,1\}^{\mathbb{N}}$ and $v \in \{-1,1\}^{\mathbb{N}}$,

$$|f(u) - f(v)| \le \max_{1 \le j \le N} \{|w_j(u-v) - 1|\} \sum_{j=0}^{N} |a_j|. \tag{5.18}$$

Therefore, if

$$V = \{t \in \{-1,1\}^{\mathbb{N}} : t(j) = 1 \text{ for } j \in [K]\}, \tag{5.19}$$

then $\nu(V) \ge 1/N$, where ν is the Haar measure on $\{-1,1\}^{\mathbb{N}}$, and $f(x) = f(x+t)$ for all $t \in V$. In particular,

$$\rho(P_N, 1) \ge 1/N. \tag{5.20}$$

Apply Lemma 7 with $u = 1$, $x = C(\ln N)^{1/2}$, and $C > 0$ chosen appropriately. □

Corollary 9 (cf. [Kah3, pp. 84–5]). *Let $\{X_j : j \in \mathbb{N}\} \subset L^2(\Omega, \mathbb{P})$ be an orthonormal sub-Gaussian system, and let*

$$B_k = \{2^k, 2^k + 1, \ldots, 2^{k+1} - 1\}, \quad k = 0, 1, \ldots.$$

Suppose $(a_j : j \in \mathbb{N})$ is a scalar sequence such that

$$s_k = \left(\sum_{j \in B_k} |a_j|^2 \right)^{\frac{1}{2}} \quad (k = 0, 1, \ldots) \textit{ is a decreasing sequence}, \tag{5.21}$$

and

$$\sum_k s_k < \infty. \tag{5.22}$$

Then, $\{\sum_{j=1}^{2^{2^k}} a_j X_j \sin 2\pi jt : k = 0, \ldots\}$ converges uniformly on $[0,1]$ almost surely. In particular, the random series $\sum_{j=1}^{\infty} a_j X_j \sin 2\pi jt$ represents almost surely a continuous function on $[0,1]$.

Proof: Consider the random trigonometric polynomials

$$p_k(t) = \sum_{j \in B_{2^k}} a_j X_j \sin 2\pi jt, \tag{5.23}$$

and the events

$$E_k = \left\{ \|p_k\|_{L^\infty} \geq 2C \left(2^k \sum_{j \in B_{2^k}} |a_j|^2 \right)^{\frac{1}{2}} \right\}, \quad k = 0, 1, \ldots, \qquad (5.24)$$

where $C > 0$ is the numerical constant in (5.11). By Theorem 8,

$$\Sigma_k \mathbb{P}(E_k) < \infty,$$

and therefore (by the Borel–Cantelli lemma), $\mathbb{P}(\overline{\lim}\, E_k) = 0$. Therefore,

$$(\|p_k\|_{L^\infty} : k \in \mathbb{N}) \text{ is } \mathscr{O}\left(2^{\frac{k}{2}} \left(\sum_{j \in C_k} |a_j|^2 \right)^{\frac{1}{2}} \right) \text{ almost surely } (\mathbb{P}). \quad (5.25)$$

Because $B_{2^k} = \bigcup\{B_j : j = 2^k, \ldots, 2^{k+1} - 1\}$, and (s_k) is a decreasing sequence, we have

$$\left(\sum_{j \in B_{2^k}} |a_j|^2 \right)^{\frac{1}{2}} = \left(\sum_{n=2^k}^{2^{k+1}-1} |s_n|^2 \right)^{\frac{1}{2}} \leq 2^{\frac{k}{2}} s_{2^k}, \quad k \in \mathbb{N}. \qquad (5.26)$$

By (5.22), $\Sigma_k 2^k s_{2^k} < \infty$, which implies, via (5.25) and (5.26), that $\Sigma_k \|p_k\|_{L^\infty} < \infty$ almost surely $(\mathbb{P})$, and hence the corollary. □

Corollary 10 *The sample-paths of a Wiener process are almost surely continuous.*

Proof: Apply Corollary 9 to the Fourier–Wiener series (stated in (3.30)). □

Remarks:

i **(another proof that Theorem VII.36 is sharp).** In Chapter VII §11, we verified by an indirect argument that the Littlewood 2n/(n+1)-inequalities are best possible. Here we give a direct proof based on Corollary 9.

Fix $n \in \mathbb{N}$. Let $m \geq n$ be an arbitrary integer. Consider the W_n-polynomial

$$f = \sum_{0<j_1<\cdots<j_n\leq m} r_{j_1}\cdots r_{j_n}, \tag{5.27}$$

whose degree (relative to the Paley ordering in (VII.4.1)) is 2^m. Randomize $\hat{f}$ by a Rademacher system indexed by $\mathbb{N}^n$ (a sub-Gaussian system),

$$f_\omega = \sum_{0<j_1<\cdots<j_n\leq m} r_{j_1\cdots j_n}(\omega)\ r_{j_1}\cdots r_{j_n}, \quad \omega \in \{-1,1\}^{\mathbb{N}^n}, \tag{5.28}$$

and then deduce from (5.11) the existence of $\omega \in \{-1,1\}^{\mathbb{N}^n}$ such that

$$\|f_\omega\|_{\mathrm{L}^\infty} \leq C\ \|f\|_{\mathrm{L}^2}(\ln 2^m)^{\frac{1}{2}} \leq C(\ln 2)^{\frac{1}{2}}|\mathrm{spect}\ f|^{\frac{1}{2}}m^{\frac{1}{2}}. \tag{5.29}$$

Then for $t \in [1,2)$,

$$\|\hat{f}_\omega\|_t/\|f_\omega\|_{\mathrm{L}^\infty} \geq |\mathrm{spect} f|^{\frac{2-t}{2t}}/C(\ln 2)^{\frac{1}{2}}m^{\frac{1}{2}} \geq K_n\ m^{\frac{2n-tn-t}{2t}}, \tag{5.30}$$

where $K_n > 0$ is a numerical constant that depends only on n (cf. (VII.11.16)). By (5.30), if $t < 2n/(n+1)$, then $\|\hat{f}_\omega\|_t/\|f_\omega\|_{\mathrm{L}^\infty}$ can be made arbitrarily large by increasing m. Therefore, $\zeta_{W_n}(t) = \infty$. (Cf. Corollary VII.42; see also Exercise 19.)

ii **(there is more. . .).** That W(t), $t \in$ [0,1], is almost surely continuous is among the most basic observations about sample-paths of a Wiener process. For various fine properties of these paths – and much more! – I refer the reader to any of several books about 'Brownian motion'; e.g., [Dur], [Hi], [KarSh], [Lé3], [Ne], [Pet], [RevYor]. (Throughout an extensive mathematical literature, with some notable exceptions (e.g., [K2], [Ne]), 'Brownian motion' has become an alias for a Wiener process. In all other scientific writing, 'Brownian motion' and 'Brownian movement' generically refer to a physical phenomenon. So far as I can determine, Paul Lévy was first to refer to Wiener's model for Brownian motion as 'Brownian motion'.)

6 Variations of the Wiener F_2-measure

Let μ_W be the Wiener F_2-measure on $\mathscr{A}\times\mathscr{B}$ defined in Chapter VI §2 and recalled in Remark iv §3 (in this chapter). In Chapter VI, we used μ_W as an example of an F_2-measure that cannot be extended to a scalar measure on $\sigma(\mathscr{A}\times\mathscr{B})$. Specifically, we verified in (VI.2.14) that the *total variation* of μ_W is infinite:

$$\|\mu_W\|_{F_1}$$

$$:= \sup\left\{\sum_{j,k}|\mu_W(A_j,B_k)| : \Sigma_j\mathbf{1}_{A_j}\le 1,\ \Sigma_k\mathbf{1}_{B_k}\le 1,\ \{A_j\}\subset\mathscr{A},\ \{B_k\}\subset\mathscr{B}\right\}$$

$$= \infty. \tag{6.1}$$

On the other hand, because $\|\mu_W\|_{F_2}$ is finite ($\|\mu_W\|_{F_2} = \sqrt{2/\pi}$; Exercise 9), Littlewood's 4/3-inequality (e.g., (II.5.1)) implies that the '4/3-variation' of μ_W is finite. Specifically, consider

$$\|\mu_W\|_{(p)} := \sup\left\{\left(\sum_{j,k}|\mu_W(A_j,B_k)|^p\right)^{\frac{1}{p}} :\right.$$

$$\left.\Sigma_j\mathbf{1}_{A_j}\le 1,\ \Sigma_k\mathbf{1}_{B_k}\le 1,\ \{A_j\}\subset\mathscr{A},\ \{B_k\}\subset\mathscr{B}\right\}, \tag{6.2}$$

and conclude that $\|\mu_W\|_{(4/3)} \le 2/\sqrt{\pi}$. A question arises: can 4/3 be replaced by a smaller exponent? (A similar question motivated Littlewood's 1930 paper [Lit4]; see Chapter I §2.) In this section we obtain sharp estimates on the variation of μ_W, which imply

$$\ell_W := \inf\{p : \|\mu_W\|_{(p)} < \infty\} = 1 \tag{6.3}$$

(while, according to (6.1), $\|\mu_W\|_{(1)} = \infty$).

To start, we fix a continuous, non-decreasing convex function φ on $[0,\infty)$, such that $\varphi(0)=0, \lim_{t\to\infty}\varphi(t)=\infty$, and

$$\varphi(t) = \exp(-1/t^2),\quad t\in\left(0,\sqrt{\frac{2}{3}}\right] \tag{6.4}$$

($\sqrt{2/3}$ is the inflection point of $\exp(-1/t^2)$). In particular, φ is an *Orlicz function.* Denote by O_φ the set of finitely supported scalar arrays (b_{jk}) such that

$$\sum_{j,k} \varphi(|b_{jk}|) \leq 1. \tag{6.5}$$

Theorem 11

$$\sup\left\{\left|\sum_{j,k} \mu_W(A_j, B_k)\ b_{jk}\right| : \Sigma_j \mathbf{1}_{A_j} \leq 1,\right.$$

$$\left.\Sigma_k \mathbf{1}_{B_k} \leq 1, \{A_j\} \subset \mathscr{A}, \{B_k\} \subset \mathscr{B}, \{b_{jk}\} \in O_\varphi\right\} < \infty. \tag{6.6}$$

Lemma 12 *Let $\{Y_k\}$ be an orthonormal sub-Gaussian system. There exists $B > 0$ (depending only on $\{Y_k\}$) such that if (c_{jk}) is a scalar array satisfying*

$$\sum_k |c_{jk}|^2 \leq 1/\log j, \quad j = 2, \ldots, \tag{6.7}$$

then

$$\mathbf{E} \sup_{j\geq 2} \left|\sum_k c_{jk}\ Y_k\right| \leq B. \tag{6.8}$$

Proof: By assumption, $c_{\{Y_j\}}$ (defined in (4.20)) is positive. We fix $0 < C < c_{\{Y_j\}}$. Then, for sufficiently large $K > 1/\sqrt{C}$, we have

$$\begin{aligned}
\mathbf{E} \sup_{j\geq 2} \left|\sum_k c_{jk}\ Y_k\right| &= \int_0^\infty \mathbb{P}(\cup_j\{|\Sigma_k\ c_{jk}\ Y_k| > t\})\ dt \\
&\leq K + \int_K^\infty \mathbb{P}(\cup_j\{|\Sigma_k\ c_{jk}\ Y_k| > t\})\ dt \\
&\leq K + \sum_{j=2}^\infty \int_K^\infty \exp(-Ct^2 \log j)\ dt := B < \infty. \qquad (6.9)
\end{aligned}$$

The equality above is routine; the second estimate follows from (6.7) and the definition of $c_{\{Y_j\}}$, and the third estimate is a calculus exercise (Exercise 20). □

Proof of Theorem 11: Let $\{A_j\} \subset \mathscr{A}$ and $\{B_k\} \subset \mathscr{B}$ each be a collection of pairwise disjoint measurable sets. Denote $d_k = \sqrt{\mathfrak{m}(B_k)}$ and $Y_k = (1/d_k)\Delta\mathrm{W}(B_k)$, $k \in \mathbb{N}$. Then, $\{Y_k\}$ is an orthonormal sub-Gaussian system. (ΔW is the white noise defined in (3.18).) Note that

$$\mu_{\mathrm{W}}(A_j, B_k) = d_k \mathbf{E} \mathbf{1}_{A_j} Y_k. \tag{6.10}$$

Fix an arbitrary $(b_{jk}) \in O_\varphi$. By rearranging the j, we can assume that for $j = 1, 2, \ldots$,

$$\sum_k \varphi(|b_{jk}|) \leq 1/j. \tag{6.11}$$

Without loss of generality we can assume $|b_{jk}| \leq \sqrt{2/3}$ (Exercise 21), and then obtain from (6.11) and the definition of φ that

$$\sup_k |b_{jk}| \leq (1/\log j)^{\frac{1}{2}}, \quad j = 2, \ldots. \tag{6.12}$$

Therefore (because $\Sigma_k |d_k|^2 = \Sigma_k \mathfrak{m}(B_k) = 1$), we have

$$\sum_k |b_{jk}\ d_k|^2 \leq 1/\log j, \quad j = 2, \ldots. \tag{6.13}$$

We rewrite, and then estimate:

$$\left|\sum_{j,k} \mu_{\mathrm{W}}(A_j, B_k)\ b_{jk}\right| = \left|\sum_{j,k} (d_k \mathbf{E} \mathbf{1}_{A_j} Y_k)\ b_{jk}\right|$$

$$= \left|\sum_j \mathbf{E} \mathbf{1}_{A_j} \sum_k b_{jk}\ d_k\ Y_k\right|$$

$$\leq \mathbf{E} \sup \left|\sum_k b_{jk}\ d_k\ Y_k\right| \leq B, \tag{6.14}$$

where the equalities follow from (6.10), and the inequality from (6.13) and Lemma 12. □

Define

$$\theta(x) = x/\{\log(1/x)\}^{\frac{1}{2}} \quad \text{for } x \in (0,1), \text{ and } \theta(0) = 0. \tag{6.15}$$

Then, by a computation (Exercise 22), there exists $0 < \delta < 1$ such that

$$\theta(x) \leq 2\ \varphi^*(x), \quad x \in (0, \delta), \tag{6.16}$$

where φ^* is the *complementary function* to φ. (See [LiTz, Vol. I, p. 147].)

Corollary 13

$$\sup\left\{\sum_{j,k}\theta(|\mu_{\mathrm{W}}(A_j,B_k)|):\Sigma_j\mathbf{1}_{A_j}\le 1,\right.$$

$$\left.\Sigma_k\mathbf{1}_{B_k}\le 1,\ \{A_j\}\subset\mathscr{A},\ \{B_k\}\subset\mathscr{B}\right\}<\infty. \tag{6.17}$$

In particular, $\|\mu_{\mathrm{W}}\|_{(p)}<\infty$ *for all* $p>1$*, and hence*

$$\ell_{\mathrm{W}}=1. \tag{6.18}$$

Proof: Apply Orlicz space-duality, (6.16), and Theorem 11. □

Next we verify that (6.17) is best possible. To this end, fix an arbitrary integer $k>0$, and consider the intervals $J_i=[(i-1)/k,i/k)$, $i\in[k]$. Define

$$E_i=\{\Delta\mathrm{W}(J_i)>0\},\ i\in[k]. \tag{6.19}$$

For $s=(s_1,\dots,s_k)\in\{-1,1\}^k$, let

$$A_s=E_1^{s_1}\cap E_2^{s_2}\cdots\cap E_k^{s_k}, \tag{6.20}$$

where $E_i^{s_i}=E_i$ if $s_i=1$, and $E_i^{s_i}=(E_i)^{\mathrm{c}}$ if $s_i=-1$. Clearly,

$$\{A_s:s\in\{-1,1\}^k\}$$

is a partition of Ω. Also observe (Exercise 23)

$$|\mu_{\mathrm{W}}(A_s,J_i)|=\sqrt{\frac{2}{\pi}}(1/2^k\sqrt{k}). \tag{6.21}$$

For $\gamma>0$, define

$$\theta_\gamma(x)=x/\{\log\,(1/x)\}^{\gamma/2},\quad x\in(0,1), \tag{6.22}$$

whence $\theta_1=\theta$. Using (6.21), we estimate

$$\sum_{s,i}\theta_\gamma(|\mu_{\mathrm{W}}(A_s,J_i)|)=\sum_{s,i}\sqrt{\frac{2}{\pi}}(1/k^{\frac12}\,2^k)\Big/\left\{\log\sqrt{\frac{\pi}{2}}k^{\frac12}\,2^k\right\}^{\gamma/2}$$

$$\ge Ck^{(1-\gamma)/2}, \tag{6.23}$$

where $C>0$ is a numerical constant. We summarize:

Theorem 14

$$\sup\left\{\sum_{i,j}\theta_\gamma(|\mu_{\mathrm{W}}(A_i,B_j)|) : \Sigma_j\,\mathbf{1}_{A_i} \le 1,\right.$$

$$\left.\Sigma_i\,\mathbf{1}_{B_j} \le 1,\ \{A_j\} \subset \mathscr{A},\ \{B_j\} \subset \mathscr{B}\right\} < \infty \tag{6.24}$$

if and only if $\gamma \ge 1$.

Remark (a measurement of complexity). The sharp result in Theorem 14 marks an extremal instance on a scale of like sharp results dealing with (that which we call) α-chaos [B1Kah]; we shall come to these later in the chapter. Indeed, I view Theorem 14 as a quantitative statement expressing precisely that the Wiener process is 'least complex' among stochastic processes within a certain class; or, that it manifests the simplest form of 'randomness'. This measurement, conveying 'least complexity' in a stochastic framework, is analogous to 1-Sidonicity, which conveys 'least complexity' in a harmonic analysis framework. (Cf. Chapter VII §11.)

7 A Multiple Wiener Integral

In this section we define a multiple integral with respect to the Wiener process, an n-dimensional construct that had first appeared – albeit disguised – in Wiener's 1938 article *The homogeneous chaos* [Wi5, pp. 917–18], and reappeared – redefined and clarified – in the works of Cameron and Martin [CaM], Itô [I2], and Kakutani [Kak2]. (See [Wi3, pp. 612–13].) I will follow Itô's construction of this integral, which is sometimes called the *multiple Itô integral*, and sometimes the *multiple Wiener–Itô integral* (e.g., [KarSh, p. 167], [Nu, p. 7]). I shall refer to it here as the *multiple Wiener integral*, and to its computation by iterated integrations (described in the next section) as the *iterated Itô integral.* To facilitate the exposition, I first will consider the archetypal case $n = 2$.

We start with the symmetric functions in $\mathrm{L}^2([0,1]^2, \mathfrak{m}^2)$,

$$\begin{aligned}\mathrm{L}^2_\sigma &= \mathrm{L}^2_\sigma([0,1]^2, \mathfrak{m}^2)\\ &= \{f \in \mathrm{L}^2([0,1]^2, \mathfrak{m}^2) : f(s,t) = f(t,s),\ (s,t) \in [0,1]^2\}.\end{aligned} \tag{7.1}$$

Let $S_\sigma = S_\sigma([0,1]^2)$ denote the space of the *standard symmetric step functions* on $[0,1]^2$ *that vanish on the diagonal;* these are 'step functions' of the form

$$f = \sum_{\substack{i=1\\j=1}}^{N} a_{ij}\, \mathbf{1}_{J_i}\, \mathbf{1}_{J_j}, \tag{7.2}$$

where

$$a_{ij} = a_{ji}, \quad a_{ii} = 0, \quad \text{and } J_i = \left[\frac{i-1}{N}, \frac{i}{N}\right) \text{ for } (i,j) \in [N]^2.$$

Because $\mathfrak{m}^2\{(s,s) : s \in [0,1]\} = 0$, we can assume without loss of generality that every $f \in \mathrm{L}^2_\sigma$ (an equivalence class of functions) is represented by a function that vanishes on the diagonal $\{(s,s): s \in [0,1]\}$, and obtain that S_σ is norm-dense in L^2_σ (Exercise 25). For $f \in S_\sigma$, define

$$I_{\mathrm{W}_2}(f) := \sum_{\substack{i=1\\j=1}}^{N} a_{ij}\, \Delta\mathrm{W}(J_i)\, \Delta\mathrm{W}(J_j). \tag{7.3}$$

Lemma 15 (Exercise 24). *The definition in* (7.3) *does not depend on the representation of f by* (7.2). *Moreover,*

$$\mathbf{E}|I_{\mathrm{W}_2}(f)|^2 = 2\|f\|^2_{\mathrm{L}^2(\mathfrak{m}^2)}. \tag{7.4}$$

This lemma and the norm-density of S_σ in L^2_σ imply that the map I_{W_2} from S_σ into $\mathrm{L}^2(\Omega, \mathbb{P})$ is uniquely extendible to a bounded linear map I_{W_2} from $\mathrm{L}^2_\sigma([0,1]^2, \mathfrak{m}^2)$ into $\mathrm{L}^2(\Omega, \mathbb{P})$. We refer to I_{W_2} as the two-fold Wiener integral.

To obtain the two-fold Wiener integral of any $f \in \mathrm{L}^2([0,1]^2, \mathfrak{m}^2)$, we consider its symmetrization

$$\tilde{f}(s,t) := \left(\frac{1}{2}\right)(f(s,t) + f(t,s)), \tag{7.5}$$

which is obviously in L^2_σ, and then define

$$I_{\mathrm{W}_2}(f) := I_{\mathrm{W}_2}(\tilde{f}). \tag{7.6}$$

Proposition 16 (cf. [I2, (I.3)], Proposition 3). *For all f and g in* $\mathrm{L}^2([0,1], \mathfrak{m}^2)$,

$$2\int_{[0,1]^2} \tilde{f}(s)\, \tilde{g}(t)\, \mathfrak{m}(ds)\, \mathfrak{m}(dt) = \mathbf{E} I_{\mathrm{W}_2}(f) I_{\mathrm{W}_2}(g). \tag{7.7}$$

Proof: If f and g are in S_σ,

$$f = \sum_{\substack{i=1\\j=1}}^{N} a_{ij} \mathbf{1}_{J_i} \mathbf{1}_{J_j}, \quad g = \sum_{\substack{i=1\\j=1}}^{N} b_{ij} \mathbf{1}_{J_i} \mathbf{1}_{J_j}, \tag{7.8}$$

then,

$$\begin{aligned}
&\mathbf{E} I_{\mathrm{W}_2}(f) I_{\mathrm{W}_2}(g) \\
&= \mathbf{E}\left(\sum_{\substack{i=1\\j=1}}^{N} a_{ij}\ \Delta\mathrm{W}(J_i)\ \Delta\mathrm{W}(J_j)\right)\left(\sum_{\substack{i=1\\j=1}}^{N} b_{ij}\ \Delta\mathrm{W}(J_i)\ \Delta\mathrm{W}(J_j)\right) \\
&= \mathbf{E}\left(2\sum_{j>i} a_{ij}\ \Delta\mathrm{W}(J_i)\ \Delta\mathrm{W}(J_j)\right)\left(2\sum_{j>i} b_{ij}\ \Delta\mathrm{W}(J_i)\ \Delta\mathrm{W}(J_j)\right) \\
&= 2\sum_{\substack{i=1\\j=1}}^{N} a_{ij}\ b_{ij}/N^2 = 2\int_{[0,1]^2} fg\ \mathrm{d}s\mathrm{d}t.
\end{aligned} \tag{7.9}$$

This implies, by the definitions in (7.5) and (7.6), that (7.7) holds for all f and g in L^2. □

The construction of a multiple Wiener integral in the general case $n \geq 2$ is similar. Let $S_{\sigma,n} = S_\sigma([0,1]^n)$ denote the linear subspace of $\mathrm{L}^2([0,1]^n, \mathfrak{m}^n)$ consisting of the standard symmetric step functions on $[0,1]^n$ that vanish on the $\binom{n}{2}$ 'hyper-diagonals'. These are the functions f on $[0,1]^n$ given by

$$f = \sum_{i_1,\ldots,i_n=1}^{N} a_{i_1\ldots i_n} \mathbf{1}_{J_{i_1}} \cdots \mathbf{1}_{J_{i_n}}, \tag{7.10}$$

where $J_i = [(i-1)/N, i/N)$ for $i \in [N]$, such that $a_{i_1\ldots i_n} = a_{i_{\pi 1}\ldots i_{\pi n}}$ for all $\pi \in \mathrm{per}[n]$, and $a_{i_1\ldots i_n} = 0$ whenever $|\{i_1, \ldots, i_n\}| < n$. For $f \in S_{\sigma,n}$, define

$$I_{\mathrm{W}_n}(f) = \sum_{i_1,\ldots,i_n=1}^{N} a_{i_1\ldots i_n} \Delta\mathrm{W}(J_{i_1}) \cdots \Delta\mathrm{W}(J_{i_n}). \tag{7.11}$$

By symmetry, if $f \in S_{\sigma,n}$, then

$$I_{\mathrm{W}_n}(f) = n!\left(\sum_{1\leq i_1<\cdots<i_n\leq N} a_{i_1\ldots i_n} \Delta\mathrm{W}(J_{i_1}) \cdots \Delta\mathrm{W}(J_{i_n})\right). \tag{7.12}$$

Because $\{\Delta W(J_{i_1}) \cdots \Delta W(J_{i_n}) : 1 \le i_1 < \cdots < i_n \le N\}$ is an orthogonal system in $L^2(\Omega, \mathbb{P})$, and $\mathbf{E}|\Delta W(J_{i_1}) \cdots \Delta W(J_{i_n})|^2 = 1/N^n$ for $1 \le i_1 < \cdots < i_n \le N$, we obtain

$$\mathbf{E}|I_{W_n}(f)|^2 = (n!)^2 \left(\sum_{1 \le i_1 < \cdots < i_n \le N} |a_{i_1 \dots i_n}|^2 / N^n \right)$$
$$= n! \, \|f\|^2_{L^2(\mathfrak{m}^n)}. \tag{7.13}$$

Now consider the symmetric functions in $L^2([0,1]^n, \mathfrak{m}^n)$,

$$L^2_\sigma([0,1]^n, \mathfrak{m}^n)$$
$$= \{f \in L^2([0,1]^n, \mathfrak{m}^n) : f(s_1, \dots, s_n) = f(s_{\pi 1}, \dots, s_{\pi n}),\ \pi \in \text{per}[n]\}, \tag{7.14}$$

and obtain, from (7.13) and the norm-density of $S_{\sigma,n}$ in $L^2_\sigma([0,1]^n, \mathfrak{m}^n)$ (Exercise 25), that I_{W_n} is uniquely extendible to $L^2_\sigma([0,1]^n, \mathfrak{m}^n)$.

Given an arbitrary $f \in L^2([0,1]^n, \mathfrak{m}^n)$, we take its symmetrization

$$\tilde{f}(s_1, \dots, s_n) = \frac{1}{n!} \sum_\pi f(s_{\pi 1}, \dots, s_{\pi n}), \tag{7.15}$$

where Σ_π is the sum over all permutations of $[n]$, and define

$$I_{W_n}(f) := I_{W_n}(\tilde{f}). \tag{7.16}$$

Proposition 17

i. (*cf. Proposition* 3). *For all f and g in* $L^2([0,1]^n, \mathfrak{m}^n)$,

$$n! \int_{[0,1]^n} \tilde{f}(\mathrm{s})\, \tilde{g}(\mathrm{s})\, \mathfrak{m}^n(d\mathrm{s}) = \mathbf{E}\, I_{W_n}(f)\, I_{W_n}(g). \tag{7.17}$$

ii. *For all integers $k > n > 0$, and for all $f \in L^2([0,1]^n, \mathfrak{m}^n)$ and $g \in L^2([0,1]^k, \mathfrak{m}^k)$,*

$$\mathbf{E}\, I_{W_n}(f)\, I_{W_k}(g) = 0. \tag{7.18}$$

Proof: The proof of Part i is similar to the proof of Proposition 3. To verify Part ii, note that for every $N > 0$, the spans of $\{\Delta W(J_{i_1}) \cdots \Delta W(J_{i_n}) : 1 \leq i_1 < \cdots < i_n \leq N\}$ and $\{\Delta W(J_{i_1}) \cdots \Delta W(J_{i_k}) : 1 \leq i_1 < \cdots < i_k \leq N\}$ are orthogonal. Therefore, (7.18) holds for all $f \in S_{\sigma,n}$ and $g \in S_{\sigma,k}$, which implies (7.18) in the general case. □

Remark (about symmetry). The role of symmetry in the definition of the multiple Wiener integral is explained by the observation that for all arrays (a_{ij}),

$$\mathbf{E}\left|\sum_{\substack{i=1\\j=1}}^{N} a_{ij}\ \Delta W(J_i)\ \Delta W(J_j)\right|^2$$

$$= \sum_{1 \leq i < j \leq N} (a_{ij} + a_{ji})^{2/N^2} + 3\sum_{j=1}^{N} a_{jj}^2/N^2 + 2 \sum_{1 \leq i < j \leq N} a_{ii}\ a_{jj}/N^2. \tag{7.19}$$

(Notice the symmetric summands.) Indeed, in the course of the definition, attention can be restricted with the same effect to symmetric functions only. To make this comment precise, we consider the equivalence relation in $[0,1]^n$ defined by

$$(t_1, \ldots, t_n) \sim (t_1', \ldots, t_n') \tag{7.20}$$

if and only if there exists a permutation π of $[n]$ such that

$$t_{\pi 1} = t_1', \ldots, t_{\pi n} = t_n'. \tag{7.21}$$

Denote the set of equivalence class representatives by $D_n := [0,1]^n/\sim$, which for convenience we take to be

$$D_n = \{(t_1, \ldots, t_n) : 0 \leq t_1 \leq \cdots \leq t_n \leq 1\}. \tag{7.22}$$

We equip D_n with structures inherited from $[0,1]^n$ via the canonical quotient map; denote the Borel field in D_n by $B_{\sigma n}$, and the Lebesgue measure restricted to it by $\mathfrak{m}_\sigma^n$. Let ϕ be the quotient map, $\phi : [0,1]^n \to D_n$. If f is a function on D_n, then let f_ϕ be the function on $[0,1]^n$ defined by

$$f_\phi(\mathbf{t}) = (f \circ \phi)(\mathbf{t}), \quad \mathbf{t} \in [0,1]^n; \tag{7.23}$$

f_ϕ is clearly symmetric.

Observe that $L^2(D_n, \mathfrak{m}_\sigma^n)$ is a natural domain of the multiple Wiener integral: for $f \in L^2(D_n, \mathfrak{m}_\sigma^n)$, define

$$I_{W_n}^{(\sigma)}(f) := (1/n!)\, I_{W_n}(f_\phi). \tag{7.24}$$

If $f \in L^2(D_n, \mathfrak{m}_\sigma^n)$ and $g \in L^2(D_n, \mathfrak{m}_\sigma^n)$, then

$$n! \int_{D_n} f(\mathbf{s})\, g(\mathbf{s})\, \mathfrak{m}_\sigma^n(d\mathbf{s}) = \int_{[0,1]^n} f_\phi(\mathbf{s})\, g_\phi(\mathbf{s})\, \mathfrak{m}^n(d\mathbf{s}), \tag{7.25}$$

and therefore (cf. (7.17)),

$$\int_{D_n} f(\mathbf{s})\, g(\mathbf{s})\, \mathfrak{m}_\sigma^n(d\mathbf{s}) = \mathbf{E}\, I_{W_n}^{(\sigma)}(f)\, I_{W_n}^{(\sigma)}(g). \tag{7.26}$$

That is, the multiple Wiener integral $I_{W_n}^{(\sigma)}$ is a unitary map from $L^2(D_n, \mathfrak{m}_\sigma^n)$ into $L^2(\Omega, \mathbb{P})$ (while $I_{W_n} : L^2([0,1]^n, \mathfrak{m}^n) \to L^2(\Omega, \mathbb{P})$ is not!).

8 The Beginning of Adaptive Stochastic Integration

Itô concluded his landmark paper [I2] with the observation that the multiple integral $I_{W_n}(f)$ could be computed iteratively. Specifically, he noted that for all f in $L_\sigma^2([0,1]^n, \mathfrak{m}^n)$,

$$I_{W_n}(f) = n! \int_0^1 \left(\int_0^{t_n} \left(\cdots \left(\int_0^{t_2} f(t_1, \ldots, t_n)\, dW(dt_1) \right) \cdots \right) dW(dt_{n-1}) \right) dW(dt_n), \tag{8.1}$$

where the iterated integrals on the right side of (8.1) had been defined in his previous paper [I1] – also a landmark. We refer to the right side of (8.1) as the *iterated Itô integral.* The gist of Itô's observation was that if $f \in L_\sigma^2([0,1]^2, \mathfrak{m}^2)$, then one could meaningfully define the iterated stochastic integral

$$\int_{[0,1]} \left(\int_{[0,1]} \mathbf{1}_{[0,t]}(s)\, f(s,t)\, dW(ds) \right) dW(dt), \tag{8.2}$$

where the integrator is a Wiener process and the integrand is the Wiener integral-valued process

$$I_{\mathrm{W}}(\mathbf{1}_{[0,t]}(\cdot)f(\cdot,t)), \quad t \in [0,1]. \tag{8.3}$$

The integral in (8.2) (defined in [I1]) marked the start of the subject of *adaptive stochastic integration* – or, as it is sometimes called, *non-anticipative stochastic integration.*

To illustrate ideas, we will verify the equality in (8.1) in a simple but archetypal case. For $t \in [0,1]$, let $f = \mathbf{1}_{[0,t]\times[0,t]}$, obviously a symmetric element in $\mathrm{L}^2([0,1]^2, \mathfrak{m}^2)$. For an integer $N > 0$, let $K = K(N,t)$ be the integer such that

$$\frac{K-1}{N} < t \leq \frac{K}{N}. \tag{8.4}$$

Denote $J_i = [(i-1)/N, i/N)$. As $N \to \infty$, the sequence (indexed by N)

$$\begin{aligned} 2 &\sum_{1\leq i<j\leq K} \Delta\mathrm{W}(J_i)\,\Delta\mathrm{W}(J_j) \\ &= \sum_{\substack{i=1\\ j=1}}^{K} \Delta\mathrm{W}(J_i)\,\Delta\mathrm{W}(J_j) - \sum_{i=1}^{K}(\Delta\mathrm{W}(J_i))^2 \\ &= \left(\mathrm{W}\left(\frac{K}{N}\right)\right)^2 - \sum_{i=1}^{K}(\Delta\mathrm{W}(J_i))^2, \quad N \in \mathbb{N}, \end{aligned} \tag{8.5}$$

converges in the $\mathrm{L}^2(\Omega, \mathbb{P})$-norm to $I_{\mathrm{W}_2}(f)$. But, we also have

$$\left(\mathrm{W}\left(\frac{K}{N}\right)\right)^2 \xrightarrow[N\to\infty]{} (\mathrm{W}(t))^2 \tag{8.6}$$

and

$$\sum_{i=1}^{K}(\Delta\mathrm{W}(J_i))^2 \xrightarrow[N\to\infty]{} t, \tag{8.7}$$

where both are L^2-norm limits. The first limit follows from

$$\lim_{N\to\infty} \mathrm{W}(K/N) = \mathrm{W}(t) \ \text{ in } \mathrm{L}^2(\Omega, \mathbb{P}) \ \text{ (Exercise 26)},$$

and the second is the quadratic variation (Remark iii §3 and Exercise 6 i). Therefore, the 2-fold Wiener integral of $\mathbf{1}_{[0,t]^2}$ is

$$I_{\mathrm{W}_2}(\mathbf{1}_{[0,t]^2}) = (\mathrm{W}(t))^2 - t. \tag{8.8}$$

On the other hand, the iterated integral of $\mathbf{1}_{[0,t]\times[0,t]}$ is

$$\begin{aligned}
&\int_0^1 \left(\int_0^v \mathbf{1}_{[0,t]}(u)\, \mathbf{1}_{[0,t]}(v)\, \mathrm{dW}(du) \right) \mathrm{dW}(dv) \\
&= \int_0^1 \mathbf{1}_{[0,t]}(v) \left(\int_0^v \mathbf{1}_{[0,t]}(u)\, \mathrm{dW}(du) \right) \mathrm{dW}(dv) \\
&= \int_0^t \mathrm{W}(v)\, \mathrm{dW}(dv) = \left(\frac{1}{2}\right) (\mathrm{W}(t))^2 - \frac{t}{2}, \qquad (8.9)
\end{aligned}$$

where the last equality is an instance of Itô's formula [I1, pp. 523–4] (Exercise 27).

These computations illustrate a fundamental distinction between stochastic and non-stochastic integration. Specifically, it is the respective presence of the second terms on the right sides of (8.8) and (8.9). In (8.8), the second term appears because the 'diagonal', a null set in $([0,1]^2, \mathtt{m}^2)$, has a non-negligible effect in $(\Omega \times [0,1]^2, \mathbb{P} \times \mathtt{m}^2)$ (cf. (8.5) and (8.7)). In (8.9), the second term appears because second order differences, negligible in the ordinary calculus, give rise to the *quadratic variation* in the stochastic calculus (e.g., (3.1)). In the next chapter, we further explain this distinction in terms of the Grothendieck factorization theorem.

The multiple Wiener integral gives rise to a stochastic process parameterized by D_n,

$$I^{(\sigma)}_{\mathrm{W}_n}(\mathbf{1}_{[0,t_1]} \cdots \mathbf{1}_{[0,t_n]}), \quad (t_1, \ldots, t_n) \in D_n, \qquad (8.10)$$

which we revisit in §11, §14, and again in the next chapter. In the case $n = 2$, (8.8) implies that for $(s,t) \in D_2$,

$$\begin{aligned}
I^{(\upsilon)}_{\mathrm{W}_2}(\mathbf{1}_{[0,s]}\mathbf{1}_{[0,t]}) &= \left(\frac{1}{2}\right) (I_{\mathrm{W}_2}(\mathbf{1}_{[0,t]^2}) - I_{\mathrm{W}_2}(\mathbf{1}_{[s,t]^2})) \\
&= \left(\frac{1}{2}\right) (\mathrm{W}(t)^2 - t - (\mathrm{W}(t) - \mathrm{W}(s))^2 - s + t) \\
&= \mathrm{W}(t)\, \mathrm{W}(s) - \left(\frac{1}{2}\right) \mathrm{W}(s)^2 - \frac{s}{2}. \qquad (8.11)
\end{aligned}$$

An interesting problem is to derive a similar formula for the process in (8.10) in the case $n > 2$ (Exercise 28).

Remarks:

i **(n-dimensional white noise?).** Recall the definition (by induction) of the standard n-fold difference operation. If f is a function on [0,1], and J is an interval with end-points $0 \leq u < v \leq 1$, then $\Delta f(J) = f(v) - f(u)$. For $n > 1$, if f is a function on $[0,1]^n, J_1, \ldots, J_n$ are intervals in [0,1], and

$$f_{J_n}(t_1, \ldots, t_{n-1}) = \Delta[f(t_1, \ldots, t_{n-1}, \cdot)](J_n),$$
$$(t_1, \ldots, t_{n-1}) \in [0,1]^{n-1}, \tag{8.12}$$

then define

$$\Delta^n f(J_1, \ldots, J_n) = \Delta^{n-1} f_{J_n}(J_1, \ldots, J_{n-1}). \tag{8.13}$$

We view $\Delta^n f$ as a function on n-cubes $Q = J_1 \times \cdots \times J_n \subset [0,1]^n$, and (slightly abusing notation) sometimes write

$$\Delta^n f(Q) = \Delta^n f(J_1, \ldots, J_n). \tag{8.14}$$

Now recall the white noise $\Delta \mathrm{W}$,

$$\Delta \mathrm{W}(B) := I_{\mathrm{W}}(\mathbf{1}_B), \quad B \in \mathscr{B}, \tag{8.15}$$

which is a random set-function whose definition extends that of the usual increment $\Delta \mathrm{W}(J)$ over an interval $J \subset [0,1]$. In the n-dimensional case, if $J_i \subset [0,1]$, $i = 1, \ldots, n$, are pairwise disjoint intervals, then

$$I_{\mathrm{W}_n}(\mathbf{1}_{J_1 \times \cdots \times J_n}) = \Delta \mathrm{W}(J_1) \cdots \Delta \mathrm{W}(J_n)$$
$$= \Delta^n \mathrm{W}^{(n)}(J_1 \times \cdots \times J_n) \tag{8.16}$$

where $\mathrm{W}^{(n)}(t_1, \ldots, t_n) := \mathrm{W}(t_1) \cdots \mathrm{W}(t_n)$. In general, however, notice that

$$I_{\mathrm{W}_n}(\mathbf{1}_{[0,s_1] \times \cdots \times [0,s_n]}), \quad (s_1, \ldots, s_n) \in [0,1]^n, \tag{8.17}$$

is not(!) the same as $\Delta^n \mathrm{W}^{(n)}([0, s_1] \times \cdots \times [0, s_n]) = \mathrm{W}(s_1) \cdots \mathrm{W}(s_n)$. This indeed is a basic feature of the Itô calculus. (For example, see (8.11) above.) A question arises: can an n-dimensional integral with respect to '$\mathrm{dW}(\mathrm{d}t_1) \ldots \mathrm{dW}(\mathrm{d}t_n)$' be defined – extending the usual one-dimensional Wiener integral – so that the resulting 'n-dimensional white noise' extends $\Delta^n \mathrm{W}^{(n)}(Q)$, where Q is an n-fold Cartesian product of intervals? We answer this question (in the affirmative) in the next chapter.

ii **(the Wiener Chaos).** Consider the subspaces of $\mathrm{L}^2(\Omega, \mathbb{P})$

$$H_n = \{I_{\mathrm{W}_n}(f) : f \in \mathrm{L}^2([0,1]^n, \mathfrak{m}^n)\}, \quad n \geq 1, \tag{8.18}$$

which are known as the nth *Wiener Chaos*, or the *Wiener Chaos of order* n (e.g., [Nu]).

The first part of Proposition 17 implies that H_n is norm-closed in $\mathrm{L}^2(\Omega, \mathbb{P})$, and the second part states that the H_n are mutually orthogonal. The complete orthogonal decomposition

$$\oplus_n H_n = \mathrm{L}^2(\Omega, \mathbb{P}) \quad \text{(the Wiener Chaos decomposition)}, \tag{8.19}$$

where $H_0 = \{\text{constants}\}$, follows from the norm-density of $\bigcup_{n=0}^{\infty} H_n$ in $\mathrm{L}^2(\Omega, \mathbb{P})$. The latter had been established first by Wiener [Wi5, Section 12], and subsequently clarified in [CaM] and [I2]. A detailed proof of (8.19) can be found in [Nu, Chapter 1].

The Wiener Chaos decomposition is analogous to the decomposition

$$\oplus_n \mathrm{L}^2_{R_n}(\Omega, \mathbb{P}) = \mathrm{L}^2(\Omega, \mathbb{P}), \tag{8.20}$$

where $\Omega = \{-1, 1\}^{\mathbb{N}}, \mathbb{P} =$ the normalized Haar measure on Ω, and R_n is the set comprising all products of n distinct Rademacher characters; see Chapter VII. Analogies between the summands $\mathrm{L}^2_{R_n}$ and H_n will be made precise in the next two sections.

9 Sub-α-systems

In this and the next section we formalize a scale calibrated by tail-probability types, which mark degrees of 'interdependence'. (See Remark ii §4.) We start with an extension of Lemma 6.

Lemma 18 (Exercise 29). *Suppose* $X \in \mathrm{L}^2(\Omega, \mathbb{P})$, $\|X\|_{\mathrm{L}^2} \leq 1$. *Let* $\alpha > 0$. *The following are equivalent:*

i. *there exists* $0 < A < \infty$ *such that*

$$\overline{\lim_{x \to \infty}} \exp(A x^{\frac{2}{\alpha}})\, \mathbb{P}(|X| > x \|X\|_{\mathrm{L}^2}) < \infty; \tag{9.1}$$

ii. *there exists* $0 < B < \infty$ *such that*

$$\sup\{\|X\|_{\mathrm{L}^p} / p^{\alpha/2} : p > 2\} \leq B; \tag{9.2}$$

iii. *there exists* $0 < C < \infty$ *such that*

$$\overline{\lim_{t \to \infty}} \mathbf{E} \exp(t|X|^{\frac{1}{\alpha}} - Ct^2) < \infty; \tag{9.3}$$

iv. *there exists* $0 < D < \infty$ *such that*

$$\mathbf{E} \exp(D|X|^{\frac{2}{\alpha}}) < \infty. \tag{9.4}$$

Definition 19 (cf. Definition 5).

i. $X \in \mathrm{L}^2(\Omega, \mathbb{P})$ is a *sub-α-variable* if X satisfies (any of) the statements in Lemma 18.
ii. $F \subset \mathrm{L}^2(\Omega, \mathbb{P})$ is a *sub-α-system* if every $X \in \mathrm{L}^2_F(\Omega, \mathbb{P})$ is a sub-α-variable. ($\alpha = 1$ is the sub-Gaussian case in Definition 5.)

Remarks:

i **(an Orlicz norm).** Consider the Orlicz function

$$\phi_\alpha(x) = \exp(x^{\frac{2}{\alpha}}) - 1, \quad x > 0, \tag{9.5}$$

and the corresponding Orlicz norm $\|Y\|_{\phi_\alpha}, Y \in \mathrm{L}^0(\Omega, \mathscr{A})$. ($\phi_1$ was defined in (4.11).) Denote the corresponding Orlicz space

$$\{Y \in \mathrm{L}^0(\Omega, \mathscr{A}) : \|Y\|_{\phi_\alpha} < \infty\}$$

by $\mathrm{L}_{\phi_\alpha}(\Omega, \mathbb{P})$. By Lemma 18, $F \subset \mathrm{L}^2(\Omega, \mathbb{P})$ is a sub-α-system if and only if

$$\mathrm{L}^2_F(\Omega, \mathbb{P}) \subset \mathrm{L}_{\phi_\alpha}(\Omega, \mathbb{P}). \tag{9.6}$$

ii **(examples).**

1. If $\{X_j : j \in \mathbb{N}\}$ is an orthonormal sub-α-system, then

$$\{X_{j_1} \otimes \cdots \otimes X_{j_n} : (j_1, \ldots, j_n) \in \mathbb{N}^n\} \subset \mathrm{L}^2(\Omega^n, \mathbb{P}^n) \tag{9.7}$$

is a sub-$n\alpha$-system (Exercise 30).
2. The Walsh system W_n is a sub-α-system if and only if $\alpha \geq n$ (Theorem VII.32).
3. Let $\{X_j : j \in \mathbb{N}\}$ be a system of statistically independent standard Gaussian variables, and define (the 'Gaussian' analog of R_n)

$$G_n = \{X_{j_1} \cdots X_{j_n} : 1 \leq j_1 < \cdots < j_n\}. \tag{9.8}$$

Then, G_n is a sub-α-system if and only if $\alpha \geq n$.

That G_n is not a sub-α-system for $\alpha < n$ is easy: merely observe that there exists a numerical constant K_n such that for all $p > 2$,

$$\|X_1 \cdots X_n\|_{\mathrm{L}^p} \geq K_n \, p^{\frac{n}{2}}. \tag{9.9}$$

The non-trivial part of the assertion, that G_n is a sub-n-system, was proved first by M. Schreiber [Sch] via a combinatorial argument similar to the proof used independently by A. Bonami in the case of the Walsh system of order n. (See [Bon2, p. 367].) That G_n is a sub-n-system can be proved also by applying (so-called) *decoupling inequalities* to $\{X_{j_1} \otimes \cdots \otimes X_{j_n} : (j_1, \ldots, j_n) \in \mathbb{N}^n\}$; cf. Example 1 above. (A detailed treatment of decoupling inequalities can be found in [dlPG].) Yet another proof (shown to me by E. Giné) makes use of Bonami's inequalities and the Central Limit Theorem (Exercise 31).

10 Measurements of Stochastic Complexity

In this section we define an index of stochastic complexity, a measurement that we have already seen in a different but equivalent form in a context of harmonic analysis. (See Chapter VII §9.)

For $Y \in \mathrm{L}^2(\Omega, \mathbb{P})$ and $s > 0$, denote

$$c(Y; s) = \varliminf_{y \to \infty} \left((-1/y^s) \log \mathbb{P}(|Y| > y)\right), \tag{10.1}$$

and define (cf. (4.20))

$$\delta(Y) = 1/\sup\{s : c(Y; s) > 0\}. \tag{10.2}$$

Equivalently,

$$\begin{aligned} 1/\delta(Y) &= \varliminf_{y \to \infty} \log(-\log \mathbb{P}\,(|Y| > y))/\log y \\ &= \sup\{s : \|Y\|_{\phi_{\frac{2}{s}}} < \infty\}. \end{aligned} \tag{10.3}$$

(The Orlicz function $\phi_{2/s}$ is defined in (9.5).) Observe that Y is a sub-γ-variable for all $\gamma > 2\delta(Y)$ and for no $\gamma < 2\delta(Y)$.

Here are some canonical examples: if $\|Y\|_{\mathrm{L}^\infty} < \infty$, then $\delta(Y) = 0$; if $\|Y\|_{\mathrm{L}^p} = \infty$ for some $p > 2$, then $\delta(Y) = \infty$; if $Y = X_1 \cdots X_n$, where the X_i are standard independent Gaussian variables, then $\delta(Y) = n/2$.

For a subspace H of $\mathrm{L}^2(\Omega, \mathbb{P})$, define

$$c_H(s) = \inf\{c(Y; s) : Y \in H,\ \|Y\|_{\mathrm{L}^2} \leq 1\}, \tag{10.4}$$

and then define the index (cf. (VII.9.29))

$$\delta_H = 1/\sup\{s : c_H(s) > 0\}. \tag{10.5}$$

Equivalently,

$$\delta_H = \inf\{\delta(Y) : Y \in H,\ \|Y\|_{\mathrm{L}^2} \leq 1\}. \tag{10.6}$$

If H is infinite-dimensional, then $1/2 \leq \delta_H \leq \infty$; see [Ru1, Theorem 3.4]. We distinguish between $c_H(\delta_H^{-1}) > 0$, in which case δ_H is said to be *exact*, and $c_H(\delta_H^{-1}) = 0$, in which case δ_H is said to be *asymptotic*.

If $H_n = \mathrm{L}^2_{W_n}(\Omega, \mathbb{P})$ (W_n is the Walsh system of order n), then

$$\delta_{H_n} = \frac{n}{2} \text{ exactly.} \tag{10.7}$$

(See (VII. 9.31)) This is analogous to

Proposition 20 *If H_n is the nth Wiener Chaos (defined in (8.18)), then*

$$\delta_{H_n} = \frac{n}{2} \text{ exactly.} \tag{10.8}$$

Proof: If $f \in S_{\sigma,n}$ and $f = \sum_{i_1,\ldots,i_n=1}^{N} a_{i_1\ldots i_n} \mathbf{1}_{J_{i_1}} \cdots \mathbf{1}_{J_{i_n}}$, then

$$I_{\mathrm{W}_n}(f) = n!\left(\sum_{1\leq i_1<\cdots<i_n\leq N} a_{i_1\ldots i_n} \Delta\mathrm{W}(J_{i_1})\cdots \mathrm{W}(J_{i_n})\right). \tag{10.9}$$

Because G_n is a sub-n-system,

$$\begin{aligned}\|I_{\mathrm{W}_n}(f)\|_{\mathrm{L}^p} &\leq K\, p^{\frac{n}{2}} \left(\sum_{i_1,\ldots,i_n=1}^{N} |a_{i_1\ldots i_n}|^2/N^n\right)^{\frac{1}{2}} \\ &\leq K\, p^{\frac{n}{2}} \|f\|_{\mathrm{L}^2}\end{aligned} \tag{10.10}$$

for all $p > 2$. This verifies (by Lemma 18 and the definition of the Orlicz norm)

$$\|I_{\mathrm{W}_n}(f)\|_{\phi_{\frac{2}{n}}} \leq K'\, \|f\|_{\mathrm{L}^2}. \tag{10.11}$$

(K and K' depend only on n.) That $I_{W_n}(f)$ is a sub-n-variable for every $f \in L^2([0,1]^n, \mathfrak{m}^n)$ follows from the norm-density of $S_{\sigma,n}$ in $L^2_\sigma([0,1]^n, \mathfrak{m}^n)$, and (10.10). In particular,

$$\delta_{H_n} \geq \frac{n}{2}. \tag{10.12}$$

To verify the opposite inequality, let $J_i \subset [0,1], i = 1, \ldots, n$, be pairwise disjoint intervals, and define $f = \mathbf{1}_{J_1 \times \cdots \times J_n}$. Then,

$$I_{W_n}(f) = \Delta W(J_1) \cdots \Delta W(J_n), \tag{10.13}$$

and $c(I_{W_n}(f); s) = 0$ for all $s > 2/n$ (Exercise 32). □

Corollary 21 *For every integer $n \geq 1$,*

i. *H_n is norm-closed in the Orlicz space $L_{\phi_{\frac{2}{n}}}(\Omega, \mathbb{P})$; in particular, H_n is norm-closed in $L^p(\Omega, \mathbb{P})$ for all $p > 2$;*
ii. *$\{|X|^p : X \in B_{H_n}\}$ is uniformly integrable for all $p \geq 1$;*
iii. *H_n is a $\Lambda(2)$-space (Definition III.6, Lemma III.6), i.e., there exists $K_n > 0$ such that*

$$\|X\|_{L^2} \leq K_n \, \|X\|_{L^1} \quad \textit{for all } X \in H_n; \tag{10.14}$$

iv. *H_n is closed in probability.*

Proof:

i. The assertions follow from the definition of the Orlicz norm, (10.10), and the equivalence ii ⇔ iv in Lemma 18 (Exercise 33).
ii. By Hölder's inequality, (10.10), and Lemma 18,

$$\begin{aligned} \mathbf{E}\,|X|^p\,\mathbf{1}_{\{|X|^p > m\}} &\leq (\mathbf{E}|X|^{2p})^{1/2}(\mathbb{P}\{|X|^p > m\})^{1/2} \\ &\leq K(2p)^p \exp\left(-\frac{1}{2}C\, m^{2/pn}\right), \end{aligned} \tag{10.15}$$

which implies the assertion.
iii. The L^1–L^2 inequality follows from Part i (e.g., Theorem II.1).
iv. It suffices to verify that if $X_j \in H_n$ for $j \in \mathbb{N}$, and $X_j \to 0$ in probability, then $X_j \to 0$ in $L^2(\Omega, \mathbb{P})$. To this end, observe that if $Y \in L^2(\Omega, \mathbb{P})$ and $\lambda \in (0,1)$, then

$$\mathbb{P}\{Y \geq \lambda\, \mathbf{E}Y\} \geq (1 - \lambda^2)\frac{(\mathbf{E}Y)^2}{\mathbf{E}(Y)^2} \tag{10.16}$$

([Kah3, p. 8]). By putting $Y = |X_j|$ in (10.16), and then applying (10.14), we obtain

$$\mathbb{P}\{|X_j| \geq \lambda\, K_n^{-1} \|X_j\|_{\mathrm{L}^2}\} \geq (1-\lambda^2) K_n^{-2}, \quad j \in \mathbb{N}, \tag{10.17}$$

which implies $\|X_j\|_{\mathrm{L}^2} \to 0$.

□

Remark (about the inequality in (10.16)). Each of the properties of H_n stated in Corollary 21 was first noted by M. Schreiber in [Sch]. However, the crucial application of (10.16) (or a variant of it) was only implicit in Schreiber's argument verifying that H_n is closed in probability. (See [Sch, p. 861] and Exercise 34.)

The importance of the inequality in (10.16) is underscored in Kahane's book [Kah3, p. 281]. A special case of it – an instance of (10.17) – appeared first in Paley's and Zygmund's 1932 paper [PaZy2, Lemma 19]; the general inequality was stated and proved in Salem's and Zygmund's 1954 paper [SaZy2]. (See [SaZy2, Lemma 4.2.4].)

11 The nth Wiener Chaos Process and its Associated F-measure

Let $\tilde{D}_n = \{(t_1, \ldots, t_n) : 0 \leq t_1 < \cdots < t_n \leq 1\}$, and denote by $\tilde{B}_{\sigma n}$ the Borel field in $\tilde{D}_n$. Consider the nth *Wiener Chaos process* (cf. (8.10))

$$\mathrm{W}_n(t_1, \ldots, t_n) = I^{(\sigma)}_{\mathrm{W}_n}(\mathbf{1}_{[0,t_1]} \cdots \mathbf{1}_{[0,t_n]}), \quad (t_1, \ldots, t_n) \in \tilde{D}_n. \tag{11.1}$$

For $A \in \mathscr{A}$ and $B \in \tilde{B}_{\sigma n}$, define

$$\mu_{\mathrm{W}_n}(A, B) := \mathbf{E}\mathbf{1}_A I^{(\sigma)}_{\mathrm{W}_n}(\mathbf{1}_B), \tag{11.2}$$

and note that μ_{W_n} is an F_2-measure on $\mathscr{A} \times \tilde{B}_{\sigma n}$ (Exercise 35).

That μ_{W_n} cannot be extended to an F_1-measure on $\sigma(\mathscr{A} \times \tilde{B}_{\sigma n})$ follows from the case $n = 1$, which was verified in Chapter VI §2 and also in §6 in this chapter. In particular, we have $\|\mu_{\mathrm{W}_n}\|_{(1)} = \infty$. (The definition of $\|\mu_{\mathrm{W}_n}\|_{(p)}$ is practically the same as in (6.2): replace $\mathscr{B}$ in (6.2) by $\tilde{B}_{\sigma n}$.) On the other hand, because $\|\mu_{\mathrm{W}_n}\|_{F_2} < \infty$, Littlewood's 4/3-inequality implies $\|\mu_{\mathrm{W}_n}\|_{(4/3)} < \infty$. As in the case $n = 1$, a question arises: what is $\ell_{\mathrm{W}_n} := \inf\{p\colon \|\mu_{\mathrm{W}_n}\|_{(p)} < \infty\}$?

The analysis is similar to that of μ_{W} in §6. Let φ_n be an Orlicz function such that

$$\varphi_n(t) = \exp(-t^{-2/n}), \quad t \in \left(0, \left(\frac{2}{2+n}\right)^{\frac{n}{2}}\right]; \tag{11.3}$$

$(2/(2+n))^{n/2}$ is the inflection point of $\exp(-t^{-2/n})$. Let $l_{\varphi_n}(\mathbb{N}^2)$ be the corresponding Orlicz space, and denote its unit ball by O_{φ_n}.

Theorem 22 (cf. Theorem 11).

$$\sup\left\{\left|\sum_{j,k} \mu_{\mathrm{W}_n}(A_j, B_k)\, b_{jk}\right| : \Sigma_j \mathbf{1}_{A_j} \le 1,\ \Sigma_k \mathbf{1}_{B_k} \le 1,\right.$$

$$\left.\{A_j\} \subset \mathscr{A},\ \{B_k\} \subset \tilde{B}_{\sigma n},\ \{b_{jk}\} \in O_{\varphi n}\right\} < \infty. \tag{11.4}$$

Proof: (Exercise 37). Fix $(b_{jk}) \in O_{\varphi n}$, and let $\{A_j\} \subset \mathscr{A}$ and $\{B_k\} \subset \tilde{B}_{\sigma n}$ be finite collections of pairwise disjoint measurable sets. Without loss of generality, assume $|b_{jk}| \le (2/(2+n))^{n/2}$, and, by rearranging the js, also

$$\sum_k \varphi_n(|b_{jk}|) \le 1/j, \quad j = 1, \ldots. \tag{11.5}$$

Then,

$$\sup_k |b_{jk}| \le (1/\log j)^{\frac{n}{2}}, \quad j = 2, \ldots. \tag{11.6}$$

By Proposition 20, there exist $L > 0$ and $K > 0$ such that for all scalar sequences $d = (d_k)$,

$$\mathbb{P}\left(\left|\sum_k d_k I^{(\sigma)}_{\mathrm{W}_n}(\mathbf{1}_{B_k})\right| > x\right) \le \exp(-K(x/\|d\|_\infty)^{\frac{2}{n}}), \quad x > L. \tag{11.7}$$

This implies

$$\left|\sum_{j,k} \mu_{\mathrm{W}_n}(A_j, B_k)\, b_{jk}\right| = \left|\sum_{j,k} b_{jk}\, \mathbf{E}\mathbf{1}_{A_j} I^{(\sigma)}_{\mathrm{W}_n}(\mathbf{1}_{B_k})\right|$$

$$\le \mathbf{E}\left|\sum_j \mathbf{1}_{A_j} \sum_k b_{jk}\, I^{(\sigma)}_{\mathrm{W}_n}(\mathbf{1}_{B_k})\right|$$

$$\leq \mathbf{E}\sup_j \left|\sum_k b_{jk}\, I^{(\sigma)}_{\mathrm{W}_n}(\mathbf{1}_{B_k})\right|$$

$$= \int_0^\infty \mathbb{P}\left(\bigcup_j \left\{\left|\sum_k b_{jk}\, I^{(\sigma)}_{\mathrm{W}_n}(\mathbf{1}_{B_k})\right| > x\right\}\right) dx$$

$$\leq c + \sum_j \int_c^\infty \mathbb{P}\left(\left|\sum_k b_{jk}\, I^{(\sigma)}_{\mathrm{W}_n}(\mathbf{1}_{B_k})\right| > x\right) dx$$

$$\leq c + \sum_j \int_c^\infty \exp(-K\, x^{\frac{2}{n}} \log j)\, dx < \infty, \tag{11.8}$$

for an appropriate choice of $c > 0$. □

Consider the instance $\gamma = n$ in (6.22),

$$\theta_n(x) = x/\{\log\,(1/x)\}^{\frac{n}{2}} \text{ for } x \in (0,1), \quad \text{and } \theta_n(0) = 0, \tag{11.9}$$

and obtain $0 < \eta < 1$ such that

$$\theta_n(x) \leq 2\,\varphi_n^*(x), \quad x \in (0,\eta), \tag{11.10}$$

where φ_n^* is the *complementary function* of φ_n. Then, by (11.10) and the Orlicz space-duality $l_{\varphi_n^*} = (l_{\varphi_n})^*$, we deduce

Corollary 23 (cf. Corollary 13).

$$\sup\left\{\sum_{j,k}\theta_n(|\mu_{\mathrm{W}_n}(A_j,B_k)|) : \Sigma_j \mathbf{1}_{A_j} \leq 1,\ \Sigma_k \mathbf{1}_{B_k} \leq 1,\right.$$
$$\left.\{A_j\} \subset \mathscr{A},\ \{B_k\} \subset \tilde{B}_{\sigma n}\right\} < \infty. \tag{11.11}$$

In particular, $\|\mu_{\mathrm{W}_n}\|_{(p)} < \infty$ *for all* $p > 1$.

To verify that θ_n is optimal, let k be an arbitrary positive integer, define $J_i = [(i-1)/k, i/k),\ i \in [k]$, and

$$Q_{i_1\ldots i_n} = J_{i_1} \times \cdots \times J_{i_n}, \quad 0 < i_1 < \cdots < i_n \leq k. \tag{11.12}$$

Then $Q_{i_1\dots i_n} \in \tilde{B}_{\sigma n}$, and

$$I^{(\sigma)}_{W_n}(\mathbf{1}_{Q_{i_1\dots i_n}}) = \Delta W(J_{i_1}) \cdots \Delta W(J_{i_n}). \tag{11.13}$$

Let $E_i \subset \Omega$ ($i \in [k]$) be defined by (6.19), and the corresponding partition $\{A_s\colon s \in \{-1,1\}^k\}$ of Ω be defined by (6.20). Then (Exercise 36),

$$\begin{aligned} |\mu_{W_n}(A_s, Q_{i_1\dots i_n})| &= |\mathbf{E}\mathbf{1}_{A_s}\Delta W(J_{i_1}) \cdots \Delta W(J_{i_n})| \\ &= \left(\frac{2}{\pi}\right)^{\frac{n}{2}} \left(\frac{1}{k}\right)^{\frac{n}{2}} \left(\frac{1}{2}\right)^{k}, \quad 0 < i_1 < \cdots < i_n \leq k. \end{aligned} \tag{11.14}$$

Therefore, if $\gamma \geq 1$, then

$$\begin{aligned} &\sum_{s\in\{-1,1\}^k, 1<i_1<\cdots<i_n\leq k} \theta_\gamma(|\mu_{W_n}(A_s, Q_{i_1\dots i_n})|) \\ &= \sum_{s\in\{-1,1\}^k, 1<i_1<\cdots<i_n\leq k} \left(\frac{2}{\pi}\right)^{\frac{n}{2}} \left(\frac{1}{k}\right)^{\frac{n}{2}} \left(\frac{1}{2}\right)^{k} \Big/ \{\log(\pi/2)^{\frac{n}{2}} k^{\frac{n}{2}} 2^k\}^{\gamma/2} \\ &\geq C_n\, k^{\frac{n-\gamma}{2}}, \end{aligned} \tag{11.15}$$

where $C_n > 0$ depends only on n. We summarize:

Theorem 24 (cf. Theorem 14).

$$\sup\left\{ \sum_{j,k} \theta_\gamma(|\mu_{W_n}(A_j, B_k)|) : \Sigma_j \mathbf{1}_{A_j} \leq 1,\ \Sigma_i \mathbf{1}_{B_k} \leq 1, \right. \\ \left. \{A_j\} \subset \mathscr{A},\ \{B_k\} \subset \tilde{B}_{\sigma n} \right\} < \infty \tag{11.16}$$

if and only if $\gamma \geq n$.

Remark (measurements of complexity). I view the estimates on variations of μ_{W_n} in Theorem 24 as measurements of stochastic complexity in W_n, measurements that are analogous to those involving the Littlewood $2n/(n+1)$-inequalities in a framework of harmonic analysis. (See Chapter VII §10, §11.) The extremal case $n = 1$ in Theorem 14 conveys that a Wiener process is stochastically 'least complex'.

Theorem 24 is an instance of the result in [B1Kah], that variations of F-measures associated with 'α-chaos' processes are controlled by the Orlicz functions $\theta_\alpha, \alpha \in [1, \infty)$. The '$\alpha$-chaos' processes, which are general continuous-time models for random walks whose steps have prescribed degrees of interdependence, will be motivated and introduced in the next three sections; their analysis will be continued in the next and the last chapters.

12 Mise en Scène (§1 continued): Further Approximations of Brownian Motion

In designing a model for movements of a Brownian particle, Wiener intended to

> treat the Brownian movement, *in a first approximation* [my italics], as an effect due to ... a series of impacts received by a particle, dependent only on the time during which the particle is exposed to collisions... [Wi1, p. 295].

He imagined Brownian movement (*in a first approximation*) to be a limit of simple random walks, a view that calls for the intervention of the Central Limit Theorem. (A view of Brownian movement as a simple random walk was subsequently reinterpreted, and further studied by others; e.g., see [K1].) In Wiener's construct, the x-coordinate of a Brownian path was a random function $\omega \in \mathrm{C}_{\mathbb{R}}([0,1])$, sampled according to a prescribed probability measure (the Wiener measure), so that for every partition $\{t_0 = 0 < t_1 < \cdots < t_n = 1\}$ of the time interval [0,1] the Brownian displacements

$$\omega(t_i) - \omega(t_{i-1}), \quad \omega \in \mathrm{C}([0,1]), i = 1, \ldots, n, \tag{12.1}$$

were statistically independent Gaussian variables with mean 0 and variance $t_i - t_{i-1}$, $i \in [n]$. The construction of the Wiener measure on $\mathrm{C}_{\mathbb{R}}([0,1])$ – the essence of the model – reflected hypotheses that *Brownian displacements were independent of one another, and that on average they were the same over time intervals of equal length.* (Review §1 and §2.)

Let us reexamine these assumptions. Loosely put, the notion that two events are independent means that the two events appear unrelated: the occurrence of one appears to have no bearing on the occurrence of the other. It was this intuitive sense of independence, and no more, that was expressed in Einstein's proposed explanation of Brownian movement [Ei1, pp. 12–13]:

> Evidently it must be assumed that each single particle executes a movement which is independent of the movement of all other particles; the movements of one and the same particle after different intervals of time must be considered as mutually independent processes, so long as we think of these intervals of time as being chosen not too small.

Notably, Einstein did not explicitly apply an assumption of independence in his analysis, but invoked an analogous 'idealized' physical principle that led to Gaussian distributions of Brownian displacements. (See Remark i §1.)

Wiener, too, considered independence first in an intuitive, undefined sense. In his introduction to 'differential-space', Wiener recalled that [Wi1, p. 296]

> Einstein's ... assumption is that the displacement of a particle in some interval of time small in comparison with those which we can observe is independent, to all intents and purposes, of its entire antecedent history... [and that]... we may regard the Brownian movement as made up ... of a large number of very brief, independent impulses acting on each particle...

Brownian displacements were subsequently viewed in Wiener's framework as *statistically independent* random variables.

The second important assumption underlying Wiener's model is that Brownian movement is homogeneous in time; this means – broadly put – that whatever happens on average in a given time interval happens on average in *every* time interval of the same length. A supposition of time-homogeneity, which can be justified in a statistical-mechanical context by an 'ergodic' argument (e.g. [Re, p. 584], [Bo, pp. 49–51]), is in effect an admission that we have no knowledge about molecular collisions. In Wiener's mathematical setup, time-homogeneity is expressed precisely by the assumption that displacements over time intervals of equal length are *identically distributed* random variables. This and the assumption of statistical independence, together with the (physically plausible) assumption that Brownian trajectories are continuous, lead via the Central Limit Theorem to Gaussian Brownian displacements – and hence the Wiener process.

The assertion that Brownian displacements are statistically independent symmetric random variables with distributions homogeneous in time conveys an idealized view: the assertion does *not* convey an intrinsic property of Brownian movement, but, rather, an observer's perception of the movement. The Wiener process is a model based on extremal assumptions of 'least stochastic complexity' and 'zero knowledge'. Still, the model has been consistent with empirical observations. To wit, while molecules colliding with a Brownian particle could not be separately tracked in sub-microscopic regions, their cumulativez effects could be detected on microscopic levels [Bo, pp. 49–51]. A classical example of such an effect is the average distance λ traveled by a Brownian particle in time t. In his 1905 'Brownian Movement' paper, after obtaining a Gaussian distribution of Brownian displacements, Einstein predicted

$$\lambda = c\sqrt{t}, \tag{12.2}$$

where c was a composite of physical constants [Ei1, pp. 12–18].* This relation, which was experimentally confirmed by J. Perrin [Pe1, §29, §30], is of course implied also by Wiener's model:

$$\mathbf{E}|\mathrm{W}(t)| = \sqrt{\frac{2t}{\pi}}. \tag{12.3}$$

But this relation follows also from more general, stochastically more complex models (e.g., see Exercise 2). A Wiener process is but an extremal idealized construct: a limit of *simple* random walks. Indeed, when modeling 'random walks' observed in the real world, an independence-of-steps hypothesis does not convey an intrinsic, physical feature; only that observers of 'random walks' are baffled by them. No one believes (I trust...) that walks in the real world have independent steps, but rather, that steps depend somehow on 'hidden variables', and are somehow *interdependent.* The problem is: how can *interdependence-of-steps* be modeled, gauged, and detected?

13 Random Walks and Decision Making Machines

Let us change the paradigm from movements of a generic Brownian particle to a stroll of a drunk. Here is an old tale commonly told to illustrate a simple random walk. A drunk leaves a pub, and walks along a road for an hour. Every $1/N$ hours, he takes a step of length $1/\sqrt{N}$ to the right or to the left with equal probability, and, a simple-minded fellow, he takes the steps independently in time. (The step's length $1/\sqrt{N}$ – *normalization* – is the result of rescaling based on (1.3).) With the pub as the origin at time $t = 0$, the drunk's position at time $t = 1/N, 2/N, \dots, 1$ is a random variable

$$X_t(\omega) = \frac{1}{\sqrt{N}} \sum_{j=1}^{tN} r_j(\omega), \quad \omega \in \{-1, +1\}^N, \tag{13.1}$$

* To derive his formula (12.2), Einstein computed the 'square root of the arithmetic mean of the squares of displacements...'; see equation (11) in [Ei1, p. 17]. He should have evaluated directly the absolute value of the mean (cf. (12.3)), thereby obtaining a slightly different evaluation of Avogadro's number; see [Ei1, p. 18].

where ω is any one of the possible 2^N paths which occurs with probability $1/2^N$, and $r_j(\omega)/\sqrt{N} := \omega(j)/\sqrt{N}$ is the step at time $(j-1)/N$.

Let us embellish the story. In this version, the drunk possesses a 'Decision Making Machine' (DMM), and steps to the right or to the left according to it. The machine consists of N labeled switches $1, \ldots, N$ inside a sealed box, and N labeled light bulbs $1, \ldots, N$ on a panel. A randomizing device inside the sealed box turns each of the N switches 'on' or 'off' independently with probability 1/2. The randomizing device, activated by a push of a blue button on the panel, produces a state $\omega \in \{-1, +1\}^N$ of the DMM: $\omega(j) = +1$ means that switch j is 'on', and $\omega(j) = -1$ means that it is 'off'; the DMM's state remains fixed until the next push of the button. The light bulbs are wired to switches so that bulb j is lit precisely when switch j is 'on'. Equipped with this DMM, our drunk starts from the pub, pushes the blue button, and walks: at time $(j-1)/N$, $j = 1, \ldots, N$, he takes a step to the right if bulb j is lit, and to the left if it is not. His position X_t at time $t = j/N$, $j = 1, \ldots, N$, is again given by (13.1), but in this tale, $\omega \in \{-1, +1\}^N$ is a state of the DMM, produced by a push of a button and fixed throughout the walk.

We now retell the story with a more complex 'Decision Making Machine'. This model has N switches labeled $1, \ldots, N$, and $N_k := \binom{N}{1} + \binom{N}{2} + \cdots + \binom{N}{k}$ light bulbs labeled $1, \ldots, N_k$, where $k > 1$ is a given integer. We fix a one–one correspondence between the light bulbs and all subsets of switches of cardinality $\leq k$,

$$A_j \subset [N],\ 0 < |A_j| \leq k,\ j = 1, \ldots, N_k, \tag{13.2}$$

and define

$$\chi_j = \prod_{i \in A_j} r_i, \quad j = 1, \ldots, N_k. \tag{13.3}$$

Switches are wired to the bulbs so that bulb j is lit if and only if $\chi_j = 1$. (Such circuits are always feasible; e.g., see [En, §1.6].) Equipped with this DMM, the drunk pushes the blue button, and leaves the pub. He takes a step of length $1/\sqrt{N_k}$ every $1/N_k$ hours (normalization); at time $(j-1)/N_k$, $j = 1, \ldots, N_k$, to the right if bulb j is lit, and to the left if it is not. The drunk's position at time t, no longer a result of a simple random walk, is given by

$$X_t(\omega) = 1/\sqrt{N_k}\sum_{j=1}^{tN_k}\chi_j(\omega), \quad \omega \in \{-1,+1\}^N,$$

$$t = 1/N_k, 2/N_k, \ldots, 1, \tag{13.4}$$

where $\omega \in \{-1,+1\}^N$ denotes the DMM's state, produced by a push of a button and fixed throughout the walk. In particular, after one hour, the drunk's position is

$$X_1(\omega) = 1/\sqrt{N_k}\sum_{m=1}^{k}\sum_{1\le l_1<\cdots<l_m\le N} r_{l_1}(\omega)\cdots r_{l_m}(\omega), \quad \omega \in \{-1,1\}^N. \tag{13.5}$$

In the general tale, the drunk walks with a DMM that has N switches labeled $1,\ldots,N$, and n light bulbs labeled $1,\ldots,n$, where $0 < N \le n$ are arbitrary integers. Distinct non-empty sets $A_1,\ldots,A_n$ of switches are designated, and the n light bulbs are wired to the N switches so that bulb j is lit if and only if

$$\chi_j(\omega) = \prod_{i\in A_j} r_i(\omega) = 1, \quad j = 1,\ldots,n. \tag{13.6}$$

The clock is calibrated by $1/n$ hours. At time j/n, $j = 0,\ldots,n-1$, the drunk takes a step of length $1/\sqrt{n}$ to the right if bulb j is lit, and to the left if not. His position at time $t = 1/n, 2/n, \ldots, 1$ is

$$X_t(\omega) = \frac{1}{\sqrt{n}}\sum_{j=1}^{tn}\chi_j(\omega), \quad \omega \in \{-1,+1\}^N. \tag{13.7}$$

I refer to (13.7) as a random *F-walk*, where $F = \{A_1,\ldots,A_n\} \subset 2^{[N]}$.

We note two extremal scenarios.

1. $A_j = \{j\}$, $j = 1,\ldots,N$. This is the standard simple random walk, the simplest possible.
2. F consists of all non-empty subsets of $\{1,\ldots,N\}$, i.e., $k = N$ in (13.5). This is the most complex F-walk possible. At the end of this stroll, the drunk will 'almost always' be found near the pub (Exercise 38).

General F-walks, falling anywhere between these two cases, manifest varying degrees of 'stochastic complexity'. We expect that this 'complexity' will be related, somehow, to the 'combinatorial complexity' of the underlying $F \subset 2^{[N]}$. In the next section, we consider the instance $F = \{A \subset [N] : 0 < |A| \leq k\}$, the degree of whose 'combinatorial complexity' is marked by k. In the last chapter, after having introduced the notion of combinatorial dimension in Chapter XIII, we will consider the more general case $F \subset \{A \subset [N] : 0 < |A| \leq k\}$, where k is a fixed positive integer.

Remarks:

i **(is it realistic. . . ?).** In a more feasible tale, the DMM is a small, hand-held computer with a screen that flashes 'right' or 'left' in a time sequence. This replaces the light bulbs of the prototype. Switches and wires are replaced by logic boards, and the 'on/off' states are determined by pressing a blue 'random' key. An added feature, making the tale more poignant, is that the number N of 'switches' as well as the collection F of sets of 'switches' can be preset (say, by the bartender). Otherwise, the story is the same: the drunk presses the blue key, and walks.

Still, *is it realistic*? The walk is determined by a DMM's randomly selected state, which is fixed throughout the walk. The drunk adheres to a pre-selected plan, ostensibly unaffected by unforseen events (barking dogs, honking cars, taunting passers-by, neurons misfiring. . .). But therein lies the paradigm. We, observers of random walks, have records only of paths, time-frame by time-frame, and no knowledge of the many events that affected them. We thus imagine a DMM's sealed compartment containing interdependent variables – subsets of switches – whose values determine the walk. All is hidden from view. A practical problem is: can we gauge, using only data about the drunk's path, the underlying 'hidden' complexity of the drunk's walk? This question is largely open-ended. We shall return to it in the last chapter.

ii **(a continuous time-model?).** Returning to the classical paradigm, we ask: if a Brownian particle executes a linear random F-walk, then what is a corresponding stochastic process that models the walk in continuous time? The Wiener process is such a model in the case $k = 1$. In the next section we will answer the question for $F = \{A \subset [N] : 0 < |A| \leq k\}$, and, in particular, will note that the kth Wiener Chaos gives rise to a continuous-time model for the corresponding random F-walk. In the last chapter, we will answer the question for $F \subset \{A \subset [N] : 0 < |A| \leq k\}$, where k is a fixed integer. The question in the general case is open.

14 α-Chaos: A Definition, a Limit Theorem, and Some Examples

Wiener's view of Brownian motion is based on an application of the Central Limit Theorem to simple random walks – a key application that motivates the definition and construction of the Wiener process. To imagine, in the same way, similar stochastic constructs that would model random F-walks, we require limit theorems.

First, let us formalize the measurements defined in (10.1) through (10.6):

Definition 25 (cf. Definition 19). $X \in \mathrm{L}^2(\Omega, \mathbb{P})$ is an α-variable, $\alpha \in [0, \infty)$, if

$$2\delta(X) = \alpha. \tag{14.1}$$

If $2\delta(X) = \alpha$ and $\|X\|_{\phi_\alpha} < \infty$, then X is an *exact* α-variable; otherwise, if $\|X\|_{\phi_\alpha} = \infty$, then X is an *asymptotic* α-variable. In either case, if $2\delta(X) = \alpha$, $\mathbf{E}|X|^2 = 1$, and $\mathbf{E}X = 0$, then X is said to be a *standard* α-variable.

For example, a bounded variable is a 0-variable; a Gaussian variable is an exact 1-variable, and a k-fold product of independent Gaussian variables is an exact k-variable.

Theorem 26 *For every integer $k \geq 1$, there exists an exact standard k-variable $Y_{(k)}$ such that*

$$\sqrt{1 \Big/ \binom{N}{k}} \sum_{1 \leq l_1 < \cdots < l_k \leq N} r_{l_1} \cdots r_{l_k} \xrightarrow[N \to \infty]{} Y_{(k)} \text{ in distribution.} \tag{14.2}$$

Proof: We denote (for convenience)

$$Z_{N,k} = \sqrt{1 \Big/ \binom{N}{k}} \sum_{1 \leq l_1 < \cdots < l_k \leq N} r_{l_1} \cdots r_{l_k}, \tag{14.3}$$

and, by induction on k, verify that $Z_{N,k}$ converges in distribution to an exact standard k-variable. The case $k = 1$ is the Central Limit Theorem. For $k = 2$, write

$$Z_{N,2} = \frac{1}{2}\sqrt{1\Big/\binom{N}{2}}\left(\sum_{j=1}^{N} r_j\right)^2 - N\Big/\sqrt{\binom{N}{2}}, \tag{14.4}$$

and then apply the Central Limit Theorem to the right side. For $k > 2$, write

$$Z_{N,k} = (1/k!)\sqrt{1\Big/\binom{N}{k}}\left(\sum_{j=1}^{N} r_j\right)^k + \sum_{i=2}^{k-1} d_{N,i}\, Z_{N,i-1}, \tag{14.5}$$

where $(d_{N,i})_{N\in\mathbb{N}}$ is a sequence of real numbers and $\lim_{N\to\infty} d_{N,i} = 0$, $i = 2, \ldots, k-1$. Apply the Central Limit Theorem to the first term on the right side of (14.5), and the induction hypothesis to each of the other $k - 2$ terms (Exercise 39). □

We return now to Brownian trajectories, and, as in previous discussions (cf. §1), imagine them to be random walks. We keep the three (heuristic) assumptions stated in §1, but do not presume that the 'walk' is simple. We fix $t > 0$, and calibrate the time interval $[0, t]$ by n subintervals of equal length t/n. If $X(t)$ is the particle's position at time t, then

$$X(t) = \sum_{j=1}^{n} X\left(\frac{j}{n}\, t\right) - X\left(\frac{j-1}{n}\, t\right) = \sqrt{\frac{c\,t}{n}}\sum_{j=1}^{n} Y_j, \tag{14.6}$$

where $c > 0$ is an absolute constant, and the Y_j are orthonormal. (See (1.3) and Exercise 2.) Now assume that for some fixed integer $k > 0$, (14.6) is a random F-walk, whose underlying 'combinatorial complexity' is k on every scale. (See comments following (13.7).) That is, for all $n = \binom{N}{k}$,

$$X(t) = \sqrt{c\,t\Big/\binom{N}{k}}\sum_{1\le l_1<\cdots<l_k\le N} r_{l_1}\cdots r_{l_k} \tag{14.7}$$

(cf. (1.4)). Then, by letting $N \to \infty$, we obtain from Theorem 26 that $X(t)/\sqrt{ct}$ is an exact standard k-variable. This motivates

Definition 27 (cf. Definition 1). A stochastic process

$$X = \{X(t) : t \in [0,1]\}$$

on a probability space $(\Omega, \mathscr{A}, \mathbb{P})$ is an α-*chaos* process, $\alpha \in [1, \infty)$, if X satisfies the following:

i. $\mathbf{E}|X(t)|^2 = Kt$ (constant $K > 0$) and $\mathbf{E}X(t) = 0$ for all $t \in [0,1]$;
ii. X has orthogonal increments, i.e., if J_1, J_2 are disjoint intervals in [0,1], then $\mathbf{E}\Delta X(J_1)\Delta X(J_2)=0$ ($\Delta X(J):=X(t) - X(s)$, where $0 \leq s < t \leq 1$ are the end-points of an interval J);
iii.

$$\delta_{H(X)} := \sup\{\delta_{\text{span}\{\Delta X(J_i)\}} : \text{intervals } J_i \subset [0,1],\ \Sigma_i \mathbf{1}_{J_i} \leq 1\} = \alpha/2,$$

where

$$H(X) = \mathrm{L}^2(\Omega, \mathbb{P})\text{-closure of} \bigcup \text{span}\{\Delta X(J_i) : \Sigma_i \mathbf{1}_{J_i} \leq 1\}.$$

An α-chaos X is *exact* if $c_{H(X)}(2/\alpha) < \infty$, and is *asymptotic* if $c_{H(X)}(2/\alpha) = \infty$.

In the next chapter we will observe that if a process X is an α-chaos, then

$$\mu_X(A, J) = \mathbf{E}\mathbf{1}_A \Delta X(J), \quad A \in \mathscr{A}, \text{ interval } J \subset [0,1], \tag{14.8}$$

determines an F_2-measure on $\mathscr{A} \times \mathscr{B}$. (This is easy. You can verify it now, or note it in the next chapter.) In particular, α-chaos processes are 'integrators' in the same sense that the Wiener process (an exact 1-chaos) is an 'integrator'. We will also note, by transcribing arguments in §5, that almost all sample-paths of an α-chaos process are continuous.

The total variation of μ_X is infinite (proof similar to the argument that the total variation of μ_W is infinite), but

$$\|\mu_X\|_{\theta_\gamma} := \sup \left\{ \sum_{j,k} \theta_\gamma(|\mu_X(A_j, B_k)|) : \Sigma_j \mathbf{1}_{A_j} \leq 1, \right.$$

$$\left. \Sigma_k \mathbf{1}_{B_k} \leq 1,\ \{A_j\} \subset \mathscr{A},\ \{B_k\} \subset \mathscr{B} \right\} < \infty \tag{14.9}$$

for all $\gamma > \alpha$, which implies $\|\mu_X\|_{(p)} < \infty$ for all $p > 1$ [BlKah]. The proof of (14.9) is similar to that of $\|\mu_{\mathrm{W}_n}\|_{\theta_n} < \infty$, where W_n is the nth Wiener Chaos process. (See Corollary 23.) I do not know whether for every α-chaos X, $\|\mu_X\|_{\theta_\gamma} = \infty$ for all $\gamma < \alpha$ (cf. Example 1 below).

Examples

1. Let τ be a measure-preserving one–one map from $([0,1], \mathscr{B}, \mathfrak{m})$ onto $(\tilde{D}_n, \tilde{B}_{\sigma n}, \mathfrak{m}_\sigma^n)$, and define

$$\mu(A, B) = \mu_{W_n}(A, \tau[B]), \quad A \in \mathscr{A},\ B \in \mathscr{B}, \tag{14.10}$$

where μ_{W_n} is the F_2-measure on $\mathscr{A} \times B_{\sigma n}$ determined by (11.2). Then, μ is an F_2-measure on $\mathscr{A} \times \mathscr{B}$ such that $\mu(\cdot, B) \ll \mathbb{P}$ for all $B \in \mathscr{B}$, and $\|\mu\|_{\theta_\gamma} < \infty$ if and only if $\gamma \geq n$ (Theorem 24). Define

$$\tilde{W}_n(t) = \frac{d}{d\mathbb{P}}(\mu(\cdot, [0,t]), \quad t \in [0,1]. \tag{14.11}$$

$\tilde{W}_n$ is an exact n-chaos, and $\mu_{\tilde{W}_n} = \mu$ where $\mu_{\tilde{W}_n}$ is determined by (14.8). (See next chapter for details.)

2. Consider a unitary map U from $L^2([0,1], \mathfrak{m})$ onto $L^2_{G_n}(\Omega, \mathbb{P}.)$ (See (9.8) for the definition of G_n.) Define

$$X(t) = U\mathbf{1}_{[0,t]}, \quad t \in [0,1]. \tag{14.12}$$

Then, X is an exact n-chaos (cf. Kakutani's realization of W in §2).

We can replace the system G_n by W_n (the Walsh system of order n), and, similarly, obtain yet another example of an exact n-chaos.

Questions

1. *Do α-chaos processes exist for non-integer α?* This problem should by now have a familiar ring. It is related to the question whether we can meaningfully define a non-integer complexity of a random walk, and obtain a limit theorem (like Theorem 26). The question is related to problems in Chapter VII concerning p-Sidon sets and Bonami's inequalities, and will be resolved in the last chapter.
2. *How do we gauge stochastic complexity of a random walk corresponding to a general collection of subsets $\mathfrak{A} \subset 2^{[N]}$?* (See Remark ii §13.) The problems that arise in connection with this question are only partly solved. They are closely related to Rudin's $\Lambda(p)$-set problem [Ru1], and to Bourgain's solution of it [Bour]. (See Chapter III §6 iv.) These issues will be discussed in the last chapter.
3. *How can we detect the stochastic complexity of Brownian movement?* This is a practical question, evoking a recurring theme of the last three sections: that Wiener's model, based on a simple random walk, is only a 'first approximation' to Brownian motion in the 'real world'.

The problem is this: given a large sample of Brownian paths – tracks of physical particles, or foreign currency fluctuations, or walks of drunks – how can we estimate the degree of complexity of the underlying process? I believe that such estimates could, somehow, be tied to the variations of the associated Fréchet measures, say in the spirit of Theorem 23. (More of this will be said in the next and the last chapter.)

Exercises

1. Let $(\Omega, \mathbb{P})$ be a probability space. If $X \in \mathrm{L}^2(\Omega, \mathbb{P})$ and $Y \in \mathrm{L}^2(\Omega, \mathbb{P})$ are uncorrelated, then find a sense in which X and Y are *functionally independent.* (For a definition and discussion of functional independence, see Remark iv in Chapter VII §8.)
2. Suppose a process $X = \{X(t) : t \in [0, \infty)\}$ has the following properties:
   (1) $\mathbf{E}X(t) = 0$ for all $t \in [0, \infty)$;
   (2) if $s_1 \leq t_1 \leq s_2 \leq t_2$, then $\mathbf{E}(X(t_1) - X(s_1))(X(t_2) - X(s_2)) = 0$;
   (3) there exists a non-negative function v on $[0, \infty)$ such that
   $$\mathbf{E}|X(t) - X(s)|^2 = v(t - s), \quad 0 \leq s \leq t < \infty.$$
   Prove that $v(t) = ct$ for all $t \in [0, \infty)$, where $c \geq 0$ is a numerical constant.
3. Prove from first principles that under the strongest possible interpretation applied to assumptions i, ii, and iii in §1, and the assumption that Brownian trajectories are continuous, the probability distribution of the displacement $X(t)$ of a Brownian particle at time t is Gaussian with mean 0 and variance $c\,t$, where $c \geq 0$ is a numerical constant.
4. i. Let H be the $\mathrm{L}^2(\Omega, \mathbb{P})$-closure of the linear span of a system of independent standard Gaussian variables. Prove that every $X \in H$ is Gaussian with mean 0 and variance $\|X\|^2_{\mathrm{L}^2}$.
   ii. Prove that if $Y_1, \ldots, Y_n$ are mutually orthogonal Gaussian variables with mean 0 such that every element in the linear span of $\{Y_1, \ldots, Y_n\}$ is Gaussian, then $Y_1, \ldots, Y_n$ are statistically independent.
5. Complete the proof of Proposition 2; that is, show that
   $$\sum_{j=1}^{n} (\mathrm{W}_{n,j})^2 \xrightarrow[n \to \infty]{} 1 \ \text{ in } \mathrm{L}^2(\Omega, \mathbb{P}) \tag{E.1}$$

implies that almost all sample-paths of W have infinite variation over [0,1].

6. i. For $t \in (0,1]$, let $\pi_k = \{0 = t_{0,k} \leq t_{1,k} < \cdots < t_{n_k,k} = t\}$, $k \in \mathbb{N}$, be partitions whose mesh goes to 0 as $k \to \infty$. By a modification of the proof of Proposition 2, establish that

$$\sum_{i=1}^{n_k} [\mathrm{W}(t_{i,k}) - \mathrm{W}(t_{i-1,k})]^2 \xrightarrow[k\to\infty]{} t \ \text{ in } \mathrm{L}^2(\Omega, \mathbb{P}).$$

(It can be shown with additional effort that the convergence above is almost sure ($\mathbb{P}$). This is a theorem due to Paul Lévy; e.g., [Doo, p. 395].)

ii. Prove that on every subinterval of [0,1], almost all ($\mathbb{P}$) sample-paths of W have infinite variation.

7. For the purpose of this exercise, you can assume Paul Lévy's result concerning the almost sure convergence of the quadratic variation.

Before we state the problems, we formalize few (neo)- classical notions. Let $\boldsymbol{\rho} = \{\rho_1, \ldots, \rho_k\}$ be a k-partition of $\mathbb{N}, |\rho_1| = \cdots = |\rho_k| = \infty$, and enumerate $\rho_i = \{n_{ij} : j \in \mathbb{N}\}, i = 1, \ldots, k$. Consider the binary digit expansion of $x \in [0,1]$, $x = \sum_{j=1}^{\infty} b_j(x)/2^j$, and define the mappings from [0,1] onto [0,1]

$$\tau_{\rho_i}(x) = \sum_{j=1}^{\infty} b_{n_{ij}}(x)/2^j, \quad i = 1, \ldots, k,$$

which are bimeasurable with respect to algebras generated by the dyadic intervals. (In this exercise, if x is a dyadic rational, then $b_j(x) = 0$ for all but finitely many j.) Consider $\tau_{\boldsymbol{\rho}} = (\tau_{\rho_1}, \ldots, \tau_{\rho_k})$, which is a one–one map from [0,1] onto $[0,1]^k$.

We say that a continuous function f on [0,1] has *type* F_k if there exists a k-partition $\boldsymbol{\rho}$ of $\mathbb{N}$ such that $f \circ \tau_{\boldsymbol{\rho}}^{-1}$ is a function with bounded F_k-variation on $[0,1]^k$; this means that the k-fold difference $\Delta^k(f \circ \tau_{\boldsymbol{\rho}}^{-1})$ determines an F_k-measure on $\mathscr{B}^k$, where $\mathscr{B}$ is the usual Borel field in [0,1]. (Review definitions in Chapter VI.).

i. Prove that if a continuous function f on [0,1] has *type* F_k, then

$$\sup\left\{\sum_{J\in\pi} |\Delta f(J)|^p : \text{ standard partition } \pi \text{ of } [0,1]\right\} < \infty,$$

$$p \leq \frac{2k}{k+1}.$$

(By a *standard partition* we mean here a finite partition that consists of contiguous intervals.)

ii. Prove that almost all sample paths of a Wiener process W do not have *type* F_k for all $k \geq 1$. What does this say about sample paths of a Wiener process?

8. Verify that if $f \in L^2([0,1], \mathfrak{m})$, then its Wiener integral $I_W(f)$ is a Gaussian random variable with mean 0 and variance $\|f\|_{L^2}^2$, and conclude that

$$\int_0^1 f(t)\, g(t)\, dt = \mathbf{E}\, I_W(f)\, I_W(g).$$

9. Let $\mu_W \in F_2(\mathscr{A}, \mathscr{B})$ be the Wiener F_2-measure defined in (3.16). Verify that

$$\|\mu_W\|_{F_2} := \sup\left\{\left\|\sum_{j,k} \mu_W(A_j, B_k)\, r_j \otimes r_k\right\|_{L^\infty} : \Sigma_j \mathbf{1}_{A_j} \leq 1,\right.$$

$$\left.\Sigma_k \mathbf{1}_{B_k} \leq 1,\ \{A_j\} \subset \mathscr{A},\ \{B_k\} \subset \mathscr{B}\right\} = \sqrt{\frac{2}{\pi}}.$$

10. Let μ_W be the Wiener F_2-measure defined in (3.16). Verify that for all bounded Borel-measurable functions f on [0,1],

$$I_W(f) = \frac{d}{d\mathbb{P}} \int_{[0,1]} f(t)\, \mu_W(\cdot, dt).$$

(This exercise is a preview of the next chapter.)

11. Let μ be a (generalized) Wiener F_2-measure defined by (3.25) and (3.26). Prove that if G is an infinite locally compact Abelian group, then μ_W cannot be extended to a scalar measure.

12. Prove that the stochastic series of W stated in (3.28) converges in $L^2(\Omega, \mathbb{P})$ uniformly in $t \in [0,1]$.

13. i. (*the Walsh–Wiener series*). Consider the stochastic series representation of W given by

$$W(t) = \sum_{j=1}^{\infty} \left(\int_{[0,t]} w_j(x)\, dx\right) X_j, \quad t \in [0,1],$$

where $\{w_j\}$ is the Paley enumeration of the Walsh system (Chapter VII §4), and $\{X_j\}$ is a statistically independent system of standard Gaussian variables. Prove that $\int_{[0,t]} w_j(x)\, dx$ is

$\mathscr{O}(1/j)$ (can you compute $\int_{[0,t]} w_j(x)\, \mathrm{d}x$ explicitly?), and obtain another proof that almost all sample-paths of a Wiener process are continuous.

ii. (*the Haar–Wiener series*). State explicitly the stochastic series of a Wiener process based on the normalized Haar system in $\mathrm{L}^2([0,1], \mathfrak{m})$ (e.g., [LiTz, Vol. I, p. 3]). (Paul Lévy used the *Haar* system to give his own construction of the Wiener process; see [Lé3, Chapter I]. So far as I can determine, a representation of the Wiener process by the Haar–Wiener series appeared first in [Ci].)

14. Here is an 'exercise' that may seem out of place in a mathematical monograph, but try it anyway!

    Ponder the difference between these two statements: (1) two events are *truly* independent, in the sense that one *really* has nothing to do with the other; (2) two events are *perceived* independent because we are unable to process, deterministically, information about interdependencies between them.
15. (Re)prove the classical Khintchin inequalities by showing that the Rademacher system is sub-Gaussian. (This proof of the Khintchin inequalities is due to E. Stein [St, Appendix D].) (See (4.25).)
16. Prove that every statistically independent system of variables that are uniformly bounded in the Orlicz space $\mathrm{L}_{\phi_1}(\Omega, \mathbb{P})$ is sub-Gaussian, where ϕ_1 is defined in (4.11).
17. Show that every Wiener integral can be represented by a series of statistically independent standard Gaussian variables.
18. Verify that all the sets in the proof of Lemma 7 are measurable.
19. In Remark i in §5 we proved Corollary VII.42. Verify the stronger statement: that Theorem VII.36 is best possible.
20. Verify (6.9).
21. Justify the assumption prior to (6.12) in the proof of Theorem 11.
22. Verify (6.16).
23. Verify (6.21).
24. Verify that the integrals in (7.3) and (7.11) do not depend on the representations of the step functions f in (7.2) and (7.10), respectively. Then verify (7.13) for standard symmetric step functions f that vanish on 'hyper-diagonals'.
25. Prove that $S_{\sigma,n}$ (standard symmetric step functions vanishing on 'hyper-diagonals') is norm-dense in $\mathrm{L}^2_\sigma([0,1]^n, \mathfrak{m}^n)$.
26. Verify (8.6).

27. (Itô's formula [I1, p. 523]) Let g be a twice-differentiable real-valued function on $\mathbb{R}$ with a continuous second derivative. Let

$$\pi = \{0 = t_0 < t_1 < \cdots < t_{n-1} < t_n = 1\}$$

be a partition of [0,1], denote $\|\pi\| = \max\{|t_j - t_{j-1}| : j = 1, \ldots, n\}$, and consider the Riemann sum

$$R_{\mathrm{W}}(g';\pi) = \sum_{j=1}^{n} g'(\mathrm{W}(t_{j-1}))(\mathrm{W}(t_j) - \mathrm{W}(t_{j-1})).$$

Prove that

$$\int_0^1 g'(\mathrm{W})\,\mathrm{dW} := \lim_{\|\pi\|\to 0} R_{\mathrm{W}}(g;\pi)$$

$$= g(\mathrm{W}(1)) - g(\mathrm{W}(0)) - \frac{1}{2}\int_0^1 g''(\mathrm{W}(t))\,\mathrm{d}t,$$

where the limit above is in $\mathrm{L}^2(\Omega, \mathbb{P})$.

28.* i. Obtain a general formula, analogous to (8.11), for the n-process (a process indexed by n parameters) defined in (8.10).

ii. For $f \in \mathrm{L}^2_\sigma([0,1]^2, \mathfrak{m}^2)$, represent the process

$$I_{\mathrm{W}}(\mathbf{1}_{[0,t]}\, f(\cdot, t)), \quad t \in [0,1],$$

by a stochastic series, and then represent the (iterated) Itô integral

$$\int_{[0,1]} I_{\mathrm{W}}(\mathbf{1}_{[0,t]}\, f(\cdot, t))\,\mathrm{dW}(\mathrm{d}t)$$

by a series whose summands involve Wiener integrals. In like fashion, state the n-process in (8.10) in terms of stochastic series.

iii. Verify explicitly that the two formulae obtained in i and ii – for the multiple integral and the iterated integral – agree in the case $f = \mathbf{1}_{[0,t_1]} \cdots \mathbf{1}_{[0,t_n]}$.

29. Prove Lemma 18.

30. Verify that if $\{X_j : j \in \mathbb{N}\}$ is a sub-α-system, then $\{X_{j_1} \otimes \cdots \otimes X_{j_n} : (j_1, \ldots, j_n) \in \mathbb{N}^n\}$ is a sub-$n\alpha$-system.

31. Let $\{X_j : j \in \mathbb{N}\}$ be a system of statistically independent standard Gaussian variables. By use of the Central Limit Theorem and Bonami's inequalities, prove that

$$G_n = \{X_{j_1} \ldots X_{j_n} : 1 \le j_1 < \cdots < j_n\}$$

is a sub-n-system.

32. In the last step of the proof of Proposition 20, show that for pairwise disjoint intervals $J_i \subset [0,1]$, $i \in [n]$, and $f = \mathbf{1}_{J_1\times\cdots\times J_n}$,

$$I_{\mathrm{W}_n}(f) = \Delta\mathrm{W}(J_1)\cdots\Delta\mathrm{W}(J_n),$$

and then verify that $c(I_{\mathrm{W}_n}(f); s) = 0$ for all $s > 2/n$.

33. Supply the missing details in the argument verifying Part i of Corollary 21.

34. i. Show that for $1 < p < \infty$, $Y \in \mathrm{L}^p(\Omega, \mathbb{P})$, and $\lambda \in (0,1)$,

$$\mathbb{P}\{Y \geq \lambda\, \mathbf{E}Y\} \geq (1-\lambda^2)\frac{(\mathbf{E}\,Y)^p}{\mathbf{E}\,Y^p}.$$

ii. Suppose $H \subset \mathrm{L}^p(\Omega, \mathbb{P})$ is a $\Lambda(p)$-space, $1 < p < \infty$; that is, H is a closed subspace of $\mathrm{L}^p(\Omega, \mathbb{P})$, wherein the L^p-norm and the L^1-norm are equivalent. (See Chapter III.) Prove that H is closed in probability.

35. Verify that the set-function μ_{W_n} defined in (11.2) is an F_2-measure on $\mathscr{A}\times \tilde{B}_{\sigma n}$.

36. Verify (11.5).

37. Fill in the missing details in the proof of Theorem 22.

38. Prove that if the drunk's walk in (13.8) is generated by $\mathfrak{A} = 2^{[N]}$, then for all but one state of the DMM, the drunk returns to the pub at the end of the walk.

39. Fill in the missing details in the proof of Theorem 26.

Hints for Exercises in Chapter X

1. The two functions defined on $\mathrm{L}^2(\Omega, \mathbb{P})$ by

$$Z \mapsto \mathbf{E}XZ \;\text{ and }\; Z \mapsto \mathbf{E}YZ, \quad Z \in \mathrm{L}^2(\Omega, \mathbb{P}),$$

are functionally independent.

2. First prove that if ϕ is a real-valued Lebesgue-measurable function on $\mathbb{R}$, and $\phi(x+y) = \phi(x) + \phi(y)$ for all $x \in \mathbb{R}$ and $y \in \mathbb{R}$, then $\phi(x) = cx$ for all $x \in \mathbb{R}$, where c is a numerical constant. Show that v is monotone and that $v(t-s) = v(t) - v(s)$.

3. This is an opportunity to review the Central Limit Theorem.
4. i. Use characteristic functions.
   ii. Show that the multidimensional characteristic function of the joint distribution of $Y_1, \ldots, Y_n$ is the product of the characteristic functions of the Y_j.
5. Assume that almost all sample-paths are continuous. The assertion in (E.1) implies

$$\sum_{j=1}^{n_k} (\mathrm{W}_{n_k,j})^2 \xrightarrow[k\to\infty]{} 1 \text{ almost surely } (\mathbb{P}) \text{ for some } n_k \uparrow \infty.$$

Now apply

$$\sum_{j=1}^{n_k} |\mathrm{W}_{n_k,j}| \geq \frac{1}{\max_j |\Delta_{n_k,j}\mathrm{W}|} \sum_{j=1}^{n_k} (\mathrm{W}_{n_k,j})^2,$$

and almost sure sample-path continuity.
7. i. Here you need to review notions concerning the Fréchet variation in Chapter VI, and the Littlewood $2k/(k+1)$-inequalities in Chapter VII.
   ii. Use the quadratic variation. A meaning of 'F_k type' is proposed in Chapter XII §4, Remark iii.
8. If $(Z_j : j \in \mathbb{N})$ is a sequence of Gaussian variables with mean 0 converging to Z in $\mathrm{L}^2(\Omega, \mathbb{P})$, then Z is Gaussian with mean 0 and variance $\lim_{j\to\infty} \|Z_j\|_{\mathrm{L}^2}$. Use characteristic functions.
9. Compute the L^1-norm of a standard Gaussian variable.
11. Cf. (VI.2.14).
13. i. This is a computation; e.g., see [Fi1, §3].
    ii. I recommend that you (at least) browse through [Ci].
14. The point of the exercise is that there is indeed a difference between (1) and (2), which all too often is blurred in scientific writing.
15. See (4.25).
16. You can assume that $\{X_j\}$ is a system of symmetric, statistically independent variables uniformly bounded in $\mathrm{L}_{\phi_1}(\Omega, \mathbb{P})$. It suffices to prove that there exist $A > 0$ and $L > 0$ with the following property: if $X = \Sigma_j a_j X_j \in \operatorname{span}\{X_j\}$ and $\Sigma_j |a_j|^2 = 1$, then

$$\mathbf{E}\exp(t|X|) \leq \exp(At^2) \quad \text{for all} \quad t > \mathrm{L}.$$

Using symmetry, statistical independence, and uniform boundedness in $L_{\phi_1}(\Omega, \mathbb{P})$, show that

$$\mathbf{E}\exp(t|\Sigma_j a_j X_j|) \leq 2\,\mathbf{E}\exp(t\Sigma_j a_j X_j)$$
$$= \prod_j \mathbf{E}\exp(ta_j X_j) \leq \prod_j \exp(At^2 a_j^2) = \exp(At^2),$$

for some $A > 0$.

22. For computations involving complementary functions, see [LiTz, p. 147]. Like Exercise 20, this too requires a small amount of calculus.
24. $\{\Delta W(I_{i_1}) \cdots \Delta W(I_{i_n}) : 1 \leq i_1 < \cdots < i_n \leq N\}$ is an orthogonal system in $L^2(\Omega, \mathbb{P})$.
25. Because elements in L^2_σ are equivalence classes determined by the $\mathfrak{m}^n$-null sets in $[0,1]^n$, and because $\mathfrak{m}^n(D) = 0$ for every 'hyper-diagonal' D, we can assume without loss of generality that every $f \in L^2_\sigma$ vanishes on all diagonals.
29. Extend the proof of Lemma 6.
30. Use induction and the generalized Minkowski inequality.
31. It suffices to show that there exists $K > 0$ such that for all $N > 0$ and all scalar n-arrays $a = (a_{j_1 \ldots j_n} : 1 \leq j_1 < \cdots < j_n \leq N)$

$$\mathbf{E}\left|\sum_{1 \leq j_1 < \cdots < j_n \leq N} a_{j_1 \ldots j_n} X_{j_1} \cdots X_{j_n}\right|^p \leq K p^{pn/2}\, \|a\|_2^p, \quad p > 2.$$

To this end, first partition the Rademacher system into N pairwise disjoint subsystems $\{r_i^{(j)} : i \in \mathbb{N}\}$, $j = 1, \ldots, N$. By the Central Limit Theorem, for each $j \in [N]$,

$$\frac{1}{\sqrt{k}} \sum_{i=1}^{k} r_i^{(j)} \xrightarrow[k \to \infty]{} X_j \quad \text{in distribution.}$$

Denote $Z_N = \sum_{1 \leq j_1 < \cdots < j_n \leq N} a_{j_1 \ldots j_n} X_{j_1} \cdots X_{j_n}$, and prove that

$$\lim_{k \to \infty} \frac{1}{k^{n/2}} \sum_{1 \leq j_1 < \cdots < j_n \leq N} a_{j_1 \ldots j_n} \left(\sum_{i=1}^{k} r_i^{(j_1)}\right) \cdots \left(\sum_{i=1}^{k} r_i^{(j_n)}\right)$$
$$\xrightarrow[k \to \infty]{} Z_N \quad \text{in distribution.}$$

Let T_k denote the kth element of the sequence above. Verify, by Bonami's inequalities, that $\{T_k\}$ is uniformly integrable, and therefore

$$\lim_{k\to\infty} \mathbf{E}|T_k|^p = \mathbf{E}|Z_N|^p.$$

Now apply Bonami's inequalities to T_k. Consult [Du1, Chapter 9] for justification of these steps.

34. i. Modify slightly the argument in [Kah3, p. 8].
    ii. Review the proof of Corollary 21 iv.
35. Matters relating to this exercise are explained in the next chapter.
36. Cf. Exercise 23.
37. Review the proofs of Lemma 12 and Theorem 11.
38. Use elementary harmonic analysis.

XI
Integrators

1 Mise en Scène: A General View

In Chapter X §1, we started with three assumptions – three perceptions – about a Brownian particle's trajectory: (i) its direction at any instant cannot be determined; (ii) displacements over disjoint time intervals are unrelated; (iii) 'statistics' of displacements over time intervals of equal length are the same. In a framework of probability theory, the strongest interpretation of these perceptions implies that a Brownian particle's position $X(t)$ at time $t \in [0,1]$ is Gaussian with mean 0 and variance ct. Specifically, we argued in Chapter X §1 that if Brownian displacements are statistically independent, symmetrically distributed random variables with distributions homogeneous in time, then $\{X(t) : t \in [0,1]\}$ is necessarily a Wiener process (Definition X.1). A Wiener process, however, conveys an idealized view: while haphazard and difficult to predict, Brownian displacements are not, in reality, independent of one another. At the end of Chapter X, imagining Brownian motion to be a random walk, we departed from the classical model, and viewed statistical independence as the first and indeed simplest instance on a scale of stochastic complexity. This view – under assumptions of time-homogeneity, finite variance, and prescribed 'randomness' – led us to *α-chaos processes*. The case $\alpha = 1$, exemplified by a Wiener process, is a continuous-time model for the simple random walk, and the case $\alpha > 1$, exemplified for integer α by the Wiener homogeneous chaos, is a continuous-time model for walks that manifest greater levels of 'randomness'.

In this chapter we study a general class of stochastic processes, which includes the α-chaos, but also much more. We make no *a priori* assumptions about time-homogeneity, finite variance, or levels of 'randomness'. We require only that processes be *integrators*:

Definition 1 A real-valued process $X = \{X(t): t \in [0,1]\}$ on a probability space $(\Omega, \mathscr{A}, \mathbb{P})$ is an *integrator* if

$$\mathbf{E}|X(t)| < \infty, \quad t \in [0,1], \tag{1.1}$$

and for all $A \in \mathscr{A}$, the variations of the functions

$$g_A(t) = \mathbf{E1}_A X(t), \quad t \in [0,1], \tag{1.2}$$

are bounded uniformly in $\mathscr{A}$, i.e.,

$$\sup\{\|g_A\|_{\mathrm{BV}} : A \in \mathscr{A}\} < \infty. \tag{1.3}$$

If for each $A \in \mathscr{A}$, g_A is continuous on [0,1], then X is said to be a *continuous integrator*, and if g_A is right-continuous, then X is said to be a *right-continuous integrator*.

The motivation for the definition is this. Imagine that a Brownian particle's position $X(t)$, $t \in [0,1]$, is obtained as a sum of displacements over successive time intervals,

$$X(t) = \sum_{i=1}^{N} \Delta X(J_i), \tag{1.4}$$

where J_i, $i \in [N]$, are pairwise disjoint time intervals whose union is $[0,t]$, and $\Delta X(J) := X(u) - X(v)$ is the displacement over a time interval J with end-points $0 \le u < v \le t$. This realization of Brownian movement as a 'random walk' – the synthesis of X from its increments – is at the heart of the matter. At the very least, we expect this realization to be consistent: that $X(t)$ be the same for any choice of time intervals J_i, $i \in [N]$, whose union is $[0,t]$. More generally, we expect realizations by finite sums of displacements to be the same as realizations by infinite sums. This leads to a basic question: can $X(t)$ be realized as an 'integral'

$$X(t) = \int_{[0,t]} \mathrm{d}X? \tag{1.5}$$

The gist of Definition 1 is that if X is an *integrator*, then the integral on the right side of (1.5) is well-defined, and if X is a *right-continuous* integrator, then (1.5) holds. Indeed, if X is a right-continuous integrator, then for every $t \in [0,1]$, $X(t)$ can be consistently *and* independently synthesized from its displacements, which means: if $\{J_i : i \in \mathbb{N}\}$ is any collection of pairwise disjoint intervals such that $\bigcup_{i=1}^{\infty} J_i = [0,t]$, then

$$X(t) = \sum_{i=1}^{\infty} \Delta X(J_i), \tag{1.6}$$

where the series on the right side converges weakly in $\mathrm{L}^1(\Omega, \mathbb{P})$. Note that this expresses a notion of 'time-independence': if $\{J_i : i \in \mathbb{N}\}$ is any collection of pairwise disjoint intervals, $\bigcup_{i=1}^{\infty} J_i = [0,t]$, $Y \in \mathrm{L}^{\infty}(\Omega, \mathbb{P})$ is arbitrary, and τ is any permutation of $\mathbb{N}$, then

$$\mathbf{E}YX(t) = \sum_{i=1}^{\infty} \mathbf{E}Y\Delta X(J_{\tau i}), \tag{1.7}$$

i.e., position at time t depends (weakly in $\mathrm{L}^1(\Omega, \mathbb{P})$) only on the set of displacements $\{\Delta X(J_i)\colon i \in \mathbb{N}\}$, and not on the time-sequence of the displacements. Eventually we shall view this particular notion of independence as a left end-point on a scale of interdependence that is calibrated by 'dimension'. But for the time being, and for a long while, we will be focusing on the 'one-dimensional' integrators of Definition 1.

2 Integrators and Integrals

Work in this chapter will be carried out in the setting of the multi-dimensional measure theory that was developed in previous chapters. We start by rephrasing Definition 1 in the terminology of this setting. (See Chapter IV.)

Lemma 2 (Exercise 1). *A process $X = \{X(t) \colon t \in [0,1]\}$ on a probability space $(\Omega, \mathscr{A}, \mathbb{P})$ is an integrator if and only if*

$$\begin{aligned} \|X\| &:= \sup\{\|\{\mathbf{E}\mathbf{1}_{A_j}\Delta X(J_k)\}_{j,k}\|_{F_2(\mathbb{N},\mathbb{N})} : \Sigma_j \mathbf{1}_{A_j} \le 1, \\ &\qquad \Sigma_k \mathbf{1}_{J_k} \le 1,\ A_j \in \mathscr{A},\ \textit{intervals}\ J_k \subset [0,1]\} \\ &= \sup\left\{ \mathbf{E}\left| \sum_{n=1}^{k} \Delta X(J_n)\, r_n(s) \right| : \right. \\ &\qquad \left. k \in \mathbb{N},\ \Sigma_{n=1}^{k} \mathbf{1}_{J_n} \le 1,\ s \in \{-1,1\}^{\mathbb{N}} \right\} < \infty. \end{aligned} \tag{2.1}$$

We consider the question: *how are functions on* [0,1] *integrated with respect to an integrator* X*?*

Approach 1 (functional-analytic). For step functions

$$f = \sum_i a_i \, \mathbf{1}_{J_i}, \tag{2.2}$$

define

$$I_X(f) = \sum_i a_i \, \Delta X(J_i). \tag{2.3}$$

Then, by Lemma 2,

$$\mathbf{E}|I_X(f)| \le 2 \, \|f\|_\infty \|X\|, \tag{2.4}$$

which implies that I_X is uniquely extendible to an $\mathrm{L}^1(\Omega, \mathbb{P})$-valued bounded linear map defined on the sup-norm closure of the space of step functions. We view this map as integration with respect to X.

Notice that, while $I_X(f)$ is well-defined for all continuous functions f, it is not obvious how to integrate by this approach an indicator function $\mathbf{1}_B$, where $B \in \mathscr{B}$ is an arbitrary Borel subset of [0,1].

Approach 2 (measure-theoretic). For $A \in \mathscr{A}$, consider the function $G_X(A)$ on [0,1] defined by

$$\begin{aligned} G_X(A)(1) &= \mathbf{E}\mathbf{1}_A X(1), \\ G_X(A)(t) &= \lim_{s \to t+} \mathbf{E}\mathbf{1}_A X(s) \ \text{ for } t \in [0,1), \end{aligned} \tag{2.5}$$

and then let

$$\mu_X(A, I) = G_X(A)(t) - G_X(A)(s), \quad I = (s,t] \subset [0,1]. \tag{2.6}$$

($\mathbf{E}\mathbf{1}_A X(\cdot)$ has bounded variation, and hence $G_X(A)(t)$ exists for every $t \in [0,1)$.) Let $\mathscr{O}$ be the algebra generated by $\{(s,t] : 0 \le s < t \le 1\}$, and extend μ_X by linearity to $\mathscr{A} \times \mathscr{O}$. Henceforth, if X is a process such that $G_X(A)(t)$ exists for all $A \in \mathscr{A}$ and $t \in [0,1]$, then μ_X will denote the corresponding set-function defined by (2.6).

Lemma 3 *If* X *is an integrator, then* $\mu_X \in F_2(\mathscr{A}, \mathscr{O})$, *and*

$$\|\mu_X\|_{F_2(\mathscr{A},\mathscr{O})} = \|X\|.$$

Proof: For each $A \in \mathscr{A}$, $G_X(A)(\cdot)$ is right-continuous and has bounded variation. This implies that $\mu_X(A, \cdot)$ is a measure on $\mathscr{O}$.

To establish that $\mu_X(\cdot, O)$ is a measure on $(\Omega, \mathscr{A})$ for every $O \in \mathscr{O}$, it suffices to verify that $G_X(\cdot)(t)$ is a measure on $(\Omega, \mathscr{A})$ for every $t \in [0, 1)$. Let $\{A_j : j \in \mathbb{N}\}$ be a collection of pairwise disjoint elements in $\mathscr{A}$, and denote $A = \cup_j A_j$. We need to verify

$$\sum_{j=1}^{\infty} G_X(A_j)(t) = G_X(A)(t). \tag{2.7}$$

Let $(s_j : j \in \mathbb{N}) \subset (t, 1)$ be a decreasing sequence converging to t, and let $J_k = (s_k, s_{k-1}]$. Then,

$$\begin{aligned}\sum_{j=1}^{\infty} G_X(A_j)(t) &= \sum_{j=1}^{\infty} \lim_{k\to\infty} \mathbf{E1}_{A_j} X(s_k) \\ &= \sum_{j=1}^{\infty} \left\{ \mathbf{E1}_{A_j} X(s_1) - \left(\sum_{k=2}^{\infty} \mathbf{E1}_{A_j} \Delta X(J_k) \right) \right\} \\ &= \mathbf{E1}_A X(s_1) - \sum_{j=1}^{\infty} \left(\sum_{k=2}^{\infty} \mathbf{E1}_{A_j} \Delta X(J_k) \right). \end{aligned} \tag{2.8}$$

By Lemma 2,

$$\{\mathbf{E1}_{A_j} \Delta X(I_k)\} \in F_2(\mathbb{N}, \mathbb{N}). \tag{2.9}$$

Therefore, we can interchange summations (Corollary IV.7),

$$\begin{aligned}\sum_{j=1}^{\infty} \left(\sum_{k=2}^{\infty} \mathbf{E1}_{A_j} \Delta X(J_k) \right) &= \sum_{k=2}^{\infty} \left(\sum_{j=1}^{\infty} \mathbf{E1}_{A_j} \Delta X(J_k) \right) \\ &= \sum_{k=2}^{\infty} \mathbf{E1}_A \Delta X(J_k) = \mathbf{E1}_A X(s_1) - G_X(A)(t), \end{aligned} \tag{2.10}$$

and thus obtain (2.7) from (2.8).

The statement $\|\mu_X\|_{F_2(\mathscr{A},\mathscr{O})} = \|X\|$ follows from (2.1) (Exercise 2). □

Corollary 4 *X is an integrator if and only if μ_X determines an F_2-measure on $\mathscr{A} \times \mathscr{B}$, and $\|\mu_X\|_{F_2(\mathscr{A},\mathscr{B})} = \|X\|$.*

Proof: Apply Theorem VI.8. □

Corollary 5 (Exercise 3). *If X is an integrator and μ_X denotes the associated F_2-measure on $\mathscr{A} \times \mathscr{B}$ determined by (2.6), then for all $B \in \mathscr{B}$, $\mu_X(\cdot, B) \ll \mathbb{P}$.*

Conversely, if μ is an F_2-measure on $\mathscr{A} \times \mathscr{B}$ such that $\mu(\cdot, B) \ll \mathbb{P}$ for all $B \in \mathscr{B}$, and $X(t) = \mathrm{d}\mu(\cdot, [0, t])/\mathrm{d}\mathbb{P}$ for $t \in [0,1]$, then X is an integrator and $\mu_X = \mu$.

Corollary 5 leads to

Definition 6 For an integrator X and $f \in \mathrm{L}^\infty([0,1], \mathscr{B})$ (:= Banach algebra of all bounded Borel measurable functions on [0,1]),

$$\int_{[0,1]} f \; \mathrm{d}X := \frac{\mathrm{d}}{\mathrm{d}\mathbb{P}} \int_{[0,1]} f(t) \; \mu_X(\cdot, \mathrm{d}t). \tag{2.11}$$

This definition is made possible by Lemma VI.9, which implies that

$$\int_{[0,1]} f(t) \; \mu_X(A, \mathrm{d}t), \quad A \in \mathscr{A}, \tag{2.12}$$

is a measure on $(\Omega, \mathscr{A})$, and by Corollary 5, which implies that this measure is absolutely continuous with respect to $\mathbb{P}$. By Lemma VI.9,

$$\mathbf{E}\left|\int_{[0,1]} f \; \mathrm{d}X\right| \leq 2\|X\| \; \|f\|_\infty, \tag{2.13}$$

i.e., $f \mapsto \int_{[0,1]} f \; \mathrm{d}X$ is a bounded linear map from $\mathrm{L}^\infty([0,1], \mathscr{B})$ into $\mathrm{L}^1(\Omega, \mathbb{P})$.

Proposition 7 *If X is an integrator and $f \in \mathrm{C}([0,1])$, then*

$$I_X(f) = \int_{[0,1]} f \; \mathrm{d}X.$$

Proof: Let $A \in \mathscr{A}$, and note that if φ is a step function,

$$\varphi = \sum_{i=1}^{n} a_i \; \mathbf{1}_{J_i}, \tag{2.14}$$

whose discontinuity points are continuity points of $G_X(A)$, then

$$\int_{[0,1]} \varphi(t) \; \mu_X(A, \mathrm{d}t) = \mathbf{E}\mathbf{1}_A \sum_{i=1}^{n} a_i \; \Delta X(J_i) = \mathbf{E}\mathbf{1}_A I_X(\varphi). \tag{2.15}$$

Let $f \in \mathrm{C}([0,1])$, and let $(\varphi_j : j \in \mathbb{N})$ be a sequence of step functions converging uniformly to f, such that discontinuity points of each φ_j are continuity points of $G_X(A)$. Then,

$$\int_{[0,1]} \varphi_j(t)\, \mu_X(A, \mathrm{d}t) \xrightarrow[j\to\infty]{} \int_{[0,1]} f(t)\, \mu_X(A, \mathrm{d}t). \tag{2.16}$$

Also,

$$I_X(\varphi_j) \xrightarrow[j\to\infty]{} I_X(f) \quad \text{(convergence in } \mathrm{L}^1(\Omega, \mathbb{P})\text{)}. \tag{2.17}$$

Therefore,

$$\mathbf{E}\mathbf{1}_A I_X(f) = \int_{[0,1]} f(t)\, \mu_X(A, \mathrm{d}t), \tag{2.18}$$

and by taking Radon–Nikodym derivatives, we obtain

$$I_X(f) = \int_{[0,1]} f\, \mathrm{d}X. \tag{2.19}$$

□

If X is an integrator and $f \in \mathrm{L}^\infty([0,1], \mathscr{B})$, then we define the *indefinite integral* $\int f \mathrm{d}X$ to be the process

$$\left(\int f\, \mathrm{d}X\right)(t) := \int_{[0,1]} f\, \mathbf{1}_{[0,t]}\, \mathrm{d}X := \int_{[0,t]} f\, \mathrm{d}X, \quad t \in [0,1]. \tag{2.20}$$

Proposition 8 (Exercise 4). *If X is an integrator and*

$$f \in \mathrm{L}^\infty([0,1], \mathscr{B}),$$

then $\int f \mathrm{d}X$ is an integrator, and $\| \int f \mathrm{d}X\| \le 2\|X\|\|f\|_\infty$.

Remarks:

i **(Riemann v. Lebesgue).** The distinction between the two integrals in Approach 1 and Approach 2 is analogous to the distinction between Riemann–Stieltjes integration with respect to a monotone function, and Lebesgue–Stieltjes integration with respect to its right-continuous 'version'. Observe that if X is a right-continuous integrator, then

$$\begin{aligned} I_X(\mathbf{1}_{(s,t]}) = X(t) - X(s) &= \frac{\mathrm{d}}{\mathrm{d}\mathbb{P}} \mu_X(\cdot, (s,t]) \\ &= \int_{[0,1]} \mathbf{1}_{(s,t]}\, \mathrm{d}X; \end{aligned} \tag{2.21}$$

in particular (cf. (1.5)),

$$X(t) - X(0) = I_X(\mathbf{1}_{[0,t]}) = \int_{[0,1]} \mathbf{1}_{[0,t]} \, \mathrm{d}X. \tag{2.22}$$

ii **('white noise').** Approach 2 leads to the 'white noise' associated with an integrator X (cf. (X.3.18)), which extends (2.22):

$$\text{'}\Delta\text{'}X(B) = \int_{[0,1]} \mathbf{1}_B \, \mathrm{d}X, \quad B \in \mathscr{B}. \tag{2.23}$$

If X is right-continuous, then 'Δ'$X(J) = \Delta X(J)$ for every interval $J \subset [0,1]$, and we drop the quotation marks around Δ. A question arises: in what sense does 'Δ'$X(\cdot)$ determine a measure on $\mathscr{B}$? For every $A \in \mathscr{A}$, $\mathbf{E}\mathbf{1}_A$'Δ'$X(\cdot)$ is a scalar measure on $\mathscr{B}$, which is immediate from definitions, but more can be said. We will return to this question later in the chapter.

iii **('randomness').** Given our objectives (and biases...), we deem an integrator X interesting when μ_X cannot be extended to a bona fide scalar measure on $\sigma(\mathscr{A}\times\mathscr{O})$. For, if μ_X does determine an F_1-measure on $\sigma(\mathscr{A}\times\mathscr{O})$, then stochastic integration with respect to X proceeds, more or less routinely, in the usual 'one-dimensional' framework of measure theory. A simple example of such a process is $X = Z \otimes f$, where $Z \in \mathrm{L}^1(\Omega, \mathbb{P})$ and f is a function of bounded variation on [0,1]. In this case,

$$\mu_X = Z\mathrm{d}\mathbb{P} \times \mathrm{d}f. \tag{2.24}$$

In general, the verification that μ_X cannot be extended to a scalar measure – the only practical way I know – is a check that its *total variation* is infinite; i.e., that

$$\begin{aligned}
&\sup\{\|\{\mathbf{E}\mathbf{1}_{A_j}\Delta X(J_k)\}_{j,k}\|_{l^1(\mathbb{N}\times\mathbb{N})} : \Sigma_j \mathbf{1}_{A_j} \le 1,\\
&\quad \Sigma_k \mathbf{1}_{J_k} \le 1,\ A_j \in \mathscr{A},\ J_k \in \mathscr{O}\} := \|\mu_X\|_{(1)}\\
&= \sup\left\{\mathbf{E}\sum_k |\Delta X(J_k)| : \Sigma_k \mathbf{1}_{J_k} \le 1,\ J_k \in \mathscr{O}\right\} = \infty. \qquad (2.25)
\end{aligned}$$

If $\|\mu_X\|_{(1)} = \infty$, then the expected variation of sample-paths of X is infinite; i.e.,

$$\begin{aligned}\infty &= \sup\left\{\mathbf{E}\sum_k |\Delta X(J_k)| : \Sigma_k \mathbf{1}_{J_k} \leq 1,\ J_k \in \mathscr{O}\right\}\\ &\leq \mathbf{E}\sup\left\{\sum_k |\Delta X(J_k)| : \Sigma_k \mathbf{1}_{J_k} \leq 1,\ J_k \in \mathscr{O}\right\}, \qquad (2.26)\end{aligned}$$

which conveys haphazardness (cf. Chapter X §3, Remark i). This suggests

Definition 9 A process $X = \{X(t) : t \in [0,1]\}$ on a probability space $(\Omega, \mathscr{A}, \mathbb{P})$ is a *random integrator* if $\mu_X \in F_2(\mathscr{A} \times \mathscr{B})$ and $\|\mu_X\|_{(1)} = \infty$.

A recurring theme in this chapter is that measurements involving variations of μ_X reflect levels of 'randomness' in X; or levels of 'stochastic complexity'. To be precise, we define

$$\ell_X := \inf\{p : \|\mu_X\|_{(p)} < \infty\}, \qquad (2.27)$$

where

$$\|\mu_X\|_{(p)} := \sup\left\{\left(\sum_{j,k} |\mu_X(A_j, B_k)|^p\right)^{\frac{1}{p}} : \right.$$

$$\left. \Sigma_j \mathbf{1}_{A_j} \leq 1,\ \Sigma_k \mathbf{1}_{B_k} \leq 1,\ \{A_j\} \subset \mathscr{A},\ \{B_k\} \subset \mathscr{B}\right\}. \qquad (2.28)$$

We call ℓ_X the *Littlewood index*, and view it as a gauge of 'randomness' in X. By Littlewood's inequality, $\ell_X \leq 4/3$, and if X is a random integrator (according to Definition 9), then $1 \leq \ell_X \leq 4/3$.

Integrators X for which $\ell_X = 1$ can be brought into sharper focus. For example, if X is an α-chaos, $\alpha \geq 1$ (Definition X.27), then the variations of μ_X are controlled by the Orlicz functions θ_γ defined in (X.6.22): $\|\mu_X\|_{\theta_\gamma} < \infty$ for all $\gamma > \alpha$, and $\|\mu_X\|_{\theta_\gamma} = \infty$ for all $\gamma < \alpha$. (See (X.14.9), and also §4 in this chapter.) Specifically,

Theorem X.14 – the assertion that $\|\mu_{\mathrm{W}}\|_{\theta_2} < \infty$ and $\|\mu_{\mathrm{W}}\|_{\theta_\gamma} = \infty$ for all $\gamma < 1$ – conveys precisely that a Wiener process is stochastically the 'least complex' among continuous-time models of random walks. Observe that for every α-chaos $X, \ell_X = 1$ (while $\|\mu_X\|_{(1)} = \infty$). We will note in §4 that for every $p \in (1, 4/3]$ there exist integrators X such that $\ell_X = p$.

3 Examples

i. L^2***-bounded processes with orthogonal increments.*** These are processes X such that

$$\mathbf{E}|X(t)|^2 := F_X(t) < \infty, \quad t \in [0,1], \tag{3.1}$$

and

$$\mathbf{E}\Delta X(I)\,\Delta X(J) = 0,$$

$$\text{intervals } I \subset [0,1],\ J \subset [0,1],\ I \cap J = \emptyset. \tag{3.2}$$

For convenience, we assume $F_X(0) = 0$. For such X, if J is an interval with end-points $0 \leq s < t \leq 1$, then $\mathbf{E}|\Delta X(J)|^2 = F_X(t) - F_X(s)$; in particular, F_X is monotonically increasing on [0,1] (Exercise 5). Let λ_X be the positive regular Borel measure on [0,1] determined by F_X, i.e.,

$$\lambda_X(J) = F_X(t^+) - F_X(s^+), \quad J = (s,t] \subset [0,1]. \tag{3.3}$$

Proposition 10 *If X is an* L^2*-bounded process with orthogonal increments, then X is an integrator, and*

$$\mu_X \ll \mathbb{P} \times \lambda_X. \tag{3.4}$$

((3.4) *means that if* $\mathbb{P}(A)=0$ *or* $\lambda_X(B)=0$, *then* $\mu_X(A,B) = 0$.)

Proof: To verify that X is an integrator, let $\{J_n\}$ be a finite collection of pairwise disjoint intervals in [0,1], let $u \in \{-1,1\}^{\mathbb{N}}$, and estimate

$$\mathbf{E}\left|\sum_n \Delta X(J_n)\, r_n(u)\right| \leq \left\|\sum_n \Delta X(J_n)\, r_n(u)\right\|_{\mathrm{L}^2} \leq \|X(1)\|_{\mathrm{L}^2}. \tag{3.5}$$

To verify (3.4), note that for each $A \in \mathscr{A}$,

$$g \mapsto \mathbf{E}\mathbf{1}_A \int_{[0,1]} g \, \mathrm{d}X, \quad g \in \mathrm{C}([0,1]), \tag{3.6}$$

is a bounded linear functional on C([0,1]), which we represent by a regular Borel measure β_A on [0,1]. Then, $\beta_A(\cdot) = \mu_X(A, \cdot)$ (Notice that for all $g \in \mathrm{C}([0,1])$,

$$\int_{[0,1]} g(t) \; \beta_A(\mathrm{d}t) = \int_{[0,1]} g(t) \; \mu_X(A, \mathrm{d}t).)$$

The linear action

$$g \mapsto \int_{[0,1]} g(t) \; \beta_A(\mathrm{d}t), \quad g \in \mathrm{C}([0,1]), \tag{3.7}$$

is uniquely extendible to a bounded linear functional on $\mathrm{L}^2([0,1], \lambda_X)$, which we denote by $\tilde{\beta}_A$ (Exercise 6 i). Then,

$$\tilde{\beta}_A(\mathbf{1}_B), \quad B \in \mathscr{B}, \tag{3.8}$$

defines a measure on $\mathscr{B}$, which we denote also by $\tilde{\beta}_A$. This measure equals β_A (Exercise 6 ii). If $B \in \mathscr{B}$, then,

$$|\mu_X(A, B)| = |\beta_A(B)| = |\tilde{\beta}_A(\mathbf{1}_B)| \leq \|\tilde{\beta}_A\| \sqrt{\lambda_X(B)}, \tag{3.9}$$

which proves $\mu_X(A, \cdot) \ll \lambda_X$.

To verify $\mu_X(\cdot, B) \ll \mathbb{P}$, observe that $\{B \in \mathscr{B} : \mu_X(\cdot, B) \ll \mathbb{P}\}$ is a σ-algebra that contains $\{(s,t] : 0 \leq s < t \leq 1\}$. □

Because X is an integrator, we obtain by the measure-theoretic approach an integral $\int_{[0,1]} f \, \mathrm{d}X$ for every bounded Borel-measurable function f on [0,1].

An integral with respect to X can be obtained also by a functional-analytic approach that mimics the construction of the Wiener integral (in Chapter X §3). Let $S_{F_X}[0,1]$ denote the space of step functions whose points of discontinuity are points of continuity of F_X. If $f \in S_{F_X}[0,1]$, $f = \sum_{i=1}^n a_i \mathbf{1}_{J_i}$, then define (Exercise 7)

$$I_X(f) = \sum_{i=1}^n a_i \; \Delta X(J_i). \tag{3.10}$$

For all $f \in S_{F_X}[0,1]$,

$$\mathbf{E}|I_X(f)|^2 = \|f\|^2_{\mathrm{L}^2(\lambda_X)}, \tag{3.11}$$

i.e., (3.10) determines a linear isometry from $S_{F_X}[0,1]$ into $\mathrm{L}^2(\Omega, \mathbb{P})$, and, because $S_{F_X}[0,1]$ is dense in $\mathrm{L}^2([0,1], \lambda_X)$, this isometry is uniquely extendible to a linear isometry from $\mathrm{L}^2([0,1], \lambda_X)$ into $\mathrm{L}^2(\Omega, \mathbb{P})$. We view I_X as an integral with respect to X. (It is the Wiener integral in the case $X = \mathrm{W}$.)

Proposition 11 (Exercise 8; cf. Proposition X.3). *Let X be an L^2-bounded process with orthogonal increments.*

i. *For $f \in \mathrm{L}^2([0,1], \lambda_X)$ and $g \in \mathrm{L}^2([0,1], \lambda_X)$,*

$$\int_{[0,1]} f(t)g(t)\lambda_X(\mathrm{d}t) = \mathbf{E} I_X(f) I_X(g). \tag{3.12}$$

ii. *Let $\{\mathbf{e}_j : j \in \mathbb{N}\}$ be an orthonormal basis of $\mathrm{L}^2([0,1], \lambda_X)$, and define*

$$\hat{X}(j) = I_X(\mathbf{e}_j) \quad (cf.\ (\mathrm{X}.3.10)). \tag{3.13}$$

If $f \in \mathrm{L}^2([0,1], \lambda_X)$ and $\hat{f}(j) = \int_{[0,1]} f(s)\mathbf{e}_j(s)\lambda_X(\mathrm{d}s)$, then

$$\sum_{j=1}^{k} \hat{f}(j)\, \hat{X}(j) \xrightarrow[k\to\infty]{} I_X(f) \quad \text{in } \mathrm{L}^2(\Omega, \mathbb{P}). \tag{3.14}$$

Proposition 12 *For an L^2-bounded process X with orthogonal increments, define*

$$H(X) = \{I_X(f) : f \in \mathrm{L}^2([0,1], \lambda_X)\}. \tag{3.15}$$

i. *$H(X)$ is a norm-closed subspace of $\mathrm{L}^2(\Omega, \mathbb{P})$.*
ii. *If $H(X)$ is a $\Lambda(2)$-space (i.e., $\|Z\|_{\mathrm{L}^2} \le \kappa \|Z\|_{\mathrm{L}^1}$ for $Z \in H(X)$), then $H(X)$ is closed in probability.*
iii. *If $H(X)$ is a $\Lambda(2)$-space and F_X is continuous, then X is a random integrator.*

Proof: Parts i and ii are exercises (Exercise 9).

To verify iii, suppose $\Sigma_k \mathbf{1}_{J_k} = \mathbf{1}_{[0,1]}$, and estimate

$$\sum_i \mathbf{E}|\Delta X(J_i)| \geq \frac{1}{\max_i \mathbf{E}|\Delta X(J_i)|} \sum_i (\mathbf{E}|\Delta X(J_i)|)^2$$

$$\geq \frac{\kappa}{\max_i \mathbf{E}|\Delta X(J_i)|} \sum_i \mathbf{E}|\Delta X(J_i)|^2$$

$$= \frac{\kappa\, \mathbf{E}|X(1)|^2}{\max_i \mathbf{E}|\Delta X(J_i)|}. \tag{3.16}$$

Because F_X is continuous, we can choose $\{J_k\}$ such that

$$\max_k \mathbf{E}|\Delta X(J_k)|$$

is as small as we like, and conclude $\|\mu_X\|_{(1)} = \infty$. □

Proposition 13 *Let X be an L^2-bounded process with orthogonal increments.*

i. *If $f \in \mathrm{C}([0,1])$, then*

$$I_X(f) = \int_{[0,1]} f \, \mathrm{d}X. \tag{3.17}$$

ii. *If F_X is right-continuous and $f \in \mathrm{L}^\infty([0,1], \mathscr{B})$, then*

$$I_X(f) = \int_{[0,1]} f \, \mathrm{d}X.$$

Proof:

i. This follows from Proposition 7 (Exercise 10).
ii. If F_X is a right-continuous function, then X is a right-continuous integrator. Therefore, if $\varphi \in S_{F_X}[0,1]$, then

$$I_X(\varphi) = \int_{[0,1]} \varphi \, \mathrm{d}X. \tag{3.18}$$

Let $f \in \mathrm{L}^\infty([0,1], \mathscr{B})$. Let $(\varphi_j: j \in \mathbb{N})$ be a sequence of step functions in $S_{F_X}[0,1]$ converging to f in the $\mathrm{L}^2(\lambda_X)$-norm. Then,

$$\int_{[0,1]} \varphi_j \, \mathrm{d}X \to I_X(f) \quad \text{(convergence in } \mathrm{L}^2(\Omega, \mathbb{P})). \tag{3.19}$$

We can assume that $\varphi_j \to f$ a.e. (λ_X), and also that $\|\varphi_j\|_\infty \leq 2\|f\|_\infty$ for $j \in \mathbb{N}$. Let $A \in \mathscr{A}$ be arbitrary. By Proposition 10, for every $\epsilon > 0$ there exist $\delta > 0$ such that if $\lambda_X(B) < \delta$ then $|\mu_X|(A, B) < \epsilon$, where $|\mu_X|(A, \cdot)$ is the total variation measure of $\mu_X(A, \cdot)$. By Egoroff's theorem, there exists $B \in \mathscr{B}$ such that $\lambda_X(B) < \delta$, and $\varphi_j \to f$ uniformly on $[0,1]\backslash B$. Therefore, there exists $N > 0$ such that for all $k \geq N$,

$$\begin{aligned}
&\int_{[0,1]} |\varphi_k(t) - f(t)| \, |\mu_X|(A, \mathrm{d}t) \\
&= \int_{[0,1]\backslash B} |\varphi_k(t) - f(t)| \, |\mu_X|(A, \mathrm{d}t) \\
&\quad + \int_B |\varphi_k(t) - f(t)| \, |\mu_X|(A, \mathrm{d}t) \\
&< \epsilon + 3\epsilon\|f\|_\infty.
\end{aligned} \tag{3.20}$$

Therefore, by (3.19),

$$\mathbf{E}\mathbf{1}_A \int_{[0,1]} \varphi_j \, \mathrm{d}X \longrightarrow \mathbf{E}\mathbf{1}_A \int_{[0,1]} f \, \mathrm{d}X = \mathbf{E}\mathbf{1}_A I_X(f). \tag{3.21}$$

The assertion follows by taking Radon–Nikodym derivatives.

□

Remarks:

i **(when is X random?).** We briefly comment on the conditions in Proposition 12 iii that imply randomness. Let $\{X_k : k \in \mathbb{N}\} \subset \mathrm{L}^2(\Omega, \mathbb{P})$ be an orthogonal set such that $\sum_{k=1}^\infty \|X_k\|_{\mathrm{L}^2}^2 < \infty$, and let $J_k = (1/(k+1), 1/k]$, $k \in \mathbb{N}$. Define

$$X(t) = \sum_{k=1}^{\infty} \mathbf{1}_{J_k}(t) X_k = \sum_{kt \geq 1} X_k, \quad t \in [0,1], \tag{3.22}$$

in which case $\Delta X(J_k) = X_k$, and

$$F_X(t) = \sum_{kt \geq 1} \|X_k\|_{\mathrm{L}^2}^2, \quad t \in [0,1]. \tag{3.23}$$

Then, X is L^2-bounded with orthogonal increments, F_X is not continuous, and

$$\|\mu_X\|_{(1)} = \sum_{k=1}^{\infty} \|X_k\|_{L^1}. \tag{3.24}$$

Now observe that there exist orthogonal sequences $(X_k : k \in \mathbb{N})$ such that $\|X_k\|_{L^2} = \|X_k\|_{L^1} = 1/k$ for $k \in \mathbb{N}$, and $H(X)$ is not a $\Lambda(2)$-space. This illustrates that neither continuity of F_X nor the condition that $H(X)$ is a $\Lambda(2)$-space are necessary for X to be random. On the other hand, $H(X)$ a $\Lambda(2)$-space does not, by itself, imply that X is random: if $\{X_k\}$ is finite, then $H(X)$ is (trivially!) a $\Lambda(2)$-space, but X is not random (Exercise 11).

ii **(sample-path properties).** If X is an L^2-bounded process with orthogonal increments, and $t \in [0,1]$ is a continuity point of F_X, then

$$X(t) = I_X(\mathbf{1}_{[0,t]}) = \sum_{j=1}^{\infty} \hat{\mathbf{1}}_{[0,t]}(j)\ \hat{X}(j)$$

$$\text{(convergence in } L^2(\Omega, \mathbb{P})). \tag{3.25}$$

To learn about sample-paths of X from series representations, we need to know more about X. For example, consider L^2-bounded processes X with orthogonal increments such that

$$F_X(t) = ct, \quad t \in [0,1]. \tag{3.26}$$

We refer to such X as *homogeneous* integrators (see Exercise X.2). For a homogeneous X, we can take $\{\cos \pi js : j = 0, \ldots\}$ to be a basis for $L^2([0,1], \lambda_X)$, and obtain

$$X(t) = btX_0 + k \sum_{j=1}^{\infty} \frac{\sin \pi jt}{j}\ \hat{X}(j), \tag{3.27}$$

where $b > 0$ and $k > 0$ are numerical constants, and $\hat{X}(j) = I_X(\cos\ \pi jt)$, $j = 0, \ldots$ (cf. (X.3.30)). In this case, if more is known about 'interactions' between displacements of X (more than orthogonality), then more can be learnt about sample-paths of X from the series representations in (3.27). For instance, if $H(X)$ is a $\Lambda(q)$-space for some $q > 2$ (L^q- and L^2-norms are equivalent in $H(X)$), then stochastic series representations imply that sample-paths of X are almost surely continuous. (See next section.)

ii. L^1***-bounded additive processes.*** A process X is L^p*-bounded* if $\mathbf{E}|X(t)|^p < \infty$ for all $t \in [0,1]$, *centered* if $\mathbf{E}X(t) = 0$ for all $t \in [0,1]$, and *additive* if $\{\Delta X(J_i)\}$ is statistically independent for every finite collection $\{J_i\}$ of pairwise disjoint intervals in $[0,1]$.

Proposition 14 *If X is a centered L^1-bounded additive process, then X is an integrator.*

Proof (Exercise 12): First note that if Y and Z are independent random variables and $\mathbf{E}Z = 0$, then $\mathbf{E}|Y + Z| \geq \mathbf{E}|Y|$. Next, by replacing the process X with $X - \tilde{X}$, where $\tilde{X}$ is a statistically independent copy of X, we can assume that X is a symmetric process. If $\{J_i\}$ is a finite collection of pairwise disjoint intervals, and $u \in \{-1,1\}^{\mathbb{N}}$, then

$$\mathbf{E}\left|\sum_i r_i(u)\, \Delta X(J_i)\right| = \mathbf{E}\left|\sum_i \Delta X(J_i)\right| \leq \mathbf{E}|X(1) - X(0)|. \tag{3.28}$$

□

iii. L^p***-bounded martingales.***

Proposition 15 *If $X = \{X(t) : t \in [0,1]\}$ is an L^p-bounded martingale process, $1 < p \leq 2$, then X is an integrator.*

Proof (Exercise 13): If $\{J_i\}$ is a finite collection of pairwise disjoint intervals, and $u \in \{-1,1\}^{\mathbb{N}}$, then by the Burkholder–Gundy inequalities (e.g., [Bu, (3.3)]),

$$\left(\mathbf{E}\left|\sum_i r_i(u)\, \Delta X(J_i)\right|\right)^p \leq \mathbf{E}\left|\sum_i r_i(u)\, \Delta X(J_i)\right|^p$$

$$\leq C_p\, \mathbf{E}\left(\sum_i |\Delta X(J_i)|^2\right)^{\frac{p}{2}} \leq c_p\, \mathbf{E}\left|\sum_i \Delta X(J_i)\right|^p . \tag{3.29}$$

□

4 More Examples: α-chaos, $\Lambda(q)$-processes, p-stable Motions

If X is homogeneous, then for all $Z \in H(X)$,

$$\mathbb{P}(|Z| > x) \leq \|Z\|^2_{\mathrm{L}^2}/x^2, \quad x > 0, \tag{4.1}$$

and if X is an *exact* 1-chaos, then for all $Z \in H(X)$,

$$\mathbb{P}(|Z| > x) \leq \exp(Kx^2/\|Z\|^2_{\mathrm{L}^2}), \quad \text{for sufficiently large } x > 0. \tag{4.2}$$

The gap between the estimates in (4.1) and (4.2) can be calibrated on two scales: one scale marked by exponential tail-probability estimates, which start at (4.2), and another scale marked by polynomial tail-probability estimates, which start at (4.1). Both calibrations are discussed below.

α-chaos, $\alpha \in [1, \infty)$

These are the processes proposed in the previous chapter as models for random walks with prescribed degrees of combinatorial complexity. (We have argued for the integer α case in Theorem X.26, and will deal with the non-integer case in the last chapter.) Recall that if X is α-chaos, $\alpha \in [1, \infty)$, then for all $Z \in H(X)$ and all $\gamma > \alpha$, there exists $K_\gamma > 0$ such that for sufficiently large $x > 0$

$$\mathbb{P}(|Z| > x) \leq \exp(K_\gamma x^{2/\gamma}/\|Z\|_{\mathrm{L}^2}^\gamma) \tag{4.3}$$

for all $Z \in H(X)$, and these estimates are best possible, i.e., there exist $Z \in H(X)$ such that (4.3) fails for all $\gamma < \alpha$ (Definition X.27). Following an interpretation of tail-probability estimates as measurements of interdependence (Chapter X §4, Remark ii), we view the α-chaos processes as random integrators whose stochastic complexity is marked precisely by α.

Every α-chaos X is a homogeneous integrator, $H(X)$ in Definition X.27 is the same as $H(X)$ in (3.15), and the definition can be restated thus (Exercise 14): A homogeneous process X is α-chaos if and only if

$$\delta_{H(X)} = \alpha/2. \tag{4.4}$$

Estimates on variations of Fréchet measures associated with α-chaos X are given in (X.14.9); in particular, $\ell_X = 1$. (See (2.27).)

We say $E \subset \mathrm{L}^2(\Omega, \mathbb{P})$ is an α-system if $\delta_{\mathrm{span}(E)} = \alpha/2$; we call E *exact* if $c_{\mathrm{span}(E)}(\delta_{\mathrm{span}(E)}^{-1}) > 0$, and asymptotic if $c_{\mathrm{span}(E)}(\delta_{\mathrm{span}(E)}^{-1}) = 0$. If X is α-chaos, then $\{\hat{X}(j): j \in \mathbb{N}\}$ (defined via (3.13)) is an α-system, which is exact if and only if X is an exact α-chaos. (For definitions of aforementioned indices, see (X.10.1)–(X.10.6).) Conversely, if $\{X_j : j \in \mathbb{N}\} \subset \mathrm{L}^2(\Omega, \mathbb{P})$ is an orthonormal α-system, and U is a unitary map from $\mathrm{L}^2([0,1], \mathfrak{m})$ onto the $\mathrm{L}^2(\Omega, \mathbb{P})$-closure of span $\{X_j\}$, then $\{U\mathbf{1}_{[0,t)} : t \in [0,1]\}$ is an α-chaos (Exercise 17). (Cf. construction of a Wiener process in Chapter X §2.)

Sample-path continuity of α-chaos can be verified by an argument similar to the one used in Chapter X, in the case of the Wiener process (1-chaos). It can also be obtained as an instance of a more general theorem asserting sample-path continuity of $\Lambda(q)$-processes. To underscore ideas, however, we will first outline the proof in the case of α-chaos – useful in its own right – and then sketch the analogous argument in the case of $\Lambda(q)$-processes.

Lemma 16 (cf. Lemma X.7; Exercise 15). *Let (S,ν) be a finite measure space, and let T be a subspace of $\mathrm{L}^\infty(S,\nu)$. Assume that for some $u \in (0,1]$,*

$$\rho(T,u) = \rho := \inf\{\nu\{|f| \geq u\|f\|_{\mathrm{L}^\infty}\} : f \in T\} > 0. \tag{4.5}$$

Let $\{Y_j : j \in \mathbb{N}\} \subset \mathrm{L}^2(\Omega,\mathbb{P})$ be an orthonormal sub-α-system, $\alpha \in [1,\infty)$. If $\{f_j\} \subset T$ satisfies

$$\left\| \sum_j |f_j|^2 \right\|_{\mathrm{L}^\infty} \leq 1, \tag{4.6}$$

then $\|p\|_{\mathrm{L}^\infty} := \operatorname{ess\,sup}_{s\in S} |\sum_j f_j(s)\; Y_j|$ is a sub-α-variable. In particular, for all $A < c_{\mathrm{span}\{Y_j\}}(2/\alpha)$, there exists $\mathrm{L} > 0$ such that

$$\mathbb{P}(\|p\|_{\mathrm{L}^\infty} > x) < \frac{\nu(S)}{\rho}\exp(-A(ux)^{2/\alpha}), \quad x > L. \tag{4.7}$$

Corollary 17 (Exercise 15). *If $\{Y_j : j\in\mathbb{N}\} \subset \mathrm{L}^2(\Omega,\mathbb{P})$ is an orthonormal sub-α-system, and $\{f_j\}$ is a finite collection of trigonometric polynomials of degree N, then*

$$\mathbb{P}\left\{ \left\| \sum_j f_j \otimes Y_j \right\|_{\mathrm{L}^\infty([0,1])} > A \left\| \sum_j |f_j|^2 \right\|_{\mathrm{L}^\infty}^{\frac{1}{2}} (\ln N)^{\frac{1}{2}} \right\} \leq 1/N, \tag{4.8}$$

where $A > 0$ depends only on $c_{\mathrm{span}\{Y_j\}}(2/\alpha)$.

Corollary 18 (Exercise 15). *Let $\{Y_j : j\in\mathbb{N}\} \subset \mathrm{L}^2(\Omega,\mathbb{P})$ be an orthonormal sub-α-system, and define blocks of integers*

$$B_k = \{[2^{k^{\frac{1}{\alpha}}}], [2^{k^{\frac{1}{\alpha}}}]+1, \ldots, [2^{(k+1)^{\frac{1}{\alpha}}}]-1\}, \quad k = 0,1,\ldots. \tag{4.9}$$

(Here $[\cdot]$ is the 'closest integer' function.) If a $\mathbb{C}$-valued sequence (a_j) satisfies

$$s_k = \left(\sum_{j \subset B_k} |a_j|^2 \right)^{\frac{1}{2}}, \quad (k = 0, 1, \ldots) \text{ is decreasing}, \tag{4.10}$$

and

$$\sum_k s_k < \infty, \tag{4.11}$$

then, $\sum_{j=1}^{\infty} a_j Y_j \sin 2\pi j t$ represents almost surely $(\mathbb{P})$ a continuous function on $[0,1]$.

Corollary 19 (Exercise 15). *For every $\alpha \in [1, \infty)$, sample-paths of α-chaos are almost surely continuous.*

$\Lambda(q)$-processes, $q \in (2, \infty)$

We begin with definitions, some old and some new (cf. Chapter III and Chapter VII). A subspace $H \subset \mathrm{L}^2(\Omega, \mathbb{P})$ is a $\Lambda(q)$-space ($q > 2$) if

$$\xi_H(q) := \sup\{\|Z\|_{\mathrm{L}^q} : Z \in H, \ \|Z\|_{\mathrm{L}^2} \leq 1\} < \infty. \tag{4.12}$$

If $\xi_H(s) < \infty$ for some $s > 2$, then

$$\lambda_H := \sup\{s : \xi_H(s) < \infty\}. \tag{4.13}$$

If $\lambda_H = q$, then we say that H is a $\lambda(q)^{\#}$-space ('*lambda-q-sharp space*'). As usual, we distinguish between $\xi_H(\lambda_H) < \infty$, in which case H is an *exact* $\Lambda(q)^{\#}$-space, and $\xi_H(\lambda_H) = \infty$, in which case H is an *asymptotic* $\Lambda(q)^{\#}$-space. Analogously, an orthonormal system $\{Y_j : j \in \mathbb{N}\} \subset \mathrm{L}^2(\Omega, \mathbb{P})$ is a $\Lambda(q)$-system if $\xi_{\mathrm{span}\{Y_j\}}(q) < \infty$, and a $\Lambda(q)^{\#}$-system (exact or asymptotic) if $\lambda_{\mathrm{span}\{Y_j\}} = q$.

Definition 20 A homogeneous process X is a $\Lambda(q)$-process ($q > 2$) if $H(X)$ is a $\Lambda(q)$-space. Similarly, a homogeneous X is a $\Lambda(q)^{\#}$-process (exact or asymptotic) if $H(X)$ is a $\Lambda(q)^{\#}$-space (exact or asymptotic, respectively).

If X is a $\Lambda(q)^{\#}$-process then $\delta_{H(X)} = 0$, and if X is α-chaos then $\lambda_{H(X)} = \infty$. If X is a $\Lambda(q)$-process, then $\{\hat{X}(j)\}$ is a $\Lambda(q)$-system. Conversely, if $\{X_j\}$ is a $\Lambda(q)$-system and U is a unitary map from $\mathrm{L}^2([0,1], \mathfrak{m})$ onto the $\mathrm{L}^2(\Omega, \mathbb{P})$-closure of $\mathrm{span}\{X_j\}$, then $\{U\mathbf{1}_{[0,t)} : t \in [0,1]\}$ is a $\Lambda(q)$-process (Exercise 18).

We sketch a proof that sample-paths of $\Lambda(q)$-processes are almost surely continuous.

Lemma 21 (Exercise 16). *Let (S,ν) be a finite measure space, and T a subspace of $\mathrm{L}^\infty(S,\nu)$ with $\rho(T,u)=\rho>0$ for some $u\in(0,1]$. (See (4.5).) Let $\{Y_j\colon j\in\mathbb{N}\}\subset \mathrm{L}^2(\Omega,\mathbb{P})$ be an orthonormal $\Lambda(q)$-system, and denote $\xi=\xi_{\mathrm{span}\{Y_j\}}(q)$. If $\{f_j\}\subset T$ satisfies*

$$\left\|\sum_j |f_j|^2\right\|_{\mathrm{L}^\infty}\le 1, \tag{4.14}$$

then

$$\mathbb{P}(\|p\|_{\mathrm{L}^\infty}>x)<\left(\frac{\nu(S)}{\rho u^q}\right)(\xi/x)^q,\quad x>0, \tag{4.15}$$

where $\|p\|_{\mathrm{L}^\infty}:=\operatorname{ess\,sup}_{s\in S}|\sum_j f_j(s)\,Y_j|$.

Corollary 22 (Exercise 16). *If $\{Y_j\colon j\in\mathbb{N}\}\subset \mathrm{L}^2(\Omega,\mathbb{P})$ is an orthonormal $\Lambda(q)$-system ($q>2$), and $\{f_j\}$ is a collection of trigonometric polynomials of degree N, then*

$$\mathbb{P}\left\{\left\|\sum_j f_j\otimes Y_j\right\|_{\mathrm{L}^\infty}>x\left\|\sum_j |f_j|^2\right\|_{\mathrm{L}^\infty}^{\frac{1}{2}}\right\}\le AN/x^q. \tag{4.16}$$

($A>0$ depends on $\xi_{\mathrm{span}\{Y_j\}}(q)$.)

Corollary 23 (cf. Corollary 18). *Suppose $\{Y_j\colon j\in\mathbb{N}\}\subset \mathrm{L}^2(\Omega,\mathbb{P})$ is an orthonormal $\Lambda(q)$-system ($q>2$), and*

$$B_k=\{[k^{\frac{q}{2}}],[k^{\frac{q}{2}}]+1,\ldots,[(k+1)^{\frac{q}{2}}]-1\},\quad k=0,1,\ldots. \tag{4.17}$$

Suppose $(a_j)\subset\mathbb{C}$ satisfies

$$s_k=\left(\sum_{j\in B_k}|a_j|^2\right)^{\frac{1}{2}},\quad k=0,1,\ldots,\ \textit{is decreasing}, \tag{4.18}$$

and

$$\sum_{k=1}^{\infty}(\log k)s_k<\infty. \tag{4.19}$$

Then, $\sum_{j=1}^{\infty} a_j Y_j \ \sin 2\pi jt$ *represents almost surely* ($\mathbb{P}$) *a continuous function on* [0,1].

Proof (Exercise 16): For $k = 1, \ldots,$ let

$$C_k = \{2^k, 2^k + 1, \ldots, 2^{k+1} - 1\}, \tag{4.20}$$

$$p_k(t) = \sum_{n \in C_k} a_n Y_n \sin 2\pi nt, \tag{4.21}$$

and

$$E_k = \left\{ \|p_k\|_{\mathrm{L}^\infty} \geq 2^{\frac{k}{q}} k \left(\sum_{n \in C_k} |a_n|^2 \right)^{\frac{1}{2}} \right\}. \tag{4.22}$$

By Corollary 23, $\mathbb{P}(E_k) \leq A/k^q, \ k = 1, \ldots,$ and therefore

$$\|p_k\|_{\mathrm{L}^\infty} = \mathscr{O}\left(2^{\frac{k}{q}} k \left(\sum_{n \in C_k} |a_n|^2 \right)^{\frac{1}{2}} \right) \text{ almost surely } (\mathbb{P}), \tag{4.23}$$

which implies the desired conclusion. □

Corollary 24 (Exercise 16). *For* $q > 2$, *sample-paths of* $\Lambda(q)$*-processes are almost surely continuous.*

Because an α-chaos process is a $\Lambda(q)$-process for (all) $q > 2$, Corollary 24 implies Corollary 19.

Remark i (other approaches).

In this section we have outlined one of several approaches – via stochastic series – to the question whether sample-paths of a given process are almost surely continuous. Detailed discussions of this general question, based on 'entropy' and 'majorizing measure' approaches, can be found in [LeT, Chapter 11]. A classical treatment based on Kolmogorov's 1934 theorem regarding sample-path continuity can be found in [Bil, Theorem 12.4]. Kolmogorov's theorem asserts that if $X = \{X(t) : t \in [0,1]\}$ is a separable process such that for some $\alpha > 0$ and $p > 1$, and all intervals $J \subset [0,1]$,

$$\mathbf{E}|\Delta X(J)|^\alpha \leq \text{length}(J)^p, \tag{4.24}$$

then the sample-paths of X are almost surely continuous (e.g., [LeT, Corollary 11.8]). Note that Kolmogorov's theorem implies Corollary 24.

A 'stochastic series' approach is useful in identifying other sample-path properties of α-chaos and $\Lambda(q)$-processes (e.g., [ČTow]).

Next we consider the Littlewood index of $\Lambda(q)$-processes.

Theorem 25 *If $q \in (2, \infty)$ and X is a $\Lambda(q)$-process, then*

$$\ell_X \leq (q+2)/(q+1).$$

Two mixed-norm inequalities are needed for the proof. The first is

Lemma 26 *If $q \in (2, \infty)$ and X is a $\Lambda(q)$-process, then for all*

$$p > q/(q-1),$$

$$\sup\left\{\sum_j \left(\sum_k |\mathbf{E}\mathbf{1}_{A_j}\Delta X(B_k)|\right)^p : \right.$$

$$\left. \Sigma_j \mathbf{1}_{A_j} \leq 1,\ \Sigma_k \mathbf{1}_{B_k} \leq 1,\ A_j \in \mathscr{A},\ B_k \in \mathscr{B}\right\} \leq K, \quad (4.25)$$

where $K > 0$ depends only on $\xi_{H(X)}(q)$.

Proof: Without loss of generality we assume $\mathbf{E}|X(t)|^2 = t$, i.e., $\mathbf{E}|I_X(f)|^2 = \|f\|^2_{\mathrm{L}^2}$ for all $f \in \mathrm{L}^2([0,1], \mathfrak{m})$. Fix finite partitions $\{A_j\} \subset \mathscr{A}$ and $\{B_k\} \subset \mathscr{B}$, and denote $Y_k = \Delta X(B_k)/\sqrt{\mathfrak{m}(B_k)}$. Then, $\{Y_k\}$ is a $\Lambda(q)$-system with

$$\xi_{\mathrm{span}\{Y_k\}}(q) := \xi \leq \xi_{H(X)}(q). \quad (4.26)$$

Using duality to prove (4.25), we will verify that if

$$\sum_j \left(\sup_k |b_{jk}|\right)^v \leq 1, \quad \frac{1}{p} + \frac{1}{v} = 1, \quad (4.27)$$

then

$$\left|\sum_{j,k} b_{jk}\ \mathbf{E}\mathbf{1}_{A_j}\Delta X(B_k)\right| \leq K. \quad (4.28)$$

To this end, we rearrange the js so that

$$\sup_k |b_{jk}|^v \leq 1/j, \quad j = 1, \ldots, \tag{4.29}$$

and, for convenience, denote $d_{jk} = b_{jk}\sqrt{\mathfrak{m}(B_k)}$. Then

$$\left(\sum_k |d_{jk}|^2\right)^{\frac{1}{2}} \leq 1/j^{\frac{1}{v}}, \tag{4.30}$$

and therefore,

$$\begin{aligned}
\left|\sum_{j,k} b_{jk}\, \mathbf{E}\mathbf{1}_{A_j} \Delta X(B_k)\right| &= \left|\sum_j \mathbf{E}\mathbf{1}_{A_j} \sum_k d_{jk}\ Y_k\right| \\
&\leq \mathbf{E}\sum_j \mathbf{1}_{A_j} \left|\sum_k d_{jk}\ Y_k\right| \leq \mathbf{E}\sup_j \left|\sum_k d_{jk}\ Y_k\right| \\
&= \int_0^\infty \mathbb{P}\left(\bigcup_j \left\{\left|\sum_k d_{jk}\ Y_k\right| > t\right\}\right) \mathrm{d}t \\
&\leq 1 + \sum_j \int_1^\infty \mathbb{P}\left(\left|\sum_k d_{jk}\ Y_k\right| > t\right)\ \mathrm{d}t \\
&\leq 1 + \xi^q \left(\int_1^\infty t^{-q}\ \mathrm{d}t\right) \sum_j 1/j^{\frac{q}{v}} := K < \infty.
\end{aligned} \tag{4.31}$$

□

Remark

ii **(an extension of the Orlicz (l^2, l^1)-mixed norm inequality).** If X is an integrator, then $\|\{\mathbf{E}\mathbf{1}_{A_j} \Delta X(B_k)\|_{\mathrm{F}_2(\mathbb{N},\mathbb{N})} \leq \|X\|$, and the Orlicz inequality (Theorem II.3) implies

$$\sum_j \left(\sum_k |\mathbf{E}\mathbf{1}_{A_j} \Delta X(B_k)|\right)^2 \leq \kappa_0^2\ \|X\|^2 = \ 2\|X\|^2. \tag{4.32}$$

Lemma 26 implies that if X is a $\Lambda(q)$-process, then we can do better than (4.32).

The second mixed-norm inequality is a restatement of the Littlewood (l^1, l^2)-inequality in the present context. In the setting here, we prove it directly without the use of the Khintchin L^1–L^2 inequality (cf. Theorem II.2).

Lemma 27 *If X is a homogeneous process and $\|X(1)\|_{L^2} = 1$, then*

$$\sup\left\{\sum_j \left(\sum_k |\mathbf{E}\mathbf{1}_{A_j}\Delta X(B_k)|^2\right)^{\frac{1}{2}} : \right.$$

$$\left. \Sigma_j \mathbf{1}_{A_j} \le 1,\ \Sigma_k \mathbf{1}_{B_k} \le 1,\ A_j \in \mathscr{A},\ B_k \in \mathscr{B} \right\} \le 1. \quad (4.33)$$

Proof: Fix finite partitions $(A_j) \subset \mathscr{A}$ and $\{B_k\} \subset \mathscr{B}$. Suppose $\{b_{jk}\} \subset \mathbb{C}$ satisfies

$$\sup_j \left(\sum_k |b_{jk}|^2\right)^{\frac{1}{2}} \le 1, \quad (4.34)$$

and estimate

$$\left|\sum_{j,k} b_{jk}\ \mathbf{E}\mathbf{1}_{A_j}\Delta X(B_k)\right| \le \sum_k \left|\mathbf{E}\left(\sum_j b_{jk}\ \mathbf{1}_{A_j}\right)\Delta X(B_k)\right|$$

$$\le \sum_k \sqrt{\mathfrak{m}(B_k)}\left(\sum_j |b_{jk}|^2\ \mathbb{P}(A_j)\right)^{\frac{1}{2}}$$

$$\le \left(\sum_k \sum_j |b_{jk}|^2\ \mathbb{P}(A_j)\right)^{\frac{1}{2}} \le 1. \quad (4.35)$$

□

Proof of Theorem 25: By Lemmas 26 and 27, it suffices to verify

$$\sum_{j,k} |a_{jk}|^{\frac{p+2}{p+1}} \le \left(\sum_j\left(\sum_k |a_{jk}|^2\right)^{\frac{1}{2}}\right)^{\frac{2}{p+1}} \left(\sum_j\left(\sum_k |a_{jk}|\right)^{\frac{p}{p-1}}\right)^{\frac{p-1}{p+1}} \quad (4.36)$$

for all scalar arrays (a_{jk}) and $p > 2$. We rewrite the left side, and then apply Hölder's inequality to the sum over k with exponents $p+1$ and $(p+1)/p$,

$$\sum_{j,k} |a_{jk}|^{\frac{p+2}{p+1}} = \sum_j \sum_k |a_{jk}|^{\frac{2}{p+1}} |a_{jk}|^{\frac{p}{p+1}}$$

$$\leq \sum_j \left(\sum_k |a_{jk}|^2 \right)^{\frac{1}{p+1}} \left(\sum_k |a_{jk}| \right)^{\frac{p}{p+1}}. \tag{4.37}$$

We obtain (4.36) by applying Hölder's inequality to the sum over j with exponents $(p+1)/2$ and $(p+1)/(p-1)$. □

Remarks:

iii **(continuous time-models for random walks?).** We already have noted that α-chaos for integer α are continuous-time models for random walks with prescribed combinatorial complexity (in the sense of Chapter X §13). In the last chapter of the book we will extend this observation to all $\alpha \in [1, \infty)$. (Existence of α-chaos in the integer α case is noted in Exercise 17; existence in the non-integer case will be established after *combinatorial dimension* is introduced.)

Whether $\Lambda(q)$-processes are continuous-time models for walks with a prescribed combinatorial complexity is an open (-ended) question that I will briefly discuss at the end of the book. Existence of $\Lambda(q)^{\#}$-processes, unlike existence of α-chaos, is easy to verify (Exercise 18).

iv **(is Theorem 25 sharp?).** For every $q > 2$ there exist $\Lambda(q)^{\#}$-processes X such that $\ell_X = (q+2)/(q+1)$. Constructions of these processes fundamentally depend – at present – on Bourgain's solution of Rudin's '$\Lambda(q)$-set problem'. I do not know whether $\ell_X = (q+2)/(q+1)$ for *every* $\Lambda(q)^{\#}$-process X.

v **($q \leq 2$?).** A homogeneous process X is a $\Lambda(2)$-process if $H(X)$ is a $\Lambda(2)$-space (Definition III.6, Lemma III.7), and a $\Lambda(2)^{\#}$-process if X is a $\Lambda(2)$-process and $\xi_{H(X)}(q) = \infty$ for all $q > 2$ (Exercise 18). How to extend and implement a notion of a $\Lambda(q)^{\#}$-process in the case $q \in (1, 2)$ is not obvious. One such extension is given below.

p-stable motion, $p \in (1, 2]$

The idea of p-stable laws is due to Paul Lévy [Lé2]. An introduction to stable laws, including some physical motivation for them, can be found in [La]. A more detailed exposition, including an introduction to stable processes, can be found in [Br]; a recent study appears in [SamTa].

A process X is called a *p-stable motion*, $p \in (0,2]$, if X is additive and for all $J = (s,t] \subset [0,1]$,

$$\mathbf{E} \exp iy\Delta X(J) = \exp \zeta(s-t)|y|^p, \quad y \in \mathbb{R}, \tag{4.38}$$

for some $\zeta > 0$; see [SamTa, p.113]. A construction of p-stable motion follows from Kolmogorov's extension theorem; e.g., [SamTa, 3.2] (Exercise 19). Basic features of p-stable motion in the case $p < 2$ are fundamentally different from those in the case $p = 2$ (the Wiener process); e.g., [CamMi] and [SamTa, p. 151].

Here we restrict attention to $p \in (1,2]$. In this range, a p-stable motion X is an integrator (Proposition 14), and we obtain via the measure-theoretic approach,

$$\int_{[0,1]} f \,\mathrm{d}X = \frac{\mathrm{d}}{\mathrm{d}\mathbb{P}} \int_{[0,1]} f(t)\, \mu_X(\cdot, \mathrm{d}t), \quad f \in \mathrm{L}^\infty([0,1], \mathscr{B}). \tag{4.39}$$

We can use also a functional-analytic approach. For step functions $f = \sum_{i=1}^n a_i \mathbf{1}_{J_i}$, define

$$I_X(f) = \sum_{i=1}^n a_i \,\Delta X(J_i), \tag{4.40}$$

and note that for each such f,

$$\mathbf{E} \exp iyI_X(f) = \exp(\zeta |y|^p \|f\|^p_{\mathrm{L}^p(\mathfrak{m})}), \quad y \in \mathbb{R}. \tag{4.41}$$

The linear map I_X can be extended to $\mathrm{L}^p([0,1], \mathfrak{m})$ so that (4.41) holds. In particular,

$$\|I_X(f)\|_{\mathrm{L}^r(\mathbb{P})} = c\|f\|_{\mathrm{L}^p(\mathfrak{m})}, \; f \in \mathrm{L}^p([0,1], \mathfrak{m}), \; r \in [1,p), \tag{4.42}$$

where $c > 0$ depends on ζ, p, and r (Exercise 20 i).

Proposition 28 *If X is a p-stable motion, $p \in (1,2]$, then*

$$I_X(f) = \int_{[0,1]} f \,\mathrm{d}X, \quad f \in \mathrm{L}^\infty([0,1], \mathscr{B}). \tag{4.43}$$

Proof: Because X is a continuous integrator,

$$\mu_X(A, J) = \mathbf{E}\mathbf{1}_A \,\Delta X(J), \quad A \in \mathscr{A}, \text{ interval } J \subset [0,1], \tag{4.44}$$

and by (4.42), there exists $c > 0$ such that

$$|\mu_X(A, B)| \leq c\, \mathfrak{m}(B)^{\frac{1}{p}},\ A \in \mathscr{A},\ B \in \mathscr{B}. \tag{4.45}$$

Therefore,

$$\mu_X \ll \mathbb{P} \times \mathfrak{m}, \tag{4.46}$$

which implies (4.43) (cf. Proposition 13; Exercise 20 ii). □

Proposition 29 *Let X be a p-stable motion, $p \in (1, 2]$, and define*

$$H(X) := \{I_X(f) : f \in \mathrm{L}^p([0,1], \mathfrak{m})\}. \tag{4.47}$$

Then

i. *$H(X)$ is a $\Lambda(r)$-space for every $0 < r < p$, i.e., $H(X)$ is a closed subspace of $\mathrm{L}^r(\Omega, \mathbb{P})$, and for all $q < r$, the L^r- and L^q-norms are equivalent in $H(X)$;*
ii. *$H(X)$ is closed in probability;*
iii. *X is a random integrator.*

Proof (Exercise 21): The first assertion follows from (4.42). The second is an immediate consequence of Exercise X.34. The third assertion follows from (4.42). □

We now show that variations of Fréchet measures μ_X associated with p-stable motions X, $p \in (1, 2)$, are controlled by the Orlicz norm $\|\cdot\|_{\theta_2}$, where θ_2 is defined in (X.6.22) [BlTow]. In particular, estimates on these variations imply $\ell_X = 1$ (Exercise 22).

The argument used to prove this is similar to the proof of Theorem X.11. To start, observe that there exists $k > 0$ such that

$$\mathbb{P}(|X(1)| > x) \leq k/x^p, \quad x > 0, \tag{4.48}$$

and, for convenience, assume $k = 1$. Suppose $\{B_k\} \subset \mathscr{B}$ is finite, and $\Sigma_k \mathbf{1}_{B_k} \leq 1$. Then, by (4.41), $\Delta X(B_k)/\mathfrak{m}(B_k)^{1/p}$ has the same distribution as $X(1)$. Define

$$X_k = \begin{cases} \Delta X(B_k) & \text{if } |\Delta X(B_k)| \leq 1 \\ 0 & \text{if } |\Delta X(B_k)| > 1, \end{cases} \tag{4.49}$$

and $X'_k = \Delta X(B_k) - X_k$. By (4.48),

$$\mathbf{E}|X_k|^q \leq \int_0^1 \mathbb{P}(|\Delta X(B_k)|^q > x)\ \mathrm{d}x$$

$$\leq \left(\frac{q}{q-p}\right) \mathfrak{m}(B_k), \quad q > p, \tag{4.50}$$

and

$$\mathbf{E}|X'_k| = \mathbb{P}(|\Delta X(B_k)| > 1) + \int_1^\infty \mathbb{P}(|\Delta X(B_k)| > x)\ \mathrm{d}x$$

$$\leq \left(\frac{p}{p-1}\right) \mathfrak{m}(B_k). \tag{4.51}$$

The X_k are independent and symmetric, and therefore, by the Khintchin inequalities, if (b_k) is a scalar sequence, then

$$\mathbf{E}\left|\sum_k b_k\ X_k\right|^q \leq q^{\frac{q}{2}}\ \mathbf{E}\left(\sum_k |b_k\ X_k|^2\right)^{\frac{q}{2}}. \tag{4.52}$$

Lemma 30 *There exists $\kappa_p > 0$ such that for all scalar sequences $b = (b_k)$,*

$$\left(\mathbf{E}\left|\sum_k b_k\ X_k\right|^q\right)^{\frac{1}{q}} \leq \kappa_p\ q\ \|b\|_\infty, \quad q > 2. \tag{4.53}$$

Proof: Let $m > 1$ be an integer, and estimate

$$\mathbf{E}\left(\sum_k |b_k\ X_k|^2\right)^m = \sum_{k_1,\ldots,k_m} |b_{k_1}|^2 \cdots |b_{k_m}|^2\ \mathbf{E}|X_{k_1}|^2 \cdots |X_{k_m}|^2$$

$$\leq \sum_{n=1}^m \sum_{j_1+\cdots+j_n=m} \binom{m}{j_1 \ldots j_n} \cdot$$

$$\cdot \left(\sum_{k_1,\ldots,k_n} |b_{k_1}|^{2j_1} \cdots |b_{k_n}|^{2j_n}\ \mathbf{E}|X_{k_1}|^{2j_1} \cdots \mathbf{E}|X_{k_n}|^{2j_n}\right) \text{(by independence)}$$

$$\leq \sum_{n=1}^m \sum_{j_1+\cdots+j_n=m} \binom{m}{j_1 \ldots j_n} \left(\sum_k |b_k|^{2j_1}\ \mathbf{E}|X_k|^{2j_1}\right) \cdots$$

$$\cdot \left(\sum_k |b_k|^{2j_n}\ \mathbf{E}|X_k|^{2j_n}\right)$$

$$\le \left(\frac{2}{2-p}\right)^m \sum_{n=1}^{m} \sum_{j_1+\cdots+j_n=m} \binom{m}{j_1 \dots j_n} \cdot$$

$$\cdot \left(\sum_k |b_k|^{2j_1}\, \mathfrak{m}(B_k)\right) \dots \left(\sum_k |b_k|^{2j_n}\, \mathfrak{m}(B_k)\right) \text{ (by (4.50), with } q=2)$$

$$\le \left(\frac{2}{2-p}\right)^m \sum_{n=1}^{m} \sum_{j_1+\cdots+j_n=m} \binom{m}{j_1 \dots j_n} \|b\|_\infty^{2j_1} \cdots \|b\|_\infty^{2j_n}$$

$$\le \left(\frac{2}{2-p}\right)^m \|b\|_\infty^{2m} \sum_{n=1}^{m} n^m \le m^{m+1} \left(\frac{2}{2-p}\right)^m \|b\|_\infty^{2m}. \tag{4.54}$$

Above, $\sum_{k_1,\ldots,k_d}$ denotes a free sum over $(k_1,\ldots,k_d) \in \mathbb{N}^d$, and $\sum_{j_1+\cdots+j_n=m}$ denotes the finite sum over all n-subsets $\{j_1,\ldots,j_n\}$ of positive integers such that $j_1+\cdots+j_n=m$. To deduce the second line in (4.54) from the first, we use the decomposition

$$\mathbb{N}^m = \bigcup_{n=1}^{m} \{(k_1,\ldots,k_m) \in \mathbb{N}^m : |\{k_1,\ldots,k_m\}| = n\}. \tag{4.55}$$

($|\{k_1,\ldots,k_m\}| = n$ means that there are n distinct elements in $\{k_1,\ldots,$ $k_m\}$.) Then, for $n = 1,\ldots,m$, we partition

$$\{(k_1,\ldots,k_m) \in \mathbb{N}^m : |\{k_1,\ldots,k_m\}| = n\}$$

according to the number of times j_i, $i = 1,\ldots,n$, that coordinates appear in $(k_1,\ldots,k_m)$; that is, partition the aforementioned set according to $\{j_1,\ldots,j_n\}$ such that $j_1+\cdots+j_n = m$. For each such partition, the multinomial $\binom{m}{j_1 \cdots j_n}$ is the number of ways that m (integer-valued) variables can be assigned n values with respective $j_1,\ldots,j_n$ repetitions.

For $q > 2$, let m be the integer such that $m \ge q/2 > m-1$, and then deduce (4.53) from (4.52) and (4.54). □

The lemma implies that there exists $K > 0$ such that for sufficiently large $x > 0$ and all scalar sequences $b = (b_k)$,

$$\mathbb{P}\left(\left|\sum_k b_k\, X_k\right| > x\right) \le \exp(-Kx/\|b\|_\infty) \tag{4.56}$$

(cf. Lemma X.18). Consider the Orlicz function φ_2 defined (in (X.11.3)),

$$\varphi_2(t) = \exp(-1/t), \quad t \in \left(0, \frac{1}{2}\right], \tag{4.57}$$

and the set O_{φ_2} of finitely supported scalar arrays (b_{jk}) such that

$$\sum_{j,k} \varphi_2(|b_{jk}|) \le 1. \tag{4.58}$$

By applying an argument nearly identical to the one used in Theorem X.11 (only the 'arithmetic' is different), we obtain

$$\sup\left\{\left|\sum_{j,k} b_{jk}\ \mathbf{E}\mathbf{1}_{A_j} X_k\right| : \Sigma_j \mathbf{1}_{A_j} \le 1,\right.$$
$$\left.\Sigma_k \mathbf{1}_{B_k} \le 1,\ \{A_j\} \subset \mathscr{A},\ \{B_k\} \subset \mathscr{B},\ \{b_{jk}\} \in O_\varphi\right\} < \infty. \tag{4.59}$$

Next, we consider

$$\theta_2(x) = x/\log\,(1/x), \quad x \in (0,1), \tag{4.60}$$

and note that there exists $0 < \delta < 1$ such that

$$\theta_2(x) \le 2\ \varphi_2^*(x), \quad x \in (0,\delta), \tag{4.61}$$

where φ_2^* is the *complementary Orlicz function* to φ_2. We conclude that

$$\sup\left\{\sum_{j,k} \theta_2(|\mathbf{E}\mathbf{1}_{A_j} \Delta X(B_k)|) : \Sigma_j \mathbf{1}_{A_j} \le 1,\right.$$
$$\left.\Sigma_k \mathbf{1}_{B_k} \le 1,\ \{A_j\} \subset \mathscr{A},\ \{B_k\} \subset \mathscr{B}\right\} < \infty. \tag{4.62}$$

Indeed, the left side of (4.62) is bounded by

$$\sup\left\{\sum_{j,k} \theta_2(|\mathbf{E}\mathbf{1}_{A_j} X_k|) : \Sigma_j \mathbf{1}_{A_j} \le 1,\ \Sigma_k \mathbf{1}_{B_k} \le 1,\ \{A_j\} \subset \mathscr{A},\ \{B_k\} \subset \mathscr{B}\right\}$$
$$+\sup\left\{\sum_{j,k} |\mathbf{E}\mathbf{1}_{A_j} X_k'| : \Sigma_j \mathbf{1}_{A_j} \le 1,\ \Sigma_k \mathbf{1}_{B_k} \le 1,\ \{A_j\} \subset \mathscr{A},\ \{B_k\} \subset \mathscr{B}\right\}. \tag{4.63}$$

By (4.59) (via Orlicz space duality), the first term is finite, and by (4.51), the second term is finite.

Remarks:

vi **(is (4.62) best possible?).** For an arbitrary integer $m > 0$, consider the intervals $B_k = [\frac{k-1}{m}, \frac{k}{m})$, $k \in [m]$, and the partition $\{A_s : s \in \{-1,1\}^m\}$ of Ω defined in (X.6.20). If X is a p-stable motion, $p \in (1,2]$, and $\mathbf{E}|X(1)| = 1$, then

$$|\mathbf{E}\mathbf{1}_{A_s}\Delta X(B_k)| = 1/(2^m m^{\frac{1}{p}}). \tag{4.64}$$

For $\gamma > 0$, define (as in (X.6.22))

$$\theta_\gamma(x) = x/\{\log\,(1/x)\}^{\gamma/2}, \quad x \in (0,1), \tag{4.65}$$

and estimate

$$\begin{aligned}\sum_{s,k}\theta_\gamma(|\mathbf{E}\mathbf{1}_{A_s}\Delta X(B_k)|) &= \sum_{s,k}(1/2^m m^{\frac{1}{p}})/\{\log\, 2^m\;\, m^{\frac{1}{p}}\}^{\gamma/2} \\ &\geq Cm^{-(\gamma p+2-2p)/2p},\end{aligned} \tag{4.66}$$

where $C > 0$ depends only on p. This implies

$$\sup\left\{\sum_{i,j}\theta_\gamma(|\mathbf{E}\mathbf{1}_{A_j}\Delta X(B_k)|) : \Sigma_j\mathbf{1}_{A_i} \leq 1,\right. \\ \left.\Sigma_i\mathbf{1}_{B_j} \leq 1,\ \{A_i\} \subset \mathscr{A},\ \{B_j\} \subset \mathscr{B}\right\} = \infty \tag{4.67}$$

for all $\gamma < (2p-2)/p$.

Problem: Close the gap between (4.67) and (4.62).

vii **(what does p-stable motion model?).** The main reason – as far as I can determine, the only reason – for calling additive processes that satisfy (4.38) *p-stable motions* is that the instance $p = 2$ and $\gamma = 1/2$ has been widely referred to in the mathematical community as 'Brownian motion'. (See Definition 2.1 in [RosWo] and the comment following it; also see Chapter X §5, Remark ii.) I have not seen any arguments, say based on a 'random walk' paradigm, that make a convincing case for choosing *p-stable motion* as a continuous-time model for physical Brownian movement. (Sample-paths of p-stable motion are almost surely continuous only in the case $p = 2$; otherwise, for $p < 2$, a p-stable motion is a pure jump process; e.g., [SamTa, p. 151].) Concerning epistemological issues that arise here – *what do p-stable motions model?* – I refer the reader to [La, pp. 73–5] for a physical interpretation of p-stable distributions in a context of 'inverse attraction laws' (Exercise 23).

5 Two Questions – a Preview

We return to general integrators, and consider these two questions:

1. Can more than *bounded* measurable functions be integrated with respect to an integrator?
2. How is integration carried out in dimensions greater than one?

In the next section we will verify that if X is an arbitrary integrator, then there exists a probability measure ν_X on [0,1] such that every $f \in L^2([0,1], \nu_X)$ is canonically integrable with respect to X. If X is an L^2-bounded process with orthogonal increments, then $\nu_X = \lambda_X / \|\lambda_X\|_M$, where λ_X is determined by (3.3). For example, if X is an α-chaos or a $\Lambda(q)$-process, then ν_X = Lebesgue measure. Similarly, if X is a p-stable motion ($p \in (1,2]$), then ν_X = Lebesgue measure, and every $f \in L^p([0,1], \mathfrak{m})$ is canonically integrable with respect to X. The general case will require the Grothendieck factorization theorem.

The second question, concerning feasibility of a 'multidimensional integral'

$$\int_{[0,1]^n} f(s_1, \ldots, s_n)\, \mathrm{d}X_1(\mathrm{d}s_1) \cdots \mathrm{d}X_n(\mathrm{d}s_n), \tag{5.1}$$

where $X_1, \ldots, X_n$ are integrators and f is a function on $[0,1]^n$, points in several directions. We can ask whether (5.1) is feasible as an iterated integral (e.g., (X.8.1)); or, whether (5.1) is feasible as a 'one-dimensional' integral via an F_2-measure associated with the n-process $X_1 \otimes \cdots \otimes X_n$ (a process indexed by n parameters); or, whether (5.1) is feasible as an integral iterated over Cartesian products of [0,1], the sum of whose respective dimensions is n. (Note that the latter case falls between the first two.) And there are also questions about the approach we take: do we use a 'functional-analytic' approach, a 'measure-theoretic' approach, or do we merge the two? We shall deal with these questions in later sections. In this section, we illustrate some typical issues that arise in the two-dimensional case.

We start with integrators X and Y, 'step' functions

$$f = \sum_j b_j \mathbf{1}_{R_j}, \tag{5.2}$$

where $\{R_j\}$ is a finite collection of pairwise disjoint rectangles, and a definition of a two-dimensional integral

$$I_{X\otimes Y}(f) := \sum_j b_j \, \Delta X(K_j) \, \Delta Y(J_j), \tag{5.3}$$

where $R_j = K_j \times J_j$, and K_j and J_j denote intervals. We can rewrite f as a *standard* step function on $[0,1]^2$,

$$f = \sum_{i,j} a_{ij} \mathbf{1}_{J_i} \mathbf{1}_{J_j}, \tag{5.4}$$

where $J_j = [\frac{j-1}{N}, \frac{j}{N})$, and obtain

$$I_{X\otimes Y}(f) = \sum_{i,j} a_{ij} \, \Delta X(J_i) \, \Delta Y(J_j). \tag{5.5}$$

If $X = Y = \mathrm{W}$ (the Wiener process), and f is in S_σ, the space of *standard symmetric step functions vanishing on the diagonal* ($a_{ij} = a_{ji}$ and $a_{ii} = 0$ in (5.4)), then

$$\|I_{\mathrm{W}\otimes\mathrm{W}}(f)\|_{\mathrm{L}^2(\Omega,\mathbb{P})} = \|I_{\mathrm{W}_2}(f)\|_{\mathrm{L}^2(\Omega,\mathbb{P})} = \sqrt{2}\, \|f\|_{\mathrm{L}^2([0,1],\mathfrak{m}^2)}, \tag{5.6}$$

and therefore, because S_σ is norm-dense in $\mathrm{L}^2_\sigma([0,1]^2, \mathfrak{m}^2)$, $I_{\mathrm{W}\otimes\mathrm{W}}(f)$ can be defined for every $f \in \mathrm{L}^2_\sigma([0,1]^2, \mathfrak{m}^2)$. ($\mathfrak{m}^2(\{(t,t) : t \in [0,1]\} = 0$ is essential here; see Chapter X §7.)

To integrate, similarly, with respect to general integrators X and Y, we need (at the very least) the norm estimates

$$\|I_{X\otimes Y}(f)\|_{\mathrm{L}^1} \le K \|f\|_\infty, \quad f \in S_\sigma, \tag{5.7}$$

where $K > 0$ depends only on X and Y. Notice, however, that an obvious use of (5.7) – taking *norm*-limits – does not go very far. Following a measure-theoretic approach, we define

$$\begin{aligned} \mu_{X\otimes Y}(A, K\times J) &= \mathbf{E}\mathbf{1}_A \, \Delta X(K) \, \Delta Y(J), \\ A \in \mathscr{A}, \; K \times J &\subset D_2, \end{aligned} \tag{5.8}$$

where $D_2 = \{(s,t) : 0 \le s < t \le 1\}$. We observe that the L^1–L^∞ norm-estimate in (5.7) implies that $\mu_{X\otimes Y}$ determines an F_2-measure on $\mathscr{A} \times B_{\sigma 2}$, where $B_{\sigma 2}$ denotes the Borel field in D_2, and then obtain an integral with respect to $X \otimes Y$ via integration with respect to $\mu_{X\otimes Y}$.

Let us illustrate the issues that arise in the case $X = Y$, and X an L^1-bounded additive process.

Lemma 31 *Suppose $\{X_j\colon j \in \mathbb{N}\}$ and $\{Y_j\colon j \in \mathbb{N}\}$ are mutually independent systems of independent symmetric random variables such that for all $N \in \mathbb{N}$,*

$$\left\| \sum_{j=1}^{N} X_j \right\|_{L^1} \leq 1 \quad \text{and} \quad \left\| \sum_{j=1}^{N} Y_j \right\|_{L^1} \leq 1. \tag{5.9}$$

Then, for all $(a_{ij}) \in l^{\infty}(\mathbb{N}^2)$ and $N \in \mathbb{N}$,

$$\left\| \sum_{\substack{i=1\\j=1}}^{N} a_{ij}\, X_i\, Y_j \right\|_{L^1(\Omega^2,\mathbb{P}^2)} \leq 2 \max_{i,j}\, |a_{ij}|. \tag{5.10}$$

Proof: By independence, symmetry, and (5.9), for all $s \in \{-1,+1\}^{\mathbb{N}}$ and $N \in \mathbb{N}$,

$$\left\| \sum_{j=1}^{N} r_j(s)\, X_j \right\|_{L^1(\Omega,\mathbb{P})} \leq 1 \quad \text{and} \quad \left\| \sum_{j=1}^{N} r_j(s)\, Y_j \right\|_{L^1(\Omega,\mathbb{P})} \leq 1. \tag{5.11}$$

Taking expectations over $s \in \{-1,+1\}^{\mathbb{N}}$, interchanging the order of integrations, and applying the L^1–L^2 Khintchin inequality, we obtain

$$\left\| \left(\sum_{j=1}^{N} |X_j|^2 \right)^{\frac{1}{2}} \right\|_{L^1(\Omega,\mathbb{P})} \leq \sqrt{2},$$

and

$$\left\| \left(\sum_{j=1}^{N} |Y_j|^2 \right)^{\frac{1}{2}} \right\|_{L^1(\Omega,\mathbb{P})} \leq \sqrt{2}. \tag{5.12}$$

By independence and symmetry, for all $(a_{ij}) \in l^{\infty}(\mathbb{N}^2)$, $s \in \{-1,+1\}^{\mathbb{N}}$, $t \in \{-1,+1\}^{\mathbb{N}}$, and $N \in \mathbb{N}$,

$$\left\| \sum_{\substack{i=1\\j=1}}^{N} a_{ij} X_i \otimes Y_j \right\|_{L^1(\Omega^2,\mathbb{P}^2)} = \left\| \sum_{\substack{i=1\\j=1}}^{N} a_{ij} r_i(s) r_j(t) X_i \otimes Y_j \right\|_{L^1(\Omega^2,\mathbb{P}^2)}. \tag{5.13}$$

Taking expectation over s and t on the right side of (5.13), interchanging the order of integrations, and applying Hölder's inequality, we obtain

$$\begin{aligned}\left\|\sum_{\substack{i=1\\j=1}}^{N} a_{ij}\; X_i \otimes Y_j\right\|_{\mathrm{L}^1(\Omega^2,\mathbb{P}^2)} &\leq \left\|\left(\sum_{\substack{i=1\\j=1}}^{N} |a_{ij}\; X_i \otimes Y_j|^2\right)^{\frac{1}{2}}\right\|_{\mathrm{L}^1(\Omega^2,\mathbb{P}^2)} \\ &\leq \max_{i,j}|a_{ij}| \left\|\left(\sum_{\substack{i=1\\j=1}}^{N} |X_i \otimes Y_j|^2\right)^{\frac{1}{2}}\right\|_{\mathrm{L}^1(\Omega^2,\mathbb{P}^2)} \\ &= \max_{i,j}|a_{ij}| \left\|\left(\sum_{j=1}^{N} |X_j|^2\right)^{\frac{1}{2}}\right\|_{\mathrm{L}^1(\Omega,\mathbb{P})} \left\|\left(\sum_{j=1}^{N} |Y_j|^2\right)^{\frac{1}{2}}\right\|_{\mathrm{L}^1(\Omega,\mathbb{P})}.\end{aligned} \tag{5.14}$$

An application of (5.12) to the last line implies (5.10). □

The lemma implies that if X is an L^1-bounded additive process, and $\tilde{X}$ is an independent copy of X, then for all $(a_{ij}) \in l^\infty(\mathbb{N}^2)$,

$$\left\|\sum_{\substack{i=1\\j=1}}^{N} a_{ij}\; \Delta X(J_i)\; \Delta\tilde{X}(J_j)\right\|_{\mathrm{L}^1(\Omega^2,\mathbb{P}^2)} \leq K \max_{i,j}|a_{ij}|, \tag{5.15}$$

where $K > 0$ depends only on X. To obtain (5.7) in the case $X \otimes X$, we use the *decoupling inequality*

$$\|I_{X\otimes X}(f)\|_{\mathrm{L}^1(\Omega,\mathbb{P})} \leq K \left\|\sum_{\substack{i=1\\j=1}}^{N} a_{ij}\; \Delta X(J_i)\; \Delta\tilde{X}(J_j)\right\|_{\mathrm{L}^1(\Omega^2,\mathbb{P}^2)}, \tag{5.16}$$

where $f \in S_\sigma$ is represented by (5.4) and $K > 0$ depends only X. This inequality is an instance of the following general theorem.

Theorem 32 ([McTa1], [Kw], [dlPG, Chapter 6]; Exercise 24). *Let $\{X_j : j \in \mathbb{N}\}$ be a system of independent symmetric random variables.*

Then, for all $q > 0$ and integers $n > 0$, there exist $K_1(q,n) = K_1 > 0$ and $K_2(q,n) = K_2 > 0$ such that for all finite tetrahedral n-arrays $(a_{j_1\dots j_n} : 0 < j_1 < \cdots < j_n \leq N) \subset \mathbb{C}$,

$$K_1 \left\| \sum_{0<j_1<\cdots<j_n\leq N} a_{j_1\dots j_n} X^{(1)}_{j_1} \cdots X^{(n)}_{j_n} \right\|_{\mathrm{L}^q(\Omega^n,\mathbb{P}^n)}$$

$$\leq \left\| \sum_{0<j_1<\cdots<j_n\leq N} a_{j_1\dots j_n} X_{j_1} \cdots X_{j_n} \right\|_{\mathrm{L}^q(\Omega,\mathbb{P})}$$

$$\leq K_2 \left\| \sum_{0<j_1<\cdots<j_n\leq N} a_{j_1\dots j_n} X^{(1)}_{j_1} \cdots X^{(n)}_{j_n} \right\|_{\mathrm{L}^q(\Omega^n,\mathbb{P}^n)}, \tag{5.17}$$

where $\{X^{(1)}_j\},\dots,\{X^{(n)}_j\}$ are independent copies of $\{X_j\}$.

Remarks:

i **(about 'decoupling').** The inequalities in (5.17) address in a framework of probability theory this general question:

if η is a symmetric n-linear functional on $\mathbb{R}^N$, then what relationships exist between norms involving $\eta(x_1,\dots,x_n)$ $(x_1 \in \mathbb{R}^N,\dots,x_n \in \mathbb{R}^N)$, and $\eta(x,\dots,x)$ $(x \in \mathbb{R}^N)$? (5.18)

Among the early works addressing this question (in a framework of functional analysis) are two 1935 papers of Mazur and Orlicz [MazOr1], [MazOr2], where the following was established:

Proposition 33 (The Mazur–Orlicz identity [MazOr1, p. 63]; Lemma VII.27). *If η is a symmetric n-linear functional on $\mathbb{R}^N$, then*

$$\eta(x_1,\dots,x_n) = \frac{1}{n!}\,\mathbf{E}\, r_1\cdots r_n\ \eta(\Sigma^n_{j=1} r_j\ x_j,\dots,\Sigma^n_{j=1} r_j\ x_j),$$
$$x_1 \in \mathbb{R}^N,\dots,x_n \in \mathbb{R}^N. \tag{5.19}$$

The Mazur–Orlicz identity (a polarization formula) implies that if

$$\|\eta\| = \sup\{\eta(x_1, \ldots, x_n) : (x_1, \ldots, x_n) \in B_N \times \cdots \times B_N\}, \quad (5.20)$$

and

$$\|\eta\|_* = \sup\{\eta(x, \ldots, x) : x \in B_N\}, \quad (5.21)$$

where B_N is the l^∞-unit ball in $\mathbb{R}^N$, then

$$\|\eta\|_* \leq \|\eta\| \leq \frac{n^n}{n!}\|\eta\|_* \quad \text{(Exercise 25)}. \quad (5.22)$$

We have already made good use of this norm-equivalence in Chapter VII, in a framework of harmonic analysis.

Theorem 32 falls naturally under the heading of (5.18). For, suppose η is a tetrahedral, symmetric n-linear functional on $\mathbb{R}^N$. (Here *tetrahedral* means: if $\{e_j : j \in [N]\}$ is the standard basis in $\mathbb{R}^N$, then $\eta(e_{j_1}, \ldots, e_{j_n}) = 0$ when $(j_1, \ldots, j_n) \in [N]^n$, and $|\{j_1, \ldots, j_n\}| < n$.) Let $\mathbb{P}_1, \ldots, \mathbb{P}_N$ be symmetric probability measures on $\mathbb{R}$, and consider the product measure $\mathbb{P} = \mathbb{P}_1 \times \cdots \times \mathbb{P}_N$ on $\mathbb{R}^N$. Define

$$\|\eta\|_q = (\mathbf{E}|\eta(x_1, \ldots, x_n)|^q)^{\frac{1}{q}}, \quad (5.23)$$

and

$$\|\eta\|_{*q} = (\mathbf{E}|\eta(x, \ldots, x)|^q)^{\frac{1}{q}}, \quad (5.24)$$

where the expectation in (5.23) is over $((\mathbb{R}^N)^n, \mathbb{P}^n)$, and the expectation in (5.24) is over $(\mathbb{R}^N, \mathbb{P})$; cf. (5.20) and (5.21). The inequalities in (5.17) can then be rephrased as

$$K_1\|\eta\|_q \leq \|\eta\|_{*q} \leq K_2\|\eta\|_q. \quad (5.25)$$

The norm-equivalence in (5.25) was first obtained by Bonami in the case where each of $\mathbb{P}_1, \ldots, \mathbb{P}_N$ has bounded support ([Bon2, pp. 366–7]), and by Schreiber in the Gaussian case ([Sch2, Theorem II.1]). Neither Bonami nor Schreiber were interested in general decoupling, as such. Bonami was motivated by problems concerning $\Lambda(p)$-sets, and Schreiber was motivated by questions regarding the Wiener Chaos. General decoupling inequalities were first established – and so dubbed – by McConnell and Taqqu [McTa1], who were motivated primarily by feasibility of double integrals with respect to p-stable processes [McTa2]. McConnell and Taqqu established the right side in (5.17). Subsequent proofs of the left side, as well as more general inequalities, appeared in [Kw] and [dlP]. The latest on decoupling can be found in [dlPG, Chapter 6].

ii **(a preview).** The proof of (5.7) in the case that X and Y are mutually independent, centered, L^1-bounded additive processes was based here on the L^1–L^2 Khintchin inequality (Lemma 31). The proof of (5.7) in the general case rests on the feasibility of *product* F_2*-measures*, which is inextricably tied to the Grothendieck factorization theorem and inequality (Chapter IX).

6 An Application of the Grothendieck Factorization Theorem

Theorem 34 *Let* $X = \{X(t) : t \in [0,1]\}$ *be an integrator.*

i. *There exists a probability measure* ν *on* $([0,1], \mathscr{B})$ *such that*

$$\mathbf{E}\left|\int_{[0,1]} f \, \mathrm{d}X\right| \leq \kappa_{\mathrm{G}} \|X\| \|f\|_{\mathrm{L}^2(\nu)}, \quad f \in \mathrm{C}([0,1]), \tag{6.1}$$

where κ_{G} *is the Grothendieck constant. (A probability measure* ν *for which* (6.1) *holds will be called a Grothendieck measure of* X*.)*

ii. *If* ν *is a Grothendieck measure of* X, *then*

$$f \mapsto \int_{[0,1]} f \, \mathrm{d}X, \quad f \in \mathrm{L}^\infty([0,1], \mathscr{B}), \tag{6.2}$$

is uniquely extendible to a bounded linear map from $\mathrm{L}^2([0,1], \nu)$ *into* $\mathrm{L}^1(\Omega, \mathbb{P})$.

Because $\mathrm{C}([0,1])$ is norm-dense in $\mathrm{L}^2([0,1], \nu)$, Part i of Theorem 34 implies immediately that the restriction to $\mathrm{C}([0,1])$ of the $\mathrm{L}^1(\Omega, \mathbb{P})$-valued map in (6.2) is extendible to $\mathrm{L}^2([0,1], \nu)$. To establish Part ii, however, we must also verify that (6.2) agrees with this extension's restriction to $\mathrm{L}^\infty([0,1], \mathscr{B})$. To this end we need

Lemma 35 *If* X *is an integrator and* ν *is a Grothendieck measure of* X, *then for all* $A \in \mathscr{A}$, $\mu_X(A, \cdot) \ll \nu$.

Proof: (cf. Proof of Proposition 10). For $A \in \mathscr{A}$,

$$\mathbf{E}\mathbf{1}_A \int_{[0,1]} g \, \mathrm{d}X, \quad g \in \mathrm{C}([0,1]), \tag{6.3}$$

defines a bounded linear functional on C([0,1]), i.e., a regular Borel measure on [0,1]. This measure, which we denote by β_A, is the same as $\mu_X(A,\cdot)$. Because ν is a Grothendieck measure, the linear functional

$$g \mapsto \int_{[0,1]} g(t)\ \beta_A(\mathrm{d}t), \quad g \in \mathrm{C}([0,1]), \tag{6.4}$$

is uniquely extendible to a bounded linear functional $\tilde{\beta}_A$ on $\mathrm{L}^2([0,1],\nu)$. Then,

$$\tilde{\beta}_A(\mathbf{1}_B), \quad B \in \mathscr{B}, \tag{6.5}$$

defines a measure on $\mathscr{B}$. This measure, which we temporarily denote by μ, is the same as β_A. To see this, let $g \in \mathrm{C}([0,1])$ be arbitrary, and let $(\varphi_j : j \in \mathbb{N})$ be a sequence of $\mathscr{B}$-simple functions converging uniformly to g. Then,

$$\int_{[0,1]} \varphi_j\ \mathrm{d}\mu \xrightarrow[j\to\infty]{} \int_{[0,1]} g\ \mathrm{d}\mu, \tag{6.6}$$

and (because $\varphi_j \to g$ in $\mathrm{L}^2([0,1],\nu)$)

$$\int_{[0,1]} \varphi_j\ \mathrm{d}\mu \xrightarrow[j\to\infty]{} \tilde{\beta}_A(g). \tag{6.7}$$

Therefore,

$$\int_{[0,1]} g\ \mathrm{d}\mu = \tilde{\beta}_A(g) = \int_{[0,1]} g(t)\ \beta_A(\mathrm{d}t), \tag{6.8}$$

which implies $\mu = \beta_A$.

Therefore, for all $A \in \mathscr{A}$ and $B \in \mathscr{B}$,

$$|\mu_X(A,B)| = |\beta_A(B)| = |\tilde{\beta}_A(\mathbf{1}_B)| \leq \|\tilde{\beta}_A\| \sqrt{\nu(B)}, \tag{6.9}$$

which proves the lemma. □

Proof of Theorem 34: We view the linear map defined by (2.11) as a bounded bilinear functional on $\mathrm{L}^\infty(\Omega,\mathbb{P}) \times \mathrm{C}([0,1])$,

$$(Y,f) \mapsto \mathbf{E}Y \int_{[0,1]} f\ \mathrm{d}X, \quad (Y,f) \in \mathrm{L}^\infty(\Omega,\mathbb{P}) \times \mathrm{C}([0,1]). \tag{6.10}$$

The Grothendieck factorization theorem implies existence of a probability measure ν on $\mathscr{B}$ such that

$$\left|\mathbf{E}Y \int_{[0,1]} f\ \mathrm{d}X\right| \leq \kappa_G \|X\| \|Y\|_{\mathrm{L}^\infty} \|f\|_{\mathrm{L}^2(\nu)},$$

$$(Y,f) \in \mathrm{L}^\infty(\Omega,\mathbb{P}) \times \mathrm{C}([0,1]), \tag{6.11}$$

which verifies Part i.

To verify ii, note that the bilinear functional in (6.10) is uniquely extendible to a bilinear functional on $L^\infty(\Omega, \mathbb{P}) \times L^2([0,1], \nu)$ with norm bounded by $\kappa_G \|X\|$. Denote this bilinear functional by β. Then, for $g \in L^2([0,1], \nu)$,

$$\beta(\mathbf{1}_A, g), \quad A \in \mathscr{A}, \tag{6.12}$$

defines a measure on $\mathscr{A}$ that is absolutely continuous with respect to $\mathbb{P}$; denote this measure by β_g (Exercise 26). Define

$$\int_{[0,1]} g \, \mathrm{d}X = \frac{\mathrm{d}}{\mathrm{d}\mathbb{P}} \beta_g, \quad g \in L^2([0,1], \nu). \tag{6.13}$$

By use of Lemma 35, we obtain that (6.13) in the case $g \in L^\infty([0,1], \mathscr{B})$ is consistent with the definition in (2.11) (Exercise 27). □

Remarks:

i **(examples of Grothendieck measures).** If X is an integrator, then there exists a probability measure ν on [0,1] such that $\int_{[0,1]} f \mathrm{d}X$ can be defined for every $f \in L^2([0,1], \nu)$. If X is a p-stable motion then $\nu = \mathfrak{m}$ (Lebesgue measure). If X is square-integrable with orthogonal increments then $\nu = \lambda_X$. These measures are obtained directly, without the intervention of the Grothendieck factorization theorem. In the case of L^1-bounded additive processes and L^p-bounded martingales, $p \in (1, 2)$, I know of the existence of such ν only by applying the Grothendieck theorem (Exercise 28).

ii **(stochastic series).** Let ν be a Grothendieck measure of an integrator X, and let $\{\mathbf{e}_n : n \in \mathbb{N}\}$ be an orthonormal basis of $L^2([0,1], \nu)$. Define

$$X_n = \int_{[0,1]} \mathbf{e}_n \, \mathrm{d}X, \quad n \in \mathbb{N} \ \text{(cf. (3.13))}. \tag{6.14}$$

If $f \in L^2([0,1], \nu)$ and $f = \Sigma_n \hat{f}(n) \mathbf{e}_n$, then

$$\int_{[0,1]} f \, \mathrm{d}X = \sum_{n=1}^{\infty} \hat{f}(n) \, X_n, \tag{6.15}$$

where the series on the right converges weakly in $L^1(\Omega, \mathbb{P})$ (Exercise 29).

iii **('white noise').** A set-function M from $\mathscr{B}$ into $L^0(\Omega, \mathscr{A})$ (scalar valued $\mathscr{A}$-measurable functions on Ω) is a *stochastic measure* on $([0,1], \mathscr{B})$ if

$$M\left(\bigcup_j B_j\right) = \sum_{j=1}^{\infty} M(B_j) \quad \text{almost surely on } (\Omega, \mathscr{A}, \mathbb{P}) \tag{6.16}$$

whenever $\Sigma_j \mathbf{1}_{B_j} = \mathbf{1}_B$, $B_j \in \mathscr{B}$, $j \in \mathbb{N}$ (cf. [KwWo, Chapter 7]). This set-function M is an L^p-valued stochastic measure if the series on the right side of (6.16) converges in $L^p(\Omega, \mathbb{P})$, and a weak-L^1 stochastic measure if it converges weakly in $L^1(\Omega, \mathbb{P})$. It follows from definitions that if X is an integrator, then 'Δ'X (defined in (2.23)) is a weak-L^1 stochastic measure. If X is a p-stable motion, then ΔX is an L^r-valued stochastic measure for every $r < p$, and if X is an L^2-bounded process with orthogonal increments, then ΔX is an L^2-valued stochastic measure. In the general case, Theorem 34 implies that if X is an integrator, then 'Δ'X is an L^1-valued stochastic measure (Exercise 30).

7 Integrators Indexed by n-dimensional Sets

Let $X = \{X(\mathbf{t}) \colon \mathbf{t} \in [0,1]^n\}$ be an n-process. Let

$$U = \{S_p \colon p = 1, \ldots, m\}$$

be a partition of $[n]$, and define

$$\|X\|_U = \sup\left\{ \mathbf{E}\left| \sum_{i_1,\ldots,i_m} \Delta^n X(Q^{(1)}_{i_1} \times \cdots \times Q^{(m)}_{i_m}) r_{i_1}(u_1) \cdots r_{i_m}(u_m) \right| : \right.$$

$$\text{finite collections } \{Q^{(p)}_i\}_i \text{ of pairwise disjoint boxes in} [0,1]^{S_p},$$

$$\left. u_p \in \{-1,1\}^{\mathbb{N}},\ p \in [m] \right\}. \tag{7.1}$$

If $\|X\|_U < \infty$, then X is said to be a U-integrator. (The n-fold difference Δ^n was defined in Chapter X, Remark i §8, a *box* in $[0,1]^{S_p}$ is a $|S_p|$-fold Cartesian product of intervals.)

The definition of $\|X\|_U$ can be rephrased thus. For a set Y, and $S \subset [n]$, consider the projection from Y^n onto Y^S defined by

$$\pi_{Y,S}(y_1, \ldots, y_n) := \mathbf{y}|_S = (y_i : i \in S), \quad \mathbf{y} = (y_1, \ldots, y_n) \in Y^n. \tag{7.2}$$

When Y is arbitrary, or understood from the context, we denote $\pi_{Y,S}$ by π_S, and $\pi_{\{i\}}$ by π_i. For $p = 1, \ldots, m$, we consider Rademacher systems $\{r_{\mathbf{i}} : \mathbf{i} \in \mathbb{N}^{S_p}\}$ indexed by $\mathbb{N}^{S_p}$. Then (Exercise 31),

$$\begin{aligned}
&\|X\|_U \\
&= \sup \Bigg\{ \mathbf{E} \Bigg| \sum_{\mathbf{i}=(i_1,\ldots,i_n)} \Delta^n X(J^{(1)}_{i_1} \times \cdots \times J^{(n)}_{i_n})\, r_{\pi_{S_1}(\mathbf{i})}(u_1) \cdots r_{\pi_{S_m}(\mathbf{i})}(u_m) \Bigg| : \\
&\qquad \text{finite collections } \{J^{(j)}_i\}_i \text{ of pairwise disjoint intervals,} \\
&\qquad u_p \in \{-1,1\}^{S_p}, p \in [m],\ j \in [n] \Bigg\}. \qquad (7.3)
\end{aligned}$$

There are two extremal instances. At one end, we have $U = \{\{1\}, \ldots, \{n\}\}$, in which case we write

$$\begin{aligned}
\|X\|_{[n]} := \sup \Bigg\{ \mathbf{E} \Bigg| \sum_{i_1,\ldots,i_n} \Delta^n X(J^{(1)}_{i_1} \times \cdots \times J^{(n)}_{i_n}) r_{i_1}(u_1) \cdots r_{i_n}(u_n) \Bigg| : \\
\text{finite collections } \{J^{(j)}_i\}_i \text{ of pairwise disjoint intervals,} \\
u_j \in \{-1,1\}^{\mathbb{N}},\ \ j \in [n] \Bigg\}; \qquad (7.4)
\end{aligned}$$

if $\|X\|_{[n]} < \infty$, then X is said to be an $[n]$-integrator. At the other end we have $U = \{[n]\}$, in which case we write

$$\begin{aligned}
\|X\|_U := \|X\| = \sup \Bigg\{ \mathbf{E} \Bigg| \sum_j \Delta^n X(Q_j)\, r_j(u) \Bigg| : \\
u \in \{-1,+1\}^{\mathbb{N}}, \{Q_i\}_i \text{ finite collections} \\
\{Q_i\}_i \text{ of pairwise disjoint boxes in } [0,1]^n \Bigg\}; \qquad (7.5)
\end{aligned}$$

if $\|X\| < \infty$, then X is said to be an integrator. Noticc that if an n-process is an integrator then it is *a fortiori* a U-integrator for every partition U of $[n]$.

If an n-process X is a U-integrator and $|U| = m$, then X gives rise to an F_{m+1}-measure. To see this, first define

$$G_X(A)(t_1, \dots, t_n) := \lim_{u_1 \to t_1^+, \dots, u_n \to t_n^+} \mathbf{E} \mathbf{1}_A X(u_1, \dots, u_n),$$

$$(t_1, \dots, t_n) \in [0,1)^n, \; A \in \mathscr{A}, \tag{7.6}$$

where limits on the right side are taken iteratively, in any order. At the very outset we need to check that the definition of G_X in (7.6) is indeed feasible. Let us verify this in the case $n = 2$. Fix $(s,t) \in [0,1)^2$, and $s_j \downarrow s$ and $t_j \downarrow t$ monotonically decreasing sequences converging to s and t, respectively. Define $I_j = [s_j, s_{j-1})$ and $J_j = [t_j, t_{j-1})$. Denote the first-order differences in the second and first variables, respectively, by

$$\Delta_2 X(s; J_j) := X(s, t_{j-1}) - X(s, t_j),$$

and

$$\Delta_1 X(I_i;\ t) := X(s_{i-1}, t) - X(s_i, t).$$

Then,

$$\begin{aligned}
&\sum_{j=2}^{\infty} \Delta \left(\sum_{i=2}^{\infty} \mathbf{E} \mathbf{1}_A \Delta_1 X(I_i;\ t) \right) (J_j) \\
&= \sum_{j=2}^{\infty} \Delta_2 \left(\mathbf{E} \mathbf{1}_A X(s_1, t) - \lim_{u \to s^+} \mathbf{E} \mathbf{1}_A X(u, t) \right) (J_j) \\
&= \mathbf{E} \mathbf{1}_A X(s_1, t_1) - \lim_{v \to t^+} \mathbf{E} \mathbf{1}_A X(s_1, v) - \lim_{u \to s^+} \mathbf{E} \mathbf{1}_A X(u, t_1) \\
&\quad + \lim_{u \to s^+, v \to t^+} \mathbf{E} \mathbf{1}_A X(u, v) \\
&= \sum_{i=2}^{\infty} \Delta \left(\sum_{j=2}^{\infty} \mathbf{E} \mathbf{1}_A \Delta X(s; J_j) \right) (I_i) \\
&= \sum_{j=2}^{\infty} \sum_{i=2}^{\infty} \mathbf{E} \mathbf{1}_A \Delta^2 X(I_i \times J_j).
\end{aligned} \tag{7.7}$$

The justification for interchanging limits in (7.7) is provided by Corollary IV.7. (Note that $\{\mathbf{E} \mathbf{1}_A \Delta^2 X(I_i \times J_j)\} \in F_2(\mathbb{N}, \mathbb{N})$.) The feasibility of the definition of G_X for $n > 2$ can be verified by induction on n (Exercise 32).

Next, for half-open intervals $J_j \subset [0,1]$, $j \in [n]$, we define

$$\mu_X(A, J_1, \ldots, J_n) = \Delta^n G_X(A)(J_1 \times \cdots \times J_n); \tag{7.8}$$

equivalently,

$$\mu_X(A, B_1, \ldots, B_m) = \Delta^n G_X(A)(B_1 \times \cdots \times B_m),$$

$$\text{half-open boxes } B_p \subset [0,1]^{S_p},\ p \in [m]. \tag{7.9}$$

(Half-open box := Cartesian product of half-open intervals.)

Proposition 36 (Exercise 33; cf. Corollary 4). *Let $U = \{S_p : p = 1, \ldots, m\}$ be a partition of $[n]$. An n-process X is a U-integrator if and only if μ_X determines an F_{m+1}-measure on $\mathscr{A} \times \mathscr{B}_{S_1} \times \cdots \times \mathscr{B}_{S_m}$ ($\mathscr{B}_{S_j} :=$ Borel field in $[0,1]^{S_j}$). Moreover, $\|X\|_U = \|\mu_X\|_{F_{m+1}(\mathscr{A}, \mathscr{B}_{S_1}, \ldots, \mathscr{B}_{S_m})}$.*

All that was done in §2 in the one-parameter case can be recast in the multi-parameter setting. Specifically, if an n-process X is a U-integrator, and $f_1 \in \mathrm{L}^\infty([0,1]^{S_1}, \mathscr{B}_{S_1}), \ldots, f_m \in \mathrm{L}^\infty([0,1]^{S_m}, \mathscr{B}_{S_m})$, then the integral of $f_1 \otimes \cdots \otimes f_m$ with respect to X is defined by

$$\int_{[0,1]^n} f_1 \otimes \cdots \otimes f_m \, \mathrm{d}X$$

$$:= \frac{\mathrm{d}}{\mathrm{d}\mathbb{P}} \left(\int_{[0,1]^n} f_1(t_1) \cdots f_m(t_m)\, \mu_X(\cdot, \mathrm{d}t_1, \ldots, \mathrm{d}t_m) \right), \tag{7.10}$$

and

$$\mathbf{E} \left| \int_{[0,1]^n} f_1 \otimes \cdots \otimes f_m \, \mathrm{d}X \right| \le 2^m \|X\|_U \|f_1\|_\infty \cdots \|f_m\|_\infty. \tag{7.11}$$

In the 'one-dimensional' case, if X is an integrator, then (by the Grothendieck factorization theorem) there exists a probability measure ν (a Grothendieck measure) on $\sigma(\mathscr{B}^n)$, such that

$$\mathbf{E} \left| \int_{[0,1]^n} f \, \mathrm{d}X \right| \le \kappa_{\mathrm{G}} \|X\| \|f\|_{\mathrm{L}^2(\nu)}, \quad f \in \mathrm{C}([0,1]^n). \tag{7.12}$$

Therefore,

$$f \mapsto \int_{[0,1]^n} f \, \mathrm{d}X, \quad f \in \mathrm{L}^\infty([0,1]^n, \sigma(\mathscr{B}^n)), \tag{7.13}$$

extends to a bounded linear map from $L^2([0,1]^n, \nu)$ into $L^1(\Omega, \mathbb{P})$ (cf. Theorem 34). An L^2-extension in the general multi-parameter case is quite another matter. For example,

$$\int_{[0,1]^n} \mathbf{1}_{B_1} \otimes \cdots \otimes \mathbf{1}_{B_m} \, \mathrm{d}X, \quad B_1 \in \mathscr{B}_{S_1}, \ldots, B_m \in \mathscr{B}_{S_m}, \tag{7.14}$$

is a 'weak-$L^1(\Omega, \mathbb{P})$ stochastic F_m-measure', but I do not know that more can be said (Exercise 34).

Remarks:

i **(Littlewood index of U-integrators).** Extending (2.27) and (2.28), we define the Littlewood index of a U-integrator X to be

$$\ell_X := \inf\{q : \|\mu_X\|_{(q)} < \infty\}, \tag{7.15}$$

where

$$\|\mu_X\|_{(q)} := \sup\left\{ \left(\sum_{j,k_1,\ldots,k_m} |\mu_X(A_j, B^{(1)}_{k_1}, \ldots, B^{(m)}_{k_m})|^q \right)^{\frac{1}{q}} : \right.$$

$$\left. \text{partitions } \{A_j\} \subset \mathscr{A},\ \{B^{(p)}_k\} \subset \mathscr{B}_{S_p},\ p \in [m] \right\}. \tag{7.16}$$

The Littlewood inequalities (Chapter X §10) imply

$$\ell_X \leq \frac{2(m+1)}{m+2}, \tag{7.17}$$

where $|U| = m$. Moreover, there exist U-integrators X such that $\|\mu_X\|_{(q)} = \infty$ for all $q < 2(m+1)/(m+2)$. We thus observe (Corollary 37 in the next section) that if U and V are partitions of $[n]$, and $|U| > |V|$, then there exists an n-process X which is a U integrator but not a V-integrator.

ii **(the meaning of it...).** In this chapter we address the issue: how are stochastic processes realized as sums of 'increments'? In the multi-parameter framework, if X is an n-process, $(t_1, \ldots, t_n) \in [0,1]^n$, and $\{J^{(i)}_i\}_i$ is a finite collection of pairwise disjoint intervals whose union equals $[0, t_j]$, $j = 1, \ldots, n$, then (clearly)

$$\Delta^n X([0,t_1] \times \cdots \times [0,t_n]) = \sum_{i_1,\ldots,i_n} \Delta^n X(J^{(1)}_{i_1} \times \cdots \times J^{(n)}_{i_n}). \tag{7.18}$$

The issue is: can the right side of (7.18) be replaced by infinite sums? (Cf. §1.)

If $U=\{S_1,\dots,S_m\}$ is a partition of $[n]$, and X is a right-continuous U-integrator (i.e., $\|X\|_U < \infty$ and $G_X(A) = \mathbf{E}1_A X)$, then for all countably infinite collections of intervals $\{J_i^{(j)} : i \in \mathbb{N}\}$, each of whose respective unions equals $[0,t_j]$, $j = 1,\dots,n$,

$$\Delta^n X([0,t_1] \times \cdots \times [0,t_n]) = \sum_{S_1} \cdots \sum_{S_m} \Delta^n X(J_{i_1}^{(1)} \times \cdots \times J_{i_n}^{(n)}), \tag{7.19}$$

where $\sum_S$ means $\sum_{i_j, j\in S}$, and the series converge weakly in $L^1(\Omega,\mathbb{P})$. If X is not an integrator (i.e., if $\|X\| = \infty$), then there exists a grid $\mathscr{G}$ of $[0,1]^n$,

$$\begin{aligned}\mathscr{G} &= \{B_{i_1\dots i_n} : (i_1,\dots,i_n) \in \mathbb{N}^n\}\\ &= \{J_{i_1}^{(1)} \times \cdots \times J_{i_n}^{(n)} : (i_1,\dots,i_n) \in \mathbb{N}^n\},\end{aligned} \tag{7.20}$$

and a one–one map τ from $\mathbb{N}$ onto $\mathbb{N}^n$, such that

$$\sum_{j=1}^{\infty} \Delta^n X(B_{\tau j}) \tag{7.21}$$

does not converge (weakly in $L^1(\Omega,\mathbb{P})$) to $\Delta^n X([0,1)^n)$ (Exercise 35). This means that the outcome of X at 'time' $(1,\dots,1)$, which is synthesized from increments $\Delta X(B)$, $B\in\mathscr{G}$, as per (7.19), depends, somehow, on 'time'-locations of increments.

iii **(the 'dimension' of a 1-process).** Much of what we do in this chapter is based on a natural association of stochastic processes with F-measures. In the one-parameter case, given a 1-process $X = \{X(t) : t\in[0,1]\}$ on a probability space $(\Omega,\mathscr{A},\mathbb{P})$, such that $\mathbf{E}|X(t)| < \infty$ for all $t \in [0,1]$, we consider

$$\mu_X(A,I) = \mathbf{E}1_A \Delta X(J), \quad \text{half-open interval } J,\ A \in \mathscr{A}, \tag{7.22}$$

which we extend to a set-function μ_X on $\mathscr{A}\times\mathscr{O}$ ($\mathscr{O}$= algebra generated by the half-open intervals in $[0,1]$): for each $J\in\mathscr{O}$, $\mu_X(\cdot,J)$ is a measure on $(\Omega,\mathscr{A})$, and for each $A \in \mathscr{A}$, $\mu_X(A,\cdot)$ is finitely additive on $\mathscr{O}$. A question arises: does μ_X enjoy a property stronger than finite additivity on $\mathscr{O}$?

So far in the one-parameter case, we focused on those μ_X that determine F_2-measures on $\mathscr{A}\times\mathscr{B}$. For these, a well-defined 'X-noise'

$$\Delta X(B) := \frac{\mathrm{d}}{\mathrm{d}\mathbb{P}}\, \mu_X(\cdot, B), \quad B \in \mathscr{B}, \tag{7.23}$$

extends

$$\Delta X(J) = X(t) - X(s), \quad J = (s,t] \subset [0,1], \tag{7.24}$$

and has the property that for every $B \in \mathscr{B}$ and all $\{B_i : i \in \mathbb{N}\} \subset \mathscr{B}$ such that $\Sigma_i \mathbf{1}_{B_i} = \mathbf{1}_B$,

$$\Delta X(B) = \sum_{i=1}^{\infty} \Delta X(B_i) \quad \text{weak convergence in } \mathrm{L}^1(\Omega, \mathbb{P}). \tag{7.25}$$

The latter – we ventured in §1 – is a statement of 'time-independence': every time-rearrangement of the underlying 'increments' leads to the same outcome. Conceivably, however, it may happen that μ_X does not determine an F_2-measure on $\mathscr{A}\times\mathscr{B}$, but still manifests more than finite additivity in the second coordinate.

Let $\mathscr{D}$ be the collection of all dyadic half-open intervals in [0,1], and let $\mathscr{O}(\mathscr{D}) = \mathscr{O}$ be the algebra generated by $\mathscr{D}$. Consider the binary expansion

$$t = \sum_{j=1}^{\infty} b_j(t)/2^j, \quad t \in [0,1], \tag{7.26}$$

where (for reasons that will become apparent) we choose the finite expansions of dyadic points (see Chapter II §1). Let $\boldsymbol{\rho} = \{\rho_1, \ldots, \rho_n\}$ be a partition of $\mathbb{N}$, each of whose elements is infinite. We enumerate $\rho_i = \{k_{ij} \colon j \in \mathbb{N}\}$, $i = 1, \ldots, n$. Each ρ_i gives rise to a 'dyadic projection' from [0,1] onto [0,1], which we denote also by ρ_i:

$$\rho_i(t) = \sum_{j=1}^{\infty} b_{k_{ij}}(t)/2^j, \quad t \in [0,1],\ i = 1, \ldots, n, \tag{7.27}$$

whence

$$\boldsymbol{\rho}(t) := (\rho_1(t), \ldots, \rho_n(t)), \quad t \in [0,1], \tag{7.28}$$

is an injection from [0,1] onto $[0,1]^n$ ('space-filling curve'). If $D_i \in \mathscr{D}$, then $\rho_i[D_i] \in \mathscr{D}$, $i = 1, \ldots, n$, and $\rho_1^{-1}[D_1] \cap \cdots \cap \rho_n^{-1}[D_n] \in \mathscr{O}(\mathscr{D})$. In particular, if $D \in \mathscr{D}$, then

$$\rho_1^{-1}[\rho_1[D]] \cap \cdots \cap \rho_n^{-1}[\rho_n[D]] = D. \tag{7.29}$$

(Therein lies the reason for choosing finite binary expansions of dyadic points.) We consider

$$\mu_{X,\boldsymbol{\rho}}(\cdot, J_1, \ldots, J_n) := \mu_X(\cdot, \rho_1^{-1}[J_1] \cap \cdots \cap \rho_n^{-1}[J_n]),$$

$$J_1 \in \mathscr{O}, \ldots, J_n \in \mathscr{O}, \tag{7.30}$$

and define (temporarily) the 'dimension' of X to be the smallest integer $n > 0$, such that there exists a partition $\boldsymbol{\rho} = \{\rho_1, \ldots, \rho_n\}$ of $\mathbb{N}$ for which $\mu_{X,\boldsymbol{\rho}}$ determines an F_{n+1}-measure on $\mathscr{A} \times \mathscr{B}^n$. (In the last chapter, 'dimension' will register continuously in $[1, \infty]$.)

Given a 1-process X, we define its Littlewood index to be

$$\ell_X := \inf\{q : \|\mu_X\|_{(q)} < \infty\}, \tag{7.31}$$

where

$$\|\mu_X\|_{(q)} := \sup\left\{ \left(\sum_k |\mu_X(A_j, O_k)|^q \right)^{\frac{1}{q}} : \right.$$

$$\left. \text{partitions } \{A_j\} \subset \mathscr{A},\ \{O_k\} \subset \mathscr{O}(\mathscr{D}) \right\}. \tag{7.32}$$

(This definition is consistent with the definitions stated in (7.15) and (7.16); Exercise 36.) The Littlewood inequalities (Chapter X §10) imply that if X is 'n-dimensional', then

$$\ell_X \leq \frac{2(n+1)}{n+2}, \tag{7.33}$$

and that there exist (via Theorem X.8) 'n-dimensional' 1-processes X with $\ell_X = 2(n+1)/(n+2)$ (Exercise 37).

Let X be a 1-process whose dimension equals $n > 1$. Then, there exists a partition $\boldsymbol{\rho} = \{\rho_1, \ldots, \rho_n\}$, such that $\mu_{X,\boldsymbol{\rho}} \in F_{n+1}$, but μ_X does not determine an F_2-measure on $\mathscr{A} \times \mathscr{B}$. Suppose $D \in \mathscr{D}$, and $\{J_k^{(i)} : k \in \mathbb{N}\}$ is an $\mathscr{O}$-partition of $\rho_i[D]$, $i = 1, \ldots, n$. Then,

$$\mathscr{G}(D) = \{\rho_1^{-1}[J_{k_1}^{(1)}] \cap \cdots \cap \rho_n^{-1}[J_{k_n}^{(n)}] : (k_1, \ldots, k_n) \in \mathbb{N}^n\}$$

$$= \{A_{k_1 \ldots k_n} : (k_1, \ldots, k_n) \in \mathbb{N}^n\} \tag{7.34}$$

is an $\mathscr{O}$-partition of D (an *n-grid* of D). Then (because $\mu_{X,\boldsymbol{\rho}} \in F_{n+1}$), for all rearrangements $\tau_1, \ldots, \tau_n$ of $\mathbb{N}$,

$$\Delta X(D) = \sum_{k_1=1}^{\infty} \cdots \sum_{k_n=1}^{\infty} \Delta X(A_{\tau k_1 \ldots \tau k_n}), \tag{7.35}$$

where the iterated series converge weakly in $L^1(\Omega, \mathbb{P})$. *But*, because $\mu_X \notin F_2$, there exist dyadic intervals D, n-grids

$$\mathscr{G} = \{A_{k_1 \dots k_n} : (k_1, \dots, k_n) \in \mathbb{N}^n\}$$

of D, and one–one maps τ from $\mathbb{N}$ onto $\mathbb{N}^n$, such that

$$\sum_{m=1}^{\infty} \Delta X(A_{\tau m}) \tag{7.36}$$

do not converge (weakly in $L^1(\Omega, \mathbb{P})$) to $\Delta X(D)$. This conveys, in effect, a dependence of the outcome $\Delta X(D)$ on the time-sequence of the increments $\Delta X(A)$, $A \in \mathscr{G}(D)$. (Cf. Remark ii above, and the discussion following Definition 1.)

Regarding integration with respect to 'n-dimensional' 1-processes X, note that if $\mu_{X,\boldsymbol{\rho}} \in F_{n+1}$, and f is a bounded $\mathscr{B}$-measurable function on [0,1] such that $f \circ \boldsymbol{\rho}^{-1} \in V_n(\mathscr{B}, \dots, \mathscr{B})$, then

$$\int_{[0,1]} f \, \mathrm{d}X$$

$$:= \frac{\mathrm{d}}{\mathrm{d}\mathbb{P}} \left(\int_{[0,1]^n} f \circ \rho^{-1}(t_1, \dots, t_n) \mu_{X; E_1, \dots, E_n}(\cdot, \mathrm{d}t_1, \dots, \mathrm{d}t_n) \right). \tag{7.37}$$

8 Examples: Random Constructions

If U and V are partitions of $[n]$ and $|U| < |V|$, then by applying Theorem X.8 and the Littlewood inequalities in Chapter X §10, we observe, via random constructions, n-processes X that are V-integrators, but not U-integrators. We sketch such a construction in the case $n = 2$, $U = \{\{1, 2\}\}$, and $V = \{\{1\}, \{2\}\}$, which is typical; constructions in the general case follow practically the same blueprint.

Let $\{J_{kj} : j \in \mathbb{N}, \ k \in [2^j]\}$ be a collection of pairwise disjoint intervals, whose union is [0,1]. Fix $j \in \mathbb{N}$, and let $\Omega_j = \{-1, 1\}^j$. By Theorem X.8, there exist $\{-1, 1\}$-valued arrays $(\epsilon_{k_1 k_2 \omega} : (k_1, k_2, \omega) \in [2^j] \times [2^j] \times \Omega_j)$, such that

$$\left\| \sum_{(k_1, k_2, \omega) \in [2^j] \times [2^j] \times \Omega_j} \epsilon_{k_1 k_2 \omega} \, r_{k_1} \otimes r_{k_2} \otimes r_\omega \right\|_{L^\infty} \leq K \, 2^{2j}, \tag{8.1}$$

where $K>0$ is an absolute constant. (For application of Theorem X.8, notice that Walsh polynomials with spectrum in

$$\{r_{k_1} \otimes r_{k_2} \otimes r_\omega : (k_1, k_2, \omega) \in [2^j] \times [2^j] \times \Omega_j\}$$

have degree at most $3 \cdot 2^{2^j}$.) For $(s,t) \in [0,1]^2$, let

$$E_j(s,t) = \{(k_1,k_2) : I_{k_1 j} \times I_{k_2 j} \subset [0,s] \times [0,t]\}, \tag{8.2}$$

and then define a 2-process X_j on the uniform probability space Ω_j by

$$X_j(s,t)(\omega) = (1/K\ 2^j) \sum_{(k_1,k_2) \in E_j(s,t)} \epsilon_{k_1 k_2 \omega},$$

$$(s,t) \in [0,1]^2, \ \omega \in \Omega_j. \tag{8.3}$$

If $(k_1, k_2, \omega) \in [2^j] \times [2^j] \times \Omega_j$, then

$$\mathbf{E}\mathbf{1}_{\{\omega\}}\ \Delta^n X_j(I_{k_1 j} \times I_{k_2 j}) = (1/K 2^{2j})\ \epsilon_{k_1 k_2 \omega}. \tag{8.4}$$

By (8.1),

$$\|X_j\|_V \leq 1 \text{ and } \|\mu_{X_j}\|_{(q)} \geq (1/K)\ 2^{\frac{3j-2qj}{q}}. \tag{8.5}$$

We view the X_j as independent processes on the product probability space $\Omega = \prod_{j=1}^{\infty} \Omega_j$, and then let

$$X = \sum_{j=1}^{\infty} X_j / j^2. \tag{8.6}$$

(X_j is defined on the jth factor of Ω.) By (8.5), $\|X\|_V < \infty$, and

$$\|\mu_X\|_{(q)} \geq \|\mu_{X_j}\|_{(q)}/j^2 \geq (1/Kj^2) 2^{\frac{3j-2qj}{q}}, \tag{8.7}$$

which implies $\|\mu_X\|_{(q)} = \infty$ for all $0 < q < 3/2$. In particular, this implies that X is a V-integrator, but not a U-integrator.

A similar construction implies

Theorem 37 (Exercise 38). *If U and V are partitions of $[n]$ and $|U| > |V|$, then there exists an n-process X such that $\|X\|_U < \infty$ and $\|X\|_V = \infty$.*

9 Independent Products of Integrators

Suppose an n-process X and an m-process Y are mutually independent integrators. Then, $X \otimes Y$ is (obviously) a

$$\{\{1, \ldots, n\}, \{n+1, \ldots, n+m\}\}\text{-integrator}.$$

In this section we verify the stronger conclusion, that $X \otimes Y$ is an integrator. ($X \otimes Y$ is the $(n+m)$-process defined by $(X \otimes Y)(\mathbf{s},\mathbf{t}) = X(\mathbf{s})\,Y(\mathbf{t})$, $(\mathbf{s},\mathbf{t}) \in [0,1]^n \times [0,1]^m$.)

If $Q \subset [0,1]^n$ and $P \subset [0,1]^m$ are boxes, then

$$\Delta^{n+m}(X \otimes Y)(Q \times P) = \Delta^n X(Q) \otimes \Delta^m Y(P). \tag{9.1}$$

We will prove that there exists $K > 0$ such that if $\{Q_j\}$ and $\{P_k\}$ are finite collections of pairwise disjoint boxes in $[0,1]^n$ and $[0,1]^m$, respectively, then for all $(a_{jk}) \in l^\infty(\mathbb{N}^2)$

$$\mathbf{E}\left|\sum_{j,k} a_{jk}\ \Delta^n X(Q_j) \otimes \Delta^m Y(P_k)\right| \le K\|(a_{jk})\|_\infty. \tag{9.2}$$

From this we will conclude that $\mu_{X\otimes Y}$ is a product F_2-measure on $\mathscr{A} \times \mathscr{B}_{n+m}$, where $\mathscr{B}_{n+m} = \sigma(\mathscr{B}_n \times \mathscr{B}_m)$, and thus obtain the multiple integrals

$$\begin{aligned} &\int_{[0,1]^{n+m}} f(\mathbf{s},\mathbf{t})\ \mathrm{d}X(\mathrm{d}\mathbf{s})\ \mathrm{d}Y(\mathrm{d}\mathbf{t}) \\ &\quad := \frac{\mathrm{d}}{\mathrm{d}\mathbb{P}}\left(\int_{[0,1]^{n+m}} f(\mathbf{s},\mathbf{t})\ \mu_{X\otimes Y}(\cdot,\mathrm{d}(\mathbf{s},\mathbf{t}))\right), \\ &\quad f \in \mathrm{L}^\infty([0,1]^{n+m}, \mathscr{B}_{n+m}). \end{aligned} \tag{9.3}$$

Here are two instances of (9.2) that are easy to verify.

1. Let X and Y be mutually independent L^2-bounded 1-processes with orthogonal increments. Then for all (a_{jk}) in the unit ball of $l^\infty(\mathbb{N})^2$,

$$\begin{aligned} \left(\mathbf{E}\left|\sum_{j,k} a_{jk}\ \Delta X(I_j)\Delta Y(J_k)\right|\right)^2 &\le \mathbf{E}\left|\sum_{j,k} a_{jk}\ \Delta X(I_j)\Delta Y(J_k)\right|^2 \\ &\le \sum_{j,k} \Delta F_X(I_j)\ \Delta F_Y(J_k) \\ &\le (\mathbf{E}|X(1)|^2 - \mathbf{E}|X(0)|^2)(\mathbf{E}|Y(1)|^2 - \mathbf{E}|Y(0)|^2). \end{aligned} \tag{9.4}$$

(See §3 for notation and basic facts.)

2. Let X and Y be mutually independent 1-processes that are L^1-bounded, centered, symmetric, and additive. Then, (9.2) follows from Lemma 31.

The general case requires the Grothendieck factorization theorem and inequality.

Theorem 38 *If X and Y are mutually independent integrators, then $X \otimes Y$ is an integrator, and*

$$\mu_{X\otimes Y} = \mu_X \times \mu_Y, \tag{9.5}$$

where $\mu_X \times \mu_Y$ is a product F_2-measure. Furthermore, if ν_1 is a Grothendieck measure of X and ν_2 is a Grothendieck measure of Y, then the product probability measure $\nu_1 \times \nu_2$ is a Grothendieck measure of $X \otimes Y$.

Proof: We consider (without loss of generality) 1-processes X and Y, redefined on the product probability space $(\Omega \times \Omega, \sigma(\mathscr{A} \times \mathscr{A}), \mathbb{P} \times \mathbb{P})$:

$$\begin{aligned} X(s)(\omega_1, \omega_2) &= X(s)(\omega_1), \quad Y(t)(\omega_1, \omega_2) = Y(t)(\omega_2), \\ (s,t) &\in [0,1]^2, (\omega_1, \omega_2) \in \Omega \times \Omega. \end{aligned} \tag{9.6}$$

For $A_1 \in \mathscr{A}$, $A_2 \in \mathscr{A}$, and intervals $J_1 \subset [0,1]$, $J_2 \subset [0,1]$,

$$\mu_{X\otimes Y}(A_1 \times A_2, J_1 \times J_2) = \mu_X(A_1, J_1)\ \mu_Y(A_2, J_2). \tag{9.7}$$

By Theorem IX.6 and Theorem IX.9, $\mu_{X\otimes Y}$ determines a product F_2-measure $\mu_X \times \mu_Y$ on $\sigma(\mathscr{A} \times \mathscr{A}) \times \sigma(\mathscr{B} \times \mathscr{B})$, which proves the first part of the theorem.

To verify the second part, we show that if ν_1 is a Grothendieck measure of X and ν_2 is a Grothendieck measure of Y, then there exists $K > 0$ such that for all $f \in C([0,1]^2)$,

$$\mathbf{E}\left|\int_{[0,1]^2} f\, d(X \otimes Y)\right| \leq K\ \|f\|_{L^2(\nu_1 \times \nu_2)} \|X\| \|Y\|. \tag{9.8}$$

To this end, let $\sum_{i,j} \delta_{ij}\, \mathbf{1}_{A_i}\mathbf{1}_{B_j}$ be a simple function on Ω^2, where the A_i and B_j are respectively pairwise disjoint, $|\delta_{ij}| \leq 1$, and observe for $f \in \mathrm{C}([0,1]^2)$,

$$\left|\mathbf{E}\left(\sum_{i,j} \delta_{ij}\mathbf{1}_{A_i}\mathbf{1}_{B_j}\right)\left(\int_{[0,1]^2} f\, \mathrm{d}(X \otimes Y)\right)\right|$$

$$= \left|\sum_{i,j} \delta_{ij} \int_{[0,1]^2} f(s,t)\ \mu_{X\otimes Y}(A_i \times B_j,\ \mathrm{d}(s,t))\right|$$

$$= \left|\sum_{i,j} \delta_{ij} \int_{[0,1]^2} f(s,t)\ \mu_X(A_i, \mathrm{d}s)\mu_Y(B_j, \mathrm{d}t)\right|$$

$$= \left|\int_{[0,1]} \sum_j \left(\int_{[0,1]} \sum_i \delta_{ij}\ f(s,t)\ \mu_X(A_i, \mathrm{d}s)\right) \mu_Y(B_j, \mathrm{d}t)\right|. \tag{9.9}$$

Consider the bilinear functionals β_1 and β_2 on $\mathrm{c}_0(\mathbb{N}) \times \mathrm{C}([0,1])$ defined by

$$\beta_1(\alpha, g) \ = \ \int_{[0,1]} \sum_i \alpha(i)\ g(s)\ \mu_X(A_i, \mathrm{d}s),$$

$$\beta_2(\alpha, g) \ = \ \int_{[0,1]} \sum_j \alpha(j)\ g(s)\ \mu_Y(B_j, \mathrm{d}s),$$

$$\alpha \in \mathrm{c}_0(\mathbb{N}),\ g \in \mathrm{C}([0,1]). \tag{9.10}$$

Because ν_1 is a Grothendieck measure of X, and ν_2 is a Grothendieck measure of Y, β_1 determines a bounded bilinear functional on $\mathrm{c}_0(\mathbb{N}) \times \mathrm{L}^2([0,1], \nu_1)$ with norm $\|\beta_1\|$, and β_2 determines a bounded bilinear functional on $\mathrm{c}_0(\mathbb{N}) \times \mathrm{L}^2([0,1], \nu_2)$ with norm $\|\beta_2\|$. By the Grothendieck factorization theorem, there exist probability measures λ_1 and λ_2 on $\mathbb{N}$ such that for all $\alpha \in \mathrm{c}_0(\mathbb{N})$ and $g \in \mathrm{C}([0,1])$,

$$\begin{aligned} |\beta_1(\alpha, g)| &\leq \kappa_{\mathrm{G}}\ \|\alpha\|_{\mathrm{L}^2(\lambda_1)}\|g\|_{\mathrm{L}^2(\nu_1)}\|\beta_1\|, \\ |\beta_2(\alpha, g)| &\leq \kappa_{\mathrm{G}}\ \|\alpha\|_{\mathrm{L}^2(\lambda_2)}\|g\|_{\mathrm{L}^2(\nu_2)}\|\beta_2\|. \end{aligned} \tag{9.11}$$

Fix orthonormal bases for $L^2(\mathbb{N}, \lambda_1)$ and $L^2([0,1], \nu_1)$, and let $\hat{\beta}_1$ be the representing matrix of β_1 in these bases. For $t \in [0,1]$, let

$$(f(\hat{n}, t) : n \in \mathbb{N}) \in l^2(\mathbb{N})$$

be the representing vector of $f(\cdot, t)$ relative to the basis in $L^2([0,1], \nu_1)$, and for $j \in \mathbb{N}$, let $(\delta(\hat{m}, j) : m \in \mathbb{N}) \in l^2(\mathbb{N})$ be the representing vector of $(\delta_{ij} : i \in \mathbb{N})$ in $L^2(\mathbb{N}, \lambda_1)$. Denote

$$F_n(t) = f(\hat{n}, t) \quad \text{and } G_n(j) = \sum_m \hat{\beta}_1(n, m)\delta(\hat{m}, j), \tag{9.12}$$

and write

$$\begin{aligned} &\int_{[0,1]} \sum_i \delta_{ij} f(s,t)\ \mu_X(A_i, \mathrm{d}s) \\ &\quad = \sum_{n,m} \hat{\beta}_1(n,m)\ f(\hat{n}, t)\delta(\hat{m}, j) = \sum_n F_n(t)\ G_n(j). \end{aligned} \tag{9.13}$$

Then,

$$\begin{aligned} &\left| \int_{[0,1]} \sum_j \left(\int_{[0,1]} \sum_i \delta_{ij}\ f(s,t)\ \mu_X(A_i, \mathrm{d}s) \right) \mu_Y(B_j, \mathrm{d}t) \right| \\ &\quad = \left| \beta_2 \left(\sum_n F_n \otimes G_n \right) \right| \le \sum_n |\beta_2(F_n, G_n)| \\ &\quad \le \kappa_G\ \|\beta_2\| \sum_n \|F_n\|_{L^2(\nu_1)} \|G_n\|_{L^2(\lambda_2)}. \end{aligned} \tag{9.14}$$

From the first line in (9.11) and the definitions of G_n and F_n, we obtain

$$\left(\sum_n \|F_n\|^2_{L^2(\nu_1)} \right)^{\frac{1}{2}} = \|f\|_{L^2(\nu_1 \times \nu_2)}, \tag{9.15}$$

and

$$\left(\sum_n \|G_n\|^2_{L^2(\lambda_2)} \right)^{\frac{1}{2}} \le \kappa_G\ \|\beta_1\|.$$

Therefore, by applying the Cauchy–Schwarz inequality to the right side of (9.14), we have

$$\left| \int_{[0,1]} \sum_j \left(\int_{[0,1]} \sum_i \delta_{ij} \ f(s,t) \ \mu_X(A_i, \mathrm{d}s) \right) \mu_Y(B_j, \mathrm{d}t) \right|$$

$$\leq \kappa_{\mathrm{G}} \ \|\beta_2\| \left(\sum_n \|F_n\|^2_{\mathrm{L}^2(\nu_1)} \right)^{\frac{1}{2}} \left(\sum_n \|G_n\|^2_{\mathrm{L}^2(\lambda_2)} \right)^{\frac{1}{2}}$$

$$\leq \kappa_{\mathrm{G}}^2 \ \|\beta_1\| \|\beta_2\| \|f\|_{\mathrm{L}^2(\nu_1 \times \nu_2)}, \tag{9.16}$$

which (maximized over all $\sum_{i,j} \delta_{ij} \mathbf{1}_{A_i} \mathbf{1}_{B_j}$) implies (9.8). □

10 Products of a Wiener Process

We have just proved that an *independent* product of integrators is an integrator (Theorem 38), and the question arises: what can be said about products of integrators when factors are not assumed to be independent?

In this section we consider the instance where each factor is the same Wiener process, and restrict out discussion to the case $n = 2$. Define the *product process* $\mathrm{W}^{(2)} := \mathrm{W} \otimes \mathrm{W}$ to be

$$\mathrm{W}^{(2)}(s,t) = \mathrm{W}(s)\mathrm{W}(t), \quad (s,t) \in [0,1]^2. \tag{10.1}$$

Then, for intervals $I \subset [0,1]$ and $J \subset [0,1]$,

$$\Delta^2 \mathrm{W}^{(2)}(I \times J) = \Delta \mathrm{W}(I) \Delta \mathrm{W}(J). \tag{10.2}$$

To verify that $\mathrm{W}^{(2)}$ is an integrator, observe that for all finite collections $\{J_j\}$ of pairwise disjoint intervals in $[0,1]$, and all finite scalar arrays (a_{jk}),

$$\mathbf{E} \left| \sum_{j,k} a_{jk} \ \Delta \mathrm{W}(J_j) \ \Delta \mathrm{W}(J_k) \right| \leq \left(\mathbf{E} \left| \sum_{j,k} a_{jk} \ \Delta \mathrm{W}(J_j) \ \Delta \mathrm{W}(J_k) \right|^2 \right)^{\frac{1}{2}}$$

$$\leq \left(\mathbf{E} \left| \sum_{j>k} a_{jk} \ \Delta \mathrm{W}(J_j) \ \Delta \mathrm{W}(J_k) \right|^2 \right)^{\frac{1}{2}}$$

$$+\left(\mathbf{E}\left|\sum_{k>j} a_{jk}\ \Delta \mathrm{W}(J_j)\ \Delta \mathrm{W}(J_k)\right|^2\right)^{\frac{1}{2}}$$

$$+\left(\mathbf{E}\left|\sum_{j} a_{jj}\ \Delta \mathrm{W}(J_j)^2\right|^2\right)^{\frac{1}{2}}$$

$$\leq 2\|(a_{jk})\|_\infty + \|(a_{jj})\|_\infty \sum_j (\mathbf{E}|\Delta \mathrm{W}(J_j)|^4)^{\frac{1}{2}} \leq 5\|(a_{jk})\|_\infty. \tag{10.3}$$

(We used above the following facts: $\{\Delta \mathrm{W}(J_j)\ \Delta \mathrm{W}(J_k) : j < k\}$ is an orthogonal system, $\mathbf{E}|\Delta \mathrm{W}(J_j)\ \Delta \mathrm{W}(J_k)|^2 =$ length J_j length J_k $(j < k)$, and $\mathbf{E}|\Delta \mathrm{W}(J_j)|^4 = 3$ length J_j^2; see Chapter X §7.). Therefore,

$$\begin{aligned}\mu_{\mathrm{W}^{(2)}}(A, I \times J) &= \mathbf{E}\mathbf{1}_A \Delta^2 \mathrm{W}^{(2)}(I \times J)\\ &= \mathbf{E}\mathbf{1}_A \Delta \mathrm{W}(I)\Delta \mathrm{W}(J), \quad \text{intervals } I, J,\ A \in \mathscr{A},\end{aligned} \tag{10.4}$$

determines an F_2-measure on $\mathscr{A}\times \mathscr{B}_2$, where $\mathscr{B}_2 = \sigma(\mathscr{B}\times \mathscr{B})$. Then, we obtain the integral

$$\begin{aligned}\int_{[0,1]^2} f(s,t)\mathrm{dW}(\mathrm{d}s)\mathrm{dW}(\mathrm{d}t) &:= \int_{[0,1]^2} f\mathrm{d}(\mathrm{W}^{(2)})\\ &= \frac{\mathrm{d}}{\mathrm{d}\mathbb{P}} \int_{[0,1]^2} f(s,t)\mu_{\mathrm{W}^{(2)}}(\cdot, \mathrm{d}(s,t)),\\ & f \in \mathrm{L}^\infty([0,1]^2, \mathscr{B}_2).\end{aligned} \tag{10.5}$$

Notice that $\int_{[0,1]^2} f\mathrm{d}(\mathrm{W}^{(2)})$ is not always the same as the two-fold Wiener integral $I_{\mathrm{W}_2}(f)$. Indeed, notice ((X.8.11))

$$I_{\mathrm{W}_2}(\mathbf{1}_{(0,s]}\mathbf{1}_{(0,t]}) = \mathrm{W}(s)\mathrm{W}(t) - s \wedge t, \quad (s,t) \in [0,1]^2, \tag{10.6}$$

($\wedge =$ minimum), while ((10.2))

$$\int_{[0,1]^2} \mathbf{1}_{(0,s]}\mathbf{1}_{(0,t]}\ \mathrm{d}(\mathrm{W}^{(2)}) = \mathrm{W}(s)\mathrm{W}(t). \tag{10.7}$$

The discrepancy between (10.6) and (10.7) is explained in the remark immediately following the proposition below.

Proposition 39 *Let $\mathfrak{m}_D$ denote the normalized Lebesgue measure on the diagonal $D = \{(s,s) : s \in [0,1]\}$ (i.e., $\int_{[0,1]^2} f(s,t)\ \mathfrak{m}_D(\mathrm{d}(s,t)) = \int_{[0,1]} f(s,s)\mathfrak{m}(\mathrm{d}s),\ f \in \mathrm{L}^\infty([0,1]^2, \mathscr{B}_2)$.)*

Then, $\nu_{\mathrm{W}^{(2)}} := (\mathfrak{m}^2 + \mathfrak{m}_D)/2$ is a Grothendieck measure of $\mathrm{W}^{(2)}$.

If ν is any other Grothendieck measure of $\mathrm{W}^{(2)}$, then $\nu(D) > 0$. In particular,

$$\Delta^2 \mathrm{W}^{(2)}(D) := \int_{[0,1]^2} \mathbf{1}_D \ \mathrm{d}(\mathrm{W}^{(2)}) = 1. \tag{10.8}$$

Proof: Let f be a step function on $[0,1]^2$, and write it as

$$f = \sum_{i=1}^{n} \sum_{j=1}^{n} a_{ij} \mathbf{1}_{j_i} \mathbf{1}_{J_j},$$

where $J_i = [i/n, i+1/n),\ i = 0, \ldots, n-1$. Then,

$$\mathbf{E}\left| \int_{[0,1]^2} f \ \mathrm{d}(\mathrm{W}^{(2)}) \right| = \mathbf{E}\left| \sum_{i=1}^{n} \sum_{j=1}^{n} a_{ij}\ \Delta\mathrm{W}(J_i)\ \Delta\mathrm{W}(J_j) \right|$$

$$\leq \mathbf{E}\left| \sum_{i \neq j} a_{ij}\ \Delta\mathrm{W}(J_i)\ \Delta\mathrm{W}(J_j) \right| + \mathbf{E}\left| \sum_{i=1}^{n} a_{ii}\ \Delta\mathrm{W}(J_i)^2 \right|$$

$$\leq \left(\mathbf{E}\left| \sum_{i \neq j} a_{ij}\ \Delta\mathrm{W}(J_i)\ \Delta\mathrm{W}(J_j) \right|^2 \right)^{\frac{1}{2}} + \mathbf{E}\left| \sum_{i=1}^{n} a_{ii}\ \Delta\mathrm{W}(J_i)^2 \right|$$

$$\leq \left(\sum_{i \neq j} |a_{ij}|^2/n^2 \right)^{\frac{1}{2}} + \sum_{i=1}^{n} |a_{ii}|/n \leq 2\|f\|_{\mathrm{L}^2(\nu_{\mathrm{W}^{(2)}})}, \tag{10.9}$$

which implies that $\nu_{\mathrm{W}^{(2)}}$ is a Grothendieck measure of $\mathrm{W}^{(2)}$.

To prove (10.8), consider $D_n = \bigcup_{i=1}^{n} J_i \times J_i$, and note that $D = \bigcap_{n=1}^{\infty} D_n$. Then, because $\mu_{\mathrm{W}^{(2)}} \in F_2(\mathscr{A}, \mathscr{B}_2)$ and $\mu_{\mathrm{W}^{(2)}}(\cdot, B) \ll \mathbb{P}$ for $B \in \mathscr{B}_2$,

$$\mathbf{E}\mathbf{1}_A \int_{[0,1]^2} \mathbf{1}_{D_n} \mathrm{d}(\mathrm{W}_2) \xrightarrow[n\to\infty]{} \mathbf{E}\mathbf{1}_A \int_{[0,1]^2} \mathbf{1}_D \ \mathrm{d}(\mathrm{W}^{(2)}), \quad A \in \mathscr{A}. \tag{10.10}$$

But

$$\int_{[0,1]^2} \mathbf{1}_{D_n}\, \mathrm{d}(\mathrm{W}^{(2)}) = \sum_{i=1}^{n} \Delta \mathrm{W}(J_i)^2 \xrightarrow[n\to\infty]{} 1 \text{ in } \mathrm{L}^2(\Omega, \mathbb{P}) \tag{10.11}$$

(quadratic variation; cf. (X.3.1)), which verifies (10.8), and, therefore, that $\nu(D) > 0$ for every Grothendieck measure ν of $\mathrm{W}^{(2)}$. □

Remarks:

i **($I_{\mathrm{W}_2}(f)$ *v.* $\int_{[0,1]^2} f\, \mathrm{d}(\mathrm{W}^{(2)})$).** The statement in (10.8) explains the difference between the two-fold Wiener integral and the integral defined in (10.5): the definition of I_{W_2} is based on the completion of step functions in the $\|\cdot\|_{\mathrm{L}^2(\mathrm{m}^2)}$-norm, while, according to the proposition above, the 'correct' norm is $\|\cdot\|_{\mathrm{L}^2(\nu_{\mathrm{W}^{(2)}})}$. For then, the two-dimensional white noise $\Delta^2 \mathrm{W}^{(2)}$,

$$\Delta^2 \mathrm{W}^{(2)}(B) := \int_{[0,1]^2} \mathbf{1}_B\, \mathrm{d}(\mathrm{W}^{(2)}), \quad B \in \mathscr{B}_2, \tag{10.12}$$

extends (as it should!)

$$\begin{aligned} &\Delta^2 \mathrm{W}^{(2)}(B_1 \times B_2) = \Delta \mathrm{W}(B_1)\, \Delta \mathrm{W}(B_2), \\ &B_1 \in \mathscr{B},\ B_2 \in \mathscr{B}. \end{aligned} \tag{10.13}$$

(See Remark i in Chapter X §8.)

ii **(the Itô integral via the measure-theoretic approach).** Let us consider a definition of $\int_0^1 \mathrm{W}\mathrm{d}\mathrm{W}$ as an $\mathrm{L}^2(\Omega, \mathbb{P})$-limit of Riemann sums. Let $\pi_n = \{0 = t_0 < t_1 < \cdots < t_n = 1\}$ be a partition of [0,1], and write

$$\begin{aligned} &\sum_{j=1}^{n} \mathrm{W}(t_{j-1})(\mathrm{W}(t_j) - \mathrm{W}(t_{j-1})) \\ &\quad = \sum_{j=1}^{n} (\mathrm{W}(t_j)^2 - \mathrm{W}(t_{j-1})^2) - \sum_{j=1}^{n} \mathrm{W}(t_j)\, (\mathrm{W}(t_j) - \mathrm{W}(t_{j-1})) \\ &\quad = \mathrm{W}(1)^2 - \sum_{j=1}^{n} \mathrm{W}(t_j)\, (\mathrm{W}(t_j) - \mathrm{W}(t_{j-1})). \end{aligned} \tag{10.14}$$

We also can write

$$\sum_{j=1}^{n} \mathrm{W}(t_{j-1})\,(\mathrm{W}(t_j) - \mathrm{W}(t_{j-1}))$$

$$= \frac{1}{2}\left(\mathrm{W}(1)^2 - \sum_{j=1}^{n}(\mathrm{W}(t_j) - \mathrm{W}(t_{j-1}))^2\right). \qquad (10.15)$$

If the mesh of π_n tends to 0 as $n \to \infty$, then the right side of (10.15) converges in $\mathrm{L}^2(\Omega, \mathbb{P})$ to $\mathrm{W}(1)^2/2 - 1/2$ (Chapter X §3, Remark ii). This, by definition, is the Itô integral $\int_0^1 \mathrm{W}d\mathrm{W}$. (Similarly, $\int_0^t \mathrm{W}d\mathrm{W} = \mathrm{W}(t)^2/2 - t/2$.)

Observe now that (10.14) can be rewritten as

$$\sum_{j=1}^{n} \mathrm{W}(t_{j-1})\,(\mathrm{W}(t_j) - \mathrm{W}(t_{j-1}))$$

$$= \int_{[0,1]^2} \left(\sum_{j=1}^{n} \mathbf{1}_{[0,t_{j-1}]} \otimes \mathbf{1}_{[t_{j-1},t_j]}\right) d\mathrm{W}^{(2)}$$

$$= \frac{1}{2}\left(\mathrm{W}(1)^2 - \sum_{j=1}^{n}(\mathrm{W}(t_j) - \mathrm{W}(t_{j-1}))^2\right). \qquad (10.16)$$

Therefore, if $f_n = \sum_{j=1}^{n} \mathbf{1}_{[0,t_{j-1}]} \otimes \mathbf{1}_{[t_{j-1},t_j]}$, and $f(s,t) = \mathbf{1}_{[0,t)}(s)$, then $f_n \to f$ in $\mathrm{L}^2([0,1]^2, \nu_{\mathrm{W}^{(2)}})$, and by Proposition 39,

$$\int_{[0,1]^2} \mathbf{1}_{[0,s)}(t)\, d\mathrm{W}^{(2)} = \int_0^1 \mathrm{W}\, d\mathrm{W}. \qquad (10.17)$$

The integral on the left side of (10.17), which can be written (formally!) as

$$\int_{[0,1]^2} \mathbf{1}_{[0,s)}(t)\, d\mathrm{W}^{(2)} = \int_{[0,1]} \left(\int_{[0,1]} \mathbf{1}_{[0,s)}(t)\, d\mathrm{W}(ds)\right) d\mathrm{W}(dt)$$

$$= \int_{[0,1]} \mathrm{W}(t)\, d\mathrm{W}(dt), \qquad (10.18)$$

is not the same as $\int_{[0,1]^2} \mathbf{1}_{[0,s]}(t) \mathrm{dW}^{(2)}$, although we could (again formally) think of the latter also as $\int_{[0,1]} \mathrm{W}(t)\ \mathrm{dW}(\mathrm{d}t)$. Indeed,

$$\begin{aligned}\int_{[0,1]^2} &\mathbf{1}_{[0,s]}(t)\ \mathrm{dW}^{(2)} \\ &= \frac{1}{2}\left(\int_{[0,1]^2} \mathbf{1}_{[0,1]^2}(s,t)\ \mathrm{dW}^{(2)} + \int_{[0,1]^2} \mathbf{1}_D(s,t)\ \mathrm{dW}^{(2)}\right),\end{aligned} \tag{10.19}$$

whereas

$$\begin{aligned}\int_{[0,1]^2} &\mathbf{1}_{[0,s)}(t)\ \mathrm{dW}^{(2)} \\ &= \frac{1}{2}\left(\int_{[0,1]^2} \mathbf{1}_{[0,1]}(s,t)\ \mathrm{dW}^{(2)} - \int_{[0,1]^2} \mathbf{1}_D(s,t)\ \mathrm{dW}^{(2)}\right).\end{aligned} \tag{10.20}$$

The left side of (10.20) can be evaluated also as follows. Start with

$$\begin{aligned}\sum_{j=1}^{n} &\mathrm{W}(t_j)\ (\mathrm{W}(t_j) - \mathrm{W}(t_{j-1})) \\ &= \int_{[0,1]^2} \left(\sum_{j=1}^{n} \mathbf{1}_{[0,t_j]}(s) \mathbf{1}_{[t_{j-1},t_j]}(t)\right) \mathrm{dW}(\mathrm{d}s)\ \mathrm{dW}(\mathrm{d}t) \\ &= \int_{[0,1]^2} \left(\sum_{j=1}^{n} \mathbf{1}_{[0,t_j]} \otimes \mathbf{1}_{[t_{j-1},t_j]}\right) \mathrm{dW}^{(2)} \\ &= \frac{1}{2}\left(\mathrm{W}(1)^2 + \sum_{j=1}^{n} (\mathrm{W}(t_j) - \mathrm{W}(t_{j-1}))^2\right)\end{aligned} \tag{10.21}$$

(cf. (10.16)). Define $g_n = \sum_{j=1}^{n} \mathbf{1}_{[0,t_j]} \otimes \mathbf{1}_{[t_{j-1},t_j]}$ and $g(s,t) = \mathbf{1}_{[0,t]}(s)$. Then, $g_n \to g$ in $\mathrm{L}^2([0,1]^2, \nu_{\mathrm{W}^{(2)}})$, and therefore, by Proposition 39,

$$\begin{aligned}\int_{[0,1]^2} \mathbf{1}_{[0,s]}(t)\ \mathrm{dW}^{(2)} &= \mathrm{W}(1)^2/2 + \frac{1}{2} \\ &= \int_0^1 \mathrm{W} \mathrm{dW} + 1.\end{aligned} \tag{10.22}$$

Notice the difference between the integral obtained by taking limits of Riemann sums on the left side of (10.14) (resulting in the *Itô integral*), and the integral obtained by taking limits of the Riemann sums on the right side of (10.21).

If we use the 'average' of the integrand over $[t_{j-1}, t_j]$ in the Riemann sums (instead of the evaluation at the left end point, or the right end-point), then we obtain

$$\lim_{\|\pi_n\|\to 0}\left(\sum_{j=1}^{n}(\mathrm{W}(t_j)+\mathrm{W}(t_{j-1})/2)\right.$$
$$\left.(\mathrm{W}(t_j)-\mathrm{W}(t_{j-1}))\right) = \mathrm{W}(1)^2/2. \qquad (10.23)$$

Stochastic integrals obtained by taking 'averages' of integrands in Riemann sums are known as *Stratonovich integrals* (e.g., [IkW, p. 101]).

Construction and analysis of $\int_{[0,1]^n} f d\mathrm{W}^{(n)}$ for arbitrary $n > 1$ follow similar lines (Exercise 39).

iii **(products of L^1-bounded additive processes).** Let us consider the n-fold product process $X^{(n)}$ of an L^1-bounded additive process X,

$$X^{(n)}(t_1,\ldots,t_n) = X(t_1)\cdots X(t_n),$$
$$(t_1,\ldots,t_n) \in [0,1]^n. \qquad (10.24)$$

Define the polyhedral set (cf. Chapter X §11)

$$\tilde{D}_n = \{(t_1,\ldots,t_n) : 0 \le t_1 < \cdots < t_n \le 1\},$$

and the polyhedral Borel field $\tilde{B}_{\sigma n}$ in it. To simplify matters, assume X is right-continuous. (Otherwise, start with (2.5) and proceed accordingly.) Define

$$\mu_{X^{(n)}}(A, J_1\times\cdots\times J_n) := \mathbf{E}\mathbf{1}_A \Delta X(J_1)\cdots\Delta X(J_n),$$
$$A \in \mathscr{A},\ \text{boxes } J_1\times\cdots\times J_n \in \tilde{B}_{\sigma n}, \qquad (10.25)$$

and deduce, via (the n-dimensional version of) Lemma 3, and the decoupling inequalities in Theorem 32, that $\mu_{X^{(n)}} \in F_2(\mathscr{A}, \tilde{B}_{\sigma n\sigma})$. An integral of $f \in \mathrm{L}^\infty(\tilde{D}_n, \tilde{B}_{\sigma n})$ with respect to $X^{(n)}$ is defined by integration with respect to $\mu_{X^{(n)}}$. Notice that unless more is known about higher moments of the increments of X, the polyhedral set $(\tilde{D}_n, \tilde{B}_{\sigma n})$ cannot be replaced by the 'full box' $([0,1]^n, \mathscr{B}_n)$.

We obtain, again by an application of decoupling inequalities (Theorem 32), that the restriction of the n-fold product $(\nu_X)^n$ to $\tilde{B}_{\sigma n}$, where ν_X is a Grothendieck measure of X, is a Grothendieck measure for $X^{(n)}$.

We define the integral (cf. (10.18); Exercise 40)

$$\begin{aligned}
&\int_{[0,1]} X \, \mathrm{d}X \\
&= \int_{[0,1]} \left(\int_{[0,1]} \mathbf{1}_{[0,t)}(s) \, \mathrm{d}X(\mathrm{d}s) \right) \mathrm{d}X(\mathrm{d}t) - X(0)^2 + X(0)X(1) \\
&:= \frac{\mathrm{d}}{\mathrm{d}\mathbb{P}} \left(\int_{\tilde{D}_2} \mathbf{1}_{[0,t)}(s) \, \mu_{X^{(2)}}(\cdot, \mathrm{d}(s,t)) \right) - X(0)^2 + X(0)X(1).
\end{aligned} \tag{10.26}$$

11 Random Integrands in One Parameter

In this section we touch on the large issue of *non-deterministic* integrands, which, broadly put, is the question: what processes are 'canonically' integrable with respect to an integrator X?

The instance $X = \mathrm{W}$ was considered first by K. Itô, who, motivated primarily by questions about diffusions, integrated (deterministic) functions of W with respect to dW. (E.g., see Remark ii in the previous section.) In particular, underscoring a fundamental distinction between the usual calculus and the stochastic calculus, he obtained

$$\int_0^1 g'(\mathrm{W}) \, \mathrm{dW} = g(\mathrm{W}(1)) - g(\mathrm{W}(0)) - \frac{1}{2} \int_0^1 g''(\mathrm{W}(t)) \, \mathrm{d}t$$

$$(\textit{the Itô formula}), \tag{11.1}$$

where g is real-valued, twice-differentiable with a continuous second derivative. (See Chapter X §8 and Exercise X.27).

Itô's integral led to a construct that has become the focus of *adaptive* stochastic integration: an integral $\int_0^1 Y \, \mathrm{d}M$, where M is an L^2-bounded semi-martingale, and Y is an M-*adapted* process. (That Y is M-*adapted* means that $Y(t)$ is $M(t)$-measurable for every $t \in [0,1]$; or equivalently, that for every $t \in [0,1]$ there exists a scalar-valued Borel-measurable function φ_t such that $Y(t) = \varphi_t(M(t))$ (cf. [Wil, p. 36]).) I will not dwell here on the intricacies and uses of adapted stochastic integration, and refer the reader to any one of several books (e.g., [KarSh], [M], [Pr]), where the subject is expansively developed. (An abridged treatment can be found in [ChWil].) In this section, moving away from adaptability, I will describe two 'non-adapted' stochastic integrals, where 'functional dependence' of integrands on integrators is not an *a priori* assumption. Both integrals are based on a measure-theoretic approach to stochastic integration.

Via Riemann Sums

Let X and Y be integrators. The issue whether Y can be integrated with respect to X can be viewed as the question: is the product process

$$X \otimes Y = \{X(s)Y(t) : (s,t) \in [0,1]^2\} \tag{11.2}$$

an integrator? For, if the answer is affirmative, then (cf. (10.26))

$$\begin{aligned} \int_{[0,1]} Y \mathrm{d}X &= \int_{[0,1]} \left(\int_{[0,1]} \mathbf{1}_{[0,s)}(t) \; \mathrm{d}Y(\mathrm{d}s) + Y(0) \right) \mathrm{d}X(\mathrm{d}t) \\ &:= \int_{[0,1]^2} \mathbf{1}_{[0,s)}(t) \mathrm{d}(X \otimes Y) + X(0)Y(0) - X(1)Y(0) \\ &= \frac{\mathrm{d}}{\mathrm{d}\mathbb{P}} \int_{[0,1]^2} \mathbf{1}_{[0,s)}(t) \; \mu_{X\otimes Y}(\cdot, \mathrm{d}(s,t)) + X(0)Y(0) - X(1)Y(0). \end{aligned} \tag{11.3}$$

The first equality in (11.3) is a formal statement based on the realization of Y as a 'sum' of its increments. The second equality is a definition motivated by the following. At the outset, we want $\int_{[0,1]} Y \mathrm{d}X$ to be a limit (say, weakly in $\mathrm{L}^1(\Omega, \mathbb{P})$) of Riemann sums

$$\sum_{j=1}^{n} Y(t_{j-1})(X(t_j) - X(t_{j-1})).$$

We assume (without loss of generality) that $X(0) = Y(0) = 0$, and then write these sums as (cf. (10.16), (10.22))

$$\begin{aligned}
&\sum_{j=1}^{n} Y(t_{j-1})\,(X(t_j) - X(t_{j-1})) \\
&\quad = \int_{[0,1]^2} \left(\sum_{j=1}^{n} \mathbf{1}_{[0,t_{j-1}]} \otimes \mathbf{1}_{[t_{j-1},t_j]} \right) \mathrm{d}(X \otimes Y) \\
&\quad = X(1)Y(1) - \sum_{j=1}^{n} X(t_j)\,(Y(t_j) - Y(t_{j-1})) \\
&\quad = X(1)Y(1) - \int_{[0,1]^2} \left(\sum_{j=1}^{n} \mathbf{1}_{[t_{j-1},t_j]} \otimes \mathbf{1}_{[0,t_j]} \right) \mathrm{d}(X \otimes Y).
\end{aligned} \tag{11.4}$$

(Sum by parts, or draw a picture.) As the mesh of partitions approaches 0, the integrals on the first line of (11.4) converge (weakly in $\mathrm{L}^1(\Omega, \mathbb{P})$) to $\int_{[0,1]^2} \mathbf{1}_{[0,t)}(s) \mathrm{d}(X \otimes Y)$, and integrals on the third line of (11.4) converge to $\int_{[0,1]^2} \mathbf{1}_{[0,t]}(s) \mathrm{d}(X \otimes Y)$ (Exercise 41). In particular, we obtain an 'integration by parts' formula

$$\int_{[0,1]} Y \,\mathrm{d}X = X(1)Y(1) - \Delta^2(X \otimes Y)(D) - \int_{[0,1]} X \,\mathrm{d}Y, \tag{11.5}$$

where

$$\begin{aligned}
\Delta^2(X \otimes Y)(D) &:= \int_{[0,1]^2} \mathbf{1}_D \,\mathrm{d}(X \otimes Y), \\
D &= \{(s,s) : s \in [0,1]\}.
\end{aligned} \tag{11.6}$$

We have already obtained, according to this definition, the integral $\int_{[0,1]} Y \,\mathrm{d}X$, where X and Y are mutually independent. This, in effect, is a counterpoint to the adapted Itô-type integral. Indeed, by Theorem 38, we obtain (under the assumption $X(0) = Y(0) = 0$)

$$\begin{aligned}
\int_{[0,1]} Y(t) \,\mathrm{d}X(\mathrm{d}t) &= \int_{[0,1]} \left(\int_{[0,1]} \mathbf{1}_{[0,t)}(s) \,\mathrm{d}Y(\mathrm{d}s) \right) \mathrm{d}X(\mathrm{d}t) \\
&= \int_{[0,1]^2} \mathbf{1}_{[0,t)}(s) \,\mathrm{d}Y(\mathrm{d}s)\,\mathrm{d}X(\mathrm{d}t) \\
&:= \frac{\mathrm{d}}{\mathrm{d}\mathbb{P}} \left(\int_{[0,1]^2} \mathbf{1}_{[0,t)}(s) \;\mu_X{\times}\mu_Y(\cdot, \mathrm{d}(s,t)) \right)
\end{aligned}$$

$$= \int_{[0,1]} \left(\int_{[0,1]} \mathbf{1}_{[0,t)}(s) \mathrm{d}X(\mathrm{d}t) \right) \mathrm{d}Y(\mathrm{d}s)$$

$$= \int_{[0,1]} \left(\int_{[0,1]} (1 - \mathbf{1}_{[0,s]}(t) \mathrm{d}X(\mathrm{d}t) \right) \mathrm{d}Y(\mathrm{d}s)$$

$$= X(1)Y(1) - \Delta^2(XY)(D) - \int_{[0,1]} X(s)\, \mathrm{d}Y(\mathrm{d}s). \qquad (11.7)$$

(Notice the intervention here of the Grothendieck factorization theorem and inequality.)

Via Stochastic Series

Suppose a 1-process X is an integrator with Grothendieck measure ν_X. If $Y = Z \otimes f$, where $Z \in \mathrm{L}^\infty(\Omega, \mathbb{P})$ and $f \in \mathrm{L}^2([0,1], \nu_X)$, then for $A \in \mathscr{A}$,

$$\int_{\Omega \times [0,1]} \mathbf{1}_A(\omega)\, Z(\omega)\, f(t)\, \mu_X(\mathrm{d}\omega, \mathrm{d}t)$$

$$= \int_{\Omega} \mathbf{1}_A(\omega)\, Z(\omega)\, \mathrm{d} \left(\int_{[0,1]} f(t)\, \mu_X(\cdot, \mathrm{d}t) \right)$$

$$= \int_{\Omega} \mathbf{1}_A(\omega)\, Z(\omega) \left(\int_{[0,1]} f\, \mathrm{d}X \right) \mathrm{d}\mathbb{P} := \eta(A). \qquad (11.8)$$

Then, η is a measure that is absolutely continuous with respect to $\mathbb{P}$. We define

$$\int_{[0,1]} Z \otimes f\, \mathrm{d}X := \frac{\mathrm{d}\eta}{\mathrm{d}\mathbb{P}} = Z \int_{[0,1]} f\, \mathrm{d}X. \qquad (11.9)$$

If X is L^2-bounded with orthogonal increments, then we can take $Z \in \mathrm{L}^2(\Omega, \mathbb{P})$; in this case, $\nu_X = \lambda_X$. (See §3, and Remark i in §6.) If $Y = \sum_{j=1}^n Z_j \otimes f_j$, where (Z_j) is a sequence in $\mathrm{L}^\infty(\Omega, \mathbb{P})$ and (f_j) is a sequence in $\mathrm{L}^2([0,1], \nu_X)$, then

$$\int_{[0,1]} Y\, \mathrm{d}X = \sum_{j=1}^{n} Z_j \int_{[0,1]} f_j\, \mathrm{d}X. \qquad (11.10)$$

If Y can be represented by $\sum_{j=1}^\infty Z_j \otimes f_j$, and $\sum_{j=1}^\infty Z_j \int_{[0,1]} f_j\, \mathrm{d}X$ converges in some sense (say, weakly in $\mathrm{L}^1(\Omega, \mathbb{P})$), then we can define the latter to be the integral of Y with respect to X. Of course, to make this precise, we need to specify how Y is represented as a 'sum of elementary tensors', each integrable with respect to X. For example, if Y is represented by

$$Y = \sum_{j=1}^{\infty} Z_j \otimes f_j \quad \text{pointwise a.e. } (\mathbb{P} \times \nu_X), \tag{11.11}$$

where

$$\sum_{j=1}^{\infty} \|Z_j\|_{\mathrm{L}^\infty} \|f_j\|_{\mathrm{L}^2(\nu_X)} < \infty,$$

(that is, $Y \in \mathrm{L}^\infty(\Omega, \mathbb{P}) \hat{\otimes} \mathrm{L}^2([0,1], \nu_X)$; e.g., Chapter IV §6, §7), then

$$\int_{[0,1]} Y \, \mathrm{d}X = \sum_{j=1}^{\infty} Z_j \int_{[0,1]} f_j \, \mathrm{d}X, \tag{11.12}$$

where the series on the right side of (11.12) converges in the $\mathrm{L}^1(\Omega, \mathbb{P})$-norm. Moreover, $\int_{[0,1]} Y \mathrm{d}X$ in (11.12) does not depend on the representation of Y by (11.11) (Exercise 42).

We have just defined integrals where *integrands* are represented by series, each integrable with respect to a given integrator. We can turn this definition around, and consider integrals obtained from series representations of *integrators*. Recall that if $Z \in \mathrm{L}^1(\Omega, \mathbb{P})$ and φ is a function of bounded variation on [0,1], then $Z \otimes \varphi$ is an integrator, and $\mu_{Z \otimes \varphi} = Z \mathrm{d}\mathbb{P} \times \mathrm{d}\varphi$ (Remark ii in §2). In this case, the processes that are naturally integrated with respect to $Z \otimes \varphi$ are 'functions' Y on $\Omega \times [0,1]$, such that $Y(\omega, \cdot)$ is in $\mathrm{L}^1([0,1], \mathrm{d}\varphi)$ for almost all ω ($\mathbb{P}$), and $\int_{[0,1]} Y(\cdot, t) \, \mathrm{d}\varphi(\mathrm{d}t)$ is in $\mathrm{L}^1(\Omega, |Z| \, \mathrm{d}\mathbb{P})$. For then,

$$\int_\Omega \left(\int_{[0,1]} Y(\omega, t) \, \mathrm{d}\varphi(\mathrm{d}t) \right) \mathbf{1}_A \, Z(\omega) \, \mathbb{P}(\mathrm{d}\omega) := \eta(A), \quad A \in \mathscr{A}, \tag{11.13}$$

determines a measure on $(\Omega, \mathscr{A})$, $\eta \ll \mathbb{P}$, and

$$\int_{[0,1]} Y \, \mathrm{d}(Z \otimes \varphi) := \frac{\mathrm{d}\eta}{\mathrm{d}\mathbb{P}} = Z \left(\int_{[0,1]} Y(\cdot, t) \, \mathrm{d}\varphi(\mathrm{d}t) \right). \tag{11.14}$$

In similar fashion, we can take integrators $\sum_{j=1}^n Z_j \otimes \varphi_j$, where the Z_j are in $\mathrm{L}^1(\Omega, \mathbb{P})$, and the φ_j are functions of bounded variation on [0,1].

In this case, if Y is a process such that $Y(\omega,\cdot) \in \mathrm{L}^1([0,1], \mathrm{d}\varphi_j)$ for almost all ω ($\mathbb{P}$), and $\int_{[0,1]} Y(\cdot,t)\, \mathrm{d}\varphi_j(\mathrm{d}t) \in \mathrm{L}^1(\Omega, |Z_j|\, \mathrm{d}\mathbb{P})$, $j = 1,\ldots,n$, then

$$\int_{[0,1]} Y\, \mathrm{d}(\Sigma_{j=1}^n Z_j \otimes \varphi_j) = \sum_{j=1}^n Z_j \int_{[0,1]} Y(\cdot,t)\, \mathrm{d}\varphi_j(\mathrm{d}t) \tag{11.15}$$

If $n = \infty$ above (i.e., if an integrator X can be represented by an infinite series $\sum_{j=1}^\infty Z_j \otimes \varphi_j$), and if the proposed integrand Y is canonically integrable with respect to each of the summands $Z_j \otimes \varphi_j$, *and* if the right side of (11.15) converges in some sense to an element in $\mathrm{L}^1(\Omega, \mathbb{P})$, then the latter can be viewed as an integral of Y with respect to X.

Because every integrator X can be represented by stochastic series (Remark ii in §6), we can make matters more concrete. Suppose ν_X is a Grothendieck measure of X, and $\{\mathbf{e}_j : j \in \mathbb{N}\}$ is an orthonormal basis for $\mathrm{L}^2([0,1], \nu_X)$. Then consider

$$X_j = \int_{[0,1]} \mathbf{e}_j\, \mathrm{d}X \ \ (:= Z_j, \text{ above}), \tag{11.16}$$

and

$$\varphi_j(t) := \hat{\mathbf{1}}_{[0,t]}(j) = \int_{[0,1]} \mathbf{1}_{[0,t]}(\mathrm{d}s)\, \mathbf{e}_j(s)\, \nu_X(\mathrm{d}s), \quad j \in \mathbb{N}, \tag{11.17}$$

and observe (cf. (6.15))

$$\mu_X(A,B) = \sum_{j=1}^\infty \mu_{X_j \otimes \varphi_j}(A,B), \quad A \in \mathscr{A},\ B \in \mathscr{B}. \tag{11.18}$$

If $Y = \{Y(t) : t \in [0,1]\}$ is canonically integrable with respect to each $\mu_{X_j \otimes \varphi_j}$, then the issue of integrability of Y with respect to X becomes the question: does the limit

$$\sum_{j=1}^\infty \int_{[0,1]} Y\, \mathrm{d}(X_j \otimes \varphi_j) = \sum_{j=1}^\infty X_j \int_{[0,1]} Y(\cdot,t)\, \mathrm{d}\varphi_j(\mathrm{d}t) \tag{11.19}$$

exist, say weakly in $\mathrm{L}^1(\Omega, \mathbb{P})$?

We can merge these two approaches: the first based on series representations of integrands, and the second based on series representations of integrators. First suppose X and Y are integrators with the same Grothendieck measure ν, and then write the partial sums of their respective stochastic series relative to a given orthonormal basis of $L^2([0,1],\nu)$,

$$S_n(X)(t) = \sum_{j=1}^{n} \varphi_j(t) X_j, \quad S_n(Y)(t) = \sum_{j=1}^{n} \varphi_j(t)\, Y_j. \tag{11.20}$$

Then,

$$\int_{[0,1]} S_n(Y)\, \mathrm{d}X = \sum_{j=1}^{n} Y_j \int_{[0,1]} \varphi_j\, \mathrm{d}X, \quad \text{by (11.10)}, \tag{11.21}$$

and

$$\int_{[0,1]} Y\, \mathrm{d}(S_n(X)) = \sum_{j=1}^{n} X_j \int_{[0,1]} Y(\cdot,t)\, \mathrm{d}\varphi_j(\mathrm{d}t), \quad \text{by (11.15)}. \tag{11.22}$$

By (6.15),

$$\int_{[0,1]} \varphi_j\, \mathrm{d}X = \sum_{k=1}^{\infty} \hat{\varphi}_j(k)\, X_k, \tag{11.23}$$

and

$$\int_{[0,1]} Y(\cdot,t)\, \mathrm{d}\varphi_j(\mathrm{d}t) = \sum_{k=1}^{\infty} \hat{\varphi}_k(j)\, Y_k, \tag{11.24}$$

where the respective series on the right sides converge weakly in $L^1(\Omega,\mathbb{P})$, and for $(j,k) \in \mathbb{N}^2$

$$\hat{\varphi}_j(k) = \int_{[0,1]^2} \mathbf{1}_{[0,t)}(s)\, \mathbf{e}_j(s)\, \mathbf{e}_k(t)\, \nu(\mathrm{d}s)\, \nu(\mathrm{d}t). \tag{11.25}$$

Then,

$$\int_{[0,1]} S_n(Y)\, \mathrm{d}X = \sum_{j=1}^{n} \sum_{k=1}^{\infty} \hat{\varphi}_j(k)\, Y_j\, X_k, \tag{11.26}$$

and

$$\int_{[0,1]} Y\, \mathrm{d}(S_n(X)) = \sum_{j=1}^{n} \sum_{k=1}^{\infty} \hat{\varphi}_k(j)\, Y_k\, X_j. \tag{11.27}$$

If either of the series in (11.26) and (11.27) converges weakly in $L^1(\Omega, \mathbb{P})$, then both series converge to the same limit, which we consider to be $\int_{[0,1]} Y \, dX$.

Remarks:

i **(a comparison of the two integrals).** The 'Riemann sums' approach yields

$$\int_{[0,1]} W \, dW = W(1)^2/2 - \frac{1}{2}, \tag{11.28}$$

which is the Itô integral; see §10.

Taking the 'stochastic series' approach, consider a series of W associated with a given orthonormal basis $\{\mathbf{e}_j\colon j \in \mathbb{N}\}$ of $L^2([0,1], \mathfrak{m})$ (cf. (X.3.33)),

$$W(t) \sim \sum_{j=1}^{\infty} \varphi_j(t) \, \zeta_j, \tag{11.29}$$

where

$$\zeta_j = \int_{[0,1]} \mathbf{e}_j \, dW, \quad j \in \mathbb{N}, \tag{11.30}$$

and

$$\varphi_j(t) = \int_{[0,1]} \mathbf{1}_{[0,t)}(s) \, \mathfrak{m}(ds), \quad t \in [0,1]. \tag{11.31}$$

(The Lebesgue measure $\mathfrak{m}$ is a Grothendieck measure for W.) Then, by (11.26),

$$\int_{[0,1]} S_n(W) \, dW = \sum_{j=1}^{n} \sum_{k=1}^{\infty} \hat{\varphi}_j(k) \, \zeta_j \, \zeta_k. \tag{11.32}$$

($\hat{\varphi}_j(k)$ is defined in (11.25).) We have $\hat{\varphi}_j(j) = 0$ for all $j \in \mathbb{N}$,

$$\sum_{j=1}^{\infty} \sum_{k=1}^{\infty} |\hat{\varphi}_j(k)|^2 < \infty,$$

and therefore, $\lim_{n\to\infty} \int_{[0,1]} S_n(W) dW$ exists in $L^2(\Omega, \mathbb{P})$. Applying 'integration by parts', we obtain

$$\int_{[0,1]} S_n(W) \, dW = W(1)^2 - \int_{[0,1]} W \, d(S_n(W)), \tag{11.33}$$

and (because $\lim_{n\to\infty}\int_{[0,1]} S_n(\mathrm{W})\,\mathrm{dW} = \lim_{n\to\infty}\int_{[0,1]} \mathrm{W}\,\mathrm{d}(S_n(\mathrm{W}))$ in $\mathrm{L}^2(\Omega,\mathbb{P})$),

$$\lim_{n\to\infty}\int_{[0,1]} S_n(\mathrm{W})\,\mathrm{dW} = \mathrm{W}(1)^2/2, \tag{11.34}$$

which, in this case, is a *Stratonovich integral* (cf. (10.23)).

ii **(interchange of limits?).** Observe that $\int_{[0,1]} S_n(\mathrm{W})\mathrm{dW}$ obtained via (11.21) is the same as $\int_{[0,1]} S_n(\mathrm{W})\mathrm{dW}$ obtained via 'Riemann sums' in (11.3). To see this, it suffices to verify (Exercise 43) that the 2-process $\zeta_j\ \varphi_j \otimes \mathrm{W}$ is an integrator (φ_j and ζ_j are defined in (11.30) and (11.31)), and

$$\begin{aligned}\int_{[0,1]} \varphi_j\ \zeta_j\ \mathrm{dW} &:= \int_{[0,1]^2} \mathbf{1}_{[0,s)}(t)\ \mathrm{d}(\zeta_j\ \varphi_j \otimes \mathrm{W}) \\ &= \frac{\mathrm{d}}{\mathrm{d}\mathbb{P}}\int_{[0,1]^2} \mathbf{1}_{[0,s)}(t)\ \mu_{\zeta_j\varphi_j\otimes \mathrm{W}}(\cdot,\mathrm{d}(s,t)) \\ &= \zeta_j \int_{[0,1]} \varphi_j\ \mathrm{dW}, \quad j \in \mathbb{N}. \end{aligned}\tag{11.35}$$

Specifically, this implies that limit operations and integrals cannot always be interchanged. For,

$$\int_{[0,1]} \mathrm{W}\ \mathrm{dW} = \int_{[0,1]} \lim_{n\to\infty} S_n(\mathrm{W})\mathrm{dW} = \mathrm{W}(1)^2/2 - \frac{1}{2}, \tag{11.36}$$

but

$$\lim_{n\to\infty}\int_{[0,1]} S_n(\mathrm{W})\mathrm{dW} = \mathrm{W}(1)^2/2,$$

where limits are in $\mathrm{L}^2(\Omega,\mathbb{P})$. This (again...!) is the effect of the diagonal: the quadratic variation of $S_n(\mathrm{W})$ over the diagonal $D = \{(s,s) : s \in [0,1]\}$ is zero, whereas the quadratic variation of W over D is positive. Put another way, Lebesgue measure $\mathfrak{m}^2$ is a Grothendieck measure of $S_n(\mathrm{W}) \otimes \mathrm{W}$ and $\mathfrak{m}^2(D) = 0$, but for every Grothendieck measure ν of $\mathrm{W} \otimes \mathrm{W}$, $\nu(D) > 0$ (Proposition 39).

iii **(some other 'named' non-adapted integrals).** The first construction of a non-adapted integral via series is due to Skorohod [Sk]. In the *Skorohod integral* [Sk], the integrator is a Wiener process, and integrands are in $\mathrm{L}^2([0,1]\times\Omega, \mathfrak{m}\times\mathbb{P})$. The idea is to represent the integrand $Y \in \mathrm{L}^2([0,1] \times \Omega, \mathfrak{m} \times \mathbb{P})$ by the Wiener Chaos-series (Remark in Chapter X §8),

$$Y(t) = \sum_{j-0}^{\infty} Y_j(t), \tag{11.37}$$

where $Y_j(t) \in H_j = \{I_{\mathrm{W}_j}(f) : f \in \mathrm{L}^2([0,1]^j, \mathfrak{m}^j\}$ for every $t \in [0,1]$, and then define

$$\int_0^1 Y \,\mathrm{dW} := \sum_{j=0}^{\infty} \int_0^1 Y_j(t) \,\mathrm{dW}(\mathrm{d}t), \tag{11.38}$$

where convergence is in $\mathrm{L}^2(\Omega, \mathbb{P})$. A crucial step in this definition is to consider the jth summand on the right side of (11.38) as a multiple Itô integral, iterated $j+1$ times. For example, notice that Skorohod's and Itô's $\int_0^1 \mathrm{W}\mathrm{dW}$ are the same (simply because $\mathrm{W}(t) \in H_1,\ t \in [0,1]$).

A second definition of a non-adapted integral based on series was given by Ogawa [O1], [O2], who, so far that I can determine, was unaware of Skorohod's (prior) integral. In Ogawa's construct, the integrator is also W, and the integrand is a process $Y = \{Y(t) : t \in [0,1]\}$ such that $\int_{[0,1]} |Y(t)|^2 \,\mathfrak{m}(\mathrm{d}t) < \infty$ a.s. ($\mathbb{P}$). An orthonormal basis $E = \{\mathbf{e}_j : j \in \mathbb{N}\}$ of $\mathrm{L}^2([0,1], \mathfrak{m})$ is chosen, and Y is expanded

$$Y = \sum_{j=1}^{\infty} \mathbf{e}_j \, \hat{Y}_E(\mathbf{e}_j), \tag{11.39}$$

where $\hat{Y}_E(\mathbf{e}_j) = \int_{[0,1]} Y(t) \,\mathbf{e}_j(t) \,\mathfrak{m}(\mathrm{d}t)$. If the series

$$\sum_{j=1}^{\infty} \hat{Y}_E(\mathbf{e}_j) \int_{[0,1]} \mathbf{e}_j(t) \,\mathrm{dW} \tag{11.40}$$

converges in probability, then the limit is the *Ogawa integral* relative to E. For example, in the simple case $Y = \mathrm{W}$, the Ogawa integral of W with respect to dW relative to *every* orthonormal basis E is the same as the integral obtained by the series in (11.34). In particular, Skorohod's and Ogawa's integrals need not be equal. Nualart and Zakai called a process Y *Ogawa integrable* if Y is integrable in Ogawa's original sense (above) relative to *every* orthonormal basis of $L^2([0,1], \mathfrak{m})$, and dubbed (11.40) in this case an *intrinsic Ogawa integral* [NuZ2].

Relationships between integrals of Itô, Skorohod, Ogawa, and Stratonovich were investigated in [NuZ1], [NuZ2]. Recent expositions of these and related matters can be found in [Nu] and [M].

Exercises

1. Verify Lemma 2.
2. Complete the proof of Lemma 3 by showing $\|\mu_X\|_{F_2(\mathscr{A},\mathscr{O})} = \|X\|$.
3. Verify Corollary 5.
4. i. Prove Proposition 8.

   ii. Show that if f and g are in $L^\infty([0,1])$, then

$$\int_{[0,1]} g \, \mathrm{d}\left(\int f \, \mathrm{d}X\right) = \int_{[0,1]} g \, f \, \mathrm{d}X. \tag{E.1}$$

   iii. (Do this after reading §6.) Let ν be a Grothendieck measure of an integrator X. Show that if $f \in L^2([0,1], \nu)$, then the indefinite integral

$$\left(\int f \, \mathrm{d}X\right)(t) := \int_{[0,1]} \mathbf{1}_{[0,t]} f \, \mathrm{d}X := \int_{[0,t]} f \, \mathrm{d}X, \quad t \in [0,1]$$

   is an integrator, that $\| \int f \, \mathrm{d}X\| \leq \kappa_G \, \|f\|_{L^2} \|X\|$, and that if $g \in L^\infty([0,1])$, then (E.1) holds. What is a Grothendieck measure of $\int f \, \mathrm{d}X$?

5. Prove that if X is an L^2-bounded process with orthogonal increments, and $F_X(t) = \mathbf{E}|X(t)|^2$, $t \in [0,1]$, then F_X is monotonically increasing on $[0,1]$.
6. In this exercise you will supply the details missing from the proof of Proposition 10.

   i. Prove that the linear action defined in (3.7),

$$g \mapsto \int_{[0,1]} g(t) \, \beta_A(\mathrm{d}t), \quad g \in C([0,1]),$$

is uniquely extendible to a bounded linear functional $\tilde{\beta}_A$ on $L^2([0,1], \lambda_X)$.

ii. Verify that $\tilde{\beta}_A(\mathbf{1}_B)$, $B \in \mathscr{B}$, is the measure on $\mathscr{B}$ that equals β_A.

7. Prove that (3.10) does not depend on the representation of the step function in it.
8. Prove Proposition 11.
9. Prove Proposition 12.
10. Prove Proposition 13 i.
11. Let X be a process defined by (3.22).

    i. Verify that $\|\mu_X\|_{(1)} = \sum_{k=1}^{\infty} \|X_k\|_{L^1}$.

    ii. Prove that there exist mutually orthogonal elements X_k, $k \in \mathbb{N}$, such that $\|X_k\|_{L^2} = \|X_k\|_{L^1} = 1/k$ ($k \in \mathbb{N}$), and $H(X)$ is not a $\Lambda(2)$-space.

    iii. Prove that there exists an integrator X such that $H(X)$ is an infinite-dimensional $\Lambda(2)$-space, but X is not *random* (according to Definition 9).

12. Supply the missing details in the proof of Proposition 14.
13. i. Supply the missing details in the proof of Proposition 15.

    ii. Prove that if $X = \{X(t) : t \in [0,1]\}$ is an L^1-bounded martingale process, then $\mathbf{E}|X(t)|$, $t \in [0,1]$, is a monotone function.

    iii.* Is every L^1-bounded martingale process an integrator?

14. Verify that a homogeneous process X is α-chaos if and only if (4.4) holds.
15. Prove Lemma 16, Corollaries 17, 18, and 19.
16. Prove Lemma 21, Corollaries 22, 23, and 24.
17. i. Let $\{X_j : j \in \mathbb{N}\}$ be an exact α-system (asymptotic α-system), and let U be a unitary map from $L^2([0,1], \mathfrak{m})$ onto $L^2_{\{X_j\}}(\Omega, \mathbb{P})$. Prove that $X = \{U\mathbf{1}_{[0,t]} : t \in [0,1]\}$ is an α-chaos (resp., asymptotic α-chaos). Conclude the existence of exact n-chaos.

    ii. Prove that the n-process $\tilde{W}_n$ defined in (X.14.11) is an exact n-chaos.

18. Let $\{X_j : j \in \mathbb{N}\}$ be an exact $\Lambda(q)^{\#}$-system (asymptotic $\Lambda(q)^{\#}$-system), and let U be a unitary map from $L^2([0,1], \mathfrak{m})$ onto

$$L^2_{\{X_j\}}(\Omega, \mathbb{P}).$$

Prove that $\{U\mathbf{1}_{[0,t]}: t\in[0,1]\}$ is an exact $\Lambda(q)^{\#}$-process (resp., asymptotic $\Lambda(q)^{\#}$-process). Conclude that for all $q > 2$ there exist exact and asymptotic $\Lambda(q)^{\#}$-processes.

19.* Can the Kakutani construction of a Wiener process (Chapter X §2; cf. Exercise 17 i) be modified to produce a p-stable motion for $p \in (1,2)$?

20. Let X be a p-stable motion, $p \in (1,2]$.

   i. Verify that the linear map I_X defined in (4.40) is extendible to $\mathrm{L}^p([0,1],\mathfrak{m})$ so that (4.41) holds. Conclude that for every $r \in [1,p)$, there exists $c > 0$ such that

   $$\|I_X(f)\|_{\mathrm{L}^r(\mathbb{P})} = c\|f\|_{\mathrm{L}^p(\mathfrak{m})}, \quad f \in \mathrm{L}^p([0,1],\mathfrak{m}).$$

   ii. Provide the missing details in the proof (of Proposition 28) that $I_X(f) = \int_{[0,1]} f \, \mathrm{d}X$ for all $f \in \mathrm{L}^\infty([0,1])$.

21. Provide the details in the proof of Proposition 29.

22. By using (4.53), prove 'quickly' that if X is a p-stable motion for $p \in (1,2)$, then $\ell_X = 1$.

23. In the spirit of the heuristic discussion in Chapter X §1, interpret a p-stable motion as a stochastic model indexed by a spatial parameter for a 'force field' associated with an 'inverse attraction' law.

24. Suppose $\{X_j : j \in \mathbb{N}\}$ is a system of symmetric p-stable independent variables. Prove (5.17) for $n = 2$ by use of the two-dimensional polarization identity,

$$\sum_{1\le i<j\le N} a_{ij}\, X_i\tilde{X}_j + \sum_{1\le i<j\le N} a_{ij}\, \tilde{X}_i X_j$$
$$= \sum_{1\le i<j\le N} a_{ij}(X_i + \tilde{X}_i)(X_j + \tilde{X}_j)$$
$$- \sum_{1\le i<j\le N} a_{ij}\, X_i X_j - \sum_{1\le i<j\le N} a_{ij}\, \tilde{X}_i\tilde{X}_j,$$

where $\{\tilde{X}_j\}$ is an independent copy of $\{X_j\}$.

25. Let η be a symmetric n-linear functional on $\mathbb{R}^n$. Define

$$\|\eta\| = \sup\,\{\eta(x_1,\ldots,x_n) : (x_1,\ldots,x_n) \in B_N \times \cdots \times B_N\},$$

and

$$\|\eta\|_* = \sup\{\eta(x,\ldots,x) : x \in B_N\},$$

where B_N denotes the l^∞-unit ball in $\mathbb{R}^N$. Prove that

$$\|\eta\|_* \le \|\eta\| \le \frac{n^n}{n!}\|\eta\|_*.$$

26. Let β denote the bilinear functional on $L^\infty(\Omega,\mathbb{P}) \times L^2([0,1],\nu)$ defined by (6.10) in the proof of Theorem 34. Prove that for $g \in L^2([0,1],\nu)$,

$$\beta(\mathbf{1}_A, g), \quad A \in \mathscr{A},$$

is a measure on $\mathscr{A}$ that is absolutely continuous with respect to $\mathbb{P}$.

27. Prove that the definition in (6.13) is consistent with the definition in (2.11).

28.* Produce explicit Grothendieck measures for L^1-bounded additive processes and L^p-bounded martingales, $p \in (1,2)$.

29. Let X be an integrator, ν a Grothendieck measure of X, $\{\mathbf{e}_n : n \in \mathbb{N}\}$ an orthonormal basis of $L^2([0,1],\nu)$, and $X_n = \int_{[0,1]} \mathbf{e}_n \, dX$ $(n \in \mathbb{N})$. Prove that if $f \in L^2([0,1],\nu)$ and $f = \sum_n \hat{f}(n)\mathbf{e}_n$, then

$$\int_{[0,1]} f \, dX = \sum_{n=1}^{\infty} \hat{f}(n)\, X_n,$$

where the series converges weakly in $L^1(\Omega,\mathbb{P})$.

30. i. Show that if X is a p-stable motion, then ΔX is an L^r-valued stochastic measure for every $r < p$, and that if X is an L^2-bounded process with orthogonal increments, then ΔX is an L^2-valued stochastic measure. In the general case, verify that if X is an integrator, then 'Δ'X defined by (2.23) is an L^1-valued stochastic measure.

    ii. Prove that if X is an L^1-bounded additive process, $\mathbf{E}X(t) = 0$ for all $t \in [0,1]$, and $B_1, \ldots, B_n$ are pairwise disjoint Borel sets in [0,1], then 'Δ'$X(B_1), \ldots,$'Δ'$X(B_n)$ are independent.

31. Verify that the definitions of $\|X\|_U$ given in (7.1) and (7.3) are the same.

32. Verify that the iterated right-limits in (7.6) are feasible.

33. Prove Proposition 36.

34.* If an n-process X is an U-integrator, where $U = \{S_1, \ldots, S_m\}$ is a partition of $[n]$, then for all $A \in \mathscr{A}$,

$$\mathbf{E}\mathbf{1}_A \int_{[0,1]^n} \mathbf{1}_{B_1} \otimes \cdots \otimes \mathbf{1}_{B_m} \, dX, \quad B_1 \in \mathscr{B}_{S_1}, \ldots, B_m \in \mathscr{B}_{S_m},$$

is an F_m-measure on $\mathscr{B}_{S_1} \times \cdots \times \mathscr{B}_{S_m}$. In the case $n = 1$, the Grothendieck factorization theorem implies a stronger statement, but what can be said in the case $n > 1$? (See Remark iii in §6.)

35. Verify that if an n-process X is not an integrator, then there is a grid of $[0,1]^n$

$$\begin{aligned}\mathscr{G} &= \{B_{i_1\dots i_n} : (i_1,\dots,i_n) \in \mathbb{N}^n\} \\ &= \{J^{(1)}_{i_1} \times \cdots \times J^{(n)}_{i_n} : (i_1,\dots,i_n) \in \mathbb{N}^n\},\end{aligned}$$

and a one–one map τ from $\mathbb{N}$ onto $\mathbb{N}^n$ such that $\sum_{j=1}^{\infty} \Delta^n X(B_{\tau j})$ does not converge weakly in $\mathrm{L}^1(\Omega,\mathbb{P})$ to $\Delta^n X([0,1)^n)$.

36. Verify that the definition of the Littlewood index in (7.31) is consistent with the definition in (7.15).

37. Verify that there exist 'n-dimensional' 1-processes X with $\ell_X = 2(n+1)/(n+2)$.

38. Prove Theorem 37.

39. Let $\mathrm{W}^{(n)}$ be the n-process defined by

$$\mathrm{W}^{(n)}(t_1,\dots,t_n) = \mathrm{W}(t_1)\cdots\mathrm{W}(t_n), \quad (t_1,\dots,t_n) \in [0,1]^n.$$

   i. Prove that $\mathrm{W}^{(n)}$ is an integrator.

   ii.* State explicit Grothendieck measures for $\mathrm{W}^{(n)}$.

40. Let X be an L^1-bounded additive process, and $\mathbf{E}X(t) = 0$ for all $t \in [0,1]$.

   i. Prove that $\mu_{X^{(n)}}$ defined by (10.25) gives rise to an element in $F_2(\mathscr{A}, \tilde{B}_{n\sigma})$.

   ii. Prove that if ν_X is a Grothendieck measure for X, then the restriction of the n-fold product $(\nu_X)^n$ to $\tilde{B}_{\sigma n}$ is a Grothendieck measure for $X^{(n)}$.

   iii. Verify that (10.26) can be viewed as an assertion about limits of Riemann sums.

41. Verify (11.4), and deduce the 'integration by parts' formula in (11.5).

42. Prove that the definition of $\int_{[0,1]} Y \mathrm{d}X$ by (11.12) does not depend on the representation by (11.11) of the integrand Y.

43. Prove that $(\zeta_j\, \varphi_j) \otimes W$ is an integrator, where φ_j and ζ_j are defined in (11.30) and (11.31), and that

$$\int_{[0,1]} \varphi_j\, \zeta_j \,\mathrm{dW} = \zeta_j \int_{[0,1]} \varphi_j \,\mathrm{dW}.$$

Hints for Exercises in Chapter XI

4. i. For each $A \in \mathscr{A}$, $\mathbf{E}\mathbf{1}_A \int_B f \, \mathrm{d}X$ determines a measure on $\mathscr{B}$.

   ii. If $A \in \mathscr{A}$, then $\mathbf{E}\mathbf{1}_A(\int f \, \mathrm{d}X)(\cdot)$ is right-continous.

9. ii. See Corollary X.21 iv.

11. ii. Let $\Omega = [0, 2\pi)$ with $\mathbb{P}$ = normalized Lebesgue measure. Let $X_k(\omega) = \frac{1}{k}\mathrm{e}^{ik\omega}$.

    iii. By the use of a lacunary sequence in $\mathbb{Z}$, choose $\{X_k : k \in \mathbb{N}\}$ such that $H(X)$ is an infinite-dimensional $\Lambda(2)$-space, and $\|X_k\|_{\mathrm{L}^2} = \|X_k\|_{\mathrm{L}^1} = 1/k^2$.

12. Consider Y and Z on $(\Omega, \mathbb{P}) \times (\Omega, \mathbb{P})$, such that $Y(\omega_1, \omega_2) = Y(\omega_1)$ and $Z(\omega_1, \omega_2) = Z(\omega_2)$, $(\omega_1, \omega_2) \in \Omega \times \Omega$. Fix $U \in \mathrm{L}^\infty(\Omega, \mathbb{P})$, $\|U\|_{\mathrm{L}^\infty} = 1$, such that $\mathbf{E}|Y| = |\mathbf{E}YU|$, view it on $\Omega \times \Omega$ so that $U(\omega_1, \omega_2) = U(\omega_1)$, and observe that

$$\mathbf{E}|Y + Z| \geq |\mathbf{E}U(Y + Z)|$$

$$= \left| \int_{\Omega \times \Omega} (Y(\omega_1) + Z(\omega_2)) \mathbb{P}(\mathrm{d}\omega_1) \, \mathbb{P}(\mathrm{d}\omega_2) \right| = |\mathbf{E}UY| = \mathbf{E}|Y|.$$

Verify that if X is a symmetric L^1-bounded additive process, and $\epsilon_j = \pm 1$ for $j \in \mathbb{N}$, then

$$\mathbf{E}\left|\sum_j \epsilon_j \Delta X(I_j)\right| = \mathbf{E}\left|\sum_j \Delta X(I_j)\right| \leq \mathbf{E}|X(1) - X(0)|.$$

To this end, show that $\sum_{j=1}^N \epsilon_j \Delta X(I_j)$ has the same distribution as $\sum_{j=1}^N \Delta X(I_j)$ by computing the respective characteristic functions.

If X is an L^1-bounded additive process such that $\mathbf{E}X(t) = 0$ for all $t \in [0,1]$, then consider $X - \tilde{X}$, where $\tilde{X}$ is a statistically independent copy of X.

14. The definition of $H(X)$ in Definition X.27 and the definition of $H(X)$ in this chapter are the same. If X is an L^2-bounded process with orthogonal increments, then $H(X)$ is the closure in $\mathrm{L}^2(\Omega, \mathbb{P})$ of span$\{\Delta X(J_i)$: intervals $J_i \subset [0,1]$, $\sum_i \mathbf{1}_{J_i} \leq 1\}$.

15. Review the proofs of analogous results in Chapter X.

16. Cf. previous exercise.

18. Let X be a symmetric random variable such that $\mathbf{E}|X|^q = 1$ and $\mathbf{E}|X|^p = \infty$ for all $p > q$. Observe that if $\{X_j\}$ is an independent system where each X_j has the same distribution as X, then $\{X_j\}$ is a $\Lambda(q)^\#$-system.

20. ii. Provide the missing details in the proof (of Proposition 28) that $I_X(\mathrm{f}) = \int_{[0,1]} f\mathrm{d}X$ for all $f \in \mathrm{L}^\infty([0,1])$. (Cf. Proposition 13 ii.)

21. Cf. Proposition 12.

22. Use duality. For arbitrary $q > 2$, let $(b_{jk}) \in B_{l^q}$ have finite support, and show, by two applications of Hölder's inequality, that $\left|\sum_{j,k} b_{jk}\mathbf{E}\mathbf{1}_{A_j}X_k\right| \leq K_q < \infty$, where the X_k are defined in (4.50) and K_q depends only on q.

23. Read [La, pp. 73–5].

24. Observe that $X_i + \tilde{X}_i$ has the same distribution as $2^{1/p}X_i$, and then apply the triangle inequality. This elementary argument appears in [McTa1, p. 944] , and is attributed to Pisier.

26. If $f_k \in C([0,1])$ and $f_k \to g$ in $\mathrm{L}^2([0,1],\nu)$, then $\beta(f_k,\cdot)$ converge uniformly to $\beta(g,\cdot)$ on bounded subsets of $\mathrm{L}^\infty(\Omega,\mathbb{P})$.

27. Because the two definitions are the same for $f \in \mathrm{C}([0,1])$, it is enough to show that if $f_k \to g$ in $\mathrm{L}^2([0,1],\nu)$, where $f_k \in \mathrm{C}([0,1])$ $(k \in \mathbb{N})$ and $g \in \mathrm{L}^\infty([0,1],\mathscr{B})$, then

$$\int_{[0,1]} f_k(t)\,\mu_X(A,\mathrm{d}t) \to \int_{[0,1]} g(t)\,\mu_X(A,\mathrm{d}t), \quad A \in \mathscr{A}.$$

If $f_k \to g$ in $\mathrm{L}^2([0,1],\nu)$, then $f_k \to g$ uniformly on $[0,1]\backslash B$, where $\nu(B)$ is as small as desired. Apply Lemma 35.

29. Estimate

$$\left|\mathbf{E}Y\sum_{n=k}^{\infty}\hat{f}(n)X_n\right| = \left|\mathbf{E}Y\int_{[0,1]}\left(\sum_{n=k}^{\infty}\hat{f}(n)\,\mathbf{e}_n\right)\mathrm{d}X\right|.$$

30. i. If X is an integrator, $B_j \in \mathscr{B}$, $j \in \mathbb{N}$, and $\Sigma_j \mathbf{1}_{B_j} = \mathbf{1}_B$, then, $\Sigma_{j=1}^n \mathbf{1}_{B_j} \to \mathbf{1}_B$ in $\mathrm{L}^2([0,1],\nu)$, where ν is a Grothendieck measure of X, and

$$\sum_{j=1}^{\infty} \text{‘}\Delta\text{’}X(B_j) = \text{‘}\Delta\text{’}X(B), \quad \text{convergence in } \mathrm{L}^1(\Omega,\mathbb{P}).$$

ii. Show

$$\mathbf{E}\,(\exp(\mathrm{i}s_1`\Delta\text{'}X(B_1)\cdots\exp(\mathrm{i}s_n`\Delta\text{'}X(B_n))$$

$$= \prod_{j=1}^{n} \mathbf{E}\,\exp(\mathrm{i}s_j`\Delta\text{'}X(B_j)).$$

32. Use induction. Mimic the proof in the case $n = 2$.
33. Cf. Corollary 4. Extend the proof of Lemma 3. In this regard, see Adams's and Clarkson's paper [CLA].
36. Use the fact that if μ_X determines an F_2-measure on $\mathscr{A} \times \mathscr{O}$, then it is a *regular* measure in the second coordinate.
37. Use Theorem X.8 together with the Littlewood inequalities (Chapter X §10). First carry out constructions, as in §8, in a framework of $\Omega \times [0,1]^n$, and then 'pull back', by use of $\boldsymbol{\rho}^{-1}$, to $\Omega \times [0,1]$.
42. Review the material about tensor products in Chapter IV.

XII

A '3/2-dimensional' Cartesian Product

1 Mise en Scène: Two Basic Questions

The n-fold Cartesian product of a set E,

$$E^n := E \times \cdots \times E = \{(x_1, \ldots, x_n) : x_i \in E, i \in [n]\}, \tag{1.1}$$

is commonly viewed as an n-dimensional set, where *dimension* means simply the number of unrestricted samplings from E. If m is an integer, $0 < m \leq n,\ b \in E$, and

$$F = \left\{ \left(x_1, \ldots, x_m, \underbrace{b, \ldots, b}_{n-m} \right) : x_i \in E,\ i \in [m] \right\}, \tag{1.2}$$

then F is an m-dimensional subset of E^n. Again, *dimension* here means the number of degrees of freedom in the definition of F: the first m coordinates are chosen freely, without restriction, and the last $n - m$ coordinates are fixed. Typically, if $0 < m \leq n$, and θ is a function from E^m into E^{n-m}, then its 'graph',

$$F = \{(x_1, \ldots, x_m, \theta(x_1, \ldots, x_m)) : (x_1, \ldots, x_m) \in E^m\}, \tag{1.3}$$

is an m-dimensional subset of E^n: there is no restriction on the first m coordinates, but once chosen, these coordinates determine the remaining $n - m$ coordinates. *Dimension* of F in this basic context is an index of interdependence between n samplings from E subject to the requirement that the outcome (an n-tuple) be in F.

A question arises: given an *arbitrary* subset F of E^n, can we gauge, precisely and meaningfully, an index of interdependence of coordinates in F? Rephrasing this question, can we precisely and meaningfully determine a 'dimension' of F? In a basic sense, we view the *dimension* of a set as the *number* of degrees of freedom enjoyed by points in the set, a number that has been traditionally viewed as an integer. And thus a second question arises: could *dimension* be, realistically, a fraction?

In the next chapter, we shall answer these questions in full, both in the affirmative. In this chapter, easing our way into fractional dimensions, we analyze an archetypal example that will serve as a guide, and eventually be seen as a 3/2-dimensional Cartesian product.

2 A Littlewood Inequality in 'Dimension' 3/2

For every $n \in \mathbb{N}$,

$$\|\hat{g}\|_p < \infty \text{ for all } g \in \mathrm{C}_{W_n}(\Omega) \text{ if and only if } p \geq 2n/(n+1)$$
$$\text{(Theorem VII.34).} \tag{2.1}$$

This result, which is in effect a calibration of Plancherel's theorem, suggested this definition: if F is an arbitrary spectral set, then let

$$\zeta_F(t) = \sup\{\|\hat{g}\|_t : g \in B_{\mathrm{C}_F}\}, \tag{2.2}$$

and

$$\sigma_F = \inf\{t : \zeta_F(t) < \infty\}$$

(Definition VII.40, Remark iv, Chapter VII §11).

The index σ_F is said to be *exact* if $\zeta_F(\sigma_F) < \infty$, and *asymptotic* if $\zeta_F(\sigma_F) = \infty$. If $\sigma_F = p$, then F is p-Sidon (exact or asymptotic), and p is the *Sidon exponent* of F. In this terminology, (2.1) becomes

$$\sigma_{W_n} = \frac{2n}{n+1} \text{ exactly, } n \in \mathbb{N}, \tag{2.3}$$

which, 'modulo decoupling', is the Littlewood $2n/(n+1)$-inequality

$$\sigma_{R^n} = \frac{2n}{n+1} \text{ exactly, (Theorem VII.36).} \tag{2.4}$$

The p-Sidon set problem – do p-Sidon sets exist for arbitrary $p \in [1, 2]$? – becomes the question: if $p \in [1,\ 2n/(n+1)] \setminus \{1,\ 4/3, \ldots,\ 2n/(n+1)\}$, then are there $F \subset R^n$ such that $\sigma_F = p$? (See Chapter VII, Remark vi §11.)

In this section we produce $F \subset R^3$ such that $\sigma_F = 6/5$. We use the following construction that played prominently in Chapter VIII, in the analysis of a trilinear Grothendieck-type inequality: index the Rademacher system by $\mathbb{N}^2$,

$$R = \{r_{ij} : (i,j) \in \mathbb{N}^2\} \tag{2.5}$$

($r_{ij}(\omega) = \omega(i,j),\ \omega \in \{-1,1\}^{\mathbb{N}^2} := \Omega$), and define

$$R^U := \{r_{ij} \otimes r_{jk} \otimes r_{ik} : (i,j,k) \in \mathbb{N}^3\};\ \text{cf. (VIII.5.3)}. \tag{2.6}$$

(The meaning of U, merely a superscript in (2.6), will later become clear.)

Lemma 1 *For all $f \in \mathrm{C}_{R^U}(\Omega^3)$,*

$$\sum_{i,j} \left(\sum_k |\hat{f}(r_{ij} \otimes r_{jk} \otimes r_{ik})|^2 \right)^{\frac{1}{2}} \leq c_1 \sqrt{2}\ \|f\|_\infty, \tag{2.7}$$

where $c_1 := \zeta_R(1)$ (the Sidon constant of R).

Proof: Let f be an R^U-polynomial, and estimate

$$\begin{aligned}
\|f\|_\infty &= \sup_{\omega_1,\omega_2,\omega_3} \left| \sum_{i,j,k} \hat{f}(r_{ij} \otimes r_{jk} \otimes r_{ik})\ r_{ij}(\omega_1)\ r_{jk}(\omega_2)\ r_{ik}(\omega_3) \right| \\
&\geq (1/c_1) \sup_{\omega_2,\omega_3} \sum_{i,j} \left| \sum_k \hat{f}(r_{ij} \otimes r_{jk} \otimes r_{ik})\ r_{jk}(\omega_2)\ r_{ik}(\omega_3) \right| \\
&\geq (1/c_1) \sum_{i,j} \int_{\Omega^2} \left| \sum_k \hat{f}(r_{ij} \otimes r_{jk} \otimes r_{ik})\ r_{jk}(\omega_2)\ r_{ik}(\omega_3)\ \mathbb{P}(\mathrm{d}\omega_2)\ \mathbb{P}(\mathrm{d}\omega_3) \right| \\
&\geq (1/c_1\sqrt{2}) \sum_{i,j} \left(\sum_k \left| \hat{f}(r_{ij} \otimes r_{jk} \otimes r_{ik}) \right|^2 \right)^{\frac{1}{2}}.
\end{aligned} \tag{2.8}$$

The last inequality follows from the Khintchin L^1–L^2 inequality applied to $\{r_{ij} \otimes r_{jk} : j \in \mathbb{N}\}$ for each $(i,k) \in \mathbb{N}^2$. □

Lemma 2

i. *If* $(b_{ij} : (i,j) \in \mathbb{N}^2)$ *is a* 2*-array of scalars, then*

$$\left(\sum_{i,j} |b_{ij}|^{\frac{4}{3}}\right)^{\frac{3}{4}} \leq \left(\sum_i \left(\sum_j |b_{ij}|^2\right)^{\frac{1}{2}}\right)^{\frac{1}{2}} \left(\sum_j \left(\sum_i |b_{ij}|^2\right)^{\frac{1}{2}}\right)^{\frac{1}{2}}. \tag{2.9}$$

ii. *If* $(a_{ijk} : (i,j,k) \in \mathbb{N}^3)$ *is a* 3*-array of scalars, then*

$$\left(\sum_{i,j,k} |a_{ijk}|^{\frac{6}{5}}\right)^{\frac{5}{6}} \leq \left(\sum_{i,j} \left(\sum_k |a_{ijk}|^2\right)^{\frac{1}{2}}\right)^{\frac{1}{3}} \cdot \left(\sum_{i,k} \left(\sum_j |a_{ijk}|^2\right)^{\frac{1}{2}}\right)^{\frac{1}{3}} \cdot \left(\sum_{j,k} \left(\sum_i |a_{ijk}|^2\right)^{\frac{1}{2}}\right)^{\frac{1}{3}}. \tag{2.10}$$

Proof: Part i has been verified in Chapter II, en route to Littlewood's 4/3-inequality ((II.5.3) in the proof of Theorem II.5 i).

To verify Part ii, write

$$\sum_{i,j,k} |a_{ijk}|^{\frac{6}{5}} = \sum_{i,j,k} |a_{ijk}|^{\frac{2}{5}} |a_{ijk}|^{\frac{4}{5}}. \tag{2.11}$$

On the right side, apply Hölder's inequality to the summation over k, and the two factors in the summand with exponents 5 and 5/4:

$$\sum_{i,j,k} |a_{ijk}|^{\frac{6}{5}} \leq \sum_{i,j} \left(\sum_k |a_{ijk}|^2\right)^{\frac{1}{5}} \left(\sum_k |a_{ijk}|\right)^{\frac{4}{5}}. \tag{2.12}$$

On the right side of (2.12), apply Hölder's inequality to the summation over i and j, and the two factors in the summand (each a sum over k) with exponents 5/2 and 5/3:

$$\sum_{i,j,k} |a_{ijk}|^{\frac{6}{5}} \leq \left(\sum_{i,j}\left(\sum_{k}|a_{ijk}|^2\right)^{\frac{1}{2}}\right)^{\frac{2}{5}} \left(\sum_{i,j}\left(\sum_{k}|a_{ijk}|\right)^{\frac{4}{3}}\right)^{\frac{3}{5}}. \tag{2.13}$$

Apply (2.9) to the second factor on the right side of (2.13) with $b_{ij} = \sum_k |a_{ijk}|$, obtaining

$$\begin{aligned}\sum_{i,j,k} |a_{ijk}|^{\frac{6}{5}} \leq &\left(\sum_{i,j}\left(\sum_{k}|a_{ijk}|^2\right)^{\frac{1}{2}}\right)^{\frac{2}{5}} \cdot \\ &\cdot\left(\sum_{i}\left(\sum_{j}\left(\sum_{k}|a_{ijk}|\right)^2\right)^{\frac{1}{2}}\right)^{\frac{2}{5}} \cdot \\ &\cdot\left(\sum_{j}\left(\sum_{i}\left(\sum_{k}|a_{ijk}|\right)^2\right)^{\frac{1}{2}}\right)^{\frac{2}{5}}.\end{aligned} \tag{2.14}$$

Applying Minkowski's inequality to the second and third factors on the right side of (2.14), we obtain

$$\begin{aligned}\sum_{i,j,k} |a_{ijk}|^{\frac{6}{5}} \leq &\left(\sum_{i,j}\left(\sum_{k}|a_{ijk}|^2\right)^{\frac{1}{2}}\right)^{\frac{2}{5}} \cdot \\ &\cdot\left(\sum_{i,k}\left(\sum_{j}|a_{ijk}|^2\right)^{\frac{1}{2}}\right)^{\frac{2}{5}} \cdot \\ &\cdot\left(\sum_{j,k}\left(\sum_{i}|a_{ijk}|^2\right)^{\frac{1}{2}}\right)^{\frac{2}{5}},\end{aligned} \tag{2.15}$$

which implies (2.10). □

These two lemmas imply

Corollary 3

$$\zeta_{R^U}\left(\frac{6}{5}\right) \leq \sqrt{2}\,\zeta_R(1). \tag{2.16}$$

Lemma 4

$$\zeta_{R^U}(t) = \infty \text{ for all } t < \frac{6}{5}. \tag{2.17}$$

Proof: We give two proofs: one based on random constructions, like those used in Chapter X to prove $\zeta_{R^n}(t) = \infty$ for $t < 2n/(n+1)$ (cf. Remark i in Chapter X §5), and the other based on Theorem VII.41,

$$\zeta_F(t) \geq \sup\{\|g\|_{L^q}/\sqrt{q} : g \in L^2_F,\ \|\hat{g}\|_{2t/(3t-2)} = 1,\ q > 2\}, \tag{2.18}$$

used in Chapter VII also to prove $\zeta_{R^n}(t) = \infty$ for $t < 2n/(n+1)$ (cf. Corollary VII.42).

i. Let m be a positive integer. By Theorem X.8, there exists a $\{-1,+1\}$-valued 3-array $(\epsilon_{ijk} : (i,j,k) \in [m]^3)$ such that if

$$f_m = (1/m^{\frac{5}{2}}) \sum_{(i,j,k)\in[m]^3} \epsilon_{ijk}\, r_{ij} \otimes r_{jk} \otimes r_{ik}, \tag{2.19}$$

then

$$\|f_m\|_\infty \leq K, \tag{2.20}$$

where $K > 0$ is a numerical constant. (For the application of Theorem X.8, note that the degree of polynomials with spectrum in $\{r_{ij} \otimes r_{jk} \otimes r_{ik} : (i,j,k) \in [m]^3\}$ is at most 2^{3m^2}.) Then

$$\|\hat{f}_m\|_t / \|f_m\|_\infty \geq m^{\frac{3}{t}-\frac{5}{2}}/K, \tag{2.21}$$

which is unbounded for $t < 6/5$, and (2.17) follows.

ii. Let m be a positive integer, and consider the Riesz product

$$R_m = \prod_{i,j=1}^{m} (1 + r_{ij}) \otimes \prod_{i,j=1}^{m} (1 + r_{ij}) \otimes \prod_{i,j=1}^{m} (1 + r_{ij}). \tag{2.22}$$

Then, $\|R_m\|_{L^1} = 1$, $\|R_m\|_{L^2} = 2^{3m^2/2}$, and therefore for all $q > 2$,

$$\|R_m\|_{L^p} \leq 2^{3m^2/q},\ \frac{1}{p} + \frac{1}{q} = 1 \quad \text{(cf. Lemma VII.30)}. \tag{2.23}$$

Let

$$h_m = \sum_{(i,j,k)\in[m]^3} r_{ij} \otimes r_{jk} \otimes r_{ik}, \tag{2.24}$$

and then, by Hölder's inequality and (2.23), obtain

$$m^3 = |\mathbf{E}\ R_m h_m| \le 2^{\frac{3m^2}{q}} \|h_m\|_{L^q}. \tag{2.25}$$

Put $q = m^2$ in (2.25), and conclude from (2.24) and (2.18) that

$$\zeta_{R^U}(t) \ge \|h_m\|_{\mathrm{L}^{m^2}}/m\ \|\hat{h}_m\|_{2t/(3t-2)} \ge \left(\frac{1}{8}\right) m^{\frac{6-5t}{2t}}, \tag{2.26}$$

which implies $\zeta_{R^U}(t) = \infty$ for all $t < 6/5$.

□

Combining Corollary 3 and Lemma 4, we obtain

Theorem 5

$$\sigma_{R^U} = \frac{6}{5} \text{ exactly.} \tag{2.27}$$

Remark: (a '3/2-dimensional' set). Theorem 5 signals that R^U behaves like a 3/2-fold Cartesian product of R. Indeed, viewing the Littlewood $2n/(n+1)$-inequality as a precise statement tied to the dimension of R^n, we could define the 'dimension' of any spectral set F by

$$\dim F = \sigma_F/(2 - \sigma_F). \tag{2.28}$$

Then, $\dim R^U = 3/2$. But this is cheating. For, we must define the 'dimension' of $F \subset R^n$ intrinsically, using only 'set-theoretic' information about F, and *then* establish the formula in (2.28). Notice, in this regard, that the definition in (2.6) is completely general: if E is any countably infinite set indexed by $\mathbb{N}^2$,

$$E = \{(x_{ij}) : (i,j) \in \mathbb{N}^2\}, \tag{2.29}$$

then we can define

$$E^U := \{(x_{ij}, x_{jk}, x_{ik}) : (i,j,k) \in \mathbb{N}^3\}. \tag{2.30}$$

In due course we will prove that the 'dimension' of E^U – intrinsically defined – equals 3/2.

In the next section we further bolster the case that the 'dimension' of $\{(r_{ij}, r_{jk}, r_{ik}) : (i,j,k) \in \mathbb{N}^3\}$ is 3/2.

3 A Khintchin Inequality in 'Dimension' 3/2

We consider the n-dimensional Khintchin inequalities, which were among the main motifs in Chapter VII and Chapter X. Recall that in Chapter VII we expressed these inequalities by

$$\delta_{R^n} = \frac{n}{2} \text{ exactly} \quad \text{(Proposition VII.31, (VII.9.30), (VII.9.31))}, \tag{3.1}$$

and in Chapter X, rephrasing (3.1), we dubbed $R^n = \{r_{j_1} \otimes \cdots \otimes r_{j_n} : (j_1, \ldots, j_n) \in \mathbb{N}^n\}$ an *exact n-system*, i.e.,

$$\delta_{\mathrm{L}_{R^n}} = \frac{n}{2} \text{ exactly ((X.10.6), (X.10.7), Definition X.25).} \tag{3.2}$$

(The measurements δ_F and δ_{L_F} denote the same index; the notation used is largely a typographical decision, and depends on the context.) We recall: if F is a spectral set, then

$$\eta_F(a) = \sup\{\|f\|_{\mathrm{L}^p}/p^a : p > 2,\ f \in B_{\mathrm{L}^2_F}\}, \quad a > 0, \tag{3.3}$$

and

$$\delta_F = \inf\{a : \eta_F(a) < \infty\} \quad \text{((VII.9.29) and (VII.9.30))}, \tag{3.4}$$

measurements that can be cast also in functional-analytic and probabilistic frameworks; see Chapter X §9, §10, §14. The assertion in (3.1) leads (via 'decoupling') to Bonami's inequalities,

$$\delta_{W_n} = \frac{n}{2} \text{ exactly} \quad \text{((VII.9.31))}, \tag{3.5}$$

which in turn lead to the question (Remark iii in Chapter VII §9): for *non-integer* $\alpha \in (1, \infty)$, are there spectral sets F such that $\delta_F = \alpha/2$? (Ostensibly couched in the language of harmonic analysis, the same question reappears in a probabilistic framework of random walks and α-chaos processes; see Remark ii in Chapter X §13, *Question* 1 in Chapter X §13, Chapter XI §4.) In this section we verify

$$\delta_{R^U} = \frac{3}{4} = \left(\frac{3}{2}\right) \Big/ 2,$$

evidence again that the 'dimension' of R^U is $\frac{3}{2}$.

We first do the groundwork. Consider a general discrete Abelian group Γ and its dual $\hat{\Gamma} = G$, wherein group operations are denoted multiplicatively. (See Chapter VII §12.) For $F \subset \Gamma, \gamma \in \Gamma$, and integers $s > 0$, denote

$$A_F(s,\gamma) = \{(\gamma_1, \ldots, \gamma_s) \in F^S : \gamma_1 \cdots \gamma_s = \gamma\}, \tag{3.6}$$

and define

$$r_F(s,\gamma) = |A_F(s,\gamma)|, \tag{3.7}$$

$$\rho_F(s) = \sup\{r_F(s,\gamma) : \gamma \in \Gamma\}. \tag{3.8}$$

Lemma 6 *For all $F \subset \Gamma$, all $f \in \mathrm{L}^2_F(G)$, and integers $s > 0$,*

$$\|f\|_{\mathrm{L}^{2s}} \leq \rho_F(s)^{1/2s} \; \|f\|_{\mathrm{L}^2}. \tag{3.9}$$

Proof: Write $f \sim \sum_{\gamma \in F} \hat{f}(\gamma)\, \gamma$, and fix an integer $s > 0$. Then,

$$\begin{aligned} \int_G |f|^{2s} \mathrm{d}x &= \int_G \left| \left(\sum_{\gamma \in F} \hat{f}(\gamma)\, \gamma \right)^s \right|^2 \mathrm{d}x \\ &= \sum_{\gamma \in \Gamma} \left| \sum_{(\gamma_1,\ldots,\gamma_s) \in A_F(s,\gamma)} \hat{f}(\gamma_1) \ldots \hat{f}(\gamma_s) \right|^2 . \end{aligned} \tag{3.10}$$

($\mathrm{d}x$ denotes integration with respect to the normalized Haar measure on G; the second equality follows from Plancherel's theorem.) By the Cauchy–Schwarz inequality,

$$\int_G |f|^{2s} \; \mathrm{d}x \leq \rho_F(s) \sum_{\gamma \in \Gamma} \sum_{(\gamma_1,\ldots,\gamma_s) \in A_F(s,\gamma)} |\hat{f}(\gamma_1)|^2 \cdots |\hat{f}(\gamma_s)|^2. \tag{3.11}$$

Observe

$$F^s = \bigcup \{A_F(s,\gamma) : \gamma \in \Gamma\}, \tag{3.12}$$

and that $A_F(s,\gamma) \cap A_F(s,\gamma') = \emptyset$ for $\gamma \neq \gamma'$. Therefore, by (3.11) and (3.12),

$$\int_G |f|^{2s} \; \mathrm{d}x \leq \rho_F(s) \sum_{(\gamma_1,\ldots,\gamma_s) \in F^s} |\hat{f}(\gamma_1)|^2 \cdots |\hat{f}(\gamma_s)|^2 = \rho_F(s) \|f\|^{2s}_{\mathrm{L}^2}. \tag{3.13}$$

□

Remark:

i **(about Lemma 6).** The 'harmonic-analytic' inequality (3.9), which rests on Plancherel's theorem, is at the very foundation of the subject. It is in effect a basic combinatorial estimate that eventually will point to the definition of *combinatorial dimension.*

Computations of L^{2s}-norms based on (3.10) appeared in Khintchin's original paper [Kh1, p. 112], where inequalities involving Rademacher functions were first derived. Similar computations were used independently by Littlewood [Lit3, p. 330], en route to similar inequalities involving Steinhaus functions. (See Remark i in Chapter VII §9.)

The key role of the combinatorial gauge r_F in the estimation of L^p-norms was first underscored and put to effective use by Walter Rudin in his classic [Ru1, Theorem 4.5]. Rudin's estimates, which were not optimal, were eventually sharpened by Aline Bonami; Lemma 6 above in her Théorème 3 in [Bon2, p. 356].

Recall (Chapter VII §12) that a spectral set $F \subset \Gamma$ is *algebraically independent* if

$$\prod_{\gamma \in S} \gamma^{n_\gamma} = \mathbf{1}_G \quad \text{for } S \subset F,\ |S| < \infty, \text{ and } n_\gamma \in \mathbb{Z}\ \ (\gamma \in S), \tag{3.14}$$

implies $n_\gamma = 0$ for all $\gamma \in S$.

Let S denote the Steinhaus system (Chapter II §6, Chapter VII §12), which we view here as a basis in $\Gamma = \oplus\, \mathbb{Z}$ $(\hat{\Gamma} = G = \mathbf{T}^{\mathbb{N}})$, and thus an archetypal algebraically independent spectral set. Enumerate $S = \{\chi_{ij} : (i,j) \in \mathbb{N}^2\}$, and consider the subset of Γ^3 (cf. (2.30))

$$S^U = \{\chi_{ij} \otimes \chi_{jk} \otimes \chi_{ik} : (i,j,k) \in \mathbb{N}^3\}. \tag{3.15}$$

Lemma 7 *For all integers $s > 0$,*

$$\rho_{S^U}(s) \leq s^{3s/2}. \tag{3.16}$$

Proof: Fix an integer $s > 0$, and $\gamma = (\gamma_1, \gamma_2, \gamma_3) \in \Gamma^3$ such that

$$\gamma = x_1 \cdots x_s, \quad x_1 \in S^U, \ldots, x_s \in S^U. \tag{3.17}$$

Write

$$x_u = \chi_{i_u j_u} \otimes \chi_{j_u k_u} \otimes \chi_{i_u k_u}, \quad u = 1, \ldots, s, \tag{3.18}$$

and restate (3.17),

$$\gamma = \left(\prod_{u=1}^{s} \chi_{i_u j_u}\right) \otimes \left(\prod_{u=1}^{s} \chi_{j_u k_u}\right) \otimes \left(\prod_{u=1}^{s} \chi_{i_u k_u}\right), \tag{3.19}$$

where

$$\gamma_1 = \prod_{u=1}^{s} \chi_{i_u j_u}, \quad \gamma_2 = \prod_{u=1}^{s} \chi_{j_u k_u}, \quad \gamma_3 = \prod_{u=1}^{s} \chi_{i_u k_u}. \tag{3.20}$$

Consider the underlying indexing sets

$$\begin{aligned} A_1 &= \{(i_1, j_1), \dots, (i_s, j_s)\}, \\ A_2 &= \{(j_1, k_1), \dots, (j_s, k_s)\}, \\ A_3 &= \{(i_1, k_1), \dots, (i_s, k_s)\}. \end{aligned} \tag{3.21}$$

Suppose $\gamma = y_1 \cdots y_s$, and

$$y_u = \chi_{i'_u j'_u} \otimes \chi_{j'_u k'_u} \otimes \chi_{i'_u k'_u}, \; u = 1, \dots, s, \tag{3.22}$$

i.e., $(y_1, \dots, y_s) \in A_{S^U}(s, \gamma)$. Then, by the algebraic independence of the Steinhaus system,

$$(i'_u, j'_u) \in A_1, \; (j'_u, k'_u) \in A_2, \; (i'_u, k'_u) \in A_3, \; u = 1, \dots, s. \tag{3.23}$$

Let $\pi_1, \dots, \pi_s$ denote the canonical projections from $A_{S^U}(s, \gamma)$ into S^U, and for convenience, denote $A = A_{S^U}(s, \gamma)$. (Recall that $A_{S^U}(s, \gamma)$ comprises s-tuples, each of whose coordinates is an element of S^U.) Then, by (3.23) and the definition of S^U, for every $u = 1, \dots, s$,

$$\begin{aligned} |\pi_u[A]| &\leq \sum_{i,j,k} \mathbf{1}_{A_1}(i,j)\, \mathbf{1}_{A_2}(j,k)\, \mathbf{1}_{A_3}(i,k) \\ &= \sum_{i,j} \mathbf{1}_{A_1}(i,j) \left(\sum_{k} \mathbf{1}_{A_2}(j,k)\, \mathbf{1}_{A_3}(i,k)\right) \\ &\leq \sum_{i,j} \mathbf{1}_{A_1}(i,j) \left(\sum_{k} \mathbf{1}_{A_2}(j,k)\right)^{\frac{1}{2}} \left(\sum_{k} \mathbf{1}_{A_3}(i,k)\right)^{\frac{1}{2}} \end{aligned}$$

$$
\begin{aligned}
&= \sum_i \left(\sum_k \mathbf{1}_{A_3}(i,k) \right)^{\frac{1}{2}} \sum_j \mathbf{1}_{A_1}(i,j) \left(\sum_k \mathbf{1}_{A_2}(j,k) \right)^{\frac{1}{2}} \\
&\le \sum_i \left(\sum_k \mathbf{1}_{A_3}(i,k) \right)^{\frac{1}{2}} \left(\sum_j \mathbf{1}_{A_1}(i,j) \right)^{\frac{1}{2}} \left(\sum_{j,k} \mathbf{1}_{A_2}(j,k) \right)^{\frac{1}{2}} \\
&\le |A_1|^{\frac{1}{2}} |A_2|^{\frac{1}{2}} |A_3|^{\frac{1}{2}} \le s^{\frac{3}{2}}
\end{aligned}
\tag{3.24}
$$

(three applications of the Cauchy–Schwarz inequality). Therefore, because $A \subset \pi_1[A] \times \cdots \times \pi_s[A]$, we conclude

$$|A_{S^U}(s,\gamma)| \le [s^{\frac{3}{2}}]^s, \tag{3.25}$$

and, by maximizing (3.25) over $\gamma \in \Gamma$, obtain the lemma. □

Theorem 8

$$\delta_{S^U} = \frac{3}{4} \textit{ exactly.} \tag{3.26}$$

Proof: Lemmas 6 and 7 imply

$$\eta_{S^U}\left(\frac{3}{4}\right) \le 1. \tag{3.27}$$

To verify that $\eta_{S^U}(a) = \infty$ for all $a < 3/4$, we apply the same argument used to prove Lemma 4 in the previous section. Let m be a positive integer, and consider the Riesz product

$$
\begin{aligned}
R_m &= \prod_{i,j=1}^{m} \left(1 + \frac{\chi_{ij} + \overline{\chi_{ij}}}{2}\right) \otimes \prod_{i,j=1}^{m} \left(1 + \frac{\chi_{ij} + \overline{\chi_{ij}}}{2}\right) \\
&\quad \otimes \prod_{i,j=1}^{m} \left(1 + \frac{\chi_{ij} + \overline{\chi_{ij}}}{2}\right).
\end{aligned}
\tag{3.28}
$$

Then, for $p \in (1, 2)$ and its conjugate $q \in (2, \infty)$,

$$\|R_m\|_{L^p} \le 2^{\frac{3m^2}{q}}. \tag{3.29}$$

Consider

$$h_m = \sum_{(i,j,k)\in[m]^3} \chi_{ij} \otimes \chi_{jk} \otimes \chi_{ik}, \tag{3.30}$$

and obtain

$$\left(\frac{1}{8}\right) m^3 = |\mathbf{E}\, h_m\, R_m| \leq 2^{\frac{3m^2}{q}} \|h_m\|_{L^q}, \quad q > 2. \tag{3.31}$$

Putting $q = m^2$ in (3.31), we have

$$\|h_m\|_{L^{m^2}} / \|h_m\|_{L^2} \geq (1/64)(m^2)^{\frac{3}{4}}, \tag{3.32}$$

and therefore

$$\eta_{S^U}(a) = \infty \quad \text{for all } a < \frac{3}{4}. \tag{3.33}$$

□

To verify that $\delta_{E^U} = 3/4$ exactly, for every infinite dissociate set E in an arbitrary discrete Abelian group Γ, we use the tensor-theoretic representation

$$A(E^3) = V_3(E, E, E) \quad \text{(Chapter VII §8, §12)}. \tag{3.34}$$

Proposition 9 (cf. Proposition VII.44). *Let Γ and Γ' be discrete Abelian groups with their respective duals G and G'. There exists $K_U > 0$ such that for all dissociate sets $E \subset \Gamma$ and $F \subset \Gamma'$*

$$\eta_{E^U}(a) \leq K_U\, \eta_{F^U}(a), \quad a > 0. \tag{3.35}$$

Proof: Write

$$\begin{aligned} E^U &= \{\gamma_{ij} \otimes \gamma_{jk} \otimes \gamma_{ik} : (i,j,k) \in \mathbb{N}^3\}, \\ F^U &= \{\chi_{ij} \otimes \chi_{jk} \otimes \chi_{ik} : (i,j,k) \in \mathbb{N}^3\}. \end{aligned} \tag{3.36}$$

Let f be an arbitrary E^U-polynomial,

$$f = \sum_{i,j,k} \hat{f}(\gamma_{ij} \otimes \gamma_{jk} \otimes \gamma_{ik})\, \gamma_{ij} \otimes \gamma_{jk} \otimes \gamma_{ik}. \tag{3.37}$$

For $\mathbf{x} = (x_1, x_2, x_3) \in (G')^3$, consider

$$f_{\mathbf{x}} = \sum_{i,j,k} \hat{f}(\gamma_{ij} \otimes \gamma_{jk} \otimes \gamma_{ik})\, \chi_{ij}(x_1)\, \chi_{jk}(x_2)\, \chi_{ik}(x_3)\, \gamma_{ij} \otimes \gamma_{jk} \otimes \gamma_{ik}. \tag{3.38}$$

By (3.34), there exist $\theta_{\mathbf{x}} \in L^1(G^3)$ such that for $\gamma_{ij} \otimes \gamma_{jk} \otimes \gamma_{ik} \in \text{spect}\,(f)$,

$$\hat{\theta}_{\mathbf{x}}(\gamma_{ij} \otimes \gamma_{jk} \otimes \gamma_{ik}) = \overline{\chi_{ij}(x_1)}\; \overline{\chi_{jk}(x_2)}\; \overline{\chi_{ik}(x_3)}, \tag{3.39}$$

and

$$\|\theta_{\mathbf{x}}\|_{L^1} \leq K, \tag{3.40}$$

where $K > 0$ does not depend on $\mathbf{x}$. Then, $f_{\mathbf{x}} \star \theta_{\mathbf{x}} = f$, and therefore

$$\|f\|_{L^q}^q = \|f_{\mathbf{x}} \star \theta_{\mathbf{x}}\|_{L^q}^q \leq K^q \|f_{\mathbf{x}}\|_{L^q}^q. \tag{3.41}$$

We integrate both sides of (3.41) over $(G')^3$, apply Fubini's theorem and then the definition of η_{F^U}, and obtain

$$\|f\|_{L^q} \leq K \eta_{F^U}(a)\, q^a \|f\|_{L^2}. \tag{3.42}$$

□

Corollary 10 *If $E \subset \Gamma$ is dissociate, then*

$$\delta_{E^U} = \frac{3}{4} \text{ exactly.} \tag{3.43}$$

Proof: Apply Theorem 8 and Proposition 9. □

Remarks:

ii **(a relationship between the δ_F and σ_F?).** In the next section we note that if E is any dissociate set in Γ, then $\sigma_{E^U} = 6/5$. And so, for $F = E^n$ or $F = E^U$,

$$\sigma_F = 4\delta_F/(2\delta_F + 1). \tag{3.44}$$

In the next chapter we verify (3.44) for all $F \subset E^n$, $n \in \mathbb{N}$. Whether this formula holds for all spectral sets F is an open question.

iii **(the same construction in Γ).** Suppose $E = \{\gamma_{ij} : (i,j) \in \mathbb{N}^2\}$ is a dissociate set in a discrete Abelian group Γ, and consider the subset of Γ

$$E_U = \{\gamma_{ij}\ \gamma_{jk}\ \gamma_{ik} : (i,j,k) \in D_3\}, \tag{3.45}$$

where $D_3 = \{(i,j,k) : (i,j,k) \in \mathbb{N}^3, 0 < i < j < k\}$ (cf. (VII.8.21)). Then, $\sigma_{E^U} = 6/5$ and $\delta_{E^U} = 3/4$ imply, through symmetrization (Chapter VII §8), that $\sigma_{E_U} = 6/5$ and $\delta_{E_U} = 3/4$ (Exercise 1). More about this will be said in the next chapter.

4 Tensor Products in 'Dimension' 3/2

In this section we amplify some of the analysis in Chapter VIII §5. We start with a countably infinite set X, let β be a scalar-valued function on X^3, and define

$$\begin{aligned}\|\beta\|_{F_U(X^3)} &= \|\beta\|_{F_U} \\ &:= \sup\left\{\left\|\sum_{\substack{(x,y)\varepsilon A\\(y,z)\varepsilon B\\(x,z)\varepsilon C}} \beta(x,y,z)\, r_{xy}\otimes r_{yz}\otimes r_{xz}\right\|_\infty : \right. \\ &\qquad\qquad \left.\text{finite sets } A\subset X^2, B\subset X^2, C\subset X^2\right\},\end{aligned} \tag{4.1}$$

where $\{r_{xy} : (x,y)\in X^2\}$ is the Rademacher system indexed by X^2, and the sup-norm $\|\cdot\|_\infty$ is evaluated over $\{-1,1\}^{X^2}\times\{-1,1\}^{X^2}\times\{-1,1\}^{X^2}$. The class of all scalar-valued functions β on X^3 such that $\|\beta\|_{F_U}<\infty$ is denoted by $F_U(X^3)$.

Let τ_1,τ_2, and τ_3 be bijections from X^2 onto X, and consider the subset of $X\times X\times X$

$$X^U = \{(\tau_1(x,y),\tau_2(y,z),\tau_3(x,z)) : (x,y,z)\in X^3\}. \tag{4.2}$$

This definition is essentially the same as that in (2.30). For, we can view X^U as the subset of $X^2\times X^2\times X^2$ defined by

$$X^U = \{((x,y),(y,z),(x,z)) : (x,y,z)\in X^3\}. \tag{4.3}$$

('Erase' the τ_i in (4.2); the effect is the same.) Now consider the elements in $F_3(X^2,X^2,X^2)$ with support in X^U,

$$\begin{aligned}&[F_3(X^2,X^2,X^2)]_{X^U} = [F_3]_{X^U} \\ &\quad := \{\beta\in F_3(X^2,X^2,X^2) : \beta(\mathbf{x}) = 0 \text{ for } \mathbf{x}\in (X^2)^3\backslash X^U\},\end{aligned} \tag{4.4}$$

and observe that $[F_3]_{X^U}$ can be naturally identified with $F_U(X^3)$: for $\beta\in F_U(X^3)$, define the function $\tilde\beta$ on $(X^2)^3$ by

$$\tilde\beta((x_1,x_2),(x_3,x_4),(x_5,x_6)) = \begin{cases}\beta(x,y,z) & x_1=x_5=x, \\ & x_2=x_3=y, \\ & x_4=x_6=z \\ 0 & \text{otherwise.}\end{cases} \tag{4.5}$$

Then, $\tilde\beta$ is supported in X^U, and $\|\beta\|_{F_U} = \|\tilde\beta\|_{F_3}$. Conversely, if $\tilde\beta\in [F_3]_{X^U}$, then 'reverse' the definition in (4.5).

If E is a countably infinite dissociate set in a discrete Abelian group Γ, enumerated as $E = \{\gamma_{ij} : (i,j) \in \mathbb{N}^2\}$, then

$$\mathrm{C}_{E^U}(G^3) = F_U(\mathbb{N}^3), \tag{4.6}$$

which is obtained by identifying $f \in \mathrm{C}_{E^U}(G^3)$ with $\beta_f \in F_U(\mathbb{N}^3)$:

$$\beta_f(i,j,k) = \hat{f}(\gamma_{ij} \otimes \gamma_{jk} \otimes \gamma_{ik}), \quad (i,j,k) \in \mathbb{N}^3. \tag{4.7}$$

In particular, Theorem 5 implies

$$\sigma_{E^U} = \frac{6}{5} \text{ exactly.} \tag{4.8}$$

Let us view the 'dual picture'. Let $V_U(X^3)$ denote the class of $\phi \in \mathrm{c}_0(X^3)$ such that

$$\phi(x,y,z) = \sum_\alpha f_\alpha(x,y)\, g_\alpha(y,z)\, h_\alpha(x,z), \tag{4.9}$$

and

$$\sum_\alpha \|f_\alpha\|_\infty \, \|g_\alpha\|_\infty \, \|h_\alpha\|_\infty < \infty, \tag{4.10}$$

where the f_α, g_α, and h_α are in $\mathrm{c}_0(X^2)$. The norm $\|\phi\|_{V_U}$ is the infimum of (4.10) over all representations of ϕ by (4.9). Note that $V_U(X^3)$ can be canonically identified with the algebra of restrictions to X^U of elements in $V_3(X^2, X^2, X^2)$, which we denote by $V_3(X^2, X^2, X^2)|_{X^U}$. Also note

$$V_U(X^3)^* = F_U(X^3), \tag{4.11}$$

or, in the language of harmonic analysis,

$$A(E^U)^* = \mathrm{C}_{E^U}(\hat{\Gamma}^3), \tag{4.12}$$

and

$$(\mathrm{C}_{E^U}(\hat{\Gamma}^3))^* = B(E^U) = \tilde{V}_U(E^U), \tag{4.13}$$

where E is a dissociate set in Γ. (Cf. (3.34) and Lemma VIII.15.) Moreover, if f, g, and h are in $l^\infty(X^2)$, and $(f \otimes g \otimes h)^U$ is the function (an elementary U-tensor) defined by

$$(f \otimes g \otimes h)^U(x,y,z) = f(x,y)\, g(y,z)\, h(x,z), \quad (x,y,z) \in X^3, \tag{4.14}$$

then the action of $\beta \in F_U(X^3)$ on $(f \otimes g \otimes h)^U$ is well defined,

$$\begin{aligned}
&\hat{\beta}((f \otimes g \otimes h)^U) \\
&:= \sum_{x,y} \left(\sum_z \beta(x,y,z)\; g(y,z)\; h(x,z) \right) f(x,y) \\
&= \sum_z \left(\sum_{x,y} \beta(x,y,z)\; g(y,z)\; h(x,z) f(x,y) \right), \qquad (4.15)
\end{aligned}$$

and

$$|\hat{\beta}((f \otimes g \otimes h)^U)| \leq 8\; \|\beta\|_{F_U} \|f\|_\infty\; \|g\|_\infty\; \|h\|_\infty. \qquad (4.16)$$

(See Exercises 2, 3.)

Remarks:

i **(tensor products of intermediate 'dimension').** The space $F_U(X^3)$ is situated between F_1 and F_2, in the sense that

$$l^1(X^3) := F_1(X^3) \subset F_U(X^3) \subset F_2(X^2, X). \qquad (4.17)$$

The left inclusion is obvious, and the inclusion on the right – also easy to verify – follows from (4.16). That both are *proper* follows from the measurements $\sigma_R = 1$, $\sigma_{R^U} = 6/5$, and $\sigma_{R^2} = 4/3$, which precisely 'locate' the respective spaces in (4.17). (The proper left inclusion was key to the existence of bounded trilinear functionals on a Hilbert space failing a Grothendieck-type inequality; see Theorem VIII.17.)

The statement dual to (4.17) is

$$V_2(X^2, X) \subsetneqq V_U(X^3) \subsetneqq V_1(X^3) := l^\infty(X^3). \qquad (4.18)$$

ii **(type F_U and type V_U).** A scalar-valued function β defined on X has *type* F_k, $k \in \mathbb{N}$, if there exists a bijection τ from X^k onto X such that $\beta \circ \tau \in F_k(X, \ldots, X)$. We denote the class of *type* F_k-functions by $\mathscr{F}_k = \mathscr{F}_k(X)$. Note the proper inclusion $\mathscr{F}_k \subsetneqq \mathscr{F}_{k+1}$, whose proof uses the full force of $\sigma_{R^k} = 2k/(k+1)$. On the 'dual' side, a scalar-valued function φ on X has *type* V_k, $k \in \mathbb{N}$, if for all bijections τ from X^k onto X, $\varphi \circ \tau \in V_k(X, \ldots, X)$. We denote the class of all such φ by $\mathscr{V}_k = \mathscr{V}_k(X)$, and observe $\mathscr{V}_{k+1} \subsetneqq \mathscr{V}_k$, which, again, uses the full force of $\sigma_{R^k} = 2k/(k+1)$. (See Remark iii in Chapter VII §11.)

Similarly, we say that a function β on X has *type* F_U if there exists a bijection τ from X^3 onto X such that $\beta \circ \tau \in F_U(X^3)$, and denote the class of such β by $\mathscr{F}_U$. On the 'dual' side, a scalar-valued function φ on X has *type* V_U if for all bijections τ from X^3 onto X, $\varphi \circ \tau \in V_U(X^3)$; we denote the class of all *type* V_U-functions by $\mathscr{V}_U$. Observe (Exercise 4)

$$\mathscr{F}_1 \subsetneqq \mathscr{F}_U \subsetneqq \mathscr{F}_2, \tag{4.19}$$

and

$$\mathscr{V}_2 \subsetneqq \mathscr{V}_U \subsetneqq \mathscr{V}_1.$$

This was previewed in Chapter VII §11, where $U_5 = \{\{1,2\},\{2,3\}, \{1,3\}\}$ corresponds to U in this chapter.

In the previous chapter, we proposed *'type'* as a measurement of stochastic interdependence (Chapter XI §7, Remark iii). We shall revisit a stochastic interpretation of *type* in the next remark, in the next section, and again in the next chapter.

iii **(a meaning...).** To start, we calibrate a countably infinite time-scale by $\mathbb{N}$. We let $q(n)$ denote the numerical value of a function q at time $n \in \mathbb{N}$, $q(0) = 0$, and let $\beta := \Delta_q$ be the sequence of increments of q over two consecutive points in time,

$$q(n) - q(n-1) = \Delta q(n) = \beta(n), \quad n \in \mathbb{N}. \tag{4.20}$$

Then, the 'final value' of q is the infinite sum

$$q_f := \sum_{n=1}^{\infty} \Delta_q(n) = \sum_{n=1}^{\infty} \beta(n). \tag{4.21}$$

We consider the question: what time-rearrangements of Δ_q produce the same q_f? (Compare this to questions stated in Chapter XI §1.)

If $\beta \in \mathscr{F}_1(\mathbb{N})$, then it matters not how (4.21) is summed: every time-rearrangement of increments leads to the same final value. With the weaker hypothesis $\beta \in \mathscr{F}_2(\mathbb{N})$, i.e., that there exists a bijection n from $\mathbb{N}^2$ onto $\mathbb{N}$ such that $\beta \circ n \in F_2(\mathbb{N},\mathbb{N})$, we obtain, by Theorem IV.6 and Corollary IV.7,

$$\begin{aligned} q_f &:= \lim_{N\to\infty} \sum_{j,k=1}^{N} (\beta \circ n)(j,k) \\ &= \sum_{j=1}^{\infty}\left(\sum_{k=1}^{\infty}(\beta \circ n)(j,k)\right) \\ &= \sum_{k=1}^{\infty}\left(\sum_{j=1}^{\infty}(\beta \circ n)(j,k)\right). \end{aligned} \tag{4.22}$$

In this case, q_f is unaffected by rearrangements ρ of the j-axis and τ of the k-axis:

$$\begin{aligned} q_f &= \sum_{j=1}^{\infty}\sum_{k=1}^{\infty}(\beta \circ n)(\rho j,\tau k) \\ &= \sum_{k=1}^{\infty}\sum_{j=1}^{\infty}(\beta \circ n)(\rho j,\tau k) \\ &= \lim_{N\to\infty} \sum_{j,k=1}^{N} (\beta \circ n)(\rho j,\tau k). \end{aligned} \tag{4.23}$$

Moreover, if $\beta \notin \mathscr{F}_1(\mathbb{N})$, then there exist (many!) rearrangements σ of $\mathbb{N}$, such that $\sum_{n=1}^{\infty} \beta(\sigma n)$ converge to different values. In essence, $\beta \notin \mathscr{F}_1(\mathbb{N})$ conveys that it indeed matters when increments happen, while $\beta \in \mathscr{F}_2(\mathbb{N})$ conveys a 'type' of admissible time-rearrangements of Δ_q that leave q_f unchanged. Precisely put, $\beta \in \mathscr{F}_2(\mathbb{N})\backslash\mathscr{F}_1(\mathbb{N})$ means that the underlying time-scale can be viewed as a plane spanned by two independent time-directions, along each of which increments Δ_q can be permuted freely without affecting the final outcome, but that this *time-plane* cannot be replaced by a *time-line*, along which the increments Δ_q could be freely interchanged and summed to the same q_f.

With increasing k, the assertion $\beta \in \mathscr{F}_k(\mathbb{N})\backslash\mathscr{F}_{k-1}(\mathbb{N})$ conveys increasing time-interdependence. The time-scale associated with β $(=\Delta q)$ can be realized as a k-dimensional domain spanned by k independent directions, along each of which increments can be rearranged without affecting the final outcome, *and* this is optimal: time cannot be represented by a $(k-1)$-dimensional domain with the same effect.

Given β in $l^2(\mathbb{N})$ (Exercise 5), we consider its 'optimal type'

$$\min\{k \in \mathbb{N} : \beta \in \mathscr{F}_k(\mathbb{N})\}, \tag{4.24}$$

which, so far in the discussion (as per (4.24)), is a positive integer. Indeed, could there be 'optimal types' that fall between k and $k+1$? To motivate matters, let us consider (4.17), where the left proper inclusion implies existence of β with *type* $\mathscr{F}_U$, but not *type* $\mathscr{F}_1$. For this β, there exists a bijection n from $\mathbb{N}^3$ onto $\mathbb{N}$, such that $\beta \circ n \in F_U(\mathbb{N}^3)$ and $\beta \notin F_1(\mathbb{N})$. For this bijection, by Corollary IV.7,

$$\begin{aligned} q_f &:= \sum_i \left(\sum_{j,k} \beta(n(i,j,k)) \right) \\ &= \sum_{j,k} \left(\sum_i \beta(n(i,j,k)) \right) \\ &= \sum_j \left(\sum_{i,k} \beta(n(i,j,k)) \right) \\ &= \sum_k \left(\sum_{i,j} \beta(n(i,j,k)) \right), \end{aligned} \tag{4.25}$$

which is unaffected by rearrangements of the i-axis, j-axis, k-axis, the (j,k)-plane, (i,k)-plane, and (i,j)-plane. Put precisely, the hypothesis $\beta \circ n \in F_U(\mathbb{N}^3)$ implies that for all rearrangements ρ_1, ρ_2 and ρ_3 of $\mathbb{N}^2$,

$$q_f = \lim_{N\to\infty} \sum_{i_1,j_1,i_2,j_2,i_3,j_3=1}^{N} (\beta \circ n)^\sim(\rho_1(i_1,j_1), \rho_2(i_2,j_2), \rho_3(i_3,j_3)), \tag{4.26}$$

where $(\beta \circ n)^\sim$ is defined by (4.5). (In (4.5), replace β by $\beta \circ n$, and X by $\mathbb{N}$.) This means that time can be spanned by three interdependent 'time-directions' (the aforementioned planes), along each of which increments Δ_q can be freely rearranged, without affecting the final outcome. Because $\beta \notin F_1(\mathbb{N})$, this 'time-box' cannot be replaced, with the same effect, by a one-dimensional scale: for every $q \in [-\infty, \infty]$, there exist rearrangements σ of $\mathbb{N}$ such that $q = \sum_{n=1}^{\infty} \beta(\sigma n)$. And we thus imagine that the 'optimal type' of β, an index of time-interdependence, is somewhere between 1 and 2.

Similarly, the proper inclusion on the right side of (4.17) implies there exist β that have *type* $\mathscr{F}_2$, but not *type* $\mathscr{F}_U$. For such β, there exist bijections n from $\mathbb{N}^3$ onto $\mathbb{N}$ with the property that

$$\begin{aligned} q_f &:= \sum_i \left(\sum_{j,k} \beta(n(i,j,k)) \right) \\ &= \sum_{j,k} \left(\sum_i \beta(n(i,j,k)) \right), \end{aligned} \tag{4.27}$$

and q_f is the same for all rearrangements of the i-axis and the (j,k)-plane. In this case, the time-domain is spanned by two independent directions – the i-axis and the (j,k)-plane – along which Δ_q can be freely rearranged without affecting the 'final' q. Because $\beta \notin \mathscr{F}_U$, the underlying time-scale cannot be a domain parameterized by interdependent (j,k)-, (i,k)-, and (i,j)-planes, along which increments can be freely permuted. In this case, the 'optimal type' of β falls somewhere between 2 and 3 (Exercise 6).

iv **(a question).** For a scalar-valued function β on $\mathbb{N}^3$, define

$$\begin{aligned} \|\beta\|_{F_{U_1}} &:= \sup \left\{ \left\| \sum_{(i,j)\in A, k \in B} \beta(i,j,k)\, r_{ij} \otimes r_k \right\|_\infty : \right. \\ &\qquad \left. \text{finite sets } A \subset X^2, B \subset X \right\}, \\ \|\beta\|_{F_{U_2}} &:= \sup \left\{ \left\| \sum_{(j,k)\in A, i \in B} \beta(i,j,k)\, r_{jk} \otimes r_i \right\|_\infty : \right. \\ &\qquad \left. \text{finite sets } A \subset X^2, B \subset X \right\}, \end{aligned} \tag{4.28}$$

and

$$\|\beta\|_{F_{U_3}} := \sup\left\{\left\|\sum_{(i,k)\in A, j\in B}\beta(i,j,k)\; r_{ik}\otimes r_j\right\|_\infty : \text{finite sets } A\subset X^2, B\subset X\right\}.$$

If $\|\beta\|_{F_U} < \infty$, then $\|\beta\|_{F_{U_l}} < \infty$ for each $l = 1,2,3$. Is the converse true?

5 Fréchet Measures in 'Dimension' 3/2

As in Chapter VII §11 and Chapter VIII §7, the symbol U in this chapter stands for the cover $\{S_1, S_2, S_3\}$ of [3], where $S_1 = (1,2)$, $S_2 = (2,3)$, $S_3 = (1,3)$. For sets X, Y, and Z, we consider the projections π_{S_i}, $i = 1,2,3$, from $X\times Y\times Z$ onto $X\times Y$, $Y\times Z$, and $X\times Z$,

$$\pi_{S_1}(x,y,z) = (x,y),\;\; \pi_{S_2}(x,y,z) = (y,z),$$
$$\pi_{S_3}(x,y,z) = (x,z). \tag{5.1}$$

Define

$$(X,Y,Z)^U = \{(\pi_{S_1}\mathbf{x},\; \pi_{S_2}\mathbf{x},\; \pi_{S_3}\mathbf{x}) : \mathbf{x}\in X\times Y\times Z\}, \tag{5.2}$$

which is a subset of the *three-fold* Cartesian product

$$(X\times Y)\times(Y\times Z)\times(X\times Z).$$

For $X = Y = Z$, we write $X^U = (X,X,X)^U$; see (4.3).

Let $(X,\mathfrak{A})$, $(Y,\mathfrak{B})$, and $(Z,\mathfrak{C})$ be measurable spaces, and suppose $\mu\in F_3(\mathfrak{A},\mathfrak{B},\mathfrak{C})$. Consider

$$\begin{aligned}&\mu(\pi_{S_1}^{-1}[A_1\times B_1]\cap\pi_{S_2}^{-1}[B_2\times C_2]\cap\pi_{S_3}^{-1}[A_3\times C_3])\\ &\quad= \mu((A_1\cap A_3)\times(B_1\cap B_2)\times(C_2\times C_3)),\\ &\qquad A_1\times B_1\in\mathfrak{A}\times\mathfrak{B},\;\; B_2\times C_2\in\mathfrak{B}\times\mathfrak{C},\;\; A_3\times C_3\in\mathfrak{A}\times\mathfrak{C},\end{aligned} \tag{5.3}$$

and extend it to a finitely additive set-function on the three-fold Cartesian product$a(\mathfrak{A}\times\mathfrak{B}) \times a(\mathfrak{B}\times\mathfrak{C}) \times a(\mathfrak{A}\times\mathfrak{C})$,

$$\begin{aligned}\tilde{\mu}(E,F,G) &:= \mu(\pi_{S_1}^{-1}[E] \cap \pi_{S_2}^{-1}[F] \cap \pi_{S_3}^{-1}[G]),\\ &\qquad E \in a(\mathfrak{A}\times\mathfrak{B}),\ F \in a(\mathfrak{B}\times\mathfrak{C}),\ G \in a(\mathfrak{A}\times\mathfrak{C}). \qquad (5.4)\end{aligned}$$

(In Chapter VI, the evaluation of μ at $(A,B,C) \in \mathfrak{A}\times\mathfrak{B}\times\mathfrak{C}$ was denoted by $\mu(A,B,C)$; when there is no confusion, and the context calls for it, we will write $\mu(A\times B\times C)$ for $\mu(A,B,C)$.) If $\tilde{\mu}$ determines an F_3-measure on $\sigma(\mathfrak{A}\times\mathfrak{B})\times\sigma(\mathfrak{B}\times\mathfrak{C})\times\sigma(\mathfrak{A}\times\mathfrak{C})$, then we say that μ is an F_U-measure on $\mathfrak{A}\times\mathfrak{B}\times\mathfrak{C}$, and denote the class of all such μ by $F_U(\mathfrak{A}\times\mathfrak{B}\times\mathfrak{C})$; for $\mathfrak{A} = 2^X, \mathfrak{B} = 2^Y$, and $\mathfrak{C} = 2^Z$, we write $\mu \in F_U(X\times Y\times Z)$ (Exercise 7).

Proposition 11 *Suppose $\mu \in F_3(\mathfrak{A},\mathfrak{B},\mathfrak{C})$, and $\tilde{\mu}$ is an F_3-measure on $a(\mathfrak{A}\times\mathfrak{B})\times a(\mathfrak{B}\times\mathfrak{C})\times a(\mathfrak{A}\times\mathfrak{C})$. Then, $\mu \in F_U(\mathfrak{A}\times\mathfrak{B}\times\mathfrak{C})$ if and only if*

$$\begin{aligned}\|\mu\|_{F_U(\mathfrak{A}\times\mathfrak{B}\times\mathfrak{C})} &= \|\mu\|_{F_U}\\ &:= \sup\left\{\left\|\sum_{(A,B,C)\in\mathfrak{g}} \mu(A,B,C)\ r_{AB}\otimes r_{BC}\otimes r_{AC}\right\|_\infty :\right.\\ &\qquad\qquad \left.(\mathfrak{A}\times\mathfrak{B}\times\mathfrak{C})\text{-grid } \mathfrak{g}\right\} < \infty, \qquad (5.5)\end{aligned}$$

where Rademacher systems are indexed, respectively, by $\pi_{S_1}[\mathfrak{g}]$, $\pi_{S_2}[\mathfrak{g}]$, and $\pi_{S_3}[\mathfrak{g}]$.

Proof: Note $\|\mu\|_{F_U(\mathfrak{A}\times\mathfrak{B}\times\mathfrak{C})} = \|\tilde{\mu}\|_{F_3(a(\mathfrak{A}\times\mathfrak{B}),a(\mathfrak{B}\times\mathfrak{C}),a(\mathfrak{A}\times\mathfrak{C}))}$, and apply Theorem VI.8. □

An F_3-measure μ on $\sigma(\mathfrak{A}\times\mathfrak{B})\times\sigma(\mathfrak{B}\times\mathfrak{C})\times\sigma(\mathfrak{A}\times\mathfrak{C})$ has support in $(X,Y,Z)^U$ if

$$\mu(A_1\times B_1, B_2\times C_2, A_3\times C_3) = \mu(A_1'\times B_1', B_2'\times C_2', A_3'\times C_3') \qquad (5.6)$$

whenever $A_1\cap A_3 = A_1'\cap A_3'$, $B_1\cap B_2 = B_1'\cap B_2'$, and $C_2\cap C_3 = C_2'\cap C_3'$. The class of F_3-measures on $\sigma(\mathfrak{A}\times\mathfrak{B})\times\sigma(\mathfrak{B}\times\mathfrak{C})\times\sigma(\mathfrak{A}\times\mathfrak{C})$ with support in $(X,Y,Z)^U$ is denoted by $[F_3]_{(X,Y,Z)^U}$, and is naturally identified with $F_U(\mathfrak{A}\times\mathfrak{B}\times\mathfrak{C})$ (cf. (4.4) and (4.5)). If $\mu \in F_U(\mathfrak{A}\times\mathfrak{B}\times\mathfrak{C})$, then $\tilde{\mu} \in [F_3]_{(X,Y,Z)^U}$ (practically by definition). If $\nu \in [F_3]_{(X,Y,Z)^U}$, then define

$$\mu(A,B,C) = \nu(A\times B, B\times C, A\times C),$$
$$(A,B,C)\in \mathfrak{A}\times\mathfrak{B}\times\mathfrak{C}, \tag{5.7}$$

and note that $\mu \in F_U(\mathfrak{A}\times\mathfrak{B}\times\mathfrak{C})$ (Exercise 8). We summarize:

$$[F_3]_{(X,Y,Z)^U} = \{\nu \in F_3(\sigma(\mathfrak{A}\times\mathfrak{B}), \sigma(\mathfrak{B}\times\mathfrak{C}), \sigma(\mathfrak{A}\times\mathfrak{C})) :$$
$$\exists\ \mu \in F_3(\mathfrak{A},\mathfrak{B},\mathfrak{C}) \text{ such that } \nu = \tilde{\mu}\}. \tag{5.8}$$

If $f \in \mathrm{L}^\infty(\sigma(\mathfrak{A}\times\mathfrak{B}))$, $g \in \mathrm{L}^\infty(\sigma(\mathfrak{B}\times\mathfrak{C}))$, and $h \in \mathrm{L}^\infty(\sigma(\mathfrak{A}\times\mathfrak{C}))$, then $(f\otimes g\otimes h)^U$ (an elementary U-tensor) is the function on $X\times Y\times Z$ defined by

$$(f\otimes g\otimes h)^U(x,y,z) = f(x,y)\ g(y,z)\ h(x,z),$$
$$(x,y,z)\in X\times Y\times Z. \tag{5.9}$$

The integral of $(f\otimes g\otimes h)^U$ with respect to $\mu \in F_U(\mathfrak{A}\times\mathfrak{B}\times\mathfrak{C})$ is

$$\int_{X\times Y\times Z} (f\otimes g\otimes h)^U \mathrm{d}\mu$$
$$:= \int_{(X\times Y)\times(Y\times Z)\times(X\times Z)} f\otimes g\otimes h\ \mathrm{d}\tilde{\mu}. \tag{5.10}$$

Suppose that X, Y, and Z are locally compact Hausdorff spaces, and that $\mathfrak{A}, \mathfrak{B}$, and $\mathfrak{C}$ are their respective Borel fields. Let $V_U(X\times Y\times Z)$ denote the class of functions ϕ on $X\times Y\times Z$ that can be represented pointwise on $X\times Y\times Z$ by

$$\phi = \sum_k (f_k\otimes g_k\otimes h_k)^U, \tag{5.11}$$

where $f_k \in \mathrm{C}_0(X\times Y)$, $g_k\in \mathrm{C}_0(Y\times Z)$, and $h_k \in \mathrm{C}_0(X\times Z)$, $k\in\mathbb{N}$, and

$$\sum_k \|f_k\|_\infty\ \|g_k\|_\infty\ \|h_k\|_\infty < \infty. \tag{5.12}$$

The V_U-norm of ϕ is the infimum of the left side of (5.12) over all representations of ϕ by (5.11). Equivalently, let

$$I_U = \{\phi \in V_3(X\times Y,\ Y\times Z,\ X\times Z) : \phi \equiv 0 \text{ on } (X,Y,Z)^U\}, \tag{5.13}$$

and consider the restriction algebra (quotient algebra)

$$V_3|_{(X,Y,Z)^U} = V_3(X{\times}Y,\ Y{\times}Z,\ X{\times}Z)/I_U. \tag{5.14}$$

Observe that

$$V_U(X{\times}Y{\times}Z) = V_3|_{(X,Y,Z)^U}, \tag{5.15}$$

that the annihilator of I_U in $F_3(\sigma(\mathfrak{A} \times \mathfrak{B}), \sigma(\mathfrak{B} \times \mathfrak{C}), \sigma(\mathfrak{A} \times \mathfrak{C}))$ is $[F_3]_{(X,Y,Z)^U}$, and obtain (Exercise 9)

Theorem 12 (a Riesz Representation theorem in 'dimension' 3/2; cf. (4.12)). *If X, Y, and Z are locally compact Hausdorff spaces with respective Borel fields $\mathfrak{A}, \mathfrak{B}$, and $\mathfrak{C}$, then*

$$V_U(X{\times}Y{\times}Z)^* = F_U(\mathfrak{A}{\times}\mathfrak{B}{\times}\mathfrak{C}). \tag{5.16}$$

Remark (another space?). Let $(X, \mathfrak{A})$, $(Y, \mathfrak{B})$, and $(Z, \mathfrak{C})$ be measurable spaces, and note the inclusion

$$F_U(\mathfrak{A}{\times}\mathfrak{B}{\times}\mathfrak{C}) \subset F_2(\mathfrak{A}, \sigma(\mathfrak{B}{\times}\mathfrak{C})) \cap F_2(\mathfrak{B}, \sigma(\mathfrak{A}{\times}\mathfrak{C})) \cap\ F_2(\mathfrak{C}, \sigma(\mathfrak{A}{\times}\mathfrak{B})). \tag{5.17}$$

Is this inclusion proper? The question is closely related to the problem in Remark iv in the previous section.

6 Product F-measures and Projective Boundedness in 'Dimension' 3/2

Let $(X_1, \mathfrak{A}_1), (X_2, \mathfrak{A}_2), (X_3, \mathfrak{A}_3), (Y_1, \mathfrak{B}_1), (Y_2, \mathfrak{B}_2)$, and $(Y_3, \mathfrak{B}_3)$ be measurable spaces with infinite underlying σ-algebras. Let

$$\mu \in F_U(\mathfrak{A}_1 \times \mathfrak{A}_2 \times \mathfrak{A}_3),\ \nu \in F_U(\mathfrak{B}_1{\times}\mathfrak{B}_2{\times}\mathfrak{B}_3),$$

and define the product (cf. (IX.1.2))

$$\mu \times \nu((A_1, B_1), (A_2, B_2), (A_3, B_3)) = \mu(A_1, A_2, A_3)\nu(B_1, B_2, B_3),$$
$$(A_1, A_2, A_3) \in \mathfrak{A}_1 \times \mathfrak{A}_2 \times \mathfrak{A}_3,\ (B_1, B_2, B_3) \subset \mathfrak{B}_1 \times \mathfrak{B}_2 \times \mathfrak{B}_3. \tag{6.1}$$

Question: *Does $\mu \times \nu$ determine an F_U-measure on $\sigma(\mathfrak{A}_1 \times \mathfrak{B}_1) \times \sigma(\mathfrak{A}_2 \times \mathfrak{B}_2) \times \sigma(\mathfrak{A}_3 \times \mathfrak{B}_3)$?*

Recall that in Chapter IX we proved that if $\mu \in F_m(\mathfrak{A}_1, \ldots, \mathfrak{A}_m)$, then $\mu \times \nu$ determines an F_m-measure for all $\nu \in F_m(\mathfrak{B}_1, \ldots, \mathfrak{B}_m)$ precisely when $\mu \subset PBF_m(\mathfrak{A}_1, \ldots, \mathfrak{A}_m)$ (Definition IX.5, Theorem IX.6). We also noted that $F_2(\mathfrak{A}_1, \mathfrak{A}_2) = PBF_2(\mathfrak{A}_1, \mathfrak{A}_2)$ (Theorem IX.9), and that for all $m > 2$, if $\mathfrak{A}_1, \ldots, \mathfrak{A}_m$ are infinite, then (Theorem IX.10) $PBF_m(\mathfrak{A}_1, \ldots, \mathfrak{A}_m) \subsetneqq F_m(\mathfrak{A}_1, \ldots, \mathfrak{A}_m)$.

For the question concerning product F_U-measures, we can take (with no loss of generality) $(X_1, \mathfrak{A}_1) = (X_2, \mathfrak{A}_2) = (X_3, \mathfrak{A}_3) = (X, \mathfrak{A})$, $(Y_1, \mathfrak{B}_1) = (Y_2, \mathfrak{B}_2) = (Y_3, \mathfrak{B}_3) = (Y, \mathfrak{B})$. By Proposition 11, $\mu \times \nu$ determines an F_U-measure precisely when there exists $K > 0$ such that for all finite partitions $\{A_i\}, \{B_i\}, \{C_i\}$ of $(X, \mathfrak{A})$, and finite partitions $\{A'_i\}, \{B'_i\}, \{C'_i\}$ of $(Y, \mathfrak{B})$,

$$\left\| \sum_{\substack{i_1, j_1, k_1 \\ i_2, j_2, k_2}} \mu(A_{i_1}, B_{j_1}, C_{k_1}) \nu(A'_{i_2}, B'_{j_2}, C'_{k_2}) r_{i_1 i_2 j_1 j_2} \otimes r_{j_1 j_2 k_1 k_2} \otimes r_{i_1 i_2 k_1 k_2} \right\|_\infty \tag{6.2}$$

is bounded by K. For ω_1, ω_2, and ω_3 in $\{-1,1\}^{\mathbb{N}^4}$, define simple functions on X^2

$$\begin{aligned} f_{i_1 j_1} &= \sum_{i_2, j_2} r_{i_1 i_2 j_1 j_2}(\omega_1)\, \mathbf{1}_{A_{i_2}} \otimes \mathbf{1}_{B_{j_2}}, \ (i_1, j_1) \in \mathbb{N}^2, \\ f_{j_1 k_1} &= \sum_{j_2, k_2} r_{j_1 j_2 k_1 k_2}(\omega_2)\, \mathbf{1}_{B_{j_2}} \otimes \mathbf{1}_{C_{k_2}}, \ (j_1, k_1) \in \mathbb{N}^2, \\ f_{i_1 k_1} &= \sum_{i_2, k_2} r_{i_1 i_2 k_1 k_2}(\omega_3)\, \mathbf{1}_{A_{i_2}} \otimes \mathbf{1}_{C_{k_2}}, \ (i_1, k_1) \in \mathbb{N}^2. \end{aligned} \tag{6.3}$$

Then,

$$\begin{aligned} &\sum_{i_2, j_2, k_2} \mu(A_{i_2}, B_{j_2}, C_{k_2})\, r_{i_1 i_2 j_1 j_2}(\omega_1)\, r_{j_1 j_2 k_1 k_2}(\omega_2)\, r_{i_1 i_2 k_1 k_2}(\omega_3) \\ &= \int_{X^3} f_{i_1 j_1} \otimes f_{j_1 k_1} \otimes f_{i_1 k_1} \mathrm{d}\mu \\ &:= \phi_\mu(i_1, j_1, k_1). \end{aligned} \tag{6.4}$$

If $\|\phi_\mu\|_{V_U} \leq K$, where $K > 0$ depends only on μ, then (6.2) is bounded by K. This leads to the definition: for $\mu \in F_U(\mathfrak{A}_1 \times \mathfrak{A}_2 \times \mathfrak{A}_3)$, $f_1 \in \mathrm{L}^\infty(\sigma(\mathfrak{A}_1 \times \mathfrak{A}_2) \times \mathbb{N}^2)$, $f_2 \in \mathrm{L}^\infty(\sigma(\mathfrak{A}_2 \times \mathfrak{A}_3) \times \mathbb{N}^2)$, and $f_3 \in$

$L^\infty(\sigma(\mathfrak{A}_1 \times \mathfrak{A}_3) \times \mathbb{N}^2)$, let

$$\phi_{\mu;f_1,f_2,f_3}(i,j,k) = \int_{X^3} (f_1(\cdot,(i,j)) \otimes f_2(\cdot,(j,k)) \otimes f_3(\cdot,(i,k)))^U \mathrm{d}\mu, \quad (i,j,k) \in \mathbb{N}^3 \tag{6.5}$$

(cf. (6.4)), and then define

$$\|\mu\|_{\mathrm{pb}_U} := \sup\{\|\phi_{\mu;f_1,f_2,f_3}\|_{V_U([N]^3)} : \|f_1\|_\infty \leq 1, \|f_1\|_\infty \leq 1, \|f_1\|_\infty \leq 1, N \in \mathbb{N}\}. \tag{6.6}$$

We say that μ is *U-projectively bounded* if $\|\mu\|_{\mathrm{pb}_U} < \infty$, and denote the space of all such μ by $PBF_U(\mathfrak{A}_1 \times \mathfrak{A}_2 \times \mathfrak{A}_3)$.

Theorem 13 (Cf. Theorem IX.6; Exercise 10). *If*

$$\mu \in F_U(\mathfrak{A}_1 \times \mathfrak{A}_2 \times \mathfrak{A}_3),$$

then $\mu \times \nu$ is an F_U-measure on $\sigma(\mathfrak{A}_1 \times \mathfrak{B}_1) \times \sigma(\mathfrak{A}_2 \times \mathfrak{B}_2) \times \sigma(\mathfrak{A}_3 \times \mathfrak{B}_3)$ for all $\nu \in F_U(\mathfrak{B}_1 \times \mathfrak{B}_2 \times \mathfrak{B}_3)$ if and only if $\|\mu\|_{\mathrm{pb}_U} < \infty$.

Remark (a 'fractional' Grothendieck-type inequality?).

If $\mu \in F_U(\mathfrak{A}_1 \times \mathfrak{A}_2 \times \mathfrak{A}_3)$, then μ is in

$$F_2(\mathfrak{A}_1, \sigma(\mathfrak{A}_2 \times \mathfrak{A}_3)) \cap F_2(\mathfrak{A}_2, \sigma(\mathfrak{A}_1 \times \mathfrak{A}_3)) \cap F_2(\mathfrak{A}_3, \sigma(\mathfrak{A}_1 \times \mathfrak{A}_2)).$$

By Theorem IX.9, which is in essence a two-dimensional Grothendieck-type inequality, μ is in

$$PBF_2(\mathfrak{A}_1, \sigma(\mathfrak{A}_2 \times \mathfrak{A}_3)) \cap PBF_2(\mathfrak{A}_2, \sigma(\mathfrak{A}_1 \times \mathfrak{A}_3)) \cap PBF_2(\mathfrak{A}_3, \sigma(\mathfrak{A}_1 \times \mathfrak{A}_2)).$$

Can we conclude from this $\mu \in PBF_U(\mathfrak{A}_1 \times \mathfrak{A}_2 \times \mathfrak{A}_3)$? The problem is related to the question in the last remark in the previous section (Exercise 11).

Exercises

1. Let $E = \{\gamma_{ij} : (i,j) \in \mathbb{N}^2\}$ be a dissociate set in a discrete Abelian group Γ, and define

   $$E_U = \{\gamma_{ij}\,\gamma_{jk}\,\gamma_{ik} : (i,j,k) \in D_3\},$$

   where $D_3 = \{(i,j,k) : (i,j,k) \in \mathbb{N}^3, 0 < i < j < k\}$. Prove that $\sigma_{E_U} = 6/5$ and $\delta_{E_U} = 3/4$.

2. Let X be a countably infinite set.

   i. Prove that for $\beta \in F_U(X^3)$, f, g, and h in $l^\infty(X^2)$,

$$\begin{aligned}\hat{\beta}((f \otimes g \otimes h)^U) &:= \sum_{x,y,z} \beta(x,y,z)\ f(x,y)\ g(y,z)\ h(x,z) \\ &= \sum_{x,y} \left(\sum_{z} \beta(x,y,z)\ g(y,z)\ h(x,z) f(x,y) \right) \\ &= \sum_{z} \left(\sum_{x,y} \beta(x,y,z)\ g(y,z)\ h(x,z) f(x,y) \right)\end{aligned}$$

   is well-defined, and

$$|\hat{\beta}((f \otimes g \otimes h)^U)| \leq 8\ \|\beta\|_{F_U}\ \|f\|_\infty\ \|g\|_\infty\ \|h\|_\infty.$$

   ii. Verify $V_U(X^3)^* = F_U(X^3)$ (a 'bare-bone' Riesz Representation-type theorem in 'dimension' 3/2).

3. Verify that if $E \subset \Gamma$ is a dissociate set, then

$$(\mathrm{C}_{E^U}(G^3))^* = B(E^U) = \tilde{V}_U(E^U).$$

4. Prove $\mathscr{F}_1(X) \subsetneqq \mathscr{F}_U(X) \subsetneqq \mathscr{F}_2(X)$, and $\mathscr{V}_2(X) \subsetneqq \mathscr{V}_U(X) \subsetneqq \mathscr{V}_1(X)$.
5. Prove that if X is countably infinite, and β is a scalar-valued function on X with type $\mathscr{F}_k$ for some $k \in \mathbb{N}$, then $\beta \in l^2(X)$.
6. The discussion of a 'stochastic' meaning of *'type'* in Remark iii §4, by and large a heuristic discussion, contains precise claims that require verification. Read through again, and – if you have not done so already – verify these claims. Specifically, verify (4.22), (4.23), (4.25), (4.26), and (4.27).
7. Verify that the definition of $F_U(X^3)$ in §4 is equivalent to the definition of $F_U(X^3)$ in §5.
8. Verify

$$\begin{aligned}[F_3]_{(X,Y,Z)^U} = \{\nu \in F_3\ (\sigma(\mathfrak{A} \times \mathfrak{B}), \sigma(\mathfrak{B} \times \mathfrak{C}), \sigma(\mathfrak{A} \times \mathfrak{C})) : \\ \exists\ \mu \in F_3(\mathfrak{A}, \mathfrak{B}, \mathfrak{C}) \text{ such that } \nu = \tilde{\mu}\}.\end{aligned}$$

9. Prove Theorem 12.
10. Prove Theorem 13.

11. Show that if the inclusion in (5.16) is an equality, then

$$PBF_U(\mathfrak{A}_1 \times \mathfrak{A}_2 \times \mathfrak{A}_3) = F_U(\mathfrak{A}_1 \times \mathfrak{A}_2 \times \mathfrak{A}_3).$$

Hints for Exercises in Chapter XII

1. Use $\sigma_{E^U} = 6/5$, $\delta_{E^U} = 3/4$, and decoupling; cf. Chapter VII §8.
2. Identify $F_U(X^3)$ with $[F_3(X^2, X^2, X^2)]_{X^U}$, and use duality.
3. Basic facts concerning tilde algebras are found in Chapter VII §8, Remark ii. The exercise itself is in Chapter VIII §5.
4. Use: $\mathscr{F}_U(X) \subset l^{6/5}(X)$, $\mathscr{F}_2(X) \subset l^{4/3}(X)$; their 'duals' $l^6(X) \subset \mathscr{V}_U(X)$, $l^4(X) \subset \mathscr{V}_2(X)$; and that these are best possible.
5. Use basic harmonic analysis.
6. Review Theorem IV.6 and Corollary IV.7.
9. Supply the details in the discussion leading to the statement of the theorem. (See Exercise 2.)
10. Review the analogous result in integer dimensions in Chapter IX.
11. If $\mu \in F_U(\mathfrak{A}_1 \times \mathfrak{A}_2 \times \mathfrak{A}_3)$ and $\nu \in F_U(\mathfrak{B}_1 \times \mathfrak{B}_2 \times \mathfrak{B}_3)$, then

$$\mu \in F_2(\sigma(\mathfrak{A}_1 \times \mathfrak{A}_2), \mathfrak{A}_3) \cap F_2(\sigma(\mathfrak{A}_2 \times \mathfrak{A}_3), \mathfrak{A}_1) \cap F_2(\sigma(\mathfrak{A}_1 \times \mathfrak{A}_3), \mathfrak{A}_2)$$

and

$$\nu \in F_2(\sigma(\mathfrak{B}_1 \times \mathfrak{B}_2), \mathfrak{B}_3) \cap F_2(\sigma(\mathfrak{B}_2 \times \mathfrak{B}_3), \mathfrak{B}_1) \cap F_2(\sigma(\mathfrak{B}_1 \times \mathfrak{B}_3), \mathfrak{B}_2).$$

Write $\sigma(\mathfrak{A}_i \times \mathfrak{A}_j) = \mathfrak{A}_{ij}$ and $\sigma(\mathfrak{B}_i \times \mathfrak{B}_j) = \mathfrak{B}_{ij}$. Then, by Theorem IX.6 and Theorem IX.9, $\mu \times \nu$ is in

$$F_2(\sigma(\mathfrak{A}_{12} \times \mathfrak{B}_{12}), \sigma(\mathfrak{A}_3 \times \mathfrak{B}_3)) \cap F_2(\sigma(\mathfrak{A}_{23} \times \mathfrak{B}_{23}), \sigma(\mathfrak{A}_1 \times \mathfrak{B}_1))$$
$$\cap F_2(\sigma(\mathfrak{A}_{13} \times \mathfrak{B}_{13}), \sigma(\mathfrak{A}_2 \times \mathfrak{B}_2)),$$

and therefore (by the assumption that the inclusion in (5.16) is an equality), $\mu \times \nu$ is an F_U-measure on

$$\sigma(\mathfrak{A}_1 \times \mathfrak{B}_1) \times \sigma(\mathfrak{A}_2 \times \mathfrak{B}_2) \times \sigma(\mathfrak{A}_3 \times \mathfrak{B}_3).$$

XIII

Fractional Cartesian Products and Combinatorial Dimension

1 Mise en Scène: Fractional Products

In the previous chapter we considered

$$X^U = \{((x_1, x_2), (x_2, x_3), (x_1, x_3)) : (x_1, x_2, x_3) \in X^3\}, \tag{1.1}$$

and suggested, by appealing to 'harmonic-analytic' measurements, that this set was a 'fractional' Cartesian product of 'dimension' 3/2. In our basic context we think of 'dimension' as an index of *interdependence*. Indeed, that the 'dimension' of X^U is 3/2 marks precisely the 'level' of interdependence between the three canonical projections from X^3 onto the three 'coordinate planes'

$$\pi_{S_1}(x_1, x_2, x_3) = (x_1, x_2), \quad \pi_{S_2}(x_1, x_2, x_3) = (x_2, x_3),$$

$$\pi_{S_3}(x_1, x_2, x_3) = (x_1, x_3), \quad (x_1, x_2, x_3) \in X^3. \tag{1.2}$$

Let us restate and motivate the notion that 'dimension' conveys interdependence (cf. Chapter XII §1). Suppose $f_1, \ldots, f_n$ are functions from F onto $Y_1, \ldots, Y_n$, respectively, where F, $Y_1, \ldots, Y_n$ are sets (with no *a priori* underlying structures), and consider the question: how 'interdependent' are $f_1, \ldots, f_n$? Loosely put, the question is about the extent to which evaluations of some functions yield information about evaluations of others. There are two extremal cases: (1) $f_1, \ldots, f_n$ are (functionally) *independent*, and (2) $f_1, \ldots, f_n$ are (functionally) *dependent*. In (1), we mean that for arbitrary $y_1 \in Y_1, \ldots, y_n \in Y_n$ there exists $x \in F$ such that $f_1(x) = y_1, \ldots, f_n(x) = y_n$ (Definition VII.41). That is,

$f_1, \ldots, f_n$ are independent if their respective evaluations are completely unrestricted; concisely put, if

$$\{(f_1(x), \ldots, f_n(x)) : x \in F\} = Y_1 \times \cdots \times Y_n, \tag{1.3}$$

which is n-dimensional. In (2), $f_1, \ldots, f_n$ *dependent* means there exist maps $\varphi_1 \colon Y_n \to Y_1, \ldots, \varphi_{n-1} \colon Y_n \to Y_{n-1}$ such that $f_1 = \varphi_1 \circ f_n$, $\ldots, f_{n-1} = \varphi_{n-1} \circ f_n$. That is, evaluations of one function determine the evaluations of each of the other $n-1$ functions. In this case,

$$\begin{aligned}&\{(f_1(x), \ldots, f_n(x)) : x \in F\}\\ &\quad = \{(\varphi_1(f_n(x)), \ldots, \varphi_{n-1}(f_n(x)), f_n(x)) : x \in F\}\end{aligned} \tag{1.4}$$

is a one-dimensional subset of $Y_1 \times \cdots \times Y_n$. The 'fractional' product X^U in (1.1), is the instance $n = 3, F = X^3$, $Y_1 = Y_2 = Y_3 = X^2$, $f_1 = \pi_{S_1}$, $f_2 = \pi_{S_2}$, and $f_3 = \pi_{S_3}$. In this instance the 'dimension' of

$$\{(\pi_{S_1}(\mathbf{x}), \pi_{S_2}(\mathbf{x}), \pi_{S_3}(\mathbf{x})) : \mathbf{x} \in X^3\} \quad \text{(cf. (1.3) and (1.4))} \tag{1.5}$$

turns out to be $3/2$, which we interpret as a measurement of the interdependence of π_{S_1}, π_{S_2}, and π_{S_3}. In the general case, the degree of *interdependence* of $f_1, \ldots, f_n$ will be the 'dimension' of the range of the $Y_1 \times \cdots \times Y_n$-valued map $(f_1, \ldots, f_n)$. The problem is: how do we define 'dimension' and compute it?

In this chapter we first define and analyze the *fractional Cartesian products*, which arose, ostensibly by fiat, in analysis of multi-linear Grothendieck-type inequalities in Chapter VIII. Then, after gaining some insight, we will define and analyze the *combinatorial dimension.* In particular, we will link the measurement of *combinatorial dimension* to measurements of interdependence in harmonic–analytic and probabilistic contexts.

To start, let X be a set that we (may as well) take to be infinite. For subsets S and T of $\mathbb{N}$, we let $\pi_{T,S}$ denote the projection from X^T onto X^S,

$$\pi_{T,S}(x_i : i \in T) = (x_i : i \in S). \tag{1.6}$$

Let $U = \{S_1, \ldots, S_n\}$ be a list of subsets of T whose union is T; we refer to such U as *covers* of T. Consider the subset of the n-fold Cartesian product $X^{S_1} \times \cdots \times X^{S_n}$

$$X^U = \{(\pi_{S_1}\mathbf{x}, \ldots, \pi_{S_n}\mathbf{x}) : \mathbf{x} \in X^T\}. \tag{1.7}$$

We refer to X^U as a *fractional Cartesian product* based on U, and to $X^{S_1} \times \cdots \times X^{S_n}$ as its *ambient product.* If we view $T \subset \mathbb{N}$ as the cover comprising all singletons in T, then (1.7) is consistent with the usual definition of X^T. For a positive integer m, we write X^m for $X^{[m]}$, and for $S \subset [m]$, we denote $\pi_{[m],S}$ by π_S. In almost all cases we will represent T by the generic $[m] = \{1, \ldots, |T|\}$. In this terminology, X^U in (XII.4.3) (in (XII.2.6) with $X = R$, and in (XII.2.30) with $X = E$) is the instance $T = [3]$ and $U = \{(1,2),(2,3),(1,3)\}$.

Let $m > 0$ be an integer, and let $U = \{S_1, \ldots, S_n\}$ be a cover of $[m]$. As in Chapter XII §2, §3, we will compute 'harmonic-analytic' indices associated with U. Specifically, we will take n independent copies of the Rademacher system enumerated respectively by $\mathbb{N}^{S_j}$,

$$R = \{r_i : \mathbf{i} \in \mathbb{N}^{S_j}\}, \quad j = 1, \ldots, n, \tag{1.8}$$

and consider

$$R^U = \{r_{\pi_{S_1}\mathbf{k}} \otimes \cdots \otimes r_{\pi_{S_n}\mathbf{k}} : \mathbf{k} \in \mathbb{N}^m\}, \tag{1.9}$$

whose ambient product is R^n. (This definition is, in effect, the same as the definition in (1.7): for each $j = 1, \ldots, n$, identify R with $\mathbb{N}^{S_j}$.) Our objective is to compute σ_{R^U} and δ_{R^U}, indices that register the interdependence of $\pi_{S_1}, \ldots, \pi_{S_n}$, and indeed the 'dimension' of X^U.

2 A Littlewood Inequality in Fractional 'Dimension'

Let $0 < k \leq m$ be integers. We refer to a cover U of $[m]$ as a *k-cover* if $|S| = k$ for all $S \in U$, and call it *maximal* if U consists of all the k-subsets of $[m]$.

Theorem 1 (cf. Corollary XII.3). *If U is the maximal k-cover of $[m]$, then*

$$\zeta_{R^U}\left(\frac{2m}{m+k}\right) \leq \zeta_R(1)(\sqrt{2})^{[\frac{m}{k}]-1}. \tag{2.1}$$

(Here $[\frac{m}{k}]$ denotes the largest integer less than or equal to $\frac{m}{k}$; ζ is defined in (XII.2.2).*)*

We require two lemmas. For the proof of the first, we need the m-linear Hölder and the generalized Minkowski inequalities, which we state below for convenience:

The m-linear Hölder Inequality

If $p_1 > 0, \ldots, p_m > 0$, and $1/p_1 + \cdots + 1/p_m \geq 1$, then for all scalar-valued functions $f_1, \ldots, f_m$ on a countable (finite or infinite) set X,

$$\sum_{x \in X} |f_1(x) \cdots f_m(x)| \leq \|f_1\|_{p_1} \cdots \|f_m\|_{p_m}. \tag{2.2}$$

The Generalized Minkowski Inequality

(See Exercise II.4.) For measure spaces (X, μ) and (Y, ν), scalar-valued measurable functions g on $X \times Y$, and $u \geq 1$,

$$\begin{aligned} &\int_Y \left(\int_X |g(x,y)|^u \mu(\mathrm{d}x) \right)^{\frac{1}{u}} \nu(\mathrm{d}y) \\ &\geq \left(\int_X \left(\int_Y |g(x,y)| \, \nu(\mathrm{d}y) \right)^u \mu(\mathrm{d}x) \right)^{\frac{1}{u}}. \end{aligned} \tag{2.3}$$

To ease notation, we use this convention: if

$$c = (c_{j_1 \ldots j_m} : (j_1, \ldots, j_m) \in \mathbb{N}^m)$$

is a scalar m-array, and $S = \{l_1, \ldots, l_k\} \subset [m]$ with $|S| > 1$, then we write

$$\sum_S c_{j_1 \ldots j_m} \quad \text{for} \quad \sum_{j_{l_1}} \cdots \sum_{j_{l_k}} c_{j_1 \ldots j_m}. \tag{2.4}$$

Lemma 2 (cf. Lemma XII.2). *If $U = \{S_1, \ldots, S_n\}$ is the maximal k-cover of $[m]$ ($n = \binom{m}{k}$), and $(b_{\mathbf{j}} : \mathbf{j} \in \mathbb{N}^m)$ is a scalar m-array, then*

$$\left(\sum_{j_1, \ldots, j_m} |b_{j_1 \ldots j_m}|^{\frac{2m}{k+m}} \right)^{\frac{k+m}{2m}} \leq \prod_{l=1}^{n} \left\{ \sum_{S_l} \left(\sum_{[m] \setminus S_l} |b_{j_1 \ldots j_m}|^2 \right)^{\frac{1}{2}} \right\}^{\frac{1}{n}}. \tag{2.5}$$

Proof: Assume $m > k > 0$ (the case $m = k$ is trivial). Write

$$\begin{aligned} &\sum_{j_1, \ldots, j_m} |b_{j_1 \ldots j_m}|^{\frac{2m}{k+m}} \\ &= \sum_{j_1, \ldots, j_m} |b_{j_1 \ldots j_m}|^{\frac{2}{k+m}} \ldots |b_{j_1 \ldots j_m}|^{\frac{2}{k+m}}. \end{aligned} \tag{2.6}$$

For bookkeeping purposes, we rename and index by $i \in [m]$ factors of summands on the right side of (2.6):

$$f_i(j_1, \ldots, j_m) = |b_{j_1 \cdots j_m}|, \quad i = 1, \ldots, m. \tag{2.7}$$

To ease notation we write f_i for $f_i(j_1, \ldots, j_m)$, and rewrite (2.6) as

$$\sum_{j_1,\ldots,j_m} |b_{j_1\cdots j_m}|^{\frac{2m}{k+m}} = \sum_{j_1,\ldots,j_m} |f_1|^{\frac{2}{k+m}} \cdots |f_m|^{\frac{2}{k+m}}. \tag{2.8}$$

Let

$$p = (m+k)/2 \quad \text{and} \quad q = (m-1)(m+k)/(m+k-2). \tag{2.9}$$

Apply the m-linear Hölder inequality to the sum over j_m in (2.8) with exponent $p_1 = p$ for f_1 and exponents $p_2 = \cdots = p_m = q$ for $f_2, \ldots, f_m$. This application implies that (2.6) is bounded by

$$\sum_{[m-1]} \left(\sum_{j_m} |f_1| \right)^{\frac{2}{k+m}} \left(\prod_{i=2}^{m} \sum_{j_m} |f_i|^{\frac{2(m-1)}{k+m-2}} \right)^{\frac{k+m-2}{(m-1)(k+m)}}. \tag{2.10}$$

If $m = 2$, then we stop. Otherwise, we apply the m-linear Hölder inequality to the sum in (2.10) over j_{m-1} with the exponent p for the factor containing f_2, and the exponent q for each of the remaining factors. This implies that (2.10) is bounded by

$$\sum_{[m-2]} \left\{ \sum_{j_{m-1}} \left(\sum_{j_m} |f_1| \right)^{\frac{2(m-1)}{k+m-2}} \right\}^{\frac{k+m-2}{(m-1)(k+m)}}$$

$$\cdot \left\{ \sum_{j_{m-1}} \left(\sum_{j_m} |f_2|^{\frac{2(m-1)}{k+m-2}} \right)^{\frac{k+m-2}{2(m-1)}} \right\}^{\frac{2}{k+m}}$$

$$\cdot \prod_{i=3}^{m} \left(\sum_{j_{m-1}, j_m} |f_i|^{\frac{2(m-1)}{k+m-2}} \right)^{\frac{k+m-2}{(m-1)(k+m)}}. \tag{2.11}$$

We continue thus: at the ith step for $i < m$, we apply the m-linear Hölder inequality to the sum over j_{m-i+1} with p for the factor containing f_i, and q for the remaining $m-1$ factors. After the $(m-1)$st step, we conclude that (2.6) is bounded by

$$\left\{\sum_{[m-1]}\left(\sum_{j_m}|f_1|\right)^{\frac{2(m-1)}{k+m-2}}\right\}^{\frac{k+m-2}{(m-1)(k+m)}}\cdots$$

$$\cdot\left\{\sum_{[m-i]}\left(\sum_{j_{m-i+1}}\left(\sum_{[m]\setminus[m-i+1]}|f_i|^{\frac{2(m-1)}{k+m-2}}\right)^{\frac{k+m-2}{2(m-1)}}\right)^{\frac{2(m-1)}{k+m-2}}\right\}^{\frac{k+m-2}{(m-1)(k+m)}}\cdots$$

$$\cdot\left\{\sum_{j_1}\left(\sum_{[m]\setminus\{1\}}|f_m|^{\frac{2(m-1)}{k+m-2}}\right)^{\frac{k+m-2}{2(m-1)}}\right\}^{\frac{2}{k+m}}. \qquad (2.12)$$

Next we apply the generalized Minkowski inequality to each of the first $m-1$ factors in (2.12): to the ith factor, $i = 1,\ldots,m-1$, we apply (2.3) with $u = 2(m-1)/(k+m-2)$,

$$g = \left(\sum_{[m]\setminus[m-i+1]}|f_i|^{\frac{2(m-1)}{k+m-2}}\right)^{\frac{k+m-2}{2(m-1)}},$$

$$\int_X = \sum_{[m-i]},\quad \int_Y = \sum_{j_{m-i+1}}.$$

The result is

$$\left(\sum_{[m]}|b_{j_1\ldots j_m}|^{\frac{2m}{k+m}}\right)^{\frac{k+m}{2m}}$$

$$\leq\left\{\prod_{i=1}^{m}\sum_{j_i}\left(\sum_{[m]\setminus\{i\}}|b_{j_1\ldots j_m}|^{\frac{2(m-1)}{k+m-2}}\right)^{\frac{k+m-2}{2(m-1)}}\right\}^{\frac{1}{m}}. \qquad (2.13)$$

We are now ready to prove (2.5) by induction. The assertion is trivial for $m \geq 2$ and $k = m$. Note also that (2.13) is the same as (2.5) for $m = 2$ and $k = 1$. Let $m > 2$, and assume (the induction hypothesis) that (2.5) holds for $m-1$ in place of m, and all $0 < k < m-1$. Let $0 < k < m$ be arbitrary, and assume that $k > 1$; for, if $k = 1$, then (2.5) is the same as (2.13). Let $\{S_1^i, \ldots, S^i_{\binom{m-1}{k-1}}\}$ be the maximal $(k-1)$-cover of $[m]\backslash\{i\}$ (the collection of all $(k-1)$-subsets of $[m]\backslash\{i\}$). By an application of the induction hypothesis to each of the m factors on the right side of (2.13), we obtain for every $i = 1, \ldots, m$,

$$\sum_{j_i}\left(\sum_{[m]\backslash\{i\}} |b_{j_1\ldots j_m}|^{\frac{2(m-1)}{k+m-2}}\right)^{\frac{k+m-2}{2(m-1)}}$$
$$\leq \sum_{j_i}\prod_{l=1}^{\binom{m-1}{k-1}}\left\{\sum_{S_l^i}\left(\sum_{([m]\backslash\{i\})\backslash S_l^i} |b_{j_1\ldots j_m}|^2\right)^{\frac{1}{2}}\right\}^{1/\binom{m-1}{k-1}}. \tag{2.14}$$

By an application of the $\binom{m-1}{k-1}$-linear Hölder inequality with $p_1 = \cdots = p_{\binom{m-1}{k-1}} = \binom{m-1}{k-1}$ to $\sum_{j_i}$ on the right side of (2.14), we obtain that the left side of (2.13) is bounded by

$$\prod_{i=1}^{m}\prod_{l=1}^{\binom{m-1}{k-1}}\left\{\sum_{j_i}\sum_{S_l^i}\left(\sum_{([m]\backslash\{i\})\backslash S_l^i} |b_{j_1\ldots j_m}|^2\right)^{\frac{1}{2}}\right\}^{1/m\binom{m-1}{k-1}}. \tag{2.15}$$

Observe that each member of the maximal k-cover of $[m]$ occurs in the list

$$\{i\} \cup S_l^i, \quad i = 1, \ldots, m, \ l = 1, \ldots, \binom{m-1}{k-1}, \tag{2.16}$$

k times. Therefore, the product in (2.15) can be reorganized as

$$\prod_{l=1}^{\binom{m}{k}}\left\{\sum_{S_l}\left(\sum_{[m]\backslash S_l} |b_{j_1\ldots j_m}|^2\right)^{\frac{1}{2}}\right\}^{k/m\binom{m-1}{k-1}}$$
$$= \prod_{l=1}^{\binom{m}{k}}\left\{\sum_{S_l}\left(\sum_{[m]\backslash S_l} |b_{j_1\ldots j_m}|^2\right)^{\frac{1}{2}}\right\}^{1/\binom{m}{k}}, \tag{2.17}$$

which proves the inductive step, and thus the lemma. □

Next we verify a Littlewood-type mixed-norm inequality involving functions in $\mathrm{C}_{R^U} = \mathrm{C}_{R^U}(\Omega^{S_1} \times \cdots \times \Omega^{S_n})$.

If U is a cover of $T \subset \mathbb{N}$, then define $q_U(\mathrm{T})$ to be the smallest integer q such that q elements of U cover T. For convenience, we identify functions $f \in \mathrm{C}_{R^U}$ with m-arrays β_f (cf. (XII.4.6)),

$$\beta_f(\mathbf{j}) = \hat{f}(r_{\pi_{S_1}\mathbf{j}} \otimes \cdots \otimes r_{\pi_{S_n}\mathbf{j}}), \quad \mathbf{j} = (j_1, \ldots, j_m) \in \mathbb{N}^m. \tag{2.18}$$

Lemma 3 *If U is a cover of $[m]$, then for all $f \in \mathrm{C}_{R^U}(\Omega^{S_1} \times \cdots \times \Omega^{S_n})$ and each $S \in U$,*

$$\sum_{S} \left(\sum_{[m]\setminus S} |\beta_f(\mathbf{j})|^2 \right)^{\frac{1}{2}} \leq \zeta_R(1)(\sqrt{2})^{q'} \|f\|_\infty, \tag{2.19}$$

where $q' = q_U([m]\setminus S)$.

Proof: We let $U = \{S_1, \ldots, S_n\}$, and prove the lemma in the case $S = S_1$. Assume f is an R^U-polynomial. Following a by-now-standard argument, we obtain

$$\begin{aligned} &\zeta_R(1)\ \|f\|_\infty \\ &\geq \left\| \sum_{S_1} \left| \sum_{[m]\setminus S_1} \beta_f(j_1, \ldots, j_m) r_{\pi_{S_2}\mathbf{j}} \otimes \cdots \otimes r_{\pi_{S_n}\mathbf{j}} \right| \right\|_\infty \\ &\geq \sum_{S_1} \mathbf{E} \left| \sum_{[m]\setminus S_1} \beta_f(j_1, \ldots, j_m) r_{\pi_{S_2}\mathbf{j}} \otimes \cdots \otimes r_{\pi_{S_n}\mathbf{j}} \right|, \end{aligned} \tag{2.20}$$

where expectation $\mathbf{E}$ is over $\Omega^{S_2} \times \cdots \times \Omega^{S_n}$. Suppose $\{S'_1, \ldots, S'_{q'}\} \subset U$ is a cover of $[m]\setminus S_1$. Let $\mathbf{E}'$ denote expectation over $\Omega^{S'_1} \times \cdots \times \Omega^{S'_{q'}}$, and let $\mathbf{E}''$ denote expectation over the remaining coordinates. Then, because $\{S'_1, \ldots, S'_{q'}\}$ covers $[m]\setminus S_1$, the q'-dimensional L^1–L^2 Khintchin inequality implies

$$\begin{aligned} &\sum_{S_1} \mathbf{E} \left| \sum_{[m]\setminus S_1} \beta_f(j_1, \ldots, j_m)\ r_{\pi_{S_2}\mathbf{j}} \otimes \cdots \otimes r_{\pi_{S_n}\mathbf{j}} \right| \\ &= \sum_{S_1} \mathbf{E}''\ \mathbf{E}' \left| \sum_{[m]\setminus S_1} \beta_f(j_1, \ldots, j_m)\ r_{\pi_{S_2}(\mathbf{j})} \otimes \cdots \otimes r_{\pi_{S_n}(\mathbf{j})} \right| \end{aligned}$$

$$\geq (\sqrt{2})^{-q'} \sum_{S_1} \mathbf{E}'' \left(\sum_{[m]\setminus S_1} |\beta_f(j_1, \ldots, j_m)|^2 \right)^{\frac{1}{2}}$$

$$= (\sqrt{2})^{-q'} \sum_{S_1} \left(\sum_{[m]\setminus S_1} |\beta_f(j_1, \ldots, j_m)|^2 \right)^{\frac{1}{2}}, \tag{2.21}$$

which, together with (2.20), proves the lemma. □

Proof of Theorem 1: If U is the maximal k-cover of $[m]$, then for every $S \in U$, $q_U([m]\setminus S) = [m/k] - 1$. Apply Lemma 2 and Lemma 3.

Lemma 4 (cf. Lemma XII.4). *If U is a k-cover of $[m]$, then*

$$\zeta_{R^U}(t) = \infty \quad \text{for all } t < \frac{2m}{m+k}. \tag{2.22}$$

Proof: As in case of Lemma XII.4, two proofs can be given: one based on random constructions, which we give here, and the other based on Riesz products, which we leave as an exercise (Exercise 1).

Suppose $U = \{S_1, \ldots, S_n\}$ is a k-cover of m, and let $N > 0$ be an arbitrary integer. By Theorem X.8, there exists a $\{-1, +1\}$-valued m-array $(\epsilon_{\mathbf{j}} : \mathbf{j} \in [N]^m)$, such that if

$$f_N = (1/N^{\frac{m+k}{2}}) \sum_{\mathbf{j}\in[N]^m} \epsilon_{\mathbf{j}}\, r_{\pi_{S_1}\mathbf{j}} \otimes \cdots \otimes r_{\pi_{S_n}\mathbf{j}}, \tag{2.23}$$

then

$$\|f_N\|_\infty \leq K, \tag{2.24}$$

where K is a numerical constant that depends only on m and k. (The degree of polynomials with spectrum in $\{r_{\pi_{S_1}\mathbf{j}} \otimes \cdots \otimes r_{\pi_{S_n}\mathbf{j}} : \mathbf{j} \in [N]^m\}$ is at most 2^{nN^k}.) Then,

$$\|\hat{f}_N\|_t / \|f_N\|_\infty \geq N^{\frac{2m - t(m+k)}{2t}} / K, \tag{2.25}$$

which is unbounded in N for $t < 2m/(m+k)$. □

Combining Theorem 1 and Lemma 4, we obtain

Corollary 5 *If U is the maximal k-cover of $[m]$, then*

$$\sigma_{R^U} = \frac{2m}{m+k} \quad \textit{exactly}. \tag{2.26}$$

Remarks:

i **(about the maximality assumption).** According to (2.26), if U is the maximal k-cover of $[m]$, then X^U could be viewed as an m/k-'dimensional' Cartesian product. (Cf. Remark in Chapter XII §2.) We will eventually verify (2.26) also for k-covers U of $[m]$ that are not maximal. (E.g., k-covers of $[m]$ such that every $j \in [m]$ is in k elements of U.) However, the maximality of U seems crucial for the constants' growth $\zeta_{R^U}(2m/(m+k)) = \mathscr{O}(\kappa^{m/k})$, which, in turn, is needed for subsequent constructions of p-Sidon sets for arbitrary $p \in [1, 2]$. (See the next remark.) I do not know whether

$$\zeta_{R^U}(2m/(m+k)) = \mathscr{O}(\kappa^{m/k})$$

for k-covers U of $[m]$ that are not maximal. This question will be revisited in the chapter.

ii **(existence of p-Sidon sets in W).**

Case 1: rational $p \in$ [1,2). Fix integers $0 < k \leq m$ such that $p = 2m/(m+k)$. Let $U = \{S_1, \ldots, S_n\}$ be the maximal k-cover of $[m]$. Let $E_1, \ldots, E_n$ be pairwise disjoint infinite subsets of the Rademacher system R, and enumerate each E_i by $\mathbb{N}^k$,

$$E_i = \{\gamma_{\mathbf{j}}^{(i)} : \mathbf{j} \in \mathbb{N}^k\}, \quad i = 1, \ldots, n. \tag{2.27}$$

Define the *fractional sum*

$$E^{(U)} = \{\gamma_{\pi_{S_1}\mathbf{j}}^{(1)} \cdots \gamma_{\pi_{S_n}\mathbf{j}}^{(n)} : \mathbf{j} \in \mathbb{N}^m\}. \tag{2.28}$$

Because the E_i are, in essence, mutually independent Rademacher systems, $E^{(U)}$ can be naturally identified with R^U (cf. Chapter VII §8). We thus obtain from Theorem 1

$$\zeta_{E^{(U)}}\left(\frac{2m}{m+k}\right) \leq \zeta_R(1)\, (\sqrt{2})^{[\frac{m}{k}]-1}, \tag{2.29}$$

and from Lemma 4

$$\zeta_{E^{(U)}}(t) = \infty \quad \text{for all } t < \frac{2m}{m+k}. \tag{2.30}$$

Case 2: arbitrary $p \in$ [1,2]. Let $(p_j : j \in \mathbb{N})$ be an increasing sequence of rationals in [1,2) converging to p. For each $j \in \mathbb{N}$, let F_j be a p_j-Sidon set obtained in Case 1, so that the F_j are mutually independent in W. (Decompose the Rademacher system into infinitely many pairwise disjoint infinite sets, and then construct each F_j by using Rademacher characters in each of the respective sets.) Then,

$$F = \bigcup_j F_j \tag{2.31}$$

is an *exact* p-Sidon set. Similarly, if $(p_j : j \in \mathbb{N})$ is a decreasing sequence of rationals in (1, 2) converging to p, and the F_j are mutually independent p_j-Sidon sets obtained in Case 1, then F defined by (2.31) is an *asymptotic* p-Sidon set (Exercise 2). Note the crucial use here of the estimates in (2.29).

iii **(existence of p-Sidon sets in an arbitrary discrete Abelian group Γ).** The constructions in W carried out in the previous remark can be mimicked in any Γ, wherein a countably infinite dissociate set plays the role of a Rademacher system. Suppose $p \in [1,2)$ is rational, and $0 < k < m$ are integers such that $p = 2m/(m+k)$. Decompose a countably infinite dissociate set $E \subset \Gamma$ into $\binom{m}{k}$ pairwise disjoint infinite sets E_i, $i = 1, \dots, \binom{m}{k}$, and enumerate each set by $\mathbb{N}^k$,

$$E_i = \{\gamma^{(i)}_{\mathbf{j}} : \mathbf{j} \in \mathbb{N}^k\}. \tag{2.32}$$

Let $U = \{S_1, \dots, S_n\}$ be the maximal k-cover of $[m]$, and define

$$E^{(U)} = \{\gamma^{(1)}_{\pi_{S_1}\mathbf{j}} \cdots \gamma^{(n)}_{\pi_{S_n}\mathbf{j}} : \mathbf{j} \in \mathbb{N}^m\}. \tag{2.33}$$

Then,

$$\zeta_{E^{(U)}}\left(\frac{2m}{m+k}\right) \le \kappa^{m/k}, \tag{2.34}$$

where κ is a numerical constant depending only on E, and

$$\zeta_{E^{(U)}}(t) = \infty \quad \text{for all } t < \frac{2m}{m+k}. \tag{2.35}$$

To verify (2.34) we need only to prove the analog to Lemma 3; to verify (2.35) we follow practically verbatim either one of the two proofs of Lemma 4 (Exercise 3).

For arbitrary $p \in [1,2]$, take an increasing sequence of rationals $(p_j : j \in \mathbb{N})$ in [1,2) converging to p. By previous constructions, we can choose finite spectral sets F_j such that

$$\zeta_{F_j}(p_j) \le K^{\frac{p}{(2-p)}}, \tag{2.36}$$

and

$$\zeta_{F_j}(p_{j-1}) > j, \tag{2.37}$$

where $K > 0$ does not depend on j. Moreover, the F_j can be chosen so that $\{F_j : j \in \mathbb{N}\}$ is a *sup-norm partition* of

$$F = \bigcup_j F_j, \tag{2.38}$$

which means that there exists $D > 0$ such that if $f \in \mathrm{C}_F(\hat{\Gamma})$, and f_j is the F_j-polynomial defined by $\hat{f}_j = \hat{f}\,\mathbf{1}_{F_j}$, then

$$\sum_j \|f_j\|_\infty \leq D\,\|f\|_\infty. \tag{2.39}$$

(See [Bl1].) Then, by (2.36), (2.37), and (2.39), F is an *exact* p-Sidon set.

By taking a decreasing sequence of rationals in (1,2) converging to p, we obtain an asymptotic p-Sidon set in Γ (Exercise 4).

iv **(tensor products in fractional 'dimension').** We transcribe the discussion in Chapter XII §4, in the case $U = \{(1,2),(2,3),(1,3)\}$, to the present general setting. We begin with a countably infinite set X, and a cover $U = \{S_1, \ldots, S_n\}$ of $[m]$. For a scalar-valued function β on X^m, define (cf. (XII.4.1))

$$\begin{aligned}\|\beta\|_{F_U(X^m)} &= \|\beta\|_{F_U} \\ &:= \sup\left\{\left\|\sum_{\mathbf{x}\in X^m} \beta(\mathbf{x})\mathbf{1}_{A_1}(\pi_{S_1}\mathbf{x})\cdots\mathbf{1}_{A_n}(\pi_{S_n}\mathbf{x})\; r_{\pi_{S_1}\mathbf{x}}\otimes\cdots\otimes r_{\pi_{S_n}\mathbf{x}}\right\|_\infty : \right. \\ &\qquad \left. \text{finite sets } A_i \subset X^{S_i}, i \in [n]\right\},\end{aligned} \tag{2.40}$$

where $\{r_{\mathbf{x}} : \mathbf{x} \in X^{S_i}\}$ is the Rademacher system indexed by X^{S_i}, $i \in [n]$. Denote by $F_U(X^m)$ the class of all scalar-valued functions β on X^m such that $\|\beta\|_{F_U} < \infty$, and identify it with the class of F_n-measures on $X^{S_1} \times \cdots \times X^{S_n}$ that are supported in X^U,

$$\begin{aligned}&[F_n(X^{S_1},\ldots,X^{S_n})]_{X^U} = [F_n]_{X^U} \\ &:= \{\beta \in F_n(X^{S_1},\ldots,X^{S_n}) : \beta(\mathbf{x}) = 0 \text{ for } \mathbf{x} \in X^{S_1}\times\cdots\times X^{S_n}\backslash X^U\},\end{aligned} \tag{2.41}$$

where the fractional Cartesian product X^U has been defined in (1.7). Specifically, if $\beta \in F_U(X^m)$, then identify it with the function $\tilde{\beta}$ on $X^{S_1} \times \cdots \times X^{S_n}$ defined by

$$\tilde{\beta}(\mathbf{i}_1, \ldots, \mathbf{i}_n) = \begin{cases} \beta(\mathbf{j}) & \pi_{S_1}\mathbf{j} = \mathbf{i}_1, \ldots, \pi_{S_n}\mathbf{j} = \mathbf{i}_n \\ 0 & \text{otherwise.} \end{cases} \tag{2.42}$$

Note that $\|\beta\|_{F_U} = \|\tilde{\beta}\|_{F_n}$.

Next we verify that $[F_n(X^{S_1}, \ldots, X^{S_n})]_{X^U}$ is complemented in $F_n(X^{S_1}, \ldots, X^{S_n})$. To this end, we use the systems S_m of statistically independent $\mathbf{T}_m$-valued random variables ($m \geq 2$) defined in Chapter II §6, and index them by X; i.e., take the underlying domain of S_m to be $(\mathbf{T}_m)^X$, and enumerate $S_m = \{\xi_{m,x} : x \in X\}$. Define the *incidence* of $j \in [m]$ relative to U,

$$i_U(j) = |\{S \in U : j \in S\}|. \tag{2.43}$$

For each S in U, $S = (j_1, \ldots, j_k)$, consider the random variables for $\mathbf{x} = (x_{j_1}, \ldots, x_{j_k}) \in X^S$,

$$Y_{S,\mathbf{x}} = \xi_{i_U(j_1), x_{j_1}} \otimes \cdots \otimes \xi_{i_U(j_k), x_{j_k}}, \tag{2.44}$$

whose underlying domain is the k-fold product

$$(\mathbf{T}_{i_U(j_1)})^X \times \cdots \times (\mathbf{T}_{i_U(j_k)})^X. \tag{2.45}$$

Lemma 6 (Exercise 5). *Let $\mathbf{1}_{X^U}$ denote the indicator function of X^U in $X^{S_1} \times \cdots \times X^{S_n}$. Then,*

$$\mathbf{E}\, Y_{S_1,\mathbf{x}} \cdots Y_{S_n,\mathbf{x}} = \mathbf{1}_{X^U}(\mathbf{x}), \quad \mathbf{x} \in X^{S_1} \times \cdots \times X^{S_n}, \tag{2.46}$$

where $\mathbf{E}$ is the expectation over the product probability space in (2.45).

Corollary 7 (cf. Proposition XII.11). *If $\beta \in F_n(X^{S_1}, \ldots, X^{S_n})$, then $\beta\, \mathbf{1}_{X^U} \in [F_n(X^{S_1}, \ldots, X^{S_n})]_{X^U}$.*

On the 'dual' side, let $V_U(X^m)$ denote the class of $\phi \in c_0(X^m)$ such that

$$\phi(\mathbf{x}) = \sum_\alpha f_{1\alpha}(\pi_{S_1}\mathbf{x}) \cdots f_{n\alpha}(\pi_{S_n}\mathbf{x}), \quad \mathbf{x} \in X^m, \tag{2.47}$$

and

$$\sum_\alpha \|f_{1\alpha}\|_\infty \cdots \|f_{n\alpha}\|_\infty < \infty, \tag{2.48}$$

where $\{f_{1\alpha}\}_\alpha \subset c_0(X^{S_1}), \ldots, \{f_{n\alpha}\}_\alpha \subset c_0(X^{S_n})$. The norm $\|\phi\|_{V_U}$ is the infimum of (2.48) over all representations of ϕ by (2.47). Then, $V_U(X^m)$ is the algebra of restrictions to X^U of elements in $V_n(X^{S_1}, \ldots, X^{S_n})$, and

$$V_U(X^m)^* = F_U(X^m). \tag{2.49}$$

Lemma 6 implies $\mathbf{1}_{X^U} \in \tilde{V}_n(X^{S_1}, \ldots, X^{S_n})$ (Exercise 5).

The preceding discussion can be rephrased in the language of harmonic analysis. Indeed, if Γ is a discrete Abelian group and E is a countably infinite dissociate set in Γ, then

$$A(E^U) = V_U(E^m), \tag{2.50}$$

$$C_{E^U}(\hat{\Gamma}) = F_U(E^m),$$

and

$$B(E^U) = \tilde{V}_U(E^m) = \tilde{V}_n(X^{S_1}, \ldots, X^{S_n})|_{X^U}.$$

In particular, if U is the maximal k-cover of $[m]$, then

$$\sigma_{E^U} = \frac{2m}{m+k} \quad \text{exactly}, \tag{2.51}$$

a measurement that distinguishes between F_{U_1} and F_{U_2} – equivalently, V_{U_1} and V_{U_2} – where U_1 is the maximal k_1-cover of $[m_1]$, U_2 is the maximal k_2-cover of $[m_2]$, and $m_1/k_1 \neq m_2/k_2$. A question naturally arises: can F_{U_1} and V_{U_1} be analogously distinguished from F_{U_2} and V_{U_2} when U_1 and U_2 are not maximal? We address this question later in the chapter.

v **(a 'fractional-dimensional' transfer device?).** If Γ and Γ' are discrete Abelian groups, $E \subset \Gamma$ and $F \subset \Gamma'$ are dissociate, and $U = \{S_1, \ldots, S_n\}$ is a cover of $[m]$, then there exists a constant $K > 0$ such that

$$\zeta_{E^{(U)}}(a) \leq K\zeta_{F^{(U)}}(a), \quad a > 0, \tag{2.52}$$

where $E^{(U)}$ and $F^{(U)}$ are defined in (2.33), and $K = \mathscr{O}(4^n)$. The verification is a straightforward adaptation of the proof of Proposition XII.9. Therefore, if U is the maximal k-cover of $[m]$, then we obtain (immediately, because $R^{(U)}$ is an exact $2m/(m+k)$-Sidon set) that if $E \subset \Gamma$ is any infinite dissociate set, then $E^{(U)}$ is an exact $2m/(m+k)$-Sidon set. However, the estimates in (2.52) are useless in deriving the full result (Remark iii above): that for *arbitrary* $p \in (1, 2)$ there exist exact and asymptotic p-Sidon sets in every infinite Abelian group Γ. (Cf. (2.34).)

This leads to a question that I cannot answer. Let E be a countably infinite dissociate set. Is there $c > 0$ such that for the maximal k-cover $U = \{S_1, \ldots, S_n\}$ of $[m]$, and $\theta_1, \ldots, \theta_n$ in the unit ball of $l^\infty(\mathbb{N}^k)$, there exists $\mu \in M(\hat{\Gamma})$ such that

$$\hat{\mu}(\gamma^{(1)}_{\pi_{S_1}\mathbf{j}} \cdots \gamma^{(n)}_{\pi_{S_n}\mathbf{j}}) = \theta_1(\pi_{S_1}\mathbf{j}) \cdots \theta_n(\pi_{S_n}\mathbf{j}), \quad \mathbf{j} \in \mathbb{N}^m, \tag{2.53}$$

and

$$\|\mu\|_M \leq c^{m/k}? \tag{2.54}$$

An affirmative answer would verify (2.52) with $K \leq c^{m/k}$, and provide a more efficient transfer device linking one group to another; cf. Remark i above.

3 A Khintchin Inequality in Fractional 'Dimension'

We say a cover $U = \{S_1, \ldots, S_n\}$ of $[m]$ is *uniformly incident* if there exists an integer $I := I_U$ such that $i_U(j) = I$ for all $j \in [m]$. ($i_U(j)$ is defined in (2.43).) For example, if U is the maximal k-cover of $[m]$, then U is uniformly incident with $I_U = \binom{m-1}{k-1}$. If U is any uniformly incident k-cover of $[m]$, then

$$|U|/I_U = m/k. \tag{3.1}$$

In this section we compute δ_{E^U} for arbitrary dissociate sets $E \subset \Gamma$, and uniformly incident covers U of $[m]$. (We will deal with general covers in §8.) We analyze first the instance $S :=$ *Steinhaus system*, which is conveniently free of 'algebraic' complications, and then will transport the results to any Γ; cf. Chapter XII §3.

Recall that if $U = \{S_1, \ldots, S_n\}$ is a cover of $[m]$, then the fractional Cartesian product S^U is obtained thus: first enumerate

$$S = \{\chi_\mathbf{i} : \mathbf{i} \in \mathbb{N}^{S_j}\}, \quad j = 1, \ldots, n, \tag{3.2}$$

and then define (cf. (1.6), (1.8))

$$S^U = \{\chi_{\pi_{S_1}\mathbf{j}} \otimes \cdots \otimes \chi_{\pi_{S_n}\mathbf{j}} : \mathbf{j} \in \mathbb{N}^m\}. \tag{3.3}$$

Theorem 8 (cf. Theorem XII.8). *If $U = \{S_1, \ldots, S_n\}$ is a uniformly incident k-cover of $[m]$, then*

$$\delta_{S^U} = m/2k \quad \textit{exactly}. \tag{3.4}$$

Lemma 9 (cf. Lemma XII.7). *If $U = \{S_1, \ldots, S_n\}$ is a uniformly incident k-cover of $[m]$, then for all integers $s > 0$,*

$$\rho_{S^U}(s) \leq s^{(m/k)s}. \tag{3.5}$$

Proof: Let $s > 0$ be an integer, and let $\gamma = (\gamma_1, \ldots, \gamma_s) \in \Gamma^n$ ($\Gamma = \oplus\, \mathbb{Z}$) be such that

$$\gamma = x_1 \cdots x_s, \quad x_1 \in S^U, \ldots, x_s \in S^U. \tag{3.6}$$

Identify $\{x_1, \ldots, x_s\}$ with its underlying indexing set in $\mathbb{N}^m$,

$$F_\gamma = \{\mathbf{j}_u \in \mathbb{N}^m : x_u = \chi_{\pi_{S_1}\mathbf{j}_u} \otimes \cdots \otimes \chi_{\pi_{S_n}\mathbf{j}_u}, u \in [s]\}, \tag{3.7}$$

and rewrite (3.6) as

$$\gamma = \left(\prod_{u=1}^{s} \chi_{\pi_{S_1}\mathbf{j}_u}\right) \otimes \cdots \otimes \left(\prod_{u=1}^{s} \chi_{\pi_{S_n}\mathbf{j}_u}\right), \tag{3.8}$$

where

$$\gamma_1 = \prod_{u=1}^{s} \chi_{\pi_{S_1}\mathbf{j}_u}, \ldots, \gamma_n = \prod_{u=1}^{s} \chi_{\pi_{S_n}\mathbf{j}_u}. \tag{3.9}$$

Consider the subsets of $\mathbb{N}^k$

$$A_1 = \pi_{S_1}[F_\gamma], \ldots, A_n = \pi_{S_n}[F_\gamma]. \tag{3.10}$$

By the algebraic independence of the Steinhaus system, if $\gamma = y_1 \cdots y_s$, and

$$y_u = \chi_{\pi_{S_1}\mathbf{j}'_u} \otimes \cdots \otimes \chi_{\pi_{S_n}\mathbf{j}'_u}, \quad \mathbf{j}'_u \in \mathbb{N}^m, \quad u = 1, \ldots, s \tag{3.11}$$

(i.e., $(y_1, \ldots, y_s) \in A_{S^U}(s, \gamma)$, where $A_{S^U}(s, \gamma)$ is defined by (XII.3.6)), then

$$\pi_{S_1}\mathbf{j}'_u \in A_1, \ldots, \pi_{S_n}\mathbf{j}'_u \in A_n, \quad u = 1, \ldots, s. \tag{3.12}$$

Designate the canonical projections from $A_{S^U}(s, \gamma)$ into S^U by $\pi_1, \ldots, \pi_s$, and denote $A = A_{S^U}(s, \gamma)$. Then, by (3.12) and the definition of S^U,

$$|\pi_u[A]| \leq \sum_{\mathbf{j} \in \mathbb{N}^m} \mathbf{1}_{A_1}(\pi_{S_1}\mathbf{j}_u) \cdots \mathbf{1}_{A_n}(\pi_{S_n}\mathbf{j}_u) \quad u = 1, \ldots, s. \tag{3.13}$$

We proceed to verify

$$\sum_{\mathbf{j}\in\mathbb{N}^m} \mathbf{1}_{A_1}(\pi_{S_1}\mathbf{j}_u)\cdots\mathbf{1}_{A_n}(\pi_{S_n}\mathbf{j}_u) \le |A_1|^{m/nk}\cdots|A_n|^{m/nk}. \tag{3.14}$$

For the sake of clarity (I hope...), we expand the notation in (2.4): if $S = (i_1,\ldots,i_k) \subset [m]$, and $c = (c(\mathbf{y}) : \mathbf{y} \in \mathbb{N}^m)$ is a scalar-valued m-array, then let

$$\sum_{y_i\in\mathbb{N}:i\in S} c(y_1,\ldots,y_m) \text{ denote the iterated sum}$$

$$\sum_{y_{i_1}}\cdots\sum_{y_{i_k}} c(y_1,\ldots,y_m), \tag{3.15}$$

which in (2.4) is denoted by $\sum\limits_S c(y_1,\ldots,y_m)$. In particular, if $S = \emptyset$, then

$$\sum_{y_i\in\mathbb{N}:i\in S} c(y_1,\ldots,y_m) = c(y_1,\ldots,y_m).$$

For $l = 0,\ldots,m$, denote

$$F(l) = \sum_{y_i\in\mathbb{N}:i\in[m]\setminus[l]} \prod_{j=1}^{n} \left(\sum_{y_i\in\mathbb{N}:i\in[l]\cap S_j} \mathbf{1}_{A_j}(y_i : i \in S_j) \right)^{m/nk}. \tag{3.16}$$

Then,

$$F(0) = \sum_{\mathbf{j}\in\mathbb{N}^m} \mathbf{1}_{A_1}(\pi_{S_1}\mathbf{j}_u)\cdots\mathbf{1}_{A_n}(\pi_{S_n}\mathbf{j}_u) \tag{3.17}$$

and

$$F(m) = |A_1|^{m/nk}\cdots|A_n|^{m/nk}.$$

We will verify the recursive inequality

$$F(l) \le F(l+1), \quad l = 0,\ldots,m-1, \tag{3.18}$$

and thus obtain (3.14) from (3.17). To prove (3.18), we first rewrite $F(l)$ as

$$\begin{aligned} F(l) &= \sum_{y_i\in\mathbb{N}:i\in[m]\setminus[l+1]} \prod_{\{j:l+1\notin S_j\}} \left(\sum_{y_i\in\mathbb{N}:i\in[l]\cap S_j} \mathbf{1}_{A_j}(y_i : i \in S_j) \right)^{m/nk} \\ &\cdot \sum_{y_{l+1}\in\mathbb{N}} \prod_{\{j:l+1\in S_j\}} \left(\sum_{y_i\in\mathbb{N}:i\in[l]\cap S_j} \mathbf{1}_{A_j}(y_i : i \in S_j) \right)^{m/nk}. \end{aligned} \tag{3.19}$$

Observe that $|\{j : l+1 \in S_j\}| = I_U = nk/m$ (by (3.1)), and apply the nk/m-linear Hölder inequality (stated in (2.2)) to the sum over y_{l+1} with $p_1 = \cdots = p_{nk/m} = nk/m$:

$$\begin{aligned} F(l) \;&\le \sum_{y_i\in\mathbb{N}:i\in[m]\setminus[l+1]} \; \prod_{\{j:l+1\notin S_j\}} \left(\sum_{y_i\in\mathbb{N}:i\in[l]\cap S_j} \mathbf{1}_{A_j}(y_i : i\in S_j)\right)^{m/nk} \\ &\quad\cdot \prod_{\{j:l+1\in S_j\}} \left(\sum_{y_{l+1}\in\mathbb{N}} \; \sum_{y_i\in\mathbb{N}:i\in[l]\cap S_j} \mathbf{1}_{A_j}(y_i : i\in S_j)\right)^{m/nk} \\ &= \sum_{y_i\in\mathbb{N}:i\in[m]\setminus[l+1]} \; \prod_{j=1}^{n} \left(\sum_{y_i\in\mathbb{N}:i\in[l+1]\cap S_j} \mathbf{1}_{A_j}(y_i : i\in S_j)\right)^{m/nk} \\ &= F(l+1). \end{aligned} \tag{3.20}$$

Because $|A_j| \le s$ for $j = 1, \ldots, n$, we obtain from (3.14)

$$|\pi_u[A]| \le s^{\frac{m}{k}}, \quad u = 1, \ldots, s, \tag{3.21}$$

and therefore, because $A_{S^U}(s,\gamma) \subset \pi_1[A] \times \cdots \times \pi_s[A]$, we conclude

$$|A_{S^U}(s,\gamma)| \le s^{(\frac{m}{k})s}, \tag{3.22}$$

which verifies the lemma. □

Lemma 10 *If $U = \{S_1, \ldots, S_n\}$ is a k-cover of $[m]$, then*

$$\eta_{S^U}(a) = \infty \quad \textit{for all } a < m/2k. \tag{3.23}$$

Proof: (cf. Proof of Theorem XII.8). Consider the Riesz product

$$H_N = \prod_{\mathbf{i}\in[N]^k} \left(1 + \frac{\chi_{\mathbf{i}} + \overline{\chi_{\mathbf{i}}}}{2}\right) \underbrace{\otimes\cdots\otimes}_{n \text{ factors}} \prod_{\mathbf{i}\in[N]^k} \left(1 + \frac{\chi_{\mathbf{i}} + \overline{\chi_{\mathbf{i}}}}{2}\right). \tag{3.24}$$

Then for all $p \in (1, 2)$,

$$\|H_N\|_{\mathrm{L}^p} \le 2^{nN^k/q}, \tag{3.25}$$

where $1/p + 1/q = 1$. Define

$$h_N = \sum_{\mathbf{j}\in[N]^m} \chi_{\pi_{S_1}\mathbf{j}} \otimes \cdots \otimes \chi_{\pi_{S_n}\mathbf{j}}, \tag{3.26}$$

and note

$$(1/2^n)N^m = |\mathbf{E}\, h_N H_N| \leq 2^{nN^k/q}\, \|h_N\|_{L^q}, \quad q > 2, \tag{3.27}$$

which, by taking $q = N^k$ and $N \to \infty$, implies the lemma. □

Proof of Theorem 8: Apply Lemma 9, Lemma XII.6, and Lemma 10. □

Remarks:

i **(existence of sets $F \subset \oplus\, \mathbb{Z}$ such that $\delta_F = \alpha/2$ for arbitrary $\alpha \in [1,\infty)$).** We follow the strategy used in §2 Remark ii. If $\alpha = m/k$ for integers $m \geq k > 0$, then we take a uniformly incident k-cover U of $[m]$. We let $E_1, \ldots, E_{|U|}$ be pairwise disjoint infinite subsets of the Steinhaus system, each enumerated by $\mathbb{N}^k$, and define the fractional sum $E^{(U)}$ by (2.28). Then (cf. Lemma 9 and Lemma 10),

$$\eta_{E^{(U)}}(m/2k) \leq 2 \tag{3.28}$$

and

$$\eta_{E^{(U)}}(a) = \infty \quad \text{for all } a < m/2k. \tag{3.29}$$

If $\alpha \in [1,\infty)$ is arbitrary, then we take an increasing sequence (α_j) of rationals in $[1,\infty)$ converging to α. We decompose the Steinhaus system into infinitely many, pairwise disjoint infinite subsets $P_j, j \in \mathbb{N}$, and construct by using the characters in P_j a fractional sum F_j such that $\eta_{F_j}(\alpha_j/2) \leq 2$, and $\eta_{F_j}(a) = \infty$ for $a < \alpha_j/2$. We let $F = \bigcup_j F_j$, and conclude that $\delta_F = \alpha/2$ *exactly*. Similarly, by using a decreasing sequence of rationals in $(1,\infty)$, we obtain $\delta_F = \alpha/2$ *asymptotically*.

ii **(existence of $F \subset \Gamma$ with $\delta_F = \alpha/2$ for arbitrary $\alpha \in [1,\infty)$).** Suppose U is a cover of $[m]$. Let $E \subset \Gamma$ and $E' \subset \Gamma'$ be countably infinite dissociate sets, where Γ and Γ' are discrete Abelian groups, and let $E^{(U)}$ and $E'^{(U)}$ be corresponding fractional sums defined by (2.28). Then there exists $K_U > 0$ such that (cf. Proposition XII.9)

$$\eta_{E^{(U)}}(a) \leq K_U\, \eta_{E'^{(U)}}(a), \quad a > 0. \tag{3.30}$$

The proof uses Riesz products, and yields $K_U \leq 4^{|U|}$, which, by taking $E' = S$, implies $\eta_{E^{(U)}}(m/2k) \leq 2 \cdot 4^{|U|}$ for dissociate $E \subset \Gamma$ and uniformly incident k-covers U of $[m]$. This estimate, however, is not sharp enough to produce $F \subset \Gamma$ with $\delta_F = \alpha/2$ as a limit of fractional sums of rational 'dimensions'; see Remark v in §2. Spectral sets F with $\delta_F = \alpha/2$ for arbitrary $\alpha \in [1, \infty)$ will be obtained by other methods later in the chapter (Exercise 6).

iii **(the next question).** The computation of the σ-index in §2 suggests that if U is the maximal k-cover of $[m]$, then the 'dimension' of X^U equals m/k, while the computation of the δ-index in this section suggests that if U is *any* uniformly incident k-cover of $[m]$, then the 'dimension' of X^U is also m/k. This naturally leads to the question: for uniformly incident k-covers U of $[m]$, and dissociate $E \subset \Gamma$, is

$$\sigma_{E^U} = \frac{2m}{m+k}? \tag{3.31}$$

And this, again, brings up the general problem (cf. Remark ii, Chapter XII §3): given a cover U of $[m]$, define the 'dimension' of X^U by using primal notions only, denote it by $\dim X^U$, and verify that for dissociate sets $E \subset \Gamma$

$$\sigma_{E^U} = 2 \dim E^U / (1 + \dim E^U), \tag{3.32}$$

and

$$\delta_{E^U} = \frac{1}{2} \dim E^U. \tag{3.33}$$

4 Combinatorial Dimension

Definition 11 Let $Y_1, \ldots, Y_n$ be sets, and $F \subset Y_1 \times \cdots \times Y_n$. For integers $s > 0$, define

$$\Psi_F(s) = \max\{|(A_1 \times \cdots \times A_n) \cap F| : \; A_i \subset Y_i, \; |A_i| \le s, \; i = 1, \ldots, n\}. \tag{4.1}$$

The *upper and lower combinatorial dimensions of* F are, respectively,

$$\dim F = \limsup_{s \to \infty} \log \Psi_F(s) / \log s, \tag{4.2}$$

and

$$\underline{\dim} F = \liminf_{s \to \infty} \log \Psi_F(s) / \log s. \tag{4.3}$$

If $\dim F = \underline{\dim} F = \alpha$, then F is an *α-product.*

If $F \subset Y_1 \times \cdots \times Y_n$ is finite, then $\underline{\dim} F = \dim F = 0$. If F is infinite, then

$$1 \le \underline{\dim} F \le \dim F \le n. \tag{4.4}$$

Although the indices defined in (4.2) and (4.3) appear symmetrical, we will observe an inherent asymmetry between them. We refer to dim defined in (4.2) as *combinatorial dimension*, and to $\underline{\dim}$ in (4.3) as *lower combinatorial dimension.*

A 'slower' but more transparent definition follows the measurements

$$d_F(a) = \sup\{\Psi_F(s)/s^a : s > 0\}, \tag{4.5}$$

and

$$\overline{d}_F(a) = \limsup_{s\to\infty} \Psi_F(s)/s^a, \quad a > 0. \tag{4.6}$$

Lemma 12 (Exercise 7).

$$\begin{aligned}\dim F &= \inf\{a : d_F(a) < \infty\} = \sup\{a : d_F(a) = \infty\}\\ &= \inf\{a : \overline{d}_F(a) < \infty\} = \sup\{a : \overline{d}_F(a) = \infty\}.\end{aligned} \tag{4.7}$$

We distinguish between two cases: (1) $\dim F$ is exact, which means $d_F(\dim F) < \infty$ (equivalently, $\overline{d}_F(\dim F) < \infty$); (2) $\dim F$ is asymptotic, which means $d_F(\dim F) = \infty$ (equivalently, $\overline{d}_F(\dim F) = \infty$).

Here are some basic properties.

Proposition 13 (Exercise 7).

i. *If F_1 and F_2 are subsets of $Y_1 \times \cdots \times Y_n$, then*

$$\dim(F_1 \cup F_2) = \max\{\dim F_1, \dim F_2\}$$

and

$$\underline{\dim}(F_1 \cup F_2) = \max\{\underline{\dim} F_1, \underline{\dim} F_2\}. \tag{4.8}$$

ii. *If $E \subset X_1 \times \cdots \times X_m$ and $F \subset Y_1 \times \cdots \times Y_n$, then,*

$$\dim(E \times F) \leq \dim E + \dim F, \tag{4.9}$$

and

$$\underline{\dim} E + \underline{\dim} F \leq \underline{\dim}(E \times F);$$

if F is an α-product, then

$$\dim(E \times F) = \dim E + \dim F \tag{4.10}$$

and

$$\underline{\dim}(E \times F) = \underline{\dim} E + \underline{\dim} F.$$

Remarks:

i **(about the terminology).** Our definition of *dimension* involves only counting, and hence the term 'combinatorial'. The definition is, in effect, a 'continuous' calibration of an elementary counting principle: for all integers $s > 0$, and s-subsets $A_1 \subset Y_1, \ldots, A_n \subset Y_n$, the number of *unrestricted* selections $(x_1, \ldots, x_n)$ from $A_1 \times \cdots \times A_n$ equals s^n, while, in general, for arbitrary $F \subset Y_1 \times \cdots \times Y_n$ and $\epsilon > 0$, the number of selections $(x_1, \ldots, x_n) \in (A_1 \times \cdots \times A_n) \cap F$ is $\mathscr{O}(s^{\dim F + \epsilon})$.

Combinatorial dimension is invariant under rearrangements of the 'coordinate-axes' $Y_1, \ldots, Y_n$: if $F \subset Y_1 \times \cdots \times Y_n$, and $\varphi_1 : Y_1 \mapsto Y_1, \ldots, \varphi_n : Y_n \mapsto Y_n$ are bijections, then (obviously)

$$\dim\{(\varphi_1(y_1), \ldots, \varphi_n(y_n)) : (y_1, \ldots, y_n) \in F\} = \dim F. \qquad (4.11)$$

This basic feature distinguishes dim from other notions of dimension listed and discussed, for example, in [Man] and [F].

ii **(interdependence).** *Combinatorial dimension* is a basic measurement of interdependence. Specifically, if $F \subset Y_1 \times \cdots \times Y_n$, then $\dim F$ gauges the interdependence of $\pi_1|_F, \ldots, \pi_n|_F$ (restrictions to F of the canonical projections from $Y_1 \times \cdots \times Y_n$ into the respective 'coordinate axes'). Generally, if $f_1, \ldots, f_n$ are functions from a set F into sets $Y_1, \ldots, Y_n$, respectively, then

$$\dim\{(f_1(t), \ldots, f_n(t)) : t \in F\} \qquad (4.12)$$

is an index of interdependence of these functions. (See discussions in Chapter XII §1, and §1 in this chapter.) The more precise, precursory measurements are given by (4.5) and (4.6).

iii **(a brief preview: two issues).** The first is existence: if X is infinite and $\alpha \in [1, n]$ is arbitrary, then are there $F \subset X^n$ such that $\dim F = \alpha$? The second focuses on the meaning of dim: how is *combinatorial dimension* related to other measurements?

Regarding existence, we have already shown in the course of the proof of Lemma 9 – and will confirm – that if $U = \{S_1, \ldots, S_n\}$ is a uniformly incident k-cover of $[m]$, then X^U is an m/k-product. In the next section we prove that if U is any cover of $[m]$, then X^U is a q-product for some rational number $q \geq 1$ associated with U. However, it turns out that for a fixed integer $n > 0$, there are only finitely many rationals $q \in [1, n]$, for which there exist fractional Cartesian products in X^n with dimension q. Later in the chapter, we will observe that for every $\alpha \in (1, n)$ there are *random* sets $F \subset X^n$ such that $\dim F = \alpha$. But the question concerning *deterministic* designs of α-dimensional sets in X^n for arbitrary $\alpha \in (1, n)$ is open.

The second issue is largely open-ended. In this chapter we establish relations linking the dim-scale, which measures interdependence in a primal setting (devoid of structure), to the σ- and δ-scales, which measure interdependence in harmonic-analytic and probabilistic settings. (Results in §2 and §3 are instances of these relations.) Further applications of combinatorial measurements (to random walks, for example) will be given in the next chapter.

5 Fractional Cartesian Products are q-products

In this section we show that if X is an infinite set, and $U = \{S_1, \ldots, S_n\}$ is a cover of $[m]$, then $\dim X^U$ is the optimal value of the linear programming problem

$$\begin{gathered}\text{maximize } t_1 + \cdots + t_m \text{ subject to the constraints that}\\ t_i \geq 0 \text{ for } i \in [m], \text{ and } \sum_{i \in S_j} t_i \leq 1 \quad \text{for } j \in [n].\end{gathered} \tag{5.1}$$

Because U is a cover of $[m]$, the *feasible set* associated with this problem is bounded, and the optimal value exists. We let $\alpha = \alpha(U)$ denote this optimal value.

Theorem 14

$$\dim X^U = \alpha(U) \text{ exactly.}$$

Proof: First we verify $\alpha = \alpha(U) \leq \dim X^U$. Let $(t_1, \ldots, t_m)$ be an optimal vector; that is, $t_1, \ldots, t_m$ satisfy the constraints in (5.1) and $\alpha = t_1 + \cdots + t_m$. Taking $(t_1, \ldots, t_m)$ to be an extreme point of the feasible set, we assume each of $t_1, \ldots, t_m$ is rational. Let $s > 0$ be an integer such that s^{t_i} is an integer for each $i \in [m]$. Let $D_1 \subset X, \ldots, D_m \subset X$ be such that $|D_i| = s^{t_i}$ for each $i \in [m]$, and consider $D = D_1 \times \cdots \times D_m \subset X^m$. Let $A_j = \times\{D_i : i \in S_j\} \subset X^{S_j}$, $j \in [n]$. Then, by the constraints in (5.1),

$$|A_j| = \prod_{i \in S_j} |D_i| = \prod_{i \in S_j} s^{t_i} \leq s, \quad j \in [n]. \tag{5.2}$$

Because $\mathbf{x} \mapsto (\pi_{S_1}\mathbf{x}, \ldots, \pi_{S_n}\mathbf{x})$ is a bijection from $D_1 \times \cdots \times D_m$ onto $(A_1 \times \cdots \times A_n) \cap X^U$,

$$|(A_1 \times \cdots \times A_n) \cap X^U| = |D| = \prod_{i=1}^{n} s^{t_i} = s^{\alpha}. \tag{5.3}$$

Because s can be chosen arbitrarily large, this implies

$$d_{X^U}(a) = \infty \quad \text{for } a < \alpha, \tag{5.4}$$

and hence $\alpha \leq \dim X^U$.

To prove $\alpha \geq \dim X^U$, we consider the dual problem:

$$\begin{gathered}\text{minimize } v_1 + \cdots + v_n \text{ subject to the constraints that}\\ v_j \geq 0 \text{ for } j \in [n], \text{ and } \sum_{i \in S_j} v_j \geq 1 \quad \text{for } i \in [m].\end{gathered} \tag{5.5}$$

By duality [G, p. 78], the optimal value exists and equals $\alpha = \alpha(U)$. Let $(v_1, \ldots, v_n)$ be an optimal vector. We will show that for all finite sets $A_1 \subset X^{S_1}, \ldots, A_n \subset X^{S_n}$,

$$|(A_1 \times \cdots \times A_n) \cap X^U| \leq |A_1|^{v_1} \cdots |A_n|^{v_n}. \tag{5.6}$$

For then, by taking $|A_1| \leq s, \ldots, |A_n| \leq s$, we have

$$\Psi_{X^U}(s) \leq s^{\Sigma v_j} = s^{\alpha}. \tag{5.7}$$

Therefore $d_{X^U}(\alpha) \leq 1$, and combining this with (5.4), we obtain $\dim X^U = \alpha$ exactly.

We apply an argument used in the proof of Lemma 9. For $q = 0, \ldots, m$, define

$$F(q) = \sum_{x_i \in X : i \in [m] \setminus [q]} \prod_{j=1}^{n} \left(\sum_{x_i \in X : i \in [q] \cap S_j} \mathbf{1}_{A_j}(x_i : i \in S_j) \right)^{v_j}. \tag{5.8}$$

Then,

$$\begin{aligned} F(0) &= \sum_{\mathbf{x} \in X^m} \mathbf{1}_{A_1}(\pi_{S_1}\mathbf{x}) \cdots \mathbf{1}_{A_n}(\pi_{S_n}\mathbf{x}) \\ &= |(A_1 \times \cdots \times A_n) \cap X^U|, \end{aligned} \tag{5.9}$$

and

$$F(m) = |A_1|^{v_1} \cdots |A_n|^{v_n}.$$

Therefore, to show (5.6) it suffices to show $F(q) \leq F(q+1)$ for $q = 0, \ldots, m-1$. To this end, rewrite $F(q)$ as

$$F(q) = \sum_{x_i \in X : i \in [m] \setminus [q+1]} \prod_{\{j : q+1 \notin S_j\}} \left(\sum_{x_i \in X : i \in [q] \cap S_j} \mathbf{1}_{A_j}(x_i : i \in S_j) \right)^{v_j}$$

$$\cdot \sum_{x_{q+1} \in X} \prod_{\{j : q+1 \in S_j\}} \left(\sum_{x_i \in X : i \in [q] \cap S_j} \mathbf{1}_{A_j}(x_i : i \in S_j) \right)^{v_j}. \tag{5.10}$$

Using $\sum_{q+1 \in S_j} v_j \geq 1$, we apply the $i_U(q+1)$-linear Hölder inequality (as stated in (2.2)) to the summation over x_{q+1} ($i_U(q+1) :=$ *incidence* of $q+1$ in U), with exponents

$$(p_1, \ldots, p_{i_U(q+1)}) = (1/v_j : q+1 \in S_j), \tag{5.11}$$

and the functions $(\sum_{x_i \in X : i \in [q] \cap S_j} \mathbf{1}_{A_j}(x_i \colon i \in S_j))^{v_j}$ such that $q+1 \in S_j$. The result is

$$F(q) \leq \sum_{x_i \in X : i \in [m] \setminus [q+1]} \prod_{\{j : q+1 \notin S_j\}} \left(\sum_{x_i \in X : i \in [q] \cap S_j} \mathbf{1}_{A_j}(x_i : i \in S_j) \right)^{v_j}$$

$$\cdot \prod_{\{j : q+1 \in S_j\}} \left(\sum_{x_{q+1} \in X} \sum_{x_i \in X : i \in [q] \cap S_j} \mathbf{1}_{A_j}(x_i : i \in S_j) \right)^{v_j}$$

$$= \sum_{x_i \in X : i \in [m] \setminus [q+1]} \prod_{j=1}^{n} \left(\sum_{x_i : i \in [q+1] \cap S_j} \mathbf{1}_{A_j}(x_i : i \in S_j) \right)^{v_j} = F(q+1). \tag{5.12}$$

□

Corollary 15 *Every fractional Cartesian product is a q-product for some rational* $q \geq 1$.

Proof: The optimal value α of the linear programming problem in (5.1) is a rational number. The proof of Theorem 14 yields

$$\lim_{s \to \infty} \Psi_{X^U}(s)/s^{\alpha} = 1 \quad \text{(Exercise 8)}. \tag{5.13}$$

□

Corollary 16 (cf. Lemma 9). *If U is a uniformly incident k-cover of $[m]$, then X^U is an m/k-product.*

Proof: Suppose $i_U(j) = I$ for $j \in [m]$. The optimal value of the linear programming problem in (5.1) is greater than or equal to m/k, and that of the problem in (5.5) is less than or equal to n/I. Use (3.1). □

Remarks:

i **(isoperimetric inequalities).** Broadly put, isoperimetric inequalities relate 'volumes' of bodies to their 'surface areas'. (E.g., see [Fe].) Here we observe that Theorem 14, and specifically its proof, can be recast and interpreted in a context of such inequalities.

Let us start with a general measure space (X, ν), a measurable set $A \subset X^m$, and define the 'volume' of A by

$$\mathrm{vol}(A) = \int_{X^m} \mathbf{1}_A \; \mathrm{d}(\nu^m), \tag{5.14}$$

where ν^m is the usual product measure. Suppose we have an effective and reasonable definition of $\mathrm{surf}_k(A)$, the k-dimensional 'surface area' of A. (An explicit definition is not important for the discussion.) Among properties of surf_k that we expect is that if A is a measurable subset of the k-dimensional 'coordinate plane' X^S, where S is a k-subset of $[m]$, then

$$\mathrm{surf}_k(A) = \int_{X^S} \mathbf{1}_A \; \mathrm{d}(\nu^S), \tag{5.15}$$

where $\nu^S = |S|$-fold product measure on X^S. We also expect that if A is any measurable subset of X^m, then $\mathrm{surf}_k(A)$ be greater than or equal to the area of the 'shadow' cast by A on each of the k-dimensional 'coordinate planes' of X^m. This means: if $U = \{S_1, \ldots, S_n\}$ is the maximal k-cover of $[m]$, then

$$\int_{X^{S_i}} \mathbf{1}_{\pi_{S_i}[A]} \; \mathrm{d}(\nu^{S_i}) \leq \mathrm{surf}_k(A), \; i = 1, \ldots, n \; \left(= \binom{m}{k} \right). \tag{5.16}$$

Let $A \subset X^m$ be arbitrary, and denote $A_1 = \pi_{S_1}[A], \ldots, A_n = \pi_{S_n}[A]$ (the 'shadows' cast by A on the respective k-dimensional 'coordinate planes'). Then,

$$A \subset \pi_{S_1}^{-1}[A_1] \cap \cdots \cap \pi_{S_n}^{-1}[A_n], \tag{5.17}$$

and therefore,

$$\begin{aligned}\mathrm{vol}(A) &= \int_{X^m} \mathbf{1}_A(\mathbf{x})\ \nu^m(\mathrm{d}\mathbf{x}) \\ &\leq \int_{X^m} \mathbf{1}_{A_1}(\pi_{S_1}\mathbf{x})\cdots\mathbf{1}_{A_n}(\pi_{S_n}\mathbf{x})\nu^m(\mathrm{d}\mathbf{x}). \qquad (5.18)\end{aligned}$$

Observe that the proof of (5.6) implies (Exercise 9)

$$\begin{aligned}&\int_{X^m} \mathbf{1}_{A_1}(\pi_{S_1}\mathbf{x})\cdots\mathbf{1}_{A_n}(\pi_{S_n}\mathbf{x})\ \nu^m(\mathrm{d}\mathbf{x}) \\ &\leq \prod_{i=1}^{n}\left(\int_{X^{S_i}} \mathbf{1}_{A_i}\mathrm{d}(\nu^{S_i})\right)^{m/kn}. \qquad (5.19)\end{aligned}$$

Then, by applying (5.14), (5.15), (5.16), and (5.18), we obtain the isoperimetric inequality (cf. [LoWh], [Os])

$$\begin{aligned}\mathrm{vol}(A) &\leq \prod_{i=1}^{n}(\mathrm{surf}_k(\pi_{S_i}[A]))^{m/kn} \\ &\leq (\mathrm{surf}_k(A))^{m/k}. \qquad (5.20)\end{aligned}$$

ii **(the next problem).** We take $\mathbb{N}$ to be our generic countable set, and restate the existence problem (Remark iii in §4): for arbitrary $\alpha \in [1, n]$, are there $F \subset \mathbb{N}^n$ such that $\dim F = \alpha$? Note that fractional Cartesian products provide examples of α-dimensional subsets of $\mathbb{N}^n$ for only finitely many rationals in $[1, n]$ (Exercise 10).

6 Random Constructions

Lemma 17 *Let $n \geq 2$ be an integer, and $1 \leq \alpha < \gamma < n$. For every integer $k > 0$, there exists $F \subset [k]^n$ such that*

$$\Psi_F(s) \leq Cs^\beta \quad \textit{for all } \beta \in [1, \alpha], \textit{ and } s \leq k^{(\gamma-\alpha)/(\gamma-\beta)} \qquad (6.1)$$

(in particular,

$$\Psi_F(s) \leq Cs^\alpha \quad \textit{for all } s > 0), \qquad (6.2)$$

and

$$|F| = \Psi_F(k) \geq \frac{1}{2}k^\alpha, \qquad (6.3)$$

where $C > 0$ depends only on n and γ.

Proof: Let $\{X_{\mathbf{i}}^{(k)} : \mathbf{i} \in [k]^n\}$ be the Bernoulli system of statistically independent $\{0,1\}$-valued variables on $(\Omega, \mathbb{P})$, such that

$$\mathbb{P}(X_{\mathbf{i}}^{(k)} = 1) = k^{\alpha - n}, \quad \mathbf{i} \in [k]^n. \tag{6.4}$$

Consider the random set $F = \{\mathbf{i} : X_{\mathbf{i}}^{(k)} = 1\}$, and denote

$$P_k = \mathbb{P}(F \text{ satisfies (6.1) and (6.3)}). \tag{6.5}$$

We establish the lemma by showing that $P_k \to 1$ as $k \to \infty$.

We use the following elementary fact about binomial probabilities: for $p \in (0, 1)$, and integers $m > 0$ and $i \geq 2mp$,

$$2\binom{m}{i+1} p^{i+1}\,(1-p)^{m-i-1} \leq \binom{m}{i} p^i (1-p)^{m-i}, \tag{6.6}$$

which implies

$$\sum_{i=j}^{m} \binom{m}{i} p^i (1-p)^{m-i} \leq 2\binom{m}{j} p^m (1-p)^{m-j}, \quad j \geq 2mp. \tag{6.7}$$

Fix $s \in [k]$, and let $\mathbf{A}$ be an s-hypercube in $[k]^n$ ($\mathbf{A} = A_1 \times \cdots \times A_n$, where $|A_1| = \cdots = |A_n| = s$). Denote

$$\beta_s = \max\left\{1, \gamma - \frac{(\gamma - \alpha)\log k}{\log s}\right\}, \tag{6.8}$$

and

$$C = \max\left\{2\mathrm{e}^{n+2}, \frac{n+1}{n-\gamma}\right\}. \tag{6.9}$$

Let $j = \lceil Cs^{\beta_s} \rceil$ (smallest integer $\geq Cs^{\beta_s}$). Because $j \geq 2s^n k^{\alpha-n}$ (by the definition of β_s), the inequality in (6.7) (with $m = s^n$ and $p = k^{\alpha-n}$) implies

$$\begin{aligned} \mathbb{P}\left(\sum_{\mathbf{i}\in\mathbf{A}} X_{\mathbf{i}}^{(k)} \geq j\right) &\leq 2\binom{s^n}{j} k^{j(\alpha-n)}(1-k^{\alpha-n})^{s^n-j} \\ &\leq 2\frac{s^{nj}}{j!}k^{j(\alpha-n)} \leq \frac{2s^{nj}}{(Cs^{\beta_s})^j\,\mathrm{e}^{-j}k^{j(n-\alpha)}}. \end{aligned} \tag{6.10}$$

Therefore,

$$
\begin{aligned}
&\mathbb{P}\left(\sum_{\mathbf{i}\subset\mathbf{A}} X_{\mathbf{i}}^{(k)} \geq Cs^{\beta_s} \text{ for some s-hypercube } \mathbf{A}\right)\\
&\leq \binom{k}{s}^n \frac{2s^{nj}}{(Cs^{\beta_s})^j\ \mathrm{e}^{-j}k^{j(n-\alpha)}}\\
&\leq \left(\frac{k^{ns}}{s^{ns}\ \mathrm{e}^{-ns}}\right)\left(\frac{2s^{nj}}{(Cs^{\beta_s})^j\ \mathrm{e}^{-j}k^{j(n-\alpha)}}\right)\\
&= 2\frac{\mathrm{e}^{j+ns}}{C^j}\left(\frac{s}{k}\right)^{nj-ns}(k^\alpha/s^{\beta_s})^j\\
&\leq 2\frac{\mathrm{e}^{j+ns}}{C^j}\left(\frac{s}{k}\right)^{(n-\gamma)j-ns} \qquad (\text{because } k^\alpha/s^{\beta_s} \leq (k/s)^\gamma)\\
&\leq \mathrm{e}^{-s}\left(\frac{s}{k}\right)^s \quad (\text{by}(6.9)).
\end{aligned} \tag{6.11}
$$

Therefore,

$$
\begin{aligned}
&\mathbb{P}\left(\sum_{\mathbf{i}\in\mathbf{A}} X_{\mathbf{i}}^{(k)} \geq Cs^{\beta_s} \text{ for some s-hypercube } \mathbf{A}, s\in[k]\right)\\
&\leq \sum_{s=1}^{k} \mathrm{e}^{-s}\left(\frac{s}{k}\right)^s.
\end{aligned} \tag{6.12}
$$

By Chebyshev's inequality,

$$
\lim_{k\to\infty} \mathbb{P}\left(\sum_{\mathbf{i}\in[k]^n} X_{\mathbf{i}}^{(k)} \leq \frac{k}{2}^\alpha\right) = 0. \tag{6.13}
$$

Because $\beta_s \leq \beta$ for all $\beta \in [1,\alpha]$ and $s \leq k^{(\gamma-\alpha)/(\gamma-\beta)}$, and because the right side of (6.12) tends to 0 as $k \to \infty$, it follows that $\lim_{k\to\infty} P_k = 1$, as required. □

Lemma 18 (Exercise 11). *Suppose F_j, $j \in \mathbb{N}$, is a sequence of pairwise n-disjoint subsets of $\mathbb{N}^n$, and let $F = \bigcup_j F_j$. Then, for every integer $m > 0$,*

$$
\sup\{\Psi_F(s)/s^\beta : s \in [m]\} \leq n \sup\{\Psi_{F_j}(s)/s^\beta : s \in [m], j \in \mathbb{N}\}. \tag{6.14}
$$

(*n-disjoint $F \subset \mathbb{N}^n$ and $G \subset \mathbb{N}^n$ means $\pi_l[F] \cap \pi_l[G] = \emptyset$ for every $l \in [n]$.*)

Theorem 19 *For all $1 \le \beta \le \alpha < n$, there exist $F \subset \mathbb{N}^n$ such that $\dim F = \alpha$ and $\underline{\dim} F = \beta$.*

Proof: We first verify that for all $\alpha \in (1, n)$, there exist α-products in $\mathbb{N}^n$. By Lemma 17, we produce a collection $\{F_j\}$ of pairwise n-disjoint subsets of $\mathbb{N}^n$ such that for each $j \in \mathbb{N}$, $|\pi_l(F_j)| = j$ for $l \in [n]$, $|F_j| \ge \frac{1}{2} j^\alpha$, and $\sup\{\Psi_{F_j}(s)/s^\alpha \colon s > 0\} \le C$. Let $F = \bigcup_j F_j$, and apply Lemma 18.

Next we verify the theorem in the case $\beta = 1$. Specifically, we produce $F \subset \mathbb{N}^n$ and two increasing sequences of positive integers (k_j) and (m_j) with these properties: for all $j \in \mathbb{N}$,

$$\Psi_F(k_j) \ge \frac{1}{2} k_j^\alpha, \tag{6.15}$$

and

$$\Psi_F(m_j) \le (n+1) C m_j; \tag{6.16}$$

for all $s > 0$,

$$\Psi_F(s) \le n C s^\alpha. \tag{6.17}$$

(For such F, $\dim F = \alpha$ and $\underline{\dim} F = 1$.) To this end, fix $\gamma \in (\alpha, n)$, and apply Lemma 17 recursively as follows: produce two increasing sequences of positive integers (k_j) and (m_j), and a sequence (F_j) of pairwise n-disjoint subsets of $\mathbb{N}^n$, such that for all $j \ge 1$,

$$m_j \ge \sum_{i=2}^{j-1} k_i^n \quad \text{and} \quad k_j^{(\gamma-\alpha)/(\gamma-1)} > m_j, \tag{6.18}$$

$$|\pi_l(F_j)| = m_j \quad \text{for } l \in [n], \tag{6.19}$$

and such that (6.1), (6.2) and (6.3) are satisfied for $F = F_j$ and $k = k_j$. Let $F = \bigcup_j F_j$. By (6.3) and (6.19), F satisfies (6.15). By Lemma 18 and (6.1), F satisfies (6.17). To verify (6.16), fix j, and let $\mathbf{A} \subset \mathbb{N}^n$ be an arbitrary m_j-hypercube. Denote $\mathbf{A}_i = \pi_1[\mathbf{A} \cap F_i] \times \cdots \times \pi_n[\mathbf{A} \cap F_i]$.

By (6.1) (with $\beta = 1$) and (6.18), $\Psi_{F_i}(s) \leq Cs$ for $i \geq j$ and $s \in [m_i]$. Therefore,

$$
\begin{aligned}
|F \cap \mathbf{A}| &= \sum_{i=1}^{\infty} |F_i \cap \mathbf{A}_i| \quad (F_i \text{ aro } n \text{ disjoint}) \\
&\leq \sum_{i=1}^{j-1} k_i^n + \sum_{i=j}^{\infty} |F_i \cap \mathbf{A}_i| \\
&\leq m_j + nCm_j \quad \text{(by Lemma 18).} \qquad (6.20)
\end{aligned}
$$

To obtain the theorem in the general case, let F_1 be a β-product, and let F_2 be such that $\dim F_2 = \alpha$ and $\underline{\dim} F_2 = 1$. Then, $\dim(F_1 \cup F_2) = \alpha$ and $\underline{\dim}(F_1 \cup F_2) = \beta$ (by Proposition 13). □

Remarks:

i **(fine-tuning).** Lemma 17 implies also that for every $\alpha \in [1, n)$, there exists $F \subset \mathbb{N}^n$ such that $\dim F = \alpha$ asymptotically.

With a bit more work, it can also be shown that there exists an increasing family of sets $F_x \subset \mathbb{N}^n$, $x \in (1, n]$, such that for each $x \in (1, n]$,

$$\dim F_x = x \ \text{ exactly} \qquad (6.21)$$

and

$$F_x = \bigcup \{F_u : u < x\}.$$

Similarly, there exists a decreasing family of sets $F_x \subset \mathbb{N}^n$, $x \in [1, n)$, such that for each $x \in [1, n)$,

$$\dim F_x = x \text{ asymptotically} \qquad (6.22)$$

and

$$F_x = \bigcap \{F_u : u > x\}.$$

(See [BlKo].)

ii **(deterministic constructions?).** The gist of Lemma 17 is that for large $k > 0$, a selection of k^α points from $[k]^n$ is likely to result in a 'finite version' of an α-product. To verify this (by the very definition of *combinatorial dimension*) we need to sift through all selections of k^α points from $[k]^n$, and count. We have done the sifting and counting randomly, and observed, in particular, that α-dimensional sets are abundant.

How to design deterministically α-dimensional sets for *arbitrary* $\alpha \in (1, n)$ is an open question. In a framework of extremal graph theory, this 'design' problem is closely related to the so-called Turán problem (e.g., see [Sim], [Bol, 6.2]). I will not dwell here on these issues. The little that I know about deterministic designs can be found in [BlPeSch].

iii **(how it came about).** Fractional Cartesian products, with their combinatorial features highlighted, first appeared in the solution to the p-Sidon set problem [Bl5]. The fractional products in [Bl5], which were maximal, subsequently gave rise to the notion of *combinatorial dimension.* The combinatorial dimension of an *arbitrary* fractional Cartesian product (Theorem 14) was computed in [BlSch]. The general existence question – whether for arbitrary $\alpha \in [1, n]$ there exist α-dimensional sets in $\mathbb{N}^n$ – was resolved by random constructions in [BlKo]. The problem left open in [BlKo] – whether for arbitrary $\alpha \in (1, n)$ there exist α-products in $\mathbb{N}^n$ – was resolved in [BlPeSch] by sharpening the estimates in [BlKo].

7 A Relation between the dim-scale and the σ-scale

Theorem 20 *For* $n \in \mathbb{N}$, *there exist* $C_n > 0$ *and* $D_n > 0$ *such that for all* $F \subset R^n$ *and* $t \geq 1$,

$$C_n \; d_F(t)^{1/2t} \leq \zeta_F\bigg(\frac{2t}{t+1}\bigg) \leq D_n \; d_F(t)^{1/2t}. \tag{7.1}$$

In particular, if F *is infinite, then*

$$\sigma_F = \frac{2 \dim F}{\dim F + 1}, \tag{7.2}$$

and σ_F *is exact if and only if* $\dim F$ *is exact.*

Lemma 21 *Let $X_1, \ldots, X_n$ be infinite sets, and suppose that φ is a scalar-valued function on $X_1 \times \cdots \times X_n$ with the property that there exists $D > 0$ such that for all $k \in \mathbb{N}$ and k-sets $A_1 \subset X_1, \ldots, A_n \subset X_n$,*

$$\sum_{(x_1,\ldots,x_n)\in A_1\times\cdots\times A_n} |\varphi(x_1,\ldots,x_n)| \leq Dk. \tag{7.3}$$

Then, for every $k \in \mathbb{N}$ and k-sets $A_1 \subset X_1, \ldots, A_n \subset X_n$, there exists a cover $\{G_1, \ldots, G_n\}$ of $A_1 \times \cdots \times A_n$ with the property that for every $i \in [n]$ and $x \in A_i$,

$$\sum_{(x_1,\ldots,x_n)\in\pi_i^{-1}[x]} |\varphi(x_1,\ldots,x_n)|\ \mathbf{1}_{G_i}(x_1,\ldots,x_n) \leq D. \tag{7.4}$$

Proof: (by induction on k). The case $k = 1$ is trivial. Suppose $k > 1$, and assume the assertion is true for $k-1$. Let $A_1 \subset X_1, \ldots, A_n \subset X_n$ be k-sets. By (7.3), for every $i \in [n]$ there exists $a_i \in A_i$ such that

$$\sum_{(x_1,\ldots,x_n)\in\pi_i^{-1}[a_i]\cap(A_1\times\cdots\times A_n)} |\varphi(x_1,\ldots,x_n)| \leq D. \tag{7.5}$$

Let $A_i' = A_i \backslash \{a_i\}$, $i \in [n]$. Then by the induction hypothesis, there exists a cover $\{G_1', \ldots, G_n'\}$ of $A_1' \times \cdots \times A_n'$ with the property that for every $i \in [n]$ and $x \in A_i'$

$$\sum_{(x_1,\ldots,x_n)\in\pi_i^{-1}[x]} |\varphi(x_1,\ldots,x_n)|\ \mathbf{1}_{G_i'}(x_1,\ldots,x_n) \leq D. \tag{7.6}$$

Let

$$G_i = G_i' \cup \pi_i^{-1}[a_i] \cap (A_1 \times \cdots \times A_n), \quad i \in [n]. \tag{7.7}$$

Then, $\{G_1, \ldots, G_n\}$ is the desired cover of $A_1 \times \cdots \times A_n$. (Exercise 12).

□

Remarks:

i **(about the lemma).** Lemma 21 provides a useful device for handling combinatorial estimates. Note this application: if $F \subset \mathbb{N}^n$, and $d_F(t) < \infty$, then for all $k \in \mathbb{N}$, and k-subsets $A_1 \subset \mathbb{N}, \ldots, A_n \subset \mathbb{N}$, there exists a partition $\{G_1, \ldots, G_n\}$ of $A_1 \times \cdots \times A_n$ such that for every $i \in [n]$ and $j \in A_i$,

$$|\pi_i^{-1}[j] \cap G_i| \leq d_F(t)\ k^{t-1}. \tag{7.8}$$

An instance of (7.8) (the case $t = 1$) appeared in [V3, Theorem 4.3] with an inductive proof much like the argument given here (credited by Varopoulos to Hörmander). This instance in another guise was key in the proof that every Sidon set in W is a finite union of independent sets [MM]. Lemma 21 (modulo terminology) appeared later in [V4, Lemma 2.1].

Next we recall the n-dimensional Littlewood (l^1,l^2)-mixed norm inequality (Lemma VII.35):

Lemma 22 *For all $f \in \mathrm{C}_{R^n}(\Omega^n)$, and every $i \in [n]$,*

$$\zeta_R(1)\, 2^{\frac{n-1}{2}} \|f\|_\infty \geq \sum_{j_i} \left(\sum_{[n]\setminus\{i\}} |\hat{f}(r_{j_1} \otimes \cdots \otimes r_{j_n})|^2 \right)^{\frac{1}{2}}. \tag{7.9}$$

(*We use the generic Rademacher system $R=\{r_i\colon i\in\mathbb{N}\}$, and follow the notation in* (2.4).)

Proof of Theorem 20: We fix $t \geq 1$, and proceed to prove the right side of (7.1). To this end, we assume $d_F(t) < \infty$, and establish (the stronger assertion): if $f \in \mathrm{C}_{R^n}(\Omega^n)$, then

$$\left(\sum_{(j_1,\ldots,j_n)\in F} |\hat{f}(r_{j_1} \otimes \cdots \otimes r_{j_n})|^{\frac{2t}{t+1}} \right)^{\frac{t+1}{2t}} \leq n\, \zeta_R(1)\, 2^{\frac{n-1}{2}} d_F(t)^{1/2t}\, \|f\|_\infty. \tag{7.10}$$

We verify (7.10) by duality. Let θ be in the unit ball of $l^{2t/(t-1)}(\mathbb{N}^n)$ with support in F. Let $k \in \mathbb{N}$ be arbitrary, and let $A_1, \ldots, A_n$ be arbitrary k-subsets of $\mathbb{N}$. We estimate by Hölder's inequality with exponents $t/(t-1)$ and t applied to $|\theta|^2$ and $\mathbf{1}_F$,

$$\begin{aligned} \sum_{(j_1,\ldots,j_n)\in A_1\times\cdots\times A_n} |\theta|^2 &\leq |F \cap (A_1 \times \cdots \times A_n)|^{\frac{1}{t}} \\ &\leq d_F(a)^{\frac{1}{t}} k. \end{aligned} \tag{7.11}$$

That is, $|\theta|^2$ satisfies the hypothesis of Lemma 21 with $D = d_F(t)^{1/t}$. Now suppose f is an R^n-polynomial with support in $A_1 \times \cdots \times A_n$, and that $|A_1| = \cdots = |A_n| = m$; i.e., $A_1, \ldots, A_n$ are m-subsets of $\mathbb{N}$, and

$$\mathrm{spect} f \subset \{r_{j_1} \otimes \cdots \otimes r_{j_n} : (j_1, \ldots, j_n) \in A_1 \times \cdots \times A_n\}. \tag{7.12}$$

By Lemma 21, there exists a cover $\{G_1, \ldots, G_n\}$ of $A_1 \times \cdots \times A_n$, such that for every $i \in [n]$,

$$\max_{j \in A_i} \sum_{(j_1,\ldots,j_n) \in \pi_i^{-1}[j]} |\theta(j_1, \ldots, j_n)|^2 \, \mathbf{1}_{G_i}(j_1, \ldots, j_n) \le d_F(t)^{\frac{1}{t}}. \tag{7.13}$$

By the Cauchy–Schwarz inequality, (7.13), and (7.9), we obtain for each $i \in [n]$

$$\begin{aligned}
&\left| \sum_{(j_1,\ldots,j_n) \in A_1 \times \cdots \times A_n} \hat{f}(r_{j_1} \otimes \cdots \otimes r_{j_n}) \, \theta \cdot \mathbf{1}_{G_i}(j_1, \ldots, j_n) \right| \\
&\le \sum_{j \in A_i} \left| \sum_{(j_1,\ldots,j_n) \in \pi_i^{-1}[j]} \hat{f}(r_{j_1} \otimes \cdots \otimes r_{j_n}) \, \theta \cdot \mathbf{1}_{G_i}(j_1, \ldots, j_n) \right| \\
&\le \sum_{j \in A_i} \left(\sum_{(j_1,\ldots,j_n) \in \pi_i^{-1}[j]} |\hat{f}(r_{j_1} \otimes \cdots \otimes r_{j_n})|^2 \right. \\
&\qquad \left. \cdot \sum_{(j_1,\ldots,j_n) \in \pi_i^{-1}[j]} |\theta|^2 \cdot \mathbf{1}_{G_i}(j_1, \ldots, j_n) \right)^{\frac{1}{2}} \\
&\le \max_{j \in A_i} \left(\sum_{(j_1,\ldots,j_n) \in \pi_i^{-1}[j]} |\theta|^2 \, \mathbf{1}_{G_i}(j_1, \ldots, j_n) \right)^{\frac{1}{2}} \\
&\qquad \cdot \sum_{j \in A_i} \left(\sum_{(j_1,\ldots,j_n) \in \pi_i^{-1}[j]} |\hat{f}(r_{j_1} \otimes \cdots \otimes r_{j_n})|^2 \right)^{\frac{1}{2}} \\
&\le d_F(t)^{1/2t} \sum_{j \in A_i} \left(\sum_{(j_1,\ldots,j_n) \in \pi_i^{-1}[j]} |\hat{f}(r_{j_1} \otimes \cdots \otimes r_{j_n})|^2 \right)^{\frac{1}{2}} \\
&\le d_F(t)^{1/2t} \, \zeta_R(1) \, 2^{\frac{n-1}{2}} (\sqrt{2})^{n-1} \, \|f\|_\infty.
\end{aligned} \tag{7.14}$$

Therefore,

$$\left| \sum_{(j_1,\ldots,j_n) \in A_1 \times \cdots \times A_n} \hat{f}(r_{j_1} \otimes \cdots \otimes r_{j_n}) \, \theta(j_1, \ldots, j_n) \right|$$

$$\leq \sum_{i=1}^{n} \left| \sum_{(j_1,\ldots,j_n)\in A_1\times\cdots\times A_n} \hat{f}(r_{j_1}\otimes\cdots\otimes r_{j_n})\,\theta\cdot\mathbf{1}_{G_i}(j_1,\ldots,j_n)\right|$$

$$\leq n\, d_F(t)^{1/2t}\,\zeta_R(1)\,2^{\frac{n-1}{2}}(\sqrt{2})^{n-1}\,\|f\|_\infty, \tag{7.15}$$

which implies the right side inequality in (7.1) with

$$D_n \leq d_F(t)^{1/2t}\,\zeta_R(1)\,2^{\frac{n-1}{2}}(\sqrt{2})^{n-1}. \tag{7.16}$$

As in Lemma 4 and Lemma XII.4, the left inequality in (7.1) can be proved by random constructions based on Theorem X.8, or by the fact (Theorem VII.41) that for every spectral set F (in any Γ),

$$\zeta_F\left(\frac{2t}{t+1}\right) \geq \sup\{\|g\|_{\mathrm{L}^q}/\sqrt{q} : g\in \mathrm{L}^2_F,\, \|\hat{g}\|_{2t/(2t-1)} = 1,\ q>2\}. \tag{7.17}$$

We give the proof based on (7.17), and leave the proof by random constructions as an exercise (Exercise 13).

Fix an integer $s>0$, and let $A_1\subset\mathbb{N},\ldots,A_n\subset\mathbb{N}$ be s-sets. Define the Riesz product

$$H_s = \prod_{j\in A_1}(1+r_j)\otimes\cdots\otimes\prod_{j\in A_n}(1+r_j). \tag{7.18}$$

Then, $\|H_s\|_{\mathrm{L}^1}=1$, $\|H_s\|_{\mathrm{L}^2}=2^{ns/2}$, and therefore,

$$\|H_s\|_{\mathrm{L}^p} \leq 2^{ns/q},\quad \frac{1}{p}+\frac{1}{q}=1. \tag{7.19}$$

Consider

$$h_s = \sum_{(j_1,\ldots,j_n)\in(A_1\times\cdots\times A_n)\cap F} r_{j_1}\otimes\cdots\otimes r_{j_n}. \tag{7.20}$$

By Hölder's inequality and (7.19), for all $q>2$,

$$|A_1\times\cdots\times A_n)\cap F| \leq |\mathbf{E}h_sH_s| \leq \|h_s\|_{\mathrm{L}^q}\,2^{ns/q}. \tag{7.21}$$

Putting $q=s$ in (7.21) and applying (7.17), we obtain

$$|A_1\times\cdots\times A_n)\cap F| \leq 2^n\,\zeta_F\left(\frac{2t}{t+1}\right)s^{\frac{1}{2}}\|\hat{h}_s\|_{2t/(2t-1)}$$

$$\leq 2^n\,\zeta_F\left(\frac{2t}{t+1}\right)s^{\frac{1}{2}}|A_1\times\cdots\times A_n)\cap\ F|^{(2t-1)/2t}, \tag{7.22}$$

which implies

$$|A_1 \times \cdots \times A_n) \cap F|/s^t \ \leq \ 2^{2nt} \ \zeta_F\left(\frac{2t}{t+1}\right)^{2t}, \tag{7.23}$$

and therefore,

$$2^{-n} \ d_F(t)^{1/2t} \leq \zeta_F\left(\frac{2t}{t+1}\right). \tag{7.24}$$

□

Corollary 23 *If U is a cover of $[m]$, then*

$$\sigma_{R^U} = 2\alpha(U)/(\alpha(U)+1) \quad \textit{exactly}, \tag{7.25}$$

where $\alpha(U) = \dim R^U$ is the solution to the linear programming problem in (5.1).

Remarks:

ii **(two applications to tensor products).** The results stated and proved above in a framework of harmonic analysis can be naturally recast in a framework of tensor products.

1. Let U be a cover of $[m]$, and let X be an infinite set. Then,

$$F_U(X^m) \subset l^p(X^m) \Leftrightarrow p \geq 2\alpha(U)/(\alpha(U)+1). \tag{7.26}$$

This implies that if U_1 and U_2 are covers of $[m]$ such that $\alpha(U_1) \neq \alpha(U_1)$, then $F_{U_1}(X^m) \neq F_{U_2}(X^m)$; on the dual side, $V_{U_1}(X^m) \neq V_{U_2}(X^m)$, and $\tilde{V}_{U_1}(X^m) \neq \tilde{V}_{U_2}(X^m)$. (For definitions, see Remark iv in §2.)

2. If $\beta \in F_n(X,\ldots,X)$ and $F \subset X^n$, then

$$\|\beta|_F\|_p \leq k_n \left(d_F\left(\frac{p}{2-p}\right)\right)^{(2-p)/2p} \qquad \|\beta\|_{F_n}; \tag{7.27}$$

in particular,

$$\|\beta|_F\|_p < \infty \quad \text{for all } p > \frac{2\dim F}{\dim F + 1}. \tag{7.28}$$

Moreover, (7.28) is best possible in the sense that if F is infinite, then there exists $\beta \in F_n(X,\ldots,X)$ with support in F such that $\|\beta\|_p = \infty$ for all $p < \frac{2\dim F}{\dim F+1}$.

iii **(how to distinguish between fractional Cartesian products with the same combinatorial dimension).** We have shown that fractional products can be partitioned into equivalence classes according to their *combinatorial dimension.* A question arises: how can fractional products be further distinguished within these classes?

Here is a case in point. Let $0 < k < m$ be relatively prime integers. Let U_1 be the maximal k-cover of $[m]$, and let U_2 be a uniformly incident k-cover of $[m]$ with incidence $I_{U_2} = k$; we refer to R^{U_1} as *maximal*, and to R^{U_2} as *minimal.* The combinatorial dimension of both R^{U_1} and R^{U_2} is m/k (Corollary 16), and the Sidon index of both is $2m/(m+k)$ (Corollary 23). These two products are extremal among *reduced* fractional products whose combinatorial dimension equals m/k, by which we mean the following. Covers $U = \{S_i, \dots, S_n\}$ (and their corresponding fractional products) are said to be *reduced* if $S_i \backslash S_j \neq \emptyset$ for all $i \neq j$. The ambient product of the maximal R^{U_1} is an $\binom{m}{k}$-fold Cartesian product of a Rademacher system, while the ambient product of the minimal R^{U_2} is an m-fold Cartesian product of R, *and* if R^U is a reduced fractional product such that $\dim R^U = m/k$, then $m \leq |U| \leq \binom{m}{k}$. (For this reason, we call R^{U_2} *minimal.*) In the case of the maximal product, we have shown in §2

$$\zeta_{R^{U_1}}\left(\frac{2m}{m+k}\right) \leq \zeta_R(1)\ (\sqrt{2})^{\left[\frac{m}{k}\right]-1}, \tag{7.29}$$

which enabled us to pass to 'irrational limits' (Remark ii in §2). In the case of the minimal R^{U_2}, the estimate

$$\zeta_{R^{U_2}}\left(\frac{2m}{m+k}\right) \leq \zeta_R(1)\ m(\sqrt{2})^{m-1} \tag{7.30}$$

follows from (7.10) and (5.13). Note the gap between the two estimates. Question: are (7.29) and (7.30) best possible in some precise sense? The general problem is this: does the measurement ζ_{R^U} distinguish between k-covers U of $[m]$ for which $\alpha(U) = m/k$? I guess an affirmative answer. (Only a guess...)

We turn now to the Walsh system of order n. Identify $F \subset R_n$ with

$$\{(r_{j_1}, \dots, r_{j_n}) : 0 < j_1 < \cdots < j_n,\ r_{j_1} \cdots r_{j_n} \in F\}, \tag{7.31}$$

denote it also by F, and apply our combinatorial measurements to it. If $F \subset W_n$, then we write $F = \bigcup_{k=1}^n F_k$, where $F_k \subset R_k$, and define

$$d_F(a) = \max_k \; d_{F_k}(a) \tag{7.32}$$

and

$$\dim F = \inf\{a : d_F(a) < \infty\} = \sup\{a : d_F(a) = \infty\}. \tag{7.33}$$

For $F \subset R_k$, define

$$\tilde{F} = \{r_{j_1} \otimes \cdots \otimes r_{j_k} : r_{j_1} \cdots r_{j_k} \in F\}. \tag{7.34}$$

Then,

$$\Psi_F(s) \le \Psi_{\tilde{F}}(s) \le \mathrm{e}^k \; \Psi_F(s), \quad \text{integers } s > 0, \tag{7.35}$$

and therefore,

$$d_F(a) \le d_{\tilde{F}}(a) \le \mathrm{e}^k \; d_F(a), \quad a > 0. \tag{7.36}$$

In particular, $\dim F = \dim \tilde{F}$. We denote the class of functions in $\mathrm{C}_{\tilde{F}}(\Omega^n)$ with symmetric transforms by

$$\begin{aligned} \mathrm{C}^\sigma_{\tilde{F}}(\Omega^k) := \{g \in \mathrm{C}_{\tilde{F}}(\Omega^k) &: \hat{g}(r_{j_1} \otimes \cdots \otimes r_{j_k}) \\ &= \hat{g}(r_{j_{\tau 1}} \otimes \cdots \otimes r_{j_{\tau k}}), \tau \in \text{per } [k]\} \end{aligned} \tag{7.37}$$

(per $[k]$:= permutations of $[k]$). By applying Theorem VII.26, we obtain that if $f \in \mathrm{C}_F(\Omega)$, then

$$\tilde{f} := \sum_{(j_1,\ldots,j_k)\in\mathbb{N}^k} \hat{f}(r_{j_1} \ldots r_{j_k}) \; r_{j_1} \otimes \cdots \otimes r_{j_k} \in \mathrm{C}^\sigma_{\tilde{F}}(\Omega^k), \tag{7.38}$$

and

$$\|\tilde{f}\|_\infty \le k! \; (2\mathrm{e}^k) \; \|f\|_\infty.$$

That is, the map $f \mapsto \tilde{f}$ is an isomorphism from $\mathrm{C}_F(\Omega)$ onto $\mathrm{C}^\sigma_{\tilde{F}}(\Omega^k)$. For $F \subset W_n$ write $F = \bigcup_{k=1}^n F_k$, where $F_k \subset R_k$, and for $f \in \mathrm{C}_F(\Omega)$, write $f = \sum_{k=1}^n f_k$, $f_k \in \mathrm{C}_{F_k}(\Omega)$. By Lemma VII.22 and (7.38), $f \mapsto \sum_{k=1}^n \tilde{f}_k$ is an isomorphism from $\mathrm{C}_F(\Omega)$ onto $\mathrm{C}^\sigma_{\tilde{F}_1}(\Omega) \oplus \cdots \oplus \mathrm{C}^\sigma_{\tilde{F}_n}(\Omega^n)$.

Corollary 24 *For $F \subset W_n$,*

$$\sigma_F = \frac{2 \dim F}{\dim F + 1} \quad \textit{exactly (asymptotically)} \tag{7.39}$$

if and only if $\dim F$ *is exact (asymptotic). Moreover, there exists* $C_n > 0$ *such that for all* $f \in C_{W_n}(\Omega)$*, and all* $t \geq 1$,

$$\left(\sum_{w \in F} |\hat{f}(w)|^{\frac{2t}{t+1}} \right)^{\frac{t+1}{2t}} \leq C_n \; d_F(t)^{1/2t} \; \|f\|_\infty. \tag{7.40}$$

Proof: (Exercise 14). Apply preceding discussion and Theorem 20.

□

Corollary 25 *Let* $n \in \mathbb{N}$ *be arbitrary. For all* $p \in [1, \frac{2n}{n+1}]$*, there exists* $F \subset W_n$ *such that* $\sigma_F = p$ *exactly, and if* $p \neq 2n/(n+1)$*, then there exists* $F \subset W_n$ *such that* $\sigma_F = p$ *asymptotically.*

Proof: By Theorem 19 and Remark i §6, for arbitrary $\alpha \in [1, n]$ there exists $F \subset W_n$ such that $\dim F = \alpha$ exactly, and if $\alpha \neq n$, then there exists $F \subset W_n$ such that $\dim F = \alpha$ asymptotically. Apply Corollary 24.

□

Remarks:

iv **(the assumption $f \in C_{W_n}(\Omega)$ in Corollary 24 cannot be erased).** If $F \subset W_n$ is infinite, then there exist $f \in C(\Omega)$ such that $\|\hat{f}|_F\|_p = \infty$ for all $p < 2$. (Do you see why?)

v **(general Abelian groups).** Let $E = \{\gamma_j : j \in \mathbb{N}\}$ be a countably infinite dissociate set in an infinite discrete Abelian group Γ, and denote

$$E_n = \{\gamma_{j_1} \cdots \gamma_{j_n} : 0 < j_1 < \cdots < j_n\}. \tag{7.41}$$

All that has been stated in this section in the framework of W_n can be transported via Riesz products to the framework of E_n.

8 A Relation between the dim-scale and the δ-scale

As in §3, we work first with the Steinhaus system S, an algebraically independent spectral set in $\Gamma = \bigoplus \mathbb{Z}$. Then, by use of Riesz products, we will transfer results to the general setting.

Lemma 26 (cf. Lemma 9). *For* $F \subset S^n$*, and all integers* $s > 0$,

$$\rho_F(s) \leq \Psi_F(s)^s. \tag{8.1}$$

Proof: Fix an integer $s > 0$, and suppose $\gamma = (\gamma_1, \ldots, \gamma_n) \in \Gamma^n$ is an s-fold product of elements in F,

$$\gamma = x_1 \cdots x_s, \quad (x_1, \ldots, x_s) \in F^s. \tag{8.2}$$

For $u = 1, \ldots, s$, write

$$x_u = \chi_{1u} \otimes \cdots \otimes \chi_{nu}, \quad \chi_{1u} \in S, \ldots, \chi_{nu} \in S, \tag{8.3}$$

and let

$$F_\gamma = \{\chi_{1u} \otimes \cdots \otimes \chi_{nu} : u = 1, \ldots, s\}. \tag{8.4}$$

Rewrite (8.2) as

$$\gamma = \left(\prod_{u=1}^{s} \chi_{1u}\right) \otimes \cdots \otimes \left(\prod_{u=1}^{s} \chi_{nu}\right); \tag{8.5}$$

that is,

$$\gamma_1 = \prod_{u=1}^{s} \chi_{1u}, \ldots, \gamma_n = \prod_{u=1}^{s} \chi_{nu}. \tag{8.6}$$

Project F_γ (an s-subset of S^n) on each of the n 'coordinate axes' (each 'axis' is a Steinhaus system), and denote

$$A_1 = \pi_1[F_\gamma], \ldots, A_n = \pi_n[F_\gamma] \tag{8.7}$$

($\pi_i :=$ ith canonical projection from S^n onto S), whence $|A_i| \leq s, i \in [n]$. By the algebraic independence of S, if $(y_1, \ldots, y_s) \in A_F(s, \gamma)$, then

$$\pi_1\{y_1, \ldots, y_s\} \in A_1, \ldots, \pi_n\{y_1, \ldots, y_s\} \in A_n. \tag{8.8}$$

($A_F(s, \gamma)$ has been defined in (XII.3.6).) Denote the canonical projections from $A_F(s, \gamma)$ into F by $\tau_1, \ldots, \tau_s$, and deduce from (8.8) that for each $u = 1, \ldots, s$,

$$|\tau_u[A_F(s, \gamma)]| \leq |F \cap (A_1 \times \cdots \times A_n)| \leq \Psi_F(s). \tag{8.9}$$

Therefore, because $A_F(s, \gamma) \subset \tau_1[A_F(s, \gamma)] \times \cdots \times \tau_s[A_F(s, \gamma)]$, we obtain

$$|A_F(s, \gamma)| \leq \Psi_F(s)^s, \tag{8.10}$$

which verifies (8.1). □

Theorem 27 *For $F \subset S^n$,*

$$16^{-n} d_F(2a)^{\frac{1}{2}} \leq \eta_F(a) \leq d_F(2a)^{\frac{1}{2}}, \quad a > 0. \tag{8.11}$$

In particular, for infinite $F \subset S^n$,

$$\delta_F = \frac{\dim F}{2}, \tag{8.12}$$

and δ_F *is exact if and only if* $\dim F$ *is exact.*

Proof: The right side of (8.11) follows from Lemma XII.6 and Lemma 26.

To verify the left side, let $s > 0$ be an arbitrary integer, and let $A_1, \ldots, A_n$ be arbitrary s-subsets of S. Consider the Riesz product

$$H_s = \prod_{\chi \in A_1} \left(1 + \frac{\chi + \overline{\chi}}{2}\right) \otimes \cdots \otimes \prod_{\chi \in A_n} \left(1 + \frac{\chi + \overline{\chi}}{2}\right). \tag{8.13}$$

As usual (cf. Lemma VII.30),

$$\|H_s\|_{\mathrm{L}^p} \leq 2^{ns/q}, \quad \frac{1}{p} + \frac{1}{q} = 1. \tag{8.14}$$

Let $h_s = \sum_{(\chi_1,\ldots,\chi_n) \in F \cap (A_1 \times \cdots \times A_n)} \chi_1 \otimes \cdots \otimes \chi_n$. By Hölder's inequality and (8.14) with $q = s$,

$$\begin{aligned} 2^{-n}|F \cap (A_1 \times \cdots \times A_n)| &= |\mathbf{E} H_s h_s| \leq 2^n \|h_s\|_{\mathrm{L}^s} \\ &\leq 2^n \, \|h_s\|_{\mathrm{L}^2} \, \eta_F(a) \, s^a, \end{aligned} \tag{8.15}$$

and therefore

$$4^{-n}|F \cap (A_1 \times \cdots \times A_n)|^{\frac{1}{2}} / s^a \leq \eta_F(a), \tag{8.16}$$

which implies the left side of (8.11).

The second assertion follows from (8.11) and the definitions of the respective indices. □

Corollary 28 *If* Γ *is an arbitrary discrete Abelian group, and* $E \subset \Gamma$ *is dissociate, then for all* $F \subset E^n$,

$$16^{-n} \, d_F(2a)^{\frac{1}{2}} \leq \eta_F(a) \leq 4^n \, d_F(2a)^{\frac{1}{2}}, \quad a > 0. \tag{8.17}$$

In particular, the second assertion in Theorem 27 *holds with* E *in place of* S.

Proof (cf. Proposition XII.9). Without loss of generality, we assume E is countably infinite, and enumerate the Steinhaus system by it,

$$S = \{\chi_\gamma : \gamma \in E\}. \tag{8.18}$$

Let f be an F-polynomial, and define for $\mathbf{x} = (x_1, \ldots, x_n) \in (T^{\mathbb{N}})^n$

$$f_{\mathbf{x}} = \sum_{(\gamma_1,\ldots,\gamma_n)\in F} \hat{f}(\gamma_1 \otimes \cdots \otimes \gamma_n)\chi_{\gamma_1}(x_1)\cdots\chi_{\gamma_n}(x_n)\ \gamma_1 \otimes \cdots \otimes \gamma_n. \tag{8.19}$$

For $\mathbf{x} \in (T^{\mathbb{N}})^n$, there exists $\theta_{\mathbf{x}} \in \mathrm{L}^1((T^{\mathbb{N}})^n)$ such that

$$\hat{\theta}_{\mathbf{x}}(\gamma_1 \otimes \cdots \otimes \gamma_n) = \overline{\chi_{\gamma_1}(x_1)}\cdots\overline{\chi_{\gamma_n}(x_n)}, \qquad \gamma_1 \otimes \cdots \otimes \gamma_n \in \text{ spect } f, \tag{8.20}$$

and

$$\|\theta_{\mathbf{x}}\|_{\mathrm{L}^1} \leq 4^n$$

(Chapter VII §8, §12). Then,

$$\|f\|_{\mathrm{L}^q}^q = \|f_{\mathbf{x}} \star \theta_{\mathbf{x}}\|_{\mathrm{L}^q}^q \leq 4^{nq}\|f_{\mathbf{x}}\|_{\mathrm{L}^q}^q. \tag{8.21}$$

Integrating both sides of (8.21) with respect to the Haar measure on $(T^{\mathbb{N}})^n$, applying Fubini's theorem, and then the right side of (8.11), we obtain

$$\|f\|_{\mathrm{L}^q} \leq 4^n\ d_F(2a)^{\frac{1}{2}}\ q^a\|f\|_{\mathrm{L}^2}, \tag{8.22}$$

which implies the right side of (8.17).

The proof of the left side is a transcription of the proof of the left side of (8.11). □

Let $E = \{\gamma_j\colon j{\in}\mathbb{N}\} \subset \Gamma$ be a dissociate set, and

$$E_n = \{\gamma_{j_1} \otimes \cdots \otimes \gamma_{j_n} : 0 < j_1 < \cdots < j_n\}.$$

If $F \subset E_n$, then we apply combinatorial measurements to its underlying indexing set (cf. (7.31)), define $\tilde{F} \subset E^n$ as in (7.34) (replace r with γ), and obtain

Corollary 29 *For infinite $F \subset E_n$,*

$$\delta_F = \frac{\dim F}{2}, \tag{8.23}$$

and δ_F is exact if and only if $\dim F$ is exact.

Proof (cf. proofs of Theorem VII.32, Theorem 27, and Corollary 28). We use by-now familiar arguments. Let f be an F-polynomial, and for $\mathbf{x} = (x_1, \ldots, x_n) \in \hat{\Gamma}^n$, define

$$f_{\mathbf{x}} = \sum_{\gamma_{j_1} \cdots \gamma_{j_n} \in F} \hat{f}(\gamma_{j_1} \cdots \gamma_{j_n}) \, \overline{\gamma_{j_1}(x_1)} \, \cdots \, \overline{\gamma_{j_n}(x_n)} \, \gamma_{j_1} \cdots \gamma_{j_n}. \tag{8.24}$$

By Theorem VII.26 (transcribed from W to Γ), there exists $\theta_{\mathbf{x}} \in \mathrm{L}^1(\hat{\Gamma})$ such that

$$\hat{\theta}_{\mathbf{x}}(\gamma_{j_1} \cdots \gamma_{j_n}) = \overline{\gamma_{j_1}(x_1)} \, \cdots \, \overline{\gamma_{j_n}(x_n)}, \; \gamma_{j_1} \cdots \gamma_{j_n} \in \text{spect } f, \tag{8.25}$$

and

$$\|\theta_{\mathbf{x}}\|_{\mathrm{L}^1} \leq 8^n.$$

Observe that $f = f_{\mathbf{x}} \star \theta_{\mathbf{x}}$, and, as in the proof of the previous corollary, conclude

$$\eta_F(a) \leq 64^n \; d_F(2a)^{\frac{1}{2}}, \quad a > 0. \tag{8.26}$$

The proof of

$$K^{-n} d_F(2a)^{\frac{1}{2}} \leq \eta_F(a) \tag{8.27}$$

(for some $K > 1$) is similar to the proof of the left side of (8.11). □

Corollary 30 *In every infinite Abelian group* Γ, *for every* $\alpha \geq \frac{1}{2}$ *there exist* $F \subset \Gamma$ *such that* $\delta_F = \alpha$ *exactly, and* $F' \subset \Gamma$ *such that* $\delta_{F'} = \alpha$ *asymptotically.*

Proof: Apply Theorem 19, Remark i in §6, and Corollary 29. □

Remarks:

i **(a motif revisited).** The theme in Chapter VII, that the Walsh system is synthesized from systems of integer indices, has been expanded in this chapter: the Walsh system can be synthesized *continuously* from systems of finite, continuously varying indices. Specifically, there exist W_x, $x \in [1, \infty)$, such that $W_x \subset W_y$ for $x \leq y$,

$$\dim W_x = x, \tag{8.28}$$

and

$$\bigcup \{W_x : x \in [1, \infty)\} = W.$$

(See Remark i §6.) The dim-index is precisely related to the σ-index and the δ-index, each of which conveys, separately, the evolving complexity of W_x: $\dim W_x$ gauges the 'combinatorial' complexity of W_x; σ_{W_x} gauges its 'functional' complexity, and δ_{W_x} gauges its 'statistical' complexity.

ii **(an open problem).** Corollary 24 and Corollary 29 imply that for all $n \in \mathbb{N}$, and $F \subset W_n$,

$$\sigma_F = 4\delta_F/(2\delta_F + 1). \tag{8.29}$$

Question: *Does* (8.29) *hold for all spectral sets* $F \subset W$*?*

The instance '$\sigma_F = 1$ exactly if and only if $\delta_F = \frac{1}{2}$ exactly' had been proved first for $F \subset W$ by Bonami [Bon1], and then for subsets F in general groups Γ by Pisier. (See Remark viii in Chapter VII §11.) Notice that (8.29) would imply an affirmative solution to the p-Sidon set union problem; see Chapter VII §13. I suspect also the reverse is true: that an affirmative resolution of the general union problem is inextricably tied to a proof of (8.29), much like the resolution of the 1-Sidon set union problem was tied to the aforementioned instance concerning $\sigma_F = 1$.

iii **(another index – another problem).** Theorem VII.41 suggests the following measurements. For $F \subset \Gamma$, and $u \in [1, 2]$, let

$$\theta_F(u) = \sup\{\|g\|_{L^s}/\sqrt{s}: \ g \in C_F(\hat{\Gamma}), \ \|\hat{g}\|_u \leq 1\} \tag{8.30}$$

and define the index

$$\xi_F = \sup\{u : \theta_F(u) < \infty\}. \tag{8.31}$$

Because $\zeta_F(2u/(3u-2)) \geq \theta_F(u)$ (Theorem VII.41),

$$\sigma_F \geq 2\xi_F/(3\xi_F - 2). \tag{8.32}$$

If $E \subset \Gamma$ is dissociate, and $F \subset E_n$, then

$$\sigma_F = 2\xi_F/(3\xi_F - 2) \quad \text{(Exercise 15)}. \tag{8.33}$$

Question: *Does* (8.33) *hold for all* $F \subset \Gamma$*?*

Exercises

1. Use Riesz products to prove that if U is a k-cover of $[m]$, then $\zeta_{R^U}(t) < \infty$ for all $t < 2m/(m+k)$.

2. Prove that exact and asymptotic p-Sidon sets exist in W for all $p \in [1,2)$.
3. Verify Lemmas 3 and 4 with a dissociate $E \subset \Gamma$ in place of the Rademacher system.
4. Prove that asymptotic p-Sidon sets exist in Γ for all $p \in [1,2)$.
5. i. Prove Lemma 6.
   ii. Show that $\mathbf{1}_{X^U} \in \tilde{V}_n(X^{S_1}, \ldots, X^{S_n})$.
6.* Let $E \subset \Gamma$ be dissociate, let U be a uniformly incident k-cover of $[m]$, and let $E^{(U)}$ be a fractional sum defined by (2.28). Prove that $\eta_{E^{(U)}}(m/2k) \leq K^{m/k}$, for $K > 0$ that does not depend on U.
7. Verify Lemma 12 and Proposition 13.
8. Prove that if the set X is infinite, U is a cover of $[m]$, and α is the optimal value of the linear programming problem in (5.1), then

$$\lim_{s \to \infty} \Psi_{X^U}(s)/s^\alpha = 1.$$

9. Prove the generalization of (5.6) stated in (5.19).
10.* Compute

$$|\{\alpha : \text{ there exist covers } U, \ |U| = n, \text{ and } \dim X^U = \alpha\}|.$$

11. Prove Lemma 18.
12. Verify the last step in the proof of Lemma 21.
13. Prove (7.1) by use of random constructions based on Theorem X.8.
14. Prove Corollary 24.
15. Prove the formula in (8.33) for $F \subset E_n$.

Hints for Exercises in Chapter XIII

1. See the proof of Lemma XII.4.
2. Supply details of the construction outlined in Remark ii §2.
8. Go through the proof of Theorem 14, and do the 'arithmetic'.
9. The proof of (5.6), properly adapted, should do it.
15. The exercise is effectively a review of what has been done in this chapter.

XIV

The Last Chapter: Leads and Loose Ends

1 Mise en Scène: The Last Chapter

I have come to the end, but believe it is only the beginning; there is hard work ahead, and much to be discovered. Questions I could not answer have been scattered throughout the book, and in this chapter I look back, assess what has been done, and try to point to future lines. I will recall various notions and results from previous chapters, and expect readers (if there are any left...) to be familiar with them.

In §2, we outline rudiments of measure theory in fractional dimensions. Grothendieck inequality-type issues surface naturally in this context. The relevant chapters are IV, V, VI, VIII, IX, XII, and XIII.

In §3, *combinatorial dimension* – a basic gauge of interdependence – is cast in general topological and measurable settings. There are other notions of *dimension*, and questions regarding connections between these and *combinatorial dimension* lead to interesting problems. Relevant chapters are XII and XIII.

In §4, we reexamine basic structures underlying classical harmonic analysis. Standard textbook harmonic analysis starts and continues with Borel measures and their transforms, but going further one could start with finitely additive set-functions, and, in the spirit of §2, follow a course where the space of measures is but a first stop. Relevant chapters are VII, XII, and XIII.

In §5, we revisit the random F-walks that were introduced in Chapter X §13, and consider the question: how random are the random F-walks? The message is that combinatorial and stochastic measurements of randomness are feasible and indeed meaningful.

In §6, we revisit the α-chaos processes, the continuous-time models for the random F-walks of the previous section. Relevant chapters are X, XI, XII, and XIII.

In §7, we recast the stochastic integrators of Chapter XI in a fractional-dimensional setting. We consider the 'dimension' of a 1-process, view it as a gauge of interdependence, and observe its precise link to the Littlewood index. Relevant chapters are XI, XII, and XIII.

2 Fréchet Measures in Fractional Dimensions

Let $(X_1, \mathfrak{A}_1), \ldots, (X_m, \mathfrak{A}_m)$ be measurable spaces with infinite underlying σ-algebras, and suppose that $\mu \in F_m(\mathfrak{A}_1, \ldots, \mathfrak{A}_m)$ (Definition VI.1). A question arises: does μ determine a measure in *more* than one coordinate at a time? The two obvious possibilities are: (1) the starting assumption $\mu \in F_m(\mathfrak{A}_1, \ldots, \mathfrak{A}_m)$ cannot be improved; (2) μ determines a scalar measure on $\sigma(\mathfrak{A}_1 \times \cdots \times \mathfrak{A}_m)$. To list all other possibilities we use the framework of fractional Cartesian products.

The Definition of $F_U(\mathfrak{A}_1 \times \cdots \times \mathfrak{A}_m)$

For $S \subset [m]$, let $\boldsymbol{\mathfrak{A}}|_S := \times\{\mathfrak{A}_i : i \in S\}$, and $\mathbf{X}|_S := \times\{X_i : i \in S\}$. Let $U = \{S_1, \ldots, S_n\}$ be a cover of $[m]$, and denote

$$C_U(i) = \{j : i \in S_j\}, \quad i \in [m]. \tag{2.1}$$

Given $\mu \in F_m(\mathfrak{A}_1, \ldots, \mathfrak{A}_m)$, view it as a set-function on $\boldsymbol{\mathfrak{A}}|_{S_1} \times \cdots \times \boldsymbol{\mathfrak{A}}|_{S_n}$,

$$\begin{aligned} &\mu\left(\bigcap\{A_{j1} : j \in C_U(1)\}, \ldots, \bigcap\{A_{jm} : j \in C_U(m)\}\right), \\ &(A_{1i} : i \in S_1) \in \boldsymbol{\mathfrak{A}}|_{S_1}, \ldots, (A_{ni} : i \in S_n) \in \boldsymbol{\mathfrak{A}}|_{S_n}, \end{aligned} \tag{2.2}$$

and observe

$$\begin{aligned} &\mu(\pi_{S_1}^{-1}[\times\{A_{1i} : i \in S_1\}] \cap \cdots \cap \pi_{S_n}^{-1}[\times\{A_{ni} : i \in S_n\}]) \\ &= \mu\left(\bigcap\{A_{j1} : j \in C_U(1)\}, \ldots, \bigcap\{A_{jm} : j \in C_U(m)\}\right). \end{aligned} \tag{2.3}$$

(In writing (2.3) we use the convention stated below (XII.5.4).) We extend μ to a finitely additive set-function $\tilde{\mu}$ on $a(\mathfrak{A}|_{S_1}) \times \cdots \times a(\mathfrak{A}|_{S_n})$ (cf. (XII.5.4)),

$$\tilde{\mu}(E_1, \ldots, E_n) = \mu(\pi_{S_1}^{-1}[E_1] \cap \cdots \cap \pi_{S_n}^{-1}[E_n]), \qquad E_j \in a(\mathfrak{A}|_{S_j}),\ j \in [n]. \tag{2.4}$$

If $\tilde{\mu}$ determines an F_n-measure on $\sigma(\mathfrak{A}|_{S_1}) \times \cdots \times \sigma(\mathfrak{A}|_{S_n})$, then we say that μ is an F_U-measure on $\mathfrak{A}_1 \times \cdots \times \mathfrak{A}_m$, and denote the class of all such μ by $F_U(\mathfrak{A}_1 \times \cdots \times \mathfrak{A}_m)$; for $\mathfrak{A}_i = 2^{X_i}$, we write $F_U(\cdots \times X_i \times \cdots)$. An extension of the proof of Proposition XII.11 yields

Proposition 1 *If $\mu \in F_m(\mathfrak{A}_1, \ldots, \mathfrak{A}_m)$, and*

$$\tilde{\mu} \in F_n(a(\mathfrak{A}|_{S_1}) \times \cdots \times a(\mathfrak{A}|_{S_n}),$$

then $\mu \in F_U(\mathfrak{A}_1 \times \cdots \times \mathfrak{A}_m)$ if and only if

$$\|\mu\|_{F_U} := \sup \left\{ \left\| \sum_{\mathbf{A} \in \mathfrak{g}} \mu(\mathbf{A}) r_{\pi_{S_1}\mathbf{A}} \otimes \cdots \otimes r_{\pi_{S_n}\mathbf{A}} \right\|_\infty : (\mathfrak{A}_1 \times \cdots \times \mathfrak{A}_m)\text{-grid } \mathfrak{g} \right\} < \infty. \tag{2.5}$$

(Rademacher systems in (2.5) are indexed, respectively, by $\pi_{S_1}[\mathfrak{g}], \ldots, \pi_{S_n}[\mathfrak{g}]$.)

Remarks:

i **(a decomposition of F_U?).** For each $i = 1, \ldots, n$, consider the cover of $[m] \backslash S_i$

$$U'_{S_i} = \{S_j \backslash S_i : j = 1, \ldots, n\}, \tag{2.6}$$

and the subsequent (modified) cover of $[m]$

$$U_i = \{S_i\} \cup U'_{S_i}. \tag{2.7}$$

For example, take $U = \{S_1, S_2, S_3\}$, where $S_1 = \{(1,2), S_2 = (2,3), S_3 = (1,3)\}$. Then, $U'_{S_1} = \{(3),(3)\}$, $U'_{S_2} = \{(1),(1)\}$, $U'_{S_3} = \{(2),(2)\}$, $U_1 = \{(1,2),(3),(3)\}$, $U_2 = \{(2,3),(1),(1)\}$, and $U_3 = \{(1,3),(2),(2)\}$. In this case, $\alpha(U) = 3/2$, and $\alpha(U_i) = 2,\ i = 1,2,3$. ($\alpha(U)$ = solution to linear programming problem in (XIII.5.1).)

If $\mu \in F_U(\mathfrak{A}_1 \times \cdots \times \mathfrak{A}_m)$, then $\mu \in F_{U_i}(\mathfrak{A}_1 \times \cdots \times \mathfrak{A}_m)$, $i = 1, \ldots, n$. Rephrased in 'long hand', this means that if μ is an F_U-measure on

$\mathfrak{A}_1 \times \cdots \times \mathfrak{A}_m$, then μ is an F_1-measure on $\mathfrak{A}|_{S_i}$ when coordinates indexed by $[m]\backslash S_i$ are fixed, and an $F_{U'_{S_i}}$-measure on $\mathfrak{A}|_{[m]\backslash S_i}$ when coordinates indexed by S_i are fixed (Exercise 1). I do not know the answer to the following.

Question: *Is the inclusion*

$$F_U(\mathfrak{A}_1 \times \cdots \times \mathfrak{A}_m) \subset \bigcap_{i=1}^{n} F_{U_i}(\mathfrak{A}_1 \times \cdots \times \mathfrak{A}_m) \tag{2.8}$$

an equality? (*This problem was stated in Chapter* XII §5 *for* $U = \{(1,2),(2,3),(1,3)\}$.)

The Definition of $[F_n]_{X^U}$

Let $\mathbf{X}^U$ be the fractional Cartesian product

$$\{(\pi_{S_1}\mathbf{x}, \ldots, \pi_{S_n}\mathbf{x}) : \mathbf{x} \in X_1 \times \cdots \times X_m\}, \tag{2.9}$$

whose ambient n-fold product is $\mathbf{X}|_{S_1} \times \cdots \times \mathbf{X}|_{S_n}$. An F_n-measure ν on $\mathfrak{A}|_{S_1} \times \cdots \times \mathfrak{A}|_{S_n}$ is supported in $\mathbf{X}^U$ if

$$\begin{aligned} &\nu(\times\{E_{1i} : i \in S_1\}, \ldots, \times\{E_{ni} : i \in S_n\}) \\ &\quad = \nu(\times\{E'_{1i} : i \in S_1\}, \ldots, \times\{E'_{ni} : i \in S_n\}) \end{aligned} \tag{2.10}$$

when $(E_{1i} : i \in S_1) \in \mathfrak{A}|_{S_1}, \ldots, (E_{ni} : i \in S_n) \in \mathfrak{A}|_{S_n}$,

$$(E'_{1i} : i \in S_1) \in \mathfrak{A}|_{S_1}, \ldots, (E'_{ni} : i \in S_n) \in \mathfrak{A}|_{S_n},$$

and

$$\begin{aligned} &\bigcap\{E_{j1} : j \in C_U(1)\} = \bigcap\{E'_{j1} : j \in C_U(1)\}, \\ &\vdots \\ &\bigcap\{E_{jm} : j \in C_U(m)\} = \bigcap\{E'_{jm} : j \in C_U(m)\}. \end{aligned} \tag{2.11}$$

(Cf. (XII.5.6).) The class of $\nu \in F_n(\mathfrak{A}|_{S_1}, \ldots, \mathfrak{A}|_{S_n})$ supported in $\mathbf{X}^U$ is denoted by $[F_n]_{\mathbf{X}^U}$ (cf. (XII.5.7)). For $\nu \in [F_n]_{\mathbf{X}^U}$, define

$$\begin{aligned} \mu(A_1, \ldots, A_m) &= \nu(\times\{A_i : i \in S_1\}, \ldots, \times\{A_i : i \in S_n\}), \\ &\qquad (A_1, \ldots, A_m) \in \mathfrak{A}_1 \times \cdots \times \mathfrak{A}_m. \end{aligned} \tag{2.12}$$

Then, $\mu \in F_U(\mathfrak{A}_1 \times \cdots \times \mathfrak{A}_m)$ and $\tilde{\mu} = \nu$. Conversely, if

$$\mu \in F_U(\mathfrak{A}_1 \times \cdots \times \mathfrak{A}_m),$$

then $\tilde{\mu} \in [F_n]_{\mathbf{X}^U}$. That is,

$$\begin{aligned}[F_n]_{\mathbf{X}^U} \\ = \{\nu \in F_n(\boldsymbol{\mathfrak{A}}|_{S_1}, \ldots, \boldsymbol{\mathfrak{A}}|_{S_n}) : \exists\ \mu \in F_U(\mathfrak{A}_1 \times \cdots \times \mathfrak{A}_m) \text{ and } \nu = \tilde{\mu}\},\end{aligned} \tag{2.13}$$

and we write $[F_n]_{\mathbf{X}^U} = F_U$ (cf. (XII.5.8), Exercise 2).

Integration with Respect to F_U-Measures

Let $f_1 \in \mathrm{L}^\infty(\sigma(\boldsymbol{\mathfrak{A}}|_{S_1})), \ldots, f_n \in \mathrm{L}^\infty(\sigma(\boldsymbol{\mathfrak{A}}|_{S_n}))$, and consider the elementary U-tensor

$$\begin{aligned}(f_1 \otimes \cdots \otimes f_n)^U(\mathbf{x}) \ &= \ f_1(\pi_{S_1}\mathbf{x}) \cdots f_n(\pi_{S_n}\mathbf{x}), \\ & \qquad \mathbf{x} \in X_1 \times \cdots \times X_m.\end{aligned} \tag{2.14}$$

The integral of $(f_1 \otimes \cdots \otimes f_n)^U$ with respect to $\mu \in F_U(\mathfrak{A}_1 \times \cdots \times \mathfrak{A}_m)$ is defined by

$$\begin{aligned}\int_{X_1 \times \cdots \times X_m} & (f_1 \otimes \cdots \otimes f_n)^U \mathrm{d}\mu \\ & := \int_{\mathbf{X}|_{S_1} \times \cdots \times \mathbf{X}|_{S_n}} f_1 \otimes \cdots \otimes f_n \ \mathrm{d}\tilde{\mu},\end{aligned} \tag{2.15}$$

where $\tilde{\mu}$ is the F_n-measure determined by (2.5).

Let $X_1, \ldots, X_m$ be locally compact Hausdorff spaces with respective Borel fields $\mathfrak{A}_1, \ldots, \mathfrak{A}_m$. Let $V_U(X_1 \times \cdots \times X_m)$ be the class of functions ϕ on $X_1 \times \cdots \times X_m$ represented (pointwise on $X_1 \times \cdots \times X_m$) by

$$\phi = \sum_k (f_{1k} \otimes \cdots \otimes f_{nk})^U, \tag{2.16}$$

where $f_{jk} \in \mathrm{C}_0(\boldsymbol{X}|_{S_j})$, $k \in \mathbb{N}$, $j \in [n]$, and

$$\sum_k \|f_{1k}\|_\infty \cdots \|f_{nk}\|_\infty < \infty. \tag{2.17}$$

The V_U-norm of ϕ is defined to be the infimum of sums in (2.17) over all representations of ϕ by (2.16). Equivalently, let

$$I_U = \{\phi \in V_n(\mathbf{X}|_{S_1}, \ldots, \mathbf{X}|_{S_n}) : \phi \equiv 0 \text{ on } \mathbf{X}^U\}, \tag{2.18}$$

and consider the restriction algebra

$$V_n(\mathbf{X}|_{S_1}, \ldots, \mathbf{X}|_{S_n})|_{\mathbf{X}^U} = V_n(\mathbf{X}|_{S_1}, \ldots, \mathbf{X}|_{S_n})/I_U \tag{2.19}$$

equipped with the quotient norm. Then,

$$V_U(X_1 \times \cdots \times X_m) = V_n(\mathbf{X}|_{S_1}, \ldots, \mathbf{X}|_{S_n})|_{\mathbf{X}^U}, \tag{2.20}$$

and the annihilator of I_U (a closed subspace of $F_n(\mathfrak{A}|_{S_1}, \ldots, \mathfrak{A}|_{S_n})$) is $[F_n]_{\mathbf{X}^U} (= F_U)$. If $\phi \in V_U(X_1 \times \cdots \times X_m)$ and $\mu \in F_U(X_1 \times \cdots \times X_m)$, then

$$\int_{X_1 \times \cdots \times X_m} \phi \, \mathrm{d}\mu := \sum_k \int_{X_1 \times \cdots \times X_m} (f_{1k} \otimes \cdots \otimes f_{nk})^U \mathrm{d}\mu, \tag{2.21}$$

and

$$\left| \int_{X_1 \times \cdots \times X_m} \phi \, \mathrm{d}\mu \right| \leq 2^n \|\phi\|_{V_U} \|\mu\|_{F_U}. \tag{2.22}$$

Theorem 2 (cf. Theorem VI.13, (XIII.2.49)). *If $X_1, \ldots, X_m$ are locally compact Hausdorff spaces with respective Borel fields $\mathfrak{A}_1, \ldots, \mathfrak{A}_m$, then*

$$V_U(X_1 \times \cdots \times X_m)^* = F_U(\mathfrak{A}_1 \times \cdots \times \mathfrak{A}_m). \tag{2.23}$$

p-variations and Littlewood Indices of F-measures

We define the p-variation of $\mu \in F_m(\mathfrak{A}_1, \ldots, \mathfrak{A}_m)$

$$\|\mu\|_{(p)}$$
$$= \sup \left\{ \left(\sum_{\mathbf{A} \in \mathfrak{g}} |\mu(\mathbf{A})|^p \right)^{\frac{1}{p}} : (\mathfrak{A}_1 \times \cdots \times \mathfrak{A}_m)\text{-grid } \mathfrak{g} \right\}, \quad p > 0, \tag{2.24}$$

and then its *Littlewood index*

$$\ell_\mu = \inf\{p : \|\mu\|_{(p)} < \infty\}. \tag{2.25}$$

If $\ell_\mu = p$ and $\|\mu\|_{(p)} < \infty$, then we write $\ell_\mu = p$ *exactly*; otherwise, if $\ell_\mu = p$ and $\|\mu\|_{(p)} = \infty$, then we write $\ell_\mu = p$ *asymptotically* (cf. (XI.2.27) and (XI.2.28)). The Littlewood inequalities in fractional dimensions (Chapter XIII §2, §7) imply that if $\alpha(U) = \alpha$ (as per (XIII.5.1)), then

$$\ell_\mu \leq 2\alpha/(\alpha + 1) \text{ for all } \mu \in F_U(\mathfrak{A}_1 \times \cdots \times \mathfrak{A}_m), \tag{2.26}$$

and

$$\exists\ \mu \in F_U(\mathfrak{A}_1 \times \cdots \times \mathfrak{A}_m) \text{ such that } \ell_\mu = 2\alpha/(\alpha+1) \text{ exactly.} \tag{2.27}$$

Corollary 3 (see examples below). *If U_1 and U_2 are covers of $[m]$, and $\alpha(U_1) < \alpha(U_2)$, then there exists $\mu \in F_{U_2}(\mathfrak{A}_1 \times \cdots \times \mathfrak{A}_m)$ such that $\mu \notin F_{U_1}(\mathfrak{A}_1 \times \cdots \times \mathfrak{A}_m)$.*

But I do not know the answer to

Question: *Suppose U_1 and U_2 are covers of $[m]$, and $\alpha(U_1) = \alpha(U_2)$. Are $F_{U_1}(\mathfrak{A}_1 \times \cdots \times \mathfrak{A}_m)$ and $F_{U_2}(\mathfrak{A}_1 \times \cdots \times \mathfrak{A}_m)$ isomorphic?*

Examples

We dub $\mu \in F_m(\mathfrak{A}_1, \ldots, \mathfrak{A}_m)$ a *true* F_U-measure if μ is an F_U-measure, but not an $F_{U'}$-measure for any cover U' of $[m]$ such that $\alpha(U') < \alpha(U)$. True F_U-measures can be observed by producing $\mu \in F_U$ that satisfy $\ell_\mu = 2\alpha/(\alpha+1)$. To this end, it suffices to produce for arbitrary integers $N > 0$, $\{-1,1\}$-valued arrays $(\epsilon_{\mathbf{j}} : \mathbf{j} \in [N]^m)$ such that

$$\left\| \sum_{\mathbf{j}\in[N]^m} \epsilon_{\mathbf{j}}\, r_{\pi_{S_1}\mathbf{j}} \otimes \cdots \otimes r_{\pi_{S_n}\mathbf{j}} \right\|_\infty \leq K N^{(\alpha+1)/2}, \tag{2.28}$$

where $K > 0$ depends only on U (Exercise 3). To obtain these arrays, recall the first part of the proof of Theorem XIII.14. Let $(t_1, \ldots, t_m)$ be a solution to the linear programming problem in (XIII.5.1) such that each $t_i \geq 0$ is rational. Let N be a positive integer such that $N^{t_1}, \ldots, N^{t_m}$ are integers, and $A_1 = \times\{[N^{t_i}] : i \in S_1\}, \ldots,\ A_n = \times\{[N^{t_i}] : i \in S_n\}$. Then,

$$|A_j| = \prod_{i\in S_j} N^{t_i} \leq N, \quad j = 1, \ldots, n, \tag{2.29}$$

and

$$|\{(\pi_{S_1}\mathbf{j}, \ldots, \pi_{S_n}\mathbf{j}) : \mathbf{j} \in [N^{t_1}] \times \cdots \times [N^{t_m}]\}| = N^\alpha. \tag{2.30}$$

By (2.29), the degree of W-polynomials with spectrum in $\{r_{\mathbf{k}} : \mathbf{k} \in A_j\}$ is at most 2^N. Therefore, by Theorem X.8 and (2.30), there exist $\epsilon_{\mathbf{j}} = \pm 1$, $\mathbf{j} \in [N^{t_1}] \times \cdots \times [N^{t_m}]$, such that

$$\left\| \sum_{\mathbf{j}\in[N^{t_1}]\times\cdots\times[N^{t_m}]} \epsilon_{\mathbf{j}}\, r_{\pi_{S_1}\mathbf{j}} \otimes \cdots \otimes r_{\pi_{S_n}\mathbf{j}} \right\|_\infty \leq K N^{(\alpha+1)/2}, \tag{2.31}$$

where $K > 0$ depends only on U. Let

$$\mu_N = (1/N^{(\alpha+1)/2}) \sum_{\mathbf{j}\in[N^{t_1}]\times\cdots\times[N^{t_m}]} \epsilon_{\mathbf{j}}\, \delta_{\pi_{S_1}\mathbf{j}}\otimes\cdots\otimes\delta_{\pi_{S_n}\mathbf{j}}. \tag{2.32}$$

By (2.31), $\|\mu_N\|_{F_U(\mathbb{N}^m)} \leq K$. By (2.30), $\|\mu_N\|_{(p)} = N^{(2\alpha-p\alpha-p)/2p}$, which is unbounded as $N \to \infty$ precisely when $p < 2\alpha/(\alpha+1)$. By applying the 'fractional' Littlewood inequalities, we deduce $\|\mu_N\|_{F_{U'}(\mathbb{N}^m)} \to \infty$ as $N \to \infty$ for all covers U' of $[m]$ such that $\alpha(U') < \alpha(U)$.

The proof of Theorem X.8 yields that arrays $(\epsilon_{\mathbf{j}} : \mathbf{j} \in [N^{t_1}] \times \cdots \times [N^{t_m}])$ above can be found with high probability. True F_U-measures are indeed ubiquitous.

Projective Boundedness

Let $(X_1, \mathfrak{A}_1), \ldots, (X_m, \mathfrak{A}_m)$, $(Y_1, \mathfrak{B}_1), \ldots, (Y_m, \mathfrak{B}_m)$ be measurable spaces, μ in $F_U(\mathfrak{A}_1 \times \cdots \times \mathfrak{A}_m)$, ν in $F_U(\mathfrak{B}_1 \times \cdots \times \mathfrak{B}_m)$, define (cf. (IX.1.2))

$$\mu \times \nu((A_1, B_1), \ldots, (A_m, B_m)) = \mu(A_1, \ldots, A_m)\, \nu(B_1, \ldots, B_m),$$

$$(A_1, \ldots, A_m) \in \mathfrak{A}_1 \times \cdots \times \mathfrak{A}_m,\ (B_1, \ldots, B_m) \in \mathfrak{B}_1 \times \cdots \times \mathfrak{B}_m. \tag{2.33}$$

Question: *When does $\mu \times \nu$ determine an F_U-measure on*

$$\sigma(\mathfrak{A}_1 \times \mathfrak{B}_1) \times \cdots \times \sigma(\mathfrak{A}_m \times \mathfrak{B}_m)?$$

In the case $U = \{(1), \ldots, (m)\}$, we proved in Chapter IX that if the underlying σ-algebras are infinite, then $\mu \times \nu$ determines an F_m-measure for every $\nu \in F_m(\mathfrak{B}_1, \ldots, \mathfrak{B}_m)$ precisely when $\mu \in PBF_m(\mathfrak{A}_1, \ldots, \mathfrak{A}_m)$. (See Definition IX.5 and Theorem IX.6.) The case $U = \{(1,2), (2,3), (1,3)\}$ was discussed in Chapter XII §6.

In the general 'fractional' case, for $\mu \in F_U(\mathfrak{A}_1 \times \cdots \times \mathfrak{A}_m)$, $f_1 \in \mathrm{L}^\infty(\mathfrak{A}|_{S_1} \times \mathbb{N}^{S_1}), \ldots, f_n \in \mathrm{L}^\infty(\mathfrak{A}|_{S_n} \times \mathbb{N}^{S_n})$, we consider (cf. (XII.6.5))

$$\phi_{\mu;f_1,\ldots,f_n}(l) = \int_{X^m} (f_1(\cdot, \pi_{S_1} l) \otimes \cdots \otimes f_n(\cdot, \pi_{S_n} l))^U \mathrm{d}\mu, \quad l \in \mathbb{N}^m, \tag{2.34}$$

and define (cf. (XII.6.6))

$$\|\mu\|_{\mathrm{pb}_U} := \sup\{\|\phi_{\mu;f_1,\ldots,f_n}\|_{V_U([N]^m)} : \|f_1\|_\infty \leq 1, \ldots, \|f_n\|_\infty \leq 1,\ N \in \mathbb{N}\}. \tag{2.35}$$

If $\|\mu\|_{\mathrm{pb}_U} < \infty$, then we say that μ is *U-projectively bounded*, and denote the space of all such μ by $PBF_U(\mathfrak{A}_1 \times \cdots \times \mathfrak{A}_m)$. For $U = \{(1), (2), \ldots, (m)\}$, $PBF_U(\mathfrak{A}_1 \times \cdots \times \mathfrak{A}_m)$ is $PBF_m(\mathfrak{A}_1, \ldots, \mathfrak{A}_m)$ in Chapter IX. The assertion $\|\mu\|_{\mathrm{pb}_U} < \infty$ conveys that μ satisfies a Grothendieck-type inequality in 'fractional dimensions'.

Theorem 4 (Cf. Theorem IX.6; Exercise 4). *If*

$$\mu \in F_U(\mathfrak{A}_1 \times \cdots \times \mathfrak{A}_m),$$

then

$$\mu \times \nu \in F_U(\sigma(\mathfrak{A}_1 \times \mathfrak{B}_1) \times \cdots \times \sigma(\mathfrak{A}_m \times \mathfrak{B}_m))$$

for all $\nu \in F_U(\mathfrak{B}_1 \times \cdots \times \mathfrak{B}_m)$ if and only if $\|\mu\|_{\mathrm{pb}_U} < \infty$.

Examples

We have already noted the abundance of *true* F_U-measures (Corollary 3 and *examples* following it), and now turn to the question whether 'true' elements exist in $PBF_U(\mathfrak{A}^m)$. We consider the problem in the integer-dimensional case (cf. Remark iii, Chapter VIII §9):

Question: *Does the inclusion*

$$\begin{aligned} &PBF_j(\sigma(\mathfrak{A}_1 \times \cdots \times \mathfrak{A}_{m-j+1}),\ \mathfrak{A}_{m-j+2,\cdots}, \mathfrak{A}_m) \\ &\quad \subset PBF_k(\sigma(\mathfrak{A}_1 \times \cdots \times \mathfrak{A}_{m-k+1}),\ \mathfrak{A}_{m-k+2,\cdots}, \mathfrak{A}_m) \end{aligned} \tag{2.36}$$

hold for $0 < j < k \leq m$?

(The inclusion

$$\begin{aligned} &F_j(\sigma(\mathfrak{A}_1 \times \cdots \times \mathfrak{A}_{m-j+1}),\ \mathfrak{A}_{m-j+2,\cdots}, \mathfrak{A}_m) \\ &\quad \subset F_k(\sigma(\mathfrak{A}_1 \times \cdots \times \mathfrak{A}_{m-k+1}), \mathfrak{A}_{m-k+2,\cdots}, \mathfrak{A}_m) \end{aligned}$$

easily follows from definitions, but (2.36) is quite another matter.) The answer is affirmative for $j = 1$ and $1 < k \leq m$ (Exercise 5, cf. (IX.7.17)), and negative for $1 < j < k \leq m$. We indicate a proof in the case $m \geq 3$, $j = m - 1$, and $k = m$.

The case $m = 3$ was worked out in Exercise IX.23 [BlCag]. For $m > 3$, we use results from Chapter VIII, Chapter IX, and Chapter XIII. Let

$$X_1 = \cdots = X_{m-2} = \Omega \times \Omega, \quad X_{m-1} = X_m = \Omega,$$

where Ω is the compact Abelian group $\{-1, 1\}^{\mathbb{N}}$. In this case, we take $\mathfrak{A}_1 = \cdots = \mathfrak{A}_{m-2} = \sigma(\mathfrak{B} \times \mathfrak{B})$, and $\mathfrak{A}_{m-1} = \mathfrak{A}_m = \mathfrak{B}$, where $\mathfrak{B}$ is the

Borel field in Ω. Let $\varphi \in l^\infty(W^{m-1})$, where $W = \hat{\Omega}$, and consider the F_m-measure μ_φ on $\mathfrak{A}_1 \times \cdots \times \mathfrak{A}_m$ defined by

$$
\begin{aligned}
&\mu_\varphi(A_1, \dots, A_m) \\
&\quad = \sum_{(w_1,\dots,w_{m-1}) \in W^{m-1}} \varphi(w_1, \dots, w_{m-1}) \cdot \\
&\qquad \cdot \hat{\mathbf{1}}_{A_1}(w_1, w_2) \cdots \hat{\mathbf{1}}_{A_{m-2}}(w_{m-2}, w_{m-1})\, \hat{\mathbf{1}}_{A_{m-1}}(w_1) \hat{\mathbf{1}}_{A_m}(w_{m-1}), \\
&\qquad A_1 \in \mathfrak{A}_1,\ A_2 \in \mathfrak{A}_2, \dots, A_m \in \mathfrak{A}_m.
\end{aligned} \tag{2.37}
$$

The F_m-measure μ_φ naturally determines an F_{m-1}-measure on $\mathfrak{A}_1 \times \cdots \times \mathfrak{A}_{m-2} \times \sigma(\mathfrak{B} \times \mathfrak{B})$,

$$
\begin{aligned}
&\mu_\varphi(A_1, \dots, A_{m-1}) \\
&\quad = \sum_{(w_1,\dots,w_{m-1}) \in W^{m-1}} \varphi(w_1, \dots, w_{m-1}) \cdot \\
&\qquad \cdot \hat{\mathbf{1}}_{A_1}(w_1, w_2) \cdots \hat{\mathbf{1}}_{A_{m-2}}(w_{m-2}, w_{m-1}) \hat{\mathbf{1}}_{A_{m-1}}(w_1, w_{m-1}), \\
&\qquad A_1 \in \mathfrak{A}_1,\ A_2 \in \mathfrak{A}_2, \dots, A_{m-1} \in \sigma(\mathfrak{B} \times \mathfrak{B}).
\end{aligned} \tag{2.38}
$$

Observe that μ_φ in (2.37) is (essentially) the m-linear functional in (VIII.7.5) with the cover

$$U = \{(1,2), (2,3), \dots, (m-2, m-1), (1), (m-1)\},$$

and, similarly, (2.38) is the $(m-1)$-linear functional with the cover $U' = \{(1,2), (2,3), \dots, (m-2, m-1), (1, m-1)\}$. The solutions to the two respective linear programming problems associated with these two covers are $\alpha(U) = (m+1)/2$, and $\alpha(U') = (m-1)/2$. Therefore, by Corollary XIII.23, there exist $\tilde{\varphi} \in B(R^U)$ such that $\tilde{\varphi} \notin B(R^{U'})$. Therefore, by applying Theorem VIII.19, we conclude that there exists $\varphi \in l^\infty(W^{m-1})$ such that $\mu_\varphi \in PBF_{m-1}(\mathfrak{A}_1, \dots, \mathfrak{A}_{m-2}, \sigma(\mathfrak{B} \times \mathfrak{B}))$, and $\mu_\varphi \notin PBF_m(\mathfrak{A}_1, \dots, \mathfrak{A}_{m-2}, \mathfrak{B}, \mathfrak{B})$.

A similar argument, based on *minimal* $(m-j+1)$-covers, resolves the general integer case $1 < j < k \leq m$ (Exercise 6 i).

Remarks:

ii **(questions I did not answer).**

1. Consider the 'fractional' version of the question stated above: in (2.36), replace F_j by F_U and F_k by F_V, where U and V are covers such that $\alpha(U) < \alpha(V)$ (Exercise 6 ii*).

2. A related question is this. Every F_U-measure μ is, by definition, an F_n-measure $\tilde{\mu}$ on $\sigma(\mathfrak{A}^{S_1}) \times \cdots \times \sigma(\mathfrak{A}^{S_n})$. The assertion that μ is U-projectively bounded appears weaker than the assertion $\tilde{\mu} \in PBF_n(\sigma(\mathfrak{A}^{S_1}), \ldots, \sigma(\mathfrak{A}^{S_n}))$. Are there $\mu \in PBF_U(\mathfrak{A}, \ldots, \mathfrak{A})$ such that $\tilde{\mu} \notin PBF_n(\sigma(\mathfrak{A}^{S_1}), \ldots, \sigma(\mathfrak{A}^{S_n}))$?
3. The Grothendieck inequality and factorization theorem imply $F_2(\mathfrak{A}, \mathfrak{A}) = PBF_2(\mathfrak{A}, \mathfrak{A})$ (Theorem IX.9). Is
$$F_U(\mathfrak{A}^m) = PBF_U(\mathfrak{A}^m)$$
for covers U of $[m]$ such that $\alpha(U) \leq 2$? This question, an instance of which was stated in the Remark in Chapter XII §6, is related to the decomposition problem stated in Remark i in this section (Exercise 7).

Types F_U and V_U (Continuation of Discussions in Chapter XII §4, Remarks ii, iii)

Let X be an infinite set, and let U be a cover of $[m]$.

Definition 5

i. $\beta \in l^2(X)$ has *type* F_U if $\beta \circ \tau \in F_U(X^m)$ for *some* bijection τ from X^m onto X; the class of such a β is denoted by $\mathscr{F}_U(X)$.
ii. $f \in c_0(X)$ has *type* V_U if $f \circ \tau \in V_U(X^m)$ for *all* bijections τ from X^m onto X; the class of such f is denoted by $\mathscr{V}_U(X)$.

Let $\alpha = \alpha(U)$, $p_\alpha = 2\alpha/(\alpha+1)$, and $q_\alpha = 2\alpha/(\alpha-1)$. Then
$$\mathscr{F}_U(X) \subset l^{p_\alpha}(X), \tag{2.39}$$
and
$$\exists\, \beta \in \mathscr{F}_U(X) \text{ such that } \|\beta\|_p = \infty \text{ for all } p < p_\alpha.$$
On the 'dual' side,
$$l^{q_\alpha}(X) \subset \mathscr{V}_U(X), \tag{2.40}$$
and
$$\forall\, q < q_\alpha\ \exists\, f \in l^q(X) \text{ such that } f \notin \mathscr{V}_U(X).$$
If U_1 and U_2 cover $[m]$, and for every $S \in U_2$ there exists $T \in U_1$ such that $S \subset T$, then we write $U_1 \prec U_2$. If $U_1 \prec U_2$, then $\alpha(U_1) \leq \alpha(U_2)$, $\mathscr{F}_{U_1}(X) \subset \mathscr{F}_{U_2}(X)$, and $\mathscr{V}_{U_2}(X) \subset \mathscr{V}_{U_1}(X)$. The assertions in (2.39) and (2.40) imply that if $U_1 \prec U_2$ and $\alpha(U_2) < \alpha(U_1)$, then $\mathscr{F}_{U_2}(X) \subsetneqq \mathscr{F}_{U_1}(X)$

and $\mathscr{V}_{U_1}(X) \subsetneqq \mathscr{V}_{U_2}(X)$ (Exercise 8). I do not know whether $U_1 \prec U_2$ and $\alpha(U_2) = \alpha(U_1)$ implies $\mathscr{F}_{U_2}(X) = \mathscr{F}_{U_1}(X)$. (This problem is related to questions stated in Chapter XIII §7, Remark iii; see also the question following Corollary 3 in this chapter.)

For rational $\alpha \geq 1$, we let $\mathfrak{F}_\alpha(X)$ denote the class of $\beta \in l^2(X)$ such that $\beta \in \mathscr{F}_U(X)$ for *some* cover U with $\alpha(U) = \alpha$. On the 'dual' side, we consider $f \in \mathrm{c}_0(X)$ such that $f \in \mathscr{V}_U(X)$ for *all* covers U with $\alpha(U) = \alpha$, and denote by $\mathfrak{V}_\alpha(X)$ the class comprising all such f. If U is an arbitrary cover with $\alpha(U) = \alpha$, then

$$\mathscr{F}_U(X) \subset \mathfrak{F}_\alpha(X) \subset l^{p_\alpha}(X) \text{ and } l^{q_\alpha}(X) \subset \mathfrak{V}_\alpha(X) \subset \mathscr{V}_U(X). \quad (2.41)$$

We define the *optimal* $\mathfrak{F}$*-type* of $\beta \in l^2(X)$ to be

$$\inf\{t : \beta \in \mathfrak{F}_t(X)\}. \quad (2.42)$$

Analogously, the *optimal* $\mathfrak{V}$*-type* of $f \in \mathrm{c}_0(X)$ is

$$\sup\{s : f \in \mathfrak{V}_s(X)\}. \quad (2.43)$$

The *optimal* $\mathfrak{F}$*-type* of $\beta \in l^2(X)$ has a 'stochastic' interpretation, discussed in Remark iii, Chapter XII §4. (See (XII.4.24), and the question following it.)

Theorem 6 (Exercise 9). *For all $\alpha \in [1, \infty]$, there exist $\beta \in l^2(X)$ and $f \in \mathrm{c}_0(X)$ whose optimal $\mathfrak{F}$-type and optimal $\mathfrak{V}$-type equal α.*

Proof: (sketch). We use results in Chapter XIII §2. If $\alpha \geq 1$ is rational, then let U be a cover such that $\alpha(U) = \alpha$, and produce $\beta \in C_{R^U}(\Omega^{|U|})$ with $\|\hat{\beta}\|_p = \infty$ for all $p < 2\alpha/(\alpha+1)$. Then (by (2.41)), the *optimal* $\mathfrak{F}$*-type* of β equals α. Similarly, let $f \in l^{q_\alpha}(X)$ such that $\|f\|_q = \infty$ for all $q < 2\alpha/(\alpha - 1)$, and conclude that its *optimal* $\mathfrak{V}$*-type* equals α.

For irrational $\alpha > 1$, let (α_j) be an increasing sequence of rationals converging to α, and apply previous productions for each α_j, $j \in \mathbb{N}$. □

These notions, cast in a structure-free setting, can be naturally recast in a framework of multidimensional measure theory. Let X be a set, $\mathscr{A}$ an algebra of subsets of X, and μ a finitely additive set-function on $\mathscr{A}$. The first main issue in classical (one-dimensional) measure theory is: does μ determine a scalar measure on $\sigma(\mathscr{A})$? Evoking a by-now-familiar theme, we view this property – $\mu \in F_1(\sigma(\mathscr{A}))$ – as an extremal instance on a continuously calibrated scale.

A finitely additive set-function μ on $\mathscr{A}$ is said to be of *type* F_U, where U covers $[m]$, if the following holds. There exist sets $X_1, \ldots, X_m$ with algebras $\mathscr{A}_1, \ldots, \mathscr{A}_m$ therein, and functionally independent bimeasurable surjections

$$\begin{gathered} \rho_1 : (X, \mathscr{A}) \mapsto (X_1, \mathscr{A}_1) \\ \vdots \\ \rho_m : (X, \mathscr{A}) \mapsto (X_m, \mathscr{A}_m), \end{gathered} \tag{2.44}$$

such that $(\rho_1, \ldots, \rho_m) : X \mapsto X_1 \times \cdots \times X_m$ is a bijection, *and*

$$\mu(\rho_1^{-1}[A_1] \cap \cdots \cap \rho_m^{-1}[A_m]), \quad A_1 \in \mathscr{A}_1, \ldots, A_m \in \mathscr{A}_m, \tag{2.45}$$

determines an F_U-measure on $\sigma(\mathscr{A}_1) \times \cdots \times \sigma(\mathscr{A}_m)$. (Bimeasurable ρ_i means: $\rho_i[A] \in \mathscr{A}_i$ if and only if $A \in \mathscr{A}$.) In this case, we say μ is of *type* F_U relative to $(\rho_1, \ldots, \rho_m)$, and note that μ can be extended to a set-function on

$$\mathscr{A}_{\{\rho_j\}} := \{\rho_1^{-1}[A_1] \cap \cdots \cap \rho_m^{-1}[A_m] : A_1 \in \sigma(\mathscr{A}_1), \ldots, A_m \in \sigma(\mathscr{A}_m)\}. \tag{2.46}$$

We define the *optimal* $\mathfrak{F}$*-type* of μ to be

$$\inf\{\alpha : \exists\, U \text{ with } \alpha(U) = \alpha, \text{and } \mu \text{ is of type } F_U\} \quad (\text{cf. } (2.42)). \tag{2.47}$$

For arbitrary $x \in [1,\infty)$, examples of finitely additive set-functions μ whose *optimal* $\mathfrak{F}$*-type* equals x can be observed by producing true F_U-measures on $\sigma(\mathscr{A}_1) \times \cdots \times \sigma(\mathscr{A}_m)$ and 'pulling' them back to $\mathscr{A}_{\{\rho_j\}}$. As usual such examples can be produced by applying Theorem X.8 and the 'fractional' Littlewood inequalities.

On the dual side, a (point-) function f on X is of type V_U if the following holds: for all $(X_1, \mathscr{A}_1), \ldots, (X_m, \mathscr{A}_m)$, and for all functionally independent bimeasurable maps $\rho_j : (X, \mathscr{A}) \mapsto (X_j, \mathscr{A}_j)$, $j = 1, \ldots, m$ such that $\boldsymbol{\rho} = (\rho_1, \ldots, \rho_m)$, is an injection,

$$f \circ \boldsymbol{\rho}^{-1} \in V_U(\mathscr{A}_1, \ldots, \mathscr{A}_m). \tag{2.48}$$

The *optimal* $\mathfrak{V}$*-type* of f is

$$\sup\{\alpha : \forall\, U \text{ with } \alpha(U) = \alpha, f \text{ is of } V_U\text{-type}\}. \tag{2.49}$$

Examples of f with arbitrarily prescribed *optimal* $\mathfrak{V}$*-types* can be produced by applying (the 'dual' version of) the 'fractional' Littlewood inequalities.

Functions on [0,1] *of Bounded* F_U*-Variation*

Let f be a scalar-valued function on [0,1]. For $p > 0$ and $\epsilon > 0$, let

$$\begin{aligned} V_\epsilon^{(p)}(f) := \sup \Bigg\{ & \sum_{j=1}^{n} |f(x_j) - f(x_{j-1})|^p : \\ & 0 = x_0 \leq x_1 \leq \cdots \leq x_n = 1, \\ & \max_j |x_j - x_{j-1}| \leq \epsilon,\ n = 1, \ldots \Bigg\}, \end{aligned} \tag{2.50}$$

and then define the *p*th-*variation* of f to be

$$V^{(p)}(f) = \lim_{\epsilon \to 0} V_\epsilon^{(p)}(f) \quad \text{(cf. (2.24))}. \tag{2.51}$$

This notion of *p*th-*variation*, which naturally extends the classical *total variation*, had been proposed and studied first by Wiener [Wi3], and then, a decade later, further studied by L.C. Young [Yo1], [Yo2]. After a long hiatus following Young's work, *p*th-*variations* were reconsidered in a probabilistic context by R. Dudley [Du2], and then, in yet another probabilistic context, by N. Towghi [Tow].

There is a precise relation between *p*th-*variations* and F_U*-variations*, which extends the observation that if $\alpha(U) = 1$, then the *total variation* and F_U*-variation* are the same. Let $x = \sum_{j=1}^{\infty} b_j(x)/2^j$ be the dyadic expansion of $x \in$ [0,1]. Let $\{\rho_1, \ldots, \rho_m\}$ be a partition of $\mathbb{N}$, each of whose elements is infinite, and enumerate $\rho_i = \{n_{ij} : j \in \mathbb{N}\}$. Consider the corresponding maps from [0,1] onto [0,1], which we denote also by ρ_i,

$$\rho_i(x) = \sum_{j=1}^{\infty} b_{n_{ij}}(x)/2^j, \quad i \in [m],\ x \in [0,1]. \tag{2.52}$$

Observe that $(\rho_1, \ldots, \rho_m)$ is an injection from [0,1] onto $[0,1]^m$. Consider the algebra $a(\mathscr{D})$ generated by the dyadic intervals in [0,1]. Note that $\rho_1, \ldots, \rho_m$ are $a(\mathscr{D})$-bimeasurable, and that every interval $[a, b] \subset [0,1]$ is in $a(\mathscr{D})_{\{\rho_j\}}$ (defined by (2.46)). The Littlewood inequalities in fractional dimensions imply

Theorem 7 *Let $U = \{S_1, \ldots, S_n\}$ be a cover of $[m]$, and f a scalar-valued function on* [0,1]*. If Δf has type F_U relative to $(\rho_1, \ldots, \rho_m)$, then $V^{(p_\alpha)}(f) < \infty$, where $p_\alpha = 2\alpha(U)/(\alpha(U)+1)$.*

3 Combinatorial Dimension in Topological and Measurable Settings

Consider grids $\mathfrak{g} = \sigma_1 \times \cdots \times \sigma_n$ of $[0,1]^n$, where $\sigma_1, \ldots, \sigma_n$ are partitions of [0,1] consisting of continuous intervals, and define

$$\|\mathfrak{g}\| := \max\{\|\sigma_1\|, \ldots, \|\sigma_n\|\}, \tag{3.1}$$

where $\|\cdot\|$ on the right side denotes the usual mesh. For $F \subset [0,1]^n$, we let

$$F_{\mathfrak{g}} = \{c : c \subset \mathfrak{g},\ F \cap c \neq \emptyset\}, \quad \text{grid } \mathfrak{g}, \tag{3.2}$$

$$D_F(a) = \lim_{\epsilon \to 0} \inf\{d_{F_{\mathfrak{g}}}(a) : \ \text{grid } \mathfrak{g},\ \|\mathfrak{g}\| < \epsilon\}, \quad a > 0, \tag{3.3}$$

and define

$$\mathrm{Dim} F = \inf\{a : D_F(a) < \infty\}. \tag{3.4}$$

(Cf. Chapter XIII §4.) The index in (3.4) is the combinatorial dimension of $F \subset [0,1]^n$ relative to the topological structure in [0,1].

In general, we consider a locally compact Hausdorff space X, and let $F \subset X^n$ be a closed set. For open covers $u_1, \ldots, u_n$ of X, and $\mathfrak{g} = u_1 \times \cdots \times u_n$ (open grid of X^n), we define $F_{\mathfrak{g}}$ by (3.2), and then

$$D_F(a) = \sup_{\mathfrak{g}} \inf_{\mathfrak{r} \prec \mathfrak{g}} d_{F_{\mathfrak{r}}}(a), \tag{3.5}$$

where $\mathfrak{g}$ and $\mathfrak{r}$ above are open grids of X^n, and $\mathfrak{r} \prec \mathfrak{g}$ means that for every $O \in \mathfrak{r}$ there is a $U \in \mathfrak{g}$ such that $O \subset U$. We define $\mathrm{Dim}F$ by (3.4). For $X = [0,1]$, $D_F(a)$ in (3.3) is finite if and only if $D_F(a)$ in (3.5) is finite (Exercise 10), and hence $\mathrm{Dim}F$ based on (3.3) is the same as $\mathrm{Dim}F$ based on (3.5).

If $(X, \mathfrak{A})$ is a measurable space, and F is a measurable subset of X^n, then in the preceding discussion, replace open grids by measurable grids, define $D_F(a)$ by (3.5), and $\mathrm{Dim}F$ by (3.4). We refer to $\mathrm{Dim}_{\mathrm{t}}F$ when using *open* grids in (3.4), and to $\mathrm{Dim}_{\mathrm{m}}F$ when using *measurable* grids in (3.4); when referring to both, either $\mathrm{Dim}_{\mathrm{t}}F$ or $\mathrm{Dim}_{\mathrm{m}}F$, we write $\mathrm{Dim}\, F$.

The combinatorial dimension of $F \subset X^n$, where X is a locally compact Hausdorff space, can be gauged by:

(i) $\dim F$ ('oblivious' to any structure in X);
(ii) $\mathrm{Dim}_{\mathrm{t}} F$ (based on the topological structure in X);
(iii) $\mathrm{Dim}_{\mathrm{m}} F$ (based on the Borel structure in X).

For closed $F \subset X^n$, $\dim F \leq \mathrm{Dim}_{\mathrm{t}}\, F \leq \mathrm{Dim}_{\mathrm{m}}\, F$. T. Körner indicated to me (in a private communication) that there exist closed sets $F \subset [0,1]^2$ such that $\dim F = 1$ and $\mathrm{Dim}_{\mathrm{t}} F = 2$, but I still do not know whether there exist closed sets F, say in $[0,1]^2$, such that $\mathrm{Dim}_{\mathrm{t}} F < \mathrm{Dim}_{\mathrm{m}} F$.

Remarks:

i **(other 'dimensions').** Various indices in diverse contexts have been dubbed *dimension* (e.g., [DSe], [F], [KolTi], [Man], [Pes], [vN3]). Arguably, the most popular and best known is the Hausdorff dimension $\mathrm{Dim}_{\mathfrak{h}}$, which marks in effect the 'correct' exponent in computations of volume. This exponent is fundamentally different from the combinatorial index Dim, which marks a degree of interdependence between coordinates. Indeed, there exist sets $F \subset [0,1]^n$ such that $\mathrm{Dim}_{\mathfrak{h}} F > \mathrm{Dim} F$, and sets $F \subset [0,1]^n$ such that $\mathrm{Dim}_{\mathfrak{h}} F < \mathrm{Dim} F$. To illustrate the first inequality, observe that if $f : [0,1] \to [0,1]$ is a continuous function, then $\mathrm{Dim}\,\mathrm{graph}(f) = 1$ (Exercise 11), and that for every $\alpha \in (1,2)$, there exist continuous functions f such that $\mathrm{Dim}_{\mathfrak{h}}\ \mathrm{graph}(f) = \alpha$ [F, Theorem 8.2]. To illustrate the second inequality, note that if $F \subset [0,1]^n$ is countably infinite, then $\mathrm{Dim}\, F (= \dim F) \geq 1$, whereas $\mathrm{Dim}_{\mathfrak{h}}\ F = 0$. The Cantor 'middle-1/3' set C provides another example:

$$\mathrm{Dim}_{\mathfrak{h}}\ C \times C = 2\ (\ln 2/\ln 3), \tag{3.6}$$

whereas

$$\mathrm{Dim}\ C \times C = 2.$$

A basic feature distinguishing Dim from $\mathrm{Dim}_{\mathfrak{h}}$ is that $\mathrm{Dim}_{\mathrm{t}}$ and $\mathrm{Dim}_{\mathrm{m}}$ are invariant, respectively, under homeomorphisms and measurable 'rearrangements' of the coordinate axes, whereas $\mathrm{Dim}_{\mathfrak{h}}$ is not.

Questions concerning relations between *combinatorial dimension* and other notions of dimension lead to interesting problems. For example, consider von Neumann's *continuous geometries* [vN3], [vN4], [vN5], which were motivated primarily by Murray's and von Neumann's studies of *factors* (e.g., [MuvN1]). (See Chapter IV §8.) A central concept underlying *continuous geometries* is a dimension function, axiomatically defined, with an ordered space as its domain, and (after normalization) the interval [0,1] as its range. In [vN4], von Neumann had produced explicit examples of such geometries by the use of projective spaces, finite analogs of which were used also in deterministic designs of fractionally dimensioned sets in [BlPeSch]. Precise connections between von Neumann's dimension function and the combinatorial dimension are otherwise unknown.

ii **(variations of Fréchet measures).** Let $(X, \mathfrak{A})$ be a measurable space, and $\mu \in \mathcal{F}_n(\mathfrak{A}, \dots, \mathfrak{A})$. For $F \in \mathfrak{A}$, define

$$|\mu|_p(F) = \sup_{\mathfrak{g}} \inf_{\mathfrak{r} \prec \mathfrak{g}} \left(\sum_{c \in F_{\mathfrak{r}}} |\mu(c)|^p \right)^{\frac{1}{p}} \quad \text{(cf. (2.24) and (2.50)).} \tag{3.7}$$

Theorem XIII.20 implies

Theorem 8 *If* $\mu \in \mathcal{F}_n(\mathfrak{A}, \dots, \mathfrak{A})$ *and* $F \in \mathfrak{A}$, *then*

$$|\mu|_p(F) < \infty, \quad p > 2\ \mathrm{Dim_m} F / (\mathrm{Dim_m} F + 1). \tag{3.8}$$

If $\mathrm{Dim_m} F = \dim F$, *then there exists* $\mu \in \mathcal{F}_n(\mathfrak{A}, \dots, \mathfrak{A})$ *such that* $|\mu|_p(F) = \infty$ *for all* $p < 2\ \mathrm{Dim_m} F / (\mathrm{Dim_m} F + 1)$.

I do not know whether the hypothesis $\mathrm{Dim_m} F = \dim F$ can be removed from the second part of the theorem.

iii **(examples).** An abundance of perfect sets $F \subset [0,1]^n$ such that $\mathrm{Dim_m} F = \mathrm{Dim_t} F = \alpha$, for arbitrary $\alpha \in (1, n)$, can be produced by random selections (Lemma XIII.17) applied iteratively in Cantor set-type constructions. These productions are detailed in [Bl7].

4 Harmonic Analysis

The setting is the archetypal compact Abelian group $\Omega = \{-1, 1\}^{\mathbb{N}}$, together with the space of finitely additive set-functions on the class $\mathfrak{C}$ of cylinders in Ω. (Cylinders are $\{(\omega_j) \in \Omega : \omega_j = \epsilon_j,\ j \in F\}$, where $F \subset \mathbb{N}$ and $\epsilon_j = \pm 1$ for $j \in F$.) For our purposes here, the setting is generic because structures in Ω and $\hat{\Omega}$ can be viewed naturally as product structures of varying dimension (cf. Chapter VII). This notion, that ambient structures can be viewed as product structures of arbitrary dimension, appears (though sometimes disguised) in more general harmonic-analytic settings.

The Space $F(\Omega)$

We rephrase a previous discussion (in §2) of *type* F_U. Let $\boldsymbol{\rho} = \{\rho_1, \dots, \rho_m\}$ be a partition of $\mathbb{N}$, whose respective elements give rise to the projections

$$\rho_1 : \{-1,1\}^{\mathbb{N}} \mapsto \{-1,1\}^{\rho_1}, \dots, \ \rho_m : \{-1,1\}^{\mathbb{N}} \mapsto \{-1,1\}^{\rho_m}, \tag{4.1}$$

where $\rho_i(\omega) := (\omega_j : j \in \rho_i)$ for $\omega = (\omega_j : j \in \mathbb{N}) \in \Omega$ and $i = 1, \dots, m$. We denote $\Omega_{\rho_i} = \{-1,1\}^{\rho_i}$, and $W_{\rho_i} = \hat{\Omega}_{\rho_i}, i = 1, \dots, m$. Every finitely additive set-function μ on $\mathfrak{C}$ can be viewed as a set-function on $\mathfrak{C}_{\rho_1} \times \cdots \times \mathfrak{C}_{\rho_m}$, where $\mathfrak{C}_{\rho_i}$ is the class of cylinders in Ω_{ρ_i}:

$$\mu(A_1, \dots, A_m) := \mu(\rho_1^{-1}[A_1] \cap \cdots \cap \rho_m^{-1}[A_m]),$$
$$A_1 \in \mathfrak{C}_{\rho_1}, \dots, A_m \in \mathfrak{C}_{\rho_m} \text{ (cf. (2.45))}. \tag{4.2}$$

Given a cover $U = \{S_1, \dots, S_n\}$ of $[m]$, we consider those μ that determine F_U-measures on $\mathfrak{B}_{\rho_1} \times \cdots \times \mathfrak{B}_{\rho_m}$, where $\mathfrak{B}_{\rho_i} = \sigma(\mathfrak{C}_{\rho_i})$, $i = 1, \dots, m$, and denote the space of all such μ by $F_{U,\boldsymbol{\rho}}(\Omega)$. If $\mu \in F_{U,\boldsymbol{\rho}}(\Omega)$, then μ (defined at the outset on $\mathfrak{C}$) is extendible to a set-function on

$$\{\pi_{S_1}^{-1}(B_1) \cap \cdots \cap \pi_{S_n}^{-1}(B_n) : \ \ B_j \in \sigma(\times\{\mathfrak{B}_{\rho_i} : i \in S_j\}), \ \ j = 1, \dots, n\}, \tag{4.3}$$

a class of sets situated between $\mathfrak{C}$ and $\mathfrak{B}$. (For $S \subset [m]$, π_S denotes the projection from $\{-1,1\}^{\mathbb{N}} = \Omega_{\rho_1} \times \cdots \times \Omega_{\rho_m}$ onto $\times\{\Omega_{\rho_i} : i \in S\}$.) If $\alpha(U) > 1$, and each member of $\boldsymbol{\rho}$ is infinite, then this class is a proper sub-class of $\mathfrak{B}$, and $M(\Omega) \subsetneqq F_{U,\boldsymbol{\rho}}(\Omega)$. Note that $F_{U,\boldsymbol{\rho}}(\Omega)$ depends on the choice of $\boldsymbol{\rho}$.

If U_1 and U_2 cover $[m]$, and $U_1 \prec U_2$, then $F_{U_2,\boldsymbol{\rho}}(\Omega) \subset F_{U_1,\boldsymbol{\rho}}(\Omega)$, and if $\alpha(U_2) < \alpha(U_1)$, then $F_{U_2,\boldsymbol{\rho}}(\Omega) \subsetneqq F_{U_1,\boldsymbol{\rho}}(\Omega)$. However, I do not know whether $\alpha(U_2) = \alpha(U_1)$ implies $F_{U_2,\boldsymbol{\rho}}(\Omega) = F_{U_1,\boldsymbol{\rho}}(\Omega)$. (See the discussion below (2.40).) In the case $U = \{(1), \dots, (m)\}$, we write $F_{\boldsymbol{\rho}}(\Omega)$ for $F_{U,\boldsymbol{\rho}}(\Omega)$. We dub μ an F-measure if μ is an $F_{\boldsymbol{\rho}}(\Omega)$-measure for some partition $\boldsymbol{\rho}$ of $\mathbb{N}$, and denote the space of such μ by $F(\Omega)$.

All this should be by now a familiar story.

Transforms

For $w \in W$, $w = r_{j_1} \cdots r_{j_k}$, write

$$w_{\rho_i} = \prod_{j \in \rho_i \cap \{j_1, \dots, j_k\}} r_j, \quad i = 1, \dots, m, \tag{4.4}$$

and thus $w = w_{\rho_1} \otimes \cdots \otimes w_{\rho_m}$. If $\mu \in F_{\boldsymbol{\rho}}(\Omega)$, then its *transform* $\hat{\mu}$ is

$$\hat{\mu}(w) = \int_{\Omega_1 \times \cdots \times \Omega_m} w_{\rho_1} \otimes \cdots \otimes w_{\rho_m} \, \mathrm{d}\mu, \quad w \in W. \tag{4.5}$$

Convolution

Convolution (considered in Chapter IX §7, 8) can be adapted to the present setting as follows: if $\mu \in F(\Omega)$, $\nu \in F(\Omega)$, *and* there exists $\lambda \in F(\Omega)$ such that $\hat{\lambda} = \hat{\mu}\,\hat{\nu}$, then $\lambda := \mu \star \nu$.

Here are three basic facts. The first is straightforward.

(1) If $\mu \in M(\Omega)$, then for all positive integers m, covers U of $[m]$, partitions $\boldsymbol{\rho} = \{\rho_1, \ldots, \rho_m\}$ of $\mathbb{N}$, and $\nu \in F_{U,\boldsymbol{\rho}}(\Omega)$, the convolution $\mu \star \nu$ is in $F_{U,\boldsymbol{\rho}}(\Omega)$ (Exercise 12 i).

Matters are more involved in higher dimensions. The next two facts, which were (essentially) verified in Chapter IX, are not quite as obvious.

(2) If $\boldsymbol{\rho} = \{\rho_1, \rho_2\}$ is a partition of $\mathbb{N}$, then for all μ and ν in $F_{\boldsymbol{\rho}}(\Omega)$, $\mu \star \nu$ is in $F_{\boldsymbol{\rho}}(\Omega)$ (Exercise 12 ii).

(3) If $\boldsymbol{\rho} = \{\rho_1, \rho_2, \rho_3\}$ is a partition of $\mathbb{N}$ such that $|\rho_1| = |\rho_2| = |\rho_3| = \infty$, then there exist $\mu \in F_{\boldsymbol{\rho}}(\Omega)$ and $\nu \in F_{\boldsymbol{\rho}}(\Omega)$, and $\mu \star \nu \notin F_{\boldsymbol{\rho}}(\Omega)$ (Exercise 12 iii).

Among the (many) unanswered questions are:

i. Suppose $\boldsymbol{\rho} = \{\rho_1, \ldots, \rho_m\}$ is a partition of $\mathbb{N}$, and U is a cover of $[m]$ such that $\alpha(U) \leq 2$, and $m > 2$. If μ and ν are in $F_{U,\boldsymbol{\rho}}(\Omega)$, then $\mu \star \nu \in F(\Omega)$ (Exercise 12 iv), but can we conclude $\mu \star \nu \in F_{U,\boldsymbol{\rho}}(\Omega)$? (See Remark ii in §2.)

ii. Can convolution be defined on the entire space $F(\Omega)$?

Examples

True $F_{U,\boldsymbol{\rho}}$-measures are observed via random constructions; see *examples* following Corollary 3. Below we describe an explicit construction in integer dimensions.

Let $\boldsymbol{\rho} = \{\rho_1, \ldots, \rho_n\}$ be a partition of $\mathbb{N}$ such that

$$|\rho_i| = \infty \text{ for } i = 1, \ldots, n.$$

Let $U = \{S_1, \ldots, S_n\}$ be a cover of $[m]$, such that

$$i_U(j) := |\{j : i \in S_j\}| \geq 2, \quad i \in [m]. \tag{4.6}$$

Choose infinite sets $E_1 \subset W_{\rho_1}, \ldots, E_n \subset W_{\rho_n}$, and index E_i by $\mathbb{N}^{S_i}$,

$$E_i = \{\chi_{\mathbf{j}}^{(i)} : \mathbf{j} \in \mathbb{N}^{S_i}\}, \quad i = 1, \ldots, n. \tag{4.7}$$

Let $\varphi \in l^\infty(\mathbb{N}^m)$, and define

$$\begin{aligned} &\mu_\varphi(\pi_1^{-1}[A_1] \cap \cdots \cap \pi_n^{-1}[A_n]) \\ &\qquad = \sum_{\mathbf{j} \in \mathbb{N}^m} \varphi(\mathbf{j})\ \hat{\mathbf{1}}_{A_1}(\chi^{(1)}_{\pi_{S_1}\mathbf{j}}) \cdots \hat{\mathbf{1}}_{A_n}(\chi^{(n)}_{\pi_{S_n}\mathbf{j}}), \\ &\qquad A_1 \in \mathfrak{B}_{\rho_1}, \ldots, A_n \in \mathfrak{B}_{\rho_n}. \end{aligned} \tag{4.8}$$

Then, Lemma VIII.18, Plancherel's theorem, and Proposition 1 imply that $\mu_\varphi \in F_{\boldsymbol{\rho}}(\Omega)$, and

$$\hat{\mu}_\varphi(w) = \begin{cases} \varphi(\mathbf{j}), & w = \chi^{(1)}_{\pi_{S_1}\mathbf{j}} \otimes \cdots \otimes \chi^{(n)}_{\pi_{S_1}\mathbf{j}}, \quad \mathbf{j} \in \mathbb{N}^m \\ 0 & \text{otherwise.} \end{cases} \tag{4.9}$$

Note that $\mu_\varphi \star \nu \in F_{\boldsymbol{\rho}}(\Omega)$ for all $\nu \in F(\Omega)$ (Exercise 13).

5 Random Walks

The random F-walks (introduced in Chapter X §13) are models for walks whose steps are caused by interdependent 'hidden variables'. In the paradigm in Chapter X §13, interdependent 'hidden variables' were represented by hidden circuits controlled by random, interdependent switching, and circuits were represented by Walsh characters.

Let $N \in \mathbb{N},\ k \in [N]$, and

$$W_k(N) = \left\{ \prod_{j \in u} r_j : u \in 2^{[N]}, 0 < |u| \leq k \right\}. \tag{5.1}$$

We apply to subsets F of $W_k(N)$ the combinatorial measurement d_F defined in (XIII.7.32). Specifically, we use the relation between d_F and the stochastic measurement η_F defined in (XII.3.3),

$$16^{-k}\ d_F\ (\alpha) \leq \eta_F(\alpha/2) \leq 4^k\ d_F(\alpha) \tag{5.2}$$

(Corollary XIII.28), and obtain the following 'limit theorem'.

Proposition 9 (cf. Theorem X.26). *Let (N_j) be a sequence of positive integers, and $F_j \subset W_k(N_j)$, $j \in \mathbb{N}$. If*

$$\sup_j d_{F_j}(\alpha) := d_{\{F_j\}}(\alpha) < \infty, \tag{5.3}$$

then there exists a subsequence (N_{j_l}) and a standard sub-α-variable Y (a sub-α-variable Y such that $\mathbf{E}\,Y = 0$ and $\mathbf{E}\,|Y|^2 = 1$) such that

$$(1/|F_{j_l}|^{\frac{1}{2}}) \sum_{\chi \in F_{j_l}} \chi \xrightarrow[l\to\infty]{} Y \text{ in distribution.} \tag{5.4}$$

Proof: Denote $Y_j = (1/|F_j|^{\frac{1}{2}}) \sum_{\chi \in F_j} \chi$. Then, $\|Y_j\|_{\mathrm{L}^2} = 1$ for $j \in \mathbb{N}$, and, by the assumption in (5.3) and the right-side inequality in (5.2), the sequence $(Y_j : j \in \mathbb{N})$ is uniformly integrable. Therefore, there exists a subsequence (Y_{j_l}) that converges in distribution to a random variable Y, such that $\|Y\|_{\mathrm{L}^2} = 1$ and $\mathbf{E}Y = 0$. By the right side inequality in (5.2),

$$\|Y_j\|_{\mathrm{L}^p} \leq 4^k\ d_{\{F_j\}}(\alpha) p^{\alpha/2}, \quad p > 2,\ j \in \mathbb{N}, \tag{5.5}$$

which implies (e.g., via [Ch, Theorem 4.5.2]) that Y is a sub-α-variable. (See Definition X.19 and discussion around it.) □

Let $F \subset W_k(N)$, $|F| = n$. We enumerate $F = (w_1, \ldots, w_n)$, and let $w_j/\sqrt{n}$ be the jth step in a random walk clocked by discrete time $t = j/n$, $j = 1, \ldots, n$. The resulting random F-walk is

$$X_t = \frac{1}{\sqrt{n}} \sum_{j=1}^{tn} w_j, \quad t = 1/n, 2/n, \ldots, 1. \tag{5.6}$$

Whereas different enumerations of F give rise to different walks in (5.6), their combinatorial complexity (gauged by $d_F(\alpha)$), and their stochastic complexity (gauged by $\eta_F(\alpha/2)$) are the same. The random walk in (5.6) can be viewed as a 'simple' version of an α-chaos process clocked by continuous time $t \in [0,1]$,

$$X(t) = \frac{1}{\sqrt{n}} \sum_{j=1}^{n} w_j \otimes \mathbf{1}_{[0,t]} \left(\frac{j}{n}\right). \tag{5.7}$$

The complexity of this process X is gauged by variations of the Fréchet measure associated with X. Precisely, by adapting arguments in

Chapter X §11, we conclude that if θ_α is the Orlicz function defined in (X.6.22), then

$$\|\mu_X\|_{\theta_\alpha} := \sup\left\{\sum_{j,k}\theta_\alpha(|\mu_X(A_j,B_k)|) : \Sigma_j\mathbf{1}_{A_j}\leq 1,\ \Sigma_k\mathbf{1}_{B_k}\leq 1,\right.$$

$$\left.\{A_j\}\subset\mathscr{A},\ \{B_k\}\subset\mathscr{B}\right\} < K_{X,\alpha}, \tag{5.8}$$

where $K_{X,\alpha}>0$ depends only on $\eta_F(\alpha/2)$. Estimating $\|\mu_X\|_{\theta_\alpha}$ upwards, we take $A_\omega=\{\omega\}\times\{-1,1\}^{\mathbb{N}\setminus[N]}$ for $\omega\in\{-1,1\}^N$, $B_k=[(k-1)/n,k/n)$ for $k=1,\dots,n$, and obtain

$$\sum_{\omega,k}\theta_\alpha(|\mu_X(A_\omega,B_k)|) = (\log 2^N\sqrt{n})^{\alpha/2}/\sqrt{n}. \tag{5.9}$$

We summarize:

Theorem 10 *Let (N_j) be a sequence of positive integers, $F_j\subset W_k(N_j)$, and let X_j be the process in* (5.7) *with $F=F_j$, $j\in\mathbb{N}$. If $d_{\{F_j\}}(\alpha)<\infty$, then $\sup_j\|\mu_{X_j}\|_{\theta_\alpha}<\infty$. If $\sup_j|F_j|/(N_j)^\alpha=\infty$, then*

$$\sup_j\|\mu_{X_j}\|_{\theta_\alpha}=\infty.$$

Remarks:

i **(time-inhomogeneity).** Under the hypotheses in Proposition 9, if we consider arbitrary $W_k(N_j)$-polynomials $\sum_{\chi\in F_j}\mathbf{v}_j(\chi)\chi$, where $\mathbf{v}_j\in B_{l^2(F_j)}$, $j\in\mathbb{N}$, then we obtain (by the same proof) a subsequence (N_{j_l}) and a standard sub-α-variable Y such that

$$\sum_{\chi\in F_{j_l}}\mathbf{v}_j(\chi)\chi\xrightarrow[l\to\infty]{}Y\quad\text{in distribution.} \tag{5.10}$$

Insofar that our goal is to model a perception of Brownian movement, these 'walks' convey a sense of time-inhomogeneity, which, for example, could be the result of information – albeit partial – about movements over different time intervals. (See the discussion of *heuristics* in Chapter 1 §1.) In the tale about the drunk and his decision making machine (in Chapter X §13), this corresponds to a hidden circuit wired to *several* light bulbs. (In the original tale, based on time-homogeneity, there were no repetitions in the enumeration of the hidden circuits.)

ii **(is Proposition 9 sharp?).** Suppose, in addition to the hypotheses in Proposition 9, we also assume $N_j \uparrow \infty$ and $|F_j| \geq KN_j^\alpha$ for all $j \in \mathbb{N}$. Is the sub-sequential limit Y a standard α-variable?

A closely related problem is this. Suppose $F \subset W_k$, and for $\alpha \geq 1$, $d_F(\alpha) < \infty$. Is there $0 < K < \infty$ depending only on F and α, such that for all $p \geq 2$, $f \in \mathrm{L}_F^p$, and $q > p$,

$$\|f\|_{\mathrm{L}^q} \leq K\left(\frac{q}{p}\right)^{\alpha/2} \|f\|_{\mathrm{L}^p}? \tag{5.11}$$

An affirmative answer would imply an affirmative answer to the previous question (Exercise 14). Note that (5.11) holds in the case $F = W_k$ (e.g., [dlPG, Theorem 3.2.1]).

iii **(another scale?).** So far we have used combinatorial measurements tied to exponential tail-probabilities. A question arises: what combinatorial measurements are linked, analogously, to statistical measurements that mark polynomial tail-probabilities? This question is related to Rudin's $\Lambda(p)$-set problem [Ru1], which had been solved first by Bourgain [Bour], and later, via a different method, by Talagrand [T].

The question can be illustrated in a context of F-walks. In the combinatorial analysis of the drunk's stroll in Chapter X §13, we have assumed that the number of switches in any hidden circuit in the drunk's decision making machine is uniformly bounded. To wit, we have considered subsets of $W_k(N)$, where N can be arbitrarily large, but k is fixed. We have obtained, accordingly, a measurement of combinatorial complexity (*combinatorial dimension*) that starts with the simplest complexity marked by $k = 1$ (the simple random walk), and increases continuously. The problem is: design a scale that begins at the *other* end, at $W_N(N) := W(N)$, and is calibrated by 'combinatorial' measurements linked to polynomial tail-probability estimates.

We propose the following scale that starts with the full Walsh system W, and moves downwards in the direction of decreasing complexity. (*Combinatorial dimension* begins at the other end, with the Rademacher system R, and moves upwards in the direction of increasing complexity.) For a finite set $A \subset \mathbb{N}$, denote

$$W(A) = \left\{ \prod_{j \in u} r_j : u \in 2^A \right\}, \tag{5.12}$$

and for $F \subset W$, define for integers $s > 0$

$$\theta_F(s) = \max\{|F \cap W(A)| : A \subset \mathbb{N}, 0 < |A| \leq s\}. \tag{5.13}$$

For $a > 0$, we let

$$e_F(a) = \max\{\theta_F(s)/2^{as} : s \geq 1\}, \tag{5.14}$$

and then define the index

$$\Lambda\text{-dim}F := \sup\{a : e_F(a) < \infty\}. \tag{5.15}$$

The measurement e_F is analogous to d_F, and Λ-dim F is analogous to dim F. The range of Λ-dim F is [0,1]. Extremal instances are: Λ-dim $W = 1$ (dim $W = \infty$), and Λ-dim $R = 0$ (dim $R = 1$).

Establishing precise links between these combinatorial measurements and polynomial tail-probability estimates is an open problem (Exercise 15*).

6 α-chaos

We view α-chaos processes (Definition X.27) as continuous-time models for random F-walks whose combinatorial complexities, marked by $d_F(\alpha)$, are bounded uniformly in F (cf. Proposition 9). For integer $n > 0$, the Wiener nth homogeneous chaos provides a canonical example of an n-chaos (noted at the end of Chapter X). For non-integer α, α-dimensional lattice sets lead, similarly, to α-chaos processes.

Existence

We mimic the Kakutani realization of a Wiener process (Chapter X §2). Fix $\alpha \in [1, n)$, and let $F \subset R^n$ be α-dimensional (Chapter XIII). Then, following results in Chapter XIII, F is an α-system, and is *exact* if and only if dim $F = \alpha$ *exactly.* (See Definition X.19.) We consider a unitary equivalence

$$U : \mathrm{L}^2([0,1], \mathfrak{m}) \mapsto \mathrm{L}^2_F(\Omega^n, \mathbb{P}^n) \tag{6.1}$$

(any unitary equivalence will do), and define

$$X = \{U\mathbf{1}_{[0,t]} : t \in [0,1]\}. \tag{6.2}$$

Then X is an α-chaos that is *exact* if and only if dim $F = \alpha$ *exactly* (Exercise 16). (Cf. Proposition XI.16.)

Detection

Variations of Fréchet measures associated with α-chaos

$$X = \{X(t) : t \in [0,1]\}$$

are controlled by the Orlicz functions θ_γ defined in (X.6.22),

$$\|\mu_X\|_{\theta_\gamma} := \sup\left\{\sum_{j,k} \theta_\gamma(|\mu_X(A_j, B_k)|) : \Sigma_j \mathbf{1}_{A_j} \leq 1,\ \Sigma_k \mathbf{1}_{B_k} \leq 1,\right.$$
$$\left. \{A_j\} \subset \mathscr{A},\ \{B_k\} \subset \mathscr{B}\right\} < \infty, \quad \gamma > \alpha. \tag{6.3}$$

(See Chapter X §14, and Proposition 10 in the previous section.) A question arises: are these estimates best possible?

Let us verify that if an α-chaos process is the result of a unitary equivalence based on an α-dimensional lattice set, as per (6.1) and (6.2), then (6.3) is optimal. Suppose $F \subset R^n$ is α-dimensional, $\alpha \geq 1$. We can assume

$$F = \bigcup_{k=0}^{\infty} F_k, \tag{6.4}$$

where $F_k \subset C_k \times \cdots \times C_k$ (n-fold Cartesian product), $C_k \subset R$, and $F_0 = \{r_1 \otimes \cdots \otimes r_1\}$; the C_k are pairwise disjoint, $|F_k| \geq K|C_k|^\alpha$, and $|C_k| \uparrow \infty$ as $k \to \infty$. Moreover, we can assume also that $|F_k| = 2^k$, and that $|C_k| \approx 2^{k/\alpha}$. For convenience, we enumerate

$$F_k = \{w_l^{(k)} : l = 1, \ldots, 2^k\}, \quad k = 0, \ldots. \tag{6.5}$$

Next we consider the orthonormal Haar basis $\mathscr{H}$ of $\mathrm{L}^2([0,1], \mathfrak{m})$,

$$\mathscr{H} = \bigcup_{k=1}^{\infty} \{2^{\frac{k}{2}} \mathfrak{h}_{k,l}\}_{l=1}^{2^k} \cup \{\mathfrak{h}_0\}, \tag{6.6}$$

where $\mathfrak{h}_0 \equiv 1$ on $[0, 1]$, and

$$\mathfrak{h}_{k,l} = \mathbf{1}_{[(2l-2)/2^{k+1}, (2l-1)/2^{k+1}]} - \mathbf{1}_{[(2l-1)/2^{k+1}, l/2^k]},$$
$$l = 1, \ldots, 2^k, k = 0, \ldots. \tag{6.7}$$

We consider $\mathscr{H}$-expansions of indicator functions,

$$\mathbf{1}_{[(l-1)/2^{k+1}, l/2^{k+1}]} = \pm\frac{1}{2}\, \mathfrak{h}_{k,[l/2]} + \sum_{i=1}^{k-1} c_{i_l} \mathfrak{h}_{i,j_i} + (1/2^{k+1})\mathfrak{h}_0,$$
$$k \geq 2,\ l = 1, \ldots, 2^{k+1},\ k \in \mathbb{N}, \tag{6.8}$$

where $[l/2]$ is the smallest integer $\geq l/2$, $j_i \in [2^i]$, and the c_{i_l} are constants whose precise values are not relevant for the estimates below. Define a unitary map U from $\mathrm{L}^2([0,1], m)$ onto $\mathrm{L}^2_F(\Omega^n, \mathbb{P}^n)$ by

$$U 2^{\frac{k}{2}} \mathfrak{h}_{k,l} = w_l^{(k)}, \quad l = 1, \ldots, 2^{k+1}, \ k \in \mathbb{N}, \tag{6.9}$$

and

$$U \mathfrak{h}_0 = w_1^{(0)} \ (= r_1 \otimes \cdots \otimes r_1).$$

Then, the process X given by (6.2) is an α-chaos.

Fix $k \geq 2$, and consider the following partition of Ω^n indexed by $\{-1,1\}^{C_k} \times \cdots \times \{-1,1\}^{C_k}$,

$$\begin{aligned} A_{\omega_1 \ldots \omega_n} &= (\omega_1, \ldots, \omega_n) \times \{-1,1\}^{\mathbb{N} \setminus C_k} \times \cdots \times \{-1,1\}^{\mathbb{N} \setminus C_k}, \\ & \quad (\omega_1, \ldots, \omega_n) \in \{-1,1\}^{C_k} \times \cdots \times \{-1,1\}^{C_k}. \end{aligned} \tag{6.10}$$

We consider also the intervals

$$J_l = [(l-1)/2^{k+1}, l/2^{k+1}], \quad l = 1, \ldots, 2^{k+1}, \tag{6.11}$$

and denote

$$a_{\omega l} = \mathbf{E} \mathbf{1}_{A_{\omega_1 \ldots \omega_n}} \Delta X(J_l). \tag{6.12}$$

The expansion in (6.8), the definition of U in (6.9), and the (statistical) independence of the F_k (the C_k are pairwise disjoint) imply

$$|a_{\omega l}| \approx 2^{-\left(\frac{k+1}{2} + n 2^{k/\alpha}\right)}. \tag{6.13}$$

Therefore, there exists $0 < K < \infty$ such that

$$\sum_{(\omega_1, \ldots, \omega_n) \in \{-1,1\}^{C_k} \times \cdots \times \{-1,1\}^{C_k}} \ \sum_{l=1}^{2^{k+1}} \theta_\gamma(|a_{\omega l}|) \geq K \ 2^{\frac{k+1}{2}} / (2^{k/\alpha})^{\frac{\gamma}{2}}. \tag{6.14}$$

We arrive, via (6.14) and (6.3) (cf. Theorem 10), at

Theorem 11 ([BlKah]). *Suppose $F \subset R^n$ and $\dim F = \alpha \geq 1$, and let X be the resulting α-chaos defined in (6.2). If $\dim F = \alpha$ exactly, then $\|\mu_X\|_{\theta_\gamma} < \infty$ if and only if $\gamma \geq \alpha$. Otherwise, if $\dim F = \alpha$ asymptotically, then $\|\mu_X\|_{\theta_\gamma} < \infty$ if and only if $\gamma > \alpha$.*

Remarks:

i **(general α-chaos?).** I believe that estimates of 'randomness' manifested by a 'real world'-process are meaningful, and indeed feasible through estimates of the variations of the Fréchet measure associated with the process. (See Remark iii in Chapter XI §2.)

Specifically, detecting α-chaos in the 'real world' could be feasible through downward estimates of $\|\mu_X\|_{\theta_\gamma}$. For a given process X, these estimates imply a lower bound on the parameter γ, such that X is a sub-α-chaos for $\alpha \geq \gamma$. In the upward direction however, whereas these measurements characterize the α-chaos associated with the unitary maps in (6.1) and (6.2), I do not know whether there exists an α-chaos process X such that $\|\mu_X\|_{\theta_\gamma} < \infty$ for $\gamma < \alpha$.

ii **($\Lambda(q)$-processes?).** Following the α-chaos, a tempting guess is that $\Lambda(q)$-processes (Definition XI.20) are, analogously, the continuous-time models for random F-walks whose combinatorial complexity is marked by the combinatorial measurements in (5.14) and (5.15). The verification of this guess is, in effect, the open problem stated in the previous section in Remark iii (Exercise 15*).

7 Integrators in Fractional Dimensions

The notion of a U-integrator (Chapter XI §7) can be recast essentially verbatim in the 'fractional' framework. In Chapter XI §7, in the integer-dimensional setting, U stood for a *partition*, and here U stands for a *cover*.

Given an integer $m \geq 1$, an m-process $X = \{X(\mathbf{t}) : \mathbf{t} \in [0,1]^m\}$ on a probability space $(\Omega, \mathscr{A}, \mathbb{P})$, and a cover $U = \{S_1, \ldots, S_n\}$ of $[m]$, we define $\|X\|_U$ by (XI.7.3). If X is an $[m]$-integrator, then X is a U-integrator ($\|X\|_U < \infty$) if and only if the set-function μ_X (defined by (XI.7.8)), determines an $F_{U\cup\{m+1\}}$-measure on $\mathscr{A} \times \mathscr{B}_{S_1} \times \cdots \times \mathscr{B}_{S_n}$. (Cf. Proposition XI.36.)

Integration with respect to U-integrators can be carried out in the 'fractional' framework of general multidimensional measure theory (§2 of this chapter). I shall not dwell further on this.

The distinguishing feature of a U-integrator X is its Littlewood index ℓ_X (defined in (XI.7.15) and (XI.7.16)). The 'fractional' Littlewood inequalities imply

$$\ell_X \leq (2\alpha(U) + 2)/(\alpha(U) + 1), \tag{7.1}$$

which is optimal: there exist U-integrators X such that

$$\ell_X = (2\alpha(U) + 2)/(\alpha(U) + 1).$$

We illustrate this below, in the case $U = \{(1,2),(2,3),(1,3)\}$, via an adaptation of a construction in Chapter XI §8.

An Example

We construct on a certain probability space $(\Omega, \mathscr{A}, \mathbb{P})$ the desired 3-process X by constructing an $F_{U\cup\{4\}}$-measure on

$$\mathscr{A} \times \mathscr{B}_{(1,2)} \times \mathscr{B}_{(2,3)} \times \mathscr{B}_{(1,3)},$$

which is absolutely continuous in the first coordinate with respect to $\mathbb{P}$. Fix an integer $N > 0$, and an N-subset $E = \{x_i : i \in [N]\} \subset [0,1]$. An application of Theorem X.8 implies that there exist $\epsilon_{sijk} = \pm 1$, for $s \in [N^2]$, and $(i,j,k) \in [N]^3$, such that

$$\left\| \sum_{s\in[N^2],(i,j,k)\in[N]^3} \epsilon_{sijk}\, r_{ij} \otimes r_{jk} \otimes r_{ik} \otimes r_s \right\|_\infty \leq K\, N^{\frac{7}{4}}, \tag{7.2}$$

where $K > 0$ does not depend on N. (For the application of Theorem X.8, note that degree of W-polynomials with spectrum in $\{r_{ij} : (i,j) \in [N]^2\}$ as well as $\{r_s : s \in [N^2]\}$ is bounded by 2^{N^2}.) We consider $\Omega_N = [N^2]$ as a uniform probability space. For $A \subset [N^2]$, $B_1 \in \mathscr{B}$, $B_2 \in \mathscr{B}$, $B_3 \in \mathscr{B}$, define

$$\begin{aligned}&\mu(A, B_1, B_2, B_3)\\ &\quad = N^{-7/4} \sum_{s\in A(i,j,k)\in[N]^3} \epsilon_{sijk}\delta_{x_i}(B_1)\, \delta_{x_j}(B_2)\, \delta_{x_k}(B_3).\end{aligned} \tag{7.3}$$

We view the discrete measure μ as an $F_{U\cup\{4\}}$-measure on

$$2^{[N^2]} \times \mathscr{B}_{(1,2)} \times \mathscr{B}_{(2,3)} \times \mathscr{B}_{(1,3)},$$

whose norm, by (7.2), satisfies

$$\|\mu\|_{F_{U\cup\{4\}}} \leq K. \tag{7.4}$$

Also note

$$\|\mu\|_{(p)} = N^{-(\frac{7}{4}-\frac{5}{p})}. \tag{7.5}$$

For each $m \in \mathbb{N}$, choose $E_m = \{x_{mi} : i \in [2^m]\} \subset [0,1]$ so that the E_m are mutually disjoint, and view $\Omega_{2^m} = [4^m]$ as a uniform probability space, as above. For each $m \in \mathbb{N}$, let μ_m be the $F_{U\cup\{4\}}$-measure on $2^{[4^m]} \times \mathscr{B}_{(1,2)} \times \mathscr{B}_{(2,3)} \times \mathscr{B}_{(1,3)}$ obtained by the procedure above, with $N = 2^m$, and $E = E_m$. Let $\Omega = \prod \Omega_{2^m}$ be the product probability space, with the product probability measure $\mathbb{P}$, the product σ-field $\mathscr{A}$, and denote the projection from Ω onto its jth-coordinate by π_j. We consider the set-function

$$\mu_m(\pi_m[A], B_1, B_2, B_3), A \in \mathscr{A},\ B_1 \in \mathscr{B},\ B_2 \in \mathscr{B},\ B_3 \in \mathscr{B}. \tag{7.6}$$

Observe that μ_m is an $F_{U\cup\{4\}}$-measure on $\mathscr{A} \times \mathscr{B}_{(1,2)} \times \mathscr{B}_{(2,3)} \times \mathscr{B}_{(1,3)}$, that $\|\mu\|_{F_{U\cup\{4\}}} \leq K$, and that it is absolutely continuous in the first coordinate with respect to $\mathbb{P}$. Define

$$\mu = \sum_{m=1}^{\infty} \mu_m/m^2, \tag{7.7}$$

which is an $F_{U\cup\{4\}}$-measure, and the corresponding 3-process

$$X(t_1,t_2,t_3) = \frac{\mathrm{d}}{\mathrm{d}\mathbb{P}}\,\mu(\cdot,[0,t_1]\times[0,t_2]\times[0,t_3]), \tag{7.8}$$

which is a U-integrator. We obtain from (7.5) that

$$\|\mu\|_{(p)} \geq 2^{m\left(\frac{7}{4}-\frac{5}{p}\right)}/m^2, \tag{7.9}$$

which is unbounded in m for all $p < 10/7$, and therefore $\ell_X = 10/7$.

We complete the discussion in Remark iii, Chapter XI §7, adding the final ingredient: that the 'dimension' of a 1-process X is the infimum of $\alpha(U)$ over all covers $U = \{S_1,\ldots,S_n\}$ of $[m]$ with the property that there exists a partition $\boldsymbol{\rho} = \{\rho_1,\ldots,\rho_m\}$ of $\mathbb{N}$ such that

$$\mu_X(A, \rho_1^{-1}[J_1] \cap \cdots \cap \rho_n^{-1}[J_n]), \quad J_1 \in \mathscr{O},\ldots, J_n \in \mathscr{O}, \tag{7.10}$$

determines an $F_{U\cup\{m+1\}}$-measure on $\mathscr{A} \times \mathscr{B}_{S_1} \times \cdots \times \mathscr{B}_{S_n}$. (Compare this to the notion of optimal $\mathfrak{F}$-*type* of a finitely additive set-function μ on a measurable space – defined in §2 of this chapter.) The 'dimension' of X gauges how a 1-process X is synthesized from its increments, and, in this sense, it gauges a degree of interdependence between its increments. (See Chapter X §1, Chapter XI §1, Remarks iii and iv in Chapter XII §4.)

The Littlewood inequalities in fractional dimensions imply that if X is 'd-dimensional', then

$$\ell_X \leq 2(d+1)/(d+2). \tag{7.11}$$

These inequalities are key in observing 'd-dimensional' 1-processes. The example described above naturally leads to a '3/2-dimensional' 1-process, and for arbitrary $d \in [1, \infty)$, similar examples can be constructed by following essentially the same blueprint (Exercise 16).

Exercises

1. Prove that if $U = \{S_1, \ldots, S_n\}$ is a cover of $[m]$, and μ is an F_U-measure on $\mathfrak{A}_1 \times \cdots \times \mathfrak{A}_m$, then for each $i = 1, \ldots, n$, μ is an F_1-measure on $\mathfrak{A}|_{S_i}$ when coordinates indexed by $[m] \setminus S_i$ are fixed, and an $F_{U'_{S_i}}$-measure on $\mathfrak{A}|_{[m]\setminus S_i}$ when coordinates indexed by S_i are fixed. (See Remark i in §2.)
2. Verify the details in the proof that
$$\begin{aligned}[F_n]_{\mathbf{X}^U} = \{\nu \in F_n(\mathfrak{A}|_{S_1}, \ldots, \mathfrak{A}|_{S_n}) : \\ \exists\, \mu \in F_U(\mathfrak{A}_1 \times \cdots \times \mathfrak{A}_m) \text{ such that } \nu = \tilde{\mu}\}\end{aligned}$$
(See *definition of* $[F_n]_{\mathbf{X}^U}$ in §2.)
3. For arbitrary integers $N > 0$, verify that there exist $\epsilon_{\mathbf{j}} = \pm 1$, for $\mathbf{j} \in [N]^m$, such that
$$\left\| \sum_{\mathbf{j}\in[N]^m} \epsilon_{\mathbf{j}}\, r_{\pi_{S_1}\mathbf{j}} \otimes \cdots \otimes r_{\pi_{S_n}\mathbf{j}} \right\|_\infty \leq K\, N^{(\alpha+1)/2},$$
where $K > 0$ depends only on U.
4. Prove (Theorem 4) that if $\mu \in F_U(\mathfrak{A}_1 \times \cdots \times \mathfrak{A}_m)$, then $\mu \times \nu \in F_U(\sigma(\mathfrak{A}_1 \times \mathfrak{B}_1) \times \cdots \times \sigma(\mathfrak{A}_m \times \mathfrak{B}_m))$ for all $\nu \in F_U(\mathfrak{B}_1 \times \cdots \times \mathfrak{B}_m)$ if and only if $\|\mu\|_{\mathrm{pb}_U} < \infty$.
5. Prove the instance $j = 1$ in (2.36). That is, verify
$$\begin{aligned}&F_1(\sigma(\mathfrak{A}_1 \times \cdots \times \mathfrak{A}_m)) \\ &\quad\subset PBF_k(\sigma(\mathfrak{A}_1 \times \cdots \times \mathfrak{A}_{m-k+1}), \mathfrak{A}_{m-k+2}, \ldots, \mathfrak{A}_m), \\ &\quad k = 2, \ldots, m.\end{aligned}$$
6. i. Prove that (2.36) fails in the general case $1 < j < k \leq m$.

   ii.* In (2.36), replace F_j by F_U and F_k by F_V, where U and V are covers such that $\alpha(U) < \alpha(V)$. Does the inclusion hold?
7. Prove that if the decomposition problem in Remark i is affirmatively resolved, then $F_U(\mathfrak{A}^m) = PBF_U(\mathfrak{A}^m)$ for all covers U of $[m]$ such that $\alpha(U) \leq 2$.

8. Prove that if $U_1 \prec U_2$ and $\alpha(U_2) < \alpha(U_1)$, then $\mathscr{F}_{U_2}(X) \subsetneqq \mathscr{F}_{U_1}(X)$ and $\mathscr{V}_{U_1}(X) \subsetneqq \mathscr{V}_{U_2}(X)$. (For notation and terminology, see Definition 5, and the discussion around it.)
9. Fill in the details of the proof of Theorem 6.
10. In the case $X = [0,1]$, prove that $D_F(a)$ in (3.3) is finite if and only if $D_F(a)$ defined in (3.5) is finite.
11. Prove that if $f : [0,1] \to [0,1]$ is a continuous function, then

$$\text{Dim graph}(f) = 1.$$

12. i. If $\mu \in M(\Omega)$, then for all positive integers m, all covers U of $[m]$, all partitions $\rho = \{\rho_1, \ldots, \rho_m\}$ of $\mathbb{N}$, and all $\nu \in F_{U,\rho}(\Omega)$, the convolution $\mu \star \nu$ is in $F_{U,\rho}(\Omega)$ (Cf. Exercise 7).
    ii. Prove that if $\rho = \{\rho_1, \rho_2\}$ is a partition of $\mathbb{N}$, then for all μ and ν in $F_\rho(\Omega)$, $\mu \star \nu \in F_\rho(\Omega)$.
    iii. Prove that if $\rho = \{\rho_1, \rho_2, \rho_3\}$ is a partition of $\mathbb{N}$ such that $|\rho_1| = |\rho_2| = |\rho_3| = \infty$, then there exist $\mu \in F_\rho(\Omega)$ and $\nu \in F_\rho(\Omega)$, and $\mu \star \nu \notin F_\rho(\Omega)$.
    iv. Suppose $\rho = \{\rho_1, \ldots, \rho_m\}$ is a partition of $\mathbb{N}$, and U is a cover of $[m]$ such that $\alpha(U) \leq 2$, and $m > 2$. If μ and ν are in $F_{U,\rho}(\Omega)$, then $\mu \star \nu \in F(\Omega)$.
13. Prove that if μ_φ is defined by (4.8), then $\mu_\varphi \star \nu \in F_\rho(\Omega)$ for all $\nu \in F(\Omega)$.
14. Suppose the second question in Remark ii §5 is answered in the affirmative. That is, if $F \subset W_k$ and $d_F(\alpha) < \infty$, then there exists $0 < K < \infty$ such that for all $p \geq 2$, $f \in \mathrm{L}^p_F$, and $q > p$,

$$\|f\|_{\mathrm{L}^q} \leq K \left(\frac{q}{p}\right)^{\alpha/2} \|f\|_{\mathrm{L}^p}.$$

Under this supposition, prove that if in addition to the hypotheses in Proposition 9, we assume $N_j \uparrow \infty$ and $|F_j| \geq K N_j^\alpha$ for all $j \in \mathbb{N}$, then the sub-sequential limit Y in (5.4) is a standard α-variable.
15.* Let $F \subset W$. Prove that for all $0 < a < 1/2$, $e_F(a) < \infty$ if and only if $\zeta_{\mathrm{L}^2_F}(4a) < \infty$. ($e_F$ was defined in (5.14), and $\zeta_{\mathrm{L}^2_F}$ was defined in (XI.4.12).) Observe that necessity follows from classical results in [Ru1]. (Reading through [Ru1] is part of the exercise.)
16. Prove that if $F \subset R^n$ is α-dimensional, then the process X defined in (6.2) is an α-chaos, which is *exact* if and only if $\dim F = \alpha$ *exactly*.
17. For arbitrary $d \in [1, \infty)$, produce 'd-dimensional' 1-processes.

Hints for Exercises in Chapter XIV

4. See Theorem IX.6.
6. i. Review the proof for $m \geq 3$, $j = m - 1$, and $k = m$. Review also results in Chapter VIII.
7. Use Theorem IX.9.
12. Review results in Chapter IX.

References

[ACl1] C.R. Adams and J.A. Clarkson, Properties of functions $f(x, y)$ of bounded variation, *Trans. Amer. Math. Soc.* **36** (1934), 711–30.

[ACl2] C.R. Adams and J.A. Clarkson, A correction to 'Properties of functions $f(x, y)$ of bounded variation', *Trans. Amer. Math. Soc.* **46** (1939), 468.

[B1] L. Bachelier, Theory of Speculation, in: *The Random Character of Stock Market Prices*, revised edition, ed. P.H. Cootner, The MIT Press, 1969.

[B2] L. Bachelier, *Calcul des Probabilités*, Gauthier-Villars, 1912.

[BaE] G. Bachelis and S. Ebenstein, On $\Lambda(p)$ sets, *Pacific J. Math.* **54** (1974), 35–8.

[Ban] S. Banach, Über einige Eigenschaften der lakünaren trigonometrischen Reihen, *Studia Math.* **2** (1930), 207–30.

[Be] D. Bernoulli, Réflexions et éclaircissements sur les nouvelles vibrations des cordes..., *Histoires Acad. r. Sci., Mémoires* **9** (1755), 173–95.

[Ber] J. Bernstein, *Einstein*, The Viking Press, 1973.

[Bi] P. Billard, Sur la convergence presque partout des séries de Fourier Walsh des fonctions de l'espace $\mathrm{L}^2(0, 1)$, *Studia Math.* **28** (1967), 363–88.

[Bil] P. Billingsley, *Convergence of Probability Measures*, John Wiley & Sons, 1968.

[Bl1] R. Blei, On trigonometric series associated with separable, translation invariant subspaces of $\mathrm{L}^\infty(G)$, *Trans. Amer. Math. Soc.* **173** (1972), 491–9.

[Bl2] R. Blei, Sidon partitions and p-Sidon sets, *Pac. J. Math.* **65** (1976), 307–13.

[Bl3] R. Blei, A uniformity property for $\Lambda(2)$ sets and Grothendieck's inequality, *Symposia Mathematica* **XXII**, Academic Press, 1977, 321–36.

[Bl4] R. Blei, Multi-dimensional extensions of the Grothendieck inequality and applications, *Arkiv För. Mat.* **17** (1979), 51–68.

[Bl5] R. Blei, Fractional Cartesian products of sets, *Ann. Inst. Fourier, Grenoble*, **29** (1979), 79–105.

[Bl6] R. Blei, Multilinear measure theory and multiple stochastic integration, *Prob. Th. Related Fields* **81** (1989), 569–84.

[Bl7] R. Blei, Combinatorial dimensions and fractional Cartesian products in Euclidean spaces, *J. London Math. Soc.* **42** (1990), 373–84.

[Bl8] R. Blei, Projectively bounded Fréchet measures, *Trans. Amer. Math. Soc.* **348** (1995), 4409–32.

[BlKo] R. Blei and T.W. Körner, Combinatorial dimension and random sets, *Israel J. Math.* **47** (1984), 65–74.

[BlKah] R. Blei and J.-P. Kahane, A computation of the Littlewood exponent of stochastic processes, *Math. Proc. Camb. Phil. Soc.* **103** (1988), 367–70.

[BlSch] R. Blei and J.H. Schmerl, Combinatorial dimension of fractional Cartesian products, *Proc. Amer. Math. Soc.* **120** (1994), 73–7.

[BlPeSch] R. Blei, Y. Peres and J.H. Schmerl, Fractional products of sets, *Random Structures and Algorithms* **6** (1995), 113–19.

[BlCag] R. Blei and J. Caggiano, Projective boundedness and convolution of Fréchet measures, *Proc. Amer. Math. Soc.* **128** (2000), 3523–28.

[BlTow] R. Blei and N. Towghi, Variations of Fréchet measures associated with p-stable motion (preprint).

[Bo] D. Bohm, *Causality and Chance in Modern Physics*, D. Van Nostrand Company, Inc., Princeton, New Jersey, New York, 1957.

[Bol] B. Bollobás, *Extremal Graph Theory*, Academic Press, London, 1978.

[Bon1] A. Bonami, Ensembles $\Lambda(p)$ dans le dual de D^∞, *Ann. Inst. Fourier, Grenoble*, **18** (1968), 193–204.

[Bon2] A. Bonami, Étude des coefficients de Fourier des fonctions de $L^p(G)$, *Ann. Inst. Fourier, Grenoble*, **20** (1970), 335–402.

[Bor] E. Borel, Les probabilités denombrablés et leurs applications arithmétiques, *Rend. del Circ. Mat. di Palermo*, **27** (1909), 247–71.

[Bou] N. Bourbaki, *Éléments de Mathématique*, VII, 1, *Les Structures Fondamentales de l' Analyse*, Livre II, Actualité Scient. et Indust., 1044, Hermann, Paris, 1948.

[Bour] J. Bourgain, Bounded orthogonal systems and the $\Lambda(p)$-set problem, *Acta. Math.* **162** (1989), 227–45.

[BozPy] M. Bożejko and T. Pytlik, Some types of lacunary Fourier series, *Colloq. Math.* **XXV** (1972), 117–24.

[Br] L. Breiman, *Probability*, Addison-Wesley Publishing Co. Inc., 1968.

[Bro] R. Brown, On the existence of active molecules in organic and inorganic bodies, *Phil. Mag.* [*New Series*] **4** (Sept. 1828), 161–73.

[Bru] S.G. Brush, *The Kind of Motion We Call Heat*, Book 2, North Holland Publishing Co., Amsterdam, New York and Oxford, 1976.

[Bu] D. Burkholder, Distribution function inequalities for martingales, *Ann. Prob.* **1** (1973).

[CW] R. Cairoli and J. Walsh, Stochastic integrals in the plane, *Acta. Math.* **134** (1975), 111–83.

[CamMi] S. Cambanis and G. Miller, Some path properties of pth order and symmetric stable processes, *Ann. Prob.* **8** (1989), 227–45.

[CaMar] R.H. Cameron and W.T. Martin, The orthogonal development of non-linear functionals in series of Fourier–Hermite functionals, *Ann. Math.* **48** (1947), 385–92.

[Car] L. Carleson, On convergence and growth of partial sums of Fourier series, *Acta. Math.* **116** (1966), 135–57.

[Cars] H.S. Carslaw, *Introduction to the Theory of Fourier's Series and Integrals*, third revised edition, Dover Publications, Inc., 1930.

[Ch] K. Chung, *A Course in Probability Theory*, Academic Press, 1974.

[ChWil] K.L. Chung and R.J. Williams, *Introduction to Stochastic Integration*, Birkhäuser, 1983.

[ChrEfSin] E. Christensen, E.G. Effros and A.M. Sinclair, Completely bounded multilinear maps and C^* algebraic cohomology, *Invent. Math.* **90** (1987), 279–96.

[Ci] Z. Ciesielski, Hölder condition for realizations of Gaussian processes, *Trans. Amer. Math. Soc.* **99** (1961), 403–13.

[Cl] J.A. Clarkson, On double Riemann–Stieltjes integrals, *Bull. Amer. Math. Soc.* **39** (1933), 929–36.

[ClA] J.A. Clarkson and C.R. Adams, On definitions of bounded variation for functions of two variables, *Trans. Amer. Math. Soc.* **35** (1933), 824–54.

[ČuTow] Z. Čučković and N. Towghi, Sample paths of α-chaos and absolute convergence of random Fourier series (to appear in *J. of Math. Analysis and Applications*).

[DSe] G. David and S. Semmes, *Fractured Fractals and Broken Dreams*, Oxford Lecture Series in Mathematics and its Applications, **7**, 1997.

[Da] A.M. Davie, Quotient algebras of uniform algebras, *J. London Math.* **7** (1973), 31–40.

[dlP] V. de la Peña, Decoupling and Khintchin's inequalities for U-statistics, *Ann. Prob.* **20** (1992), 1877–92.

[dlPG] V. de la Peña and E. Giné, *Decoupling: From Dependence to Independence*, Springer-Verlag, New York, 1999.

[DiU] J. Diestel and J.J. Uhl, *Vector Measures*, Math. Surveys, **15**, American Math. Soc., Providence, 1977.

[DiJTon] J. Diestel, H. Jarchow and A. Tonge, *Absolutely Summing Operators*, Cambridge University Press, 1995.

[Die1] J. Dieudonné, Recent developments in the theory of locally convex spaces, *Bull. Amer. Math. Soc.* **59** (1953), 495–512.

[Die2] J. Dieudonné and von Neumann, in: *Dictionary of Scientific Biography*, Editor in Chief C.C. Gillispie, Vol. XIV, Charles Scribner's Sons, New York, 1976.

[Do] I. Dobrakov, On extension of Vector Polymeasures, *Czech. Math. J.* **38** (1988), 88–94.

[Doo] J. Doob, *Stochastic Processes*, John Wiley & Sons, Inc., New York, 1953.

[DrP] R.E. Dressler and L. Pigno, Rosenthal sets and Riesz sets, *Duke J. Math.* **41** (1974), 675–7.

[Dru] S.W. Drury, Sur les ensembles de Sidon, *C. R. Acad. Sc. Paris*, t. **271** (1970), 162–3.

[Du1] R.M. Dudley, *Real Analysis and Probability*, Wadsworth & Brooks/Cole, 1989.

[Du2] R.M. Dudley, Fréchet differentiability, p-variation and uniform donsker classes, *Ann. Prob.* **20** (1992), 1968–82.

[Dug] P. Dugac, Euler, d'Alembert et les fondements de l'analyse, in: *Leonhard Euler, 1707–1783, Beiträge sur Leben und Werk*, Birkhäuser-Verlag, Basel, 1983.

[DunSchw] N. Dunford and J.T. Schwartz, *Linear Operators I*, Interscience Publishers, Inc., New York, 1964.

[Dur] R. Durrett, *Brownian Motion and Martingales in Analysis*, Wadsworth, 1984.

[DvErKak] A. Dvoretzky, P. Erdös and S. Kakutani, Nonincrease everywhere of the Brownian motion process, *Proceedings of the Fourth Berkeley Symposium*, **II** (1961), 103–16.

[ERos] R.E. Edwards and K.A. Ross, p-Sidon sets, *J. Func. Anal.* **15** (1974), 404–27.

[Ei1] A. Einstein, *Investigations On the Theory of Brownian Movement*, edited by R. Fürth, translated by A.D. Cowper, Dover Publications, Inc., 1956.

[Ei2] A. Einstein, Autobiographical notes, in: *Albert Einstein: Philosopher–Scientist*, ed. P.A. Schillpp, The Library of Living Philosophers, Tudor Pub., Co., New York, 1951.

[Eis] L.P. Eisenhart, *An Introduction to Differential Geometry with Use of the Tensor Calculus*, Princeton University Press, Princeton, 1947.

[En] H. Enderton, *A Mathematical Introduction to Logic*, Academic Press, New York and London, 1972.

[F] K.J. Falconer, *The Geometry of Fractal Sets*, Cambridge Tracts in Mathematics, Cambridge University Press, 1989.

[Fe] H. Federer, *Geometric Measure Theory*, Springer-Verlag, New York, 1969.

[Fey] R. Feynman, *The Feynman Lectures on Physics*, v. 1, Addison-Wesley, Inc., 1963.

[Fig] A. Figà-Talamanca, An example in the theory of lacunary Fourier series, *Boll. U.M.I.* **3** (1970), 1–4.

[FigRid] A. Figà-Talamanca and D. Rider, A theorem of Littlewood and lacunary series for compact groups, *Pac. J. Math.* **16** (1966), 505–14.

[Fi1] N.J. Fine, On the Walsh functions, *Trans. Amer. Math. Soc.* **65** (1949), 372–414.

[Fi2] N.J. Fine, The generalized Walsh functions, *Trans. Amer. Math. Soc.* **69** (1950), 66–77.

[Fr] M. Fréchet, Sur les fonctionnelles bilinéaires, *Trans. Amer. Math. Soc.* **16** (1915), 215–34.

[G] D. Gale, *The Theory of Linear Economic Models*, McGraw-Hill, New York, 1960.

[Ga] A. Garsia, *Martingale Inequalities, Seminar Notes on Recent Progress*, W.A. Benjamin, Inc., 1973.

[GiISch] J.E. Gilbert, T. Itô and B.M. Schreiber, Bimeasure algebras on locally compact groups, *J. Funct. Anal.* **64** (1985), 134–62.

[GrMc] C.C. Graham and O.C. McGehee, *Essays in Commutative Harmonic Analysis*, Springer-Verlag, 1979.

[GrSch1] C.C. Graham and B.M. Schreiber, Bimeasure algebras on LCA groups, *Pac. J. Math.* **115** (1984), 91–127.

[GrSch2] C.C. Graham and B.M. Schreiber, Projections in spaces of bimeasures, *Canad. Math. Bull.* **31** (1988), 19–25.

[GrY] C.C. Graham and K. Ylinen, Classes of trimeasures: applications of harmonic analysis, *Probability Measures on Groups, X* (ed. H. Heyer), Plenum Press, New York and London, 1991, 169–76.

[GraR] I. Grattan-Guiness and J.R. Ravetz, *Joseph Fourier, 1768–1830*, MIT Press, 1972.

[Gro1] A. Grothendieck, Produits tensoriels topologiques et espaces nucléaires, *Mem. Amer. Math. Soc.* **16**, 1955.

[Gro2] A. Grothendieck, Resumé de la théorie metrique des produits tensoriels, *Bol. Soc. Matem. Saõ Paulo*, **8** (1956), 1–79.

[H1] U. Haagerup, Les meilleures constantes de l'inégalité de Khintchine, *C. R. Acad. Sc. Paris*, t. **286** (1978), 259–62.

[H2] U. Haagerup, The best constants in the Khintchin inequality, *Studia Math.* **70** (1982), 231–83.

[Ha1] A. Haar, Zur Theorie des orthogonalen Funktionensysteme, *Math. Ann.* **69** (1910), 331–71.

[Ha2] H. Haar, Der Massbegrief in der Theorie der kontinuierlichen Gruppen, *Ann. Math.* **34** (1933), 147–69.

[Hal] R.R. Hall, On a conjecture of Littlewood, *Proc. Camb. Phil. Soc.* **78** (1975), 443–5.

[Har] G.H. Hardy, Raymond Edward Alan Christopher Paley, *J. London Math. Soc.* **9** (1934), 76–80.

[He] S.J. Heims, *John von Neumann and Norbert Wiener*, The MIT Press, Cambridge, 1980.

[Hel] H. Helson, *Harmonic Analysis*, Hindustan Book Agency, second edition, 1995.

[Her] J. Herivel, *Joseph Fourier, The Man and the Physicist*, Clarendon Press, 1975.

[Herz] C. Herz, Sur les ensembles parfaits et séries trigonométriques by N. Varopoulos, *AMS Math. Reviews*, **31**, 2567 (1966), 462–63.

[HewRos] E. Hewitt and K. Ross, *Abstract Harmonic Analysis, V. II*, Springer-Verlag, 1970.

[Hi] T. Hida, *Brownian Motion*, Springer-Verlag, 1980.

[Hu] R.A. Hunt, On the convergence of Fourier series, *Proceedings of the Conference on Orthogonal Expansions and Their Continuous Analogues*, Southern Illinois University Press, 1968, 234–55.

[IkW] N. Ikeda and S. Watanabe, *Stochastic Differential Equations and Diffusion Processes*, North-Holland, 1981.

[I1] K. Itô, Stochastic integral, *Proc. Imp. Acad. Tokyo*, **20** (1944), 519–24.

[I2] K. Itô, Multiple Wiener integral, *J. Math. Soc. Japan*, **3** (1951), 157–67.

[I3] K. Itô, *Introduction to Probability Theory*, Cambridge University Press, 1984.

[JWo] G. Johnson and G. Woodward, On p-Sidon sets, *Indiana Univ. Math. J.* **24** (1974), 161–7.

[K1] M. Kac, Random walk and the theory of Brownian motion, *Amer. Math. Month.* **54** (1947), 369–91.

[K2] M. Kac, *Probability and Related Topics in Physical Sciences*, Interscience Publishers, 1959.

[Ka] S. Kaczmarz, Über ein Orthogonalsystem, *Comptes rendus du premier congrès des math. des pays slaves*, Warsaw, 1929, 189–92.

[KaSte] S. Kaczmarz and H. Steinhaus, Le système orthogonal de M. Rademacher, *Studia Math.* **II** (1930), 231–47.

[Kad] R.V. Kadison, Operator Algebras – an overview, in: *The Legacy of John von Neumann, Proceedings of Symposia in Pure Mathematics*, **50**, American Math. Soc., Providence, 1990.

[Kah1] J.-P. Kahane, Propriétés locales des fonctions à séries aléatoires, *Studia Math.* **19** (1960), 1–25.

[Kah2] J.-P. Kahane, Brownian motion and classical analysis, *Bull. London Math. Soc.* **8** (1976), 145–55.

[Kah3] J.-P. Kahane, *Some Random Series of Functions*, second edition, Cambridge University Press, Cambridge, 1985.

[Kak1] S. Kakutani, Brownian motion and duality of locally compact Abelian groups, *Proceedings of the International Congress of Mathematicians*, 1950, 456–7.

[Kak2] S. Kakutani, Determination of the spectrum of the flow of Brownian motion, *Proc. Nat. Acad. Sci. USA*, **36** (1950), 319–23.

[Kak3] S. Kakutani, Spectral analysis of stationary Gaussian processes, *Proceedings of the Fourth Berkeley Symposium, Vol. II*, University of California Press, 1961.

[KarSh] I. Karatzas and S. Shreve, *Brownian Motion and Stochastic Calculus*, Springer-Verlag, 1988.

[Kat] Y. Katznelson, *An Introduction to Harmonic Analysis*, John Wiley & Sons, Inc., 1968, second edition, 1976.

[KatMc] Y. Katznelson and O.C. McGehee, Measures and pseudomeasures on compact subsets of the line, *Math. Scand.* **23** (1968), 57–68.

[Kh1] A. Khintchin, Über dyadische Brüche, *Math. Zeit.* **18** (1923), 109–16.

[Kh2] A. Khintchin, Sur une théorème général relatif aux probabilités dénombrables, *C. R. Acad. Sci. Paris*, **178** (1924), 617–19.

[KolTi] A.N. Kolmogorov and V.M. Tihomirov, ϵ-entropy and ϵ-capacity of sets in function spaces, *American Math. Soc. Translations*, **17** (1961), 277–364.

[Ko] T.W. Körner, *Fourier Analysis*, Cambridge University Press, 1988.

[Kw] S. Kwapień, Decoupling inequalities and polynomial chaos, *Ann. Prob.* **15** (1987), 1062–72.

[KwWo] S. Kwapień and W. Woyczyński, *Random Series and Stochastic Integrals: Single and Multiple*, Birkhäuser, 1992.

[L] P. Lancaster, *The Theory of Matrices*, second edition with applications, Academic Press, 1985.

[La] J. Lamperti, *Probability*, W.A. Benjamin, Inc., 1966.

[LatO] R. Latała and K. Oleszkiewicz, On the best constants in the Khinchin-Kahane inequality, *Studia Math.* **109** (1994), 101–4.

[LeT] M. Ledoux and M. Talagrand, *Probability in Banach Spaces*, Springer-Verlag, New York, 1991.

[Lé1] P. Lévy, *Calcul des Probabilités*, Gauthier-Villars, Paris, 1925.

[Lé2] P. Lévy, Théorie des erreurs. La loi de Gauss et les lois exceptionnelles, *Soc. Math. France Bull.* **52** (1924), 49–85.

[Lé3] P. Lévy, *Processes Stochastiques et Mouvement Brownien*, Gauthier-Villars, Paris, 1948.

[Lé4] P. Lévy, *Quelques Aspects de la Pensées d'un Mathématicien*, Librairie Scientifique et Technique, Albert Blanchard, Paris, 1970.

[LiPe] J. Lindenstrauss and A. Pelczynski, Absolutely summing operators in $\mathscr{L}_p$-spaces and their applications, *Studia Math.* **29** (1968), 275–326.

[LiTz] J. Lindenstrauss and L. Tzafriri, *Classical Banach Spaces I*, Springer-Verlag, Berlin, Heidelberg and New York, 1977.

[Lit1] J.E. Littlewood, A theorem on power series, *Proc. London Math. Soc.* **23** (1925), 94–103.

[Lit2] J.E. Littlewood, On inequalities in the theory of functions, *Proc. London Math. Soc.* **23** (1925), 481–519.

[Lit3] J.E. Littlewood, On the mean values of power series, *Proc. London Math. Soc.* **25** (1926), 328–37.

[Lit4] J.E. Littlewood, On bounded bilinear forms in an infinite number of variables, *Quart. J. Math. Oxford*, **1** (1930), 164–74.

[LoWh] L.H. Loomis and H. Whitney, An inequality related to the isoperimetric inequality, *Bull. Amer. Math. Soc.* **55** (1949), 961–2.

[Lu1] F. Lust, Produits tensoriels injectifs d'espaces de Sidon, *Colloq. Math.* **XXXII** (1975), 285–9.

[Lu2] F. Lust, Produits tensoriels injectifs d'espaces faiblement séquentiellement complets, *Colloq. Math.* **XXXIII** (1975), 289–90.

[McTa1] T. McConnell and M. Taqqu, Decoupling inequalities for multilinear forms in independent symmetric random variables, *Ann. Prob.* **14** (1986), 943–54.

[McTa2] T. McConnell and M. Taqqu, Dyadic approximation of double integrals with respect to symmetric stable processes, *Stoch. Proc. App.* **22** (1986), 323–31.

[Mac] G.W. Mackey, Harmonic analysis as the exploitation of symmetry – a historical survey, *Bull. Amer. Math. Soc.* **3** (1980), 543–699.

[Ma] N. Macrae, *John von Neumann*, Pantheon Books, New York, 1992.

[M] P. Malliavin, *Stochastic Analysis*, Springer-Verlag, 1997.

[MM] M. Malliavin-Brameret and P. Malliavin, Cararactérisation arithmétique d'une classe d'ensembles de Helson, *C. R. Acad. Sc. Paris*, t. **264** (1967), 192–3.

[Man] B. Mandelbrot, *The Fractal Geometry of Nature*, W.H. Freeman and Co., New York, 1983.

[MarPi] M. Marcus and G. Pisier, *Random Fourier Series with Applications to Harmonic Analysis*, Princeton University Press, Princeton, N. J., 1981.

[MazOr1] S. Mazur and W. Orlicz, Grundlegende Eigenschaften der polynomischen Operationen, *Studia Math.* **5** (1934), 50–68.

[MazOr2] S. Mazur and W. Orlicz, Grundlegende Eigenschaften der polynomischen Operationen, zweite Mitteilung, *Studia Math.* **5** (1935), 179–89.

[Mé] J.-F. Méla, Approximation diophantienne et ensembles lacunaires, *Bull. Soc. Math. Fr.* **19** (1969), 26–54.

[Mey1] Y. Meyer, Spectres des mesures et mesures absolutment continues, *Studia Math.* **XXX** (1968), 87–99.

[Mey2] Y. Meyer, Recent advances in spectral synthesis, in: *Conference on Harmonic Analysis, College Park Maryland, 1971*, Springer-Verlag Lecture Notes in Mathematics, 266.

[Mo] G. Morgenthaler, On Walsh-Fourier series, *Trans. Amer. Math. Soc.* **84** (1957), 472–507.

[Mor] M. Morse, Bimeasures and their integral extensions, *Ann. Mat. Pura App.* **39** (1955), 345–56.

[MorTr1] M. Morse and W. Transue, Functionals of bounded Fréchet variation, *Canadian J. Math.* **1** (1949), 153–65.

[MorTr2] M. Morse and W. Transue, Functionals F bilinear over the product $A \times B \ldots$, I, *Ann. Math.* **50** (1949), 777–815.

[MorTr3] M. Morse and W. Transue, Functionals F bilinear over the product $A \times B \ldots$, II, *Ann. Math.* **51** (1950), 576–614.

[MorTr4] M. Morse and W. Transue, Norms of distribution functions associated with bilinear functionals, in: *Contributions to Fourier Analysis*, Princeton University Press, **25** (1950), 104–44.

[MorTr5] M. Morse and W. Transue, Semi-normed vector spaces with duals of integral type, *J. d'Analyse Math.* **4** (1955), 149–86.

[MorTr6] M. Morse and W. Transue, C-bimeasures Λ and their superior integrals Λ^*, *Circolo Mat. Palermo* **4** (1955), 270–300.

[MorTr7] M. Morse and W. Transue, C-bimeasures Λ and their integral extensions, *Ann. Math.* **64** (1956), 480–504.

[Mu1] F.J. Murray, On complementary manifolds and projections in spaces L_p and l_p, *Trans. Amer. Math. Soc.* **41** (1937), 138–52.

[Mu2] F.J. Murray, Bilinear transformations in Hilbert space, *Trans. Amer. Math. Soc.* **45** (1939), 474–507.

[MuvN1] F.J. Murray and J. von Neumann, On rings of operators, *Ann. Math.* **37** (1936), 116–229.

[MuvN2] F.J. Murray and J. von Neumann, On rings of operators IV, *Ann. Math.* **44** (1943), 716–808.

[NF] B. Sz.-Nagy and C. Foias, *Harmonic Analysis of Operators on Hilbert Spaces*, North-Holland, 1970.

[Ne] E. Nelson, *Dynamical Theories of Brownian Motion*, Mathematical Notes, Princeton University Press, 1967.

[Ni] O.M. Nikodym, Sur les familles bornées des fonctions parfaitement additive d'ensemble abstrait, *Montash. Math. Phys.* **40** (1933), 418–26.

[Nu] D. Nualart, *The Malliavin Calculus and Related Topics*, Springer-Verlag, 1995.

[NuZ1] D. Nualart and M. Zakai, Generalized stochastic integrals and the Malliavin calculus, *Prob. Th. Rel. Fields*, **73** (1986), 255–80.

[NuZ2] D. Nualart and M. Zakai, On the relation between the Stratonovich and Ogawa integrals, *Ann. Prob.* **17** (1989), 1536–40.

[Ny] M.J. Nye, *Molecular reality, A Perspective on the Scientific Work of Jean Perrin*, MacDonald, London and American Elsevier Inc., New York, 1972.

[O1] S. Ogawa, Une remarque sur l'approximation de l'intégrale stochastique du type noncausal par une suite des intégrales de Stieltjes, *Tôhoku Math. J.* **36** (1984), 41–8.

[O2] S. Ogawa, Quelques propriétés de l'intégrale stochastique du type noncausal, *Japan J. Appl. Math.* **1** (1984), 405–16.

[Or] W. Orlicz, Über unbedingte Konvergenz in Funktionenraumen (I), *Studia Math.* **IV** (1933), 33–7.

[Os] R. Osserman, The isoperimetric inequality, *Bull. Amer. Math. Soc.* **84** (1978), 1182–1238.

[Pa] R.E.A.C. Paley, A remarkable series of orthogonal functions (I), *Proc. London Math. Soc.* **34** (1932), 241–79.

[PaZy1] R.E.A.C. Paley and A. Zygmund, On some series of functions (1), *Proc. Camb. Phil. Soc.* **26** (1930), 337–57.

[PaZy2] R.E.A.C. Paley and A. Zygmund, On some series of functions (3), *Proc. Camb. Phil. Soc.* **28** (1932), 190–205.

[PaWi1] R.E.A.C. Paley and N. Wiener, *Fourier Transforms in the Complex Domain*, Amer. Math. Soc., Colloq. Pub. XIX, Providence, 1934.

[PaWi2] R.E.A.C. Paley and N. Wiener, Characters of Abelian groups, *Proc. Nat. Acad. Sc.* **19** (1933), 253–7.

[PaWiZy] R.E.A.C. Paley, N. Wiener and A. Zygmund, Notes on random functions, *Math. Z.* **37** (1933), 647–68.

[Pe1] J. Perrin, Mouvement Brownien et réalité moléculaire, *Ann. Chimie et Physique,* 8ᵉ série, **18** (1909), 1–114.

[Pe2] J. Perrin, *Atoms*, trans. by D.L. Hammick, second edition, Constable, London, 1923.

[Pes] Y.B. Pesin, *Dimension Theory in Dynamical Systems: Contemporary Views and Applications*, The University of Chicago Press, Chicago and London, 1997.

[Pet] K.E. Petersen, *Brownian Motion, Hardy Spaces and Bounded Mean Oscillation*, Cambridge University Press, 1977.

[Pie] A. Pietsch, Absolute *p*-summierende Abbildungen in normierten Räumen, *Studia Math.* **XXVIII** (1967), 333–53.

[Pi1] G. Pisier, Sur l'éspace de Banach des séries de Fourier aléatoires presque surement continues, *Séminaire sur la Géometrie des espaces de Banach*, 1977–1978, Exposés No. XVII–XVIII.

[Pi2] G. Pisier, Arithmetic characterizations of Sidon sets, *Bull. Amer. Math. Soc.* **8** (1983), 87–9.

[Pi3] G. Pisier, *Factorization of Linear Operators and Geometry of Banach Spaces, Amer. Math. Soc.* CBMS **60**, 1986.

[Pr] P. Protter, *Stochastic Integration and Differential Equations*, Springer-Verlag, New York, 1990.

[R] H. Rademacher, Einige Sätze über Reihen von allgemeinen Orthogonalfunktionen, *Math. Ann.* **87** (1922), 112–38.

[Re] F. Reif, *Fundamentals of Statistical and Thermal Physics*, McGraw-Hill, 1965.

[RevYor] D. Revuz and M. Yor, *Continuous Martingales and Brownian Motion*, second edition, Springer-Verlag, 1994.

[RiLe] G. Ricci-Curbastro and T. Levi-Civita, Méthodes de calcul différential absolu et leurs applications, *Math. Ann.* **54** (1901), 125–201.

[Rid] D. Rider, Randomly continuous functions and Sidon sets, *Duke Math. J.* **42** (1975), 759–64.

[$\mathrm{Ri_f}$1] F. Riesz, Sur les opérations fonctionnelles linéaires, *C. R. Acad. Sci. Paris*, **149** (1909), 974–7.

[$\mathrm{Ri_f}$2] F. Riesz, Über die Fourierkoeffizienten einer stetigen Funktion von beschränkter Schwankung, *Math. Zeit.* **2** (1918), 312–15.

[$\mathrm{Ri_m}$] M. Riesz, Sur les fonctions conjuguées, *Math. Zeit.* **27** (1928), 218–44.

[Ro] H.P. Rosenthal, On trigonometric series associated with weak* closed subspaces of continuous functions, *J. Math. Mech.* **17** (1967), 485–90.

[RosWo] J. Rosiński and W.A. Woyczyński, On Itô stochastic integration with respect to *p*-stable motion: inner clock, integrability of sample paths, double and multiple integrals, *Ann. Prob.* **14** (1986), 271–86.

[Roy] H.L. Royden, *Real Analysis*, third edition, Macmillan Pub. Co., New York, 1988.

[Ru1] W. Rudin, Trigonometric series with gaps, *J. Math. Mech.* **9** (1960), 203–27.

[Ru2] W. Rudin, *Principles of Mathematical Analysis*, third edition, McGraw-Hill, 1976.

[Ru3] W. Rudin, *Fourier Analysis on Groups*, Interscience, New York, 1967.

[Ru4] W. Rudin, *The Way I Remember It*, History of Mathematics Vol. 12, Amer. Math. Soc. 1997.

[S] S. Saeki, The ranges of certain isometries of tensor products of Banach spaces, *J. Math. Soc. Japan*, **23** (1971), 27–39.

[SaZy1] R. Salem and A. Zygmund, On a theorem of Banach, *Proc. Nat. Acad. U.S.A.* **33** (1947), 293–5.

[SaZy2] R. Salem and A. Zygmund, Some properties of trigonometric series whose terms have random signs, *Acta Math.* **91** (1954), 245–301.

[SamTa] G. Samorodnitsky and M. Taqqu, *Stable Non-Gaussian Random Processes*, Chapman & Hall, 1994.

[Sap] P. Saphar, Applications á puissance nucléaire et applications de Hilbert-Schmidt dans les espaces de Banach, *C. R. Acad. Sci. Paris,* **261** (1965), 867–70.

[Saw] J. Sawa, The best constant in the Khintchin inequality for complex Steinhaus variables, the case $p = 1$, *Studia Math.* **81** (1985), 107–26.

[Sc1] R. Schatten, On the direct product of Banach spaces, *Trans. Amer. Math. Soc.* **53** (1943), 195–217.

[Sc2] R. Schatten, On reflexive norms for the direct product of Banach spaces, *Trans. Amer. Math. Soc.* **54** (1943), 498–506.

[Sc3] R. Schatten, The cross-space of linear transformations, *Ann. Math.* **47** (1946), 73–84.

[Sc4] R. Schatten, *A Theory of Cross-Spaces*, Annals of Math. Studies, **26**, Princeton University Press, 1950.

[ScvN1] R. Schatten and J. von Neumann, The cross-space of linear transformations II, *Ann. Math.* **47** (1946), 608–30.

[ScvN2] R. Schatten and J. von Neumann, The cross-space of linear transformations III, *Ann. Math.* **49** (1948), 557–82.

[Sch1] M. Schreiber, Fermeture en probabilité des chaos de Wiener, *C. R. Acad. Sc. Paris, Série A*, t. **265** (1967), 859–61.

[Sch2] M. Schreiber, Fermeture en probabilité de certains sous-espaces d'un espace L^2, *Z. W. Geb.* **14** (1969), 36–48.

[Schu1] J. Schur, Bemerkungen zur Theorie der beschränkten Bilinearformen mit unendlich vielen Veränderlichen, *J. Reine angew. Math.* **140** (1911), 1–32.

[Schu2] J. Schur, Über lineare Transformationen in der Theorie der unendlichen Reihen, *J. Reine angew. Math.* **151** (1920), 79–111.

[Schw] L. Schwartz, *Théorie des Distributions*, Hermann, Paris, 1966.

[Sh] C.E. Shannon, A mathematical theory of communication, in: *Claude Elwood Shannon Collected Papers*; edited by N.J.A. Sloane and Aaron D. Wyner, IEEE Press, 1993.

[Si1] S. Sidon, Ein Satz über die absolute Konvergenz von Fourierreihen in denen sehr viele Glieder fehlen, *Math. Ann.* **96** (1926), 418–19.

[Si2] S. Sidon, Verallgemeinerung eines Satzes über die absolute Konvergenz von Fourierreihen mit Lücken, *Math. Ann.* **97** (1927), 675–6.

[Si3] S. Sidon, Ein Satz über trigonometrische Polynome mit Lücken und seine Anwendungen in der Theorie der Fourier-Reihen, *J. Reine angew. Math.* **163** (1930), 251–2.

[Sim] M. Simonovits, Extremal graph theory, in: *Selected Topics in Graph theory 2*, L.W. Beineke and R.J. Wilson, eds., Academic Press, London, 1983, 161–200.

[Sj] P. Sjölin, An inequality of Paley and convergence a.e. of Walsh–Fourier series, *Arkiv För. Mat.* **7** (1968), 551–70.

[Sk] V. Skorohod, On a generalization of a stochastic integral, *Theory Prob. Appl.* **XX** (1975), 219–33.

[Sm] R.R. Smith, Completely bounded multilinear maps and Grothendieck's inequality, *Bull. London Math. Soc.* **20** (1988), 606–12.

[St] E.M. Stein, *Singular Integrals and Differentiability Properties of Functions*, Princeton University Press, 1970.

[Ste1] H. Steinhaus, Über die Wahrscheinlichkeit dafür, daß der Konvergenzkreis einer Potenzreihe ihre natürliche Grenze ist, *Math. Z.* **31** (1930), 408–16.

[Ste2] H. Steinhaus, Sur la probabilité de la convergence de séries, *Studia Math.* **2** (1930), 21–39.

[Sz] S.J. Szarek, On the best constants in the Khinchin inequality, *Studia Math.* **58** (1976), 197–208.

[T] M. Talagrand, Sections of smooth convex bodies via majorizing measures, *Acta. Math.* **175** (1995), 273–300.

[To] B. Tomaszewski, Two remarks on the Khintchin-Kahane inequality, *Colloq. Math.* **46** (1982), 283–8.

[Tow] N. Towghi, Stochastic integration of processes of finite generalized variations I, *Ann. Prob.* **23** (1995), 629–67.

[Tr] F. Treves, *Topological Vector Spaces, Distributions and Kernels*, Academic Press, New York and London, 1967.

[Tu] H. Tucker, *A Graduate Course in Probability*, Academic Press, New York and London, 1967.

[V1] N. Varopolous, Tensor algebras and harmonic analysis, *Acta. Math.* **119** (1967), 51–112.

[V2] N. Varopolous, On a problem of A. Beurling, *J. Func. Anal.* **2** (1968), 24–30.

[V3] N. Varopolous, Tensor algebras over discrete spaces, *J. Func. Anal.* **3** (1969), 321–35.

[V4] N. Varopolous, On an inequality of von Neumann and an application of the metric theory of tensor products to operator theory, *J. Funct. Anal.* **16** (1974), 83–100.

[vN1] J. von Neumann, *Mathematical Foundations of Quantum Mechanics*, trans. by Robert T. Beyer, Princeton University Press, Princeton, 1955.

[vN2] J. von Neumann, On infinite direct products, *Compositio Math.* **6** (1938), 1–77.

[vN3] J. von Neumann, Continuous geometry, *Proc. Nat. Acad. Sci.* **22** (1936), 92–100.

[vN4] J. von Neumann, Examples of continuous geometries, *Proc. Nat. Acad. Sci.* **22** (1936), 101–8.

[vN5] J. von Neumann, *Continuous Geometry*, Princeton University Press, 1960.

[W] J. Walsh, A closed set of normal orthogonal functions, *Am. J. Math.* **45** (1923), 5–24.

[We] A. Weil, *L'integration dans les groupes topologiques et ses applications*, Actualités Sci. et Ind. 869, Paris, Hermann & Cie, 1938.

[Wey1] H. Weyl, *Space–Time–Matter*, fourth edition, trans. from German by H. Brose, Dover Publications, Inc., 1952.

[Wey2] H. Weyl, *The Theory of Groups and Quantum-Mechanics*, trans. from German by H.P. Robertson, London, Methuen & Co., 1931.

[Wi1] N. Wiener, The average of an analytic functional and the Brownian movement, *Proc. Nat. Acad. Sci.* **7** (1921), 294–8.

[Wi2] N. Wiener, Differential-Space, *J. Math. Phys.* **2** (1923), 131–74.

[Wi3] N. Wiener, The quadratic variation of a function and its Fourier series, *J. Math. Phys.* **3** (1924), 73–94.

[Wi4] N. Wiener, Un problème de probabilitès denombrables, *Bull. Soc. Math. France*, **11** (1924), 569–78.

[Wi5] N. Wiener, The homogeneous chaos, *Amer. J. Math.* **60** (1938), 897–936.

[Wi6] N. Wiener, *Collected Works, Vol. I*, ed. P. Masani, The MIT Press, 1976.

[Wil] D. Williams, *Probability with Martingales*, Cambridge University Press, 1991.

[Y1] K. Ylinen, On vector bimeasures, *Ann. Mat. Pura Appl.* **4** (1978), 119–38.

[Y2] K. Ylinen, Noncommutative Fourier transforms of bounded bilinear forms and completely bounded multilinear operators, *J. Funct. Anal.* **79** (1988), 144–65.

[Yo1] L.C. Young, An inequality of the Hölder type, connected with Stieltjes integration, *Acta. Math.* **67** (1936), 251–82.

[Yo2] L.C. Young, General inequalities for Stieltjes integrals and the convergence of Fourier series, *Math. Ann.* **115** (1938), 581–612.

[ZS] G. Zhao and B.M. Schreiber, Algebras of multilinear forms on groups, *Contemp. Math.* **189** (1995), 497–511.

[Zy1] A. Zygmund, Sur les séries trigonométriques lacunaires, *J. London Math. Soc.* **5** (1930), 138–45.

[Zy2] A. Zygmund, *Trigonometric Series*, reprinted second edition, Cambridge University Press, 1968.

Index